About Island Press

Since 1984, the nonprofit Island Press has been stimulating, shaping, and communicating the ideas that are essential for solving environmental problems worldwide. With more than 800 titles in print and some 40 new releases each year, we are the nation's leading publisher on environmental issues. We identify innovative thinkers and emerging trends in the environmental field. We work with world-renowned experts and authors to develop cross-disciplinary solutions to environmental challenges.

Island Press designs and implements coordinated book publication campaigns in order to communicate our critical messages in print, in person, and online using the latest technologies, programs, and the media. Our goal: to reach targeted audiences—scientists, policymakers, environmental advocates, the media, and concerned citizens—who can and will take action to protect the plants and animals that enrich our world, the ecosystems we need to survive, the water we drink, and the air we breathe.

Island Press gratefully acknowledges the support of its work by the Agua Fund, Inc., The Margaret A. Cargill Foundation, Betsy and Jesse Fink Foundation, The William and Flora Hewlett Foundation, The Kresge Foundation, The Forrest and Frances Lattner Foundation, The Andrew W. Mellon Foundation, The Curtis and Edith Munson Foundation, The Overbrook Foundation, The David and Lucile Packard Foundation, The Summit Foundation, Trust for Architectural Easements, The Winslow Foundation, and other generous donors.

Assessment of Climate Change in the Southwest United States

A Report Prepared for the National Climate Assessment

Printed on recycled, acid-free paper

Manufactured in the United States of America

Citation: Garfin, G., A. Jardine, R. Merideth, M. Black, and S. LeRoy, eds. 2013. *Assessment of Climate Change in the Southwest United States: A Report Prepared for the National Climate Assessment*. A report by the Southwest Climate Alliance. Washington, DC: Island Press.

Keywords: Adaptation, agriculture, air quality, assessment, atmospheric river, biodiversity, climate change, climate impacts, climate modeling, climate variability, coastal, Colorado River, decision making, drought, electric power generation, extreme events, flooding, forest mortality, Great Basin, heat related illness, heat wave, land-use change, mitigation, Native American tribes, natural resource management, ocean acidification, phenology, public health, ranching, Rio Grande, Sacramento-San Joaquin, sea-level rise, social vulnerability, Southwest, stationarity, uncertainty, urban metabolism, U.S.Mexico border, vector-borne disease, water resources, wildfire

This technical input document in its current form does not represent a Federal document of any kind and should not be interpreted as the position or policy of any Federal, State, Local, or Tribal Government or Non-governmental entity.

Front Cover Images: Eagle Dancers courtesy of the New Mexico Tourism Department. All other images courtesy of iStock.

About This Series

This report is published as one of a series of technical inputs to the National Climate Assessment (NCA) 2013 report. The NCA is being conducted under the auspices of the Global Change Research Act of 1990, which requires a report to the President and Congress every four years on the status of climate change science and impacts. The NCA informs the nation about already observed changes, the current status of the climate, and anticipated trends for the future. The NCA report process integrates scientific information from multiple sources and sectors to highlight key findings and significant gaps in our knowledge. Findings from the NCA provide input to federal science priorities and are used by U.S. citizens, communities and businesses as they create more sustainable and environmentally sound plans for the nation's future.

In fall of 2011, the NCA requested technical input from a broad range of experts in academia, private industry, state and local governments, non-governmental organizations, professional societies, and impacted communities, with the intent of producing a better informed and more useful report in 2013. In particular, the eight NCA regions, as well as the Coastal and the Ocean biogeographical regions, were asked to contribute technical input reports highlighting past climate trends, projected climate change, and impacts to specific sectors in their regions. Each region established its own process for developing this technical input. The lead authors for related chapters in the 2013 NCA report, which will include a much shorter synthesis of climate change for each region, are using these technical input reports as important source material. By publishing this series of regional technical input reports, Island Press hopes to make this rich collection of information more widely available.

This series includes the following reports:

Climate Change and Pacific Islands: Indicators and Impacts
Coastal Impacts, Adaptation, and Vulnerabilities
Great Plains Regional Technical Input Report
Climate Change in the Midwest: A Synthesis Report for the National Climate Assessment
Climate Change in the Northeast: A Sourcebook
Climate Change in the Northwest: Implications for Our Landscapes, Waters, and Communities
Oceans and Marine Resources in a Changing Climate
Climate of the Southeast United States: Variability, Change, Impacts, and Vulnerability
Assessment of Climate Change in the Southwest United States

Electronic copies of all reports can be accessed on the Climate Adaptation Knowledge Exchange (CAKE) website at *www.cakex.org/NCAreports*. Printed copies are available for sale on the Island Press website at *www.islandpress.org/NCAreports.*

Assessment of Climate Change in the Southwest United States

A Report Prepared for the National Climate Assessment

EXECUTIVE EDITOR

Gregg Garfin (University of Arizona)

ASSOCIATE EDITOR

Angela Jardine (University of Arizona)

CONTRIBUTING EDITORS

Robert Merideth (University of Arizona)

Mary Black (University of Arizona)

Sarah LeRoy (University of Arizona)

ISLANDPRESS

Washington | Covelo | London

Acknowledgements

The authors thank the members of the Southwest Climate Alliance Executive Committee, who developed the original Expression of Interest for this technical report. Members included: David Busch (USGS), Dan Cayan (Scripps Institution of Oceanography/UCSD), Michael Dettinger (USGS), Erica Fleishman (University of California, Davis), Gregg Garfin (University of Arizona), Alexander Sasha Gershunov (Scripps Institution of Oceanography/UCSD), Glen MacDonald (UCLA), Jonathan Overpeck (University of Arizona), Kelly Redmond (Desert Research Institute), Mark Schwartz (University of California, Davis), William Travis (University of Colorado), and Brad Udall (University of Colorado). We acknowledge the contributions of the three NOAA Regional Integrated Sciences and Assessments (RISA) projects in the region, and their program managers, through the generous donation of staff time and assessment advice: California-Nevada Applications Program (CNAP; Dan Cayan), Climate Assessment for the Southwest (CLIMAS; Dan Ferguson), and Western Water Assessment (WWA; Brad Udall). We also acknowledge the contributions of the Southwest Climate Science Center (SWCSC). The NOAA Climate Program Office and the USGS provided funding and development support.

For guidance and input in writing this report, we thank the National Climate Assessment staff: Emily Cloyd, Katharine Jacobs, Fred Lipschultz, Sheila O'Brien, and Anne Waple. We thank the National Climate Assessment Development and Advisory Committee for guidance and input in writing this report. This includes the following people: Jim Buizer (University of Arizona), Guido Franco (California Energy Commission), Nancy Grimm (Arizona State University), Diana Liverman (University of Arizona), Susanne Moser (Susanne Moser Research and Consulting, Stanford University), Richard Moss (Pacific Northwest National Laboratory), and Gary Yohe (Wesleyan University).

The authors are grateful to Jaimie Galayda (University of Arizona) for exceptional assistance in organizing the August 2011 prospective authors' workshop. Eric Gordon (University of Colorado), Sarah Guthrie (CIRES), William Travis (University of Colorado), and Suzanne van Drunick (CIRES) hosted the 2011 workshop, providing rooms, technical support, and hospitality. Mary Floyd (Zantech IT Services) assisted with travel arrangements for some of the workshop participants. Tamara Wall (Desert Research Institute) also provided assistance. Jennifer Paolini (Scripps Institution of Oceanography) provided exceptional logistical support and hospitality for our August workshop and our January 2012 Coordinating Lead Authors' workshop. Lesa Langan DuBerry (University of Arizona) and Anh Le (University of Arizona) provided logistical support for both workshops.

The *Assessment of Climate Change in the Southwest United States* was developed with the benefit of a scientifically rigorous first draft expert review and we thank the review editors and expert reviewers. We also thank everyone who contributed in the public review, as well as the decision makers, resource managers, citizens, and experts who provided feedback on the content of the report, in addition to valuable insights about the form of the report and ancillary materials.

We acknowledge the modeling groups, the Program for Climate Model Diagnosis and Intercomparison (PCMDI) and the World Climate Research Programme (WCRP) Working Group on Coupled Modeling (WGCM) for their roles in making available the WCRP CMIP3 multi-model dataset. Support of this dataset is provided by the Office of Science, U.S. Department of Energy. The GCM projections from the CMIP3 dataset are part of the World Climate Research Programme's (WCRP's) Coupled Model Intercomparison Project phase 3 (CMIP3) multi-model dataset. Bias Corrected and Downscaled (BCSD) and VIC hydrological model projections were obtained from the http://gdo-dcp.ucllnl.org/downscaled_cmip3_projections/ archive provided by the U.S. Bureau of Reclamation and partners (see Reclamation, 2011, "West-Wide Climate Risk Assessments: Bias-Corrected and Spatially Downscaled Surface Water Projections," Technical Memorandum No. 86-68210-2011-01, prepared by the U.S. Department of the Interior, Bureau of Reclamation, Technical Services Center, Denver, Colorado, 138pp.). We also thank the North American Regional Climate Change Assessment Program (NARCCAP) for providing climate change projection data used in several analyses presented in this report.

We would like to extend our greatest appreciations to the members of the report technical support team. Betsy Woodhouse (University of Arizona) advised us throughout the creation of this report. The University of Colorado (CIRES) members (Eric Gordon, Jeff Lukas, and Bill Travis) provided guidance and editing expertise throughout this report both in text and graphics. Andrea Ray (NOAA Earth System Research Laboratory) provided valuable graphic and figure caption edits. The University of Arizona Institute of the Environment communications team members provided support in numerous ways (Matt Price and Rey Granillo developed the review and report websites; Melissa Kerr created the SW Climate Summit draft of chapter 1; Stephanie Doster provided editing assistance; and Gigi Owen created video media). Melissa Espindola's attention to detail in assembling various lists and ancillary materials was also essential. Annisa Tangreen, Jaimie Galayda, Sarah LeRoy, Katherine Waser, and Gretel Hakanson spent countless hours ensuring report-wide standardization and formatting, and checking bibliographic citations and information related to figures and tables. Special thanks to Jaimie Galayda and Sarah LeRoy for overseeing the manuscript coordination. Lastly, words cannot express our gratitude to Ami Nacu-Schmidt (CIRES, University of Colorado) for creating the majority of the imagery, which she wove into a consistent graphical story throughout this report.

Abbreviations and Acronyms

AR – atmospheric river
BAFC – Border Area Fire Council
BTU - British thermal unit (a measure of energy consumption)
BCCA – bias correction and constructed analogues downscaling method
BCDC – San Francisco Bay Conservation and Development Commission
BCSD – bias correction and spatial downscaling method
BIA – Bureau of Indian Affairs of the U.S. Dept. of the Interior
CAFE – corporate average fuel economy
CAM – crassulacean acid metabolism; a type of plant metabolism
CARB – California Air Resources Board
CCAWWG – Climate Change and Western Water Group
CIAT – Center for International Earth Science Information Network
CICESE – Centro de Investigación Científica de Educación Superior de Ensenada
CICS – Cooperative Institute for Climate and Satellites, North Carolina State University
CIESIN – Center for International Earth Science Information
CILA – Comisión Internacional de Límites y Agua, the Mexican counterpart to IBWC
CIMIS – California Irrigation Management Information System (of the California Dept. of Water Resources)
CIRES – Cooperative Institute for Research in Environmental Sciences
CLIMAS – Climate Assessment for the Southwest
CMIP3 – phase 3 of the World Climate Research Programme's Coupled Model Intercomparison Project
CNAP – California-Nevada Applications Program
CO_2 – carbon dioxide
COLEF – El Colegio de la Frontera Norte
CONAGUA – Comisión Nacional del Agua, the Mexican National Water Commission
COOP – National Weather Service Cooperative Observing Network
CRJCP – Colorado River Joint Cooperative Process
CSC – Climate Science Center
CSI – cold spell index
CSP – concentrated solar power
CWCB – Colorado Water Conservation Board
DWP – Los Angeles Department of Water and Power
EC – eddy covariance
EERS – Energy Efficiency Resource Standard

EIA – U.S. Energy Information Agency

ENSO – El Niño-Southern Oscillation

ET – evapotranspiration

FEMA – Federal Emergency Management Agency

GCM – global climate model

GCRA – Global Change Research Act of 1990

GDP – gross domestic product

GNEB – Good Neighbor Environmental Board

GHG – greenhouse gas

Gt – gigaton

HWI – heat wave index

IBWC – International Boundary and Water Commission, the U.S. counterpart to CILA

ICLEI – Local Governments for Sustainability (originally founded in 1990 as "International Council for Local Environmental Initiatives")

IPCC – Intergovernmental Panel on Climate Change

LCC – Landscape Conservation Cooperatives (of the Dept. of the Interior)

LCP – local coastal plan

LFP – Livestock Forage Disaster Program

mac – million acres

maf – million acre-feet

MARAD – Maritime Administration

MCA – Medieval Climate Anomaly

MJO – Madden–Julian Oscillation

MMT – million metric tons

MRGCD – Middle Rio Grande Conservancy District

MWh – megawatt hours

NAAQS – National Ambient Air Quality Standards (of the EPA)

NADB – North American Development Bank

NALC – Native American Land Conservancy

NAM – North American monsoon

NARCCAP – North American Regional Climate Change Assessment Program

NAS – National Academy of Sciences

NCA – National Climate Assessment

NCADAC – National Climate Assessment and Development Advisory Committee (of NOAA)

NCDC – NOAA's National Climate Data Center

NETL – National Energy Technology Laboratory

NFIP – National Flood Insurance Program (of FEMA)

NGO – non-governmental organization

NIDIS – National Integrated Drought Information System
NOAA – National Oceanic and Atmospheric Administration
NOEP – National Ocean Economic Program
NPS – National Park Service
NRC – National Research Council (of the National Academies)
NREL – National Renewable Energy Laboratory
PDF – probability density function
PDO –Pacific Decadal Oscillation
PDSI – Palmer Drought Severity Index
PEAC-BC – State of Baja California Climate Action Plan
PIA – practicably irrigable acreage
PLPT – Pyramid Lake Paiute Tribe
PM – particulate matter
POT – peaks over threshhold
ppm – parts per million
PV – photovoltaic
QDO – Pacific Quasi-Decadal Oscillation
RCM – regional climate model
Reclamation – U.S. Bureau of Reclamation
RISA – Regional Integrated Sciences and Assessments (a NOAA Climate Program)
RPS – renewable portfolio standards
RTEP – Regional Transmission Expansion Project
SECURE – Science and Engineering to Comprehensively Understand and Responsibly Enhance
SFTP – Supplemental Federal Test Procedure (of the EPA)
SIP – state implementation plan
SRES – Special Report on Emissions Scenarios (of the IPCC Fourth Assessment Report)
SURE – Supplemental Revenue Assistance Payments Program
SWCA – Southwest Climate Alliance
SWCSC – Southwest Climate Science Center
SWE – snow-water equivalent
TAAP – Transboundary Aquifer Assessment Program, a U.S.-Mexico program that assesses shared aquifers
Tmax – maximum daily temperature
Tmin – minimum daily temperature
TRNERR – Tijuana River National Estuarine Research Reserve
UABC – Universidad Autónoma de Baja California
UBEES – Utah Building Energy Efficiency Strategies
USFS – U.S. Forest Service (Department of Agriculture)

USFWS – U.S. Fish and Wildlife Service (Dept. of the Interior)
USGCRP – U.S. Global Change Research Program
USGS – U.S. Geological Survey
VMT – vehicle miles travelled
WestFAST – Western Federal Agency Support Team
WGA – Western Governors' Association
WHO – World Health Organization
W/m^2 – watts per square meter
WNV – West Nile virus
WRF – Weather Research and Forecasting model
WSWC – Western States Water Council of the Western Governors Association
WUCA – Water Utility Climate Alliance
WWA – Western Water Assessment

Contents

WEATHER AND CLIMATE OF THE SOUTHWEST

Chapter 1

Summary for Decision Makers

COORDINATING LEAD AUTHOR

Jonathan Overpeck (University of Arizona)

LEAD AUTHORS

Gregg Garfin (University of Arizona), Angela Jardine (University of Arizona), David E. Busch (U.S. Geological Survey), Dan Cayan (Scripps Institution of Oceanography), Michael Dettinger (U.S. Geological Survey), Erica Fleishman (University of California, Davis), Alexander Gershunov (Scripps Institution of Oceanography), Glen MacDonald (University of California, Los Angeles), Kelly T. Redmond (Western Regional Climate Center and Desert Research Institute), William R. Travis (University of Colorado), Bradley Udall (University of Colorado)

With contributions from the authors of this assessment report

1.1 Introduction

Natural climate variability is a prominent factor that affects many aspects of life, livelihoods, landscapes, and decision-making across the Southwestern U.S. (Arizona, California, Colorado, Nevada, New Mexico, and Utah; included are the adjacent United States-Mexico border and Southwest Native Nations land). These natural fluctuations have caused droughts, floods, heat waves, cold snaps, heavy snow falls, severe winds, intense storms, the battering of coastal areas, and acute air-quality conditions. And as a region that has experienced—within the relatively short time span of several decades—rapid increases in human population (Figure 1.1), significant alterations in land use and land cover, limits on the supplies of water, long-term drought, and other climatic

Chapter citation: Overpeck, J., G. Garfin, A. Jardine, D. E. Busch, D. Cayan, M. Dettinger, E. Fleishman, A. Gershunov, G. MacDonald, K. T. Redmond, W. R. Travis, and B. Udall. 2013. "Summary for Decision Makers." In *Assessment of Climate Change in the Southwest United States: A Report Prepared for the National Climate Assessment*, edited by G. Garfin, A. Jardine, R. Merideth, M. Black, and S. LeRoy, 1–20. A report by the Southwest Climate Alliance. Washington, DC: Island Press.

changes, the Southwest can be considered to be one of the most "climate-challenged" regions of North America. This document summarizes current understanding of climate variability, climate change, climate impacts, and possible solution choices for the climate challenge, all issues that are covered in greater depth in *Assessment of Climate Change in the Southwest United States.*[i]

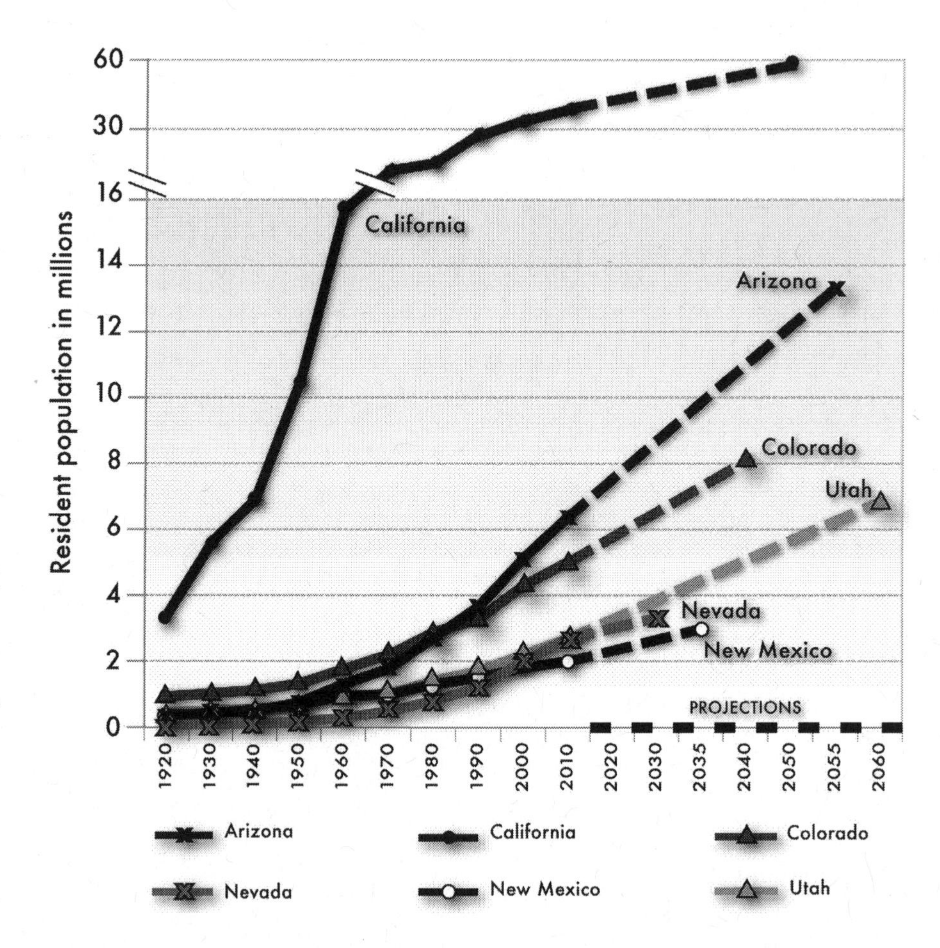

Figure 1.1 Rapid population growth in the Southwest is expected to continue. The current (2010) population is 56 million, and an additional 19 million people are projected to be living in the region by 2030. Source: US Department of Commerce, Bureau of Economic Analysis http://www.bea.gov/regional/index.htm. [Chapter 3]

The juxtaposition of the Southwest's many landscapes—mountains, valleys, plateaus, canyons, and plains—affect both the region's climate and its response to climate change. Whether human and natural systems are able to adapt to changes in climate will be influenced by many factors, including the complex topographic pattern of land ownership and the associated policies and management goals. Moreover, the human population in the region will likely grow, primarily in urban areas, from a population of about 56 million in 2010 to an estimated 94 million by 2050 (Figure 1.1). [Chapter 3]

The Southwest climate is highly variable across space and over time related to such factors as ocean-land contrasts, mountains and valleys, the position of jet streams, the North American monsoon, and proximity to the Pacific Ocean, Gulf of California, and

Gulf of Mexico. The Mojave and Sonoran Deserts of Southern California, Nevada, and Arizona are the hottest (based on July maximum temperatures), driest regions of the contiguous United States. Coastal zones of California and northwestern Mexico have large temperature gradients and other properties from the shore to inland. Mountain regions are much cooler and usually much wetter regions of the Southwest, with the Sierra Nevada and mountains of Utah and Colorado receiving nearly half of their annual precipitation in the form of snow. The resulting mountain snowpack provides much of the surface water for the region, in the form of spring runoff. [Chapter 4]

There is mounting scientific evidence that climate is changing and will continue to change. There is also considerable agreement—at varying levels of confidence sufficient to support decision making—regarding why the climate is changing, or will change [Chapter 19]. Readers of this summary may wish to review all or parts of the complete report, *Assessment of Climate Change in the Southwest United States*, to learn more about the region's climate, and its likely changes and effects.

1.2 Observed Recent Climatic Change in the Southwest

The climate of the Southwest is already changing in ways that can be attributed to human-caused emissions of greenhouse gases, or that are outcomes or expressions consistent with such emissions—with these notable observations:

- ***The Southwest is warming.*** Average daily temperatures for the 2001–2010 decade were the highest (Figure 1.2) in the Southwest from 1901 through 2010. Fewer cold waves and more heat waves occurred over the Southwest during 2001–2010 compared to average decadal occurrences in the twentieth century. The period since 1950 has been warmer than any period of comparable length in at least 600 years, as estimated on the basis of paleoclimatic tree-ring reconstructions of past temperatures. [Chapter 5]
- ***Recent drought has been unusually severe relative to droughts of the last century, but some droughts in the paleoclimate record were much more severe.*** The areal extent of drought over the Southwest during 2001–2010 was the second largest observed for any decade from 1901 to 2010. However, the most severe and sustained droughts during 1901–2010 were exceeded in severity and duration by multiple drought events in the preceding 2,000 years (Figure 1.3). [Chapter 5]
- ***Recent flows in the four major drainage basins of the Southwest have been lower than their twentieth century averages.*** Streamflow totals in the Sacramento-San Joaquin Rivers, Upper Colorado, Rio Grande, and Great Basin were 5% to 37% lower during 2001–2010 than their twentieth century average flows. Moreover, streamflow and snowmelt in many snowmelt-fed streams of the Southwest tended to arrive earlier in the year during the late twentieth century than earlier in the twentieth century, and up to 60% of the change in arrival time has been attributed to increasing greenhouse-gas concentrations in the atmosphere (Figure 1.4). [Chapter 5]

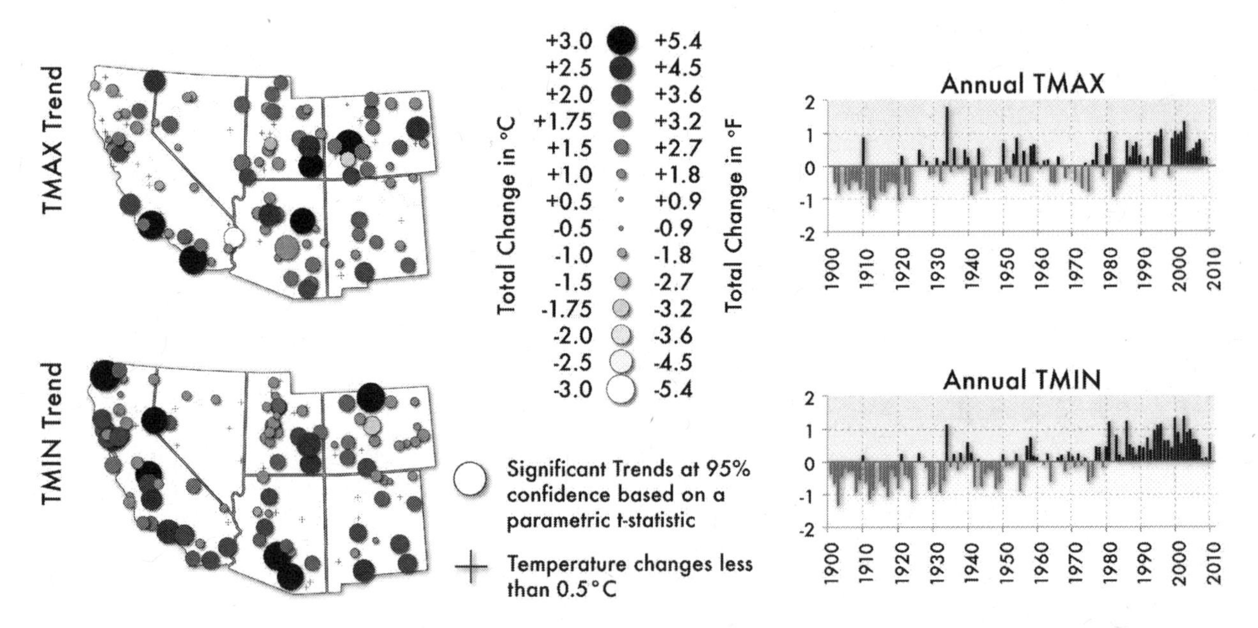

Figure 1.2 Temperature trends in the twentieth century. The 1901–2010 trends in annually averaged daily maximum temperature (TMAX, top) and daily minimum temperature (TMIN, bottom). Units are the change in °C/110yrs. Trends computed from 251 stations for precipitation analysis and 180 stations for temperature analysis using GHCN V3 data. Source: Menne and Williams (2009). [Chapter 5]

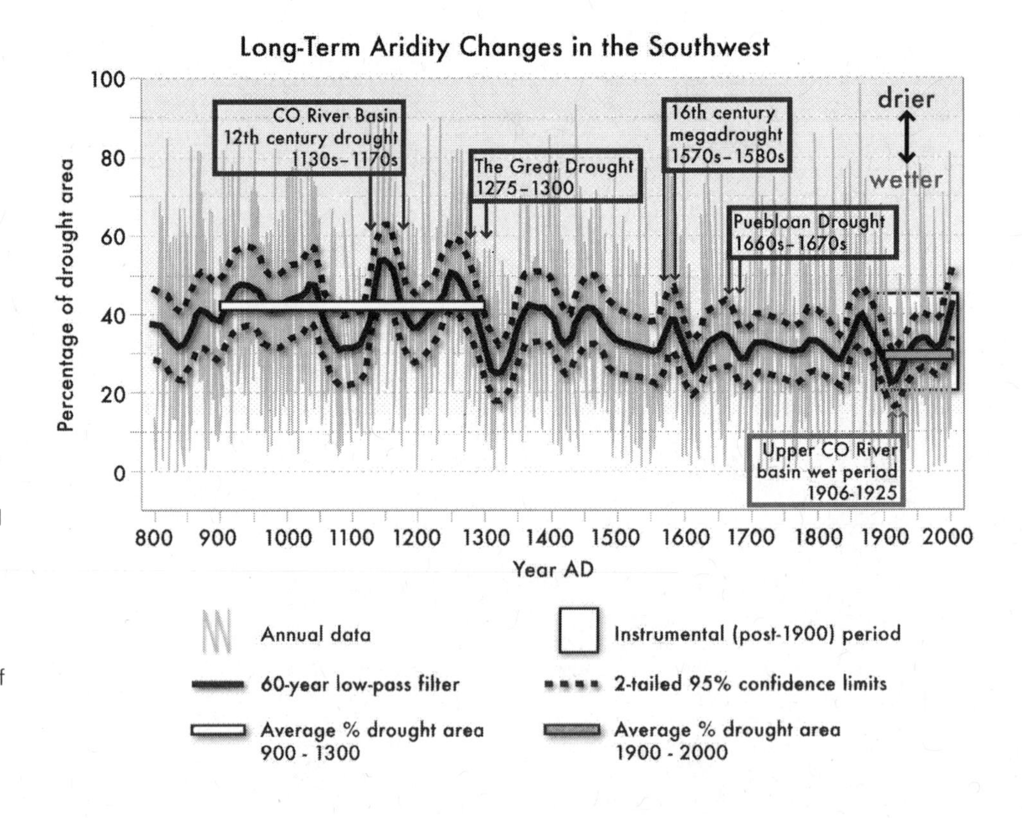

Figure 1.3 History of drought in the West. Percent area affected by drought (PDSI<–1) across the western United States, as reconstructed from tree-ring data. Modified from Cook et al. (2004), reprinted with permission from the American Association for the Advancement of Science. [Chapter 5]

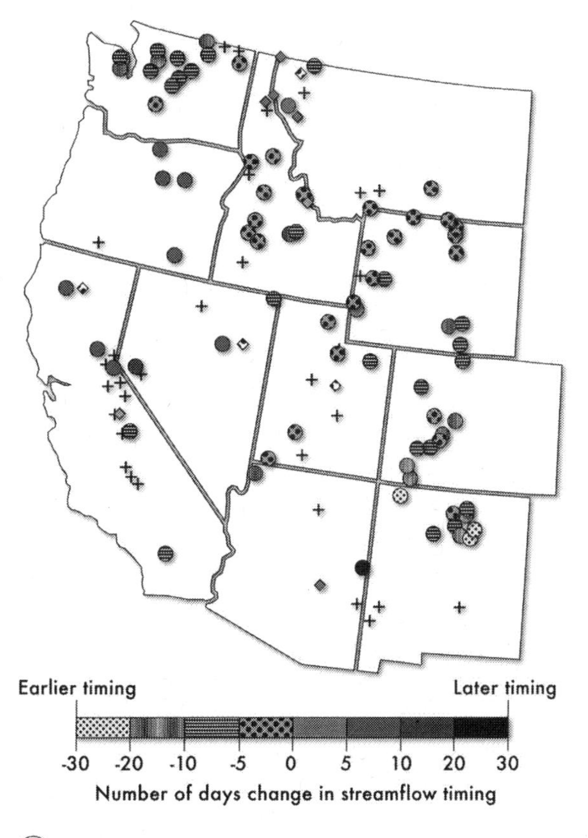

Figure 1.4 Changing streamflow timing 2001–2010 compared to 1950–2000. Differences between 2001–2010 and 1950–2000 average date when half of the annual streamflow has been discharged (center of mass) for snowmelt-dominated streams (Stewart, Cayan and Dettinger 2005). [Chapter 5]

1.3 Projected Future Climatic Change in the Southwest

Climate scientists have high confidence that the climate of the Southwest will continue to change through the twenty-first century and beyond, in response to human-generated greenhouse gas emissions, and will continue to vary in ways that can be observed in historic and paleoclimate records (Table 1.1). However, not all aspects of the climate change or variation can be projected with equal confidence.[ii] The highest confidence is associated with projections that are consistent among climate models and with observed changes, such as those described in the previous section. The magnitude and duration of future change depends most on the amount of greenhouse gases emitted to the atmosphere, particularly carbon dioxide emitted by the burning of coal, oil, and natural gas. Much of the future change will be irreversible for centuries after substantial anthropogenic carbon dioxide emissions have ceased.

- ***Warming will continue, with longer and hotter heat waves in summer.*** Surface temperatures in the Southwest will continue to increase substantially over the twenty-first century (high confidence), with more warming in summer and fall than winter and spring (medium-high confidence) (Figures 1.5 and 1.6). Summer heat waves will become longer and hotter (high confidence). Winter cold snaps will become less frequent but not necessarily less severe (medium-high confidence). [Chapter 6 and 7]

- ***Average precipitation will decrease in the southern Southwest and perhaps increase in northern Southwest.*** Precipitation will decline in the southern portion of the Southwest region, and change little or increase in the northern portion (medium-low confidence) (Figure 1.6). [Chapter 6]
- ***Precipitation extremes in winter will become more frequent and more intense (i.e., more precipitation per hour)*** (medium-high confidence). Precipitation extremes in summer have not been adequately studied. [Chapter 7]
- ***Late-season snowpack will continue to decrease.*** Late winter-spring mountain snowpack in the Southwest will continue to decline over the twenty-first century, mostly because temperature will increase (high confidence) (Figure 1.7). [Chapter 6]
- ***Declines in river flow and soil moisture will continue.*** Substantial portions of the Southwest will experience reductions in runoff, streamflow, and soil moisture in the mid- to late-twenty-first century (medium-high confidence) (Figure 1.7). [Chapter 6]
- ***Flooding will become more frequent and intense in some seasons and some parts of the Southwest, and less frequent and intense in other seasons and locations.*** More frequent and intense flooding in winter is projected for the western slopes of the Sierra Nevada (medium-high confidence), whereas snowmelt-driven spring and summer flooding could diminish in that mountain range (high confidence). [Chapter 7]
- ***Droughts in parts of the Southwest will become hotter, more severe, and more frequent*** (high confidence). Drought, as defined by Colorado River flow amount, is projected to become more frequent, more intense, and more prolonged, resulting in water deficits in excess of those during the last 110 years (high confidence). However, northern Sierra Nevada watersheds may become wetter with climate change (low confidence). [Chapter 7]

1.4 Recent and Future Effects of Climatic Change in the Southwest

Terrestrial and freshwater ecosystems

Natural ecosystems are being affected by climate change in noticeable ways, which may lead to their inhabitants needing to adapt, change, or move:

- ***The distributions of plant and animal species will be affected by climate change.*** Observed changes in climate are associated strongly with some observed changes in geographic distributions of species in the Southwest (high confidence). [Chapter 8]

Table 1.1 Current and predicted climate phenomena trends discussed in this report

Projected Change Parameter	Direction of Change	Is it Occurring?	Remarks	Confidence	Chapter
Average annual temperature	Increase	Yes. Southwest temperatures increased 1.6°F +/- 0.5°F, between 1901-2010.	Depending on the emissions scenario, model projections show average annual temperature increases of 1-4°F in the period 2021-2050, 1-6°F in 2041-2070, and 2-9°F in 2070-2099. Changes along the coastal zone are smaller than inland areas.	High	5; 6
Seasonal temperatures	Increase	Yes, in all seasons. Studies conclusively demonstrate partial human causation of winter/spring minimum temperature increases.	Model projections show the largest increases in summer and fall. The largest projected increases range from 3.5°F in the period 2021-2050 to 9.9°F in 2070-2099.	High	5; 6
Freeze-free season length	Increase	Yes, the freeze-free season for the Southwest increased about 7% (17 days) during 2001–2010 compared to the average season length for 1901–2000.	Model projections using a high emissions scenario (A2) show that by 2041–2070, most of the region exhibits increases of at least 17 freeze-free days, with some parts of the interior showing 38-day increases.	High	5; 6
Heat waves	Increase	Yes. More heat waves occurred over the Southwest during 2001–2010 compared to average occurrences in the twentieth century.	Model projections show an increase in summer heat wave frequency and intensity.	High	5; 7
Cold snaps	Decrease	Fewer cold waves occurred over the Southwest during 2001–2010 compared to average occurrences in the twentieth century.	Winter time cold snaps are projected to diminish their frequency but not necessarily their intensity into late century. Interannual and decadal variability will modulate occurrences across the region.	Medium-high	5; 7

Table 1.1 Current and predicted climate phenomena trends discussed in this report (Continued)

Projected Change Parameter	Direction of Change	Is it Occurring?	Remarks	Confidence	Chapter
Average annual precipitation	Decrease	Not yet detectable. During 1901–2010 there was little regional change in annual precipitation.	For all periods and both scenarios, model simulations show both increases and decreases in precipitation. For the region as a whole, most of the median values are negative, but not by much, whereas the range of changes, among different models, is high. Annual precipitation projections generally show decreases in the southern part of the region and increases in the northern part.	Medium-low	6
Spring precipitation	Decrease	Not yet detectable.	By mid-century, all but one model projects spring regional precipitation decreases. By 2070-2099, the median projected decrease is 9-29%, depending on the emissions scenario.	Medium-high	6
Extreme daily precipitation	Increase	Maybe. Studies indicate the frequency of extreme daily precipitation events over the Southwest during 1901–2010 had little regional change in extreme daily precipitation events.	Models project more intense atmospheric river precipitation; some studies project more frequent intense precipitation during the last half of the twenty-first century, especially in the northern part of the region.	Medium-low	5; 7
Mountain snowpack	Decrease	Yes, in parts of the Southwest.	Model projections from this report and other studies project a reduction of late winter-spring mountain snowpack in the Southwest over the twenty-first century, mostly because of the effects of warmer temperature.	High	6

Table 1.1 Current and predicted climate phenomena trends discussed in this report (Continued)

Projected Change Parameter	Direction of Change	Is it Occurring?	Remarks	Confidence	Chapter
Snowmelt and streamflow timing	Earlier	Yes, snowmelt and snowmelt-fed streamflow in many streams of the Southwest trended towards earlier arrivals in the late-twentieth century and early twenty-first century.	Not analyzed in this report, but implied by projections of diminished April 1 snow water equivalent in most Southwest river basins.	High	5, 6
Flooding	Increase	No. Annual peak streamflow rates declined from 1901 to 2008 in the Southwest.	More frequent and intense flooding in winter is projected for the western slopes of the Sierra Nevada range; Colorado Front Range flooding in summer is projected to increase.	Low	5; 7
Drought severity	Increase	Yes. During the period 1901-2010. However, the most severe and sustained droughts during 1901–2010 were exceeded in severity and duration by drought events in the preceding 2000 years.	Observed Southwest droughts have been exacerbated by anomalously warm summer temperatures. Model projections of increased summer temperatures would exacerbate future droughts. Model projections show depletion of June 1 soil moisture and lower total streamflow.	Medium-high	5; 6

- ***Ecosystem function and the functional roles of resident species will be affected.*** Observed changes in climate are associated strongly with some observed changes in the timing of seasonal events in the life cycles of species in the region (high confidence). [Chapter 8]
- ***Changes in land cover will be substantial.*** Observed changes in climate are affecting vegetation and ecosystem disturbance (Figure 1.8). Among those disturbances are increases in wildfire and outbreak of forest pests and disease. Death of plants in some areas of the Southwest also is associated with increases in temperature and decreases in precipitation (high confidence). [Chapter 8]
- ***Climate change will affect ecosystems on the U.S.-Mexico border.*** Potential changes to ecosystems that transect the international border are often not explicitly considered in the public policy exposing these sensitive ecosystems to climate change impacts (high confidence). [Chapter 16]

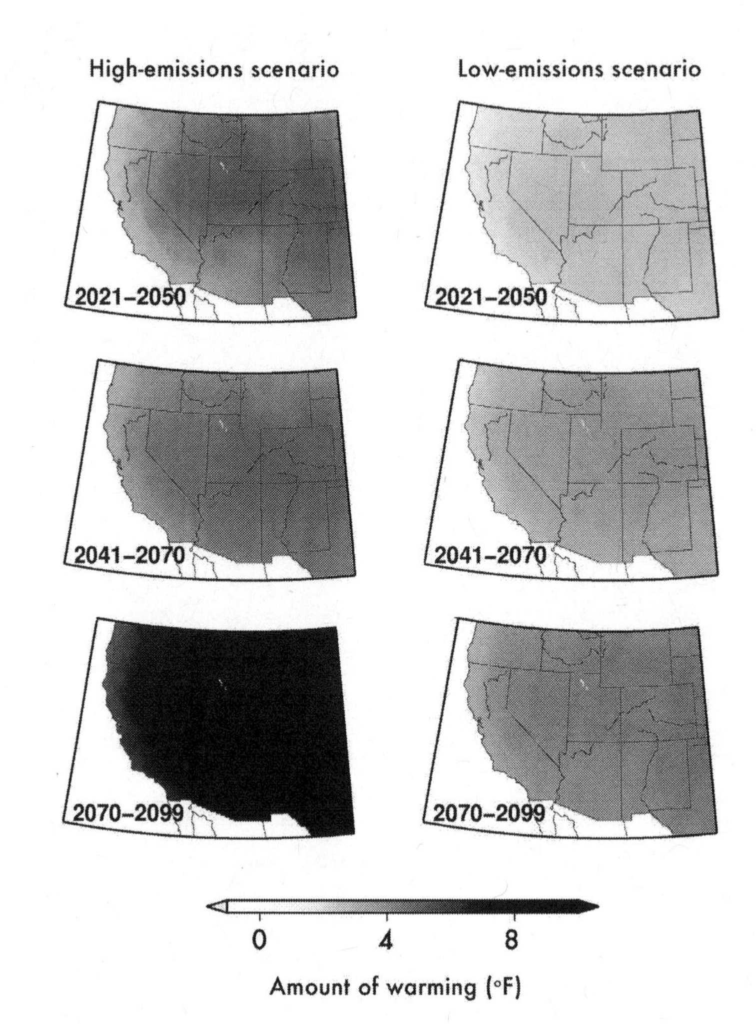

Figure 1.5 Projected temperature changes for the high (A2) and low (B1) GHG emission scenario models. Annual temperature change (oF) from historical (1971–2000) for early- (2021–2050; top), mid- (2041–2070; middle) and late- (2070–2099; bottom) twenty-first century periods. Results are the average of the sixteen statistically downscaled CMIP3 climate models. Source: Nakićenović and Swart (2000), Mearns et al. (2009). [Chapter 6]

Coastal systems

Coastal California is already being affected by climate change, and future climate-related change will become more notable if greenhouse-gas emissions are not substantially reduced:

- ***Coastal hazards, including coastal erosion, flooding, storm surges and other changes to the shoreline will increase in magnitude as sea level continues to rise*** (high confidence). Sea levels along the California coast have risen less than a foot since 1900, but could rise another two feet (high confidence), three feet (medium-high confidence), or possibly more (medium-low confidence) by the end of the twenty-first century (Figure 1.9). [Chapter 9]
- ***Effects of coastal storms will increase.*** Increased intensity (medium-low confidence) and frequency (medium-low confidence) of storm events will further change shorelines, near-shore ecosystems, and runoff. In many regions along the

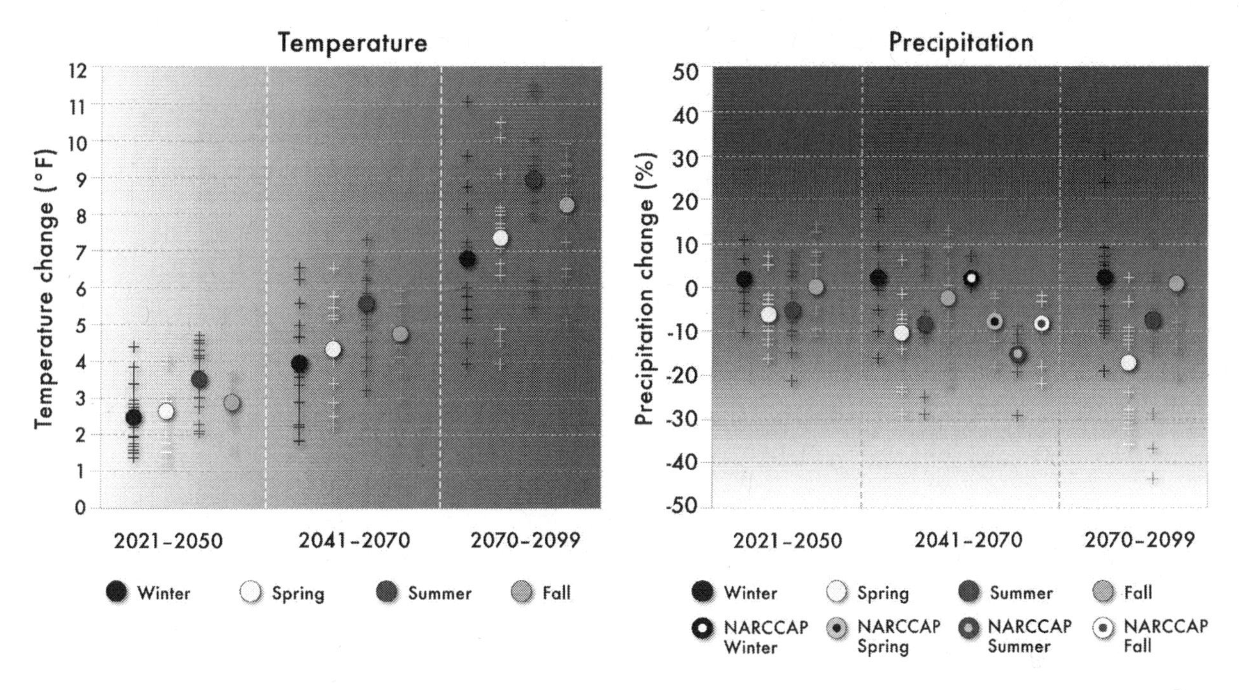

Figure 1.6 Projected change in average seasonal temperatures (°F, left) and precipitation (% change, right) for the Southwest region for the high-emissions (A2) scenario. A fifteen-model average of mean seasonal temperature and precipitation changes for early-, mid-, and late-twenty-first century with respect to the simulations' reference period of 1971–2000. Changes in precipitation also show the averaged 2041–2070 NARCCAP four global climate model simulations. The seasons are December–February (winter), March–May (spring), June–August (summer), and September–November (fall). Plus signs are projected values for each individual model and circles depict overall means. Source: Mearns et al. (2009). [Chapter 6]

coast, storms coupled with rising sea levels will increase the exposure to waves and storm surges (medium-high confidence). [Chapter 9]

- ***Economic effects of coastal climate change will be large.*** Between 2050 and 2100, or when sea levels are approximately 14–16 inches higher than in 2000, the combined effects of sea-level rise and large waves will result in property damage, erosion, and economic losses far greater than currently experienced (high confidence). [Chapter 9]
- ***Coastal ecosystems and their benefits to society will be affected.*** Ocean warming, reduced oxygen content, and sea-level rise will affect marine ecosystems, abundances of fishes, wetlands, and coastal communities (medium-high confidence). However, there is uncertainty in how and by how much coastal ecosystems will be affected. [Chapter 9]
- ***Ocean acidification is taking place.*** Many marine ecosystems will be negatively affected by ocean acidification that is driven by increased levels of atmospheric carbon dioxide (high confidence). But there is substantial uncertainty about the effects of acidification on specific coastal fisheries and marine food webs. [Chapter 9]

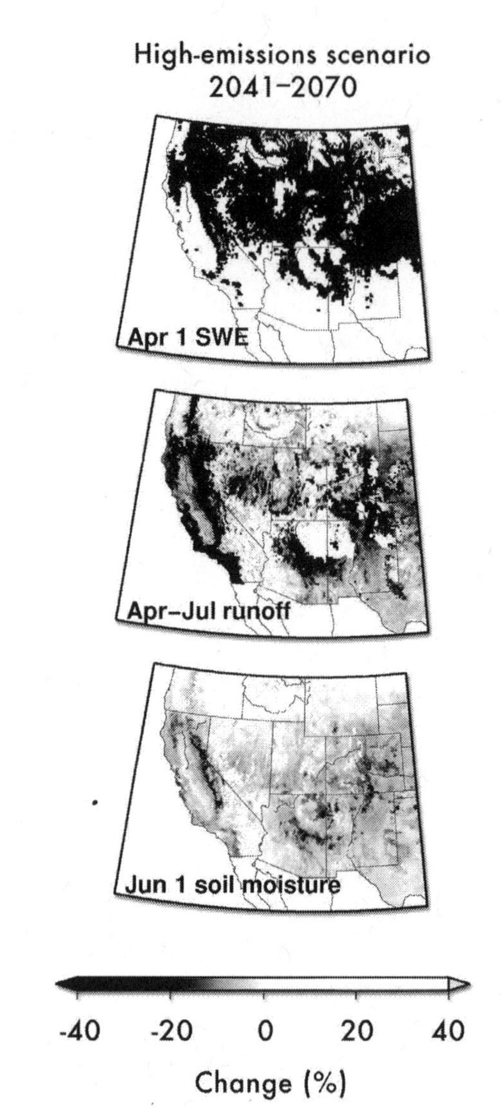

Figure 1.7 Predicted changes in the water cycle. Mid-century (2041–2070) percent changes from the simulated historical median values from 1971–2000 for April 1 snow water equivalent (SWE, top), April–July runoff (middle) and June 1 soil moisture content (bottom), as obtained from median of sixteen VIC simulations under the high-emissions (A2) scenario. Source: Bias Corrected and Downscaled World Climate Research Programme's CMIP3 Climate Projections archive at http://gdo-dcp.ucllnl.org/downscaled_cmip3_projections/#Projections:%20Complete%20Archives. [Chapter 6]

Water

Water is the limiting resource in the Southwest, and climate variability and change will continue to have substantial effects on water across much of the region. Reduction in water supplies can lead to undesirable changes in almost all human and natural systems including agriculture, energy, industry, forestry, and recreation. In particular:

- ***Climate change could further limit water availability in much of the Southwest.*** A large portion of the Southwest, including most of the region's major river systems (e.g., Rio Grande, Colorado, and San Joaquin), is expected to experience reductions in streamflows and other limitations on water availability in the twenty-first century (medium-high confidence) (Figure 1.7). [Chapters 5, 6, 7, and 10]

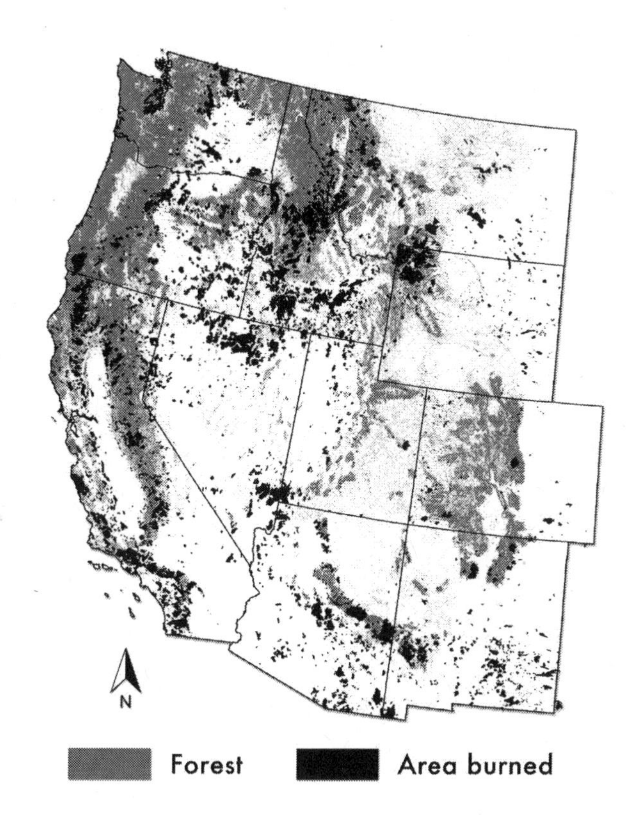

Figure 1.8 Areas of the western United States burned in large (> 1000 acres [400 ha]) fires, 1984–2011. Dark shading shows fires in areas classified as forest or woodland at 98-feet (30-meter) resolution by the LANDFIRE project (http://www.landfire.gov/). Fire data from 1984–2007 are from the Monitoring Trends in Burn Severity project (http://www.mtbs.gov/) and fire data from 2008–2011 are from the Geospatial Multi-Agency Coordination Group (http://www.geomac.gov/). [Chapter 8]

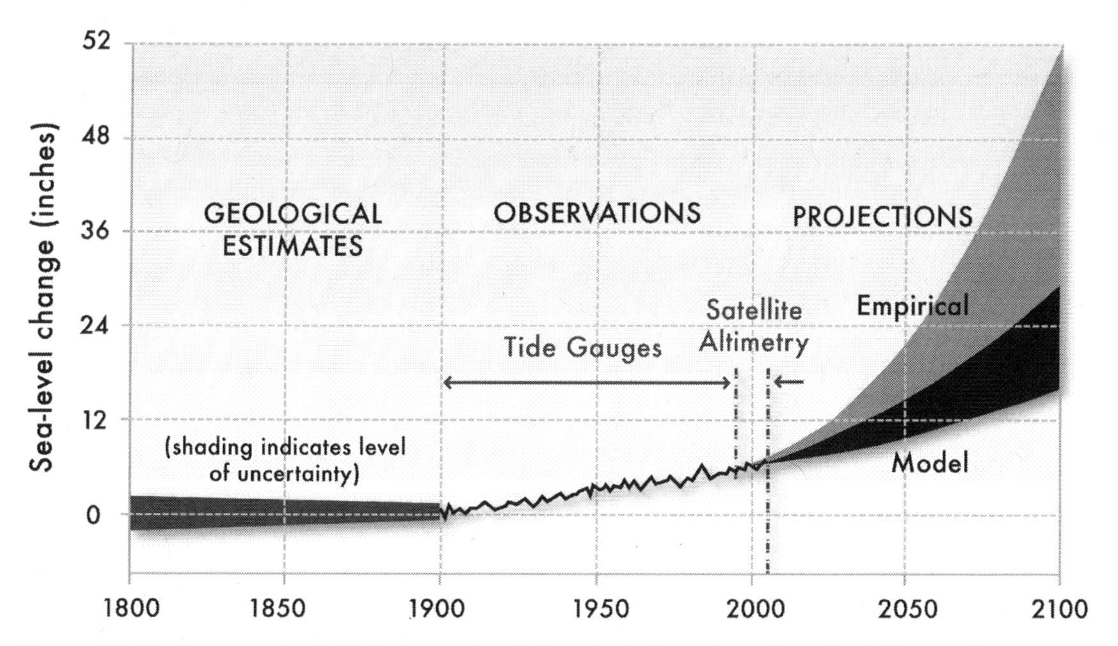

Figure 1.9 Past, present, and future sea-level rise. Geologic and recent sea-level histories (from tide gauges and satellite altimetry) are combined with projections to 2100 based on climate models and empirical data. Modified with permission from Russell and Griggs (2012), Figure 2.1. [Chapter 9]

- ***Water availability could be decreased even more by unusually warm, decades-long periods of drought.*** Much of the Southwest, including major river systems such as the Colorado and Rio Grande, has experienced decades-long drought repeatedly over the last 1,000 to 2,000 years. Similar exceptional droughts could occur in the future, but temperatures are expected to be substantially hotter than in the past (high confidence) (Figure 1.3). [Chapters 5, 6, 7, and 10]
- ***The past will no longer provide an adequate guide to project the future.*** Twentieth-century water management has traditionally been based in part on the principle of "stationarity," which assumes that future climate variations are similar to past variations. As climate changes, temperature will increase substantially and some areas of the Southwest will become more arid than in the past (high confidence). [Chapters 6 and 10]
- ***Surface water quality will be affected by climate change.*** In some areas, surface water quality will be affected by scarcity of water, higher rates of evaporation, higher runoff due to increased precipitation intensity, flooding, and wildfire (high confidence). [Chapter 10]

Human health

The Southwest's highly complex and often extreme geography and climate increase the probability that climate change will affect public health. Several potential drivers of increased health risk exist only or primarily in the Southwest, and there is substantial variation in the sensitivity, exposure, and adaptive capacity of individuals and groups of people within the Southwest to climate change-related increases in health risks:

- ***Climate change will drive a wide range of changes in illness and mortality.*** In particular, climate change will exacerbate heat-related human morbidity and mortality, and lead to increased concentrations of airborne particulates and pollutants from wildfires and dust storms. Climate change may affect the extent to which organisms such as mosquitoes and rodents can carry pathogens (e.g., bacteria and viruses) and transmit disease from one host to another (medium-high confidence). [Chapter 15]
- ***Allergies and asthma will increase in some areas.*** On the basis of data showing earlier and longer spring flower bloom, allergies and asthma may worsen for individual sufferers or become more widespread through the human population as temperature increases (medium-low confidence). [Chapter 15]
- ***Disadvantaged populations will probably suffer most.*** The health of individuals who are elderly, infirm, or economically disadvantaged is expected to decrease disproportionately to that of the general population (high confidence), due to their increased exposure to extreme heat and other climate hazards. [Chapter 15]

Additional effects of climate change

Climate change has the potential to affect many other sectors and populations within the Southwest. For example:

- ***Agriculture will be affected by climate change.*** Effects of climate change and associated variability on production of both crops and livestock could be long-lasting, with short-term reductions in profitability (medium-low confidence). [Chapter 11]

- ***Energy supplies will become less reliable as climate changes and climate change will drive increasing energy demand in some areas.*** Delivery of electricity may become more vulnerable to disruption due to extreme heat and drought events that increase demand for home and commercial cooling, reduce thermal power plant efficiency or ability to operate, reduce hydropower production, or reduce or disrupt transmission of energy (medium-high confidence) (Figure 1.10). [Chapter 12]
- ***Climate change will affect urban areas in differing ways depending on their locations and on their response or adaptive capacities.*** Climate change will affect cities in the Southwest in different ways depending on their geographic locations. Local capacity to address effects of climate change will also vary depending on governmental, institutional, and fiscal factors. Incidences of air pollution related to increased heat are likely to increase, and water supplies will become less reliable (medium-high confidence). [Chapter 13]
- ***Reliability of transportation systems will decrease.*** Climate change will affect transportation systems in different ways depending on their geographic location (e.g., changing sea level and storm surge affect coastal roads and airports), potentially impeding the movement of passengers and goods (medium-high confidence). [Chapter 14]
- ***Climate change may disproportionately affect human populations along the U.S.-Mexico border.*** Climate changes will stress on already severely limited water systems, reducing the reliability of energy infrastructure, agricultural production, food security, and ability to maintain traditional ways of life in the border region (medium-high confidence). [Chapter 16]
- ***Native American lands, people, and culture are likely to be disproportionately affected by climate change.*** Effects of climate change on the lands and people of Southwestern Native nations are likely to be greater than elsewhere because of endangered cultural practices, limited water rights, and social, economic, and political marginalization, all of which are relatively common among indigenous people (high confidence). [Chapter 17]

1.5 Choices for Adjusting to Climate and Climate Change

A century of economic and population growth in the Southwest has already placed pressures on water resources, energy supplies, and ecosystems. Yet the Southwest also has a long legacy of human adaptation to climate variability that has enabled society to live within environmental constraints and to support multiple-use management and conservation across large parts of the region. Governments, for-profit and non-profit organizations, and individuals in the Southwest have already taken a variety of steps to respond to climate change. A wide range of options are available for entities and individuals choosing to reduce greenhouse gas emissions or to prepare and adapt to climate

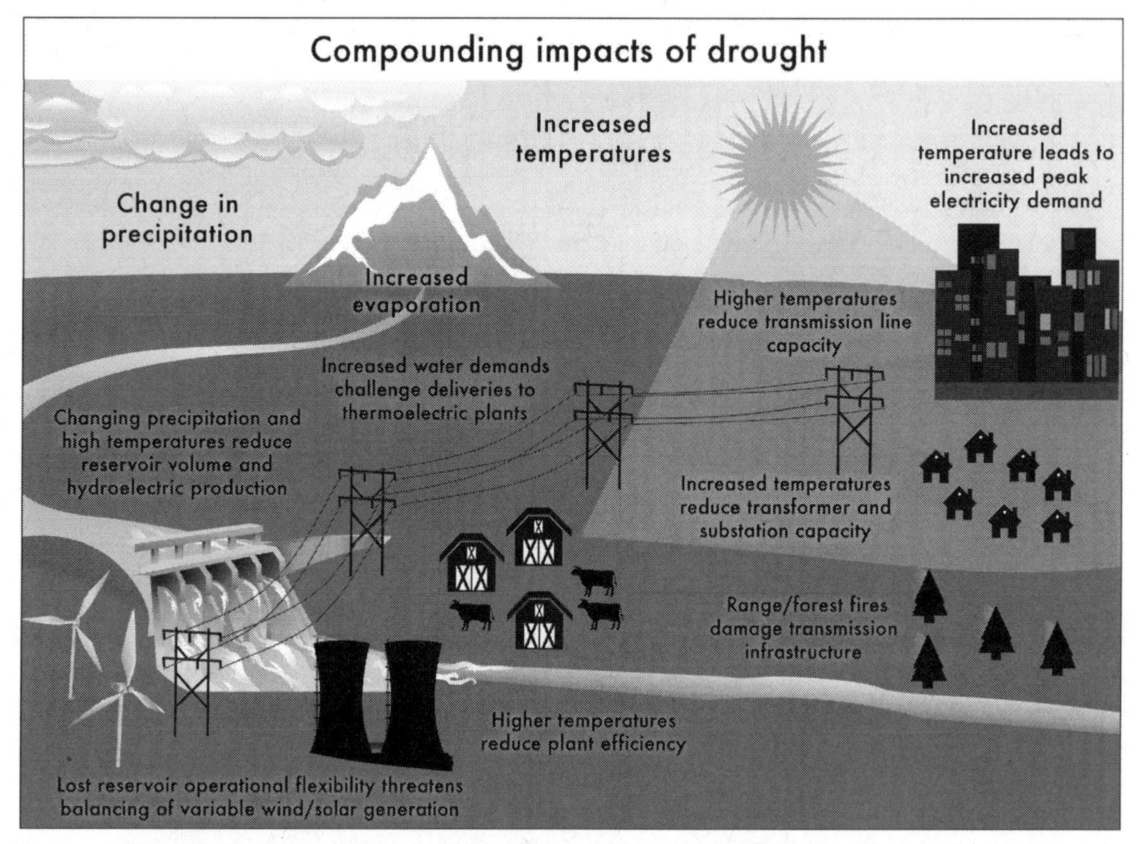

Figure 1.10 Compounding impacts of drought on energy. [Chapter 12]

variability and change (Table 1.2). Others who have not yet begun to respond to climate change directly are choosing to reduce energy and water use for immediate economic benefit or as ways of enhancing the sustainability of water supply, energy, and food production [Chapter 18]. Many options for responding to climate change in the Southwest have been, or are being, investigated, and are assessed in the full report, *Assessment of Climate Change in the Southwest United States.* Notable examples include:

- ***Reducing greenhouse gas emissions.*** Governments, for-profit and non-profit organizations, and individuals are already taking many steps to reduce the causes of climate change in the Southwest, and there are lessons to learn from the successes and failures of these early efforts, such as the first U.S. implementation of cap-and-trade legislation in California. There have been few systematic studies, however, that evaluate the effectiveness of the choices made in the Southwest to reduce greenhouse gas emissions (medium-low confidence). California has established targets and the National Research Council has recommended targets for reduction in emissions of greenhouse gases. Meeting these targets will be challenging. However, there are many low-cost or revenue-generating opportunities for emissions reductions in the Southwest, especially those related to energy efficiency and to the development of renewable sources of energy (medium-high confidence). [Chapter 18]

Table 1.2 Adaptation options relevant for the Southwest [Chapter 18]

Sector	Adaptation Strategies
Agriculture	Improved seeds and stock for new and varying climates (and pests, diseases), increase water use efficiency, no-till agriculture for carbon and water conservation, flood management, improved pest and weed management, create cooler livestock environments, adjust stocking densities, insurance, diversify or change production.
Coasts	Plan for sea level rise—infrastructure, planned retreat, natural buffers, land use control. Build resilience to coastal storms—building standards, evacuation plans. Conserve and manage for alterations in coastal ecosystems and fisheries.
Conservation	Information and research to identify risks and vulnerabilities, secure water rights, protect migration corridors and buffer zones, facilitate natural adaptations, manage relocation of species, reduce other stresses (e.g., invasives)
Energy	Increase energy supplies (especially for cooling) through new supplies and efficiency. Use sustainable urban design, including buildings for warmer and variable climate. Reduce water use. Climate-proof or relocate infrastructure.
Fire management	Use improved climate information in planning. Manage urban-wild land interface.
Forestry	Plan for shifts in varieties, altered fire regimes, protection of watersheds and species.
Health and emergencies	Include climate in monitoring and warning systems for air pollution, allergies, heat waves, disease vectors, fires. Improve disaster management. Cooling, insulation for human comfort. Manage landscape to reduce disease vectors (e.g. mosquitos), Public health education and training of professionals.
Transport	Adjust or relocate infrastructure (coastal and flood protection, urban runoff), plan for higher temperatures and extremes.
Urban	Urban redesign and retrofit for shade, energy, and water savings. Adjust infrastructure for extreme events, sea-level rise.
Water management	Enhance supplies through storage, transfers, watershed protection, efficiencies and reuse, incentives or regulation to reduce demand and protect quality, reform or trade water allocations, drought plans, floodplain management. Use climate information and maintain monitoring networks, desalinate. Manage flexibly for new climates not stationarity.

Source: Smith, Horrocks et al. (2011); Smith, Vogel et al. (2011).

- ***Planning and implementing adaptation programs.*** There is a wide range of options in most sectors for adapting to climate variability and extreme events, including many that have ecological, economic, or social benefits (medium-high confidence). [Chapter 18]

- ***Lowering or removing barriers to optimize capacity for adaptation.*** A number of relatively low-cost and easily implemented options for adapting to climate variability and change are available in the Southwest, including some "no-regrets" options with immediate benefits that could foster economic growth. Lowering or removing financial, institutional, informational, and attitudinal barriers will increase society's ability to prepare for and respond to climate change (medium-high confidence). [Chapter 18]
- ***Connecting adaption and mitigation efforts.*** Many options exist to implement both adaptation and mitigation, i.e. options that reduce some of the causes of climate change while also increasing the readiness and resilience of different sectors to reduce the impacts of climate change (high confidence). The significant probability of severe and sustained drought in the drought-prone Southwest makes some adaptation options applicable even in the absence of significant climate change (high confidence). [Chapters 5, 7, 10 and 18]
- ***Planning in coastal areas.*** Coastal communities are increasingly interested in and have begun planning for adaptation. There are opportunities to increase use of policy and management tools and to implement adaptive policies (high confidence). [Chapter 9]
- ***Changing water management.*** Considerable resources are now being allocated by the water-management sector to understand how to adapt to a changing water cycle. A full range of options involving both supplies and demands are being examined. Large utilities have been more active in assessing such options than relatively small utilities (high confidence). [Chapter 10]
- ***The large amounts of water currently used for irrigated agriculture can buffer urban supplies.*** Assuming water allocations to agriculture remain substantial, short-term agricultural-urban water transfers can greatly reduce the total cost of water shortages and limit effects on urban water users during climate- or weather-induced water shortages (medium-high confidence). [Chapter 11]
- ***Changing energy policy.*** A shift from the traditional fossil fuel economy to one rich in renewable energy will have substantial effects on water use, land use, air quality, national security, and the economy. The reliability of the energy supply in the Southwest as climate changes depends on how the energy system evolves over this century (medium-high confidence). [Chapter 12]
- ***Adaptation and mitigation on federal and tribal land.*** The Southwest has the highest proportion of federal and tribal land in the nation (Figure 1.11). Native nations are taking action to address climate change by actively seeking additional resources for adaptation, and by initiating climate-change mitigation (medium-low confidence). Federal land and resource management agencies are beginning to plan with the assumption that climate is changing, although efforts are not consistent across agencies (high confidence). [Chapters 17 and 18]

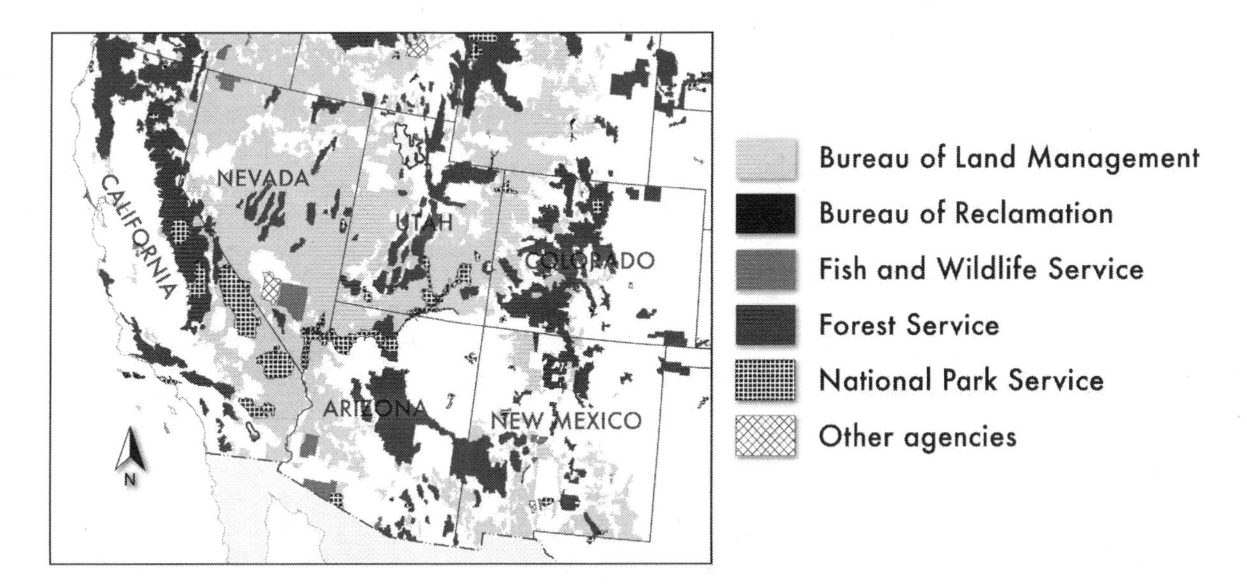

Figure 1.11 Extensive federal lands in the Southwest: A legacy for the future. This map illustrates the legacy of federal land ownership in the Southwest, covering nearly 30% of the entire United States. Protected habitat and ecosystem services ensure sustainable management of resources and may be the greatest insurance policy against losses in the future, because natural resource use and biological species can more easily adapt to rapidly changing climatic conditions. Modified from *The National Atlas of the United States of America* (http://www.nationalatlas.gov; see also http://nationalatlas.gov/printable/images/pdf/fedlands/fedlands3.pdf; accessed October 8, 2012). [Chapter 18]

1.6 Key Unknowns

Although there has been a substantial increase in the understanding of how Southwest climate is changing and will change and how this change will affect the human and natural systems of the region, much remains to be learned. The full report, *Assessment of Climate Change in the Southwest United States*, identifies many key unknowns, and assesses the data, monitoring, modeling, and other types of research needed to increase knowledge [Chapters 19 and 20]. Yet, current knowledge and experience is sufficient to support climate change adaptation and mitigation actions, such as reducing greenhouse gas emissions or adapting to the changes that cannot be avoided, minimized, or mitigated. Many of these potential actions represent "no-regrets" options that are already either cost-effective in the immediate or short-term. [Chapter 18]

References

Cook, E. R., C. Woodhouse, C. M. Eakin, D. M. Meko, and D. W. Stahle. 2004. Long-term aridity changes in the western United States. *Science* 306:1015–1018.

Mearns, L. O., W. Gutowski, R. Jones, R. Leung, S. McGinnis, A. Nunes, and Y. Qian. 2009. A regional climate change assessment program for North America. *Eos Transactions AGU* *90*:311.

Menne, M. J., and C. N. Williams, Jr. 2009. Homogenization of temperature series via pairwise comparisons. *Journal of Climate* 22:1700–1717, doi:10.1175/2008JCLI2263.1.

Nakićenović, N., and R. Swart, eds. 2000. *Special report on emissions scenarios: A special report of Working Group III of the Intergovernmental Panel on Climate Change.* Cambridge: Cambridge University Press.

Russell, N. L., and G. B. Griggs. 2012. *Adapting to sea-level rise: A guide for California's coastal communities.* Sacramento: California Energy Commission, Public Interest Environmental Research Program.

Smith, M. S., L. Horrocks, A. Harvey, and C. Hamilton. 2011. Rethinking adaptation for a 4°C world. Philosophical Transactions of the Royal Society A 369:196–216.

Smith, J., Vogel, J., Carney, K. and C. Donovan. 2011. Adaptation case studies in the western United States: Intersection of federal and state authority for conserving the greater sage grouse and the Colorado River water supply. Washington, DC: Georgetown Climate Center.

Stewart, I., D. Cayan, and M. Dettinger. 2005. Changes towards earlier streamflow timing across western North America. *Journal of Climate* 18:1136–1155.

Endnotes

i Much of the text in this summary is taken directly, or with minor modification, from the full report, *Assessment of Climate Change in the Southwest United States*, and where this is the case, chapter citations appear in brackets at the end of each paragraph or bullet.

ii Confidence estimates cited in this document (high, medium-high, medium-low, or low) are explained in more detail in the main report. Confidence was assessed by authors of the main report on the basis of the quality of the evidence and the level of agreement among experts with relevant knowledge and experience. [Chapters 2 and 19]

Chapter 2

Overview

COORDINATING LEAD AUTHOR

Gregg Garfin (University of Arizona)

LEAD AUTHOR

Angela Jardine (University of Arizona)

EXPERT REVIEW EDITOR

David L. Feldman (University of California, Irvine)

2.1 Introduction

The first comprehensive analysis of the implications of climate variability and change[i] stated that, "the influence of climate permeates life throughout the United States" (Sprigg and Hinkley 2000, 2).

Since the report was issued, the scientific evidence, the concerns of decision makers, and demonstrated temperature trends and multi-year and decadal variability show that *climate change* also permeates life throughout the Southwestern United States. Since 2000, the region has experienced episodes of severe and sustained drought, declines in water supplies, notable floods, the widespread die-off of conifer trees, increasing temperatures, and severe wildland fires of record extent. These occurrences are related in part to climate change. They also are related to the ways in which climate interacts with other drivers (or forces) of change across the region, such as population growth, economic development, urban expansion, food production, and the extraction and consumption of natural resources, including water, timber, minerals, and energy fuels. Therefore, regular assessment of the state of climate knowledge—and of the climate-related vulnerabilities and risks to citizens and the economy—is vital to clearly define choices available to those who make decisions about the quality of human life

Chapter citation: Garfin, G. and A. Jardine. 2013. "Overview." In *Assessment of Climate Change in the Southwest United States: A Report Prepared for the National Climate Assessment*, edited by G. Garfin, A. Jardine, R. Merideth, M. Black, and S. LeRoy, 21–36. A report by the Southwest Climate Alliance. Washington, DC: Island Press.

and livelihoods, the well-being of communities, or the management of resources and landscapes across the Southwest.

The *Assessment of Climate Change in the Southwest United States* is a summary and synthesis of the past, present, and projected future of the region's climate, examining what this means for the health and well-being of human populations and the environment throughout the six Southwestern states—Arizona, California, Colorado, Nevada, New Mexico, and Utah—an area of about 700,000 square miles that includes vast stretches of coastline, an international border, and the jurisdictions of 182 federally recognized Native American tribes.

The report looks at climate and its effects on scales ranging from states to watersheds and across ecosystems and regions; at links between climate and resource supply and demand; at effects on sectors—such as water, agriculture, energy, and transportation—that are critical to the well-being of the region's inhabitants; the vulnerabilities to climate changes of all facets of the region, and the responses, or adaptations, that society may choose to make.

What is an assessment?[ii]

This report is an *assessment* of climate change for the Southwest region of the United States and as such is *not* a research project, review paper, or advocacy piece. We define scientific assessment as a critical evaluation of information for purposes of guiding decisions on a complex issue: climate change and its interactions with other aspects of natural systems and society. Stakeholders, who are typically decision makers, have been actively engaged in defining the scope of this report and in reviewing the document. This assessment is intended to be relevant to public policy and resource management, but our findings, judgments, and recommendations are not prescriptive; we do not present findings as "must-do's," but as options. We have summarized complexity by synthesizing and sorting what is known and widely accepted from what is not known (or not agreed upon). Written chiefly during late 2011, with revisions through mid-2012, this assessment provides a snapshot of the current state of climate change information and knowledge related to the region.

We have synthesized, through evaluation and judgment, information from a range of sources, including data sets of observations, simulations and projections from computer modeling, peer-reviewed scientific papers, case studies, and other sources. This assessment represents the consensus findings of nearly 200 authors and reviewers. In this assessment, experts and decision makers representing a variety of disciplines have discussed and made judgments about the importance and quality of information and about ways to characterize uncertainty and confidence.

Data evaluated in this assessment were collected previously (in some cases by the authors of this report) and are publicly available. Some new understanding results from synthesis. Part of our charge was to identify important gaps in knowledge about climate change and the type of research that would reduce or better define areas of uncertainty. This report focuses on the implications of the science results for management and policy and so is not limited to previously published ideas. Thus, we have clearly labeled and consistently judged the importance of information and our level of confidence in its accuracy or validity. This report is evidence-based as verified by multiple reviews.

2.2 Context and Scope

The U.S. National Climate Assessment (NCA; http://www.globalchange.gov/what-we-do/assessment; see Box 2.1) for 2013, a national report on climate change and impacts, provided the motivation to produce this regional report.

Previously, the first National Climate Assessment (National Assessment Synthesis Team 2000) received technical input from multiple geographic regions in the United States. That assessment's sixty-page Southwest region report (Sprigg and Hinkley 2000) examined the effects of climate variability and change (including projections of the future) on water resources, ranching, natural ecosystems, extractive industries (oil, gas, mining), human health, urban areas, energy, and planning for the future. The 2000 report emphasized observed climate trends and phenomena and identified potential vulnerabilities related to climate, yet gave relatively little attention to adaptation planning and risk management.

The second National Climate Assessment in 2009 (Karl, Melillo, and Peterson 2009) was summarized from twenty-one synthesis and assessment products produced by the U.S. Climate Change Science Program (CCSP). The CCSP did not solicit technical input from regions, instead focusing on key sectors (e.g., transportation, agriculture, and water resources) and problems (e.g., strengths and limitations of climate models, temperature trends, model reliability, and adaptation options for ecosystems). The five-page Southwest section of the 2009 National Climate Assessment gave increased attention to projected climate changes, impacts to vulnerable water and ecosystem resources and, to a lesser degree, agriculture and urban areas. For these topics, the second assessment built a strong foundation for this report.

The present *Assessment of Climate Change in the Southwest United States*, part of the third National Climate Assessment, emphasizes new information and understandings since publication of the 2009 National Climate Assessment and expands the scope of previous regional assessments by analyzing the effects of climate change on Native American lands and the U.S.-Mexico border area, by presenting key uncertainties associated with each topic discussed in the report, and by providing a compendium of research needed to address these uncertainties. With its regional perspective, this report also provides the basis for similar assessments to be made at state, watershed, municipal, tribal, or other local levels for decision making at finer scales.

The report uses the established Intergovernmental Panel on Climate Change (IPCC) greenhouse gas (GHG) emissions scenarios, A2 (high) and B1 (low).[iii] These scenarios were used as inputs into global climate models to project climate changes in the IPCC Fourth Assessment Report and are fully described in the Special Report on Emissions Scenarios (Nakićenović and Swart 2000). Increases in the accumulation of GHGs in the atmosphere are thought to be the main cause of twenty-first century climate change stemming from human economic development choices. While GHGs are not the only influence on climate change considered by the IPCC, estimating the amount of GHGs in the future atmosphere is probably the largest uncertainty in projecting future climate. The estimation depends on predicting such factors as the state of the future global economy, global population growth, public policies and regulations, and the rate of adoption of technologies that reduce GHG emissions. While it is unrealistic to expect to know with certainty the future variations in these factors, scientists are able to use plausible scenarios to project likely ranges of future GHG emissions. Other published scenarios and approaches are also incorporated in this report.

The report is guided, in part, by issues identified by stakeholders[iv] within the region, solicited through a workshop convened in June 2011, three teleconferences conducted during the second half of 2011, and review of reports from other climate change workshops and needs assessments. Early in the process, regional stakeholders mentioned that they would have little incentive to read a long report. Thus, we have limited the length of the report and have provided brief summaries online (http://www.swcarr.arizona.edu), which stakeholders suggested would be useful.

Box 2.1

National Climate Assessment

The National Climate Assessment (NCA) is being conducted under the auspices of the Global Change Research Act of 1990 (GCRA). The GCRA requires a report to the President and Congress every four years that analyzes the effects of global change on the natural environment, agriculture, energy production and use, land and water resources, transportation, human health and welfare, human social systems, and biological diversity. The report examines current trends in global change (both human-induced and natural) and projects major trends for the next 25 to 100 years.

National climate assessments serve as status reports on climate change and its impacts. The assessments rely on observations made across the country and compare these observations to projections from climate-system models. As with previous assessments, the third NCA (2013) evaluates the current state of scientific knowledge relative to climate impacts and trends. But it additionally evaluates the effectiveness of U.S. activities to mitigate and adapt to climate change and identify economic opportunities and challenges that arise as the climate changes.

The objectives of the NCA are to provide information and reports in the context of a continuing, inclusive national process that will:

- synthesize relevant science and information;
- increase understanding of what is known and not known;
- identify needs for information related to preparing for climate variability and change and reducing climate impacts and vulnerability;
- evaluate progress of adaptation and mitigation activities;
- inform science priorities;
- build assessment capacity in regions and sectors;
- build societal understanding and skilled use of assessment findings; and
- recognize the global and international context of climate trends and connections between climate risk and impacts in the United States and elsewhere.

The 2013 NCA differs from previous climate assessments in that it: (1) is a continuing effort rather than a periodic report-writing activity; (2) fosters partnerships with non-governmental entities; and (3) provides web-based data and information. For a list of the U.S. assessments, see http://globalchange.gov/publications/reports.

2.3 Other Southwest Region Climate Assessments

Many other climate-change reports and assessments have been produced by federal and state agencies, non-governmental organizations, and municipalities. These documents (some of which are listed in Table 2.1) relate in whole or in part to the Southwest region. For instance, the U.S. Forest Service assessed the state of knowledge about climate-change trends and associated impacts on U.S. forests (Joyce and Birdsey 2000). The report focused on plant productivity in response to elevated atmospheric carbon dioxide levels, and the authors turned to models to explore potential changes to ecosystem succession and forest productivity. The Southwest is included, implicitly, in maps and text on changes to forest ecosystems. In a more recent federal effort, the U.S. Bureau of Reclamation (2011) examined climate variability and trends and used projections of future climate and hydrology to assess risks to water resources in the Western United States. The assessment reported on selected river basins in the Southwest: Sacramento-San Joaquin, Truckee-Carson, Colorado, and Upper Rio Grande.

Table 2.1 Selected climate change assessments and reports pertaining to the Southwest region

Year	Institution	Report Name
2000	USDA-Forest Service	The Impact of Climate Change on America's Forests: A Technical Document Supporting the 2000 USDA Forest Service RPA Assessment (Joyce and Birdsey 2000) http://www.fs.fed.us/rm/pubs/rmrs_gtr059.pdf
2006	The Nature Conservancy	Ecoregion-Based Conservation Assessments of the Southwestern United States and Northwestern Mexico (Marshall, List, and Enquist 2006) http://azconservation.org/dl/TNCAZ_Ecoregions_SW_Ecoregional_Summary.pdf
2007	National Academy of Sciences	Colorado River Basin Water Management: Evaluating and Adjusting to Hydroclimatic Variability (NRC 2007) http://www.nap.edu/catalog.php?record_id=11857
2007	City of Denver	City of Denver Climate Action Plan (Mayor's Greenprint Denver Advisory Council 2007) http://www.greenprintdenver.org/about/climate-action-plan-reports/
2008	Colorado Water Conservation Board	Climate Change in Colorado: A Synthesis to Support Water Resource Management and Adaptation (Ray et al. 2008) http://cwcb.state.co.us/public-information/publications/Documents/ReportsStudies/ClimateChangeReportFull.pdf

Table 2.1 Selected climate change assessments and reports pertaining to the Southwest region (Continued)

Year	Institution	Report Name
2009	National Audubon Society	Birds and Climate Change: Ecological Disruption in Motion (National Audubon Society 2009) http://birds.audubon.org/sites/default/files/documents/birds_and_climate_report.pdf
2010	EPA	Climate Change Indicators in the United States (EPA 2010) http://www.epa.gov/climatechange/indicators/pdfs/ClimateIndicators_full.pdf
2010	State of California	2010 Climate Action Team Report to Governor Schwarzenegger and the California Legislature (California Climate Action Team 2010) http://www.energy.ca.gov/2010publications/CAT-1000-2010-005/CAT-1000-2010-005.PDF
2011	NOAA	State of the Climate in 2010 (Blunden, Arndt, and Beringer 2011) http://www.ncdc.noaa.gov/bams-state-of-the-climate/2010.php
2011	Bureau of Reclamation	SECURE Water Act Section 9503(c) - Reclamation Climate Change and Water 2011 (Reclamation 2011) http://www.usbr.gov/climate/SECURE/docs/SECUREWaterReport.pdf
2011	National Wildlife Federation	Scanning the Conservation Horizon: A Guide to Climate Change Vulnerability Assessment (Glick, Stein, and Edelson 2011) http://www.nwf.org/~/media/PDFs/Global-Warming/Climate-Smart-Conservation/NWFScanningtheConservationHorizonFINAL92311.ashx

States and cities have also produced climate change assessments for parts of the Southwest. A landmark executive order in California triggered a series of assessments and Climate Action Team reports to the governor (http://www.climatechange.ca.gov/climate_action_team/reports/#2010), beginning in 2006. This extensive series of reports formed the basis for numerous implementation plans. Colorado's Water Conservation Board commissioned a study (Ray et al. 2008) to determine the state of knowledge about Colorado's climate and the implications of projected future variations on the state's water resources. Several Colorado cities and municipalities inventoried GHG emissions and existing programs for emissions reduction as a foundation for climate-change planning (e.g., Mayor's Greenprint Denver Advisory Council 2007). Such assessments provide valuable local data and assessment at levels of analysis that regional and national reports cannot encompass.

Finally, non-governmental organizations have produced assessment reports for the region. Many of these assess a combination of peer-reviewed materials, new and existing data, and internal reports. For example, The Nature Conservancy aggregated and standardized data across multiple ecoregions (large areas of land and water that are characterized by plant and animal communities and other environmental factors) and assessed the status and condition of native species, ecological systems, and natural resources such as water (Marshall, List, and Enquist 2006).

2.4 Sponsors and Authors of this Report

In July 2011, the Southwest Climate Alliance (SWCA)[v] submitted an expression of interest (EOI)[vi] to produce this regional technical input report for the NCA.[vii] The SWCA institutions and their partners have individually contributed to previous national assessments and to state-level assessments for California and Colorado. They also recently convened a Colorado River Basin workshop to assess regional capacity to perform ongoing assessments.

The SWCA team obtained funding from the National Oceanic and Atmospheric Administration (NOAA) and the Department of the Interior to convene a workshop for potential regional assessment authors and hire temporary staff to coordinate production of the report. Experts in the report subject areas were recruited to serve as assessment chapter lead authors. Report authors are primarily from university and federal research labs, with some contributors from state agencies, non-governmental organizations, and the private sector (see Table 2.2 and Figure 2.1). They have donated their time to write this report.

Table 2.2 Author affiliations for *Assessment of Climate Change in the Southwest United States*

Sector	Total Number	Number of Unique Institutions
Federal	23	13
State	5	5
University	86**	25
NGO	3	2
Private	3	3
Tribal*	1	1
TOTAL	**121**	**49**

* Authors with only tribal affiliation. Some federal and university authors also have tribal affiliations.

** Some authors with university affiliations also have affiliations with federal agencies.

Participating Institutions

1. Lawrence Berkeley National Laboratory; The Nature Conservancy; NOAA Coastal Services Center; University of California, Berkeley
2. NASA Ames Research Center; Stanford University; Susanne Moser Research & Consulting; U.S. Geological Survey
3. California Coastal Commission; University of California, Santa Cruz
4. Lawrence Livermore National Laboratory
5. California Air Resources Board; California Dept. of Water Resources; California Energy Commission; California Office of Env. Health Hazard Assessment; University of California, Davis
6. University of California, Merced
7. NASA Jet Propulsion Laboratory; University of California, Los Angeles; University of Southern California
8. Nossaman, Inc.
9. San Diego State University; Scripps Institution of Oceanography; U.S. Geological Survey
10. Colegio de la Frontera Norte
11. Centro de Investigación Científica y de Educación Superior de Ensenada
12. Desert Research Institute
13. Arizona State University
14. Northern Arizona University; U.S. Geological Survey
15. National Phenology Network; U.S. Geological Survey; University of Arizona
16. University of Utah
17. Utah State University
18. University of Wyoming
19. Colorado State University; National Park Service; U.S. Geological Survey
20. National Center for Atmospheric Research; NOAA Earth System Research Laboratory; University of Colorado
21. Bureau of Land Management; Bureau of Reclamation; Colorado Governor's Energy Office; Denver Water
22. Zuni Tribe Water Resources Program
23. Sandia National Laboratories
24. University of Washington
25. Bureau of Land Management
26. University of Illinois, Chicago
27. NOAA Cooperative Institute for Climate and Satellites
28. Woods Hole Research Center
29. Universidad Nacional Autónoma de México

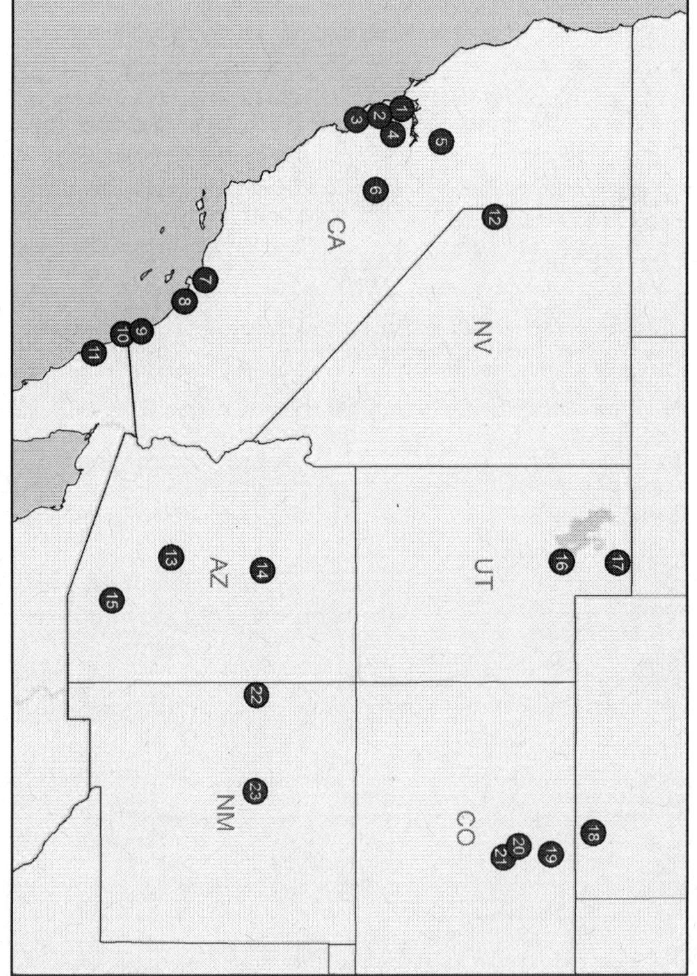

Figure 2.1 Locations of authors and their institutions contributing to this report. A total of 121 authors volunteered their time to writing this report. Map by Christine Albano.

2.5 Characterizing and Communicating Uncertainty

While climate changes have effects on human populations, human activities likewise affect the atmosphere and climate. As mentioned earlier, predicting the effects of future climate changes on the environment and society will always require estimating a range of social realities, such as population growth, economic development, new technology development, and enactment of new laws and regulations. These and other factors ultimately impact GHG concentrations in the atmosphere, for example, which in turn affect climate.

We also are limited by our present incomplete understanding of some biophysical processes that feed into global climate models to project an outcome. Consequently, these processes must be expressed mathematically in computer models and statistically in terms of *ranges*, with commentary on the confidence of the estimates. We refer to evaluation of the ranges of estimates of possible future climate and impacts—accounting for the scenarios used to drive the climate models, the information used to construct the models, and the interpretation and use of the models' data for planning and decision making—as characterization of *uncertainty*. Scientific research and assessments can provide information and characterize uncertainty in a way that facilitates choices that are *risk-based* (see Box 2.2).

In this report, we have adopted guidance from the National Climate Assessment (http://www.globalchange.gov/images/NCA/Draft-Uncertainty-Guidance_2011-11-9.pdf) to characterize and communicate uncertainty[viii]. We have attempted to frame questions or problems to allow appraisal of the level of knowledge or understanding in the context of the question or problem. We review the range of scientific information for each question and describe the information used, the standards of evidence applied, and the confidence of the authors in their results. In reporting key findings, we have followed these steps to communicate our level of confidence in key conclusions:

1. We framed a manageable number (three or four) of key questions or issues that address the most important information needs of stakeholders.
2. We evaluated the available information, considering the type, amount, quality, and consistency of evidence, summarizing the level of evidence as strong, fair, or weak.
3. We formulated well-posed conclusions that can be confirmed or falsified.
4. We identified key uncertainties and briefly describe what monitoring, research, or other work is needed to improve the information base.
5. We assessed the levels of confidence (high, medium-high, medium-low, or low) by considering (a) the quality of the evidence and (b) the level of agreement among experts with relevant knowledge and experience. Confidence is a subjective judgment, but it is based on systematic evaluation of the type, amount, quality, and consistency of evidence, and the degree of agreement among experts (Table 2.3).
6. Especially for findings that identify potential high-consequence outcomes, we estimated uncertainty probabilistically (i.e., provided a likelihood that the outcome could occur under a stipulated scenario or conditions). Likelihoods can be based on quantitative methods, such as model results or statistical sampling, or on expert judgment. Some authors may use standardized ranges (<5% likely, <33% likely, 33–66% likely, >66% likely, or >95% likely).

7. To ensure transparency in reporting uncertainty and confidence, we prepared brief traceable accounts that describe the main factors that contributed to a particular conclusion and level of confidence.[ix]

Box 2.2

Risk-based Framing[x]

The National Climate Assessment and the *Assessment of Climate Change in the Southwest United States* use a risk-based management approach to describe statements about key vulnerabilities to climate change and how they may change over time. A key vulnerability has:

- a large magnitude;
- early onset of impacts;
- a high degree of persistence and irreversibility;
- a wide distribution (e.g., across levels of society, or spatially);
- a high likelihood of occurrence; or
- great importance (based on perceptions).

The motivation for using this risk-based approach is based on research (IPCC 2007; NRC 2010a, 2010b) and interaction with climate information user communities, such as the Department of Defense.

Risk is defined as the product of likelihood of occurrence of an event or condition and the consequence of that occurrence, where:

- consequence ("importance") can be assessed using metrics ranging from physical impacts to vulnerability;
- vulnerability depends on exposure to a climate phenomenon or stimulus, sensitivity (the degree to which a vulnerable system responds to the climate phenomenon or stimulus), and adaptive capacity, or ability to adapt, respond, or rebound to the climate phenomenon or stimulus;
- likelihood depends on sensitivity to the climate phenomenon or stimulus, and the associated climate variability.

Risk management can be based on either quantitative or qualitative representations of likelihood and consequence. For qualitative information, the report authors use rigorous methods to describe likelihood and consequence. The authors have submitted traceable accounts of the sources used and the rationale behind the quantitative and qualitative judgments regarding risk. Qualitative techniques are useful in circumstances in which there may be a range of future likelihoods or consequences. This report focuses attention on highly likely impacts and vulnerabilities, but also on lower likelihood impacts and vulnerabilities that carry high consequences. The latter is in recognition of stakeholder concerns about climate extremes and rare events that may have significant impacts on infrastructure and investments.

2.6 Accountability and Review

To ensure transparency in developing this regional report's conclusions and key findings, we also have cited all sources of information, as is common peer-review practice, and sources of data for all graphics and tables.

For the key findings in each chapter's Executive Summary, the respective authors have submitted traceable accounts, as suggested in guidance from the National

Climate Assessment (http://www.globalchange.gov/what-we-do/assessment/nca-activities/guidance).

In addition, the report received two independent reviews. The first review was by three experts, who were nominated in late 2011 by the chapter lead authors and the report editors. The second was at an open review in spring 2012. For both, independent review editors evaluated the review comments to ensure that authors adequately addressed the review comments.

Table 2.3 Factors contributing to assessment confidence associated with key findings

Confidence Level	Examples of Combinations of Factors that Could Contribute to this Confidence Evaluation
High	Strong evidence (established theory, multiple sources, consistent results, well documented and accepted methods, etc.), high consensus
Medium-High	Moderate evidence (several sources, some consistency, methods vary and/or documentation limited, etc.), medium consensus
Medium-Low	Suggestive evidence (a few sources, limited consistency, models incomplete, methods emerging, etc.), competing schools of thought
Low	Inconclusive evidence (limited sources, extrapolations, inconsistent findings, poor documentation and/or methods not tested, etc.), disagreement or lack of opinions among experts

Source: Moss and Yohe (2011).

2.7 Organization of This Report

The report comprises twenty chapters:

Chapter 1: Summary for Decision Makers describes the key issues found in Chapters 3–20.

Chapter 2: Overview describes the basis for, and methods used to create, this report.

Chapter 3: The Changing Southwest describes the important characteristics that affect exposure and sensitivity of the Southwest to climate change. Chapter 3 examines general socio-economic and land-use patterns and trends for the region. These include a brief examination of the physical context, human demographics and population trends, key laws relevant to resource management, and institutions conducting climate-assessment or policy initiatives.

Chapter 4: Present Weather and Climate: Average Conditions describes baseline characteristics of current climate and hydrologic parameters, such as temperature, precipitation, and snowpack, as well as the factors that contribute to the unique climates of the region. Chapter 4 discusses the main factors contributing to regional climate variability, and describes important climate hazards and impacts, such as droughts, floods, wildland fires, air quality and extreme temperatures.

Chapter 5: Present Weather and Climate: Evolving Conditions assesses weather and climate variability and trends in the Southwest, using observed climate and paleoclimate records. Chapter 5 analyzes the last 100 years of climate variability in comparison to the last 1,000 years, and links the important features of evolving climate conditions to river flow variability in four of the region's major drainage basins. The chapter closes with an assessment of the monitoring and scientific research needed to increase confidence in understanding when climate episodes, events, and phenomena are attributable to human-caused climate change.

Chapter 6: Future Climate: Projected Average presents climate-model projections of future temperature, precipitation, and atmospheric circulation (long-term weather patterns) for the Southwest. Chapter 6 also examines projections of hydrologic parameters, such as snow water equivalent, soil moisture, and runoff for a subset of basins in the region, including the Colorado River Basin.

Chapter 7: Future Climate: Projected Extremes summarizes current scientific understanding about how specific weather and climate extremes are expected to change in the Southwest as global and regional temperatures increase. Chapter 7 examines heat waves, cold snaps, drought, floods, and weather related to wildland fires. The chapter also examines possible changes in weather patterns associated with climate extremes, such as atmospheric rivers and Santa Ana winds.

Chapter 8: Natural Ecosystems addresses the observed changes in climate that are associated strongly with observed changes in geographic distributions and phenology (recurring phenomena of biological species such as timing of blossoms or migrations of birds) in Southwestern ecosystems. Chapter 8 also examines disturbances such as wildfires and outbreaks of forest pathogens and discusses issues associated with how carbon is stored and released in Southwestern ecosystems, in relation to climate-change threats.

Chapter 9: Coastal Issues examines climate-change threats to coastal ecosystems and human habitats, as well as available management and adaptation options such as insurance incentives. The chapter describes and evaluates key climate-induced impacts, including sea-level rise, erosion, storm surges, and oceanographic factors, including nutrient upwelling, ocean acidification, and oxygen-depleted zones. Chapter 9 also describes interactions between existing vulnerabilities (such as human development in coastal ecosystems).

Chapter 10: Water: Impacts, Risks, and Adaptation focuses on societal vulnerabilities to impacts from changes in sources, timing, quantity, and quality of the Southwest's water supply. The chapter addresses both vulnerabilities related to environmental factors (such as wildfire risk and increased stream temperatures) and issues related to water management (such as water and energy demand, and reservoir operation). Chapter 10 describes water management strategies for the coming century, including federal, regional, state, and municipal adaptation initiatives. (Note: Surface hydrology is addressed in Chapters 4–7.)

Chapter 11: Agriculture and Ranching reviews the climate factors that influence crop production and agricultural water use. The chapter discusses modeling studies that use climate-change model projections to examine effects on agricultural water allocation and scenario studies that investigate economic impacts and the potential for using adaptation strategies to accommodate changing water supplies, crop yields, and pricing. Chapter 11 concludes with sections on ranching and drought and on disaster-relief programs.

Chapter 12: Energy: Supply, Demand, and Impacts describes the potential effects of climate change on the production, demand, and delivery of energy. Chapter 12 describes climate effects on peak energy production and examines the vulnerability of infrastructure to climate change. The chapter describes direct and indirect climate effects on the generation of electricity, with analyses of different methods of generation, such as natural gas turbines, hydropower, and thermoelectric. The chapter concludes with an assessment of the evolution of fuel mixes for energy generation and transportation, and offers mitigation strategies for the present and future.

Chapter 13: Urban Areas describes the unique characteristics of Southwest cities and the ways they will be affected by and contribute to future climate changes. The chapter draws particular attention to six large urban areas: Albuquerque, Denver, Las Vegas, Los Angeles, Phoenix, and Salt Lake City. Chapter 13 addresses ways in which cities may contribute to climate change through their urban metabolisms—flows of water, energy, materials, nutrients, air, water, and soil impacts. The chapter also examines key pathways through which cities will be affected, including fire, water resources, flooding, urban infrastructure, and sea-level rise.

Chapter 14: Transportation examines climate change issues across a broad range of transportation sectors in the Southwest, including land transportation (passenger and freight), marine transportation, and air transportation, beginning with current trends. Chapter 14 analyzes possible direct and indirect impacts to transportation infrastructure and to the economy. The chapter concludes by examining vulnerabilities and uncertainties with respect to potential disruptions to the transportation system.

Chapter 15: Human Health reviews the state of knowledge with regard to climate-related public health threats, including those related to extreme heat, air quality (including respiratory ailments, dust, and fire-related particulate matter), and changes to disease vectors (such as mosquito populations). Chapter 15 examines factors that interact with and complicate disease transmission and risk. The chapter concludes by discussing public health planning and adaptation planning.

Chapter 16: Climate Change and U.S.-Mexico Border Communities evaluates some factors unique to the U.S.-Mexico border that affect the vulnerability of human populations to climate change, including border demographic changes, urban expansion, and socio-economic issues. Chapter 16 also addresses border climate and ecosystem issues, such as climate extremes, wildfires, and potential climate effects on the Colorado River estuary. The chapter includes a discussion of border adaptation measures, with an emphasis on the role of cross-border collaboration.

Chapter 17: Unique Challenges Facing Southwestern Tribes evaluates observed climate effects on Native American lands, and discusses the intersection of climate and the unique cultural, socioeconomic, legal and governance contexts for addressing these issues in Indian Country. Chapter 17 highlights some preparedness-, mitigation-, and adaptation-planning initiatives currently underway in the Southwest.

Chapter 18: Climate Choices for a Sustainable Southwest describes challenges to implementing mitigation and adaptation plans, given specific governance issues related to states, municipalities, and regional institutions. The chapter discusses new environmental management initiatives in the region, and gives examples of current climate-change mitigation and adaptation initiatives and successes. Chapter 18 analyzes the barriers to implementing solutions, and highlights the practical opportunities afforded through maximizing the co-benefits of mitigation and adaptation, and minimizing costs and environmental and social harms.

Chapter 19: Moving Forward with Imperfect Information builds on information from previous chapters, focusing on uncertainties, monitoring deficiencies, and data challenges. Chapter 19 summarizes the scope of what we do and do not know about climate in the Southwestern United States, and outlines those uncertainties that hamper scientific understanding of the climate system and potentially impede successful adaptation to the impacts of climate change. The chapter emphasizes issues related to climate and impact models, and scenarios of the future.

Chapter 20: Research Strategies for Addressing Uncertainties builds on descriptions of research and research needs articulated in earlier chapters. The chapter describes current research efforts and the challenges and opportunities for reducing uncertainties. It explores strategies to improve characterization of changes in climate and hydrology, and emphasizes the application of research strategies to decisions, including methods such as scenario planning.

References

Blunden, J., D. S. Arndt, and M. O. Baringer. 2011. State of the climate in 2010. *Bulletin of the American Meteorological Society* 92: S1–S236, doi:10.1175/1520-0477-92.6.S1. http://www.ncdc.noaa.gov/bams-state-of-the-climate/2010.php.

California Climate Action Team (CCAT). 2010. *2010 Climate Action Team report to Governor Schwarzenegger and the California Legislature.* N.p.: CCAT. http://www.energy.ca.gov/2010publications/CAT-1000-2010-005/CAT-1000-2010-005.PDF.

Cayan, D. R., A. L. Luers, G. Franco, M. Hanemann, B. Croes, and E. Vine. 2008. Overview of the California climate change scenarios project. *Climatic Change* 87 (Suppl. 1): S1–S6, doi:10.1007/s10584-007-9352-2.

Glick, P., B. A. Stein, and N. A. Edelson, eds. 2011. *Scanning the conservation horizon: A guide to climate change vulnerability assessment.* Washington, DC: National Wildlife Federation. http://www.nwf.org/~/media/PDFs/Global-Warming/Climate-Smart-Conservation/NWFScanningtheConservationHorizonFINAL92331.ashx.

Hayhoe, K., D. Cayan, C. B. Field, P. C. Frumhoff, E. P. Maurer, N. L. Miller, S. C. Moser, et al. 2004. Emissions pathways, climate change, and impacts on California. *Proceedings of the National Academy of Sciences* 101:12422–12427.

Intergovernmental Panel on Climate Change (IPCC). 2007. *Climate change 2007: Synthesis report. Contribution of Working Groups I, II and III to the Fourth Assessment Report of the Intergovernmental Panel on Climate Change,* eds. R. K. Pachauri and A. Reisinger. Geneva: IPCC. http://www.ipcc.ch/pdf/assessment-report/ar4/syr/ar4_syr.pdf.

Joyce, L. A., and R. Birdsey, eds. 2000. *The impact of climate change on America's forests: A Technical Document supporting the 2000 USDA Forest Service RPA Assessment*. Gen. Tech. Rep. RMRS-GTR-59. Fort Collins, CO: U.S. Forest Service, Rocky Mountain Research Station. http://www.fs.fed.us/rm/pubs/rmrs_gtr059.pdf.

Karl, T. R., J. M. Melillo, and T. C. Peterson, eds. 2009. *Global climate change impacts in the United States*. Cambridge: Cambridge University Press. http://downloads.globalchange.gov/usimpacts/pdfs/climate-impacts-report.pdf.

Marshall, R., M. List, and C. Enquist. 2006. *Ecoregion-based conservation assessments of the southwestern United States and northwestern Mexico: A geodatabase for six ecoregions, including the Apache Highlands, Arizona-New Mexico mountains, Colorado Plateau, Mojave Desert, Sonoran Desert, and southern Rocky Mountains*. Tucson, AZ: The Nature Conservancy. http://azconservation.org/dl/TNCAZ_Ecoregions_SW_Ecoregional_Summary.pdf.

Mayor's Greenprint Denver Advisory Council. 2007. *City of Denver Climate Action Plan: Final recommendations to Mayor Hickenlooper*. Denver, CO: City of Denver. http://www.greenprintdenver.org/about/climate-action-plan-reports/.

Moss, R. H., and G. Yohe. 2011. Assessing and communicating confidence levels and uncertainties in the main conclusions of the NCA 2013 Report: Guidance for authors and contributors. N.p.: National Climate Assessment Development and Advisory Committee (NCADAC). http://www.globalchange.gov/images/NCA/Draft-Uncertainty-Guidance_2011-11-9.pdf.

Nakićenović, N., and R. Swart, eds. 2000. *Special report on emissions scenarios: A special report of Working Group III of the Intergovernmental Panel on Climate Change*. Cambridge: Cambridge University Press.

National Assessment Synthesis Team. 2000. *Climate change impacts on the United States: The potential consequences of climate variability and change*. Report for the U.S. Global Change Research Program. Cambridge: Cambridge University Press.

National Audubon Society. 2009. Birds and climate change: Ecological disruption in motion. http://birds.audubon.org/sites/default/files/documents/birds_and_climate_report.pdf.

National Research Council (NRC). 2007. *Colorado River Basin water management: Evaluating and adjusting to hydroclimatic variability*. Washington, DC: National Academies Press.

—. 2010a. *Adapting to the impacts of climate change*. Washington, DC: National Academies Press.

—. 2010b. *Informing an effective response to climate change*. Washington, DC: National Academies Press.

Ray, A., J. Barsugli, K. Averyt, K. Wolter, M. Hoerling, N. Doesken, B. Udall, and R. S. Webb. 2008. *Climate change in Colorado: A synthesis to support water resources management and adaptation*. Boulder, CO: Western Water Assessment. http://cwcb.state.co.us/public-information/publications/Documents/ReportsStudies/ClimateChangeReportFull.pdf.

Sprigg, W. A., and T. Hinkley, eds. 2000. *Preparing for a changing climate: The potential consequences of climate variability and change, Southwest*. A report of the Southwest Regional Assessment Group for the U.S. Global Change Research Program. Tucson, AZ: Institute for the Study of Planet Earth.

U.S. Bureau of Reclamation (Reclamation). 2011. *SECURE Water Act Section 9503(c) – Reclamation Climate Change and Water, Report to Congress*. Denver: U.S. Bureau of Reclamation. http://www.usbr.gov/climate/SECURE/docs/SECUREWaterReport.pdf.

U.S. Environmental Protection Agency (EPA). 2010. *Climate change indicators in the United States*. http://epa.gov/climatechange/science/indicators/download.html.

Endnotes

i The phrase "climate variability and change" is used many times in this document. Climate variability refers to the inherent variability of climate, for instance, from year to year or decade to decade; climate change refers to ways in which systematic trends in some climate factors, such as increases in heat-trapping gases in the atmosphere (greenhouse gases) and associated increases in temperature, alter the climate system and its variations.

ii This section is based, in part, on remarks from Dr. David Stephenson (University of Exeter, UK) and text from the National Climate Assessment.

iii Taken together, the A2 and B1 scenarios provide reasonable estimates of what "high" and "low" global GHG emissions might be throughout the remainder of the twenty-first century. For example, scenario A2 (referred to in this report as the "high-emissions scenario") assumes a future with a high global population growth rate, slow global economic development rate, slow global technological change, and global fossil fuels use at rates slightly lower than observed in historical records. This combination of conditions would result in relatively high GHG emissions that continue to rise throughout the twenty-first century at an increasing rate (to a concentration of approximately 900 parts per million (ppm) in 2100), and substantially increased global temperatures. In contrast, scenario B1 (referred to in this report as the "low-emissions scenario") assumes a future in which global population peaks in the year 2050 and economies shift rapidly toward the introduction of clean and resource-efficient technologies, with an emphasis on global solutions to economic, social, and environmental sustainability. In the B1 scenario, greenhouse gas (GHG) emissions reach a peak in the mid-twenty-first century and then decline, resulting in carbon dioxide (CO_2) concentration of approximately 540 ppm in 2100, and smaller increases in global temperatures than those resulting from the A2 scenario. As has been emphasized in the IPCC study results and in prior regional climate change assessments, the outcomes of different mitigation strategies (as expressed by the A2 and B1 scenarios), in terms of the cumulative GHG concentrations and resultant climate changes, do not become very clear until after the middle of the twenty-first century, when the warming and other impacts from the B1 low-emissions scenario begin to be clearly exceeded by those of the A2 (and other) high-emissions scenarios (Nakićenović and Swart 2000; Hayhoe et al. 2004; IPCC 2007; Cayan et al. 2008).

iv Stakeholders are natural resource managers whose decision making relies in part on understanding how climate related variables impact their domains.

v The Southwest Climate Alliance consists chiefly of three NOAA-funded Regional Integrated Sciences and Assessments projects (California-Nevada Applications Program [http://meteora.ucsd.edu/cap/], Climate Assessment for the Southwest [http://www.climas.arizona.edu/], and Western Water Assessment [http://wwa.colorado.edu/]) and the Department of the Interior-funded Southwest Climate Science Center (http://www.doi.gov/csc/southwest/index.cfm), as well as a number of partner universities and federal research laboratories.

vi Federal Register / Vol. 76, No. 134 / Wednesday, July 13, 2011 / Notices http://www.gpo.gov/fdsys/pkg/FR-2011-07-13/pdf/2011-17379.pdf.

vii This technical report is only one of several for which EOIs were submitted. As far as the editors of this report know, this was the only EOI intending to produce a comprehensive regional report.

viii These were the October 27, 2011, NCA pre-decisional draft guidelines, which were the guidelines available to the authors of this report during the report draft and review periods.

ix Traceable accounts consist of (1) the reasoning behind the conclusion, (2) the sources of data and information contributing to the conclusion, (3) an assessment of the amount of evidence and degree of agreement among sources of evidence, (4) an assessment of confidence in the finding, and (5) an assessment of uncertainty associated with the finding.

x Many of the remarks here are drawn from National Climate Assessment guidance documents and insights from National Climate Assessment and Development Advisory Committee co-chair, Gary Yohe.

Chapter 3

The Changing Southwest

COORDINATING LEAD AUTHOR

David M. Theobald (National Park Service)

LEAD AUTHORS

William R. Travis (University of Colorado), Mark A. Drummond (U.S. Geological Survey), Eric S. Gordon (University of Colorado)

EXPERT REVIEW EDITOR

Michele Betsill (Colorado State University)

Executive Summary

This chapter describes important geographical and socio-economic characteristics and trends in the Southwest—such as population and economic growth and changes in land ownership, land use, and land cover—that provide the context for how climate change will likely affect the Southwest. The chapter also describes key laws and institutions relevant to adaptive management of resources.

- The Southwest is home to a variety of unique, natural landscapes—mountains, valleys, plateaus, canyons, and plains—that are both important to the region's climate and respond uniquely to changes in climate. Potential adaptation of human and natural systems will face challenges due to a complex pattern of land ownership, which crosses political and management jurisdictions and transverses significant elevational gradients. This decreases the adaptive capacity of the region because it makes it more difficult to coordinate decision making across landscapes. (medium-low confidence)

Chapter citation: Theobald, D. M., W. R. Travis, M. A. Drummond, and E. S. Gordon. 2013. "The Changing Southwest." In *Assessment of Climate Change in the Southwest United States: A Report Prepared for the National Climate Assessment*, edited by G. Garfin, A. Jardine, R. Merideth, M. Black, and S. LeRoy, 37–55. A report by the Southwest Climate Alliance. Washington, DC: Island Press.

- The Southwest has experienced rapid population increases and urban expansion for the past 150 years or so, and rapid population growth will likely continue to be an enduring feature, especially in urban areas. Indeed, the region will likely grow by an additional 19 million people by 2030 (from 2010). These changes will make it more difficult to manage natural resources because of the additional demand for and reliance on natural resources (e.g., water supply). (medium-high confidence)
- The coordination of climate-change adaptation strategies will be challenging because environmental management decisions will be made at many geographic scales, over different time frames, and by multiple agencies pursuing numerous associated policies and management goals. Adaptive capacity may be bolstered through lessons learned from emerging assessment projects (see Chapter 18). (medium-high confidence)

3.1 Lay of the Land: Geographical Themes and Features

Regions can be defined in many ways, but an important lesson from decades of geographical research is that the definition depends on the theme or topic being studied, the manner in which it is being studied, and the intended outcome of such a study. An assemblage of states provides the National Climate Assessment a way to divide assessment activities regionally. The "Southwest"—defined as the six contiguous states of Arizona, California, Colorado, Nevada, New Mexico, and Utah—rests on the certain logic of proximity and on the fact that states are important governmental units that must respond to the effects of climate variation and change. Beyond this basic political geography are several "critical zones" that are important to highlight because of their vulnerability to climate change, such as the coastal zone (see Chapter 9), the wildland-urban interface (see Chapter 13), the U.S.-Mexico borderlands (see Chapter 16), and the lands of Native nations (see Chapter 17).

Natural features

Two common geographical features tie the six states together. First, the states collectively span the most extensive arid and semi-arid climates and lands in the United States. Each state also touches and makes use of the waters of the Colorado River Basin. On the other hand, the six-state region, covering nearly 700,000 square miles, encompasses a variety of topography and landscapes, from the highest mountains in the conterminous United States (Mt. Whitney at 14,505 feet in California and Mt. Elbert at 14,440 feet in Colorado) to the lowest terrestrial point in the western hemisphere (Bad Water Basin in Death Valley at 282 feet below sea level). Significant physiographic and hydrologic features (Figure 3.1) include: a 3,400-mile shoreline along the Pacific Ocean that varies from cliff and rocky headlands to low-gradient coastal and brackish marshlands; the Central Valley of California; the Sierra Nevada; a southern reach of the Cascade Range; the extensive Basin and Range province; the Colorado Plateau; the Southern Rocky Mountains; and the western Great Plains (or "high plains") that skirt the region's eastern edge in Colorado and New Mexico (Hunt 1974). This natural landscape is also broken into hydrological basins, most notably the Sacramento-San Joaquin, Colorado, and Rio Grande, as well as a large (260,000-square-mile) interior drainage—the Great Basin—which covers nearly one-fifth of the six-state region.

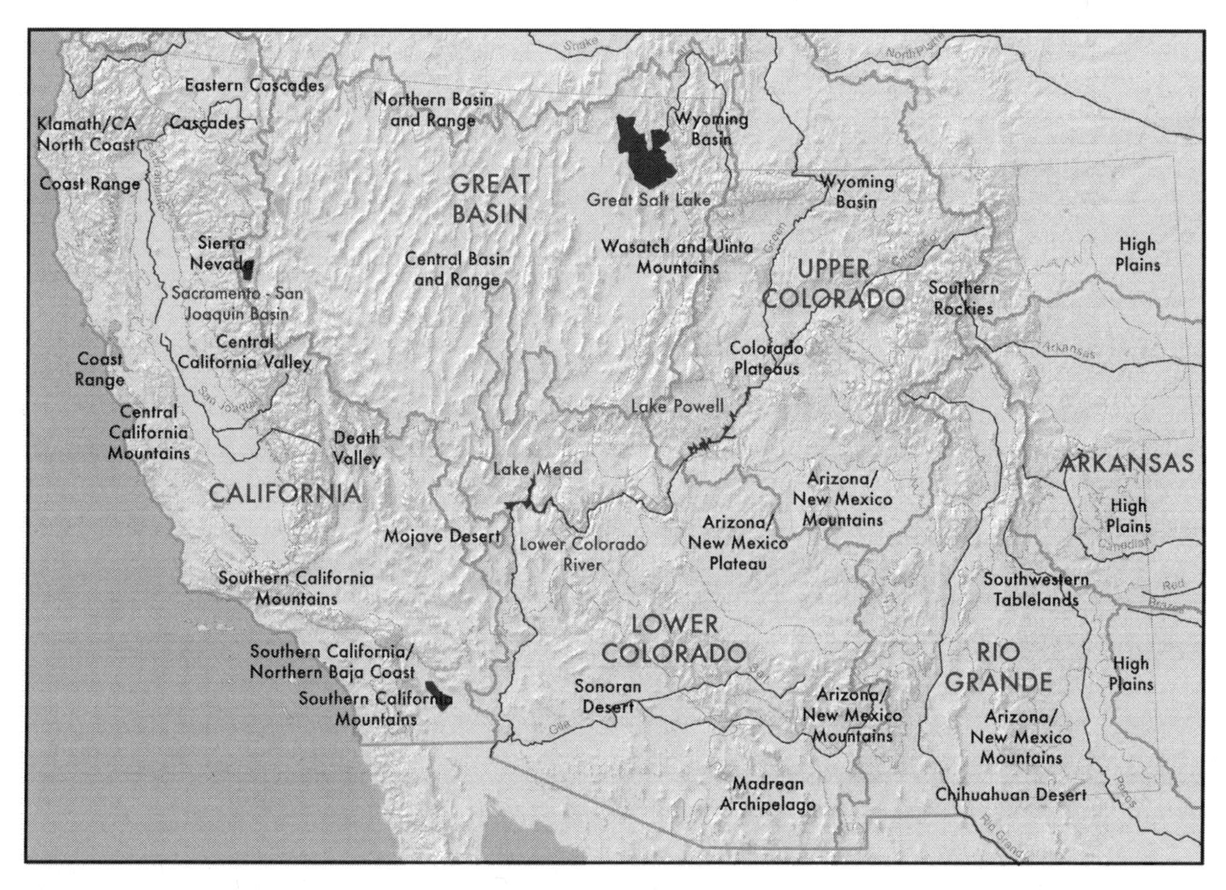

Figure 3.1 Important physiographic and ecoregional features of the Southwest. Water basin names are in upper-case, ecoregional names in lower-case. Source: ESRI ArcDate v10.

The juxtaposition of mountains, valleys, plateaus, canyons, and plains increases the degree to which the region will be affected by climate change. For example, the higher elevations produce the net annual runoff that provides water resources to the drier valleys, piedmonts, and plains where most of the region's human settlements are located. As a result, important sources of water for many urban areas are often quite far away (Southern California partially relies on water from the Colorado River, for example). As a result, potential feedbacks in the water resources system (in this case between the water users and their water sources) may be fairly weak or even "decoupled." Also, at a local scale the topographic variability of the Southwest is important because it may provide a buffer to climate change by conserving biodiversity (Ackerly et al. 2010). Yet many public and private land ownership boundaries occur in areas of steep elevational changes (Travis 2007), coinciding with boundaries between ecological systems (i.e., ecotones).

The land covers (Figures 3.2 and 3.3, and Table A3.1) draped on this topography are principally grassland and shrubland (55.3% of the region's land cover), marked by California chaparral and Great Plains grasslands as well as by extensive sagebrush and desert shrub and cacti mixes (such as found in the Sonoran and Mojave Deserts). Nearly one-quarter of the Southwest is covered by forests in a diverse array of mountain and high-plateau settings, including: extensive lodgepole pine in the Southern Rockies (notable for experiencing a significant die-off in recent years; see Bentz et al. 2010); topographically controlled forest islands in otherwise desert landscapes (the "Sky Islands" of southern New Mexico and Arizona); moist coastal and redwood and inland sequoia forests in California; park-like forests of ponderosa pine skirting the southern Colorado Plateau and eastern slopes of the Southern Rockies; and extensive pinyon-juniper at middle elevations in the Colorado Plateau and Great Basin (with pinyon also experiencing a significant die-off early this century; see Chapter 8). At the highest elevations are mountain peaks and alpine tundra (0.7%). About 6.6% of the Southwest has been converted to cropland agriculture, and another 2.3% has been developed as urban areas.

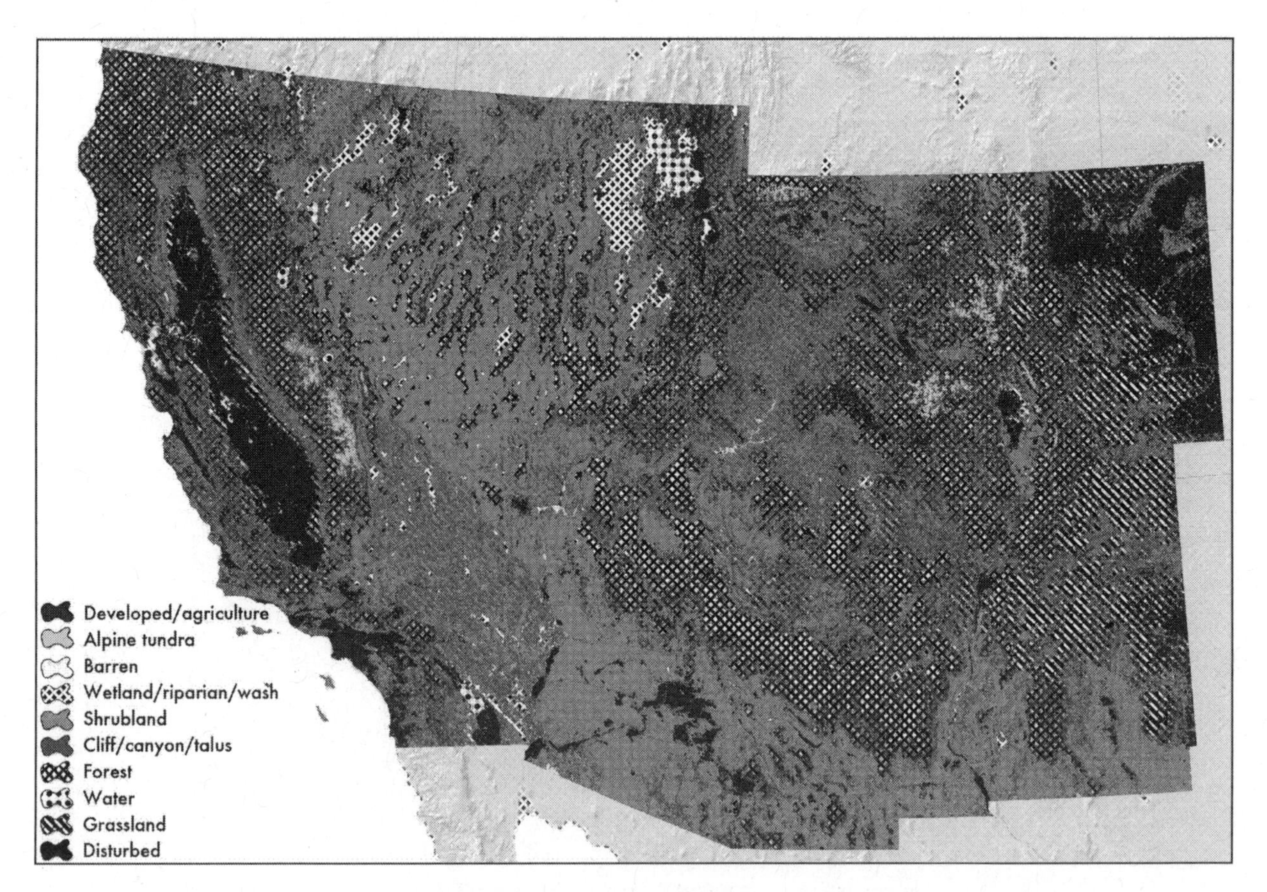

Figure 3.2 Land cover types in the Southwest. See Appendix Table A3.1 for classifications. Source: USGS (2010).

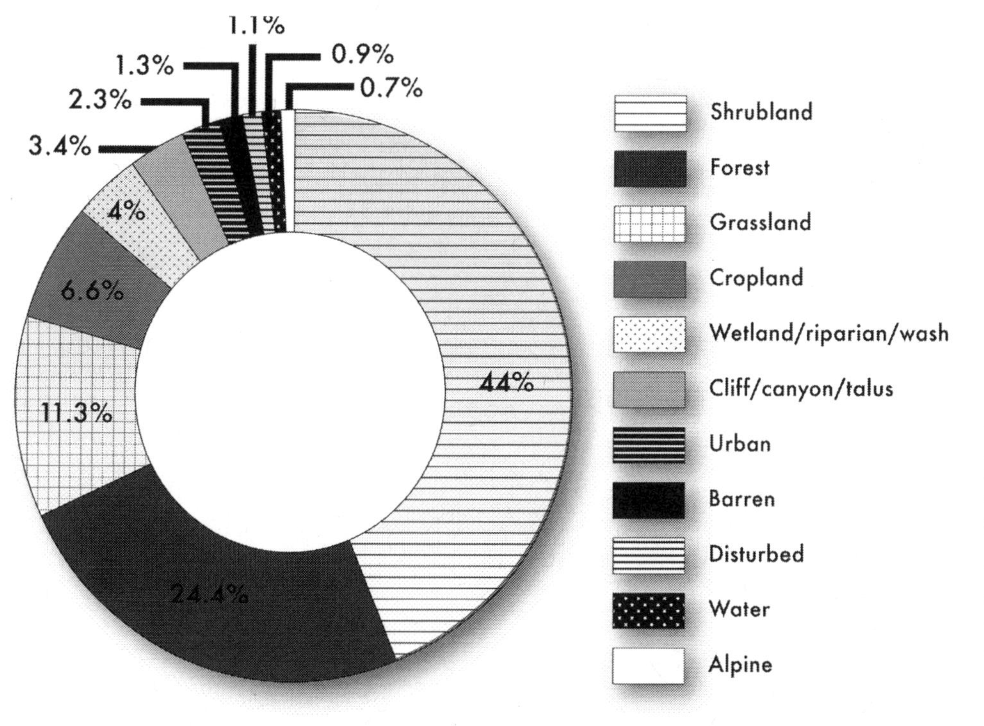

Figure 3.3 The proportion of land cover types found in the Southwest. See Appendix Table A3.1 for classifications. Source: USGS (2010).

Human geography

The human landscape of the Southwest is marked by a few large cities, some comprising sprawling metropolitan swaths, embedded in a predominantly rural landscape and in some places wilderness (Theobald 2001; Lang and Nelson 2007). The most notable metropolitan footprints include the Southern California conurbation around Los Angeles and San Diego; the San Francisco Bay Area; the string of cities marking California's Central Valley (from Redding to Bakersfield); Phoenix to Tucson; the Wasatch Front (anchored in Salt Lake City); and the Colorado Front Range centered on Denver. Smaller urban-suburban footprints in the region include Las Vegas, Reno, and Albuquerque. All told, there are thirty-nine metropolitan planning organizations centered on urban areas in the Southwest (these are described more fully in Chapter 13). Nearly all of these urban areas have grown significantly in the last few decades in both population and extent (Theobald 2001; Theobald 2005; Travis 2007) and many are surrounded by exurban development, much of which can be described as the "wildland-urban interface" (Radeloff et al. 2005; Theobald and Romme 2007). Beyond the exurban fringe, the region's rural landscapes include areas of dryland and irrigated agriculture, extensive rangelands (see Chapter 11), and isolated small towns and resorts. Although infrastructure is rather thinly dispersed across this rural landscape, areas of intense energy development and pockets of earth-transforming hard-rock mining also mark the landscapes.

A dominant feature of the region's rural geography is its extensive public lands, mostly federal, that encompass fully 59% of the six-state region's land surface (Figures 3.4 and 18.1). The federal lands are divided among agencies with different management mandates and goals, chiefly the Bureau of Land Management (BLM), Forest Service

(USFS), National Park Service (NPS), and Fish and Wildlife Service (USFWS). Each agency has efforts underway to plan for and adapt to climate change (Smith and Travis 2010). The lands of Native nations occupy another 7% of the region. Nearly five million acres of privately owned lands have been conserved in the past decade through land trust conservation (a 65% increase over that period). Especially relevant to climate vulnerability and adaptation in the Southwest is the mixture of ownership that occurs along the elevational gradients (Figure 3.4), which hints at the complexities of managing and cooperating for possible latitudinal and upward shifts of climates and migration of species. (For further discussion about the potential responses of plant and animal species to climate change, see Chapter 8).

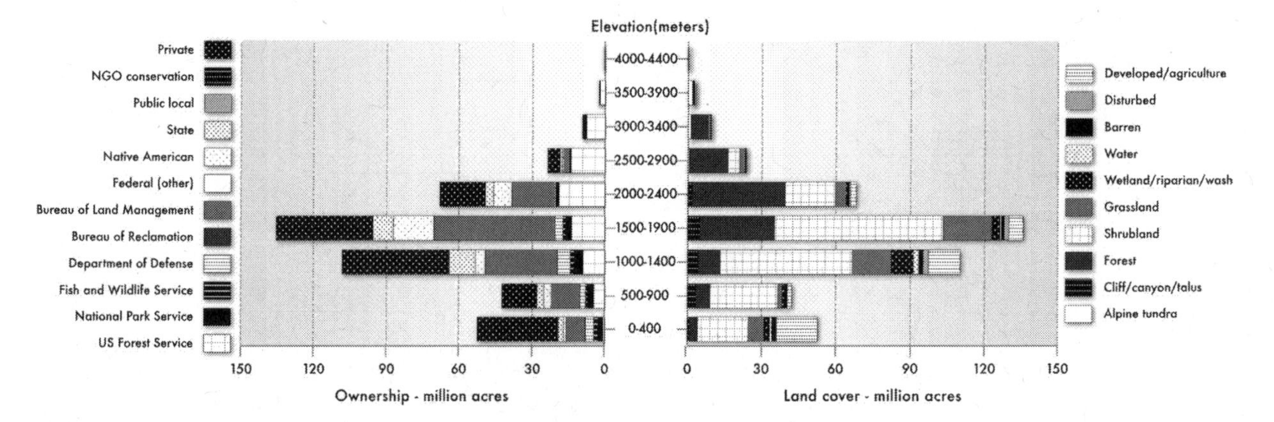

Figure 3.4 Spatial patterns of ownership and land cover types, arrayed along elevation gradients, are two critical aspects that hint at the complexities of coordinating adaptation strategies in the Southwest. All data up to 2010 taken from the US Census Bureau, with state specific projections from: AZ Dept. of Economic Security, CA Dept. of Finance, CO State Demographer's Office, NV State Demographer's Office, NM Bureau of Business and Economic Research, UT Governor's Office of Planning and Budget, and UT State Demographer's Office.

Public lands

The federal lands in the Southwest comprise 22 national parks, 74 national wildlife refuges, nearly 66 million acres of national forests, and 120 million acres under the jurisdiction of the BLM (see Figure 18.12). A patchwork of federal laws governs resource management policies on these lands (Table 3.1). For example, BLM policies are set under the Federal Land Policy and Management Act of 1976, which codified public ownership of BLM-managed lands and prescribed "multiple-use" management intended to direct resource use to "best meet the present and future needs of the American people" (Public Law 94-579). BLM lands are often managed for grazing, mineral and hydrocarbon extraction, and recreation, among other uses. The Department of Agriculture's USFS oversees National Forests through policies developed in accordance with the 1976 National Forest Management Act. This law requires National Forest System managers to develop integrated management plans intended to balance multiple intended uses while maintaining forest resources for future generations. Primary uses of National Forests include timber harvesting, grazing, mineral extraction, and recreation. National Wildlife Refuges are administered by the USFWS, under the Department of the Interior. Refuges are managed under the National Wildlife Refuge System Administration Act of 1966 with the stated goal of establishing a network of lands for conservation, management, and restoration of fish and wildlife resources. Although primarily managed for species conservation and restoration, refuges may also host extractive industries and recreation, including hunting and fishing. The NPS was created under the 1916 National Park Service Act, which instructed NPS to manage scenery and natural and historic resources "unimpaired for the enjoyment of future generations." Individual units are managed under the terms of specific laws establishing each park.

A number of federal laws prescribe policies relevant to federal and other lands. The Wilderness Act of 1964, the Antiquities Act of 1906, and the National Landscape Conservation System Act of 2009 all provide additional legal authority to protect public lands. The National Environmental Policy Act of 1969 requires agencies to review environmental impacts of major environmental actions, while the Endangered Species Act of 1973 prohibits government and private actors from destroying habitat critical to the survival of threatened and endangered species.

Extractive resource use on federal lands is further guided by a number of laws, including the Surface Mining Control and Reclamation Act of 1977 (regarding coal extraction), the General Mining Act of 1872 (regarding hardrock mining), the Mineral Leasing Act of 1920 (regarding oil and gas resources) and the Taylor Grazing Act of 1934 (regarding sheep and cattle grazing).

A central difficulty of the patchwork of laws, policies, and regulatory agencies is that it poses a significant challenge to coordinate adaptation to climate change (although Landscape Conservation Cooperatives have recently been developed to address this issue under the auspices of the USFWS). The problem is further compounded by the relatively high levels of uncertainty associated with climate model predictions. A key is to develop proactive strategies to anticipate change and to adaptively manage resources throughout changing circumstances.

Population

The Southwest hosted a permanent resident population of 56.2 million in 2010 (Table 3.2). It has been the fastest-growing region of the nation for several decades as part of the so-called Sun-Belt Migration that began in earnest in the 1970s. The Interior West topped the national charts of population growth over the last two decades (1990–2010), with Nevada, Arizona, Utah, and Colorado comprising the four fastest-growing states in the country. The Southwest grew by 37%, from 41.2 to 56.2 million residents, during 1990–2010, compared to a national growth rate of 24% (1.2% annualized).

Growth in the region is concentrated in the metropolitan areas, and several Southwestern cities (most notably Las Vegas and Phoenix) have been among the fastest growing in the United States over the past two decades. The region is slightly more urbanized than much of the nation, with 82% of the population residing in urban areas compared to a national average of 78%. (See further discussion of the Southwest's urban areas in Chapter 13.)

Table 3.1 Federal laws and policies relevant to federal and other lands in the Southwest

Federal Law (Year Enacted)	Land Base or Resource Covered	Relevant Agency	Overarching Goal
POLICIES GUIDING FEDERAL LAND MANAGEMENT AGENCIES			
National Park Service Organic Act (1916)	National Parks and other park units	NPS	"Conserve the scenery and natural and historic objects and wild life ... unimpaired for the enjoyment of future generations"
National Wildlife Refuge System Administration Act (1966)	National Wildlife Refuges	USFWS (on-shore resources); NOAA (offshore resources)	Conservation, management, and restoration of species
Federal Land Policy and Land Management Act (1976)	BLM Lands	BLM	Multiple use to best meet the present and future needs of the American people
National Forest Management Act (1976)	National Forests	USFS	Integrated planning for sustained multiple uses of renewable resources
ADDITIONAL LAWS PROTECTING PUBLIC LANDS			
Antiquities Act (1906)	National Monuments	Primarily NPS, also including USFS and BLM	Preservation of resources of "historic or scientific interest"
Wilderness Act (1964)	Specified federal public lands	Primarily USFS, BLM, and NPS	Preservation of lands with wilderness characteristics
National Landscape Conservation System Act (2009)	Specified federal public lands	NPS, USFS, and BLM	Conservation, protection, and restoration of nationally significant western public lands with outstanding natural, cultural, or scientific values
LAWS PROTECTING WILDLIFE AND RESOURCE MANAGEMENT POLICIES			
National Environmental Policy Act (1969)	Any major federal action	All federal agencies	Requires review of environmental impacts resulting from any major federal action
Endangered Species Act (1973)	Threatened and endangered species	USFWS, although applies to all federal agencies	Conservation, protection, and recovery of threatened and endangered species

Table 3.1 Federal laws and policies relevant to federal and other lands in the Southwest (Continued)

Federal Law (Year Enacted)	Land Base or Resource Covered	Relevant Agency	Overarching Goal
LAWS GOVERNING RESOURCE EXTRACTION AND GRAZING			
General Mining Act (1872)	Minerals found on federal lands	All federal land management agencies	Set policies for the discovery, claim, and recovery of hardrock resources under federal lands
Mineral Leasing Act (1920)	Oil and gas extraction	All federal land management agencies	Set policies for the extraction of oil, gas, phosphate, sodium, and coal on federal lands
Taylor Grazing Act (1934)	Rangeland	Federal agencies that manage grazing (primarily BLM and USFS)	Prevent overgrazing and provide for the permitting of grazing on public lands
Surface Mining Control and Reclamation Act (1977)	Coal on federal lands	All federal land management agencies	Ensure appropriate regulation of mining and reclamation on federal lands

Most analysts expect the West, and especially the Southwest, to continue growing in population faster than the nation as a whole for the foreseeable future (Travis 2007). This prediction is based on positive trends in all of the demographic components of population change: natural growth (births over deaths), domestic net in-migration, and international net in-migration. The Census Bureau's population projections to 2030 (Table 3.2) reflect this scenario. Arizona and Utah likely will grow by about 50% of their 2010 populations, and Colorado and New Mexico are expected to add another third to their populations (Figure 3.5). Even California, building on a large base (37.2 million in 2010), is projected to grow by nearly a third. In all, some 18.8 million more people likely will live in the West by 2030 than did in 2010. Most states extend their projections even further in time; linear extrapolation to each state's extended population projection suggests a regional population in 2050 of around 94.8 million, a 69% (1.37% annualized) increase over the 2010 census.

Natural resource economy

Two trends are clear with respect to the Southwest's natural resource-based economy (Figure 3.6). First, the iconic Western economies of agriculture, ranching, fishing, hunting, and mining have lost ground, and now contribute only a small fraction of the overall gross domestic product or GDP of the region, averaging around 4.5% for the past three decades and never reaching higher than 7% per year during that period. Second,

Table 3.2 Trends in population growth in the Southwest (in thousands of people)

State	1990	2000	2010	Total Growth 1990–2010	% Growth 1990-2010	Projected Pop. 2030	% Growth 2010–2030	Total Growth 2010–2030
Arizona	3,665	5,130	6,392	2,726	74	9,480	48	3,088
California	29,760	33,871	37,253	7,493	25	48,380	30	11,127
Colorado	3,294	4,301	5,029	1,734	53	6,564	31	1,535
Nevada	1,201	1,998	2,700	1,498	125	3,363	25	663
New Mexico	1,515	1,819	2,059	544	36	2,825	37	767
Utah	1,722	2,233	2,763	1,041	60	4,394	59	1,631
TOTAL	**41,159**	**49,353**	**56,198**	**15,039**	**37**	**75,010**	**33**	**18,811**

Sources: U.S. Census sources [for pre-2010] and state demographer's projections [for 2010 and beyond].

after a period of relative stability or small increases from the 1970s to the mid-1980s (averaging 5.4% over that 15-year period), the contribution of these natural resource sectors has declined by a third in the past 15 years (now averaging 2.9% per year). Finance, professional services, and the like now contribute a large majority of GDP, followed by construction and manufacturing.

3.2 Land Use and Land Cover

The pace and types of land-use and land-cover change (from one type of land use or land cover to another) from 1973 to 2000 varied across the Southwestern states (Figure 3.7, and Table A3.2). The average annual rate of the combined changes ranged from <0.1% of the total area of Nevada to 0.4% of neighboring California. Annual rates of change were consistently higher in Colorado and California, although the amount of change in New Mexico tripled beginning in the mid-1980s. Numerous factors contributed to the state-by-state variability, including the mix of land ownership, population changes, government policies and regulations, and climate variability.

The arid states with extensive public lands that limit land use options—Arizona, Utah, and Nevada—have some of the lowest rates of land-use and land-cover change in the nation. These states and other areas of warm deserts (i.e., the Chihuahuan, Sonoran, and Mojave) also lack the large extent of agricultural land cover fluctuation (such as occurs in the Great Plains of eastern Colorado, New Mexico, and California's Central Valley) and intensive forest harvesting that contribute to higher rates of land-use and

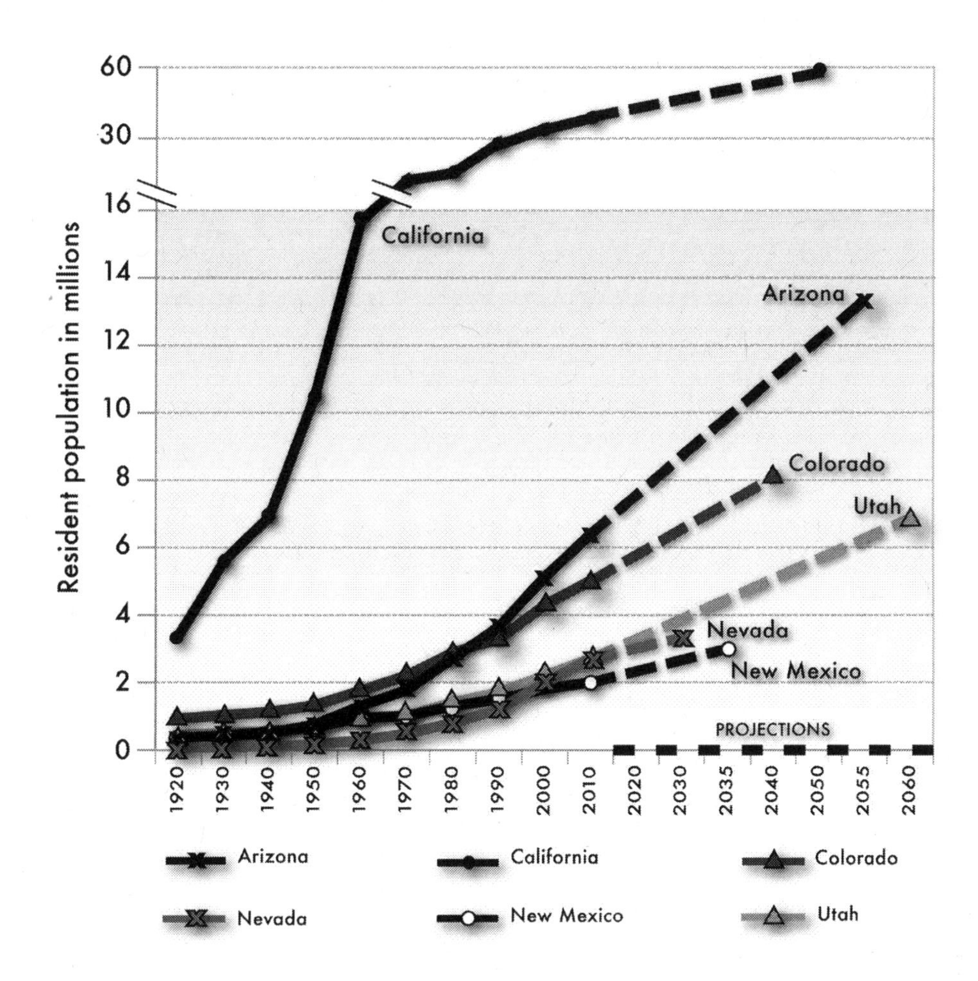

Figure 3.5 Rapid population growth in the Southwest is expected to continue. The current (2010) population is 56 million, and an additional 19 million people are projected to be living in the region by 2030. Source: US Department of Commerce, Bureau of Economic Analysis (http://www.bea.gov/regional/index.htm).

land-cover changes in other U.S. regions. However, lower rates of land-use change do not preclude important change-related effects, such as irreversible or slow recovery of disturbed lands. For example, in Nevada, although low rates of change occurred, disturbed forested areas were slow to recover and grasslands/shrublands converted for urban development and mining contributed to the net decline of natural cover types.

Other trends between 1973 and 2000 are notable. The extent of urban development, mining, fire, and other natural land disturbance increased across all Southwestern states. Urban land cover increased by an estimated 45%, affecting 0.5% of the total area. Most of the growth in urban and other developed lands occurred on grassland/shrubland (56%), although more than one-third of the expansion was at the expense of cropland agriculture and maintained pasture (34%). Nearly 90% of the agricultural land converted to urban areas was in California and Colorado. The loss of agriculture to development and other causes in California's Central Valley is offset by expansion of new cultivated areas; however, other types of conversions cumulatively resulted in a small net loss of agricultural land cover in the state. California's developed lands increased overall by an estimated 40% between 1973 and 2000. This increased land-use conversion and development in the Southwest generates increased pressure and need for a coordinated land management approach for successful adaptation to climate change.

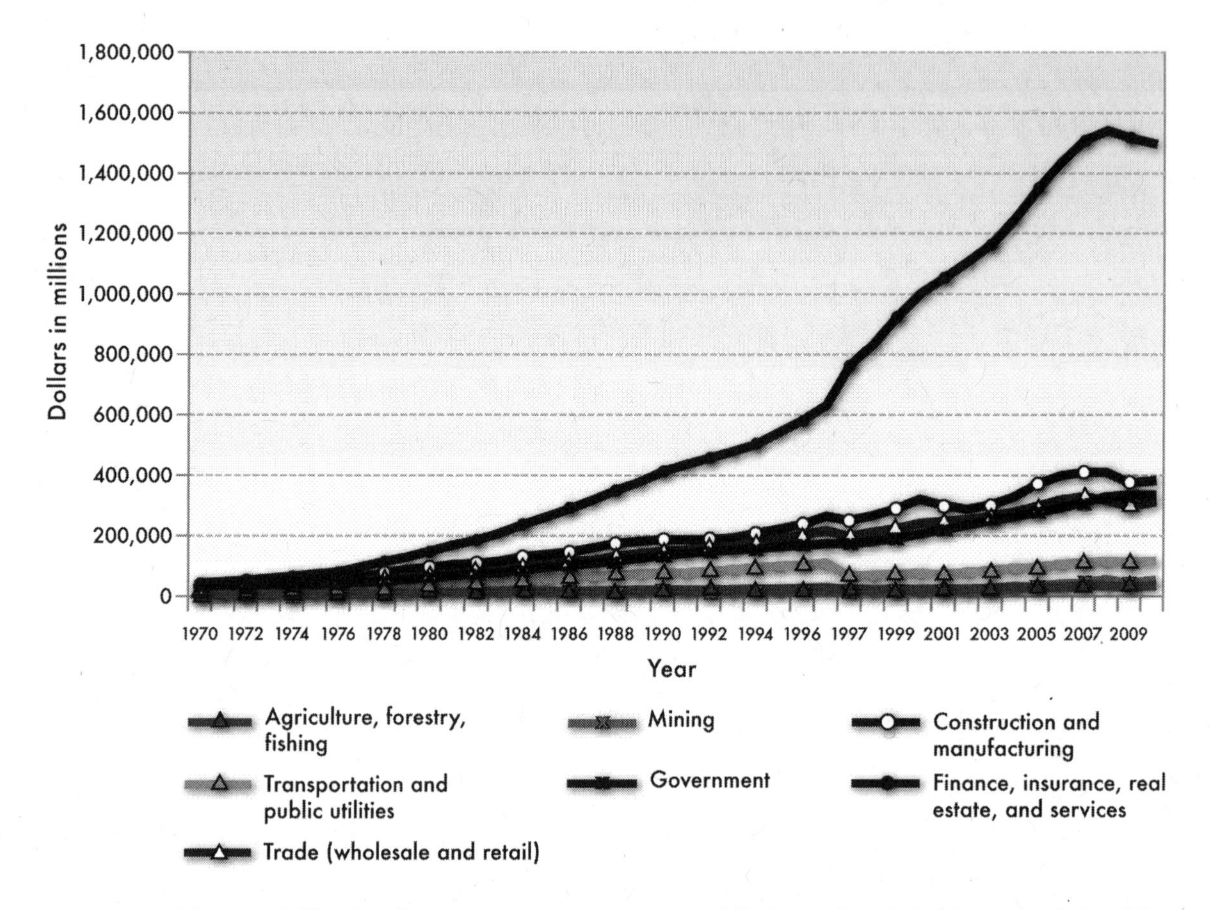

Figure 3.6 The Southwestern economy grew rapidly from the 1970s through 2008, with a decline commencing with the recession. The strongest economic sectors were finance, insurance, real estate, and services, followed by construction and manufacturing, trade, and government. The more traditional natural resource economies remain important but provide only a small portion of the GDP of the region (shown in millions of dollars). Note that previous to 1998 income by industry were defined using the Standard Industrial Classification, and in 1998 and after were defined using the North American Industry Classification System. This definitional change resulted in a slight down-tick in Construction and manufacturing, Agriculture, forestry, and fishing, and Transportation and public utilities, and the up-tick in Finance, insurance, real estate, and services. Longer-term trends (>5-10 years) remain robust to this definitional change. Source: U.S. Department of Commerce, Bureau of Economic Analysis (http://www.bea.gov/regional/index.htm).

Forest cover declined in all states by a combined 2.2% (0.5% of the region) due primarily to mechanical disturbance (e.g., timber harvest) and fire, although some of the decrease occurred on land with potential for eventual tree regrowth following fire or post-harvest replanting. The extent of mechanical disturbance was highest in the mountains and foothills of California, central Arizona, and New Mexico. However, Colorado and other states may see an increase in timber harvest related to insect-related forest die-off exacerbated by changing climatic conditions.

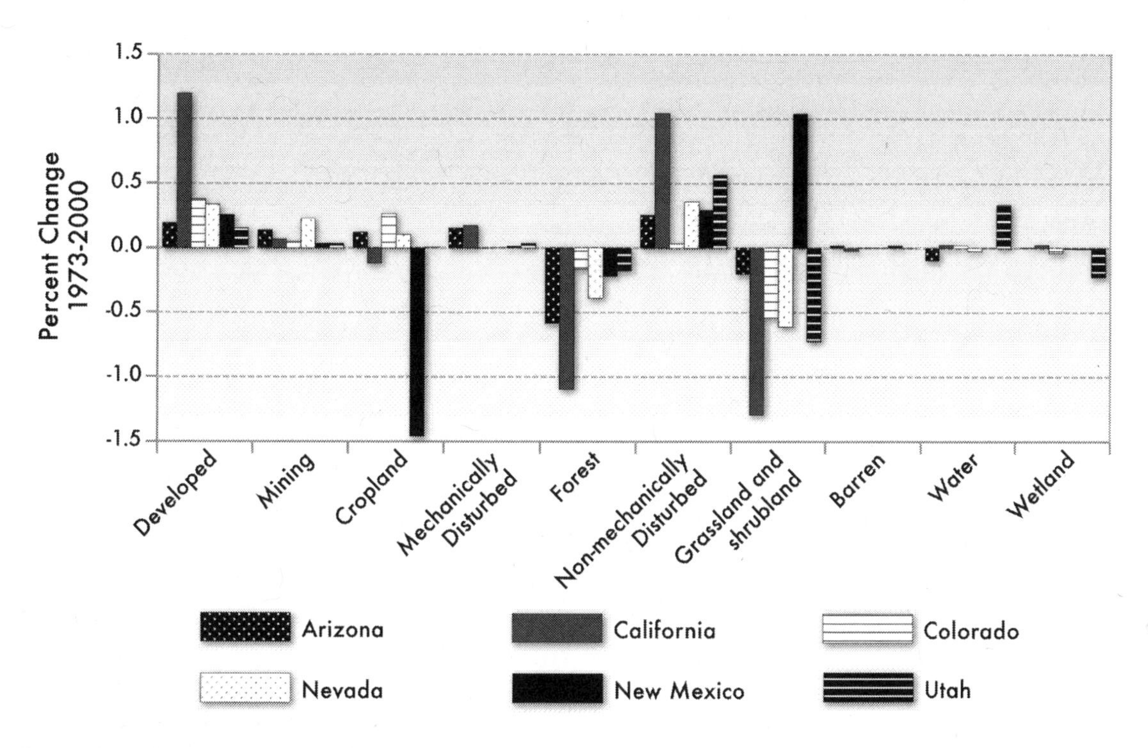

Figure 3.7 Percent of total state area affected by net change in land use and land cover types from 1973 to 2000 for the six Southwestern states. See Appendix Table A3.2 for class descriptions. Source: USGS land cover trends project (http://landcovertrends.usgs.gov); Loveland et al. (2002).

Changes in agricultural land cover, which declined by 3.5% (0.2% of the region), often show a reciprocal relationship with grassland/shrubland changes (0.6% decline, 0.4% of the region), although the extent of exchange between the two types of cover is often uneven. Conversions from grassland and shrubland to agriculture were more extensive in Colorado, Arizona, and Nevada, resulting in small net increases in agriculture. A substantial net decline in agricultural land cover occurred in New Mexico, where a significant amount of cropland was returned to grassland cover in response to incentives of the Conservation Reserve Program to set aside environmentally sensitive land. The overall decline in grassland/shrubland (except in New Mexico) is tied to agricultural expansion, as well as to urban growth and development, expansion of mining, and other disturbance.

Trends for urban and exurban development

Associated with rapid population growth in the Southwest, the extent of urban land (housing density greater than one unit per 2.5 acres) and exurban land (one unit per 2.5–40 acres) will continue to increase (Figure 3.8; Table 3.3). The extent of urban land is forecast to double (from 4.1 to 8.1–9.3 million acres) by 2050, while lower-density exurban lands will expand by 33% to 41% (from 13.6 to 18.2–19.1 million acres) (Bierwagen et al. 2010).

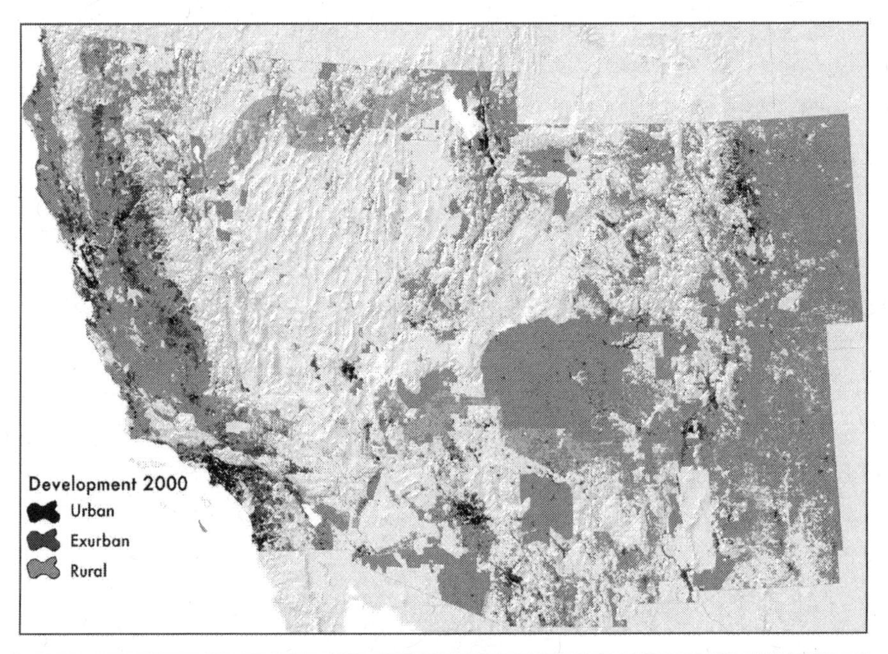

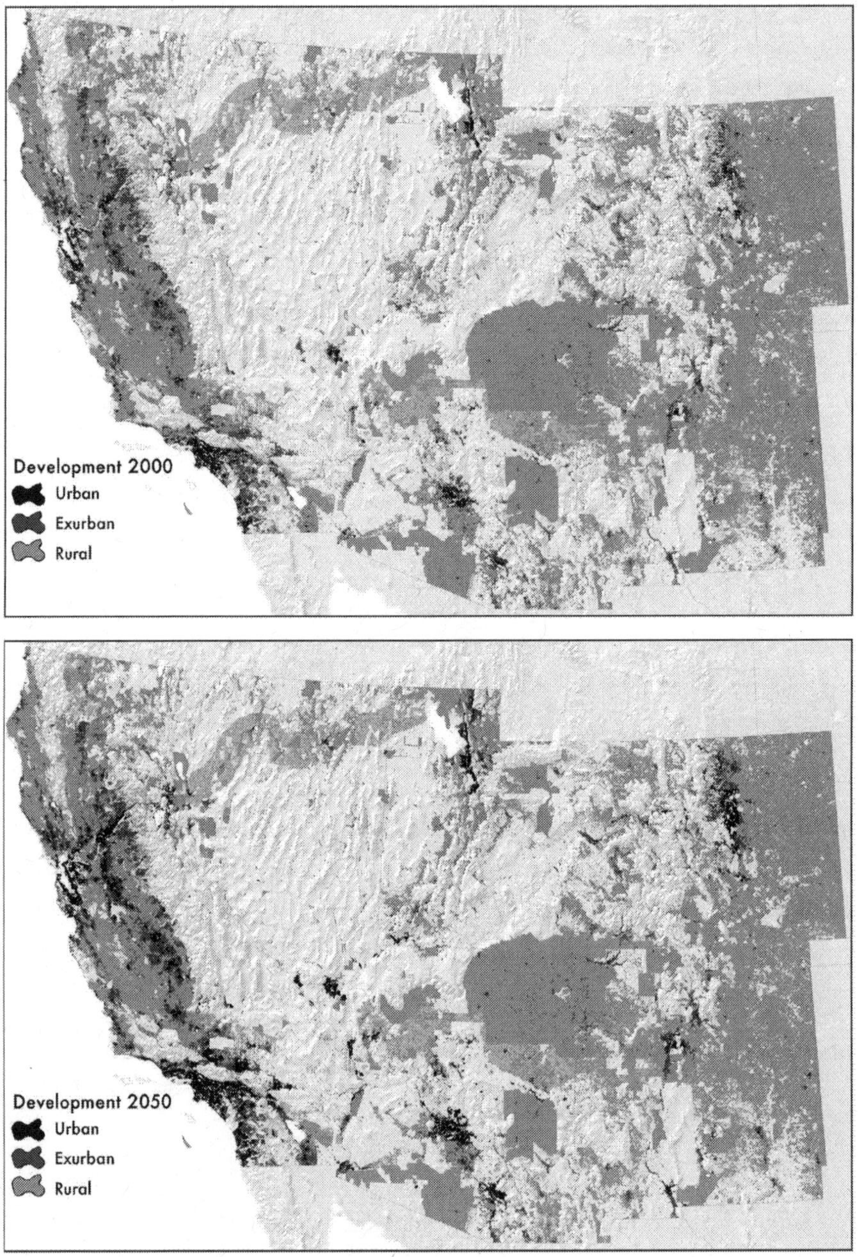

Figure 3.8 The pattern of urban, exurban, and rural residential development for 2000 and forecast for 2050. Source: Bierwagen et al. (2010).

Both the rapid pace and patterns of population growth and ensuing land use change provide both challenges and opportunities for adapting to climate change. A near-doubling of population from 2000 to 2050 will increase already stressed water resources in particular. Although most of the population in the Southwest lives in urban areas, the footprint of these areas is likely to more than double, from about 4 million acres to 8–9 million acres. An additional 10–11 million acres of low-density (exurban) housing density (see Table 3.3) is likely to contribute significantly to the number of miles travelled in vehicles.

Table 3.3 Historical, current, and forecasted expansion of urban and exurban lands in the Southwest (data expressed as thousands of acres [kac])

Geography	Historical (1950)		Current (~2000)		Forecast (~2050)	
	Urban developed (kac)	Exurban developed (kac)	Urban developed (kac)	Exurban developed (kac)	Urban developed (kac)	Exurban developed (kac)
STATES						
Arizona	30	224	544	1,441	1,255	1,967
					1,448	1,818
					1,054	1,928
					1,163	1,884
California	597	2,334	2,516	7,962	4,995	11,727
					5,349	11,374
					4,874	10,555
					5,058	10,384
Colorado	68	355	402	1,690	1,024	2,204
					1,003	2,202
					785	2,311
					766	2,333
New Mexico	24	237	191	1,328	324	1,730
					348	1,812
					277	1,841
					287	1,925
Nevada	8	65	179	428	562	510
					500	521
					463	516
					419	521
Utah	33	235	224	655	696	863
					580	965
					599	938
					485	1058
Southwest	767	3,477	4,083	13,563	8,894	19,074
					9,270	18,762
					8,092	18,158
					8,218	18,174
WATER RESOURCE REGIONS						
Rio Grande	21	208	161	1,035	291	1,361
					314	1,386
					241	1,407
					250	1,456

Table 3.3 Historical, current, and forecasted expansion of urban and exurban lands in the Southwest (data expressed as thousands of acres [kac]) (Continued)

	Historical (1950)		Current (~2000)		Forecast (~2050)	
Geography	Urban developed (kac)	Exurban developed (kac)	Urban developed (kac)	Exurban developed (kac)	Urban developed (kac)	Exurban developed (kac)
Upper Colorado	7	112	79	889	150	1,118
					144	1,146
					118	1,183
					114	1,198
Lower Colorado	34	269	677	1,665	1,691	2,161
					1,879	2,006
					1,393	2,182
					1,496	2,144
Great Basin	37	259	283	835	899	1,199
					738	1,306
					769	1,222
					629	1,335
California	596	2311	2495	7874	4908	11,582
					5250	11,248
					4818	10,413
					4991	10,254

Source: Bierwagen et al. (2010).

Note: These reflect the storylines used in the IPCC's Special Report on Emissions Scenarios (Nakićenović and Swart 2000): A1, A2, B1, and B2, from top to bottom.

A remaining question of importance for this region is how well emerging "green design" strategies will be able to diminish or reduce resource demands for energy and water. Much of the development of alternative resources such as wind and solar energy has occurred remotely from the urban areas to be served, as has water-supply infrastructure. This geographical decoupling can be useful in some settings, but further removes social systems from natural system feedbacks. This can be a positive thing, but also can hinder the development of adaptive strategies because of a perceived lack of need to change behavior.

With environmental management decisions taking place at many geographic scales, over different time frames, and by multiple agencies, the coordination of climate change adaptation strategies will be a particular challenge.

References

Ackerly, D. D., S. R. Loarie, W. K. Cornwell, S. B. Weiss, H. Hamilton, R. Branciforte, and N. J. B. Kraft. 2010. The geography of climate change: Implications for conservation biogeography. *Diversity and Distributions* 16:476–487.

Bentz, B. J., J. Régnière, C. J. Fettig, E. M. Hansen, J. L. Hayes, J. A. Hicke, R. G. Kelsey, J. F. Negrón, and S. J. Seybold. 2010. Climate change and bark beetles of the western United States and Canada: Direct and indirect effects. *BioScience* 60:602–613.

Bierwagen, B. G., D. M. Theobald, C. R. Pyke, A. Choate, P. Groth, J. V. Thomas, and P. Morefield. 2010. National housing and impervious surface scenarios for integrated climate impact assessments. *Proceedings of the National Academy of Sciences* 107:20887–20892.

Hunt, C. B. 1974. *Natural regions of the United States and Canada*. New York: W. H. Freeman.

Lang, R. E., and A. C. Nelson. 2007. The rise of the megapolitans. *Planning* 73 (1): 7–12.

Loveland, T. R., T. L. Sohl, S. V. Stehman, A. L. Gallant, K. L. Sayler, and D. E. Napton. 2002. A strategy for estimating the rates of recent United States land-cover changes. *Photogrammetric Engineering and Remote Sensing* 68:1091–1099.

Nakićenović, N., and R. Swart, eds. 2000. *Special report on emissions scenarios: A special report of Working Group III of the Intergovernmental Panel on Climate Change*. Cambridge: Cambridge University Press.

Radeloff, V. C., R. B. Hammer, S. I. Stewart, J. S. Fried, S. S. Holcomb, and J. F. McKeefry. 2005. The wildland-urban interface in the United States. *Ecological Applications* 15:799–805.

Smith, J. B., and W. R. Travis. 2010. *Adaptation to climate change in public lands management*. Issue Brief 10-04. Washington, DC: Resources for the Future. http://www.rff.org/RFF/Documents/RFF-IB-10-04.pdf.

Theobald, D. M. 2001. Land use dynamics beyond the American urban fringe. *Geographical Review* 91:544–564.

Theobald, D. M. 2005. Landscape patterns of exurban growth in the USA from 1980 to 2020. *Ecology and Society* 10 (1): 32. http://www.ecologyandsociety.org/vol10/iss1/art32/.

Theobald, D. M., and W. H. Romme. 2007. Expansion of the US wildland-urban interface. *Landscape and Urban Planning* 83:340–354.

Travis, W. R. 2007. *New geographies of the American west: Land use and the changing patterns of place*. Washington, DC: Island Press.

U.S. Geological Survey (USGS), National Gap Analysis Program. 2010. *National land cover: Version 1*.

Appendix

See following page.

Table A3.1 List of ecological systems and groups, modified from USGS Southwest region

Group (L1)	Ecological systems
Alpine	Alpine sparse/barren
	Alpine grassland
Cliff-canyon-talus	Cliff, canyon and talus
Developed	Urban/built-up
	Cropland
Disturbed	Mining
	Recently burned
	Introduced vegetation
	Other disturbed or modified
Forest	Deciduous-dominated forest and woodland
	Mixed deciduous/coniferous forest and woodland
	Conifer-dominated forest and woodland
Grassland	Montane grassland
	Lowland grassland and prairie
	Sand prairie, coastal grasslands and lomas
	Wet meadow or prairie
Shrubland	Scrub shrubland
	Steppe
	Chaparral
	Deciduous-dominated savanna and glade
	Conifer-dominated savanna
	Sagebrush-dominated shrubland
	Deciduous-dominated shrubland
Sparse-barren	Beach, shore and sand
	Bluff and badland
	Other sparse and barren
Water	Rivers, lakes, reservoirs
Wetland-riparian	Playa, wash, and mudflat
	Salt, brackish & estuary wetland
	Freshwater herbaceous marsh
	Freshwater forested marsh or swamp
	Bog or fen
	Depressional wetland
	Floodplain and riparian

Source: USGS (2010).

Table A3.2 List of USGS land cover trends class descriptions

Land Cover Class	Description
Agriculture (cropland and pasture)	Land in either a vegetated or unvegetated state used for the production of food and fiber, including cultivated and uncultivated croplands, hay lands, pasture, orchards, vineyards, and confined livestock operations. Forest plantations are considered forests regardless of their use for wood products.
Barren	Land comprised of soils, sand, or rocks where <10% of the area is vegetated. Does not include land in transition recently cleared by disturbance.
Developed (urban and built-up)	Intensive use where much of the land is covered by structures or human-made impervious surfaces (residential, commercial, industrial, roads, etc.) and less-intensive use where the land-cover matrix includes both vegetation and structures (low-density residential, recreational facilities, cemeteries, utility corridors, etc.), and including any land functionally related to urban or built-up environments (parks, golf courses, etc.).
Forest and Woodland	Non-developed land where the tree-cover density is >10%. Note cleared forest land (i.e. clear-cuts) is mapped according to current cover (e.g. mechanically disturbed or grassland/shrubland).
Grassland/Shrubland (including rangeland)	Non-developed land where cover by grasses, forbs, or shrubs is >10%.
Mechanically Disturbed	Land in an altered, often unvegetated transitional state caused by disturbance from mechanical means, including forest clear-cutting, earthmoving, scraping, chaining, reservoir drawdown, and other human-induced clearance.
Mines and Quarries	Extractive mining activities with surface expression, including mining buildings, quarry pits, overburden, leach, evaporative features, and tailings.
Non-mechanically Disturbed	Land in an altered, often unvegetated transitional state caused by disturbance from non-mechanical means, including fire, wind, flood, and animals.
Open Water	Persistently covered with water, including streams, canals, lakes, reservoirs, bays, and ocean
Wetland	Land where water saturation is the determining factor in soil characteristics, vegetation types, and animal communities. Wetlands can contain both water and vegetated cover.

Source: USGS (2010).

Chapter 4

Present Weather and Climate: Average Conditions

COORDINATING LEAD AUTHOR

W. James Steenburgh (University of Utah)

LEAD AUTHORS

Kelly T. Redmond (Western Regional Climate Center and Desert Research Institute), Kenneth E. Kunkel (NOAA Cooperative Institute for Climate and Satellites, North Carolina State University and National Climate Data Center), Nolan Doesken (Colorado State University), Robert R. Gillies (Utah State University), John D. Horel (University of Utah)

CONTRIBUTING AUTHORS

Martin P. Hoerling (NOAA Earth System Research Laboratory), Thomas H. Painter (Jet Propulsion Laboratory)

EXPERT REVIEW EDITOR

Roy Rasmussen (National Center for Atmospheric Research)

Executive Summary

This chapter describes the weather and climate of the Southwest, which straddles the mid- and subtropical latitudes and includes the greatest range of topographic relief in the contiguous United States. The key findings are as follows:

Chapter citation: Steenburgh, W. J., K. T. Redmond, K. E. Kunkel, N. Doesken, R. R. Gillies, J. D. Horel, M. P. Hoerling, and T. H. Painter. 2013. "Present Weather and Climate: Average Conditions." In *Assessment of Climate Change in the Southwest United States: A Report Prepared for the National Climate Assessment*, edited by G. Garfin, A. Jardine, R. Merideth, M. Black, and S. LeRoy, 56–73. A report by the Southwest Climate Alliance. Washington, DC: Island Press.

- The climate of the Southwest United States is highly varied and strongly influenced by topographic and land-surface contrasts, the mid-latitude storm track, the North American monsoon, and proximity to the Pacific Ocean, Gulf of California, and Gulf of Mexico. (high confidence)
- The low-elevation Mojave and Sonoran Deserts of Southern California, Nevada, and Arizona are the hottest (based on July maximum temperatures), driest regions of the contiguous United States. The mountain and upper-elevation regions of the Southwest are much cooler, with the Sierra Nevada and mountains of Utah and Colorado receiving more than 60% of their annual precipitation in the form of snow. (high confidence)
- Storms originating over the Pacific Ocean produce most of the cool-season (November to April) precipitation, generating a mountain snowpack that provides much of the surface-water resources for the region as spring runoff. (high confidence)
- Persistent cold pools, also known as inversions, form in valleys and basins during quiescent wintertime weather periods, leading to a buildup of pollution in some areas. (high confidence)
- The North American monsoon is important during the warm season and is most prominent in Arizona and New Mexico where it produces up to half of the average annual precipitation from July to September. (high confidence)
- The Madden-Julian Oscillation (MJO), El Niño-Southern Oscillation (ENSO), Pacific Quasi-Decadal Oscillation (QDO), and Pacific Decadal Oscillation (PDO) contribute to but do not fully explain month-to-month, year-to-year, and decade-to-decade climate variability within the region. (medium-high confidence)

4.1 Introduction

The Southwest straddles the mid- and subtropical latitudes, with mountains, land-surface contrasts, and proximity to the Pacific Ocean, Gulf of California, and Gulf of Mexico having substantial impacts on climatic conditions (Sheppard et al. 2002). Much of California has a Mediterranean-like climate characterized by hot, dry summers and mild winters with episodic, but occasionally intense rainstorms. The interior, southern, low-elevation portion of the region, which includes the Mojave and Sonoran Deserts (see Chapter 3 for a geographic overview of the Southwest), contains the hottest (based on summertime maximum temperatures) and driest locations in the United States, a result of persistent subtropical high pressure and topographic effects. Interior northern and eastern portions of the Southwest have lower mean annual temperatures and see a larger seasonal temperature range, greater weather variability, and more frequent intrusions of cold air from the higher latitudes due to increased elevation and distance from the Pacific Ocean and Gulf of Mexico.

During winter, the mid-latitude storm track influences the region (Hoskins and Hodges 2002; Lareau and Horel 2012). Average cool-season precipitation is greatest in the coastal ranges and Sierra Nevada of California by virtue of their position as the first mountain barriers in the path of Pacific storms. Interior ranges receive less precipitation on average, but seasonal snow accumulations can still be quite large and are an essential source of water for the region. During the summer, a notable feature of the climate of the interior Southwest is a peak in precipitation caused by the North American monsoon, a shift in the large-scale atmospheric circulation that brings moisture originating from the Gulf of Mexico, Gulf of California, and Pacific Ocean into the Southwest (Adams and Comrie 1997; Higgins, Yao, and Wang 1997). The influence of the North American monsoon is strongest in Arizona and New Mexico, where up to 50% of the average annual precipitation falls from July to September (Douglas et al. 1993).

4.2 General Climate Characteristics

Surface-air temperature

The surface-air temperature (hereafter temperature) climatology of the Southwest varies with latitude, distance from large bodies of water, and altitude. The average annual temperature is highest (greater than 70°F) in the Mojave and Sonoran Deserts of southwestern Arizona, southeastern California, and extreme southern Nevada, including Death Valley and the lower Colorado River valley, and greater than 55°F in a swath extending from central California to southern New Mexico (Figure 4.1a). The highest average annual temperatures in the interior Southwest (greater than 50°F) occur in lower altitude regions of southern and western Nevada, the Great Salt Lake Basin, the Colorado Plateau, and the high plains of Colorado and New Mexico. Throughout the Southwest, the average annual temperature decreases with altitude and is lowest (less than 32°F) over the upper elevations of the Uinta Mountains of Utah and the Rocky Mountains of Colorado. The total spatial range in average annual temperature is very large and exceeds 40°F, which contributes to large ecosystem variations (see Chapter 8).

During July, climatologically the warmest month of the year across most of the region, average maximum temperatures exceed 100°F in the low basins and valleys of southeastern California, southern Nevada, and southern and western Arizona and exceed 85°F in lower altitude valleys, basins, and plains throughout the Southwest region (Figure 4.1b). Much lower average maximum temperatures are found along and near the California coast due to the influence of the Pacific Ocean, and at the upper elevations.

During January—climatologically the coldest month of the year across most of the region—average minimum temperatures are highest and above freezing (greater than 32°F) across lower-altitude regions of California, including the Pacific coast and Central Valley, southern Nevada, southern and western Arizona, and southern New Mexico (Figure 4.1c). Elsewhere, January minimum temperatures are below freezing. Although there is a tendency for temperature to decrease with increasing altitude, this relationship is weaker in the winter than in summer because of the tendency of cold-air pools to develop and persist over mountain valleys and basins (Wolyn and McKee 1989; Whiteman, Bian, and Zhong 1999; Reeves and Stensrud 2009). For example, the average minimum temperatures in the Uinta Basin of northeast Utah and the San Luis Valley of Colorado and New Mexico are comparable to those found in the surrounding higher terrain. Similar cold-air pools occur in valleys and basins throughout the Southwest region.

The difference between the average July maximum temperature and average January minimum temperature is smallest (less than 40°F) along and near the Pacific coast and generally smaller over much of California than the interior Southwest due to the moderating influence of the Pacific Ocean (Figure 4.1d). The largest annual temperature ranges are found in the interior mountain valleys and basins of Utah and Colorado that are prone to wintertime cold pools. Throughout the region, including California, the annual temperature range is generally larger in valleys and basins than in the surrounding higher topography.

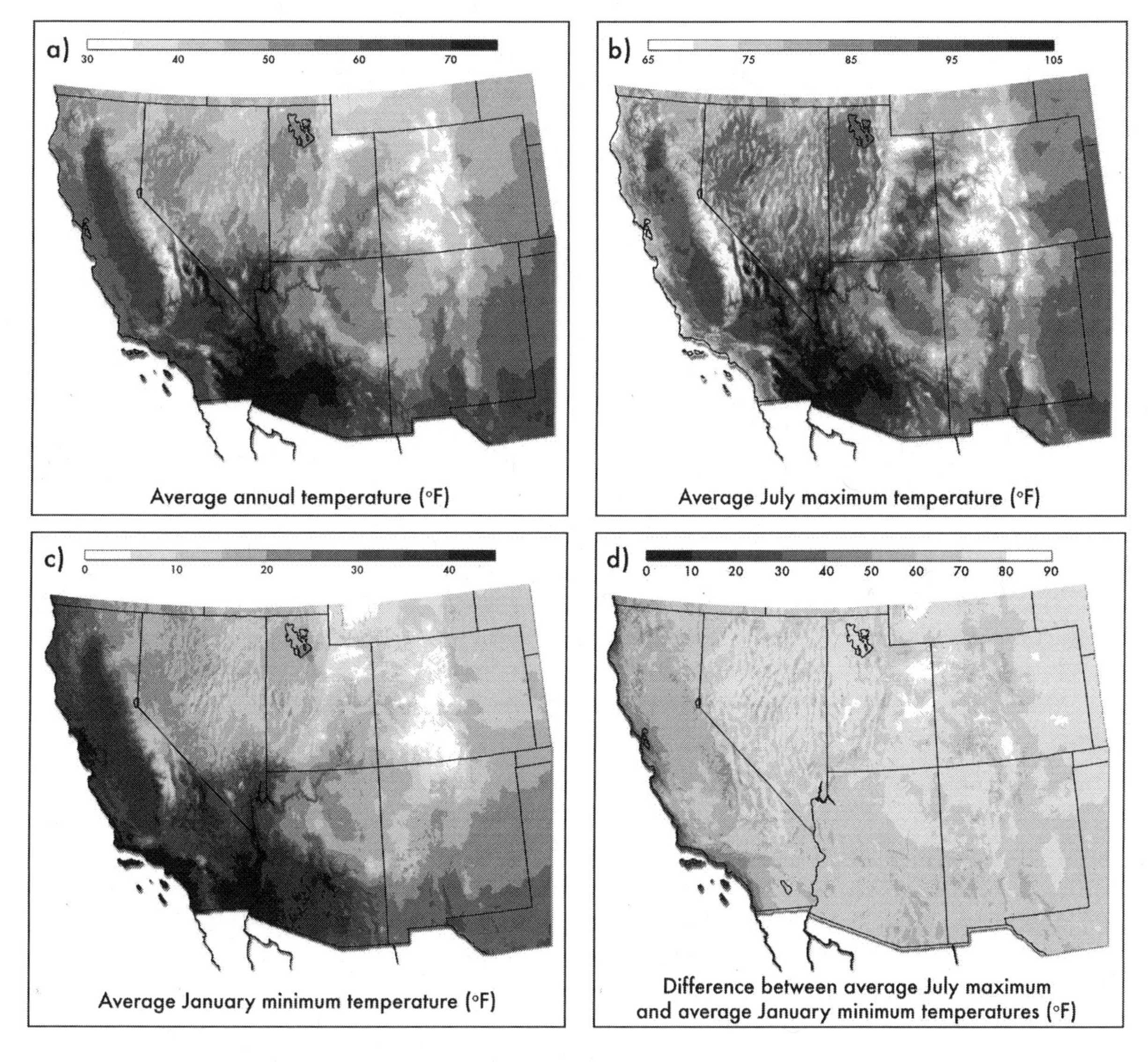

Figure 4.1 Temperature climatology of the Southwest (°F, 1971–2000). Source: PRISM Climate Group, Oregon State University (http://prism.oregonstate.edu).

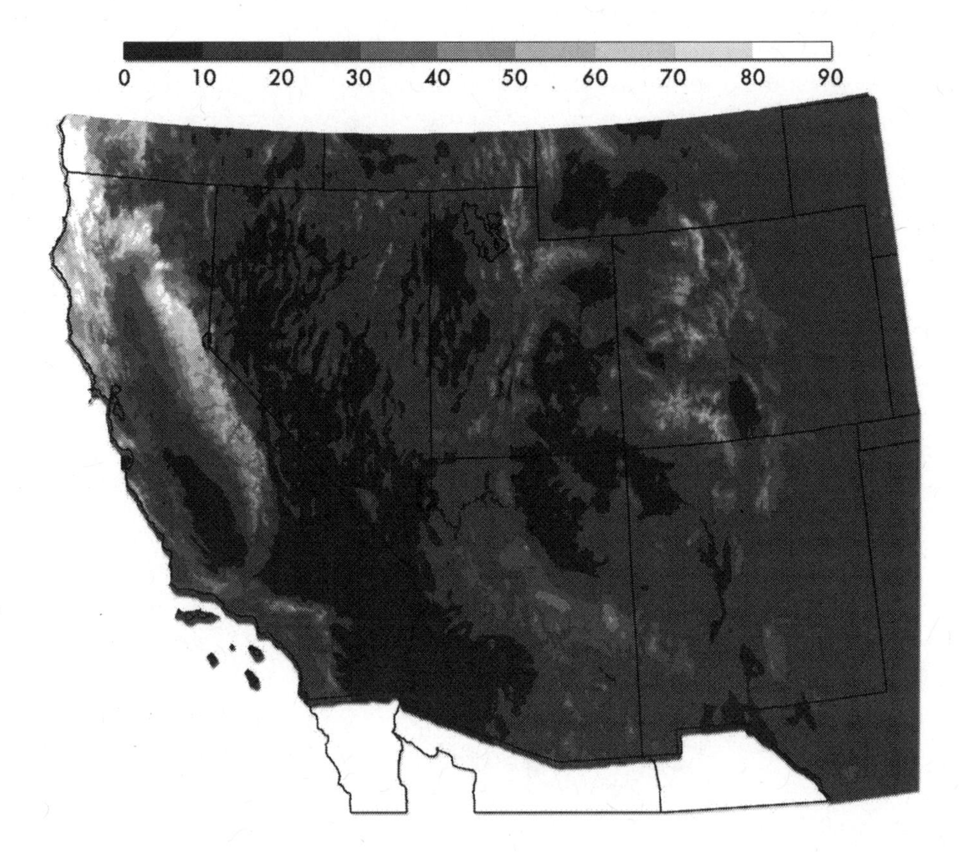

Figure 4.2 Annual average precipitation in the Southwest, 1971–2000 (in inches). Source: PRISM Climate Group, Oregon State University (http://prism.oregonstate.edu).

Precipitation

Average annual precipitation varies from less than 5 inches in the lower valleys and basins of southwestern Arizona, southeastern California, and extreme southern and western Nevada to more than 90 inches in portions of the coastal mountains, southern Cascade Mountains, and northern Sierra Nevada of Northern California (Figure 4.2). Large variations in precipitation exist throughout the region due to the influence of topography (Daly, Neilson, and Phillips 1994).

The seasonality of precipitation varies substantially across the Southwest depending on exposure to the mid-latitude westerly storm track during the cool season, the monsoon circulation during the warm season, and elevation. Most of California and portions of Nevada, Utah, and Colorado are strongly influenced by the mid-latitude storm track, with most precipitation falling during the cool season when the mid-latitude storm track is most active (Figure 4.3a). Monsoon precipitation in these areas is less abundant. Most of Arizona, western New Mexico, and portions of extreme southeast California, southern Nevada, southern Utah, and southwest Colorado observe a pronounced peak in precipitation in late summer due to the influence of the monsoon (Figure 4.3b). Since monsoon precipitation is produced primarily by thunderstorms, large spatial contrasts in seasonal precipitation can be found within these areas during individual summers. The high plains and tablelands of New Mexico and Colorado also observe a summer maximum, but with a broader peak due to frequent spring storms prior to the development of the monsoon (Figure 4.3c). This region also receives less wintertime precipitation due to the predominant westerly flow and drying influence of upstream mountain ranges. In central Nevada, Utah, and western Colorado, the seasonal cycle of precipitation is not as pronounced, but generally peaks from March to May (Figure 4.3d). Areas not identified in Figure 4.3 do not have strong seasonal precipitation variations.

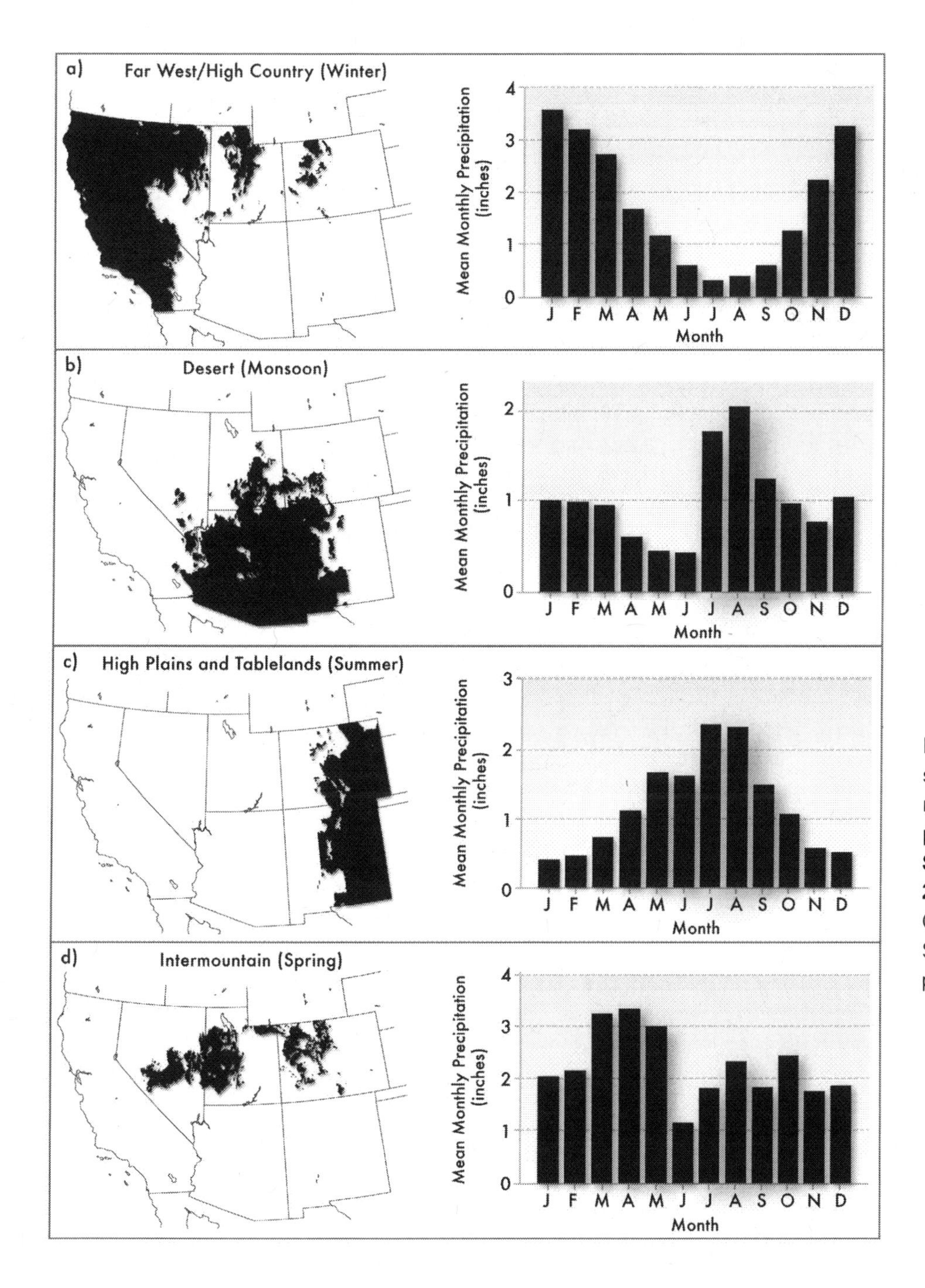

Figure 4.3 Major spatial patterns of monthly varying precipitation over the Southwest, 1901–2010. Source: PRISM Climate Group, Oregon State University (http://prism.oregonstate.edu).

Snowfall, snowpack, and water resources

A large fraction of the precipitation in the upper-elevations of the Southwest falls as snow, which serves as the primary source of water for the region and enables a winter tourism economy involving skiing, snowboarding, snowmobiling, and other recreational activities. In the Sierra Nevada and mountains of Utah and Colorado, more than 60% of the annual precipitation falls as snow (Serreze et al. 1999). Snowier locations in the Sierra Nevada, Wasatch Mountains, and Colorado Rockies average over thirty feet of snow annually, with lesser amounts in other ranges of the Southwest (Steenburgh and Alcott 2008). Snowfall provides a smaller fraction of the annual precipitation in the mountains of Arizona and southern New Mexico where winter storms are less frequent and the monsoon dominates in the summer (Serreze et al. 1999; Stewart, Cayan, and Dettinger 2004).

The mountain snowpack that develops during the winter serves as a natural water reservoir for the western United States. Snowmelt and runoff during the spring and summer provide most of the surface water resources for the region, with 50% to 90% of the total runoff occurring during the April to July snowmelt runoff season in most Southwest drainage basins (Serreze et al. 1999; Stewart, Cayan, and Dettinger 2004, 2005).

4.3 Major Climate and Weather Events

The Southwest is susceptible to hazardous and costly weather and climate events. The greatest social and environmental impacts come from drought, winter storms, floods, thunderstorms, temperature extremes, and air pollution.

Drought

The Southwest is susceptible to periods of dryness that can span months to years. The most significant and severe droughts persist for multiple years and result from a diminished frequency or intensity of winter storms (Cayan et al. 2010; Woodhouse et al. 2010). Although water storage and delivery infrastructure (such as dams, reservoirs, canals, and pipelines) helps stabilize municipal water supplies during these droughts, rural, agricultural, and recreational impacts are still sometimes substantial. Seasonal and multiyear droughts also affect wildfire severity (Westerling et al. 2003). Parts of the Southwest experienced relatively wet conditions during the 1980s and 1990s followed by reduced precipitation beginning around 2000 (Cayan et al. 2010). This, in combination with wildfire suppression and land management practices (Allen et al. 2002), contributed to wildfires of unprecedented size, with five states (Arizona in 2002 and 2011; Colorado in 2002; Utah in 2007; California in 2003; and New Mexico in 2011 and 2012) experiencing their largest fires on record at least once during the last decade. Past climatic conditions, reconstructed from tree rings, suggest that droughts lasting up to several decades have occurred in the Colorado River Basin approximately once or twice per century during the last 500 to 1,000 years (Figure 4.5) (Grissino-Mayer and Swetnam 2000; Woodhouse 2003; Meko et al. 2007; Woodhouse et al. 2010).

Drought characteristics vary across the Southwest. For example, Colorado, which has multiple sources of precipitation throughout the year, is less prone to lengthy droughts (Redmond 2003). Droughts in Arizona and New Mexico tend to be strongly related to

Box 4.1

Dust and Snow

How dust affects the timing and intensity of snowmelt and runoff in the Southwest is an area of emerging understanding. During the winter and spring, wind-blown dust from lowland regions can accumulate in the mountain snowpack (Figure 4.4) (Painter et al. 2007; Steenburgh, Massey, and Painter 2012). Because dust is darker than snow, this increases the amount of sunlight absorbed by the snow, leading to an earlier, more rapid snowmelt. Studies in Colorado's San Juan Mountains, for example, indicate that the duration of snowcover in the spring and summer can be shortened by several weeks in years with large dust accumulations (Painter et al. 2007; Painter et al. 2012; Skiles et al. 2012). Modeling studies suggest that this results in a runoff with a more rapid increase, earlier peak, and reduced volume in the upper Colorado River Basin (Painter et al. 2010).

Efforts to better understand the spatial and year-to-year variations in dust accumulation and characteristics in the mountains of the Southwest are ongoing. Better understanding of these variations will help improve the prediction of spring runoff timing and volume, as well as projections of the impacts of climate change on mountain snowpack and ecosystems.

Figure 4.4 Dust-covered snow in the Dolores River headwaters, San Juan Mountains, Colorado, 19 May 2009. Photo courtesy of T. H. Painter, Snow Optics Laboratory, JPL/Caltech.

large-scale shifts in the atmospheric circulation associated with the El Niño-Southern Oscillation (ENSO). ENSO refers collectively to episodes of warming and cooling of the equatorial Pacific Ocean and their related atmospheric circulation changes. Warm and cool ENSO episodes are known as El Niño and La Niña, respectively. La Niña years are associated with reduced cool-season precipitation over southern portions of the Southwest region (Redmond and Koch 1991; Cayan, Redmond, and Riddle 1999).

Winter storms

Winter storms in the Southwest can produce heavy snowfall, heavy rainfall, flooding, high winds, and large, abrupt temperature drops (Marwitz 1986; Marwitz and Toth 1993; Poulos et al. 2002; Schultz et al. 2002; Steenburgh 2003; White et al. 2003; Neiman et al. 2008; Shafer and Steenburgh 2008). With their close proximity to the Pacific Ocean,

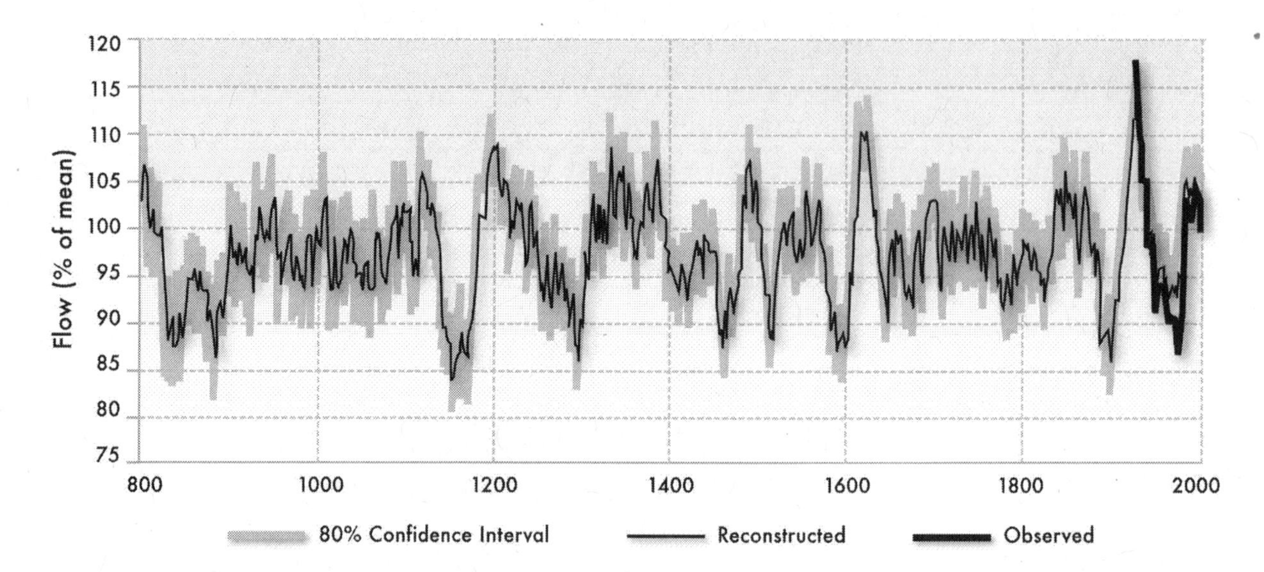

Figure 4.5 25-year running mean of reconstructed (thin line) and observed (thick line) flows at Lees Ferry, Arizona. Adapted from Meko et al. (2007) with permission from the American Geophysical Union.

the coastal ranges and the Sierra Nevada of California experience episodes of heavy precipitation as impressive as any in the United States (Dettinger, Ralph, Das et al. 2011). Narrow corridors of moisture known as atmospheric rivers, which are typically found near or ahead of cold fronts, contribute to many of these episodes (Neiman et al. 2008). Mountains throughout the Southwest, as well as lowlands in the Southwest interior, can experience large, multiday snowstorms that produce hazardous travel and avalanche conditions and play an important role in the regional precipitation, runoff, and water balance (Poulos et al. 2002; Steenburgh 2003). Arctic outbreaks can produce large snow accumulations and blizzard conditions over the high plains and Front Range of eastern Colorado and New Mexico (Marwitz and Toth 1993; Rasmussen et al. 1995; Poulos et al. 2002).

Strong winds, which in some areas are enhanced by coastal and topographic effects, also occur throughout the region. Severe downslope winds occur along several mountain ranges in the Southwest including the Sierra Nevada, Wasatch Mountains, and Front Range of Colorado. These events can produce severe aircraft turbulence and surface wind gusts that exceed 100 miles per hour (Lilly and Zipser 1972; Clark et al. 2000; Grubišić et al. 2008). In Southern California, the Santa Ana winds can produce extreme wildfire behavior and have played important roles in recent megafires in the region (Keeley et al. 2009).

Floods

Several mechanisms contribute to flooding in the Southwest (Hirschboeck 1988; Michaud, Hirshboeck, and Winchell 2001). During the winter, heavy precipitation associated with landfalling mid-latitude cyclones and concomitant atmospheric rivers can produce widespread flooding in California, Arizona, Nevada, Utah, and New Mexico. Because of its proximity to the Pacific Ocean, proclivity for slow moving, multi-day storms that tap into moisture from the tropics and subtropics, and high mountains with extensive cool-season snow cover, California is especially vulnerable to this type of flooding (Ralph et al. 2006; Dettinger 2011; Dettinger, Ralph, Das et al. 2011).

Climatologists have recently recognized that gigantic cool-season flooding events such as the California Flood of 1861–62 (Null and Hubert 2007; Porter et al. 2011) can be identified in paleoclimate evidence (such as tree rings and sediment layers), with an approximate average recurrence interval of about 300 years. California was sparsely populated during the 1861–62 winter, but recent exercises by the emergency response community suggest that damages and expenses from such a several-week sequence of major storms could today reach $0.5 trillion to $1.0 trillion (Dettinger, Ralph, Hughes et al. 2011; Porter et al. 2011).

Flooding can also occur during the spring runoff, especially in years in which the snowpack persists into the late spring and is followed by an early summer heat wave or rain-on-snow event. Such flooding occurred in Utah during the spring and summer of 2011 (FEMA 2011).

Flash floods associated with thunderstorms occur throughout the Southwest, many during the months of the North American monsoon (Hirschboeck 1987; Maddox, Canova, and Hoxit 1980). In some instances, moisture associated with the remnants of decaying tropical cyclones from the eastern Pacific contributes to the flooding (Ritchie et al. 2011). Because of heavy precipitation rates, topographic channeling, and the impervious nature of the land surface in some urban and desert areas, the flooding produced by these thunderstorms can be abrupt and severe. Along the Colorado Front Range, extensive complexes of nearly stationary heavy thunderstorms have produced very destructive flash floods, including the 1976 Big Thompson Flood that killed more than 125 people and the 1997 Fort Collins Flood that produced more than $250 million in property damage (Maddox et al. 1978; Caracena et al. 1979; Weaver, Gruntfest, and Levy 2000). Given the localized nature of the rainfall produced by these thunderstorms, however, flooding is rarely severe in larger drainage basins.

Thunderstorms

Hazardous weather produced by thunderstorms does occur in the Southwest. Lightning, a primary concern for public safety, killed 49 people in the Southwest from 2001–2010, including 26 in Colorado (NWS 2012). Lightning also ignites wildfires and contributes to the regional wildfire climatology (Swetnam and Betancourt 1998). Although damaging hail is rare in most of the Southwest, the hail intensity in eastern Colorado and New Mexico is among the highest in North America (Changnon 1977). In Colorado, the Rocky Mountain Insurance Information Association reported $3 billion in hail damage during the past 10 years (RMIIA 2012). Strong winds can also accompany thunderstorms and, in some instances, generate severe dust storms (Brazel and Nickling 1986), known as haboobs in Arizona. The dry, low-level environment commonly found over the Southwest during thunderstorms can contribute to the development of microbursts, localized areas of sinking air that generate strong straight-line winds at the surface and are a concern for public safety and aviation. Tornadoes are rare in California, Nevada, Utah, and

Arizona, but have been reported in all four states (NCDC 2012). The frequency of days with tornadoes is, however, much higher in eastern Colorado and New Mexico, which lie on the western edge of "Tornado Alley." In particular, the frequency of tornado days in northeast Colorado is among the highest in the United States (Brooks, Doswell, and Kay 2003).

Temperature extremes

Cold and heat waves occur in the Southwest (Golden et al. 2008; Grotjahn and Faure 2008; Gershunov, Cayan, and Iacobellis 2009). Much of California and portions of southern Nevada, southwest Utah, southern Arizona, and southern New Mexico experience generally mild winters, but are susceptible to hard freezes when the storm track plunges far to the south of its average position. Hard freezes damage agricultural crops, ornamental plants, and (through frozen pipes) public and household utilities. Hard freezes causing in excess of $1 billion in damages and losses in California occurred in December 1990, December 1997, and January 2006 (NCDC 2011). Although net losses are not as great as those in California, early growing-season freezes can also produce agricultural damage in interior regions of the Southwest.

The Southwest also experiences episodes of extended high temperatures that affect ecosystems, hydrology, agriculture and livestock, and human comfort, health, and mortality (Gershunov, Cayan, and Iacobellis 2009; see also Chapter 15). For example, a 2006 California heat wave contributed to more than 140 deaths, 16,000 excess emergency department visits, and 1,000 excess hospitalizations (Knowlton et al. 2009). Such human health impacts are greatly exacerbated by high humidity and high nighttime temperatures (Golden et al. 2008; Gershunov, Cayan, and Iacobellis 2009). (See also the discussions of past and projected extreme climate events in the Southwest in Chapters 5 and 7.)

Air quality

As of August 2011, the Environmental Protection Agency had designated at least one county in each Southwest state as being in nonattainment for the National Ambient Air Quality Standards for one or more pollutants (Figure 4.6; see also EPA 2012). During the cool season, persistent cold-air pools that form in mountain valleys and basins can trap emissions (from motor vehicles, wood-burning stoves, industry, etc.) that lead to the formation of secondary particulate matter. During multi-day events, elevated particulate matter levels can develop in large urban areas (such as Salt Lake City, Utah). Even smaller cities in relatively confined valleys (such as Logan, Utah) may experience elevated particulate levels from urban and agricultural emissions (Malek et al. 2006; Silva et al. 2007; Gillies, Wang, and Booth 2010).

Elevated wintertime ozone levels have also been reported during intense temperature inversions in the vicinity of rural natural gas fields (Schnell et al. 2009). In the warm season, quiescent weather conditions, high temperatures, and intense solar radiation can lead to elevated ozone levels. In portions of coastal California, pollutants are often trapped beneath the marine inversion, which is a climatological feature over the eastern Pacific Ocean. Emissions from wildfires can also contribute to particulate matter or ozone production and wind-blown dust can produce elevated particulate matter levels in some areas (Pheleria et al. 2005; Steenburgh, Massey, and Painter 2012).

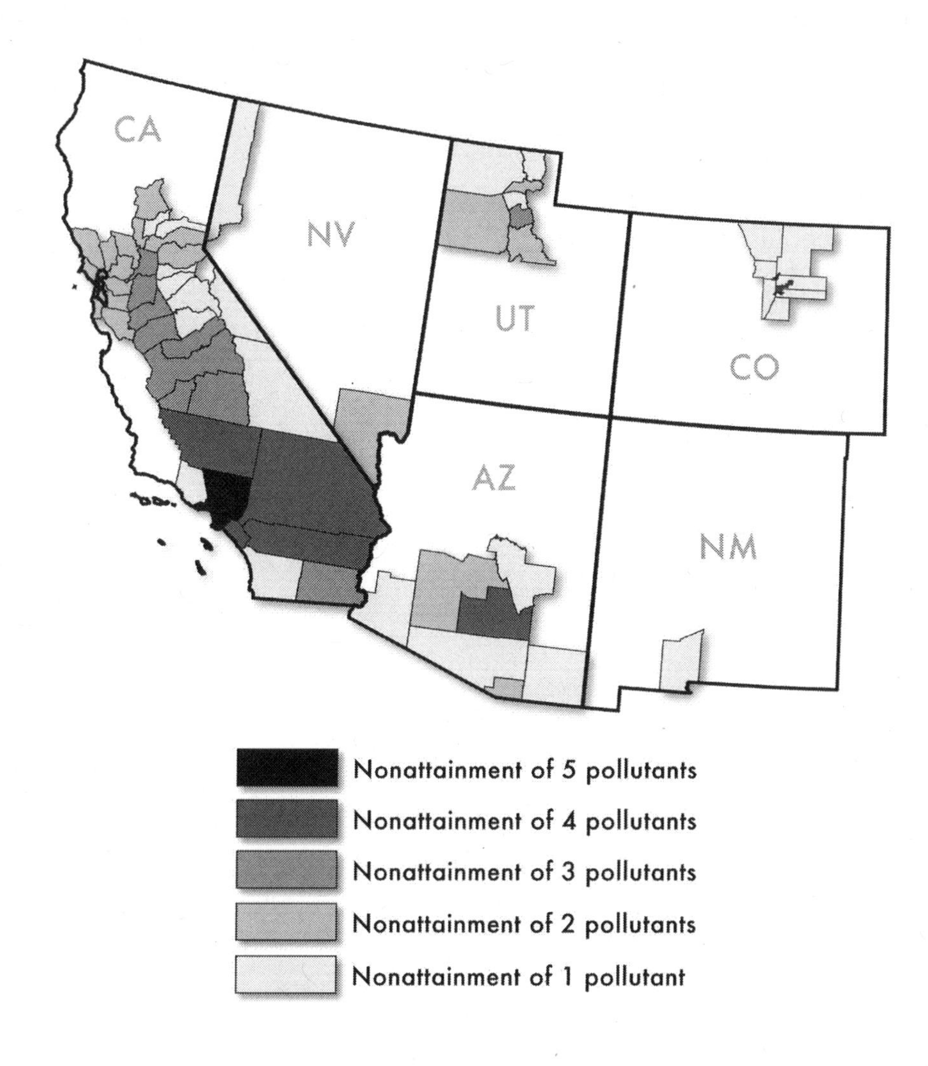

Figure 4.6 Counties designated by the Environmental Protection Agency as nonattainment areas for National Ambient Air Quality Standards in April 2010. Source: EPA (n.d.).

4.4 Climate Variability

The general climate characteristics of the Southwest described in section 4.2 reflect averages over a period of roughly 30 years. There is, however, considerable variability in the climate of the Southwest that occurs from month-to-month (intraseasonal), year-to-year (interannual), and decade-to-decade (interdecadal). In portions of the region, especially California, Nevada, and Arizona, this variability leads to the largest fluctuations relative to the mean in annual precipitation and streamflow in the contiguous United States, posing challenges for water-resource management and drought and flood mitigation (Dettinger, Ralph, Das et al. 2011).

Interannual and interdecadal climate variations in the Southwest are demonstrably related to fluctuations in Pacific sea-surface temperatures (SST) over periods of years to decades, such as the 2- to 7-year ENSO (Dettinger et al. 1998; Higgins and Shi 2001), the Pacific Quasi-Decadal Oscillation (QDO) (Tourre et al. 2001; Wang et al. 2009, 2010), and the Pacific Decadal Oscillation (PDO) (Mantua et al. 1997; Zhang, Wallace, and Battisti 1997). The occurrence and transition from long-lasting droughts to periods of above normal rainfall have been linked to these very slow changes in the Pacific Ocean (Sangoyomi 1993; Gershunov and Barnett 1998; Brown and Comrie 2004; Zhang and Mann 2005; Wang et al. 2010).

The impact of ENSO on precipitation and drought anomalies over western North America has been studied extensively. It is now well established that ENSO tends to produce the so-called North American dipole, a situation in which relative conditions of precipitation and temperature (high vs. low) occur in opposition simultaneously for the Pacific Northwest and for the Southern California-Arizona-New Mexico area (Dettinger at al. 1998), with marginal influence on conditions for areas in between (Rajagopalan and Lall 1998). This yields a tendency for above-average precipitation and temperatures over the southern Southwest during El Niño winters and below-average precipitation during La Niña winters. Despite these shifts relative to average, extreme episodic precipitation events can occur in either El Niño and La Niña winters (Feldl and Roe 2010, 2011).

Intraseasonal (that is, time scales beyond a few days and shorter than a season) variations in precipitation have been examined for the summer North American monsoon and winter season precipitation (Mo 1999, 2000; Higgins and Shi 2001; Mo and Nogues-Paegle 2005; Becker, Berbery, and Higgins 2011). Over California during winter, these variations have been linked to the Madden-Julian Oscillation (MJO), an atmospheric phenomenon that contributes to cyclical outbreaks of convection near the equator that in turn affect atmospheric circulations over the midlatitude Pacific Ocean (Mo 1999; Becker, Berbery, and Higgins 2011; Guan et al. 2012).

References

Adams, D. K., and A. C. Comrie. 1997. The North American monsoon. *Bulletin of the American Meteorological Society* 78:2197–2213.

Allen, C. D., M. Savage, D. A. Falk, K. F. Suckling, T. W. Swetnam, T. Shulke, P. B. Stacey, P. Morgan, M. Hoffman, and J. T. Klingel. 2002. Ecological restoration of southwestern ponderosa pine ecosystems: A broad perspective. *Ecological Applications* 12:1418–1433.

Becker, E. J., E. H. Berbery, and R. W. Higgins. 2011. Modulation of cold-season U.S. daily precipitation by the Madden–Julian Oscillation. *Journal of Climate* 24:5157–5166.

Brazel, A. J., and W. G. Nickling. 1986. The relationship of weather types to dust storm generation in Arizona (1965–1980). *International Journal of Climatology* 6:255–275.

Brooks, H. E., C. A. Doswell III, and M. P. Kay. 2003. Climatological estimates of local daily tornado probability for the United States. *Weather and Forecasting* 18:626–640.

Brown, D. P., and A. C. Comrie. 2004. A winter precipitation "dipole" in the western United States associated with multidecadal ENSO variability. *Geophysical Research Letters* 31: L09203.

Caracena, F., R. A. Maddox, L. R. Hoxit, and C. F. Chappell. 1979. Mesoanalysis of the Big Thompson Storm. *Monthly Weather Review* 107:1–17.

Cayan, D. R., T. Das, D. W. Pierce, T. P. Barnett, M. Tyree, and A. Gershunov. 2010. Future dryness in the southwest US and the hydrology of the early 21st century drought. *Proceedings of the National Academy of Sciences* 107:21271–21276.

Cayan, D. R., K. T. Redmond, and L. G. Riddle. 1999. ENSO and hydrologic extremes in the western United States. *Journal of Climate* 12:2881–2893.

Changnon, Jr., S. A. 1977. The scales of hail. Journal of Applied Meteorology and Climatology 16:626–648.

Clark, T. L., W. D. Hall, R. M. Kerr, D. Middleton, L. Radke, F. M. Ralph, P. J. Neiman, and D. Levinson. 2000. Origins of aircraft-damaging clear-air turbulence during the 9 December 1992 Colorado downslope windstorm: Numerical simulations and comparison with observations. *Journal of the Atmospheric Sciences* 57:1105–1131.

Daly, C., R. P. Neilson, and D. L. Phillips. 1994. A statistical–topographic model for mapping climatological precipitation over mountainous terrain. *Journal of Applied Meteorology and Climatology* 33:140–158.

Dettinger, M. D. 2011. Climate change, atmospheric rivers, and floods in California – A multimodel analysis of storm frequency and magnitude changes. *Journal of the American Water Resources Association* 47:514–523.

Dettinger, M. D., D. R. Cayan, H. F. Diaz, and D. M. Meko. 1998. North–south precipitation patterns in western North America on interannual-to-decadal timescales. *Journal of Climate* 11:3095–3111.

Dettinger, M. D., F. M. Ralph, T. Das, P. J. Neiman, and D. R. Cayan. 2011. Atmospheric rivers, floods, and the water resources of California. *Water* 3:445–478.

Dettinger, M. D., F. M. Ralph, M. Hughes, T. Das, P. Neiman, D. Cox, G. Estes, et al. 2011. Design and quantification of an extreme winter storm scenario for emergency preparedness and planning exercises in California. *Natural Hazards* 60:1085–1111, doi:10.1007/s11069-011-9894-5.

Douglas, M. W., R. A. Maddox, K. Howard, and S. Reyes. 1993. The Mexican monsoon. *Journal of Climate* 6:1665–1677.

Feldl, N., and G. H. Roe. 2010. Synoptic weather patterns associated with intense ENSO rainfall in the southwest United States. *Geophysical Research Letters* 37: L23803.

—. 2011. Climate variability and the shape of daily precipitation: A case study of ENSO and the American West. *Journal of Climate* 24:2483–2499.

Federal Emergency Management Agency (FEMA). 2011. President declares a major disaster for Utah. News release, August 8, 2011. http://www.fema.gov/news-release/president-declares-major-disaster-utah-0.

Gershunov, A., and T. P. Barnett. 1998. Interdecadal modulation of ENSO teleconnections. *Bulletin of the American Meteorological Society* 79:2715–2725.

Gershunov, A., D. R. Cayan, and S. F. Iacobellis. 2009. The great 2006 heat wave over California and Nevada: Signal of an increasing trend. *Journal of Climate* 22:6181–6203.

Gillies, R. R., S-Y. Wang, and M. R. Booth. 2010. Atmospheric scale interaction on wintertime Intermountain West low-level inversions. *Weather and Forecasting* 25:1196–1210.

Golden, J. S., D. Hartz, A. Brazel, G. Luber, and P. Phelan. 2008. A biometeorology study of climate and heat-related morbidity in Phoenix from 2001 to 2006. *International Journal of Biometeorology* 52:471–480.

Grissino-Mayer, H. D., and T.W. Swetnam. 2000. Century-scale climate forcing of fire regimes in the American Southwest. *Holocene* 10:213–220.

Grotjahn, R., and G. Faure. 2008. Composite predictor maps of extraordinary weather events in the Sacramento, California, region. *Weather and Forecasting* 23:313–335.

Grubišić, V., J. D. Doyle, J. Kuettner, R. Dirks, S. A. Cohn, L. L. Pan, S. Mobbs, et al. 2008. The terrain-induced rotor experiment. *Bulletin of the American Meteorological Society* 89:1513–1533.

Guan, B., D. E. Waliser, N. P. Molotock, E. J. Fetzer, and P. J. Neiman. 2012. Does the Madden–Julian Oscillation influence wintertime atmospheric rivers and snowpack in the Sierra Nevada? *Monthly Weather Review* 140:325–342.

Higgins, R. W., and W. Shi. 2001. Intercomparison of the principal modes of interannual and intraseasonal variability of the North American Monsoon System. *Journal of Climate* 14:403–417.

Higgins, R. W., Y. Yao, and X. L. Wang. 1997. Influence of the North American Monsoon System on the U.S. summer precipitation regime. *Journal of Climate* 10:2600–2622.

Hirshboeck, K. K. 1987. Hydroclimatically-defined mixed distributions in partial duration flood series. In *Hydrologic frequency modeling*, ed. V. P. Singh, 199–212. Dordrecht: D. Reidel.

—. 1988. Flood hydroclimatology. In *Flood geomorphology*, ed. V. R. Baker, R. C. Kochel, and P. C. Patton, 27–49. Hoboken, NJ: John Wiley and Sons.

Hoskins, B. J., and K. I. Hodges. 2002. New perspectives on the Northern Hemisphere winter storm tracks. *Journal of the Atmospheric Sciences* 59:1041–1061.

Keeley, J. E., H. Safford, C. J. Fotheringham, J. Franklin, and M. Moritz. 2009. The 2007 Southern California wildfires: Lessons in complexity. *Journal of Forestry* 107:287–296.

Knowlton, K., M. Rotkin-Ellman, G. King, H. G. Margolis, D. Smith, G. Solomon, R. Trent, and P. English. 2009. The 2006 California heat wave: Impacts on hospitalizations and emergency department visits. *Environmentl Health Perspectives* 117:61–67.

Lareau, N., and J. D. Horel. 2012. The climatology of synoptic-scale ascent over western North America: A perspective on storm tracks. *Monthly Weather Review* 140:1761–1778.

Lilly, D. K., and E. J. Zipser. 1972. The Front Range windstorm of January 11, 1972: A meteorological narrative. *Weatherwise* 25:56–63.

Maddox, R. A., F. Canova, and L. R. Hoxit. 1980. Meteorological characteristics of flash flood events over the western United States. *Monthly Weather Review* 108:1866–1877.

Maddox, R. A., L. R. Hoxit, C. F. Chappell, and F. Caracena. 1978. Comparison of meteorological aspects of the Big Thompson and Rapid City flash floods. *Monthly Weather Review* 106:375–389.

Malek, E., T. Davis, R. S. Martin, and P. J. Silva. 2006. Meteorological and environmental aspects of one of the worst national air pollution episodes (January, 2004) in Logan, Cache Valley, Utah, USA. *Atmospheric Research* 79:108–122.

Mantua, N. J., S. R. Hare, Y. Zhang, J. M. Wallace, and R. C. Francis. 1997. A Pacific decadal climate oscillation with impacts on salmon. *Bulletin of the American Meteorological Society* 78:1069–1079.

Marwitz, J. D. 1986. A comparison of winter orographic storms over the San Juan Mountains and Sierra Nevada. In *Precipitation enhancement: A scientific challenge*, ed. R. R. Braham, Jr., 109–113. Meteorological Monograph No. 43. Boston: American Meteorological Society.

Marwitz, J., and J. Toth. 1993. The Front Range blizzard of 1990. Part I: Synoptic and mesoscale structure. *Monthly Weather Review* 121:402–415.

Meko, D. M., C. A. Woodhouse, C. A. Baisan, T. Knight, J. J. Lukas, M. K. Hughes, and M. W. Salzer. 2007. Medieval drought in the upper Colorado River Basin. *Geophysical Research Letters* 34: L10705, doi:10.1029/2007GL029988.

Michaud, J. D., K. K. Hirschboeck, and M. Winchell. 2001. Regional variations in small-basin floods in the United States. *Water Resources Research* 37:1405–1416.

Mo, K. C. 1999. Alternating wet and dry episodes over California and intraseasonal oscillations. *Monthly Weather Review* 127:2759–2776.

—. 2000. Intraseasonal modulation of summer precipitation over North America. *Monthly Weather Review* 128:1490–1505.

Mo, K. C., and J. Nogues-Paegle. 2005. Pan-America. In *Intraseasonal variability in the atmosphere-ocean climate system*, ed. K. M. Lau and D. E. Waliser, 95–121. New York: Springer.

National Climatic Data Center (NCDC). 2011. Billion dollar U.S. weather/climate disasters, 1980–2011. http://www.ncdc.noaa.gov/img/reports/billion/billionz-2011.pdf.

—. 2012. U.S. tornado climatology. http://www.ncdc.noaa.gov/oa/climate/severeweather/tornadoes.html (accessed January 23, 2012).

National Weather Service (NWS). n.d. Natural hazard statistics. http://www.nws.noaa.gov/om/hazstats.shtml (accessed January 23, 2012).

Neiman, P. J., F. M. Ralph, G. A. Wick, J. D. Lundquist, and M. D. Dettinger. 2008. Meteorological characteristics and overland precipitation impacts of atmospheric rivers affecting the West Coast of North America based on eight years of SSM/I satellite observations. *Journal of Hydrometeorology* 9:22–47.

Null, J., and J. Hulbert. 2007. California washed away: The great flood of 1862. *Weatherwise* 60 (1): 27–30.

Painter, T. H., A. P. Barrett, C. C. Landry, J. C. Neff, M. P. Cassidy, C. R. Lawrence, K. E. McBride, and G. L. Farmer. 2007. Impact of disturbed desert soils on duration of mountain snow cover. *Geophysical Research Letters* 34: L12502, doi:10.1029/2007GL030284.

Painter, T. H., J. S. Deems, J. Belnap, A. F. Hamlet, C. C. Landry, and B. Udall. 2010. Response of Colorado River runoff to dust radiative forcing in snow. *Proceedings of the National Academy of Sciences* 107:17125–17130.

Painter, T. H., S. M. Skiles, J. S. Deems, A. C. Bryant, and C. C. Landry. 2012. Dust radiative forcing in snow of the Upper Colorado River Basin. Part I: A 6-year record of energy balance, radiation, and dust concentrations. *Water Resources Research* 48. W07521, doi:10.1029/2012WR011985.

Pheleria, H. C., P. M. Fine, Y. Zhu, and C. Sioutas. 2005. Air quality impacts of the October 2003 southern California wildfires. *Journal of Geophysical Research* 110: D07S20.

Porter, K., A. Wein, C. Alpers, A. Baez, P. Barnard, J. Carter, A. Corsi, et al. 2011. *Overview of the ARkStorm scenario*. U.S. Geological Survey Open-File Report 2010-1312. http://pubs.usgs.gov/of/2010/1312/.

Poulos, G. S., D. A. Wesley, J. S. Snook, and M. P. Meyers. 2002. A Rocky Mountain storm. Part I: The blizzard—Kinematic evolution and the potential for high-resolution numerical forecasting of snowfall. *Weather and Forecasting* 17:955–970.

Ralph, F. M., P. J. Neiman, G. A. Wick, S. I. Gutman, M. D. Dettinger, D. R. Cayan, and A. B. White. 2006. Flooding on California's Russian River: Role of atmospheric rivers. *Geophysical Research Letters* 33: L13801, doi:10.1029/2006GL026689.

Rajagopalan, B., and U. Lall, 1998. Interannual variability in western US precipitation. *Journal of Hydrology* 210:51–67.

Rasmussen, R. M., B. C. Bernstein, M. Murakami, G. Stossmeister, and J. Reisner. 1995. The 1990 Valentine's Day Arctic outbreak. Part I: Mesoscale and microscale structure and evolution of a Colorado Front Range shallow upslope cloud. *Journal of Applied Meteorology and Climatology* 34:1481–1511.

Redmond, K. T. 2003. Climate variability in the intermontane West: Complex spatial structure associated with topography, and observational issues. In *Water and climate in the western United States*, ed. W. M. Lewis, 29–48. Boulder, CO: University Press of Colorado.

Redmond, K. T., and R. W. Koch.1991. Surface climate and streamflow variability in the western United States and their relationship to large scale circulation indices. *Water Resources Research* 27:2381–2399.

Reeves, H. D., and D. J. Stensrud, 2009. Synoptic-scale flow and valley cold-pool evolution in the western United States. *Weather and Forecasting* 24:1625–1643.

Ritchie, E. A., K. M. Wood, D. S. Gutzler, and S. R. White. 2011. The influence of eastern Pacific tropical cyclone remnants on the southwestern United States. *Monthly Weather Review* 139:192–210.

Rocky Mountain Insurance Information Association (RMIIA). 2012. Hail. http://www.rmiia.org/Catastrophes_and_Statistics/Hail.asp.

Sangoyomi, T. B. 1993. Climatic variability and dynamics of Great Salt Lake hydrology. Ph.D. dissertation, Utah State University.

Schnell, R. C., S. J. Oltmans, R. R. Neely, M. S. Endres, J. V. Molenar, and A. B. White. 2009. Rapid photochemical production of ozone at high concentrations in a rural site during winter. *Nature Geoscience* 2:120–122.

Schultz, D. M., W. J. Steenburgh, R. J. Trapp, J. Horel, D. E. Kingsmill, L. B. Dunn, W. D. Rust, et al. 2002. Understanding Utah winter storms: The Intermountain precipitation experiment. *Bulletin of the American Meteorological Society* 83:189–210.

Serreze, M. C., M. P. Clark, R. L. Armstrong, D. A. McGinnis, and R. S. Pulwarty. 1999. Characteristics of the western United States snowpack from snowpack telemetry (SNOTEL) data. *Water Resources Research* 35:2145–2160.

Shafer, J. C., and W. J. Steenburgh. 2008. Climatology of strong Intermountain cold fronts. *Monthly Weather Review* 136:784–807.

Sheppard, P. R., A. C. Comrie, G. D. Packin, K. Angersbach, and M. K. Hughes. 2002. The climate of the US Southwest. *Climate Research* 21:219–238.

Silva, P. J., E. L. Vawdrey, M. Corbett, and M. Erupe. 2007. Fine particle concentrations and composition during wintertime inversions in Logan, Utah, USA. *Atmospheric Environment* 26:5410–5422.

Skiles, S. M., T. H. Painter, J. S. Deems, A. C. Bryant, and C. C. Landry. 2012. Dust radiative forcing in snow of the Upper Colorado River Basin. Part II: Interannual variability in radiative forcing and snowmelt rates. *Water Resources Research* 48. W07522, doi:10.1029/2012WR011986.

Steenburgh, W. J. 2003. One hundred inches in one hundred hours: Evolution of a Wasatch Mountain winter storm cycle. *Weather and Forecasting* 18:1018–1036.

Steenburgh, W. J., and T. I. Alcott. 2008. Secrets of the greatest snow on Earth. *Bulletin of the American Meteorological Society* 89:1285–1293.

Steenburgh, W. J., J. D. Massey, and T. H. Painter. Forthcoming. Episodic dust events along Utah's Wasatch Front. *Journal of Applied Meteorology and Climatology* 51.

Stewart, I. T., D. R. Cayan, and M. D. Dettinger, 2004. Changes in snowmelt runoff timing in western North America under a 'business as usual' climate change scenario. *Climatic Change* 62:217–232.

—. 2005. Changes toward earlier streamflow timing across western North America. *Journal of Climate* 18:1136–1155.

Swetnam, T. W., and J. L. Betancourt. 1998. Mesoscale disturbance and ecological response to decadal climatic variability in the American Southwest. *Journal of Climate* 11:3128–3147.

Tourre, Y., B. Rajagopalan, Y. Kushnir, M. Barlow, and W. White. 2001. Patterns of coherent decadal and interdecadal climate signals in the Pacific Basin during the 20th century. *Geophysical Research Letters* 28:2069–2072.

U.S. Environmental Protection Agency (EPA). n.d. Green book: Currently designated nonattainment areas for all criteria pollutants, as of July 20, 2012. http://epa.gov/oaqps001/greenbk/ancl3.html (last updated on July 20, 2012).

Wang, S.-Y., R. R. Gillies, J. Jin, and L. E. Hipps. 2009. Recent rainfall cycle in the Intermountain region as a quadrature amplitude modulation from the Pacific decadal oscillation. *Geophysical Research Letters* 36: L02705.

—. 2010. Coherence between the Great Salt Lake level and the Pacific quasi-decadal oscillation. *Journal of Climate* 23:2161–2177.

Weaver, J. F., E. Gruntfest, and G. M. Levy. 2000. Two floods in Fort Collins, Colorado: Learning from a natural disaster. *Bulletin of the American Meteorological Society* 81:2359–2366.

Westerling, A. L., A. Gershunov, T. J. Brown, D. R. Cayan, and M. D. Dettinger. 2003. Climate and wildfire in the western United States. *Bulletin of the American Meteorological Society* 84:595–604.

White, A. B., P. J. Neiman, F. M. Ralph, D. E. Kingsmill, and P. O. G. Persson. 2003. Coastal orographic rainfall processes observed by radar during the California Land-Falling Jets Experiment. *Journal of Hydrometeorology* 4:264–282.

Whiteman, C. D., X. Bian, and S. Zhong. 1999. Wintertime evolution of the temperature inversion in the Colorado Plateau Basin. *Journal of Applied Meteorology and Climatology* 38:1103–1117.

Wolyn, P. G., and T. B. McKee. 1989. Deep stable layers in the Intermountain western United States. *Monthly Weather Review* 117:461–472.

Woodhouse, C.A. 2003. A 431-year reconstruction of western Colorado snowpack from tree rings. *Journal of Climate* 16:1551–1561.

Woodhouse, C. A., D. M. Meko, G. M. MacDonald, D. W. Stahle, and E. R. Cook. 2010. A 1,200-year perspective of 21st century drought in southwestern North America. *Proceedings of the National Academy of Sciences* 107:21283-21288.

Zhang, Y., J. M. Wallace, and D. S. Battisti. 1997. ENSO-like interdecadal variability: 1900–93. *Journal of Climate* 10:1004–1020.

Zhang, Z., and M. E. Mann. 2005. Coupled patterns of spatiotemporal variability in Northern Hemisphere sea level pressure and conterminous U.S. drought. *Journal of Geophysical Research* 110: D03108.

Chapter 5

Present Weather and Climate: Evolving Conditions

COORDINATING LEAD AUTHOR

Martin P. Hoerling (NOAA, Earth System Research Laboratory)

LEAD AUTHORS

Michael Dettinger (U.S. Geological Survey and Scripps Institution of Oceanography), Klaus Wolter (University of Colorado, CIRES), Jeff Lukas (University of Colorado, CIRES), Jon Eischeid (University of Colorado, CIRES), Rama Nemani (NASA, Ames), Brant Liebmann (University of Colorado, CIRES), Kenneth E. Kunkel (NOAA Cooperative Institute for Climate and Satellites, North Carolina State University, and National Climate Data Center)

EXPERT REVIEW EDITOR

Arun Kumar (NOAA)

Executive Summary

This chapter assesses weather and climate variability and trends in the Southwest, using observed climate and paleoclimate records. It analyzes the last 100 years of climate variability in comparison to the last 1,000 years, and links the important features of evolving climate conditions to river flow variability in four of the region's major drainage basins. The chapter closes with an assessment of the monitoring and scientific research needed to increase confidence in understanding when climate episodes, events, and phenomena are attributable to human-caused climate change.

Chapter citation: Hoerling, M. P., M. Dettinger, K. Wolter, J. Lukas, J. Eischeid, R. Nemani, B. Liebmann, and K. E. Kunkel. 2013. "Present Weather and Climate: Evolving Conditions." In *Assessment of Climate Change in the Southwest United States: A Report Prepared for the National Climate Assessment*, edited by G. Garfin, A. Jardine, R. Merideth, M. Black, and S. LeRoy, 74–100. A report by the Southwest Climate Alliance. Washington, DC: Island Press.

- The decade 2001–2010 was the warmest and the fourth driest in the Southwest of all decades from 1901 to 2010. (high confidence)
- Average annual temperature increased 1.6°F (+/- 0.5°F) over the Southwest during 1901–2010, while annual precipitation experienced little change. (high confidence)
- Fewer cold waves and more heat waves occurred over the Southwest during 2001–2010 compared to their average occurrences in the twentieth century. (high confidence)
- The growing season for the Southwest increased about 7% (seventeen days) during 2001–2010 compared to the average season length for the twentieth century. (high confidence)
- The frequency of extreme daily precipitation events over the Southwest during 2001–2010 showed little change compared to the twentieth-century average. (medium-high confidence)
- The areal extent of drought over the Southwest during 2001–2010 was the second largest observed for any decade from 1901 to 2010. (medium-high confidence)
- Streamflow totals in the four major drainage basins of the Southwest were 5% to 37% lower during 2001–2010 than their average flows in the twentieth century. (medium-high confidence)
- Streamflow and snowmelt in many snowmelt-fed streams of the Southwest trended towards earlier arrivals from 1950–1999, and climate science has attributed up to 60% of these trends to the influence of increasing greenhouse gas concentrations in the atmosphere. (high confidence)
- Streamflow and snowmelt in many of those same streams continued these earlier arrivals during 2001–2010, likely in response to warm temperatures. (high confidence)
- The period since 1950 has been warmer in the Southwest than any comparable period in at least 600 years, based on paleoclimatic reconstructions of past temperatures. (medium-high confidence)
- The most severe and sustained droughts during 1901–2010 were exceeded in severity and duration by several drought events in the preceding 2,000 years, based on paleoclimatic reconstructions of past droughts. (high confidence)

5.1 Introduction

Wallace Stegner (1987) expressed a sentiment held by many familiar with the Southwest: "If there is such a thing as being conditioned by climate and geography, and I think there is, it is the West that has conditioned me."

As the twenty-first century unfolds, two principal concerns make it important to take stock of the region's climate. One concern is that the annual demand for water in the Southwest—especially from the Colorado River, which supplies water to each of the region's states—has risen to an amount that nearly matches the natural annual flow in the Colorado River. With only a small margin between supply and demand—both of which are sensitive to climate variability and change—the importance of reservoirs in the Colorado River Basin increases (Barnett and Pierce 2008, 2009; Rajagopalan et al. 2009). Droughts, whose periodic occurrences in the Southwest have tested the resilience of the region's indigenous populations (Liverman and Merideth 2002), have increasingly significant effects. In particular, excess (unconsumed) water supply capacity has been diminishing, virtually vanishing, as was especially evident during the region's drought that began in 2000 (Fulp 2005). The second concern is the expectation, based on a growing body of scientific evidence, that climate change in the Southwest will most likely reduce water resources, including a decline in the annual flow of the Colorado River (Milly, Dunne, and Vecchia 2005; Christensen and Lettenmaier 2007; McCabe and Wolock 2008; see also the discussion of the effects of climate change on the water supplies of the Southwest in Chapter 10.)

This chapter reviews the nature of weather and climate variability in the Southwestern United States based on recorded observations and measurements that span the last century. The chapter links how changing climate conditions affect the variability of river flows specifically in four of the region's major drainage basins: Sacramento-San Joaquin, Humboldt (in the Great Basin), Upper Colorado, and Rio Grande. To place current climatic conditions in a longer-term context, the chapter looks at the indirect evidence (*paleoclimatic reconstruction*, as from tree rings, pollen, sediment layers, and so on) of climatic conditions over the last thousand years, showing the variations that occurred before humans substantially increased emissions of greenhouse gases. The chapter concludes by appraising the data gaps and needs for monitoring the evolving climate and hydrological conditions in the Southwest.

5.2 Climate of the First Decade of the Twenty-first Century

"Exceptionally warm" aptly describes temperatures in the Southwest during the first decade of the twenty-first century. Annual temperatures for 2001–2010 were warmer than during any prior decade of the twentieth century, both for the Southwest as a whole and for each state in the region (Table 5.1).[i] Annual averaged temperatures for 2001 –2010 were 1.4°F (0.8°C) warmer than the 1901–2000 average. The intensity of warming is related to changes in temperatures at particular times of the day and in particular seasons. For example, greater warming has occurred due to increases in daily minimum temperatures than to increases in daily maximum temperatures, though the reasons for this difference are not well-known and may be related to local effects and to adjustments applied to station data (as discussed further below; see also Fall et al. 2011). The key features of a warming Southwest appear robustly across various data sets and methods of analysis, as shown further in Appendix Table A5.1.

With respect to the seasons, when looking at average seasonal temperatures for the period 2001–2010 versus those for the twentieth century, greater warming (i.e., larger differences) occurred during the spring and summer than occurred during the other seasons, especially winter. Based on results of a rigorous detection and attribution study, the recent rapid increase in late winter/early spring minimum temperatures are *very unlikely* due to natural variability alone, but are consistent with a regional sensitivity to increased greenhouse gases and aerosols (Bonfils et al. 2008). During winter, maximum

Table 5.1 Comparison of Southwest annual and seasonal surface temperatures averaged for 2001–2010

AVERAGE TEMPERATURE

Season	1901–2000 6-State Avg. Mean °F (°C)	1901–2000 6-State Avg. Std.Dev. °F (°C)	2001–2010 Decadal Mean °F (°C)	2001–2010 Decadal Anom °F (°C)	Rank	AZ	CA	CO	NV	NM	UT
DJF	35.1 (1.7)	0.7 (0.4)	36 (2.2)	+0.9 (0.5)	2	3	2	3	2	3	2
MAM	50.0 (10.0)	0.9 (0.5)	51.8 (11)	+1.8 (1.0)	1	1	1	1	1	1	1
JJA	70.0 (21.1)	0.9 (0.5)	72 (22.2)	+2.0 (1.1)	1	1	1	1	1	1	1
SON	53.2 (11.8)	0.7 (0.4)	54.7 (12.6)	+1.4 (0.8)	1	1	1	1	1	1	1
Annual	52.2 (11.2)	0.7 (0.4)	53.6 (12)	+1.4 (0.8)	1	1	1	1	1	1	1

MAXIMUM TEMPERATURE

Season	1901–2000 6-State Avg. Mean °F (°C)	1901–2000 6-State Avg. Std.Dev. °F (°C)	2001–2010 Decadal Mean °F (°C)	2001–2010 Decadal Anom °F (°C)	Rank	AZ	CA	CO	NV	NM	UT
DJF	47.7 (8.7)	0.7 (0.4)	47.8 (8.8)	+0.2 (0.1)	5	4	3	8	6	3	4
MAM	64.6 (18.1)	0.9 (0.5)	66.2 (19)	+1.6 (0.9)	1	1	2	1	2	1	1
JJA	86.0 (30.0)	0.7 (0.4)	87.4 (30.8)	+1.4 (0.8)	1	1	1	2	1	2	1
SON	68.2 (20.1)	0.7 (0.4)	68.9 (20.5)	+0.7 (0.4)	3	2	3	5	3	2	4
Annual	66.6 (19.2)	0.7 (0.4)	67.6 (19.8)	+1.1 (0.6)	1	1	1	2	1	1	1

MINIMUM TEMPERATURE

Season	1901–2000 6-State Avg. Mean °F (°C)	1901–2000 6-State Avg. Std.Dev. °F (°C)	2001–2010 Decadal Mean °F (°C)	2001–2010 Decadal Anom °F (°C)	Rank	AZ	CA	CO	NV	NM	UT
DJF	22.5 (-5.3)	0.9 (0.5)	24.1 (-4.4)	+1.6 (0.9)	2	2	2	2	1	2	2
MAM	35.6 (2.0)	0.9 (0.5)	37.4 (3.0)	+1.8 (1.0)	1	1	2	1	1	1	1
JJA	54.1 (12.3)	0.9 (0.5)	56.5 (13.6)	+2.3 (1.3)	1	1	1	1	1	1	1
SON	38.3 (3.5)	0.9 (0.5)	40.6 (4.8)	+2.3 (1.3)	1	1	1	1	1	1	1
Annual	37.6 (3.1)	0.9 (0.5)	39.7 (4.3)	+2.2 (1.2)	1	1	1	1	1	1	1

Note: Comparison of annual and seasonal surface temperatures for the six Southwestern states, averaged for 2001–2010 versus 1901–2000, and a ranking of the 2001–2010 decadal averages relative to the ten individual decades of the twentieth century. Results shown for daily averaged, maximum, and minimum temperatures.

Source: PRISM monthly gridded analysis for 1901–2010 (PRISM Climate Group, Oregon State University, http://prism.oregonstate.edu).

temperature changes have been muted, being only 0.2°F (0.1°C) above the twentieth-century average for the Southwest. These data further indicate that the winter maximum temperatures for 2001–2010 averaged over Colorado were actually colder than during the majority of decades in the twentieth century.

"Unusually dry" best describes Southwest moisture conditions during the first decade of the twenty-first century (Table 5.2). Annual precipitation, averaged across the entire Southwest, ranked 2001–2010 the fourth driest of all decades since 1901, a condition that is found to be robust across various data sets (see Appendix Table A5.2). The departure of -0.59 inches (-15 mm) represents a reduction of 4% of the twentieth-century average annual total. Much of the deficit was accumulated in the early half of the decade in association with one of the most severe droughts on (instrumental) record (Hoerling and Kumar 2003; Pielke et al. 2005). It is *likely* that most of recent dryness over the Southwest is associated with a natural, decadal coolness in tropical Pacific sea-surface temperatures, and is mostly unrelated to influences of increased greenhouse gases and aerosols (Hoerling, Eischeid, and Perlwitz 2010). The strongest percentage declines occurred during spring and summer, which were 11% and 8% below normal,[ii] respectively. The winter season, when the bulk of the region's precipitation is delivered, actually experienced a small increase relative to twentieth-century averages. Precipitation conditions during 2001–2010 varied considerably among the six Southwestern states, with Arizona experiencing its driest decade since 1901 and Utah experiencing one of its wetter decades.

Table 5.2 Comparison of Southwest annual and seasonal precipitation totals averaged for 2001–2010

PRECIPITATION

	1901–2000 6-State Avg.		2001–2010 Decadal								
Season	Mean inches (mm)	Std.Dev. inches (mm)	Mean inches (mm)	Anom inches (mm)	Rank	AZ	CA	CO	NV	NM	UT
DJF	5.1 (129.3)	0.38 (9.6)	5.2 (133.3)	+0.16 (4.0)	5	7	5	2	4	2	3
MAM	3.8 (96.6)	0.32 (8.2)	3.4 (86.8)	-0.39 (9.8)	10	11	9	10	5	7	9
JJA	3.3 (83.4)	0.26 (6.7)	3.0 (77.1)	-0.25 (6.3)	9	10	9	7	10	6	9
SON	3.4 (85.3)	0.39 (9.9)	3.2 (82.3)	-0.12 (3.0)	6	11	7	3	5	7	3
Annual	15.5 (394.6)	0.8 (20.2)	15.0 (381.0)	-0.60 (15.1)	8	11	7	6	5	5	3

Note: Comparison of annual and seasonal precipitation totals for the six Southwestern states, averaged for 2001–2010 versus 1901–2000, and a ranking of the 2001–2010 decadal averages relative to the ten individual decades of the twentieth century.

Source: PRISM monthly gridded analysis for 1901–2010 (PRISM Climate Group, Oregon State University, http://prism.oregonstate.edu).

5.3 Climate Trends for 1901–2010

The trend in surface temperature during 1901–2010 was upward over all of the stations in the Southwest that have long-term climate records (Figure 5.1, upper panel). Average annual temperature increased 1.6°F (0.9°C) over the Southwest during 1901–2010, with a range of magnitudes from 1.4°F to 2.0°F (+0.8°C to +1.1°C) based on analyses conducted with other data sets (see Appendix). The 95% confidence interval for the linear trend is +/- 0.5°F (0.3°C).

The linear warming trend continued in the first decade of the twenty-first century, the warmest over the region during the 110-year period of record (see Table 5.1). Increases have been more than 1.8°F (1°C) in many parts of the Southwest over the last 110 years, with isolated 3.6°F (2°C) increases occurring in southwestern portions of the region. Both daytime high temperatures (Figure 5.1, second panel) and nighttime low temperatures (Figure 5.1, third panel) have exhibited widespread warming trends. In this data set, which has been homogenized, minimum temperatures are found to increase at about the same rate as the maximum temperatures for the Southwest as a whole, though there is considerable variability from one station to another. Note that the trends in unadjusted raw data show somewhat higher minimum temperatures (Figure A5.1).

In light of the warming over the Southwest in all seasons, it is not surprising that the growing season duration has increased over the last century. Figure 5.2 shows the growing season departures for each year during 1901–2010, based on a 32°F (0°C) threshold, for some sixty Southwest stations that form part of the National Weather Service Cooperative Observing Network (COOP) (see Kunkel et al. 2004). The average growing season over the Southwest during 2001–2010 was seventeen days longer (about 7% longer) than the twentieth-century average and one month longer than that of the first decade of the twentieth century.[iii]

The trend in annual precipitation during 1901–2010 (expressed as percent of annual precipitation climatology) is shown for individual stations in Figure 5.1 (bottom panel). Plotted atop the regional map of precipitation trend are outlines of four major drainage basins in the Southwest (clockwise from upper left: Sacramento-San Joaquin, Great Basin, Upper Colorado, and the Rio Grande). A summary of the climate conditions and river runoff characteristics for these hydrologic regions will be subsequently provided.

Although the decade 2001–2010 has been relatively dry for the Southwest as a whole (Table 5.1), the trend in annual precipitation computed for the entire 1901–2010 period reveals little change over the 110-year period (Figure 5.1, bottom panel). Some significant local trends (filled colored circles) can be discerned, however, including wet trends at select stations in the Great Basin. It should be noted that the bulk of the region's precipitation falls at high elevations, areas not well observed. Although water supplies for the region's major drainage basins are especially dependent on the precipitation falling in these remote areas, the long-term trends in such resources are not well known.

5.4 Extreme Weather Variability During 1901–2010

Further indicators of a warming Southwest climate are provided by indices for occurrences of cold waves and heat waves, shown in Figure 5.3. Cold waves (defined as four-day periods colder than the threshold of a one-in-five-year frequency) have been

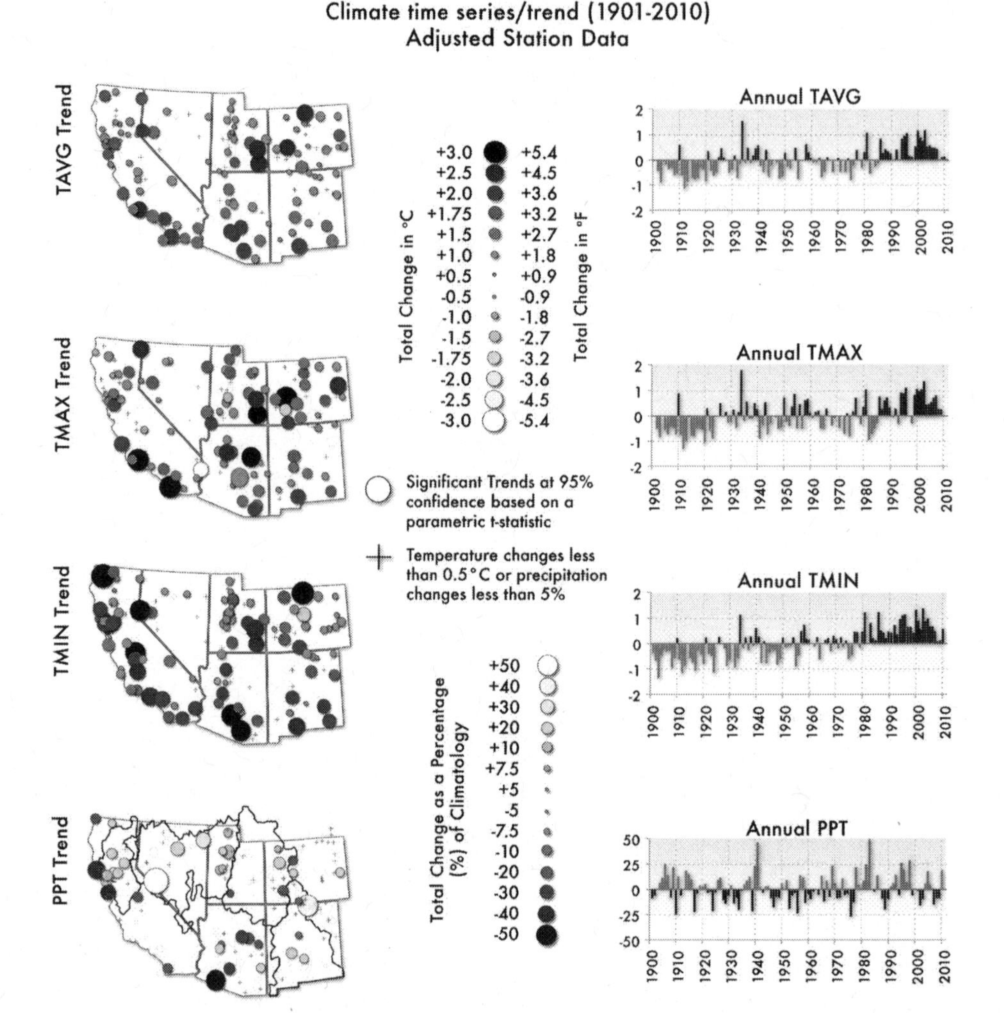

Figure 5.1 Trends in temperature and precipitation, 1901–2010. Shown in the first three (upper) panels are the pattern and intensity of 110-year trends in the Southwest in annually averaged daily temperature (TAVG), and daily maximum temperature (TMAX) and daily minimum temperature (TMIN), respectively, as estimated from station data for which there are at least 90 years of available data during the period. The magnitude of trends is indicated by a station circle's size, with warming (cooling) trends denoted by dark grey (light grey) shades. Bottom panel shows trend in annual averaged precipitation for 1901-2010, showing data for individual stations plotted with outlines of four major basins in the Southwest (Sacramento-San Joaquin, Great Basin, Upper Colorado, and Rio Grande). Units are the total change expressed as percent of annual climatology, and positive (negative) trends are shown in light grey (dark grey). Larger circle sizes denote greater magnitude trends. Filled stations denote statistically significant trends at 95% confidence based on a parametric t-statistic. Stations with temperature (precipitation) changes less than 0.5°C (5%) are denoted with a + symbol. The figure uses so-called "homogenized" data, in which adjustments to remove artificial temperature changes at a station are made by using the method of pairwise difference comparisons between monthly temperatures from a network of reporting stations. The trends calculated from the raw data that preserve various inhomogeneities in the original time series are provided in Appendix Figure A5.1. Results from both the raw and the homogenized data show substantial warming. Source: Menne and Williams (2009).

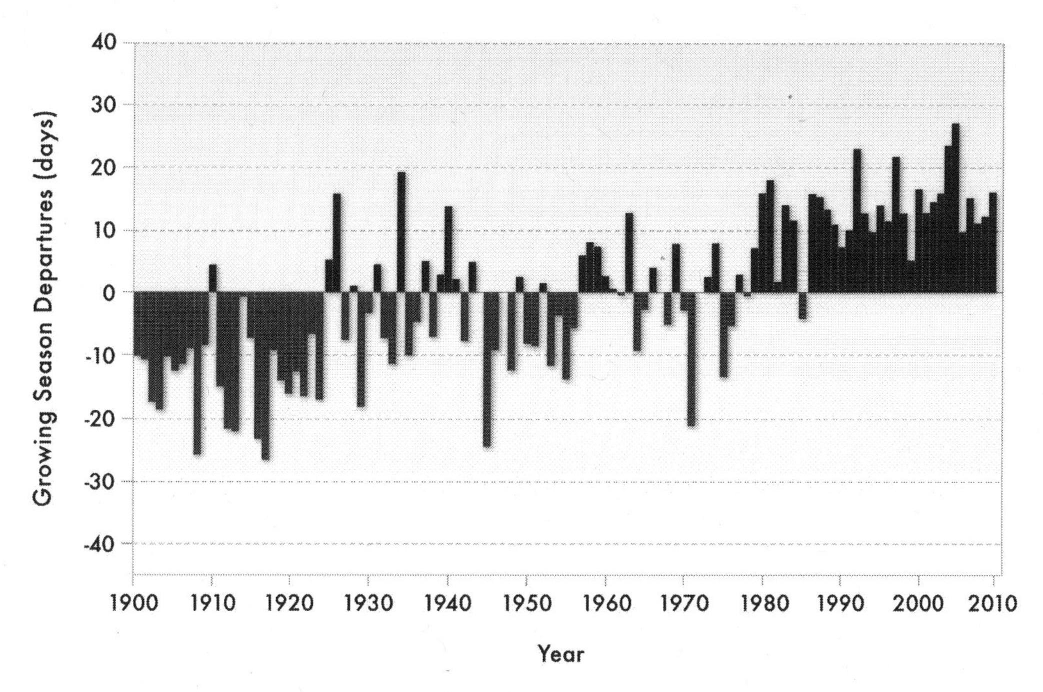

Figure 5.2 Growing season anomalies for 1900 to 2010. Departure from the normal number of days in the growing season are shown as number of days per year. Length of the growing season is defined as the period between the last freeze in spring and the first freeze in the subsequent fall. Source: NOAA National Climatic Data Center for the Cooperative Observer Network (http://www.ncdc.noaa.gov/land-based-station-data/cooperative-observer-network-coop).

especially rare since about 1990, while the frequency of heat waves (defined similarly to cold waves) increased.[iv] Heat-wave frequency during 2001–2010, however, is not appreciably different from that occurring during the 1930s. Both periods were characterized by drought, and the feedback from dry soils and clear skies likely enhanced the severity of summertime heat during both decades. The increase in heat waves in tandem with a decrease in cold waves is consistent with other research findings that showed an increase in record high maximum temperatures relative to record low minimum temperatures over the entire United States since 1950 (Meehl et al. 2009).

COOP station data have also been used to derive an extreme precipitation index (Kunkel et al. 2003); Figure 5.4 shows an index time series of five-day rainfall extremes (events wetter than a threshold for a one-in-five year-frequency). There is no discernible trend in this statistic of heavy precipitation events, with the time series characterized by appreciable decadal variability. Nor is there evidence for trends in precipitation extremes using other indices such as for extreme daily precipitation totals (not shown). Bonnin, Maitaria, and Yekta (2011) diagnosed trends in rainfall exceedences (exceeding precipitation thresholds over multiple timescales as defined in the NOAA Atlas 14) for the semiarid Southwest encompassing much of our six-state region but excluding Colorado and central and Northern California. For their 1908–2007 period of analysis, they found negative trends in rainfall exceedences for the semi-arid Southwest for all multi-day durations, though only the one-day duration negative trends were statistically

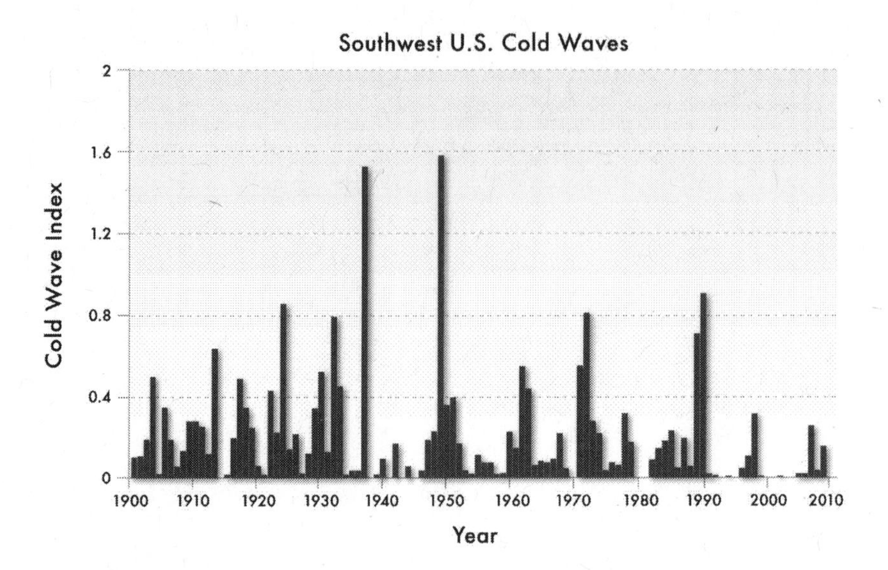

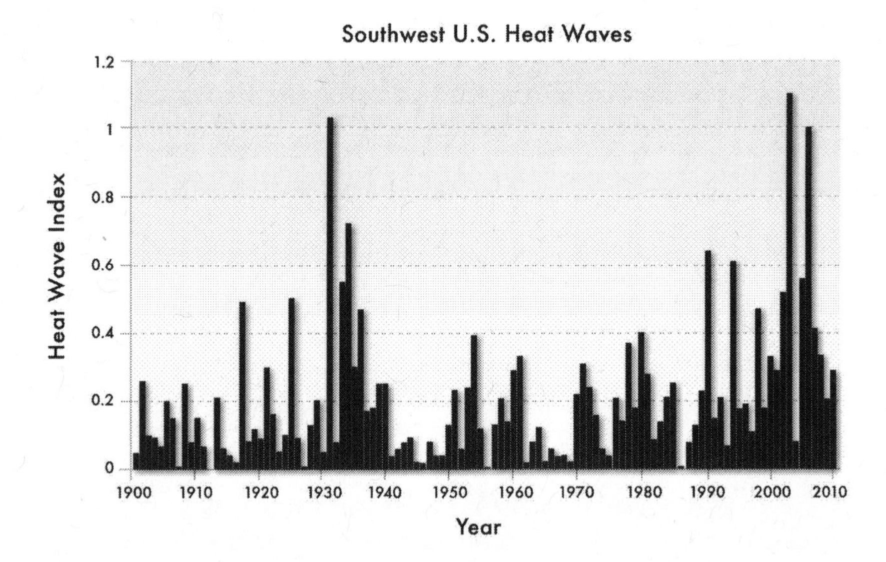

Figure 5.3 Occurrence of cold waves and heat waves, 1901–2010. Cold waves (top) are defined as four-day periods that are colder than the threshold of a one-in-five-year occurrence; heat waves (bottom) are four-day periods warmer than the same threshold. The thresholds are computed for the entire 1901–2010 period. Source: NOAA National Climatic Data Center for the Cooperative Observer Network (http://www.ncdc.noaa.gov/land-based-station-data/cooperative-observer-network-coop).

significant. This means that in recent years there were fewer events exceeding high precipitation thresholds for multi-day episodes than during earlier in the period of analysis. The Southwest is thus unlike some other areas of the United States, such as the Great Lakes and Ohio Valley regions, where significant upward trends in very heavy daily precipitation (during 1908–2000) have been noted (Groisman et al. 2004).

5.5 Summertime Drought During 1901–2010

To quantify variability in drought across the region, the Palmer Drought Severity Index (PDSI) was used, based on data from the NOAA National Climate Data Center (NCDC). The PDSI is a widely used indicator of dryness, representing the balance of water inputs

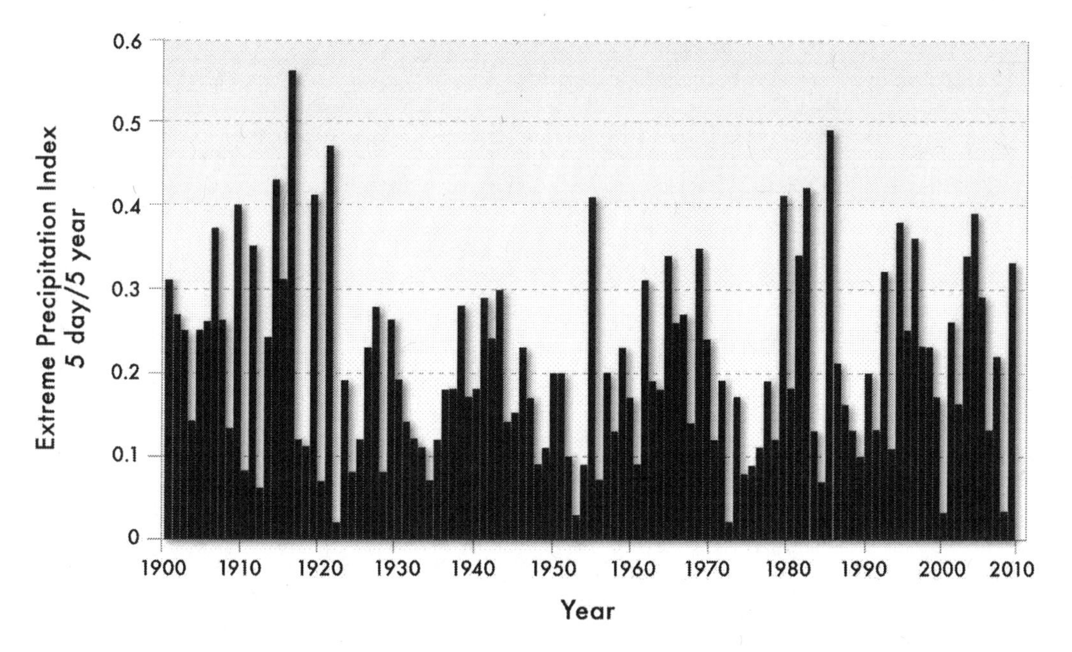

Figure 5.4 Occurrence of extreme precipitation events, 1900–2010. Extreme precipitation events are defined as five-day totals that are wetter than the threshold of a one-in-five-year occurrence. Source: NOAA National Climatic Data Center for the Cooperative Observer Network (http://www.ncdc.noaa.gov/land-based-station-data/cooperative-observer-network-coop).

to the soil (based on observed monthly precipitation) and water losses from the soil (based on observed monthly temperature). This soil-moisture balance calculation incorporates the water-holding ability of soil, and thus PDSI for a given month or season actually reflects the previous nine to twelve months of weather conditions (Palmer 1965; Alley 1984). Thus, while summer PDSI captures moisture anomalies that peak during three months of summer, it also incorporates conditions from the previous fall, winter, and spring. For this assessment, the divisional monthly data were averaged by area across the six Southwest states, then summer (June, July, August) PDSI was averaged to represent drought conditions during the height of the growing season.

Figure 5.5 shows variations in the areal extent of drought[v] in the Southwest from 1901 to 2010. Over that period, there is a trend towards increasing drought extent, in large part due to widespread drought during the 2001–2010 decade, which had the second-largest area affected by drought (after 1951–1960) and the most severe average drought conditions (average summer PDSI = -1.3) of any decade. The severity of drought in 2001–2010 reflects both the decade's low precipitation and high temperatures, since both affect the surface water balance that enter into the calculation of PDSI (see Table 5.2). An analysis of PDSI trends over the globe by Dai (2011) indicates that the warmer temperatures across the Southwest in recent decades are at least partly responsible for the increasing drought coverage. Widespread and severe drought also occurred in the Southwest from 1950 to 1956, whereas the Dust Bowl conditions of the 1930s (though having a severe impact on the eastern portions of the region and the rest of the Great Plains) did not extend significantly into the Southwest except in 1934.

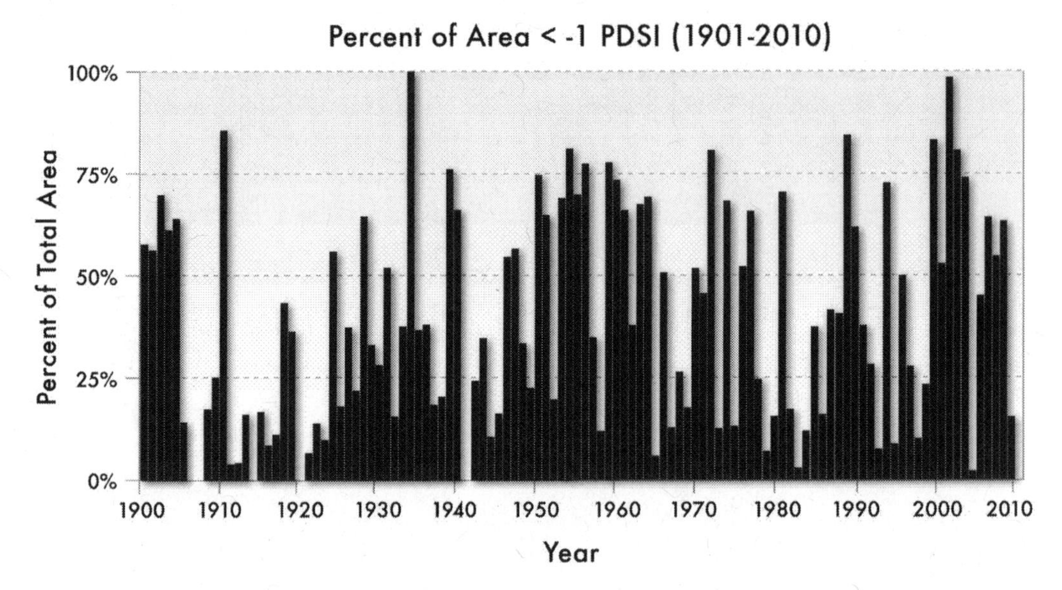

Figure 5.5 Areal coverage of drought in the Southwest United States, 1900–2010. Data is for drought during summer (June through August). Graph based on the Palmer Drought Severity Index (PDSI) and monthly data from the NOAA National Climatic Data Center for 1901–2010.

PDSI is a useful indicator of local-to-regional meteorological drought (low precipitation) and agricultural drought (low soil moisture). It is less useful for indicating hydrologic drought (low water supply), especially if the PDSI measurements do not adequately capture the specific locations where the water supply for a given area originates—often in mountain headwaters remote from the point of use. Also, while PDSI can be closely correlated with annual streamflow in basins where runoff is mainly from the melting of the winter snowpack, this is not the case in basins where streamflow mainly derives from infrequent large rainfall events or from groundwater discharge. The most consistently useful indicator of hydrologic drought across the Southwest is streamflow itself, which is the subject of the next section.

5.6 Hydroclimatic Variability During 1901–2010

The water resources of the Southwest, and especially its rivers, reflect variations of precipitation, evaporation, and transpiration (the uptake of water by plants) over the region, with evaporation and transpiration strongly modulated by temperature. They also reflect human interventions into the hydrologic system—dams, diversions, and water uses such as irrigation. The climatic conditions of the 2001–2010 period described previously thus resulted in significant deviations of the flows of rivers in the region from twentieth-century norms. Here we will focus on naturalized or near-natural flows in four major hydrologic basins—the Sacramento-San Joaquin system, central Great Basin (represented by the Humboldt River at Palisade), Upper Colorado River, and Rio Grande (see Figure 5.1, bottom panel)—to represent recent hydrologic variations. Naturalized flows are best estimates of the total streamflow that would have reached the outlet of the river system in the absence of human actions and are based on historic measured flows, corrected for human influences including diversions and other uses.

As noted earlier, the 2001–2010 period was unusually warm in the Southwest, with many areas also drier than normal (Table 5.2). The four river systems analyzed here all responded to those climatic conditions with lower-than-normal measured flows (and, where available, lower-than-normal naturalized flows) (Table 5.3). Naturalized flows in the Sacramento-San Joaquin Rivers system reflected drier than normal conditions, with a 2001–2010 average daily flow of 6.8 million acre-feet/year compared to the 1931–2000 average flow of 10.8 million acre-feet/year. This 37% overall deficit ranked 2001–2010 as the lowest-flow decade since 1931 in the Sacramento-San Joaquin system. Flows in the Humboldt River at Palisade during 2001–2010 averaged 134,000 acre-feet/year during the 2001–2010 period, or 5% below the 141,000 acre-feet/year average of the 1921–2000 period, ranking that decade as the sixth driest in nine decades of record. Warm temperatures and dry conditions reduced average naturalized flows in the Colorado River (measured at Lees Ferry) to 12.6 million acre-feet/year, compared to the 1901–2000 average of 15.0 million acre-feet/year (Cayan et al. 2010). This 16% decadal deficit (Table 5.3) made 2001–2010 the second-lowest-flow decade at Lees Ferry (among eleven) since 1901. Observed flows for 2001–2010 in the Rio Grande at El Paso (where the river leaves the Southwest region) were about 23% lower than the period from 1941 to 2000, even though overall precipitation in the basin was 3% above normal.

Overall, then, the 2001–2010 climatic conditions contributed to unusually low annual flows in major drainage systems across the Southwest (Table 5.3). The low flows resulted from less precipitation, warm temperatures, and, to some extent, water-management impacts that have not been completely accounted for in the naturalized records. The influences of these various factors are known to differ from basin to basin. The extent to which the warmth and dryness of the decade might be attributable to greenhouse-gas-fueled climate change is not currently known, as no formal detection-and-attribution study has been conducted for temperature, precipitation, or runoff of this most recent decade, nor for individual basins. Generally, though, these lower flows for the decade are beyond what would be expected from the reduced precipitation; they could be symptomatic of the Southwest hydroclimates that are projected for the latter decades of this century under scenarios of continued warming (Cayan et al. 2010). Annual peak streamflow rates declined from 1901 to 2008 in the interior Southwest, the only region in the continental U.S. that has experienced a regional-scale significant decline (Hirsch and Rhyberg 2011).

Various other hydrologic changes in the Southwest symptomatic of a warmer climate occurred between 1950 and 1999 (Barnett et al. 2008). These include declines in the late-winter snowpack in the northern Sierra Nevada (Roos 1991), trends toward earlier snowmelt runoff in California and across the West (Dettinger and Cayan 1995; Stewart, Cayan, and Dettinger 2005), earlier spring onset in the western United States as indicated by changes in the timing of plant blooms and spring snowmelt-runoff pulses (Cayan et al. 2001), declines in mountain snowpack over Western North America (Mote et al. 2005), general shifts in western hydroclimatic seasons (Regonda et al. 2005), and trends toward more precipitation falling as rain instead of snow over the West (Knowles, Dettinger, and Cayan 2006).

Table 5.3 Differences between 2001–2010 and twentieth-century averages of basin-mean precipitation, average temperature, and streamflow for four major hydrologic basins in the Southwest

River Basin	Periods Compared	Precipitation Difference	Temperature Difference	Streamflow Difference
Colorado River at Lees Ferry (naturalized)	2001–2010 vs 1901–2000	-4%	+0.7°C	-16%
Sacramento-San Joaquin Rivers (naturalized)	2001–2010 vs 1931–2000	-7%	+0.7°C	-37%
Humboldt River at Palisade, NV	2001–2010 vs 1921–2000	-3%	+0.7°C	-5%
Rio Grande at El Paso	2001–2010 vs 1941–2000	+3%	+0.6°C	-23%

Note: Different baselines reflect different periods of streamflow record.
Source: PRISM monthly gridded analysis for 1901–2010 (PRISM Climate Group, Oregon State University, http://prism.oregonstate.edu).

These various indicators have recently been studied in an integrated program of hydroclimatic trends assessment for the period 1950–1999. The research findings for a region of the Western United States[vi] demonstrated that during this period human-induced greenhouse gases began to impact: (a) wintertime minimum temperatures (Bonfils et al. 2008); (b) April 1 snowpack water content as a fraction of total precipitation (Pierce et al. 2008); (c) snow-fed streamflow timing (Hidalgo et al. 2009), and (d) a combination of (a), (b), and (c) (Barnett et al. 2008). These evaluations also indicated, with high levels of statistical confidence, that as much as 60% of the climate-related trends in these indicators were human-induced and that the changes—all of which reflect temperature influences more than precipitation effects—first rose to levels that allowed confident detection in the mid-1980s.

Analyses of differences between the average ratios of rain-to-snow in precipitation, snowpack water contents, and snowmelt-fed streamflow-timing indicators in the 2001–2010 period and available twentieth-century baselines indicate that the anomalous conditions in these indicators seen at the end of the twentieth century (Barnett et al. 2008) have persisted through the first decade of the present century. Differences between the averages of the central date of water-year hydrographs in snowmelt-fed rivers (i.e., the date when half of the annual streamflow has occurred) in the 2001–2010 period compared to twentieth-century baselines at 130 stream gauges across the Western states are shown in Figure 5.6, which corroborates recent findings of Fritze, Stewart, and Pebesma (2011). The pattern that has emerged (see Figure 5.6) is that snow-fed streamflows arrived five to twenty days earlier in the recent decade compared to twentieth-century averages across broad areas of the West. These most recent changes in snowmelt timing cannot yet be formally attributed to climate change because the necessary formal detection-and-attribution studies have not been extended to the most recent decade.

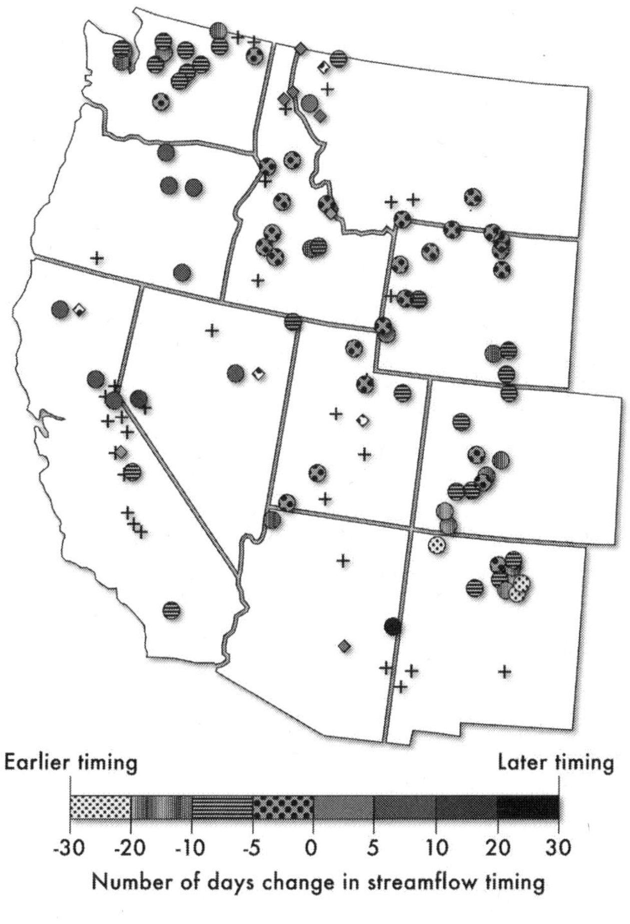

Figure 5.6 Changing streamflow timing 2001–2010 compared to 1950–2000. Differences between 2001–2010 and 1950–2000 average date when half of the annual streamflow has been discharged (center of mass) for snowmelt-dominated streams (Stewart, Cayan and Dettinger 2005).

Indeed, we must consider how temperature and precipitation have naturally varied over the region, since an important change in an apparently natural climate pattern bracketed the 2001–2010 period. There was a decisive switch from the negative to the positive PDO phase (more El Niño-like conditions) in the mid-1970s that may have favored subsequent wet and warm conditions in parts of the Southwest in the 1980s and 1990s (Gershunov and Barnett 1998). Yet, there was a transition back to negative PDO (more La Niña-like) conditions just prior to the turn of the century, which in turn might have favored a drier and cooler Southwest in 2001–2010 (Ropelewski and Halpert 1986; Yarnal and Diaz 1986; Redmond and Koch 1991; Kahya and Dracup 1993). These interannual and decadal variations in oceanic conditions to which Southwest precipitation, snowpack, and hydrology are quite sensitive are mainly the result of the natural variability in the climate system (see also Hoerling, Eischeid, and Perlwitz 2010). These natural variations could readily explain the dryness of the past decade, but its exceptional warmth would seem, for now, to run counter to the temperature changes that might be expected from the natural variations alone.

Thus far, the temperature-driven changes in snowpack volumes and streamflow timing in the Southwest have not risen to levels that have disrupted water supplies; however, continuation of such trends—to be expected if recent warming in the region continues or accelerates—would eventually challenge water-resource systems that have historically relied on the (now) delicate pairing of manmade structures for storage of runoff with seasonal storage of water in natural snowpacks to meet warm-season water demands (Rajagopalan et al. 2009).

5.7 Paleoclimate of the Southwest United States

Since the relatively brief instrumental climate records in the Southwest (covering about 100 years) are unlikely to capture the full range of natural hydroclimatic variability, environmental proxies are used to reconstruct the pre-instrumental climate, or *paleoclimate*. The most broadly useful proxies for reconstructing the past one-to-two millennia of climate in the Southwest are tree rings, which can record either temperature or moisture variability, depending on the species, elevation, and location. Many other proxies—ice cores, glacier size and movement, sand dunes, lake sediments, cave speleothems (e.g., stalactites)—provide information that complements the tree-ring data. We assess here the evidence of past Southwest hydroclimate from these paleorecords.

Paleotemperature

There are far fewer high-resolution paleotemperature records for the past 1,000 to 2,000 years for the Southwest than paleodrought records (described below). But there are sufficient data, mainly from tree rings, to broadly describe the long-term variability in regional temperature, and place the observed temperatures in the Southwest since 1900 into a much longer context (Figure 5.7).

All paleotemperature records for the region indicate that the modern period (since about 1950) has been warmer than at any time in the past 600 years. Most of these records agree that the modern period was also warmer than the Medieval Climate Anomaly (MCA, a period of warm climate in the Northern Hemisphere from ca. AD 900–1350) or any other period in the past 2,000 years (Salzer and Kipfmueller 2005; Ababneh 2008; Salzer et al. 2009). Results of global climate model (GCM) experiments using estimates of past solar variability and volcanic activity also suggest that recent warmth in the Southwest exceeds MCA conditions (Stevens, González-Rouco, and Beltrami 2008; Woodhouse et al. 2010). However, other studies point to warmer conditions in the Southwest during some (Graumlich 1993) or all (Millar et al. 2006) of the MCA compared to the past fifty years.

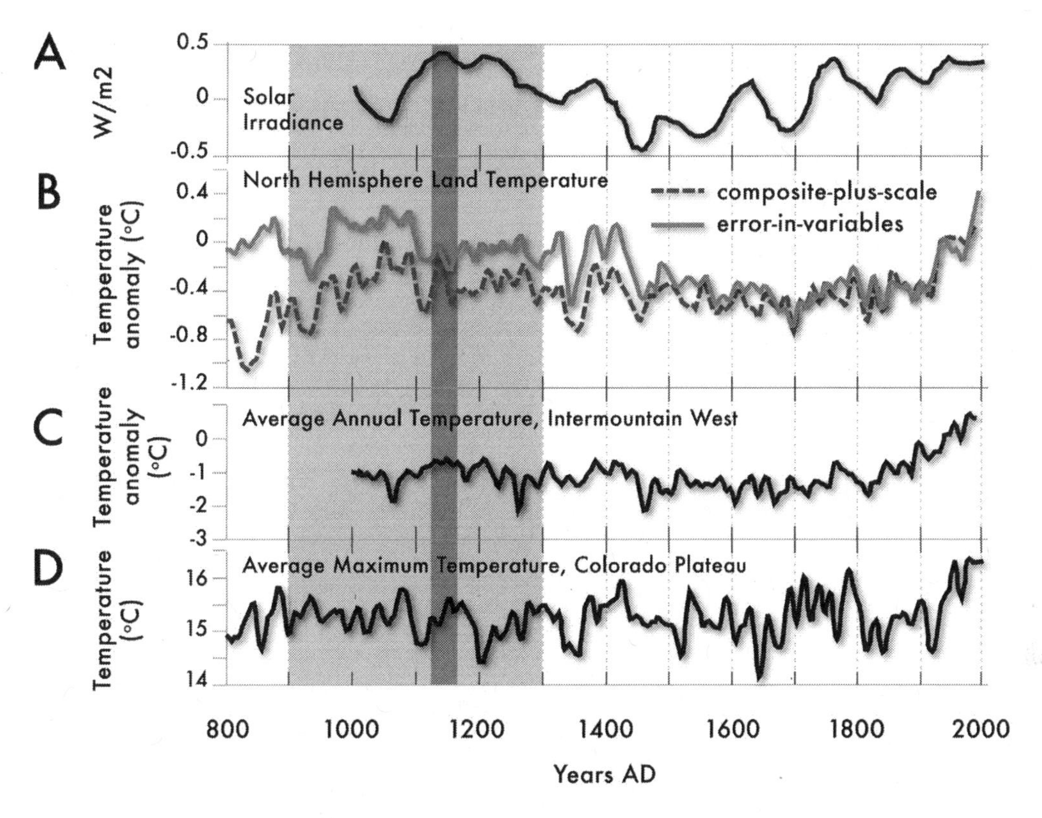

Figure 5.7 Proxy and modeled data for the past 2000 years. (A) Solar variability (solar irradiance) shown in watts per square meter (Bard et al. 2000), (B) Multi-proxy reconstructions of Northern Hemisphere temperatures, two estimates (Mann et al. 2008), (C) ECHO-G GCM simulations of temperatures for the U.S. Southwest-Intermountain region (Stevens, Gonzalez-Ruico and Beltrani 2008), (D) Temperature reconstruction for the Colorado Plateau from tree rings (Salzer and Kipfmueller 2005). The lighter shading indicates the period of the Medieval Climate Anomaly and the darker shading indicates the mid-12th century megadrought in the Colorado River Basin. Adapted from Woodhouse et al. (2010).

Regardless of how warm the MCA was, it is clear that the reconstructed warmest periods during the MCA were associated with widespread severe drought in the Southwest (MacDonald et al. 2008; Woodhouse et al. 2010; Woodhouse, Pederson, and Gray 2011), as indicated by the paleodrought records described below. More generally, the paleorecord indicates that warmer periods of the past 2,000 years have often been associated with increased aridity (Woodhouse, Pederson, and Gray 2011).

Paleodrought

The Southwest contains one of the greatest concentrations in the world of yearly paleodrought records, in the form of hundreds of highly moisture-sensitive tree-ring records. These site-level tree-ring data have been calibrated with observed records at regional to sub-regional scales to reconstruct summer and winter PDSI (see Figure 5.8; Woodhouse and Brown 2001; Cook et al. 2004; Cook et al. 2009; MacDonald 2007; MacDonald and Tingstad 2007), seasonal and annual precipitation (Grissino-Mayer 1996; Ni et al. 2002; Touchan et al. 2010), April 1 snow-water equivalent (Woodhouse 2003; Tingstad and MacDonald 2010; Pederson et al. 2011), as well as annual streamflow in many basins (see below). All of these reconstructions extend back at least 350 years, and some up to 2,200 years. The annual paleodrought information from tree rings is complemented by coarser-resolution information derived from oxygen isotopes in lake sediments (Benson 1999; Anderson 2011), renewed movement of sand-dune fields (Muhs et al. 1997), low stands of lakes (Stine 1994), and speleothems (Asmerom et al. 2007).

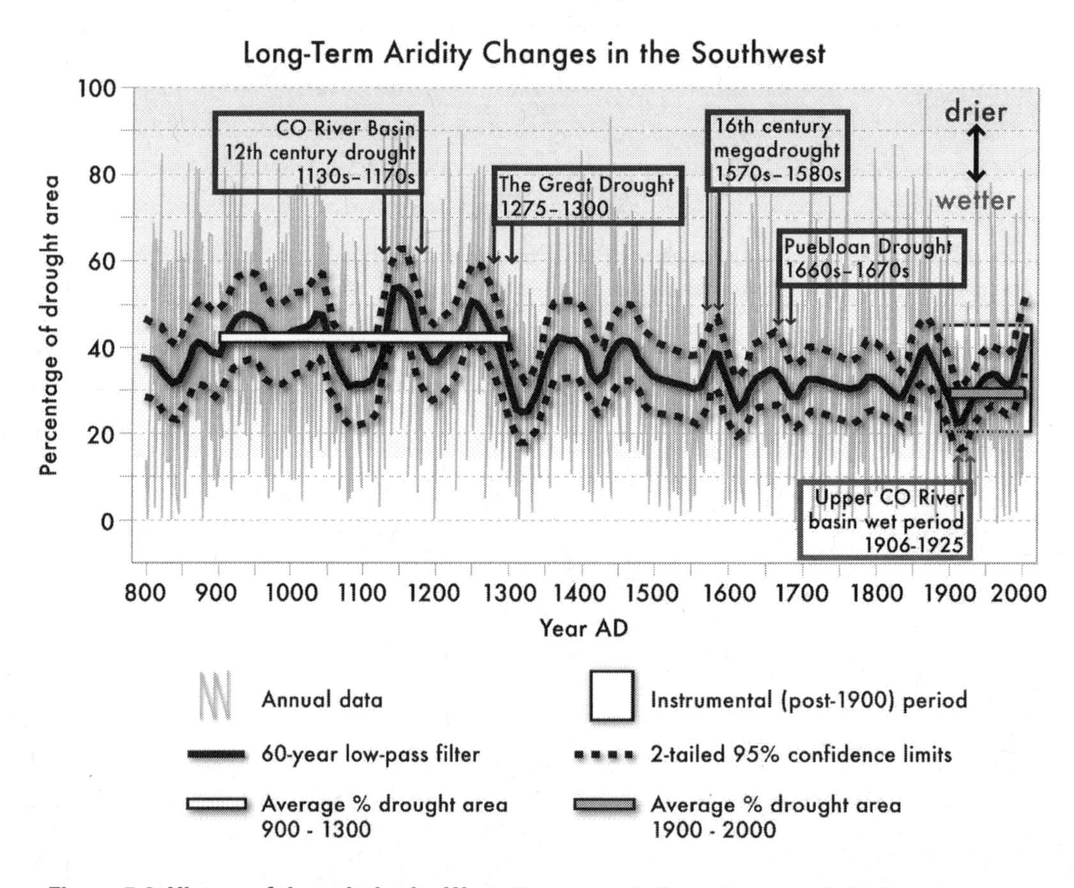

Figure 5.8 History of drought in the West. Percent area affected by drought (PDSI <–1) across the western United States, as reconstructed from tree-ring data. Modified from Cook et al. (2004), reprinted with permission from the American Association for the Advancement of Science.

Collectively, these paleodrought records for the Southwest provide unequivocal evidence that the most severe multi-year droughts observed during the past 110 years in the Southwest, such as in the 1950s and the early 2000s, were exceeded in severity and duration multiple times by droughts during the preceding 2,000 years. The most severe and sustained paleodroughts (sometimes called "megadroughts") occurred during the MCA from AD 900–1350, were associated with high temperatures in the Southwest, and were likely caused by persistently cool La Niña-like conditions in the tropical Pacific Ocean (MacDonald 2007; MacDonald et al. 2008; Seager et al. 2008). Severe paleodroughts also occurred during other times, most prominently the late 1500s megadrought (Stahle et al. 2000) and the early second-century drought (Routson, Woodhouse, and Overpeck 2011). The Southwest paleorecords also clearly indicate that overall the twentieth century experienced less drought than most of the preceding four to twenty centuries (Barnett and Pierce 2009). The reconstructed Drought Area Index (DAI) for the Western United States since AD 800 (Cook et al 2004; Figure 5.8) nicely illustrates the conclusions drawn from the broader set of paleodrought records.[vi]

Paleostreamflow

Information recorded in the growth rings of moisture-sensitive trees has been used to reconstruct annual streamflows in three of the four major Southwest river basins described in section 5.6: the Sacramento-San Joaquin (since AD 901; Meko 2001; Meko and Woodhouse 2005; MacDonald, Kremenetski, and Hidalgo 2008), the Colorado (since AD 762; Meko et al. 2007; MacDonald, Kremenetski, and Hidalgo 2008), and the Rio Grande (since AD 1450; Meko 2008). As with the other records of paleodrought, all of these streamflow reconstructions indicate periods of hydrologic drought prior to 1900 that were more severe and sustained than any since 1900. During the regional droughts in the mid-1100s and in the late 1500s, the Sacramento-San Joaquin basin experienced sustained drought simultaneously with the Colorado and Rio Grande basins. This simultaneity apparently did not occur again until the regional drought that began in 1999–2000.

5.8 Future Monitoring and Science Needs

This chapter offered an account of how weather, climate, and hydrology varied over the Southwestern United States during the last century and in the paleoclimate record. The chapter focused on conditions during the first decade of the new millennium and compared those to conditions of the twentieth century. Emphasis was also given to climate-hydrology links in recognition that growing human populations in the Southwest are placing greater demands on water resources, demands that may create new vulnerabilities in the face of a varying climate. This hydroclimate assessment is also cognizant of concerns regarding human-induced climate change. Human-induced climate change for some variables (especially temperature) has already been detected on global to continental scales, and a human-induced change appears to now be emerging in the Southwest for some hydrology-sensitive variables (such as late wintertime elevated minimum temperatures, which accelerate runoff timing). Human-induced climate change since 1950 has also been detected at a regional scale over the greater western United States. These include detectable effects on hydrology as a consequence of warmer temperatures, reduced spring snowpack, and earlier runoff for a region that includes the Colorado Rocky Mountains and Sierra Nevada, but also the Cascades, Blue Mountains of Oregon, and the northern Rockies of Idaho and Montana. However, a scientific study, using advanced methods of detection and attribution, focused directly on the particular six-state Southwest region of this report, and including data updated through 2010, has not yet been conducted.

This assessment offers a reliable broad-scale view of meteorological and hydrological conditions over the Southwest as a whole, but the reader should be aware that confidence is considerably lower with respect to how local conditions have varied. This is a consequence of sparse observational networks (in time and space), poor monitoring of some key physical processes that connect climate and hydrology (such as high-elevation conditions, soil-moisture conditions, and groundwater variation), and a limited understanding of how weather, climate, and hydrology link at such scales.

If increasing confidence in quantifying how local conditions have varied is desirable, then there is a need for enhanced and especially more abundant observations to better monitor these processes at scales that are informative with respect to terrain. Enhanced monitoring is needed throughout the Southwest, particularly with respect to the water cycle. Augmented capabilities would address the occurrence of heavy precipitation during winter storms and summer convection (thunderstorms), precipitation falling as rain versus snow, rain-on-snow events, snowpack formation and melt off, and basin-scale runoff efficiency (the ratio of precipitation that infiltrates as groundwater as compared to runoff).

In addition to such enhanced data collection efforts, ongoing scientific assessments are needed that connect these data through physically based attribution studies. These would require improved climate and hydrological modeling at scales consistent with resolving the effects of terrain on weather and climate patterns throughout the Southwest. They would also require improved representation of physical processes such as atmospheric convection, evapotranspiration, snowpack formation, and runoff production. Efforts to improve monitoring must be complemented by improvements in our ability to detect and attribute changes in climate conditions and hydrology throughout the Southwest for years to decades. Attention to variations and changes in the region's water cycle, informed by these more precise and extensive science, modeling, and monitoring efforts, will allow us to create better early warning systems for the region and to also identify effective adaptation and mitigation strategies to address climate-change impacts (e.g. Ralph et al. 2011).

References

Ababneh, L. 2008. Bristlecone pine paleoclimatic model for archeological patterns in the White Mountain of California. *Quaternary International* 188:9–78.

Alley, W. 1984. The Palmer Drought Severity Index: Limitations and assumptions. *Journal of Climatology and Applied Meteorology* 23:1100–1109.

Anderson, L. 2011. Holocene record of precipitation seasonality from lake calcite ^{18}O in the central Rocky Mountains, United States. *Geology* 39:211–214.

Asmerom, Y., V. Polyak, S. Burns, and J. Rassmussen. 2007. Solar forcing of Holocene climate: New insights from a speleothem record, southwestern United States. *Geology* 35:1–4.

Bard, E., G. Raisbeck, F. Yiou, and J. Jouzel. 2000. Solar irradiance during the last millennium based on cosmogenic nucleides. *Tellus* 52B: 985–992.

Barnett, T. P., and D. W. Pierce. 2008. When will Lake Mead go dry? *Water Resources Research* 44: W03201, doi:10.1029/2007WR006704.

—. 2009. Sustainable water deliveries from the Colorado River in a changing climate. *Proceedings of the National Academy of Sciences* 106:7334–7338.

Barnett, T. P., D. W. Pierce, H. Hidalgo, C. Bonfils, B. Santer, T. Das, G. Bala, et al. 2008. Human-induced changes in the hydrology of the western United States. *Science* 319:1080–1083, doi:10.1073/pnas.0812762106

Benson, L.V., 1999. Records of millennial-scale climate change from the Great Basin of the western United States. In *Mechanisms of global climate change at millennial time scales,* ed. P. Clark, R. Webb, and L. Keigwin, 203–225. Geophysical Monograph Series 112. Washington, DC: American Geophysical Union.

Bonfils, C., B. D. Santer, D. W. Pierce, H. G. Hidalgo, G. Bala, T. Das, T.P. Barnett, et al. 2008. Detection and attribution of temperature changes in the mountainous western United States. *Journal of Climate* 21:6404–6424.

Bonnin, G. M., K. Maitaria, and M. Yekta. 2011. Trends in rainfall exceedances in the observed record in selected areas of the United States. *Journal of the American Water Resources Association* 46:344–353, doi:10.1111/j.1752-1688.2011.00603.x.

Cayan, D. R., S. A. Kammerdiener, M. D. Dettinger, J. M. Caprio, and D. H. Peterson. 2001. Changes in the onset of spring in the western United States. *Bulletin of the American Meteorological Society* 82:399–415.

Cayan, D. R., T. Das, D. W. Pierce, T. P. Barnett, M. Tyree, and A. Gershunov. 2010. Future dryness in the southwest US and the hydrology of the early 21st century drought. *Proceedings of the National Academy of Sciences* 107:21271-21276, doi:10.1073/pnas.0912391107.

Christensen, N. S., and D. P. Lettenmaier. 2007. A multimodel ensemble approach to assessment of climate change impacts on the hydrology and water resources of the Colorado River Basin. *Hydrology and Earth System Sciences* 11:1417–1434.

Cook, E. R., C. Woodhouse, C. M. Eakin, D. M. Meko, and D. W. Stahle. 2004. Long-term aridity changes in the western United States. *Science* 306:1015–1018.

Cook, E. R., R. Seager, R. R. Heim, R. S. Vose, C. Herweijer, and C. Woodhouse. 2009. Megadroughts in North America: Placing IPCC projections of hydroclimatic change in a long-term paleoclimate context. *Journal of Quaternary Science* 25:48–61, doi: 10.1002/jqs.1303.

Dai, A. 2011. Characteristics and trends in various forms of the Palmer Drought Severity Index (PDSI) during 1900-2008. *Journal of Geophysical Research* 116: D12115, doi:10.1029/2010JD015541.

Daly, C. 2006. Guidelines for assessing the suitability of spatial climate data sets. *International Journal of Climatology* 26:707–721.

Daly, C., M. Halbleib, J. I. Smith, W. P. Gibson, M. K. Doggett, G. H. Taylor, J. Curtis, and P. A. Pasteris. 2008. Physiographically-sensitive mapping of temperature and precipitation across the conterminous United States. *International Journal of Climatology* 28:2031–2064.

Dettinger, M. D., and D. R. Cayan. 1995. Large-scale atmospheric forcing of recent trends toward early snowmelt runoff in California. *Journal of Climate* 8:606–623.

Fall, S., A. Watts, J. N. Gammon, E. Jones, D. Niyogi, J. Christy, and R. Pielke Sr. 2011. Analysis of the impacts of station exposure on the U.S. Historical Climatology Network temperatures and temperature trends. *Journal of Geophysical Research* 116: D14120.

Fritze, H., I. T. Stewart, and E. Pebesma. 2011. Shifts in western North American runoff regimes for the recent warm decades. *Journal of Hydrometeorology* 12:989–1006.

Fulp, T. 2005. How low can it go? *Southwest Hydrology* 4 (2):16–17, 28.

Gershunov, A., and T. P. Barnett. 1998. Interdecadal modulation of ENSO teleconnections. *Bulletin of the American Meteorological Society* 79:2715–2725.

Graumlich, L. J. 1993. A 1000-year record of temperature and precipitation in the Sierra Nevada. *Quaternary Research* 39:249–255.

Grissino-Mayer, H., 1996. A 2129-year reconstruction of precipitation for northwestern New

Mexico, U.S.A. In *Tree Rings, environment and humanity: Proceedings of the international conference, Tucson, Arizona, 17-21 May 1994*, ed. J. S. Dean, D. M. Meko, and T. W. Swetnam, 191-204. Tucson, AZ: Radiocarbon, Dept. of Geosciences, University of Arizona.

Groisman, P. V., R. W. Knight, T. R. Karl, D. R. Easterling, B. Sun, and J. H. Lawrimore. 2004. Contemporary changes of the hydrological cycle over the contiguous United States: Trends derived from in situ observations. *Journal of Hydrometeorology* 5:64–85, doi: 10.1175/1525–7541.

Hidalgo, H. G., T. Das, M. D. Dettinger, D. R. Cayan, D. W. Pierce, T. P. Barnett, G. Bala, et al. 2009. Detection and attribution of stream flow timing changes to climate change in the western United States. *Journal of Climate* 22:3838–3855, doi:10.1175/2009JCLI2470.1.

Hirsch, R. M., and K. R. Ryberg. 2011. Has the magnitude of floods across the USA changed with global CO_2 levels? *Hydrological Sciences Journal* 57:1–9, doi:10.1080/02626667.2011.621895.

Hoerling, M., J. Eischeid, and J. Perlwitz. 2010. Regional precipitation trends: Distinguishing natural variability from anthropogenic forcing. *Journal of Climate* 23:2131–2145.

Hoerling, M. P., and A. Kumar. 2003. The perfect ocean for drought. *Science* 299:691–694.

Kahya, E., and J. A. Dracup. 1993. U.S. streamflow patterns in relation to the El Niño/Southern Oscillation. *Water Resources Research* 29:2491–2503.

Knowles, N., M. D. Dettinger, and D. R. Cayan, 2006. Trends in snowfall versus rainfall in the western United States. *Journal of Climate* 19:4545–4559.

Kunkel, K. E., D. R. Easterling, K. Hubbard, and K. Redmond. 2004. Temporal variations in frost-free season in the United States: 1895–2000. *Geophysical Research Letters* 31: L03201, doi:10.1029/2003GL018624.

Kunkel, K. E., D. R. Easterling, K. Redmond, and K. Hubbard. 2003. Temporal variations of extreme precipitation events in the United States: 1895–2000. *Geophysical Research Letters* 30:1900-1903, doi:10.1029/2003GL018052.

Liverman, D. M., and R. W. Merideth, Jr. 2002. Climate and society in the U.S. Southwest: The context for a regional assessment. *Climate Research* 21:199–218.

MacDonald, G. M. 2007. Severe and sustained drought in Southern California and the West: Present conditions and insights from the past on causes and impacts. *Quaternary International* 173/174:87–100, doi:10.1016/j.quaint.2007.03.012.

MacDonald, G. M., and A. M. Tingstad. 2007. Multicentennial precipitation variability and drought occurrence in the Uinta Mountains region, Utah. *Arctic, Antarctic and Alpine Research* 39:549–555.

MacDonald, G.M., K. V. Kremenetski, and H. Hidalgo. 2008. Southern California and the perfect drought: Simultaneous prolonged drought in Southern California and the Sacramento and Colorado River systems. *Quaternary International* 188:11–23, doi:10.1016/j.quaint.2007.06.027.

MacDonald, G. M., D. W. Stahle, J. Villanueva Diaz, N. Beer, S. J. Busby, J. Cerano-Paredes, J. E. Cole, et al. 2008. Climate warming and twenty-first century drought in southwestern North America. *EOS Transactions AGU* 89:82.

Mann, M. E., Z. Zhang, M. K. Hughes, R. S. Bradley, S. K. Miller, S. Rutherford, and F. Ni. 2008. Proxy-based reconstructions of hemispheric and global surface temperature variations over the past two millennia. *Proceedings of the National Academy of Sciences* 105:13252-13257, doi:10.1073/pnas.0805721105.

McCabe, G. J., and D. M. Wolock. 2007. Warming may create substantial water supply shortages in the Colorado River Basin. *Geophysical Research Letters* 34: L22708, doi:10.1029/2007GL031764.

Meehl, G., C. Tibaldi, G. Walton, D. Easterling, and L. McDaniel. 2009. Relative increase of record high maximum temperature compared to record low minimum temperature in the US. *Geophysical Research Letters* 36: L23701, doi:10.1029/2009GL040736.

Meko, D. M. 2001. Reconstructed Sacramento River system runoff from tree rings. Report prepared for the California Department of Water Resources, July 2001.

—. 2008. Streamflow reconstruction, Rio Grande River at Otowi Ridge, 1450-2002. Prepared for the Rio Grande Basin Workshop, 30 May 2008, New Mexico State University Extension, Albuquerque, New Mexico. http://treeflow.info/riogr/riograndeotowinatural.txt.

Meko, D. M., and C. A. Woodhouse. 2005. Tree-ring footprint of joint hydrologic drought in Sacramento and Upper Colorado River Basins, western USA. *Journal of Hydrology* 308:196–213.

Meko, D. M., C. A. Woodhouse, C. A. Baisan, T. Knight, J. J. Lukas, M. K. Hughes, and M. W. Salzer. 2007. Medieval drought in the Upper Colorado River Basin. *Geophysical Research Letters* 34: L10705.

Menne, M. J., and C. N. Williams, Jr. 2009. Homogenization of temperature series via pairwise comparisons. *Journal of Climate* 22:1700–1717, doi:10.1175/2008JCLI2263.1.

Millar, C. I., J. C. King, R. D. Westfall, H. A. Alden, and D. L. Delany. 2006. Late Holocene forest dynamics, volcanism, and climate change at Whitewing Mountain and San Joaquin Ridge, Mono County, Sierra Nevada, CA, USA. *Quaternary Research* 66:273–287.

Milly, P. C. D., K. A. Dunne, and A. V. Vecchia. 2005. Global pattern of trends in streamflow and water availability in a changing climate. *Nature* 438:347–350.

Mote, P. W., A. F. Hamlet, M. P. Clark, and D. P. Lettenmaier. 2005. Declining mountain snowpack in western North America. *Bulletin of the American Meteorological Society* 86:39–49.

Muhs, D. R., T. W. Stafford, J. B. Swinehart, S. D. Cowherd, S. A. Mahan, C. A. Bush, R. F. Madole, and P. B. Maat. 1997. Late Holocene eolian activity in the mineralogically mature Nebraska Sand Hills. *Quaternary Research* 48:162–176.

Ni, F., T. Cavazos, M. K. Hughes, A. C. Comrie, and G. Funkhouser, 2002. Cool-season precipitation in the southwestern USA since AD 1000: Comparison of linear and nonlinear techniques for reconstruction. *International Journal of Climatology* 22:1645–1662.

Palmer, W. C. 1965. *Meteorological drought*. Research Paper No. 45. Washington, DC: U.S. Weather Bureau.

Pederson, G. T., S. T. Gray, C. A. Woodhouse, J. L. Betancourt, D. B. Fagre, J. S. Littell, E. Watson, B. H. Luckman, and L. J. Graumlich. 2011. The unusual nature of recent snowpack declines in the North American Cordillera. *Science* 333:332–335, doi: 10.1126/science.1201570.

Pielke, R. Sr., N. Doesken, O. Bliss, T. Green, C. Chaffin, J. Salas, C. Woodhouse, J. Lukas, and K. Wolter. 2005. Drought 2002 in Colorado: An unprecedented drought or a routine drought? *Pure and Applied Geophysics* 162:1455–1479.

Pierce, D., T. Barnett, H. Hidalgo, T. Das, C. Bonfils, B. Santer, G. Bala, et al. 2008. Attribution of declining western US snowpack to human effects. *Journal of Climate* 21:6425–6444, doi:10.1175/2008JCLI2405.1.

Rajagopalan, B., K. Nowak, J. Prairie, M. Hoerling, B. Harding, J. Barsugli, A. Ray, and B. Udall. 2009. Water supply risk on the Colorado River: Can management mitigate? *Water Resources Res*earch 45: W08201, doi:10.1029/2008WR007652.

Ralph, F. M., M. D. Dettinger, A. White, D. Reynolds, D. Cayan, T. Schneider, R. Cifelli, et al. 2011. A vision of future observations for western US extreme precipitation events and flooding: Monitoring, prediction and climate. Report to the Western States Water Council, Idaho Falls.

Redmond, K. T., and R. W. Koch. 1991. Surface climate and streamflow variability in the western United States and their relationship to large scale circulation indices. *Water Resources Research* 27:2381–2399, doi:10.1029/91WR00690.

Regonda, S. K., B. Rajagopalan, M. Clark, and J. Pitlick. 2005. Seasonal cycle shifts in hydroclimatology over the western United States. *Journal of Climate* 18:372–384.

Roos, M. 1991. A trend of decreasing snowmelt runoff in Northern California. In *Proceedings of the 59th Western Snow Conference, April 12-15, 1991, Juneau, Alaska,* 29–36. Juneau, AK: Western Snow Conference.

Ropelewski, C. F., and M. S. Halpert. 1986. North American precipitation and temperature patterns associated with the El Niño/Southern Oscillation (ENSO). *Monthly Weather Review* 114:2352–2362.

Routson, C. C., C. Woodhouse, and J. T. Overpeck. 2011. Second century megadrought in the Rio Grande headwaters, Colorado: How unusual was medieval drought? *Geophysical Research Letters* 38: L22703, doi:10.1029/2011GL050015.

Salzer, M. W., and K. F. Kipfmueller. 2005. Reconstructed temperature and precipitation on a millennial timescale from tree rings in the southern Colorado Plateau, USA. *Climatic Change* 70:465–487.

Salzer, M., M. Hughes, A. Bunn, and K. Kipfmueller. 2009. Recent unprecedented tree-ring growth in bristlecone pine at the highest elevations and possible causes. *Proceedings of the National Academy of Sciences* 106:20348–20353.

Seager, R., R. Burgman, Y. Kushnir, A. Clement, E. R. Cook, N. Naik, and J. Miller. 2008. Tropical Pacific forcing of North American medieval megadroughts: Testing the concept with an atmosphere model forced by coral reconstructed SSTs. *Journal of Climate* 21:6175–6190.

Stahle, D. W., E. R. Cook, M. K. Cleaveland, M. D. Therrell, D. M. Meko, H. D. Grissino-Mayer, E. Watson, and B. H. Luckman. 2000. Tree-ring data document 16th century megadrought over North America. *EOS Transactions AGU* 81:121, doi:10.1029/00EO00076.

Stegner, W. 1987. *The American West as living space.* Ann Arbor: University of Michigan Press.

Stevens, M. B., J. F. González-Rouco, and H. Beltrami. 2008. North American climate of the last millennium: Underground temperatures and model comparison. *Journal of Geophysical Research* 113: F01008, doi: 10.1029/2006JF000705.

Stewart, I., D. Cayan, and M. Dettinger. 2005. Changes towards earlier streamflow timing across western North America. *Journal of Climate* 18:1136–1155.

Stine, S. 1994. Extreme and persistent drought in California and Patagonia during mediaeval time. *Nature* 369:546–549.

Tingstad, A. H., and G. M. MacDonald. 2010. Long-term relationships between ocean variability and water resources in northeastern Utah. *Journal of the American Water Resources Association* 46:987–1002.

Touchan, R., C. Woodhouse, D. Meko, and C. Allen. 2010. Millennial precipitation reconstruction for the Jemez Mountains, New Mexico, reveals changing drought signal. *International Journal of Climatology* 31:896–906.

Woodhouse, C.A., and P. M. Brown. 2001. Tree-ring evidence for Great Plains drought. *Tree-Ring Research* 57:89–103.

Woodhouse, C. A., 2003. A 431-year reconstruction of western Colorado snowpack. *Journal of Climate* 16:1551–1561.

Woodhouse, C. A., D. M. Meko, G. M. MacDonald, D. W. Stahle, and E. R. Cook. 2010. A 1,200-year perspective of 21st century drought in southwestern North America. *Proceedings of the National Academy of Sciences* 107:21283–21288, doi:10.1073/pnas.0911197107.

Woodhouse, C. A., G. T. Pederson, and S. T. Gray. 2011. An 1800-year record of decadal-scale hydroclimatic variability in the Upper Arkansas River Basin from bristlecone pine. *Quaternary Research* 75:483–490, doi:10.1016/j.yqres.2010.12.007.

Yarnal, B., and H. F. Diaz. 1986. Relationships between extremes of the Southern Oscillation and the winter climate of the Anglo-American Pacific Coast. *Journal of Climate* 6:197–219.

Appendix

Table A5.1 Comparison of annual surface temperatures in the Southwest

NCDC AVERAGE TEMPERATURE

	1901–2000 6-State Avg.		2001–2010 Decadal		
Season	**Mean °F (°C)**	**Std.Dev. °F (°C)**	**Mean °F (°C)**	**Anom °F (°C)**	**Rank**
DJF	36.0 (2.2)	0.7 (0.4)	36.7 (2.6)	+0.7 (0.4)	2
MAM	51.3 (10.7)	0.9 (0.5)	52.9 (11.6)	+1.6 (0.9)	1
JJA	70.9 (21.6)	0.7 (0.4)	72.9 (22.7)	+2.0 (1.1)	1
SON	54.1 (12.3)	0.5 (0.3)	55.4 (13.0)	+1.3 (0.7)	1
Annual	53.0 (11.7)	0.7 (0.4)	54.5 (12.5)	+1.4 (0.8)	1

STATION-ADJUSTED AVERAGE TEMPERATURE

	1901–2000 6-State Avg.		2001–2010 Decadal		
Season	**Mean °F (°C)**	**Std.Dev. °F (°C)**	**Mean °F (°C)**	**Anom °F (°C)**	**Rank**
DJF	36.7 (2.6)	0.7 (0.4)	37.2 (2.9)	+0.5 (0.3)	2
MAM	52.2 (11.2)	0.9 (0.5)	53.6 (12)	+1.4 (0.8)	1
JJA	71.2 (21.8)	0.7 (0.4)	72.7 (22.6)	+1.4 (0.8)	1
SON	55.0 (12.8)	0.5 (0.3)	55.9 (13.3)	+0.9 (0.5)	1
Annual	53.8 (12.1)	0.5 (0.3)	54.9 (12.7)	+1.1 (0.6)	2

STATION-UNADJUSTED AVERAGE TEMPERATURE

	1901–2000 6-State Avg.		2001–2010 Decadal		
Season	**Mean °F (°C)**	**Std.Dev. °F (°C)**	**Mean °F (°C)**	**Anom °F (°C)**	**Rank**
DJF	37.2 (2.9)	0.5 (0.3)	37.4 (3)	+0.2 (0.1)	4
MAM	52.9 (11.6)	0.7 (0.4)	54.0 (12.2)	+1.1 (0.6)	1
JJA	71.8 (22.1)	0.7 (0.4)	73.0 (22.8)	+1.3 (0.7)	1
SON	55.6 (13.1)	0.4 (0.2)	56.3 (13.5)	+0.7 (0.4)	1
Annual	54.3 (12.4)	0.5 (0.3)	55.2 (12.9)	+0.9 (0.5)	2

Note: Temperatures are averaged for the decade 2001–2010 versus 1901–2000; there is also a ranking of the 2001–2010 decadal averages relative to the ten individual decades of the twentieth century. The Southwest is comprised of Arizona, California, Colorado, Nevada, New Mexico, and Utah.

Sources: National Climatic Data Center, climate division, station-based adjusted, and station-based unadjusted data (http://www.ncdc.noaa.gov/oa/ncdc.html).

Table A5.2 Comparison of annual precipitation in the Southwest

PRISM PRECIPITATION					
	1901–2000 6-State Avg.		2001–2010 Decadal		
Season	**Mean inches (mm)**	**Std.Dev.inches (mm)**	**Mean inches (mm)**	**Anom inches (mm)**	**Rank**
DJF	5.1 (129)	0.4 (10)	5.2 (133)	+0.2 (4)	5
MAM	3.8 (97)	0.3 (8)	3.4 (87)	-0.4 (10)	10
JJA	3.3 (83)	0.3 (7)	3.0 (77)	-0.2 (6)	9
SON	3.3 (85)	0.4 (10)	3.2 (82)	-0.1 (3)	6
Annual	15.6 (395)	0.8 (20)	15.0 (380)	-0.6 (15)	8
NCDC CLIMATE DIVISION PRECIPITATION					
	1901–2000 6-State Avg.		2001–2010 Decadal		
Season	**Mean inches (mm)**	**Std.Dev.inches (mm)**	**Mean inches (mm)**	**Anom inches (mm)**	**Rank**
DJF	4.7 (120)	0.4 (11)	4.8 (122)	+0.04 (1)	5
MAM	3.5 (90)	0.3(8)	3.0 (77)	-0.5 (13)	10
JJA	3.3 (85)	0.3 (7)	3.0 (77)	-0.3 (8)	9
SON	3.2 (82)	0.4 (10)	3.0 (76)	-0.2 (6)	9
Annual	14.8 (377)	0.8 (21)	14.1 (359)	-0.7 (18)	10
NCDC STATION PRECIPITATION					
	1901–2000 6-State Avg.		2001–2010 Decadal		
Season	**Mean inches (mm)**	**Std.Dev.inches (mm)**	**Mean inches (mm)**	**Anom inches (mm)**	**Rank**
DJF	6.0 (153)	0.5 (13)	6.1 (154)	+0.04 (1)	6
MAM	4.1 (104)	0.4 (9)	3.6 (91)	-0.6 (14)	10
JJA	3.3 (85)	0.4 (9)	3.0 (76)	-0.4 (9)	8
SON	3.5 (90)	0.5 (12)	3.3 (84)	-0.2 (6)	8
Annual	16.9 (429)	0.8 (20)	16.4 (417)	-0.5 (13)	7

Note: Precipitation is averaged for the decade 2001–2010 versus 1901–2000 along with a ranking of the 2001–2010 decadal averages relative to the ten individual decades of the twentieth century.

Sources: PRISM (PRISM Climate Group, Oregon State University, http://prism.oregonstate.edu); National Climatic Data Center (NCDC), climate division, and station-based data (http://www.ncdc.noaa.gov/oa/ncdc.html).

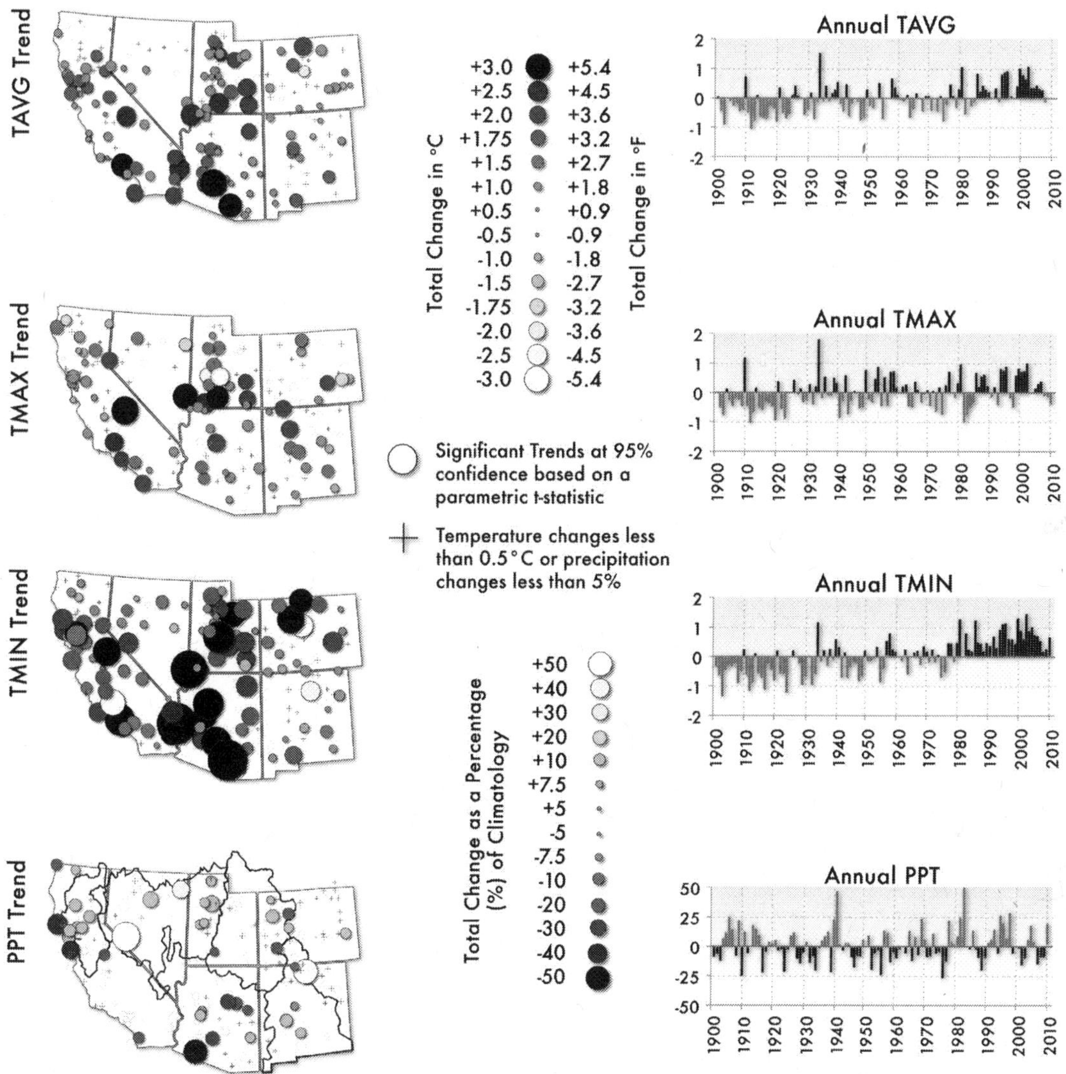

Figure A5.1 The 1901–2010 trends in annual averaged daily temperature (TAVG, top), daily maximum temperature (TMAX, second panel), and daily minimum temperature (TMIN, third panel). Units are the change in °C/110yrs, and stronger positive (negative) trends shown in dark grey (light grey). Bottom panel is the 1901–2010 trend in annual averaged precipitation. Units are the total change expressed as percent of annual climatology, and positive (negative) trends are shown in light grey (dark grey). Larger circle sizes denote greater magnitude trends. Trends are computed at station locations using the GHCN (Version 3) raw monthly (unadjusted) data. Results are shown only for locations having at least 90 years of available data. Filled stations denote statistically significant trends at 95% confidence based on a parametric t-statistic. Stations with temperature (precipitation) changes less than 0.5°C (5%) are denoted with a + symbol. The fraction of stations with significant trends is 61%, 40%, and 63% for TAVG, TMAX, and TMIN respectively. The fraction of stations with significant trends in precipitation is 16%. Source: Menne and Williams (2009).

Endnotes

i Tables 5.1 and 5.2 present temperature and precipitation conditions for the first decade of the twenty-first century, respectively, and compare those conditions against 100-year averages of the previous century. Shown also are the rankings of the recent decadal conditions relative to the ten decades of the twentieth century, both for the Southwest as a whole and for the six individual states comprising the region. The tables assess average temperature, maximum temperature, minimum temperature, and precipitation. The data are based on the monthly PRISM analysis (Daly 2006) which incorporates physiographic features (e.g., complex topography and coastal zones) in the process of generating climate grids from available in situ data, the consequence of which is to substantially improve analyses in the Western United States relative to other climate analyses (Daly et al. 2008).

ii Data used in Tables 5.1 and 5.2 are based on 2.5 mile (4km) resolution PRISM analyses (data available at: http://www.prism.oregonstate.edu/products/matrix.phtml?view=data). For purposes of long period trend estimates, we present diagnoses conducted at station locations, rather than from gridded data, and examine those sites that possess historical observations spanning most of the 1901–2010 period.

iii The trends were calculated at station locations based on Global Historical Climate Network (GHCN) Version 3 (Menne and Williams 2009; data available at http://www.ncdc.noaa.gov/ghcnm/v3.php).

iv Another measure of heat and cold waves is discussed in Chapter 7, Sections 7.2 and 7.3.

v Drought is defined here as having at least a -1 (or lower) PDSI intensity.

vi These included the Southwest as well as drainage basins in the Cascades, Blue Mountains of Oregon, and the northern Rockies of Idaho and Montana.

vii Analysis of a subset of the Cook et al. dataset, covering only the Southwest region, shows DAI variability across the Southwest over the last 1,200 years to be very similar to that across the larger area depicted in Figure 5.7.

Chapter 6

Future Climate: Projected Average

COORDINATING LEAD AUTHOR

Daniel R. Cayan (Scripps Institution of Oceanography, University of California, San Diego and U.S. Geological Survey)

LEAD AUTHORS

Mary Tyree (Scripps Institution of Oceanography, University of California, San Diego), Kenneth E. Kunkel (NOAA Cooperative Institute for Climate and Satellites, North Carolina State University and National Climate Data Center), Chris Castro (University of Arizona), Alexander Gershunov (Scripps Institution of Oceanography, University of California, San Diego), Joseph Barsugli (University of Colorado, Boulder, CIRES), Andrea J. Ray (NOAA), Jonathan Overpeck (University of Arizona), Michael Anderson (California Department of Water Resources), Joellen Russell (University of Arizona), Balaji Rajagopalan (University of Colorado), Imtiaz Rangwala (University Corporation for Atmospheric Research), Phil Duffy (Lawrence Livermore National Laboratory)

EXPERT REVIEW EDITOR

Mathew Barlow (University of Massachusetts, Lowell)

Executive Summary

This chapter describes possible climate changes projected to evolve during the twenty-first century for the Southwest United States, as compared to recent historical climate. It focuses on how climate change might affect longer-term aspects of the climate in the

Chapter citation: Cayan, D., M. Tyree, K. E. Kunkel, C. Castro, A. Gershunov, J. Barsugli, A. J. Ray, J. Overpeck, M. Anderson, J. Russell, B. Rajagopalan, I. Rangwala, and P. Duffy. 2013. "Future Climate: Projected Average." In *Assessment of Climate Change in the Southwest United States: A Report Prepared for the National Climate Assessment*, edited by G. Garfin, A. Jardine, R. Merideth, M. Black, and S. LeRoy, 101–125. A report by the Southwest Climate Alliance. Washington, DC: Island Press.

Southwest and is closely related to Chapter 7, which is concerned with the implications of climate change on shorter period phenomena, especially extreme events. The projections derive from the outcomes of several global climate models, and associated "downscaled" regional climate simulations, using two emissions scenarios ("A2" or "high-emissions," and "B1" or "low-emissions") developed by the Intergovernmental Panel on Climate Change (IPCC) *Special Report on Emissions Scenarios* (SRES; Nakićenović and Swart 2000). The key findings are:

- Temperatures at the earth's surface in the Southwest will rise substantially (by more than 3°F [1.7°C] over recent historical averages) over the twenty-first century from 2001–2100. (high confidence)
- The amount of temperature rise at the earth's surface in the Southwest will be higher in summer and fall than winter and spring. (medium-high confidence)
- Climate variations of temperature and precipitation over short periods (year-to-year and decade-to-decade) will continue to be a prominent feature of the Southwest climate. (high confidence)
- There will be lower precipitation in the southern portion of the Southwest region and little change or increasing precipitation in the northern portion. (medium-low confidence)
- There will be a reduction of Southwest mountain snowpack during February through May from 2001 through 2100, mostly because of the effects of warmer temperature. (high confidence)
- Substantial parts of the Southwest region will experience reductions in runoff and streamflow from the middle to the end of the twenty-first century. (medium-high confidence)

6.1 Global Climate Models: Statistical and Dynamical Downscaling

Global climate models (GCMs) are the fundamental drivers of regional climate-change projections (IPCC 2007). GCMs allow us to characterize changes in atmospheric circulation associated with human causes at global and continental scales. However, because of the planetary scope of the GCMs, their resolution, or level of detail, is somewhat coarse. A typical GCM grid spacing is about 62 miles (100 km) or greater, which is inadequate for creating projections and evaluating impacts of climate change at a regional scale. Thus, a "downscaling" procedure is needed to provide finer spatial detail of the model results.

Downscaling is done in two ways—statistical (or empirical) downscaling and dynamical downscaling—each with its inherent strengths and weaknesses. Statistical downscaling relates historical observations of local variables to large-scale measures. For climate modeling, this means taking the observed relationship of atmospheric circulation and regional-scale surface data of interest (temperature and precipitation) and applying those empirical relationships to GCM data for some future period (Wilby et al. 2004; Maurer et al. 2010). Many of the results shown here, involving projected temperature and precipitation, are based upon a "bias correction and spatial downscaling"

(BCSD) method (Maurer et al. 2010).[i] "Bias" is a statistical term reflecting systematic error. The main advantage of statistical downscaling is that it is computationally simple, so a relatively large number of GCMs and greenhouse gas emission scenarios may be considered for a more robust characterization of statistical uncertainty (Maurer et al. 2010).[ii]

In contrast, dynamical downscaling produces climate information by use of a limited-area, regional climate model (RCM) driven by the output from a global climate model (Mearns et al. 2003; Laprise et al. 2008). Dynamical RCMs, similar to GCMs, are numerical representations of the governing set of equations that describe the climate system and its evolution through time over a particular region. Though use of a physically based process model at a grid spacing of tens (rather than hundreds) of kilometers is substantially more computationally expensive, the method has two advantages over statistical downscaling. Climate stationarity (the concept that past climatic patterns are a reasonable representation of those in the future) is not assumed and the influence of complex terrain on the climate of the western United States is better represented (Mearns et al. 2003), improving the simulation of precipitation from winter storms (Ikeda et al. 2010) and thunderstorms during the summer monsoon (Gutzler et al. 2005). However, because of their cost, long runs (lengthy computer processing) of regional climate simulations generally are not undertaken. In addition, each regional model contains some degree of bias, so statistical adjustments are almost always required. In other words, regardless of whether statistical or dynamical downscaling is pursued, bias-correction of GCM or GCM-RCM output is a necessary part of the process.

In this chapter, we use the "variable infiltration capacity" (VIC) model (Liang et al. 1994) to derive land-surface hydrological variables that are consistent with the downscaled forcing data.[iii] The VIC model has been applied in many studies of hydrologic impacts of climate variability and change (e.g. Wood et al. 2004; Das et al. 2009).

6.2 Climate Scenarios

Following the lead of the U.S. National Climate Assessment, the *Special Report on Emissions Scenarios* A2 ("high emissions") and B1 ("low emissions") scenarios (IPCC 2007) are employed here (see endnote iii in Chapter 2).

The choice of the high (A2) and low (B1) emissions scenarios was also guided by the need to span a range of GHG emissions and the availability of a reasonably large number of GCM simulations. GCMs, to varying degrees, capture average recent historical climate and a statistical representation of its variability over the Southwest (Ruff, Kushnir, and Seager 2012) and to some extent key regions such as the tropical Pacific that are known to drive important climate variations in the Southwest region (Dai 2006; IPCC 2007; Cayan et al. 2009). Many applications employ multiple model simulations, made for one or more GHG emissions scenario using one or more GCMs, which are generally referred to as an ensemble. Beyond this, however, it has been shown (Pierce et al. 2009; Santer et al. 2009) that a model's performance in simulating characteristics of observed climate is not a very useful measure of how well it will simulate future climate under climate change. Because of this, it is better to draw upon a number of simulations to construct a possible distribution of climate change; i.e., it is important to consider results from several climate models rather than to rely on just a few.

6.3 Data Sources

Projected climate for the *Assessment of Climate Change in the Southwest United States* is based on a series of GCM and downscaled projection data sets. A set of CMIP3[iv] GCM outputs, from fifteen GCMs (used in an initial set of experiments) and sixteen GCMs (used in later experiments) that were identified in the 2009 NCA report, provided a core set of thirty and thirty-two climate simulations, respectively. Each GCM simulated one high- and one low-emissions scenario. Only one simulation for each individual GCM-emissions scenario pairing was included. The GCMs provide historical simulations in addition to projected twenty-first-century climate simulations based on the high- and low-emissions scenarios.

Statistical downscaled monthly temperature and precipitation data from the sixteen GCMs using the BCSD method were employed. These data are at a horizontal resolution of 1/8° (roughly 7.5-mile [12-km] resolution), covering the period of 1950–2099 (Maurer, Brekke et al. 2007; Gangopadhyay et al. 2011).

Dynamical downscaled simulations from the multi-institutional North American Regional Climate Change Assessment Program (NARCCAP; Mearns et al. 2009) were used. At this time, there are nine high-emissions simulations available using different combinations of a regional climate model driven by a global climate model. Each simulation includes the periods of 1971–2000 and 2041–2070 for the high-emissions scenario only, and is at a horizontal resolution of approximately 31 miles (50 km).

Peer-reviewed and publicly available hydrologic projections (Gangopadhyay et al. 2011) are associated with the same BCSD CMIP3 climate projections that are supporting evaluations in this chapter. Hydrologic simulations using the VIC model from each of the sixteen historical BCSD simulations, sixteen high-emissions BCSD simulations, and sixteen low-emissions BCSD simulations were employed.

6.4 Temperature Projections

There is high confidence that climate will warm substantially over the twenty-first century, as all of the projected GCM and associated downscaled simulations exhibit progressive warming over the Southwest United States. Within the modeled historical simulations, the model warming begins to become distinguished from the range of natural variability in the 1970s; similar warming is also found in observed records and appears, partially, to be a response to the effects of GHG increases (Barnett et al. 2008; Bonfils et al. 2008). Concerning the projected climate, in the early part of the twenty-first century, the warming produced by the high-emissions scenario is not much greater than that of the low-emissions scenario; there is considerable overlap between the high- and low-emissions scenario results. But by the mid-2000s, as GHG concentrations under the high-emissions scenario become considerably higher than those in the low-emissions scenario, warming in the high-emissions simulations becomes increasingly greater than those from the low-emissions scenario. The projected rate of warming is substantially greater than the historical rates estimated from observed temperature records in California (Bonfils et al. 2008).

Maps showing the sixteen CMIP3 ensemble mean annual temperature changes for three future time periods (2035, 2055, 2085) and two emissions scenarios (high and low) are shown in Figure 6.1. The three periods show successively higher temperatures than the model-simulated historical mean for 1971–2000. Spatial variations are relatively small, especially for the low-emissions scenario. Changes along the coastal zone are noticeably smaller than inland areas (see also Cayan et al. 2009; Pierce et al. 2012). Also, the warming tends to be slightly greater in the north, especially in the states of Nevada, Utah, and Colorado. Warming increases over time, and also increases between the high- and low-emissions scenarios for each respective period as shown in the thirty-year early-, mid-, and late-twenty-first century plots in Figure 6.2 for the aggregated six-state region that defines the Southwest. Figure 6.3 shows the mean seasonal changes for each future time period for the high-emissions scenario, averaged over the entire Southwest region for fifteen CMIP3 models. For the low-emissions scenario, the amount of annual warming ranges between 1°F and 3°F (0.6°C to 1.7°C) for the period, 2021–2050; over 1°F

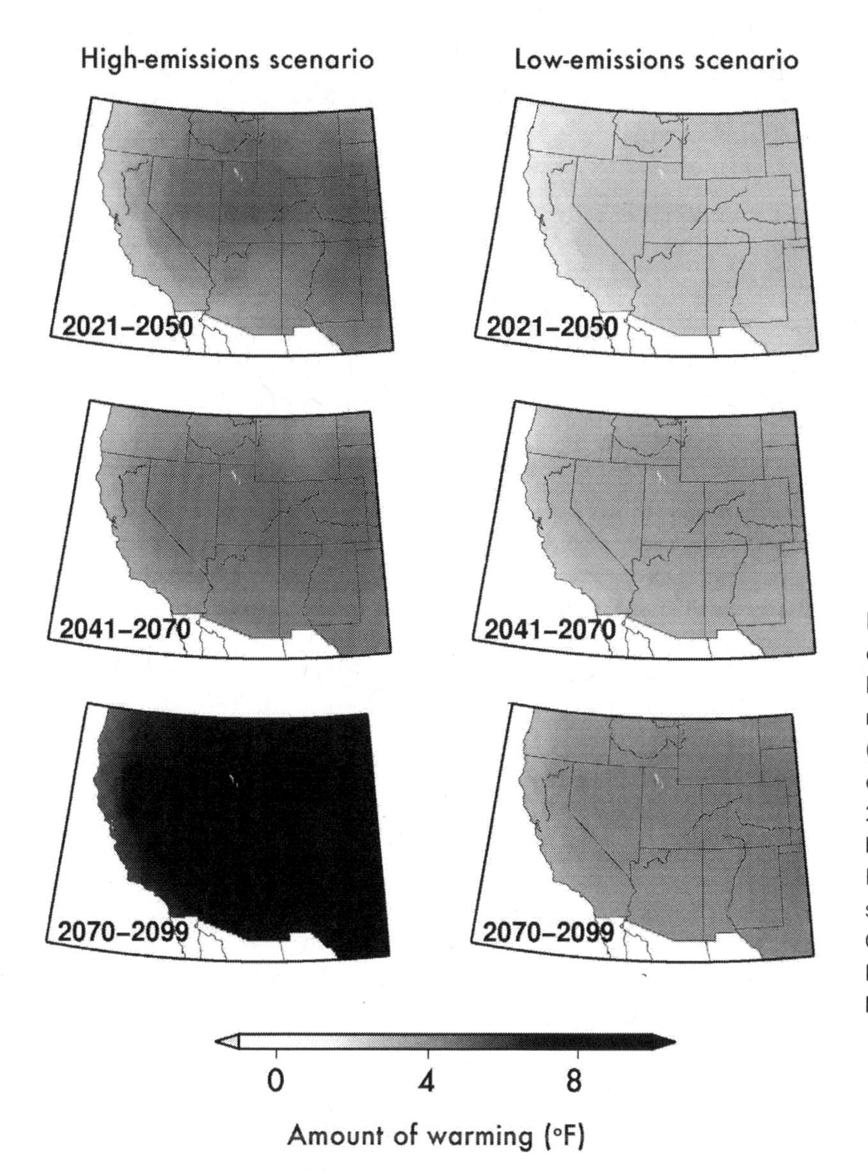

Figure 6.1 Projected temperature changes for the high (A2) and low (B1) GHG emission scenario models. Annual temperature change (°F) from historical (1971–2000) for early- (2021–2050; top), mid- (2041–2070; middle) and late- (2070–2099; bottom) twenty-first century periods. Results are the average of the sixteen statistically downscaled CMIP3 climate models. Source: Nakićenović and Swart (2000), Mearns et al. (2009).

to 4°F (0.6°C to 2.2°C) for 2041–2070; and 2°F to 6°F (1.1°C to 3.3°C) for 2070–2099. For the high-emissions scenario, values range slightly higher, from about 2°F to 4°F (1.1°C to 2.2°C) for 2021–2050; 2°F to 6°F (1.1°C to 3.3°C) for 2041–2070; and are much higher, a 5°F to 9°F (2.8°C to 5°C) range, by 2070–2099. For 2055, the average temperature change simulated by the NARCCAP models (4.5°F, or 2.5°C) is close to the mean of the CMIP3 GCMs for the high-emissions scenario.

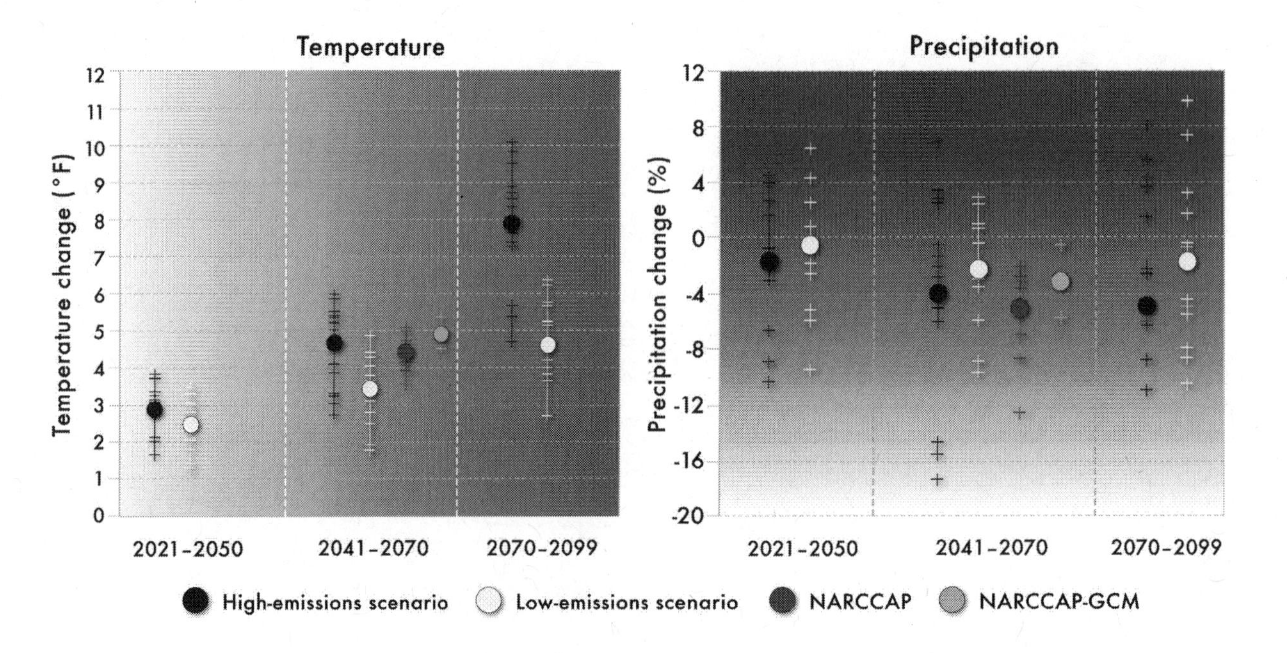

Figure 6.2 Mean annual temperature changes (°F; left) and precipitation changes (%; right) for early-, mid- and late-twenty-first-century time periods. Temperature changes and precipitation changes are with respect to the simulations' reference period of 1971–2000 for 15 CMIP3 models, averaged over the entire Southwest region for the high (A2) and low (B1) emissions scenarios. Also shown are results for the NARCCAP simulations for 2041–2070 and the four GCMs used in the NARCCAP experiment (A2 only). The small plus signs are values for each individual model and the circles depict the overall means. Source: Nakicenovic and Swart (2000), Mearns et al. (2009).

The warming, as it emerges from the variability within and across model simulations, is shown for each of three subregions in the Southwest by the ensemble time series in Figure 6.4. Temperature increases are largest in summer, with means around 3.5°F (1.9°C) in 2021–2050, 5.5°F (3.1°C) in 2041–2070, and 9°F (5°C) in 2070–2099. The least warming is in winter, starting at 2.5°F (1.4°C) in 2021–2050 and building to almost 7°F (3.9°C) in 2070–2099. However, it is important to note that differences between individual model temperature changes are relatively large. Within a given emissions scenario, differences in the resultant mean temperature of the simulations can be attributed to differences in the models (for example, the way they represent and calculate key physical processes) and from differences across simulations resulting from the inherent

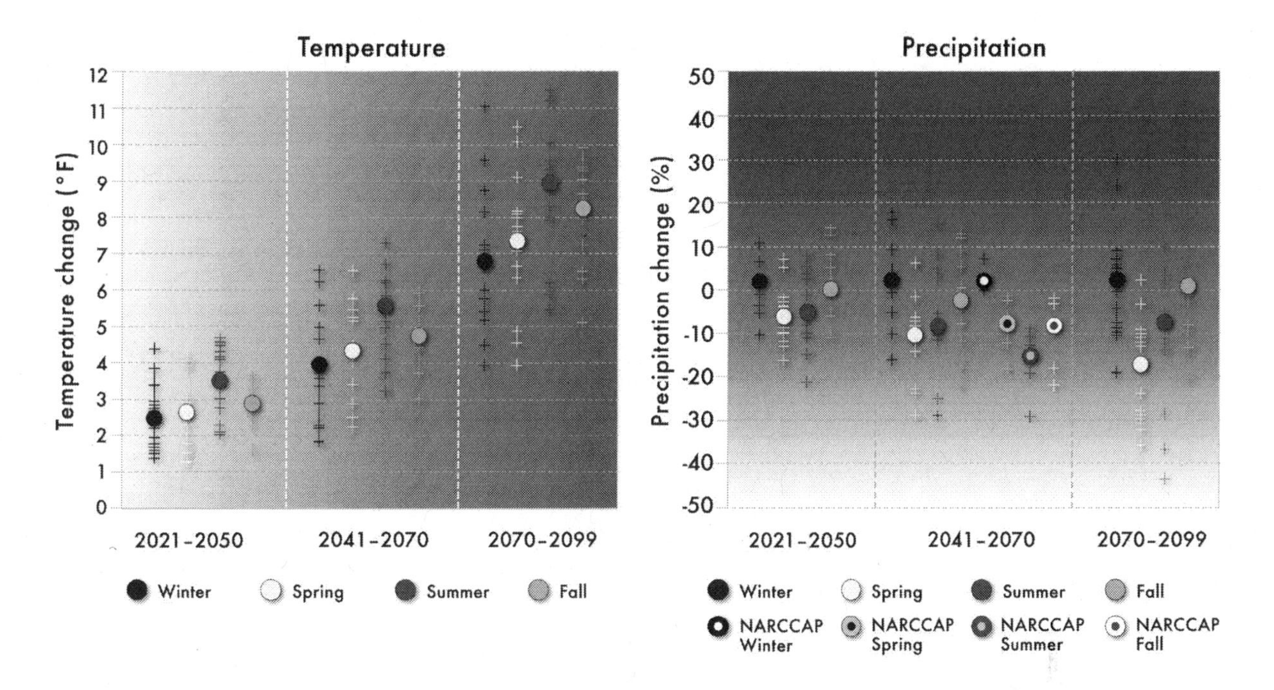

Figure 6.3 Projected change in average seasonal temperatures (°F, left) and precipitation (% change, right) for the Southwest region for the high-emissions (A2) scenario. A fifteen-model average of mean seasonal temperature and precipitation changes for early-, mid-, and late-twenty-first century with respect to the simulations' reference period of 1971–2000. Changes in precipitation also show the averaged 2041–2070 NARCCAP four global climate model simulations. The seasons are December–February (winter), March–May (spring), June–August (summer), and September–November (fall). Plus signs are projected values for each individual model and circles depict overall means. Source: Mearns et al. (2009).

variability of shorter-period climate fluctuations (Hawkins and Sutton 2009). The magnitude of the differences between models using the same emissions scenario is large compared to the difference in the change between seasons and to the difference between high- and low-emissions scenarios, and is comparable to that of the mean differences between the projections for the early- and late-twenty-first century.

6.5 Projections of Other Temperature Variables

The projected length of the annual freeze-free season increases across the region, which historically has exhibited a freeze-free season ranging from 50 to 300 days, depending on location (Figure 6.5, top). By the mid-twenty-first century (Figure 6.5, bottom, from NARCCAP simulations), the entire region exhibits increases of at least 17 additional freeze-free days, excepting parts of the California coast, which show increases of 10 to 17 days. The largest increases, more than 38 days, are in the interior far West. The freeze-free season in eastern parts of Colorado and New Mexico increases by 17 to 24 days, while in some areas along the Rocky Mountains it increases up to 30 days.

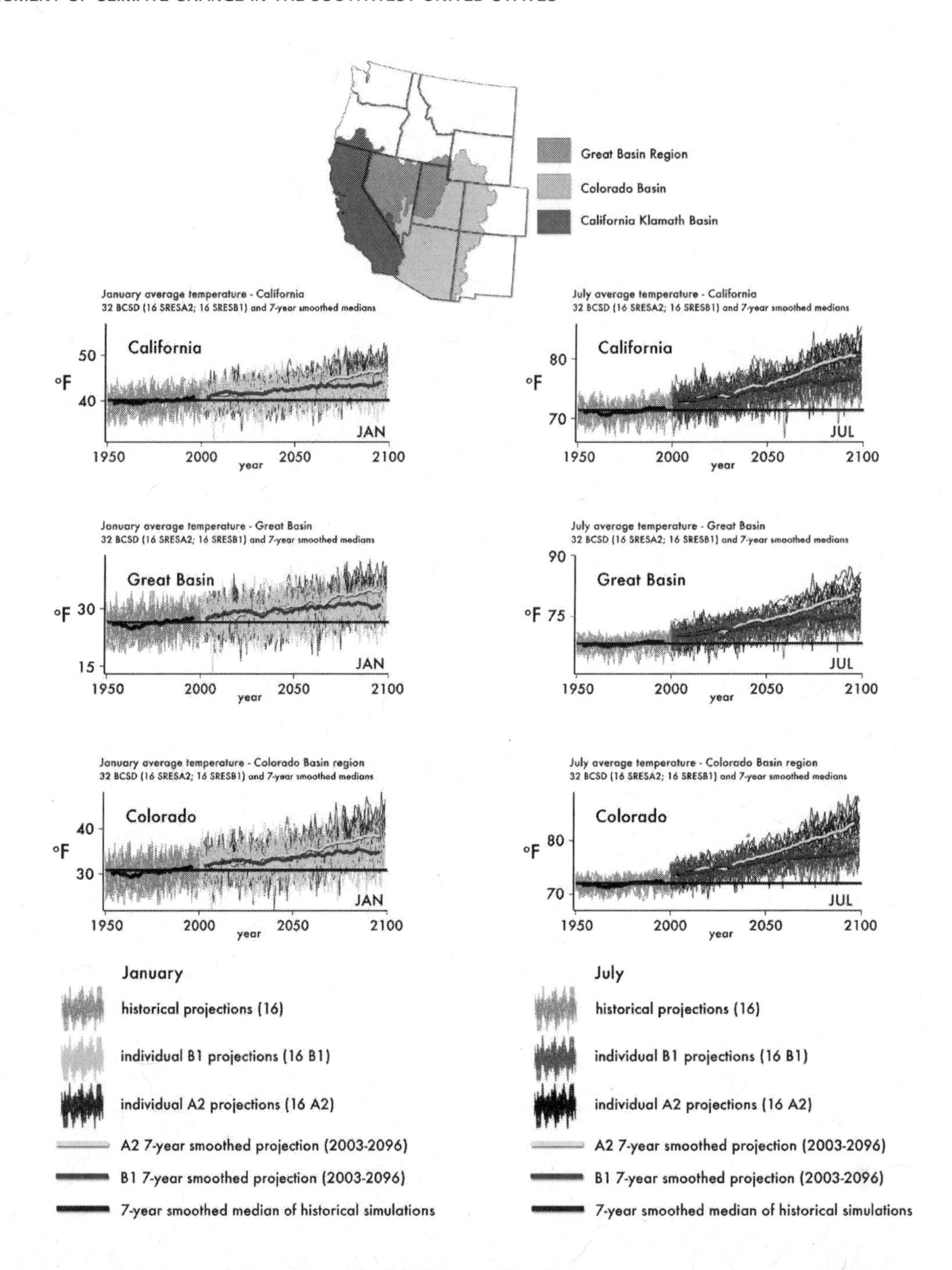

Figure 6.4 January (left) and July (right) BCSD average temperature for California (top), the Great Basin (middle), and Colorado (bottom). The maps show the three regions over which the temperatures were averaged. Source: Bias Corrected and Downscaled World Climate Research Programme's CMIP3 Climate Projections archive at http://gdo-dcp.ucllnl.org/downscaled_cmip3_projections/#Projections:%20Complete%20Archives.

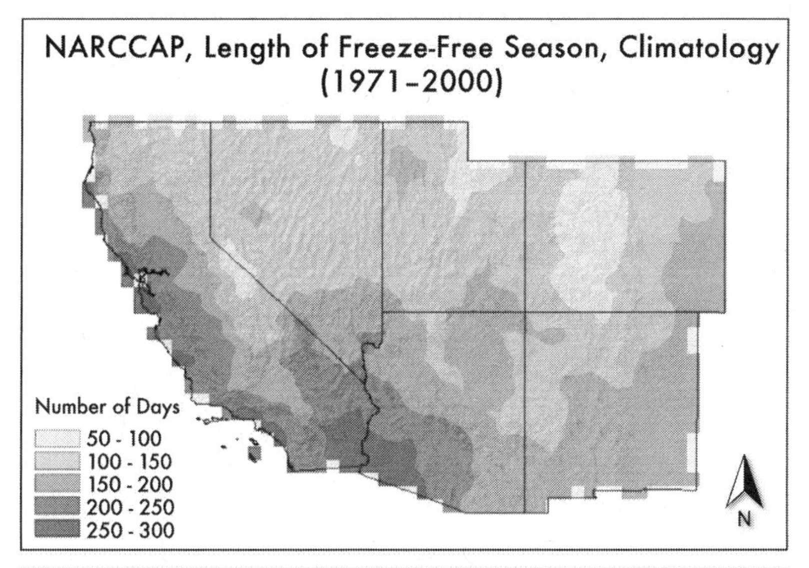

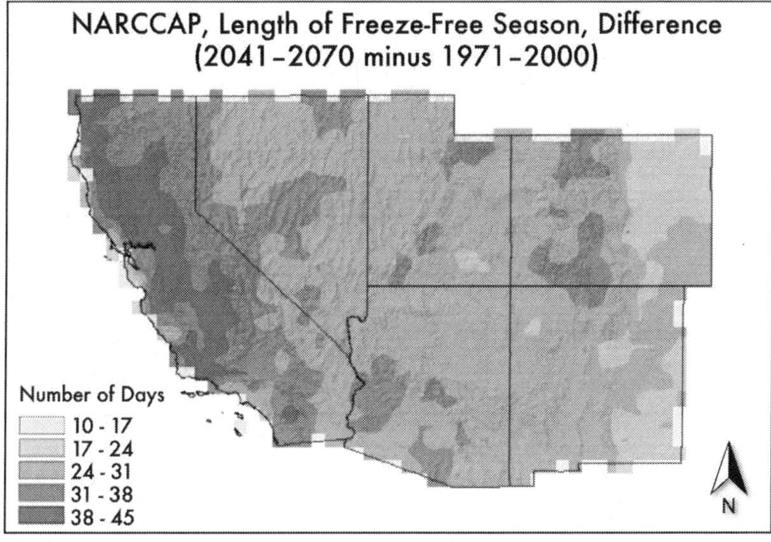

Figure 6.5 NARCCAP multi-model mean change in the length of the freeze-free season between 2041–2070 and 1971–2000 (top) and simulated NARCCAP climatology of the length of the freeze-free season (bottom). Source: Mearns et al. (2009).

Heating degree days (a measurement that reflects the amount of energy needed to heat a home or structure) decrease substantially. In general, by the mid-twenty-first century as gauged from the mean of NARCCAP simulations, the entire region is projected to experience a decrease of at least 500 heating degree days per year, using a heating degree day baseline of 65°F.[v] The largest changes occur in higher-elevation areas, where the decreases are up to 1,900 heating degree days. Areas along the coast, along with southern Arizona, are projected to experience the smallest decreases in heating degree days per year.

On the other hand, cooling degree days increase over the entire Southwest region, with the warmest areas showing the largest increases and vice versa for the coolest areas. The hottest areas, such as Southern California and southern Arizona, are simulated to have the largest increase of cooling degree days per year, up to 1,000, using a 65°F baseline. Areas east of the Rocky Mountains, as well as the California coast, show increases of 400 to 800 cooling degree days per year. Areas with the highest elevations, including the Rocky Mountains and the Sierra Nevada, have the smallest simulated increases, around 200 days or fewer. Cooling and heating requirements become acute during extreme conditions that fall into the tail of the temperature distribution—heat waves and cool outbreaks—whose future occurrences and intensity are affected by the underlying changes in the center of the distribution as described in Chapter 7.

6.6 Precipitation Projections

The precipitation climatology in the Southwest is marked by a large amount of spatial and temporal variability. Observed variability over time in parts of the Southwest, as scaled by mean precipitation, is greater than that in other regions of the United States (Dettinger et al. 2011), from few-day events (see Chapter 7) to scales of months, years, and decades (Cook et al. 2004; Woodhouse et al. 2010). The climate-model-projected simulations indicate that a high degree of variability of annual precipitation will continue during the coming century, as illustrated by the ensemble time series of annual total precipitation in inches shown in Figure 6.6. This suggests that the Southwest will remain susceptible to unusually wet spells and, on the other hand, will remain prone to occasional drought episodes.

To some degree, the model simulations also contain trends over the twenty-first century, as presented in Table 6.1. It is emphasized that these results have medium-low confidence, however, because the trends are generally small in comparison to the high level of shorter-period variability and the considerable variability that occurs among model simulations. As with the temperature projections, the difference in mean precipitation over a given epoch between simulations is due to internal variability, differences between model formations, and between emissions scenarios (Hawkins and Sutton 2009). The distribution of the CMIP3 multi-model median changes in annual precipitation, as a fraction of the modeled historical (1971–2000) annual mean, is shown in Figure 6.7 for the three future periods: 2021–2050 (referred to as "2035"), 2041–2070 (or "2055"), and 2070–2099 (or "2085") and for the low-emissions and high-emissions scenarios. Generally, the median changes shift from drier conditions (than historical climatology) in the south to somewhat wetter in the north. In the high-emissions simulations these changes increase in magnitude through the twenty-first century, but in the low-emissions scenario the differences over time are not as great and peak near the mid-twenty-first century, as seen in Figure 6.7 and at the right in Figure 6.2. The largest north-south percentage differences are for the high-emissions scenario in 2085, varying from an increase of 2% in the far north of the Southwest region to a decrease of 12% in the far south of the Southwest. The smallest difference occurs for the high-emissions scenario in 2035, with increases of 2% in the Nevada-Utah area, and a decrease of about 4% to 6% in areas such as Colorado, New Mexico, and California. However, in the high-emissions scenario in the late twenty-first century, weak increases are found in median precipitation in southeastern California and southern Nevada.

Figure 6.2, right panel, shows the mean annual changes in precipitation for each future time period and both emissions scenarios, averaged over the entire Southwest region for fifteen CMIP3 models. In addition, averages for the nine NARCCAP simulations and

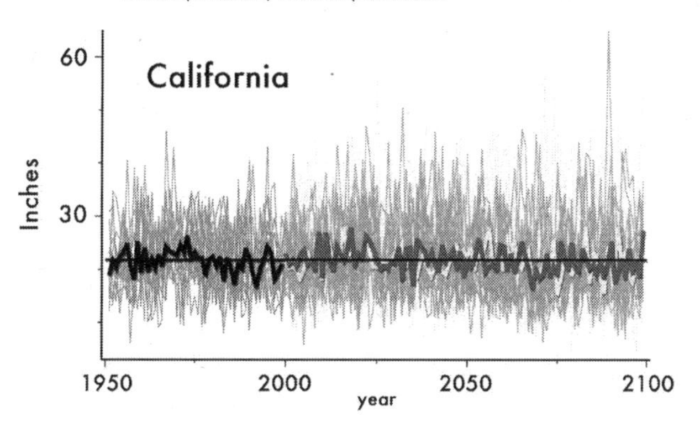

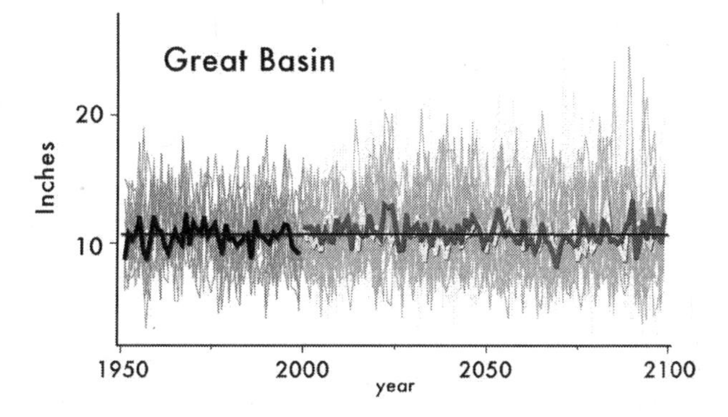

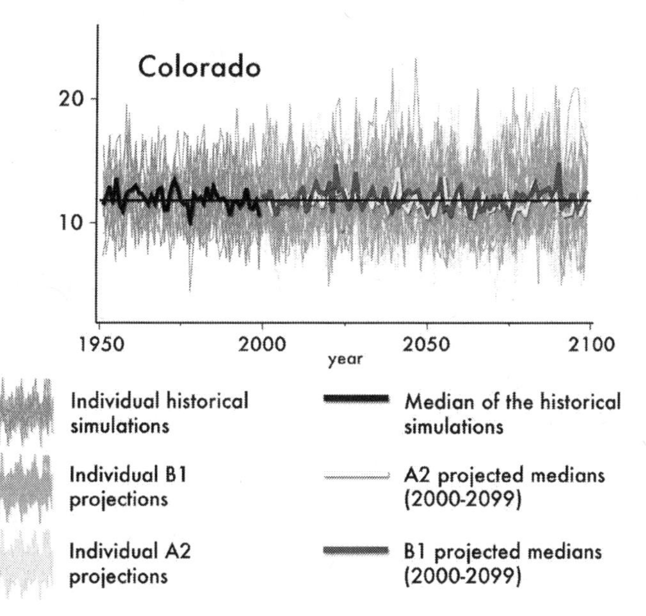

Figure 6.6 Water year precipitation (in inches) averaged over California (top), Great Basin (middle) and Colorado (bottom). Source: Bias Corrected and Downscaled World Climate Research Programme's CMIP3 Climate Projections archive at http://gdo-dcp.ucllnl.org/downscaled_cmip3_projections/#Projections:%20Complete%20Archives.

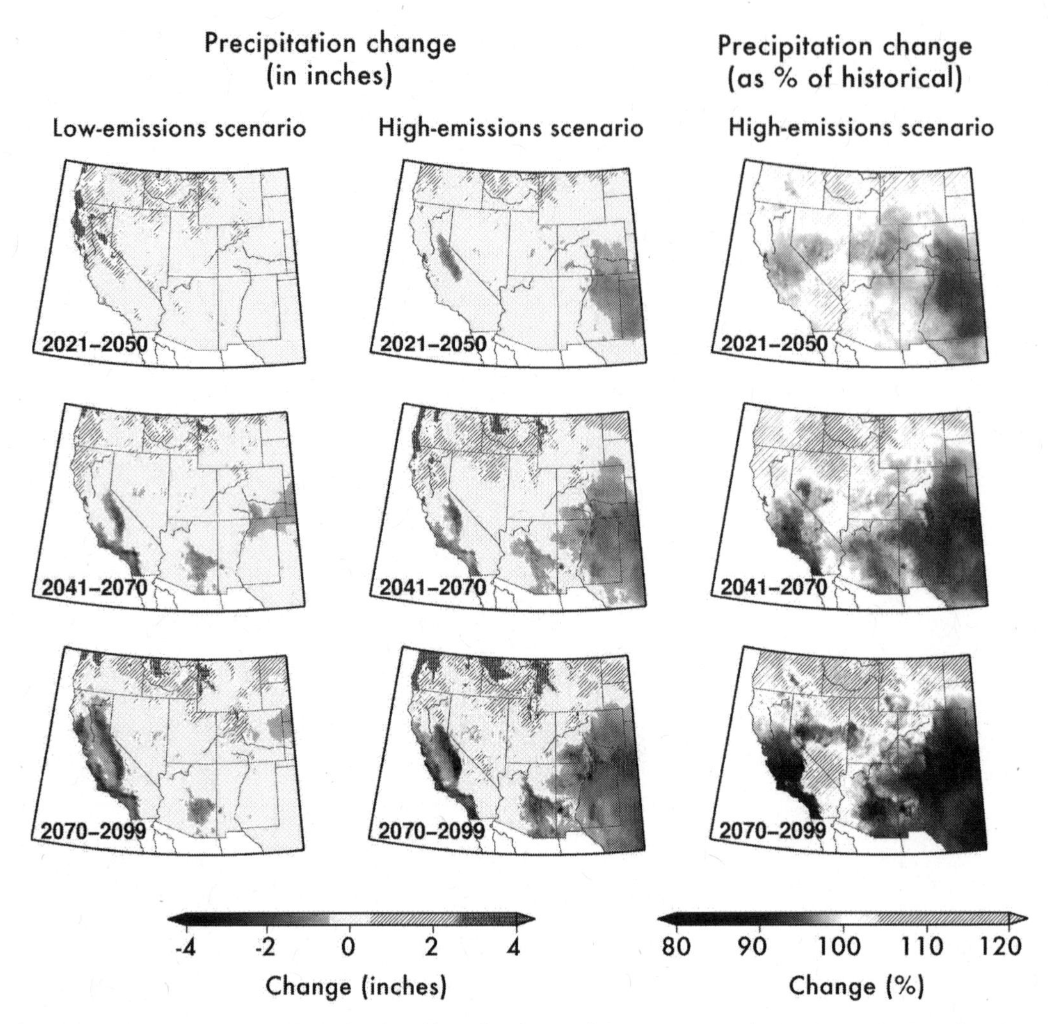

Figure 6.7 Annual precipitation change (in inches) from the historical simulation for the low-emissions scenario (left), the high-emissions scenario (middle), and percentage of historical simulation compared to the high-emissions scenario (right). Early- (2021–2050), mid- (2041–2070) and late- (2070–2099) twenty-first century periods shown in top, middle and bottom panels. Values shown are the median of sixteen simulations downscaled via BCSD. Source: Bias Corrected and Downscaled World Climate Research Programme's CMIP3 Climate Projections archive at http://gdo-dcp.ucllnl.org/downscaled_cmip3_projections/#Projections:%20Complete%20Archives.

the four GCMs used in the NARCCAP experiment are shown for 2055 (high-emissions scenario only). All the mean changes are negative, although the values are rather small overall. For the high-emissions scenario, the CMIP3 models project average decreases of around 2% in 2035, 4% by 2055, and about 5% by 2085. The decreases for the low-emissions scenario are only slightly smaller; in 2085 the decrease is 2%, compared to 5% for high-emissions. The mean of the NARCCAP simulations is more negative than the mean of the CMIP3 GCMs or the mean of the four GCMs used in the NARCCAP experiment, although the differences are small. The range of individual model changes is large compared to the differences in the ensemble means, as also illustrated by the spread of changes shown in Table 6.1. In fact, for all three future periods and for the two scenarios, the individual model range is larger than the differences in the CMIP3 ensemble mean changes, relative to the historical mean precipitation.

The distribution of changes in the Southwest region's mean annual precipitation for each future time period and both emissions scenarios across the fifteen CMIP3 models is shown in Figure 6.2, right panel, and in Table 6.1, which also shows the distribution of the NARCCAP simulations (for 2055, high-emissions scenario only) for comparison. For all periods and both scenarios, the CMIP3 model simulations include both increases and decreases in precipitation. For the region as a whole, most of the median values are negative, but not by much, with change values having magnitudes of 3.1% or less. The range of changes is between 15% and 30%. For example, in the high-emissions scenario, the precipitation change for 2055 varies from a low of -17% to a high of +7%. The NARCCAP range of changes varies from -13% to -2%.

Table 6.1 Distribution of changes in mean annual precipitation (%) for the Southwest region for the 15 CMIP3 models

Scenario	Period	Low	25%ile	Median	75%ile	High
A2	2021-2050	-10	-3	-2	2	5
	2041-2070	-17	-6	-3	1	7
	2070-2099	-20	-10	-3	3	8
	NARCCAP	-13	-7	-3	-3	-2
B1	2021-2050	-10	-2	1	2	6
	2041-2070	-10	-3	-2	0	3
	2070-2099	-10	-5	-1	1	10

Annual precipitation changes over individual regions are stronger or weaker than the aggregate Southwest, as exhibited by the median of the ensemble high-emissions and low-emissions simulations for California, the Great Basin, and the Colorado Basin regions in Figure 6.6. These reinforce the evidence from the mapped median changes in Figure 6.7, showing that the California region exhibits the greatest reduction in precipitation, while the Colorado Basin remains nearly the same as historical levels. The ensemble swarms in Figure 6.6 emphasize the importance of individual wet years in affecting longer term climatological values; at least for a few models, the wettest years grow wetter during the last half of the twenty-first century.

Considering the Southwest as a whole, a majority of the models contain different levels and even directions of change in different seasons, as shown in Table 6.2 and Figure 6.3, right panel. These include increases in winter precipitation, while for the other three seasons, most of the models simulate decreases. In the spring, all but one model simulate decreases. In both the summer and fall, a few models produced sizeable increases in precipitation. In the low-emissions scenario, the range of changes is generally smaller, with a tendency toward somewhat wetter conditions. For example, a majority of the low-emissions models simulate wetter conditions, as compared to the drier majority for the high-emissions scenario. A central feature of the results shown in Table 6.2 is the tendency for greatest precipitation reductions in spring months, albeit with large uncertainty in the seasonal changes (as expressed by the range of results across the ensemble of model simulations).

Table 6.2 Distribution of changes in mean seasonal precipitation (%) for the Southwest region for the 15 CMIP3 models

Scenario	Period	Season	Low	25%ile	Median	75%ile	High
A2	2070-2099	DJF	-19	-8	3	8	30
		MAM	-36	-29	-12	-10	2
		JJA	-44	-13	-9	3	20
		SON	-21	-8	-1	0	38
B1	2070-2099	DJF	-12	-6	2	5	17
		MAM	-27	-9	-7	-1	11
		JJA	-16	-7	0	3	18
		SON	-24	-4	-1	6	13

6.7 Atmospheric Circulation Changes

Climate changes in the Southwest are governed overwhelmingly by global influences. The IPCC Fourth Assessment Report indicated that several climate models project that the mid-latitude storm tracks in both the Southern and Northern Hemispheres will migrate poleward over the twenty-first century (e.g. Meehl et al. 2007). This result was reinforced by analyses by Salathé (2006) and Cayan et al. (2009), who showed a northward shift in the North Pacific winter storm track over the mid- and late-twenty-first century. This result is consistent with the findings of Favre and Gershunov (2009), who examined paths of mid-latitude cyclones and anticyclones[vi] in the North Pacific impinging on the North American West Coast in observations and in the CNRM-CM3 model[vii] high-emissions projection. Wintertime statistics of these trajectories indicate that the flow pattern will become less stormy in the Gulf of Alaska, with more northerly flow along the West Coast of North America. This projected trend is on par with interannual variability in this region and indicates future conditions somewhat reminiscent of today's La Niña phase of the El Niño-Southern Oscillation (ENSO) and negative-phase

Pacific Decadal Oscillation (PDO) winters. These projected circulation changes would result in less frequent winter precipitation in the Southwest United States and northwestern Mexico, and more frequent, albeit less intense, cold outbreaks moderating average wintertime warming, especially in California's low-lying valleys and east of the Front Range of the Rocky Mountains (see Chapter 7).

The shift of the mid-latitude northeastern Pacific storm track poleward may be a result of uneven warming over the earth's surface. Warming is projected to be lower at the low latitudes than the high latitudes, causing a diminished gradient of temperature from the equator to the North and South Poles (Lambert 1995; Lambert and Fyfe 2006; IPCC 2007). In a warming atmosphere, the models also predict increasing humidity. Enhanced evaporation from the ocean's surface and the resulting heating of the atmosphere from condensation of water vapor aloft could reinforce the deepest cyclones, making deep low pressure systems (cyclonic storms) more numerous while moderate events decline in frequency (Lambert 1995; Lambert and Fyfe 2006).

The storm track displacement is also consistent with the projected enhancement and poleward extension of the large descending limb of the tropical atmospheric circulation—the Hadley Cell, which impels the trade winds and jet stream in the tropics and subtropics (see, for example, Lu, Vecchi, and Reichler 2007). The enhancement and broadening of subsiding air in the subtropics and low middle latitudes of the eastern North Pacific could also result in an increase and seasonal expansion of the low-level coastal clouds, the "marine layer" along the California coast, especially in spring and summer. This important potential impact on coastal climate has so far not been investigated.

6.8 North American Monsoon

Representing the North American monsoon (see Chapter 4) in an atmospheric model is extremely challenging because it is governed by multiple factors at different spatial and temporal scales (Douglas and Englehart 2007; Bieda et al. 2009). Most important, the initiation of convection over mountains during the day and thunderstorm organization and growth must be appropriately addressed (Janowiak et al. 2007; Lang et al. 2007; Nesbitt, Gochis, and Lang 2008). GCMs cannot resolve the North American monsoon as a distinct climatological feature because they cannot resolve several key regional processes (Liang et al. 2008; Dominguez, Cañon, and Valdes 2009). Dynamical downscaling using RCMs (at a grid spacing of tens of kilometers) has simulated well at least the start of the convective process over mountains and improved the climatology of monsoon precipitation (Gutzler et al. 2005; Castro, Pielke, and Adegoke 2007). Very high resolution (1.2 mile, or 2-km grid spacing or less) is required to simulate the *evolution* of organized convection so that individual thunderstorms can be explicitly represented (Gao, Li, and Sorooshian 2007). Interannual variability of monsoon precipitation in the Southwest United States is related in part to ENSO and PDO variability (Dominguez, Kumar, and Vivoni 2008). How this natural variability may potentially change in the future is not clear in the GCMs and is a source of large uncertainty (Castro et al. 2007). The evolving pattern of sea-surface temperature is generally important for models to properly simulate key changes in surface temperature and precipitation on a regional scale (Barsugli, Shin, and Sardeshmukh 2006). Moreover, enhanced subsidence of air in the subtropical region (as described in Lu, Vecchi, and Reichler 2007) can potentially impact convection and therefore moderate the monsoon, but this possible mechanism of monsoonal change has so far been unexplored. In general, future expectations of the North American monsoon suffer from uncertainties currently common to monsoon systems around the world.

Figure 6.8 shows the simulated changes in warm-season (June through September) temperature and precipitation in the NARCCAP high-emissions simulations, shaded according to the level of model agreement. The ensemble mean change in model temperatures shows summer warming that ranges between +2°F and +6°F throughout most of the western United States by the middle of the twenty-first century (2041–2069), with the largest temperature increases in the central Rocky Mountains. Precipitation is projected to decrease overall in the Southwest by about 10% to 20%, consistent with an overall drying trend in subtropical regions. This projected precipitation decrease is relatively smaller in eastern Arizona and western New Mexico, with relatively weaker agreement among the individual models. Though NARCCAP models strongly agree as to the amount of warming during the warm season, the weak level of consensus about changes in the monsoon circulation reflects the enormous challenges of representing the monsoon in an atmospheric model. In summary, how monsoon precipitation may change is not yet clear, especially in those areas where monsoon precipitation accounts for a greater proportion of total annual precipitation.

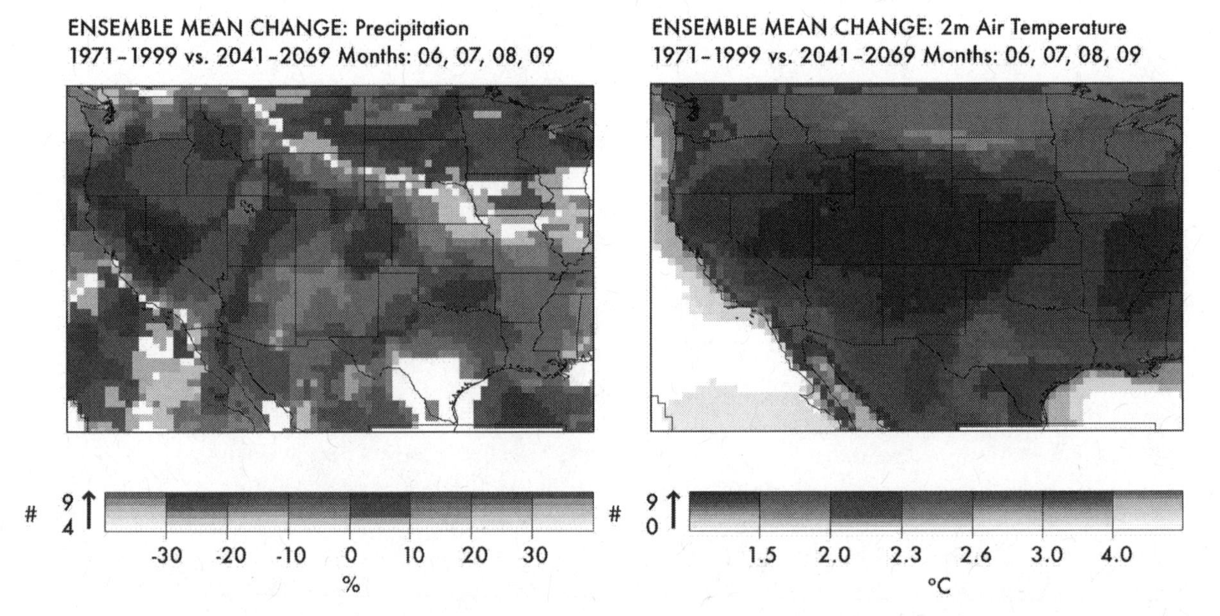

Figure 6.8 NARCCAP precipitation and temperature difference, June–September, 2041–2069 relative to simulated historical means 1971-1999. The degree of change is indicated by the color, whereas the degree of agreement among the nine RCMs is indicated by the intensity of the color. For precipitation (left), the color intensity shows the agreement among the RCMs on the direction of change (i.e., positive or negative percent change in future precipitation); for temperature (right), the color intensity shows agreement among the RCMs on areas where future temperature is projected to be at least 2°C (3.6°F) higher than the 1971-1999 average. Source: Mearns et al. (2009).

6.9 Changes in Precipitation-related Measures

The escalating effect of warming, coupled with a tendency in parts of the Southwest toward annual precipitation decreases, would amplify recent observed trends of lower spring snowpack across much of the western United States (Mote et al. 2005; Knowles, Dettinger, and Cayan 2006; Pierce et al. 2008). Additionally, the Southwest straddles both a region to the north where precipitation is projected to increase and a region to the south where precipitation is projected to decrease—as shown by a consensus of global model simulations (IPCC 2007; Seager and Vecchi 2010). The GCM projections, downscaled and run through the VIC hydrological model, show changes in hydrological measures that are consistent with the warming trend. They indicate a marked reduction in spring snow accumulation in mountain watersheds across the Southwestern United States (Figure 6.9 top panel) that becomes more pronounced over the decades of the twenty-first century. The relatively gradual decline for the California, Colorado, and Rio Grande basins shown in Figure 6.10 (top row) is consistent with several other studies (e.g. Knowles and Cayan 2002; Christensen and Lettenmaier 2007; Ray et al. 2008; Cayan et al. 2009; Das et al. 2009; Wi et al. 2012). More rain and less snow, earlier snowmelt, and, to some extent, drying tendencies cause a reduction in late-spring and summer runoff (Figure 6.9, middle panel, and 6.10, middle row). Together these effects, along with increases in evaporation, result in lower soil moisture by early summer (Figure 6.9, bottom panel, and 6.10, bottom row).

Recent studies have projected Colorado River flows to show possible reductions from climate-change impacts, ranging from about -5% to about -20% by mid-century (Hoerling and Eischeid 2007; Das et al. 2011; Reclamation 2011a; Vano, Das, and Lettenmaier 2012). Changes in streamflow are driven by changes in precipitation and also by increases in temperature. Recent estimates, from several hydrological models, suggest reductions in annual Colorado River flow at Lees Ferry, Arizona, which is the location established by the Colorado River Compact in 1922 as the dividing point between the Upper Colorado River Basin and the Lower Colorado River Basin. Estimates of the reductions in Colorado River flow range from approximately 3% to 16% decrease per 1°F (0.6°C) warming and a reduction of 1% to 2% of flow per 1% reduction of precipitation (Hoerling et al. 2009; Reclamation 2011b; Vano, Das, and Lettenmaier 2012). As estimated by the VIC hydrological model, runoff and streamflow are more sensitive to warming in the Colorado Basin than in the Columbia River watershed and are much greater than in the northern and southern drainages of the Sierra Nevada in California (Das et al. 2011). Figure 6.11, based on the ensemble of sixteen VIC simulations under the high-emissions scenario, shows the median tendency for reductions in total annual runoff over the Southwest in the mid-twenty-first century. Over the Colorado Basin, the composite of simulations from the VIC simulations exhibits reductions of runoff of approximately 5% to 18% by the middle portion of the twenty-first century, consistent with the estimates described above.

The early twenty-first-century drought in the Southwest (see Chapters 4 and 5) underscores that the Southwest climate is prone to dry spells. Such droughts have a tendency to take on large areal footprints, although both observations and climate model simulations indicate different degrees of dryness in California, the Great Basin, and the Colorado Basin. As quantified by the VIC hydrological model, the most extreme drought

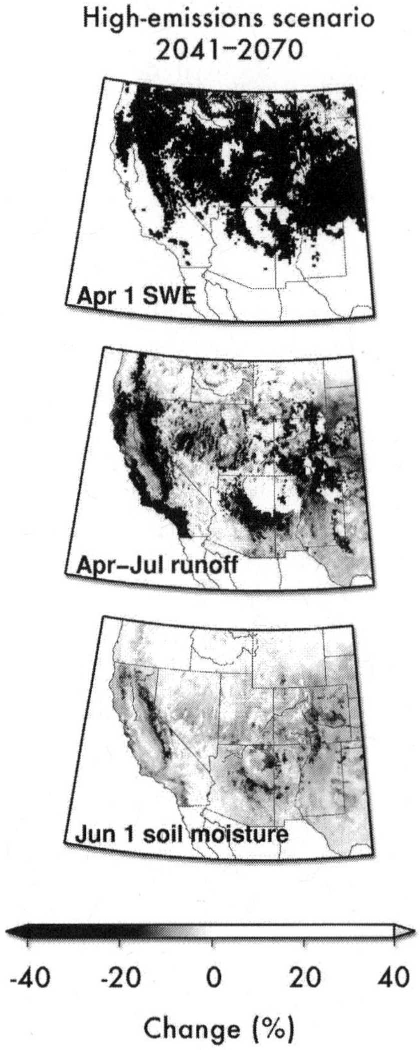

High-emissions scenario
2041–2070

Apr 1 SWE

Apr–Jul runoff

Jun 1 soil moisture

-40 -20 0 20 40
Change (%)

Figure 6.9 Predicted changes in the water cycle. Mid-century (2041–2070) percent changes from the simulated historical median values from 1971-2000 for April 1 snow water equivalent (SWE, top), April–July runoff (middle) and June 1 soil moisture content (bottom), as obtained from median of sixteen VIC simulations under the high-emissions (A2) scenario. Source: Bias Corrected and Downscaled World Climate Research Programme's CMIP3 Climate Projections archive at http://gdo-dcp.ucllnl.org/downscaled_cmip3_projections/#Projections:%20Complete%20Archives.

years throughout the instrumental record have tended to build up and finally abate over an extended multiyear period. Historically, and especially during the early twenty-first century, Southwestern droughts have been exacerbated by unusually warm summer temperatures. This tendency could worsen in future decades: several twenty-first-century climate model simulations suggest that dry years will include anomalously warm summer temperatures even above and beyond the warming trend in the Southwest (Cayan et al. 2010). During extreme droughts, the deficit in soil moisture grows larger, and also grows in comparison to the deficit in precipitation. Although projected precipitation anomalies during dry spells do not change markedly from observed past conditions, other hydrologic measures—including soil moisture—become more depleted.

Human-induced climate change impacts on temperature, snowpack, and the timing of streamflow over the western United States have already been detected (Maurer, Stewart et al. 2007; Barnett et al. 2008; Bonfils et al. 2008; Pierce et al. 2008; Hidalgo et al. 2009),

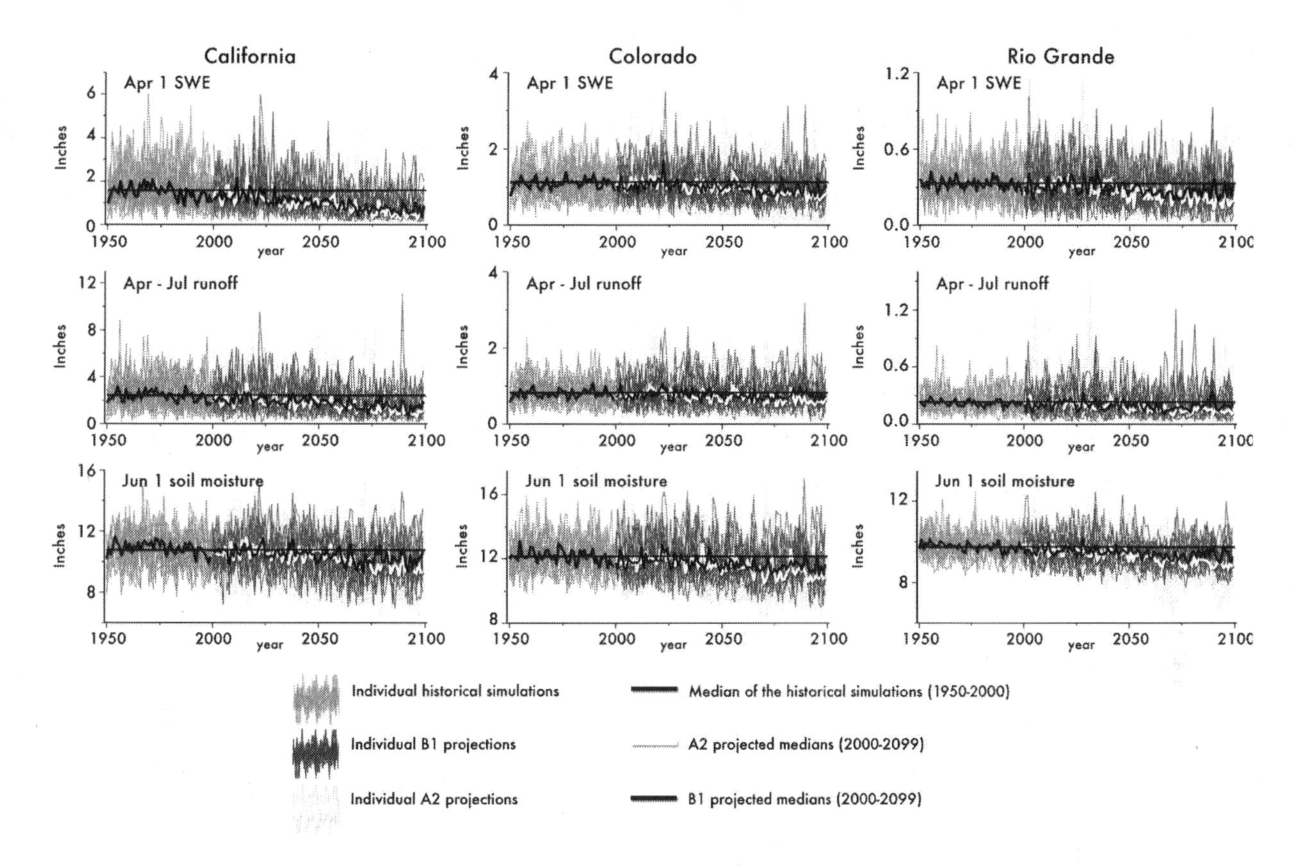

Figure 6.10 Spatially averaged values (in inches) for April 1 snow water equivalent (top), April–July runoff (middle) and June 1 soil moisture content (bottom). Averages are shown for California (left), Colorado (middle) and Rio Grande (right). Source: Bias Corrected and Downscaled World Climate Research Programme's CMIP3 Climate Projections archive at http://gdo-dcp.ucllnl.org/downscaled_cmip3_projections/#Projections:%20Complete%20Archives.

and as climate continues to warm there will be serious impacts on the hydrological cycle and water resources of the Southwest United States (Barnett et al. 2004; Seager et al. 2007). Water resource implications are described in Chapter 10. Downscaled temperature, precipitation, and modeled hydrologic measures already provide sufficient spatial detail to assess hydroclimatic effects that will be critical in planning for risks to water resources and ecosystems, risk of wildfire, and other key issues in the Southwest. A strong consensus among the model projections across the Southwest for substantially lower spring snowpack, lower spring-summer runoff, and drier summers underscores that traditional planning practices can no longer be supported (see Milly et al. 2008) or that the past can be assumed to be a reasonable representation of the future. Thus, past hydrological observations cannot sufficiently frame the risks of unfavorable future outcomes, such as an inability to meet demands for water.

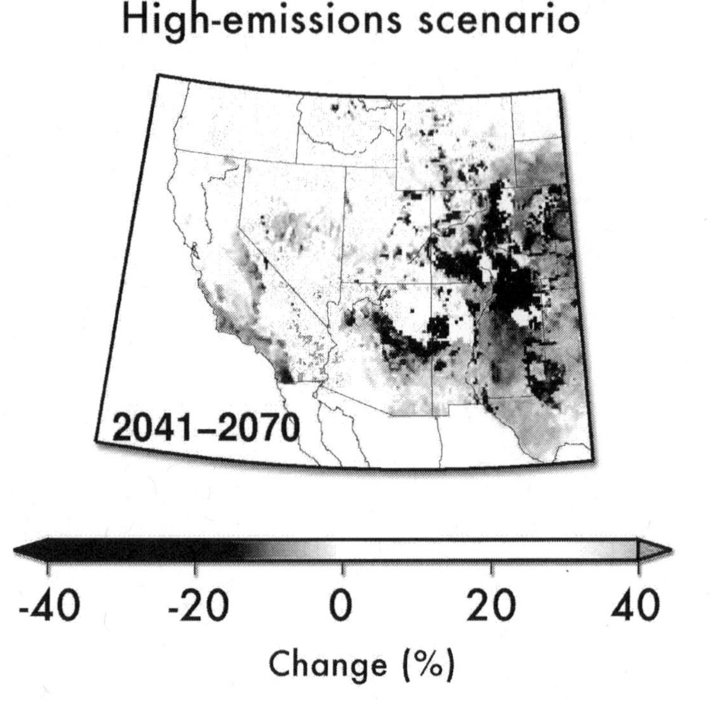

Figure 6.11 High-emissions scenario mid-century (2041–2070) percent change from the historical period (1971–2000) for annual runoff. Data obtained from median of sixteen VIC simulations. Source: Bias Corrected and Downscaled World Climate Research Programme's CMIP3 Climate Projections archive at http://gdo-dcp.ucllnl.org/downscaled_cmip3_projections/#Projections:%20Complete%20Archives.

References

Barnett, T., R. Malone, W. Pennell, D. Stammer, B. Semtner, and W. Washington. 2004. The effects of climate change on water resources in the west: Introduction and overview. *Climate Change* 62:1–11, doi:10.1023/B:CLIM.0000013695.21726.b8.

Barnett, T. P., D. W. Pierce, H. Hidalgo, C. Bonfils, B. Santer, T. Das, G. Bala, et al. 2008. Human-induced changes in the hydrology of the western United States. *Science* 316:1080–1083.

Barsugli, J. J., S-I. Shin, and P. D. Sardeshmukh. 2006. Sensitivity of global warming to the pattern of tropical ocean warming. *Climate Dynamics* 27:483–492, doi:10.1007/s00382-006-0143-7.

Bieda, S.W. III, C. L. Castro, S. L. Mullen, A. Comrie, and E. Pytlak. 2009. The relationship of transient upper-level troughs to variability of the North American monsoon system. *Journal of Climate* 22:4213–4227.

Bonfils, C., B. D. Santer, D. W. Pierce, H. G. Hidalgo, G. Bala, T. Das, T. P. Barnett, et al. 2008. Detection and attribution of temperature changes in the mountainous western United States. *Journal of Climate* 21:6404–6424, doi:10.1175/2008JCL12397.1.

Castro, C. L., R. A. Pielke, Sr., and J. O. Adegoke. 2007. Investigation of the summer climate of the contiguous U.S. and Mexico using the Regional Atmospheric Modeling System (RAMS). Part I: Model climatology (1950-2002). *Journal of Climate* **20**:3844–3865.

Castro, C. L., R. A. Pielke, Sr., J. O. Adegoke, S. D. Schubert, and P. J. Pegion. 2007. Investigation of the summer climate of the contiguous U.S. and Mexico using the Regional Atmospheric Modeling System (RAMS). Part II: Model climate variability. *Journal of Climate* 20:3866–3887.

Cayan, D. R., T. Das, D. W. Pierce, T. P. Barnett, M. Tyree and A. Gershunov. 2010. Future dryness in the southwest US and the hydrology of the early 21st century drought. *Proceedings of the National Academy of Sciences* 107:21271–21276.

Cayan, D., M. Tyree, M. Dettinger, H. Hidalgo, T. Das, E. Maurer, P. Bromirski, N. Graham and R. Flick. 2009. *Climate change scenarios and sea level rise estimates for the California 2009 climate change scenarios assessment*. Final paper CEC-500-2009-014-F. Np: California Climate Change Center. http://www.energy.ca.gov/2009publications/CEC-500-2009-014/CEC-500-2009-014-F.PDF.

Christensen, N., and D. P. Lettenmaier. 2007. A multimodel ensemble approach to assessment of climate change impacts on the hydrology and water resources of the Colorado River Basin. *Hydrology and Earth System Sciences* 11:1417–1434.

Cook, E. R., C. Woodhouse, C. M. Eakin, D. M. Meko, and D. W. Stahle. 2004. Long-term aridity changes in the western United States. *Science* 306:1015–1018.

Dai, A. 2006. Precipitation characteristics in eighteen coupled climate models. *Journal of Climate* 19:4605–4630.

Das, T., H. Hidalgo, D. Cayan, M. Dettinger, D. Pierce, C. Bonfils, T.P. Barnett, G. Bala and A. Mirin. 2009. Structure and detectability of trends in hydrological measures over the western United States. *Journal of Hydrometeorology* 10:871–892, doi:10.1175/2009JHM1095.1.

Das, T., D. W. Pierce, D. R. Cayan, J. A. Vano, and D. P. P. Lettenmaier. 2011. The importance of warm season warming to western U.S. streamflow changes. *Geophysical Research Letters* 38: L23403, doi:10.1029/2011GL049660.

Dettinger, M. D., F. M. Ralph, T. Das, P. J. Neiman, and D. R. Cayan. 2011. Atmospheric rivers, floods and the water resources of California. *Water* 3:445–478.

Dominguez, F., J. Cañon, and J. Valdes. 2009. IPCC-AR4 climate simulations for the southwestern U.S.: The importance of future ENSO projections. *Climatic Change* 99:499–514, doi:10.1007/s10584-009-9672-5.

Dominguez, F., P. Kumar, and E. R. Vivoni. 2008. Precipitation recycling variability and ecoclimatological stability—A study using NARR data. Part II: North American monsoon region. *Journal of Climate* 21:5187–5203.

Douglas, A. V., and P. J. Englehart. 2007. A climatological perspective of transient synoptic features during NAME 2004. *Journal of Climate* 20:1947–1954.

Favre, A., and A. Gershunov. 2009. North Pacific cyclonic and anticyclonic transients in a global warming context: Possible consequences for western North American daily precipitation and temperature extremes. *Climate Dynamics* 32:969–987.

Gangopadhyay, S., T. Pruitt, L. Brekke, and D. Raff. 2011. Hydrologic projections for the western United States. *Eos Transactions AGU* 92:441, doi:10.1029/2011EO480001.

Gao, X., J. Li, and S. Sorooshian. 2007. Modeling intraseasonal features of 2004 North American monsoon precipitation. *Journal of Climate* 20:1882–1896.

Gutzler, D. S., H-K. Kim, R. W. Higgins, H-M. H. Juang, M. Kanamitsu, K. Mitchell, K. Mo, et al. 2005. The North American monsoon Model Assessment Project: Integrating numerical modeling into a field-based process study. *Bulletin of the American Meteorological Society* 86:1423–1429.

Hawkins, E., and R. T. Sutton. 2009. The potential to narrow uncertainty in regional climate predictions. *Bulletin of the American Meteorological Society* 90:1095–1107.

Hidalgo, H. G., T. Das, M. D. Dettinger, D. R. Cayan, D. W. Pierce, T. P. Barnett, G. Bala, et al. 2009. Detection and attribution of stream flow timing changes to climate change in the western United States. *Journal of Climate* 22:3838–3855, doi:10.1175/2009JCLI2470.1.

Hoerling, M., and J. Eischeid. 2007. Past peak water in the Southwest. *Southwest Hydrology* 6 (1): 18–35.

Hoerling, M., D. Lettenmaier, D. Cayan, and B. Udall. 2009. Reconciling projections of Colorado River streamflow. *Southwest Hydrology* 8 (3): 20–21, 31.

Ikeda, K., R. Rasmussen, C. Liu, D. Gochis, D. Yates, F. Chen, M. Tewari, et al. 2010. Simulation of seasonal snowfall over Colorado. *Atmospheric Research* 97:462–477, doi:10.1016/j.atmosres.2010.04.010.

Intergovernmental Panel on Climate Change (IPCC). 2007. *Climate change 2007: The physical science basis. Contribution of Working Group I to the Fourth Assessment Report of the Intergovernmental Panel on Climate Change,* ed. S. Solomon, D. Qin, M. Manning, Z. Chen, M. Marquis, K. B. Averyt, M. Tignor and H.L. Miller. Cambridge: Cambridge University Press. http://www.ipcc.ch/ipccreports/ar4-wg1.htm.

Janowiak, J. E., V. J. Dagostaro, V. E. Kousky, and R. J. Joyce. 2007. An examination of precipitation in observations and model forecasts during NAME with emphasis on the diurnal cycle. *Journal of Climate* 20:1680–1692.

Knowles, N., and D. R. Cayan. 2002. Potential effects of global warming on the Sacramento/San Joaquin watershed and the San Francisco estuary. *Geophysical Research Letters* 29:1891, doi:10.1029/2001GL014339.

Knowles, N., M. D. Dettinger, and D. R. Cayan. 2006. Trends in snowfall versus rainfall in the western United States. *Journal of Climate* 19:4545–4559.

Lambert, S. J. 1995. The effect of enhanced greenhouse warming on winter cyclone frequencies and strengths. *Journal of Climate* 8:1447–14452.

Lambert, S. J., and J. C. Fyfe. 2006. Changes in winter cyclone frequencies and strengths simulated in enhanced greenhouse warming experiments: Results from the models participating in the IPCC diagnostic exercise. *Climate Dynamics* 26:713–728.

Lang, T. J., D. A. Ahijevych, S. W. Nesbitt, R. E. Carbone, S. A. Rutledge, and R. Cifelli. 2007. Radar-observed characteristics of precipitating systems during NAME 2004. *Journal of Climate* 20:1713–1733.

Laprise, R., R. de Elía, D. Caya, S. Biner, P. Lucas-Pincher, E. Diaconescu, M. Leduc, A.

Alexandru, and L. Separovic. 2008. Challenging some tenets of regional climate modelling. *Meteorology and Atmospheric Physics* 100:3–22, doi:10.1007/s00703-0080292-9.

Liang, X., D. P. Lettenmaier, E. Wood, and S. J. Burges. 1994. A simple hydrologically based model of land surface water and energy fluxes for General Circulation Models. *Journal of Geophysical Research* 99:14415–14428.

Liang, X-L., J. Zhu, K. E. Kunkel, M. Ting, and J. X. L. Wang. 2008. Do CGCMs simulate the North American monsoon precipitation seasonal-interannual variability? *Journal of Climate* 21:4424–4448.

Lu, J., G. A. Vecchi, and T. Reichler. 2007. Expansion of the Hadley cell under global warming. *Geophysical Research Let*ters 34: L06805, doi:10.1029/2006GL028443.

Maurer, E. P., L. Brekke, T. Pruitt, and P. B. Duffy. 2007. Fine-resolution climate projections enhance regional climate change impact studies. *EOS Transactions AGU* 88:504.

Maurer, E. P., H. G. Hidalgo, T. Das, M. D. Dettinger, and D. R. Cayan. 2010. The utility of daily large-scale climate data in the assessment of climate change impacts on daily streamflow in California. *Hydrology and Earth System Sciences* 14:1125–1138.

Maurer, E. P., I. T. Stewart, C. Bonfils, P. B. Duffy, and D. Cayan. 2007. Detection, attribution, and sensitivity of trends toward earlier streamflow in the Sierra Nevada. *Journal of Geophysical Research* 112: D11118, doi:10.1029/2006JD008088.

Mearns, L. O., F. Giorgi, P. Whetton, D. Pabon, M. Hulme, and M. Lal, 2003. *Guidelines for use of climate scenarios developed from Regional Climate Model experiments.* N.p.: Data Distribution Centre of the Intergovernmental Panel on Climate Change. http://www.ipcc-data.org/guidelines/RCM6.Guidelines.October03.pdf.

Mearns, L. O., W. Gutowski, R. Jones, R. Leung, S. McGinnis, A. Nunes, and Y. Qian. 2009. A regional climate change assessment program for North America. *Eos Transactions AGU 90*:311.

Meehl, G. A., T. F. Stocker, W. D. Collins, P. Friedlingstein, A. T. Gaye, J. M. Gregory, A. Kitoh, et al. 2007. Global climate projections. In *Climate change 2007: The physical science basis. Contribution of Working Group I to the Fourth Assessment Report of the Intergovernmental Panel on Climate Change*, ed. S. Solomon, D. Qin, M. Manning, Z. Chen, M. Marquis, K.B. Averyt, M. Tignor and H.L. Miller, SM.10-1 – SM.10-8. Cambridge: Cambridge University Press. http://www.ipcc.ch/pdf/assessment-report/ar4/wg1/ar4-wg1-chapter10-supp-material.pdf.

Milly, P. C. D., J. Betancourt, M. Falkenmark, R. M. Hirsch, Z. W. Kundzewicz, D. P. Lettenmaier, and R. J. Stouffer. 2008. Stationarity is dead: Whither water management? *Science* 319:573–574.

Mote, P. W., A. F. Hamlet, M. P. Clark, and D. P. Lettenmaier. 2005. Declining mountain snowpack in western North America. *Bulletin of the American Meteorological Society* 86:39–49.

Nakićenović, N., and R. Swart, eds. 2000. *Special report on emissions scenarios: A special report of Working Group III of the Intergovernmental Panel on Climate Change*. Cambridge: Cambridge University Press.

Nesbitt, S. W., D. J. Gochis, and T. J. Lang. 2008. The diurnal cycle of clouds and precipitation along the Sierra Madre Occidental observed during NAME-2004: Implications for warm season precipitation estimation in complex terrain. *Journal of Hydrometeorology* 9:728–743.

Pierce, D. W., T. P. Barnett, H. G. Hidalgo, T. Das, C. Bonfils, B. D. Santer, G. Bala, et al. 2008. Attribution of declining western U.S. snowpack to human effects. *Journal of Climate* 21:6425–6444, doi:10.1175/2008JCLI2405.1.

Pierce, D. W., T. P. Barnett, B. D. Santer, and P. J. Gleckler. 2009. Selecting global climate models for regional climate change studies. *Proceedings of the National Academy of Sciences* 106:8441–8446, doi:10.1073/pnas.0900094106.

Pierce, D. W., T. Das, D. R. Cayan, E. P. Maurer, N. L. Miller, Y. Bao, M. Kanamitsu, et al. 2012. Probabilistic estimates of future changes in California temperature and precipitation using statistical and dynamical downscaling. *Climate Dynamics* 39, published online, doi: 10.1007/s00382-012-1337-9.

Ray, A. J., J. J. Barsugli, K. B. Averyt, K. Wolter, M. Hoerling, N. Doesken, B. Udall, and R. S. Webb. 2008. *Climate change in Colorado: A synthesis to support water resources management and adaptation*. Boulder, CO: Western Water Assessment. http://cwcb.state.co.us/public-information/publications/Documents/ReportsStudies/ClimateChangeReportFull.pdf.

Ruff, T. W., Y. Kushnir, and R. Seager. 2012. Comparing 20th and 21st century patterns of interannual precipitation variability over the western United States and northern Mexico. *Journal of Hydrometeorology* 13:366–378.

Salathé, E. P. Jr. 2006. Influences of a shift in North Pacific storm tracks on western North American precipitation under global warming. *Geophysical Research Letters* 33: L19820, doi:10.1029/2006GL026882.

Santer, B. D., K. E. Taylor, P. J. Gleckler, C. Bonfils, T. P. Barnett, D. W. Pierce, T. M. L. Wigley, et al. 2009. Incorporating model quality information in climate change detection and attribution studies. *Proceedings of the National Academy of Sciences* 106:14778–14783, doi:10.1073/pnas.0901736106.

Seager, R., M. Ting, I. Held, Y. Kushnir, J. Lu, G. Vecchi, H-P. Huang, et al. 2007. Model projections of an imminent transition to a more arid climate in southwestern North America. *Science* 316:1181–1184.

Seager, R., and G. A. Vecchi. 2010. Greenhouse warming and the 21st century hydroclimate of southwestern North America. *Proceedings of the National Academy of Sciences* 107:21277–21282.

U.S. Bureau of Reclamation (Reclamation). 2011a. *Literature synthesis on climate change* implications for water and environmental resources, 2nd ed. Technical Memorandum 86-68210-2010-03. Denver: Reclamation, Research and Development Office. http://www.usbr.gov/research/docs/climatechangelitsynthesis.pdf.

— . 2011b. *West-wide climate risk assessments: Bias-corrected and spatially downscaled surface water projections.* Technical Memorandum 86-68210-2011-01. Denver: Reclamation, Technical Service Center.

Vano, J. A., T. Das, and D. P. Lettenmaier. 2012. Hydrologic sensitivities of Colorado River runoff to changes in precipitation and temperature. *Journal of Hydrometeorology* 13:932–949.

Wi, S., F. Dominguez, M. Durcik, J. Valdes, H. Diaz, and C. L. Castro. 2012. Climate change projections of snowfall in the Colorado River Basin using dynamical downscaling. *Water Resources Research* 48: W05504.

Wilby, R. L., S. P. Charles, E. Zorita, P. Whetton, and L. O. Mearns. 2004. *Guidelines for use of climate scenarios developed from statistical downscaling methods.* N.p.: IPCC Data Distribution Center. http://www.ipcc-data.org/guidelines/dgm_no2_v1_09_2004.pdf.

Wood, A. W., L. R. Leung, V. Sridhar, and D. P. Lettenmaier. 2004. Hydrologic implications of dynamical and statistical approaches to downscaling climate model outputs. *Climatic Change* 62:189–216.

Woodhouse, C.A., D. M. Meko, G. M. MacDonald, D. W. Stahle, and E. R. Cook. 2010. A 1,200-year perspective of 21st century drought in southwestern North America. *Proceedings of the National Academy of Sciences* 107:21283–21288.

Endnotes

i The BCSD method removes bias in the climate model output by mapping from the probability distribution of a current climate simulation to the probability distribution of observations on a monthly basis.

ii For downscaling simulated surface temperature from the GCMs, the BCSD methodology preserves GCM (large-scale) trends by removing them initially and adding them back after the downscaling is implemented. For downscaling simulated precipitation, no explicit step is included in BCSD to preserve the GCM trends, because trends are not so obviously present. Other inherent weaknesses of the BCSD approach are the assumption of climate "stationarity"—the idea that statistical relationships developed in a historical period are applicable to a future period—and the underestimation of variability (Wilby et al. 2004; Milly et al. 2008).

iii VIC is a macroscale, distributed, physically based hydrologic model that balances both surface energy and water over a grid mesh. For this report, VIC simulations, run from BCSD downscaled precipitation and temperature data, were employed.

iv CMIP3 is phase 3 of the World Climate Research Programme's Coupled Model Intercomparison Project.

v With a baseline of 65°F, heating degree days are the sum of the temperature differences of the daily mean temperature subtracted from 65°F, for all days when the mean temperature is less than 65°F. Cooling degree days are calculated similarly, but for when the mean temperature exceeds 65°F.

vi Cyclones are the rapid circulation of winds around a low pressure center, traveling counterclockwise in the Northern Hemisphere and clockwise in the Southern Hemisphere. Anticyclones spiral out from a high pressure area and travel clockwise in the Northern Hemisphere and counterclockwise in the Southern Hemisphere.

vii CNRM-C3 is the third version of a global ocean-atmosphere model originally developed at the Centre National de Recherches Meteorologiques, France.

Chapter 7

Future Climate: Projected Extremes

COORDINATING LEAD AUTHOR

Alexander Gershunov (Scripps Institution of Oceanography, University of California, San Diego)

LEAD AUTHORS

Balaji Rajagopalan (University of Colorado), Jonathan Overpeck (University of Arizona), Kristen Guirguis (Scripps Institution of Oceanography), Dan Cayan (Scripps Institution of Oceanography), Mimi Hughes (National Oceanographic and Atmospheric Administration [NOAA]), Michael Dettinger (U.S. Geological Survey), Chris Castro (University of Arizona), Rachel E. Schwartz (Scripps Institution of Oceanography), Michael Anderson (California State Climate Office), Andrea J. Ray (NOAA), Joe Barsugli (University of Colorado/Cooperative Institute for Research in Environmental Sciences), Tereza Cavazos (Centro de Investigación Científica y de Educación Superior de Ensenada), and Michael Alexander (NOAA)

REVIEW EDITOR

Francina Dominguez (University of Arizona)

Executive Summary

This chapter summarizes the current understanding about how and why specific weather and climate extremes are expected to change in the Southwest with climate warming over the course of the current century.

Summertime heat waves and wintertime cold snaps are among the extremes most directly affected by climate change as well as the ones with the greatest impacts.

Chapter citation: Gershunov, A., B. Rajagopalan, J. Overpeck, K. Guirguis, D. Cayan, M. Hughes, M. Dettinger, C. Castro, R. E. Schwartz, M. Anderson, A. J. Ray, J. Barsugli, T. Cavazos, and M. Alexander. 2013. "Future Climate: Projected Extremes." In *Assessment of Climate Change in the Southwest United States: A Report Prepared for the National Climate Assessment*, edited by G. Garfin, A. Jardine, R. Merideth, M. Black, and S. LeRoy, 126–147. A report by the Southwest Climate Alliance. Washington, DC: Island Press.

- Heat waves, as defined relative to current climate, are projected to increase in frequency, intensity, duration, and spatial extent. (high confidence)
- Heat waves are projected to become more humid and therefore expressed relatively more strongly in nighttime rather than daytime temperatures, with associated stronger impacts on public health, agriculture, ecosystems, and the energy sector. (medium-low confidence)
- Wintertime cold snaps are projected to diminish in their frequency, but not necessarily in their intensity, into the late twenty-first century. (medium-high confidence)

Precipitation extremes are projected to become more frequent and more intense in the wintertime. Summertime precipitation extremes have not been adequately studied.

- Enhanced precipitation extremes are generally expected due to greater moisture availability in a warming atmosphere, even if average precipitation declines. (medium-low confidence).
- Enhanced precipitation specifically associated with atmospheric rivers, a wintertime phenomenon typically yielding extreme precipitation, is projected by most current climate models. (medium-low confidence)

Flooding is expected to change in timing, frequency, and intensity, depending on season, flood type, and location.

- Floods from winter storms on the western slopes of the Sierra Nevada have been projected to increase in intensity in winter by all climate models that have been analyzed thus far, including models that otherwise project drier conditions. (medium-high confidence)
- Snowmelt-driven spring and summertime floods are expected to diminish in both frequency and intensity. (high confidence)
- Transition from hail to rain on the Front Range of the Rocky Mountains is expected to result in higher flash-flood risk specifically in eastern Colorado. (medium-low confidence)

Drought is generally expected to intensify in a warming climate, but some variation across basins can be expected, although few basins have been analyzed.

- Drought, as expressed in Colorado River flow, is projected to become more frequent, more intense, and longer-lasting, resulting in water deficits not seen during the instrumental record. (high confidence)
- Northern Sierra Nevada watersheds may become wetter, and in terms of flow, somewhat less drought-prone with climate change. (medium-low confidence)
- In terms of soil moisture, drought is expected to generally intensify in the dry season due to warming. (high confidence)

Extreme fire weather can be associated with a combination of factors, none of which need be particularly extreme. For example, dry and commonly warm Santa Ana winds (which are not necessarily themselves extreme) are frequently associated with extreme fire potential in Southern California.

- Santa Ana winds are expected to diminish in frequency, but at the same time become drier and hotter. (medium-low confidence) However, the combined effect of decreased winds and increased temperatures and dryness on Southern California's fire risk is not clear.

Beyond these projections, the region is fraught with important uncertainties regarding future extremes, as many have yet to be projected.

7.1 Introduction

Extreme events can be defined in many ways. Typical definitions of weather and climate extremes consider either the maximum value during a specified time interval (such as season or year) or exceedance of a threshold (the "peaks-over-threshold" [POT] approach), in which universal rather than local thresholds are frequently applied. For example, temperatures above 95°F (35°C) are often considered extreme in most locations across the United States, except in areas such as the low-lying deserts of Arizona and California, where such temperatures are typical in the summer. Temperatures at these levels are obviously extreme for living organisms from a non-adapted, physiological perspective, and technological adaptation for humans is required for day-to-day functioning in such temperatures. But such temperatures are not necessarily extreme from the statistical or local climate perspectives. In statistics, extremes are considered low-probability events that differ greatly from typical occurrences. The IPCC defines extremes as 1% to 10% of the largest or smallest values of a distribution (Trenberth et al. 2007). Studies over large or complex regions marked by significant climatic variation require definitions that are relevant to local climate. Across the Southwest, location-specific definitions of extreme temperature, precipitation, humidity, and wind are required if a meaningful region-wide perspective is desired.

In spite of common claims that climate change will result in past extremes becoming more commonplace, only a few scientific studies have actually considered future projections of extremes (e.g., Meehl et al. 2000; Tebaldi, Hayhoe, and Arblaster 2006; Parry et al. 2007; Trenberth et al. 2007). Even fewer have focused on regional extremes, usually in response to specific events such as the European heat wave of 2003. Studies examining projections of temperature, precipitation, and hydrological extremes typically resolve the Southwest as part of a much larger spatial domain. Hydrological drought research is the exception, as it naturally focuses on river basins. Drought in the Colorado River Basin, which encompasses a large swath of the Southwest and channels a large part of its water supply, was the focus of a recent drought projection study (Cayan et al. 2010). As a state, California has probably been the focus of more climate-change research than any other in the United States—research that has translated to state policy action. Not surprisingly, some of the first regional extreme climate projections in the nation have been carried out for some of California's weather and climate extremes (e.g. Das et al. 2011; Mastrandrea et al. 2011; Gershunov and Guirguis 2012). Results of these and other relevant studies are described in the Southwestern context below.

For climate science to inform impact assessment and policy research, it is important to define the most relevant impact-based indices of environmental extremes. This impact-driven (or "bottom-up") approach represents the current thrust of climate science striving to be relevant to society. To accomplish this goal, close collaborations among science and the public-private policy sectors must be initiated and maintained. This process is perhaps further along in the Southwest than in other regions of the nation; but even for the Southwest, the necessary cross-sector relationships are still in their infancy. One of the future goals of the Southwest Climate Alliance (SWCA) is to define extremes by first understanding their impacts in key sectors. But for now, while keeping mindful of their impacts, we define extremes based solely on climate records and models.

For meaningful projections of extremes, models must be validated with respect to the mechanisms (such as heat waves or atmospheric rivers) that produce specific extremes. Without careful validation, a multi-model approach can introduce more uncertainties into projection of extremes than when multi-model projections are averaged to study mean climate trends, an approach that is typically assumed to increase certainty. In contrast to average climate, changes in extremes cannot be assumed to be more adequately diagnosed from averaging across a set of models, or ensemble members. The rare nature of extremes demands that they be carefully analyzed in each realization of modeled climate.

7.2 Heat Waves

Background climate warming can be expected to result in increased heat wave activity as long as the thresholds used to define heat waves remain unchanged. Multi-model and downscaled projections are clear on this, globally and specifically for the Southwest United States (see, for example, Diffenbaugh and Ashfaq 2010).

Gershunov, Cayan, and Iacobellis (2009) showed that heat waves over California and Nevada are not simply increasing in frequency and intensity but are also changing their character: they are becoming more humid and therefore are expressed more strongly in nighttime rather than daytime temperatures (i.e., in minimum [Tmin] rather than maximum [Tmax] daily temperatures). These changes started in the 1980s and appear to have accelerated since 2000. Moreover, the seasonal average humidity levels have not increased; rather, rare synoptic circulations (regional pressure patterns and their associated surface winds) that bring hot air to the extreme Southwest tend to also bring increased humidity. The trend in humid heat waves was shown to be due to the warming of the Pacific Ocean surface west of Baja California, a regionally intensified part of the global ocean warming trend. Following up on these observational results, Gershunov and Guirguis (2012) first identified a global climate model (GCM) from which daily data were available and which was able to simulate both the synoptic causes of California heat waves (i.e., the observed pressure and humidity patterns associated with regional heat waves) as well as the observed trend in the flavor of regional heat waves disproportionately intensifying at night compared to the daytime. They then considered downscaled projections[i] over California and its subregions. Given the lack of heat wave projections studies for the entire Southwest and the potentially disproportionate impacts of humid heat on a region where life is acclimatized to dry heat, in the section below we expand the observational diagnosis of Gershunov, Cayan, and Iacobillis (2009) and then follow the approach of Gershunov and Guirguis (2012), extending their heat wave projections to the entire Southwest.

Heat wave index

Heat waves are hereby defined locally, but are described over the entire Southwest as a period lasting at least one day when daily temperature exceeds the 95th percentile of the local daily May-to-September climatology of maximum or minimum temperatures for 1971–2000. In other words, a local heat wave is registered when temperature rises to the level of the hottest 5% of summer days or nights.[ii] The local magnitude of the heat wave (the heat wave index, or HWI) is the difference between the actual Tmax or Tmin and its corresponding 95th percentile threshold, summed over the consecutive days of the heat wave, or over the entire season if a measure of *summertime* heat wave activity is desired. This measure is similar to the familiar *degree days*, except that the threshold temperature is defined relative to local climatology, as opposed to an absolute threshold, making the HWI consistent and comparable for all locations representing a region. The regional HWI is then constructed by taking the regional average of the local values. HWI reflects the frequency, intensity, duration, and spatial extent of heat waves across the Southwest (Figure 7.1).

Projections

Observations and modeling indicate that Southwestern heat wave activity is increasing as expected with climate change, however as in California, it is increasing disproportionately relative to minimum versus maximum temperatures (Figure 7.1). The Tmin trend is clearly visible during the historical period and it is comparable to the modeled trend (inset on panel B), in contrast, the historical modeled Tmax trend has not yet been observed. For the future, heat waves are projected to increase at an accelerating rate, with nighttime heat waves projected to increase at a faster rate than daytime heat waves. Much of the projected increase in Southwestern heat wave activity is to be expected simply from average seasonal warming driving temperatures to exceed the stationary local 95th percentile thresholds—by larger margins, more often, for more consecutive days, and over larger parts of the Southwest—driving this cumulative heat wave index dramatically upward.

Mastrandrea and others (2011) adopted a multi-model view on California heat waves to examine 100-year events.[iii] Their results also suggested higher minimum temperatures are projected to increase more than maximum temperatures, but their modeling results were not as clear on this point as those from the well-validated CNRM-CM3 model or from observations. The main result from multi-model heat wave projections is that observed 100-year return period heat waves become heat waves with a 10-year or even shorter return period during the last half of the twenty-first century.

The disproportionate increase in nighttime versus daytime projected heat wave occurrence is consistent with observations and is indicative of enhanced future impacts on health (of humans, animals, and ecosystems), agriculture, and energy infrastructure, due to the elevated humidity and diminished nighttime respite from heat. The intensifying heat thus becomes more difficult for the biota of the Southwest to tolerate, acclimatized as they are to dry daytime heat and cool nights. Given the high correspondence of

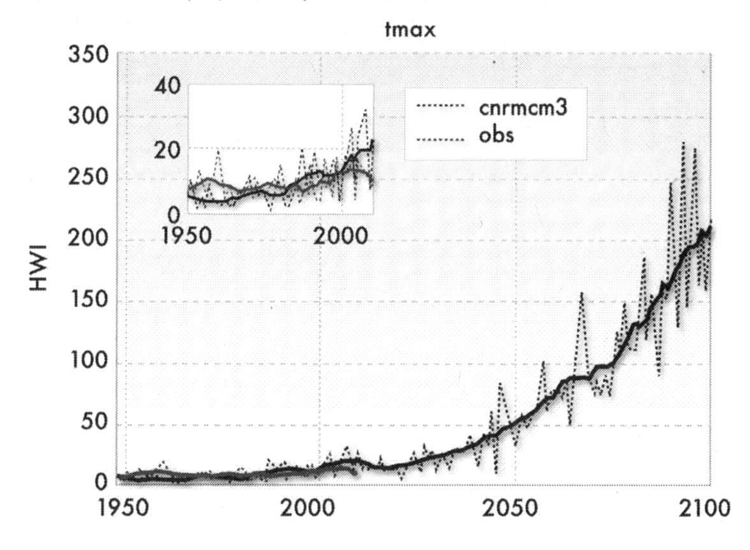

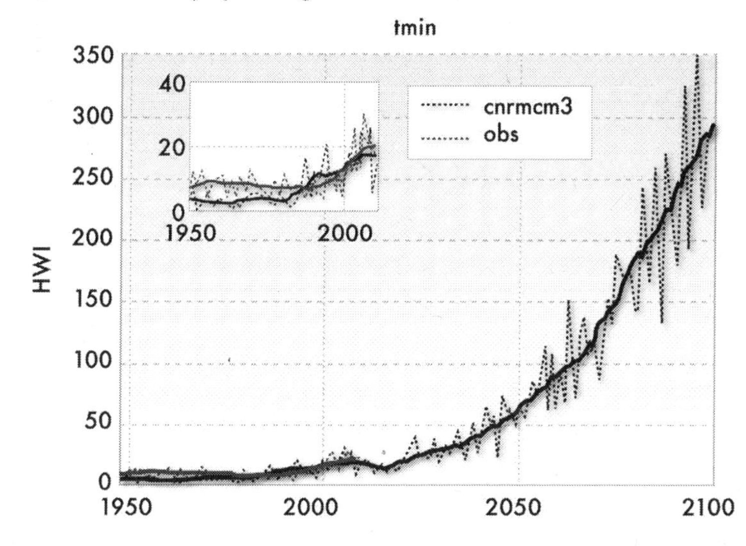

Figure 7.1 The summertime (May–September) Heat Wave Index (HWI) for Tmax (a) and Tmin (b) for the Southwest region. Solid line is the 5-year running mean. HWI values show °C above the local 95th historical percentile. Inset shows the same data on a scale appropriate for the historical period (1950-1999). Historical observed and modeled data as well as twenty-first century projections (according to the SRES-A2 "high-emissions" scenario) are shown from observations as well as from a GCM (CNRM-CM3) historical simulation and projection averaged over the Southwest. Adapted for the Southwest based on the work of Gershunov and Guirguis (2012) for California. Data source: Salas-Mélia et al. (2005); see also http://www.ipcc-data.org/ar4/model-CNRM-CM3-change.html.

observed and modeled trends, we consider this to be a medium-low confidence result. Agreement from additional models, if validated to produce realistic heat waves, will likely increase the confidence of this conclusion into the "high-confidence" category.

Sub-regional variation of heat wave trends is possible. For example, working with BCCA-downscaled data for California, Gershunov and Guirguis (2012) found intriguing patterns of change in the magnitude of coastal heat waves relative to median summertime warming of coastal regions. Although average warming is projected to be weaker at the coast than over inland areas in summer (see Chapter 6), a trend in heat waves of progressively enhanced magnitude through the twenty-first century is projected along the coast—the most highly populated and least heat-adapted of all California sub-regions. This trend is already observed over coastal Northern California.

7.3 Wintertime Cold Outbreaks

A diminishing trend in cold spell frequency and intensity has been observed over the Northern Hemisphere and its continental sub-regions (Guirguis et al. 2011). Within regions, cold outbreaks are most intense in low-lying topography that channels the cold dense air. Their frequency also responds to changes in atmospheric circulation patterns, in particular to transient high pressure systems that cause cold air outbreaks. As the subtropical high-pressure zones intensify and expand poleward (Lu, Vecchi, and Reichler 2007; Lu, Deser, and Reichler 2009) and the storm track contracts towards the pole (IPCC 2007), fewer cyclones and more anticyclones are expected to reach the Southwest United States and northwestern Mexico, resulting in less-frequent precipitation but increased frequency of atmospheric circulation conditions leading to cold outbreaks (Favre and Gershunov 2009; Chapter 6). Considered with regional seasonal warming trends, this change may affect the frequency of future cold extremes. However, because of the topographic complexity of the Southwestern United States (Chapter 3, Section 3.1), this would apply only to cold extremes in coastal low-lying valleys west of the Sierra Nevada. Winter cold outbreaks in much of the rest of the mountainous Southwest are not affected by transient anticyclones arriving from the North Pacific (see Favre and Gershunov 2009, Figure 14).

Following the heat wave index definition provided above and the approach of Guirguis and others (2011), cold outbreaks are here defined as the coldest five percent of the wintertime daily temperature distribution, aggregating degree days below the local 5th percentile thresholds over the cold season (November–March), and averaging over the region. In other words, cold outbreaks occur when temperature drops below the local levels defining the coldest 5% of winter days or nights, and are measured by how far they drop below those levels over the entire region.[iv] The resulting cold spell index (CSI), derived from observations and the CNRM-CM3 model, is presented in Figure 7.2. It reflects frequency, intensity, duration, and spatial extent of wintertime cold spells over the entire Southwest. Cold spells are clearly projected to diminish in both maximum and minimum temperatures. The trend is not projected to be steady, as the influence of natural interannual and interdecadal variability on the occurrence of cold extremes is projected to continue to strongly modulate cold outbreaks in the future. Kodra, Steinhauser, and Ganguly (2011) project that occasional extreme cold events are likely to persist across each continent under twenty-first century warming scenarios, however, and this agrees with recent results for California using multi-model downscaled projections (Pierce et al. 2012). In the states of California and Northern Baja California, the more frequent occurrence of mid-latitude anticyclones that produce the cold snaps (Favre and Gershunov 2009) may be elevating the probability that some of the future cold snaps will be nearly as cold as those in the past. However, as warming continues into the late twenty-first century while the local thresholds used to define cold extremes remain static, this probability should diminish over the entire Southwest, leading to generally less frequent if not always less severe cold outbreaks by the end of the century.

a) Observed and projected daytime Cold Spell Index

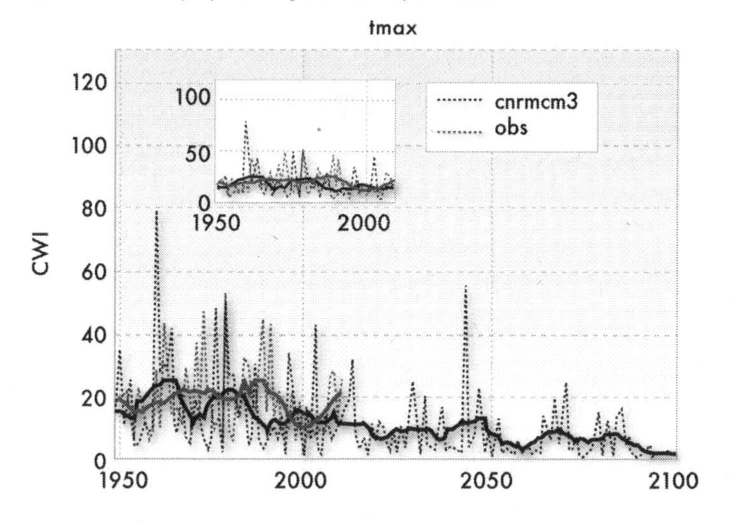

b) Observed and projected nighttime Cold Spell Index

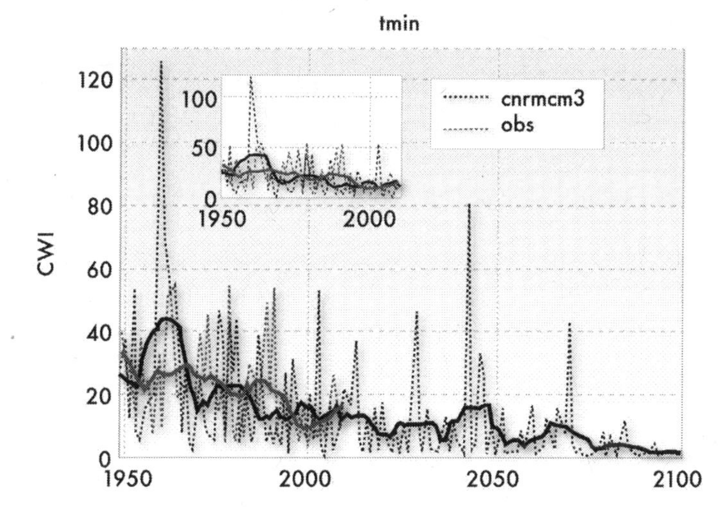

Figure 7.2 The wintertime (November–March) Cold Spell Index (CSI) for Tmax (a) and Tmin (b) for the Southwest region. Explanation of lines, values, and insets as in Figure 7.1. Adapted for the Southwest based on the work of Gershunov and Guirguis (2012) for California. Data source: Salas-Mélia et al. (2005); see also http://www.ipcc-data.org/ar4/model-CNRM-CM3-change.html.

7.4 Precipitation

General results and key uncertainties

Models project that there will be augmented extreme precipitation events even in regions where total precipitation is generally expected to decrease (Groisman et al. 2005; Wang and Zhang 2008), such as in the southern portion of the Southwest in winter (see Chapter 6). The reason for this expectation is that warmer air holds more moisture at saturation (100% relative humidity) and therefore "extreme" storms should be able to produce more precipitation than similar events in the past. Global climate models are notoriously deficient at simulating high frequency precipitation, especially its extreme values, but they generally agree on this result (Groisman et al. 2005; Kharin et al. 2007).

To circumvent these deficiencies in simulating precipitation and to rely more on model strengths, Wang and Zhang (2008) used statistical downscaling to relate large-scale atmospheric circulation and humidity to locally observed precipitation and then applied this downscaling scheme to a global climate model projection. By doing this, the effects of changes in circulation and humidity could be evaluated separately, and changes in the daily extreme values of winter precipitation could be diagnosed. They found that over much of North America during the last half of the current century, extreme precipitation events that currently occur on average once in a twenty-year period are projected to occur up to twice as frequently, even in regions that are projected to have decreased precipitation due to circulation changes. In other words, increased specific humidity in a warming atmosphere is expected to dominate future trends in extreme precipitation.

Dynamically downscaled GCM projections support this result. Multi-model dynamically downscaled simulations project a significant increase (of 13% to14% on average by mid-century) in the intensity of wintertime extremes with 20- and 50-year return periods under the high-emissions scenario (Dominguez et al. 2012), even though the same simulations project average precipitation to decrease over the Southwest. Increased water vapor content in a warming atmosphere seems a key element in such projections, and statistical downscaling schemes that do not include explicit accounting for this increasing moisture content do not necessarily support these findings. Mastrandrea and others (2011), for example, used a precipitation downscaling scheme that did not use atmospheric humidity as a predictor. When applied in a multi-model context over California, this approach did not yield clear or significant changes in precipitation extremes.

Such studies are not directly comparable due to different choices of global climate models, downscaling schemes, and definitions of extreme precipitation. However, simple physical reasoning suggests that in a warming and moistening atmosphere, greater precipitation extremes can co-evolve with generally drier conditions. This argument is consistent with observations and modeling over many of the world's regions (Groisman et al. 2004; Groisman et al. 2005), especially in summer, but increasing extreme precipitation trends have not yet been observed over the Southwest. This is in spite of the fact that the Southwest has been at the forefront of warming among regions of the contiguous United States. Although different projection results are not yet perfectly congruent and more research is clearly needed, modeling schemes that explicitly resolve increasing moisture content in a warming atmosphere consistently result in more frequent and larger future precipitation extremes over the otherwise drying Southwest and their results should be regarded with more confidence.

Next, we examine precipitation extremes due to specific storm systems and physical processes.

North American monsoon (NAM)

The North American monsoon (NAM) is the source of summertime precipitation for much of the Southwest, particularly Arizona and New Mexico. The core region of NAM is along the Sierra Madre Occidental in northwestern Mexico (Cavazos, Turrent, and Lettenmaier 2008; Arriaga-Ramírez and Cavazos 2010). These studies examined observed trends in extreme summertime rainfall, but did not find any significant precipitation trends related specifically to NAM.[v] The northern tip of NAM penetrates into the Southwestern United States. Monsoonal precipitation modeling and projections present many uncertainties (see Chapter 6), which translate to key uncertainties for the extremes. Projections of extremes have not been specifically evaluated for NAM rainfall; however, in a new study, Cavazos and Arriaga-Ramirez (2012), found that model scenarios of increased greenhouse gases (A2 "high-emissions" scenario) in conjunction with statistical downscaling, show a weakening of the monsoon rainfall due to longer and more frequent dry periods in the monsoon region by the end of this century. However, as explained in Chapter 6, the North American monsoon system is also influenced by large-scale patterns of natural variability. Great challenges in monsoon modeling prevent confident conclusions about the future of NAM, especially its extremes. Chou and Lan (2011) show a negative trend in the annual total maximum precipitation in the monsoon region during the twenty-first century associated with increased subtropical subsidence induced by global warming. Projected increase in subtropical subsidence (as discussed in Chapter 6 and in Lu, Vecchi, and Reichler 2007) could negatively impact NAM precipitation; however, its potential impact on NAM extremes is, even intuitively, less clear. The low confidence in projected decreased total NAM precipitation and the lack of understanding about its extremes make NAM an important topic for ongoing research. Future research should consider NAM in its entirety on both sides of the U.S.-Mexico border.

Atmospheric rivers

Much of the Southwest is within reach of an important class of Pacific storms that are often referred to as "atmospheric rivers" (ARs). ARs are storms in which enormous amounts of water vapor are delivered to the region from over the Pacific Ocean in corridors that are low-level (less than about 6,600 feet [2000 m] above sea level), long (greater than 1,200 miles [2000 km]), and narrow (less than about 300 miles [500 km] wide) (Ralph and Dettinger 2011). So far, atmospheric rivers have been the only extreme-precipitation-producing systems in the Southwest that are large enough to be adequately modeled by GCMs even without downscaling and that have received recent careful attention in the context of climate change.

When these ARs encounter the mountains of the Southwest—most often in California, but occasionally penetrating as far inland as Utah and New Mexico (Figure 7.3)—they produce many of the most intense precipitation events that define the storm and flood climatology of the region (Ralph et al. 2006; Ralph et al. 2011). These storms are present in climate-change simulations by the coupled atmosphere-ocean global climate models included in the IPCC Fourth Assessment Report (e.g., Bao et al. 2006; Dettinger 2011; Dettinger, Ralph, Das et al. 2011; Dettinger, Ralph, Hughes et al. 2011), and presumably will be better represented in the generally better-resolved models of the Fifth Assessment Report. A preliminary study of their occurrences and intensities in the Fourth Assessment Report projections of climate changes in response to the SRES A2 "high-emissions" scenario by seven GCMs (Dettinger 2011) indicates that in a warmer climate, ARs making landfall on the California coast will carry more water vapor in general. By the mid-twenty-first century, ARs are projected to increase by an average of about 30% per year and about twice as many years are projected to have many more than historical numbers of ARs. Also, all seven models yielded occasional twenty-first century ARs that were considerably more intense than any simulated (or observed) in the historical period. Together these results suggest that the risks of storm and flood hazards in the Southwest from AR storms may increase under the changing climate of the twenty-first century; however the analyses and even our understanding of historical ARs (Dettinger 2011; Ralph and Dettinger 2011) are still preliminary and warrant further investigation.

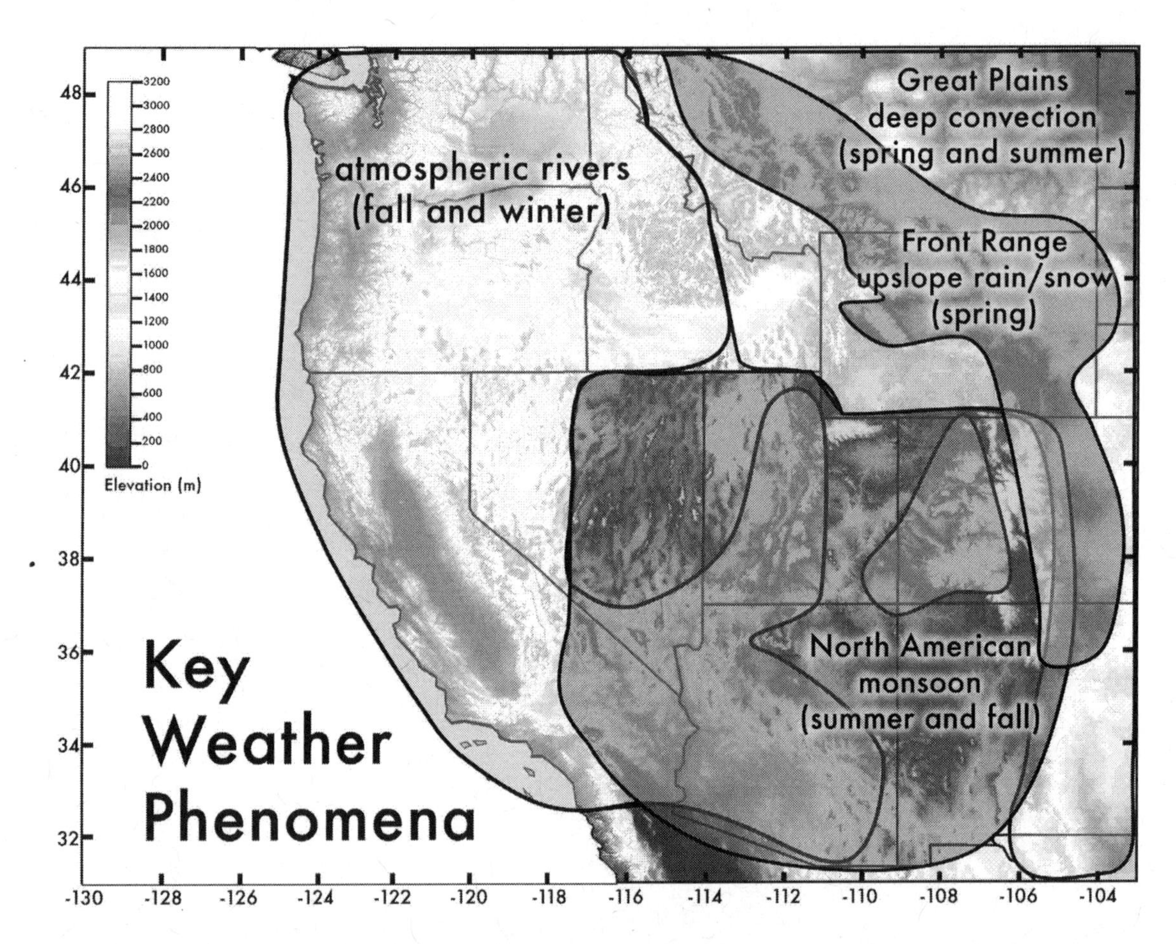

Figure 7.3 Key weather phenomena that cause extreme precipitation in the Southwest. Schematic illustration of regional patterns of the primary weather phenomena that lead to extreme precipitation and flooding, while also contributing to water supplies (Guan et al. 2010; Dettinger et al. 2011), across the western United States. Modified from Ralph et al. (2011); see http://www.westgov.org/wswc/167%20council%20meeting%20-%20id/167%20council%20mtg%20-%20oct2011.html.

Hail on the Colorado Front Range

Parts of the Southwest are prone to precipitation from intense summertime thunderstorms that fall as hail rather than heavy rain. Although it can inflict significant damage on property and agriculture, hail may help prevent or delay flash flooding. The most active region in terms of hailstorm intensity, frequency, and duration in the United

States is the leeward (eastern) side of the Rocky Mountains, especially eastern Colorado. Mahoney et al. (2012) used a dynamical downscaling framework to compare past (1971–2000) and future (2041–2070) warm-season convective storm characteristics, with a focus on hail, in the Colorado Front Range and Rocky Mountain regions.[vi] The authors found that surface hail in the 2041–2070 time period was projected to be nearly eliminated, despite an increase in in-cloud hail, due to a higher altitude level of melting (32°F [0°C] isotherm). The initial level of melting increased from about 16,400 feet (5,000 m) above sea level to about 18,000 feet (5,500 m) over the study region over time. The model simulations suggest this deeper vertical layer of above-freezing temperatures will be sufficient to melt the hailstones before they reach the surface. Additionally, across most elevations in the region, the future simulations produced greater total maximum precipitation and surface runoff. The combination of decreased surface hail and increased rainfall as well as overall precipitation intensity, implies flash flooding may become more likely in mountainous regions that currently experience hail, especially where the surfaces are relatively impervious.[vii]

7.5 Surface Hydrology

Flooding

Changes due to warming in the type of intense precipitation received will affect flooding. Just as the change of summertime precipitation on the Rockies' Front Range from hail to rain will increase the risk of flash flooding, so too will the change from wintertime snow to rain for areas of the Southwest. Projected changes in winter storms, including both intensities and temperatures, are expected and projected (Das et al. 2011) to yield increased winter floods, especially in the Sierra Nevada, where winter storms are typically warmer than those farther inland. Even in global climate model scenarios with decreased total regional precipitation, flood magnitudes are projected to increase (Das et al. 2011). More frequent and/or intense precipitation extremes are an important cause of increased flooding, but warming also plays an important role as it results in wintertime precipitation falling more as rain rather than snow. The projected late-century increase in flooding generated in the Sierra Nevada watersheds is therefore due to wintertime storm-driven runoff, while spring and early summer snowmelt-driven floods are expected to wane.

Future changes in flooding elsewhere in the Southwest will depend on the future of the storm mechanisms partially summarized in Figure 7.3. For example, where enhanced ARs drive extreme precipitation, wintertime flooding may be expected to increase, although probably not as much as on the western slopes of the Sierra Nevada, where much moisture is squeezed from these systems. In other regions and seasons (e.g. the monsoon region in the summertime), uncertainties about future changes in precipitation extremes translate directly into uncertainties about future flooding.

Drought

Global models suggest more dry days and drier soils in the future for the southern part of the Southwest (Field et al. 2011). Along with projected warming and increased evapotranspiration, this can only mean that droughts will become more severe. The crucial importance of water resources and their natural volatility in the arid, thirsty, and growing Southwest has motivated numerous hydrological studies over the decades. The Colorado River—which provides at least partial water supply to all Southwestern states—has been the natural focus of many of these studies. Recent research was motivated by a prolonged drought that afflicted the Southwest, and particularly the Colorado River, for much of the first decade of the twenty-first century (MacDonald 2008).

This contemporary drought was examined in the context of past records and future projections utilizing a hierarchy of GCM projections, statistical downscaling, and hydrologic modeling to focus on Southwestern drought and describe it in the context of past and likely future conditions (Cayan et al. 2010). The results are summarized here.

The recent drought is a perfect example of droughts that the Southwest has been prone to experience about once per century. The analysis of Cayan and others (2010), based on the high emissions scenario, suggests that the current 100-year drought will become commonplace in the second half of this century and that future droughts will be much more severe than those previously recorded. This possibility should not be surprising given the magnitude of megadroughts on the paleorecord (Cook at al. 2004; Cook et al. 2009; Chapter 5), but importantly, climate change is slowly tipping the balance in favor of more frequent, longer, and more intense droughts. Figure 7.4 shows the difference in the Colorado River flow deficit accumulated over consecutive years of observed versus projected 100-year drought. This projection of intensified drought conditions on the Colorado River is not due to changes in precipitation, but rather due directly to warming and its effect on reducing soil moisture (Figure 7.5) by reducing snowpack and increasing evapotranspiration. The projected longer, more intense droughts in the Colorado Basin will pose challenges to sustaining water supplies of the already over-allocated Colorado River (Chapter 10). Increased drought may not be expected for all river basins of the Southwest, however; the Sacramento River Basin, for example, is projected to become slightly wetter (Cayan et al. 2010).

7.6 Fire Weather

Fire weather is persistent in the Southwest most of the year and actual outbreaks of fire in the region are significantly affected by human factors such as ignition and arson, fire management, and fire suppression practices that might (in the long run) provoke stronger wild fires (Westerling et al. 2003). Hot, gusty, dry winds can greatly exacerbate the risk of extremely large wildfires if they occur when the fuels are dry and plentiful and especially where the wind itself can influence the risk of a spark (e.g., when it contacts power lines). The topographically complex Southwest is home to several regional downslope winds of the rain-shadow-type (such as the Colorado Front Range chinook) or the gravity-driven type (such as the Santa Ana winds of Southern California). The Santa Ana winds, accelerating down the west slopes of Southern California's coastal ranges, are particularly notorious for spreading uncontrollable fires, since the beginning of the Santa Ana season in the fall coincides with the end of the long dry warm season (e.g. Westerling et al. 2004).

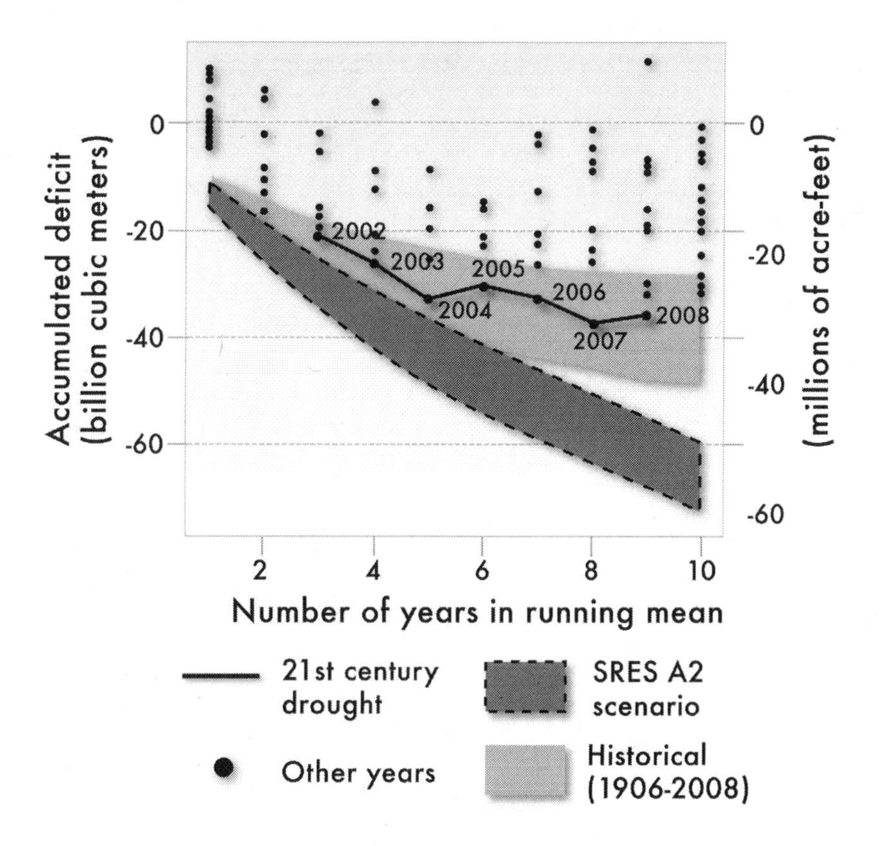

Figure 7.4 Drought associated anomalies in Colorado River streamflow. Accumulated deficit in flow (10^9 m^3, or billions of cubic meters) on the Colorado River at Lees Ferry relative to the mean flow observed over the period 1906–2008 (Y axis), as a function of drought duration in years (X axis). The twenty-first-century drought is shown as a dark grey line; other years are shown as black dots. For example, the 2007 label indicates an eight-year drought (2000–2007), with an accumulated deficit of around 40 billion cubic meters (around 30 million acre-feet). Light grey shading indicates where, two-thirds of the time, the worst drought (accumulated streamflow deficit) of the century should fall; the darker grey hatched region shows where the worst drought should fall for the end of this century, estimated from downscaled climate models. The twenty-first-century drought was consistent with the expected 100-year event, given the observed climate, whereas future 100-year droughts are expected to be much more severe in terms of accumulated flow deficit. Modified from Cayan et al. (2010).

Santa Ana winds

The cool, relatively moist fall and winter climate of Southern California is often disrupted by dry, hot days with strong winds known as the Santa Anas that blow out of the desert. The Santa Ana winds are a dominant feature of the cool-season climate of this region (Conil and Hall 2006; Hughes and Hall 2010), and they have important ecological impacts. The most familiar is their influence on wildfires: following the hot, dry Southern California summer, the extremely low relative humidities and strong, gusty winds associated with Santa Anas introduce extreme fire risk, often culminating in wildfires with large economic loss (Westerling et al. 2004; Moritz et al. 2010).

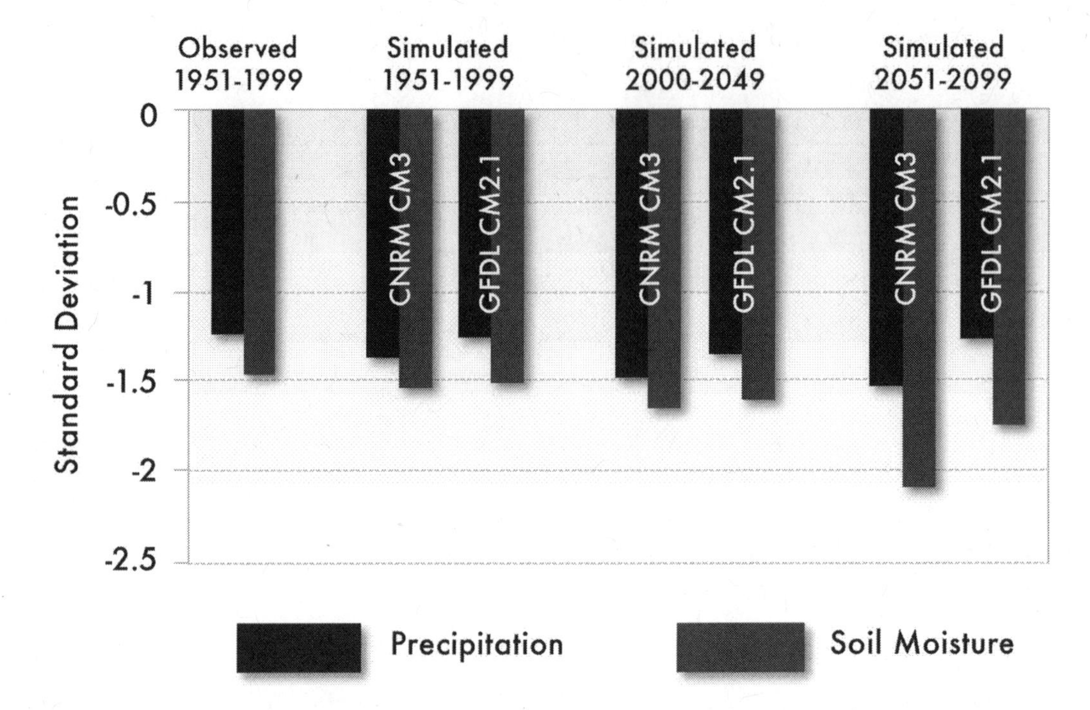

Figure 7.5 Drought-associated anomalies in precipitation and soil moisture. Composite average of water-year precipitation (dark grey) and water-year soil moisture (light grey) anomalies associated with extreme negative soil moisture anomalies for the Southwest estimated from observations and simulated climate input from CNRM CM3 and GFDL CM2.1 GCMs SRES A2 emission scenario, for 1951–1999, 2000–2049, and 2050–2099. Modified from Cayan et al. (2010).

Hughes, Hall, and Kim (2011) recently documented the potential impacts of human-generated climate change on the frequency of the Santa Anas and associated meteorological conditions with a high-resolution dynamical downscaling of a Fourth Assessment Report (NCAR CCSM3) model. They project that the number of Santa Ana days per winter season would be approximately 20% fewer in the mid twenty-first century compared to the late-twentieth century.

In addition to the change in Santa Anas' frequency, Hughes, Hall, and Kim (2011) also investigated changes during Santa Anas in two other meteorological variables known to be relevant to fire weather conditions—relative humidity and temperature—and found a decrease in the relative humidity and an increase in temperature. Both of these changes would favor fire, while the reduction in Santa Ana wind events would reduce fire risk. More work is necessary to ensure these results are robust across different climate models and emission scenarios, and to quantify the impact of these changes on fire weather. Santa Ana winds are treated as weather extremes in this report because they cause extreme fire danger conditions. However, Santa Ana winds are rather commonplace and not extreme in and of themselves. The extremes of Santa Ana winds have not been studied either in the observed or projected climate. This is an important topic for future research.

7.7 Discussion of Key Uncertainties

Large-scale climate drivers such as El Niño-Southern Oscillation (ENSO) and the Pacific Decadal Oscillation (PDO) in particular, play an important role in the winter and spring precipitation extremes (and consequently floods) over the Southwest and West. A rich body of literature shows the Southwestern hydroclimate—in particular the daily precipitation extremes—to be very sensitive to natural interannual and decadal variability (Gershunov and Barnett 1998; Cayan et al. 1999; Gershunov and Cayan 2003). Thus, flood risk can be affected by changes to moisture delivery processes (see, for example, Dettinger et al. 1998; Gershunov and Barnett 1998; Cayan, Redmond and Riddle 1999; Rajagopalan et al. 2000), temperature, and land surface conditions. An important uncertainty for the Southwest therefore is how the relevant modes of natural variability in the Pacific sector (ENSO and PDO) and their combined influences on Southwestern climate may be affected by climate change. The most predictable climate regime in the Southwest is the dry winter associated with La Niña and the negative phase of the PDO. For the first time in over a century, in spring 2011 this combination of natural forcing did not result in a dry winter; instead, great snow accumulations ended the early twenty-first century drought.[viii] The influence of climate change on the stability of teleconnections[ix] and the reliability of traditional seasonal climate forecasts should be investigated in future research.

Flooding is a result of complex interactions between the type and characteristics of moisture delivery, water catchment attributes, and land surface features. In a broad sense, the common mechanisms of moisture delivery—heavy winter rainfall, runoff, heavy winter snow followed by spring melt, rain-on-snow events, and summer convection connected with the North American monsoon system—operate in conjunction with temperature regimes and catchment and land surface features. The key to flooding outcomes are the *processes* that deliver moisture to this region: an intricate choreography of large-scale, ocean-atmospheric climate drivers and orography (the physical geography of mountains).[x]

Significant changes to flood risks during the twentieth century were observed over the entire Western United States as a result of general warming (Hamlet and Lettenmaier 2007). Winter temperature changes modify the precipitation patterns; most significantly, higher temperatures increase winter rainfall at the expense of snowfall, thus reducing spring flooding while potentially increasing winter flooding. Warm and cold phases of ENSO and PDO also strongly modulate flooding risk. For example, cold ENSO and PDO phases reduce overall precipitation in the Colorado River Basin, thereby reducing flood risk. These insights provide a template for flood-risk changes under a warmer climate in the twenty-first century, assuming climate change does not affect the nature or stability of the teleconnections climate forecasters have come to trust. This is an assumption that needs to be verified.

Floods that cause severe property damage and loss of life in populous regions are predominantly caused by severe precipitation events on already saturated soils. As discussed above, the Southwest is likely to experience increased flood risk from short-duration extreme atmospheric river precipitation events. The future role of rain-on-snow events in twenty-first-century flood regimes remains highly uncertain (Dettinger et al. 2009).

The ability of climate models to reproduce extreme high-frequency precipitation is a key uncertainty in projections. Dynamical models typically overestimate the frequency of precipitation and underestimate precipitation intensity (e.g. Gershunov et al. 2000). Given the known modeling precipitation biases, how certain are projected trends in precipitation extremes?

Behavior of co-occurring high-impact extremes such as drought and heat waves is a key uncertainty that has not so far been adequately addressed. It is likely that soil moisture anomalies predetermine a region's capacity for extreme heat waves, while heat waves, in turn, deplete soil moisture. Decadal drought cycles can therefore modify the clearly projected trends in heat-wave activity. Research on the interactions between drought cycles and heat wave activity is needed to understand possible decadal variation of heat waves in a warming Southwest projected to experience deeper and longer droughts (Cayan et al. 2010).

Coastal climate is characterized by persistent low-level clouds in summer, to which coastal ecosystems and society are adapted. This "marine layer" responds to a host of natural weather and climate influences on global and local scales. The marine layer is particularly sensitive to inland temperatures and can respond to heat waves in different ways depending on regional to large-scale atmospheric circulation, coastal upwelling of cold nutrient-rich ocean water, and the state of the PDO. It typically protects the highly populated and sparsely air-conditioned coast from heat waves, but its absence during a heat wave (such as in July 2006 along the California coast) can severely impact public health (see Chapter 15), agriculture (Chapter 11), and the energy sector (Chapter 12). Marine-layer dynamics are not well understood or modeled. Future behavior of the marine layer in general and specifically in conjunction with extreme heat is unknown.

The Southwest is demarcated by an international border that transforms the region but fails to confine the impacts of extreme weather and climate. With its core region in northwestern Mexico, the North American monsoon is an excellent example of a climate phenomenon straddling both sides of the border. Although the region experiences the same extremes on either side of the U.S.-Mexico border, the impacts of these extremes—and the ability to observe and mitigate them—are not equally shared because of socioeconomic and sociopolitical disparities (see also Chapter 16 for an extensive discussion of climate change effects and adaptation in the U.S.-Mexico border region).

Irreversible changes and tipping points

In addition to these and numerous other uncertainties, the possibilities for abrupt or irreversible changes and tipping points exist, particularly in the impacts of climate change on biological and social systems. These issues, however, are highly speculative and uncertain and we only briefly list a few considerations here.

- Warmer winters and drought can and have led to bark beetle infestations that threaten the pine forests and influence wildfire risk (see Chapter 8, Section 8.4).
- The Southwest appears prone to abrupt shifts in climate regimes as evidenced by the paleorecord (e.g., historic megadroughts, see Chapter 5) that, coupled with enhanced heat wave activity, could lead to irreversible impacts on ecology as well as human adaptation, and may reduce the productive capacity of resources such as soils and rangelands.

- Impacts that promote devastating wildfires can result in irreversible land-cover and ecological changes.
- A declining snowpack has the capacity to irreversibly change the hydrologic regime.
- Change in the atmospheric vertical temperature profile can influence atmospheric stability and precipitation.
- A shift in the jet stream, although gradual, could have irreversible consequences on human time scales.
- The massive changes in the Arctic may be impacting the Southwest in ways currently unknown.
- Asian dust and aerosols may have a lasting influence on precipitation.

These and other uncertainties, coupled with the region's unique diversity and compounding vulnerabilities, create a highly volatile landscape for climate to write its story upon, employing an evolving lexicon and extreme punctuation marks.

References

Arriaga-Ramírez, S., and T. Cavazos. 2010. Regional trends of daily precipitation indices in northwest Mexico and southwest United States. *Journal of Geophysical Research* 115: D14111, doi:10.1029/2009JD013248.

Bao, J. W., S. A. Michelson, P. J. Neiman, F. M. Ralph, and J. M. Wilczak. 2006. Interpretation of enhanced integrated water vapor bands associated with extratropical cyclones: Their formation and connection to tropical moisture. *Monthly Weather Review* 134:1063–1080.

Barlow, M. 2011. Influence of hurricane-related activity on North American extreme precipitation. *Geophysical Research Letters* 38: L04705, doi:10.1029/2010GL046258.

Cavazos, T., and S. Arriaga-Ramírez. 2012. Downscaled climate change scenarios for Baja California and the North American monsoon during the 21st century. *Journal of Climate* 25:5904–5915.

Cavazos, T., C. Turrent, and D. P. Lettenmaier. 2008. Extreme precipitation trends associated with tropical cyclones in the core of the North American Monsoon. *Geophysical Research Letters* 35: L21703, doi:10.1029/2008GL035832.

Cayan, D. R., T. Das, D. W. Pierce, T. P. Barnett, M. Tyree, and A. Gershunov. 2010. Future dryness in the southwest US and the hydrology of the early 21st century drought. *Proceedings of the National Academy of Sciences* 107:21271–21276, doi:10.1073/pnas.0912391107.

Cayan, D. R., K. T. Redmond, and L. G. Riddle. 1999. ENSO and hydrologic extremes in the western United States. *Journal of Climate* 12:2881–2893.

Chou, C., and C-W. Lan. 2011. Changes in the annual range of precipitation under global warming. *Journal of Climate* 25:222–235.

Conil, S., and A. Hall. 2006. Local regimes of atmospheric variability: A case study of southern California. *Journal of Climate* 19:4308–4325.

Cook, E. R., R. Seager, R. R. Heim, R. S. Vose, C. Herweijer, and C. Woodhouse. 2009. Megadroughts in North America: Placing IPCC projections of hydroclimatic change in a long-term paleoclimate context. *Journal of Quaternary Science* 25:48–61, doi: 10.1002/jqs.1303. http://www.ldeo.columbia.edu/res/div/ocp/pub/cook/2009_Cook_IPCC_paleo-drought.pdf

Cook, E. R., C. Woodhouse, C. M. Eakin, D. M. Meko, and D. W. Stahle. 2004. Long-term aridity changes in the western United States. *Science* 306:1015–1018.

Das, T., M. D. Dettinger, D. R. Cayan, and H. G. Hidalgo. 2011. Potential increase in floods in California's Sierra Nevada under future climate projections. *Climatic Change* 109 (Suppl. 1): S71–S94, doi:10.1007/s10584-011-0298-z.

Dettinger, M. D. 2011. Climate change, atmospheric rivers and floods in California – A multi-model analysis of storm frequency and magnitude changes. *Journal of the American Water Resources Association* 47:514–523.

Dettinger, M. D., D. R. Cayan, H. F. Diaz, and D. M. Meko. 1998. North–south precipitation patterns in western North America on interannual-to-decadal timescales. *Journal of Climate* 11:3095–3111.

Dettinger, M. D., H. Hidalgo, T. Das, D. Cayan, and N. Knowles. 2009. *Projections of potential flood regime changes in California.* Final Paper CEC-500-2009-050. Sacramento: California Climate Change Center. http://www.energy.ca.gov/2009publications/CEC-500-2009-050/CEC-500-2009-050-F.PDF.

Dettinger, M. D., F. M. Ralph, T. Das, P. J. Neiman, and D. Cayan. 2011. Atmospheric rivers, floods, and the water resources of California. *Water* 3:455-478. http://www.mdpi.com/2073-4441/3/2/445/.

Dettinger, M. D., F. M. Ralph, M. Hughes, T. Das, P. Neiman, D. Cox, G. Estes, et al. 2011. Design and quantification of an extreme winter storm scenario for emergency preparedness and planning exercises in California. *Natural Hazards* 60:1085–1111, doi:10.1007/s11069-011-9894-5.

Diffenbaugh, N. S., and M. Ashfaq. 2010. Intensification of hot extremes in the United States. *Geophysical Research Letters* 37: L15701, doi:10.1029/2010GL043888.

Dominguez, F., E. Rivera, D. P. Lettenmaier, and C.L. Castro. 2012. Changes in winter precipitation extremes for the western United States under a warmer climate as simulated by regional climate models. *Geophysical Research Letters* 39: L05803, doi: 10.1029/2011GL050762.

Favre, A., and A. Gershunov. 2009. North Pacific cyclonic and anticyclonic transients in a global warming context: Possible consequences for western North American daily precipitation and temperature extremes. *Climate Dynamics* 32:969–987.

Field, C. B., V. Barros, T. Stocker, and Q. Dahe, eds. 2011. *Managing the risks of extreme events and disasters to advance climate change adaptation: Special report of the Intergovernmental Panel on Climate Change.* Cambridge: Cambridge University Press. http://ipcc-wg2.gov/SREX/report/.

Gershunov, A., and T. P. Barnett. 1998. ENSO influence on intraseasonal extreme rainfall and temperature frequencies in the contiguous United States: Observations and model results. *Journal of Climate* 11:1575–1586.

Gershunov, A., T. Barnett, D. Cayan, T. Tubbs, and L Goddard. 2000. Predicting and downscaling ENSO impacts on intraseasonal precipitation statistics in California: The 1997–1998 event. *Journal of Hydrometeorology* 1:201–209.

Gershunov, A., and D. R. Cayan. 2003. Heavy daily precipitation frequency over the contiguous United States: Sources of climate variability and seasonal predictability. *Journal of Climate* 16:2752–2765.

Gershunov, A., D. Cayan, and S. Iacobellis. 2009. The great 2006 heat wave over California and Nevada: Signal of an increasing trend. *Journal of Climate* 22:6181–6203.

Gershunov, A., and K. Guirguis. 2012. California heat waves in the present and future. *Geophysical Research Letters* 39: L18710, doi:10.1029/2012/GL052979.

Groisman, P. V., R. W. Knight, T. R. Karl, D. R. Easterling, B. Sun, and J. H. Lawrimore. 2004. Contemporary changes of the hydrological cycle over the contiguous United States: Trends derived from in situ observations. *Journal of Hydrometeorology* 5:64–85, doi:10.1175/1525–7541.

Groisman, P. Y., R. W. Knight, D. R. Easterling, T. R. Karl, G. C. Hegerl, and V. N. Razuvaev. 2005. Trends in intense precipitation in the climate record. *Journal of Climate* 18:1326–1350.

Guan, B., N. P. Molotch, D. E. Waliser, E. J. Fetzer, and P. J. Neiman. 2010. Extreme snowfall events linked to atmospheric rivers and surface air temperature via satellite measurements. *Geophysical Research Letters* 37: L20401, doi:10.1029/2010GL044696.

Guirguis, K., A. Gershunov, R. Schwartz and S. Bennett. 2011. Recent warm and cold daily winter temperature extremes in the Northern Hemisphere. *Geophysical Research Letters* 38: L17701, doi:10.1029/2011GL048762.

Hamlet, A. F., and D. P. Lettenmaier. 2007. Effects of 20th century warming and climate variability on flood risk in the western U.S. *Water Resources Research* 43: W06427, doi:10.1029/2006WR005099.

Hirschboeck, K. K. 1991. Climate and floods. In *National water summary 1988-89: Hydrologic events and floods and droughts*, comp. R. W. Paulson, E. B. Chase, R. S. Roberts, and D. W. Moody, 99–104. U.S. Geological Survey Water-Supply Paper 2375. Washington, DC: U.S. Government Printing Office.

Hughes, M., and A. Hall. 2010. Local and synoptic mechanisms causing Southern California's Santa Ana winds. *Climate Dynamics* 34:847–857.

Hughes M., A. Hall, and J. Kim. 2011. Human-induced changes in wind, temperature and relative humidity during Santa Ana events. *Climatic Change* 109 (Suppl. 1): S119–S132, doi:10.1007/s10584-011-0300-9.

Intergovernmental Panel on Climate Change (IPCC). 2007. Summary for policymakers. In *Climate change 2007: The physical science basis. Contribution of Working Group I to the Fourth Assessment Report of the Intergovernmental Panel on Climate Change*, ed. S. Solomon, D. Qin, M. Manning, Z. Chen, M. Marquis, K. B. Averyt, M. Tignor, and H. L. Miller. Cambridge: Cambridge University Press.

Kharin, V. V., F. W. Zwiers, X. Zhang, and G. Hegerl. 2007. Changes in temperature and precipitation extremes in the IPCC ensemble of global coupled model simulations. *Journal of Climate* 20:1419–1444.

Kodra, E., K. Steinhauser, and A. R. Ganguly. 2011. Persisting cold extremes under 21st-century warming scenarios. *Geophysical Research Letters* 38: L08705. doi:10.1029/2011GL047103.

Lu, J., C. Deser, and T. Reichler. 2009. The cause for the widening of the tropical belt since 1958. *Geophysical Research Letters* 36: L03803, doi: 10.1029/GL036076.

Lu, J., G. A. Vecchi, and T. Reichler. 2007. Expansion of the Hadley cell under global warming. *Geophysical Research Letters* 34: L06805, doi:10.1029/2006GL028443.

Mahoney, K., M. A. Alexander, G. Thompson, J. J. Barsugli, and J. D. Scott. 2012. Changes in hail and flood risk in high-resolution simulations over the Colorado Mountains. *Nature Climate Change* 2:125–131, doi:10.1038/nclimate1344.

Mastrandrea, M. D., C. Tebaldi, C. W. Snyder, and S. H. Schneider. 2011. Current and future impacts of extreme events in California. *Climatic Change* 109 (Suppl. 1): S43–S70, doi 10.1007/s10584-011-0311-6.

Maurer, E. P., and H. G. Hidalgo. 2008. Utility of daily vs. monthly large-scale climate data: An intercomparison of two statistical downscaling methods. *Hydrology and Earth System Sciences* 12:551–563.

MacDonald, G. M., D. W. Stahle, J. Villanueva Diaz, N. Beer, S. J. Busby, J. Cerano-Paredes, J. E. Cole, et al. 2008. Climate warming and twenty-first century drought in southwestern North America. *Eos Transactions AGU* 89:82.

Meehl, G. A., T. Karl, D. R. Easterling, S. Changnon, R. Pielke, D. Changnon, J. Evans, et al. 2000. An introduction to trends in extreme weather and climate events: Observations, socioeconomic impacts, terrestrial ecological impacts, and model projections. *Bulletin of the American Meteorological Society* 81:413–416.

Moritz, M., T. Moody, M. Krawchuk, M. Hughes, and A. Hall. 2010. Spatial variation in extreme winds predicts large wildfire locations in chaparral ecosystems. *Geophysical Research Letters* 37: L04801.

Nakićenović, N., and R. Swart, eds. 2000. *Special report on emissions scenarios: A special report of Working Group III of the Intergovernmental Panel on Climate Change.* Cambridge: Cambridge University Press.

Pierce, D. W., T. Das, D. R. Cayan, E. P. Maurer, N. Miller, Y. Bao, M. Kanamitsu, et al. 2012. Probabilistic estimates of future changes in California temperature and precipitation using statistical and dynamical downscaling. *Climate Dynamics* published online, doi:10.1007/s00382-012-1337-9.

Rajagopalan, B., E. Cook, U. Lall, and B. Ray. 2000. Spatiotemporal variability of ENSO and SST teleconnections to summer drought over the United States during the twentieth century. *Journal of Climate* 13:4244–4255.

Ralph, F. M., and M. D. Dettinger. 2011. Storms, floods and the science of atmospheric rivers. *Eos Transactions AGU* 92:265–266.

Ralph, F. M., M. D. Dettinger, A. White, D. Reynolds, D. Cayan, T. Schneider, R. Cifelli, et al. 2011. A vision of future observations for western US extreme precipitation events and flooding: Monitoring, prediction and climate. Report to the Western States Water Council, Idaho Falls.

Ralph, F. M., P. J. Neiman, G. A. Wick, S. I. Gutman, M. D. Dettinger, D. R. Cayan, and A. B. White. 2006. Flooding on California's Russian River: Role of atmospheric rivers. *Geophysical Research Letters* 33: L13801, doi:10.1029/2006GL026689.

Salas-Mélia, D., F. Chauvin, M. Déqué, H. Douville, J. F. Guérémy, P. Marquet, S. Planton, J. F. Royer, and S. Tyteca. 2005. Description and validation of the CNRM-CM3 global coupled model. CNRM Working Note 53. Toulouse, France: Centre National de Recherches Météorologiques. http://www.cnrm.meteo.fr/scenario2004/paper_cm3.pdf.

Sheppard, P. R., A. C. Comrie, G. D. Packin, K. Angersbach, and M. K. Hughes. 2002. The climate of the US Southwest. *Climate Research* 21:219–238.

Tebaldi, C., K. Hayhoe, and J. M. Arblaster. 2006. Going to the extremes: An intercomparison of model-simulated historical and future changes in extreme events. *Climatic Change* 79:185–211.

Trenberth, K. E., P. D. Jones, P. Ambenje, R. Bojariu, D. R. Easterling, A. K. Tank, D. Parker, et al. 2007. Observations: Surface and atmospheric climate change. In *Climate change 2007: The physical science basis. Contribution of Working Group I to the Fourth Assessment Report of the Intergovernmental Panel on Climate Change*, ed. S. Solomon, D. Qin, M. Manning, Z. Chen, M. Marquis, K.B. Averyt, M. Tignor and H.L. Miller, 235–336. Cambridge: Cambridge University Press.

Wang, J., and X. Zhang. 2008. Downscaling and projection of winter extreme daily precipitation over North America. *Journal of Climate* 21:923–937.

Westerling, A. L., D. R. Cayan, T. J. Brown, B. L. Hall, and L. G. Riddle. 2004. Climate, Santa Ana winds and autumn wildfires in Southern California. *Eos Transactions AGU* 85:289, 296.

Westerling, A. L., A. Gershunov, T. Brown, D. Cayan, and M. Dettinger. 2003. Climate and wildfire in the western United States. *Bulletin of the American Meteorological Society* 84:595–604.

Endnotes

i Downscaled using the Bias Corrected Constructed Analogue (BCCA) statistical downscaling method of Maurer and Hidalgo (2008).

ii Another measure of heat waves is discussed in Chapter 5, Section 5.4.

iii Events with a probability of occurrence of 1% in any given year or 100% in 100 years.

iv Another measure of cold waves is discussed in Chapter 5, Section 5.4.

v They detected tropical cyclone-related trends in the core NAM region. However, because this is in Mexico and since Barlow (2011) showed that hurricane-related activity contributes only 1% of Southwestern precipitation extremes, we did not consider hurricanes and tropical storms in this chapter.

vi Mahoney et al. (2012) used a multi-tiered downscaling approach where first a GCM (GFDL) was downscaled to 31 mile (50-km) grid as a part of North American Regional Climate Change Assessment Program (NARCCAP), using the high-emissions scenario (Nakicenovic and Swart 2000). Then extreme precipitation events in NARCCAP were further downscaled using the Weather Research and Forecasting (WRF) model. High-resolution WRF simulations (up to 0.8 mile [1.3 km] horizontal grid), initialized using composite future and past conditions were produced.

vii Potential sensitivities of model microphysical parameterization (especially hail size distribution to melting hail) merits further investigation. Nevertheless, although based on one GCM projection only, Mahoney and colleagues (2012) claim their results are robust due to consistency with different initialized climate projections, and different WRF methodologies (event and composite).

viii In winter 2011–2012, at the time of this writing, similar forcings are producing more than expected dryness, however.

ix *Teleconnections* are persistent large-scale patterns of atmospheric circulation that reflect changes in the jet stream or atmospheric waves over very large areas. They can derive from internal atmospheric dynamics or from changes in sea-surface temperatures and convection, as in the tropical Pacific ENSO cycle resulting in, for example, anomalous precipitation patterns in the Southwest.

x These are described in detail in Hirschboeck (1991) and Sheppard et al. (2002).

Chapter 8

Natural Ecosystems

COORDINATING LEAD AUTHOR

Erica Fleishman (University of California, Davis)

LEAD AUTHORS

Jayne Belnap (U.S. Geological Survey), Neil Cobb (Northern Arizona University), Carolyn A.F. Enquist (USA National Phenology Network/The Wildlife Society), Karl Ford (Bureau of Land Management), Glen MacDonald (University of California, Los Angeles), Mike Pellant (Bureau of Land Management), Tania Schoennagel (University of Colorado), Lara M. Schmit (Northern Arizona University), Mark Schwartz (University of California, Davis), Suzanne van Drunick (University of Colorado), Anthony LeRoy Westerling (University of California, Merced)

CONTRIBUTING AUTHORS

Alisa Keyser (University of California, Merced), Ryan Lucas (University of California, Merced)

EXPERT REVIEW EDITOR

John Sabo (Arizona State University)

Executive Summary

Existing relations among land cover, species distributions, ecosystem processes (such as the flow of water and decomposition of organic matter), and human land use are the basis for projecting ranges of ecological responses to different scenarios of climate change.

Chapter citation: Fleishman, E., J. Belnap, N. Cobb, C. A. F. Enquist, K. Ford, G. MacDonald, M. Pellant, T. Schoennagel, L. M. Schmit, M. Schwartz, S. van Drunick, A. L. Westerling, A. Keyser, and R. Lucas. 2013. "Natural Ecosystems." In *Assessment of Climate Change in the Southwest United States: A Report Prepared for the National Climate Assessment*, edited by G. Garfin, A. Jardine, R. Merideth, M. Black, and S. LeRoy, 148–167. A report by the Southwest Climate Alliance. Washington, DC: Island Press.

However, because such relations evolve, projections based on current relations are likely to be inaccurate. Additionally, changes in climate, land use, species distributions, and disturbance regimes (such as fire and outbreaks of disease) will affect the ability of ecosystems to provide habitat for animals and plants that society values, to maintain ecosystem processes, and to serve as reservoirs of carbon. There is reliable evidence for the following key findings, which are true of the Southwest and many other regions.

- Observed changes in climate are associated strongly with some changes in geographic distributions of species that have been observed since the 1970s. The extent of these observed changes in geographic distribution varies considerably among species. (high confidence)
- Observed changes in climate are associated strongly with some observed changes in the timing of seasonal events in the life cycles of species. The magnitude of these changes in timing of seasonal events varies considerably among species. (high confidence)
- Some disturbance processes that result in mortality or decreases in the viability of native plants are associated strongly with observed changes in climate. Among those disturbances are wildfires and outbreaks of forest pests and pathogens. Mortality of some species of plants and of plants in some regions also is associated directly with higher temperatures and decreases in precipitation. (high confidence)
- The probability that a species will occupy and reproduce in a specified geographic area for a selected number of years may increase if the physiology or behavior of individuals of the species is able to change in response to environmental change. These changes, which often have a genetic basis, may increase probabilities of persistence (the likelihood that a species will occupy and reproduce in a certain geographic area for a certain number of years) beyond what might be expected on the basis of current associations between species and climatic variables. (high confidence)

8.1 Introduction: Climate, Climate Change, and Ecosystems of the Southwest

The Southwest's high species richness of diverse groups of plants and animals (Kier et al. 2009) in part reflects the considerable geographic and seasonal variation in climate within the region (see Figure 4.1). For example, the difference in absolute minimum and maximum temperatures at a given location within a year can be as much as 113°F (45°C) in the interior of the Southwest and as little as 59°F (15°C) near the coast. High elevations in the Sierra Nevada and Rocky Mountains receive 39 inches to 79 inches (100 cm to 200 cm) of precipitation annually, whereas low elevations receive less than 4 inches (10 cm).

Climatic variation in the Southwest, as in any region, also is reflected by variations in land cover and land use (see Chapter 3). Within the Southwest, the U.S. Gap Analysis Project (USGS 2004) mapped 209 ecological systems,[i] which are defined as groups of plant community types that tend to co-occur within landscapes with similar ecological processes, geology, soils, or ranges of environmental attributes such as elevation and precipitation (Comer et al. 2003), and twenty additional classes of land that has been disturbed or modified by humans.

Climatic variables such as actual evapotranspiration (the amount of water delivered to the atmosphere by evaporation and plant transpiration), soil water deficit (the amount of available water removed from the soil within the active root depth of plants), average temperatures of the coldest and warmest months, and different measures of precipitation are highly correlated with the geographic distributions of individual species and ecological systems (e.g., Rehfeldt et al. 2006; Parra and Monahan 2008; Franklin et al. 2009). Increasing temperatures and aridity (MacDonald 2010) and earlier snowmelt and peak streamflow (Bonfils et al. 2008) also have been linked to changes in the geographic distributions of species (e.g., Kelly and Goulden 2008; Moritz et al. 2008; Forister et al. 2010). Changes in climate have been associated with changes in phenology (the timing of seasonal events in the life cycle of plants and animals) (e.g., Bradley and Mustard 2008; Kimball et al. 2009; Miller-Rushing and Inouye 2009) and changes in the frequency, extent, duration, and severity of fires and outbreaks of forest pathogens (e.g., Westerling et al. 2006; Bentz 2008).

Projections suggest that by 2100, average annual temperatures in the Southwest may increase by 2°F to 9°F (1°C to 5°C), which will increase rates of evaporation and transpiration of surface water and soil water to the atmosphere. Annual runoff across much of the region is projected to decrease 10% to 40% by 2100, and the severity and length of droughts and soil-moisture depletion are expected to increase substantially (IPCC 2007; Cayan et al. 2010; Seager and Vecchi 2010; see also Chapters 6 and 7). Extremes in high temperatures are anticipated to increase, whereas extreme cold events are expected to become less severe and shorter in duration. Changes in temperature and water deficits in soils and plants are projected to be greatest in interior regions and least near the Pacific Coast (Pan et al. 2010).

Computer models that associate climate with the distribution of species suggest that by 2100, the locations occupied by individual species may change substantially in response to projected changes in temperature and precipitation (Lenihan et al. 2003; Archer and Predick 2008; Loarie et al. 2008). For example, increases in water temperature in rivers and streams may cause mortality of some native fish species and some of the invertebrates on which they prey, and increase the likelihood that non-native salmonid fishes (which spawn in freshwater but may spend a portion of their life in the ocean) will colonize these rivers and streams. Abundances of some native fishes may decrease and the probability of breeding among native and non-native fishes may increase. For example, the amount of habitat for a native cutthroat trout (*Oncorhynchus clarkia*) is projected to decrease as much as 58% in response to increases in temperature and competition with other species (Wenger et al. 2011).

Existing plant and animal species or their recent ancestors have persisted through substantial climatic changes. However, the anticipated rate of widespread climate change from 2010 to 2100 generally exceeds that documented in paleoenvironmental records from the recent geologic past (around 2 million years). Additionally, human land uses such as urbanization and agriculture have reduced the quantity and quality of habitat for some species and created barriers to dispersal of some species (Willis and MacDonald 2011). Patterns of human settlement and other land uses vary considerably across the Southwest. For example, human population density across most of the Great Basin is relatively low, and there is comparatively little human infrastructure that might impede dispersal of native species. By contrast, coastal Southern California is densely populated by humans and the little remaining natural land cover is highly fragmented by human activity.

Despite the clear relation of the distributions of some species to climate, the relation between changes in climate and recent changes in the geographic distribution of species is highly uncertain. Additionally, there is considerable uncertainty about how species and the communities and ecosystems they form will respond to projected changes in climate. Some shifts in species' ranges observed in the late 1900s and early 2000s likely reflect not only changes in climate but changes in land use (e.g., Thorne, Morgan, and Kennedy 2008; Forister et al. 2010). For example, local extinction and changes in the distribution of the butterfly species *Euphydryas editha* were represented as a response to climate change (Parmesan 1996). But that study did not account for geographic variation in diet of the species. Nor did it account for geographic differences in the extent of non-native plant species or urbanization, both of which affect the probability of local extinctions and changes in the butterfly's distribution (Fleishman and Murphy 2012).

Most climate-based projections of species' distributions are based on their current climatic niches (e.g., Rehfeldt et al. 2006; Parra and Monahan 2008; Franklin et al. 2009), which are assumed to be unchanging over time and uniform in space. These projections may overestimate the size of species' ranges and consequently overestimate the probability of persistence (occupancy and reproduction at a level that will not lead to local extinction) of populations that are adapted to a comparatively narrow range of climatic conditions or resources (Reed, Schindler, and Waples 2011). The projections also may underestimate ranges and probabilities of persistence of species that can adapt to changes in the living and nonliving attributes of their environment (Visser 2008; Chevin, Lande, and Mace 2010; Nicotra and Davidson 2010). Both natural environmental changes and management interventions (even those intended to mimic natural processes) may accelerate the process of evolution (Hellman and Pfrender 2011). Recent and prehistoric (Willis and MacDonald 2011) data on terrestrial and aquatic vertebrates, invertebrates, and plants demonstrate these responses may be rapid, on the order of years or decades. Moreover, temperatures and the amount of precipitation projected by 2100 may fall outside the current ranges for the region (Williams and Jackson 2007). When values of a variable used in building a predictive model (such as temperature) do not include the full range of values for which projections are being made, the uncertainty of the model's projections increases, and the accuracy of the model's projections may decrease.

In the following sections, we examine how some species, communities, and ecosystems of the Southwest may respond to changes in climate. These sections are not comprehensive treatments, but they illustrate potential responses across the Southwest. First, we explore how changes in climate may be reflected in changes in phenology of species (seasonal phenomena such as development of leaves, blooms of flowers, spawning of fish, and migrations of birds) and the resulting interactions among species. Second, we investigate how changes in precipitation and temperature may affect soils, vegetation, and carbon storage in arid regions. The response of plants and animals to increases in atmospheric concentrations of carbon dioxide and associated changes in climate also may affect the way non-native invasive species compete with native species and are distributed (Thuiller, Richardson, and Midgley 2007; Hellmann et al. 2008; Bradley et al. 2010). Third, we highlight potential changes in tree mortality and fire across the extensive forests and woodlands of the Southwest. Each of these examples highlights the uncertainty of projected ecological responses to changes in climate.

8.2 Phenology and Species Interactions

Variability in weather, climate, and hydrology largely drive phenology (Walther et al. 2002). The timing of these seasonal events in turn directly affects interactions among species and the environment (Parmesan and Yohe 2003; Cleland et al. 2007; IPCC 2007) and is likely to be a major force in shaping ecological responses to climate change. Interactions among species that shape the structure and function of ecosystems include competition, predation, consumption of plants by animals, parasitism, disease, and mutually beneficial relations (Yang and Rudolf 2010).

Organisms may adapt phenologically to environmental change through evolution or phenotypic plasticity (the ability of individuals to consciously or unconsciously increase their probability of survival and reproduction by responding to environmental cues). For example, earlier spring thaws can induce earlier opening of buds either through natural selection or through a direct physiological response of individual plants. However, recent environmental changes have led to both earlier and later timing of these phenological events and have exceeded the ability of some species to adapt to such changes. Differences in phenological responses of different species can disrupt interactions among species (Parmesan 2006; Both et al. 2009). The differences in phenological responses among interacting species in response to changing climate may increase the probability of changes in abundance, population growth rate, and local persistence of individual species (Parmesan 2007; Miller-Rushing et al. 2010; Thackeray et al. 2010). For example, in the Netherlands the peak abundance of caterpillars that feed on oak leaves has become earlier than the peak abundance of migratory birds that feed on the caterpillars, resulting in a decrease in abundance of the birds (Both et al. 2009). Species that are more capable of adapting to environmental change (such as many non-native invasive species and species with general food requirements) may have a higher probability of persisting as climate changes than species with more fixed phenotypes (such as many endemic species—species that occur only in a particular location—and species with restricted diets) (Møller, Rubolini, and Lehikoinen 2008; Willis et al. 2008; Kellermann et al. 2009). Knowing more about how the phenology of non-native invasive plants is affected by climate change may allow more effective timing of actions to eradicate these plants or minimize their spread (Marushia, Cadotte, and Holt 2010; Wolkovich and Cleland 2010).

Phenology and interactions among species in terrestrial systems

The average timing of developmental events of plants, such as bud formation and flowering, is occurring one day earlier per decade across the Northern Hemisphere and 1.5 days earlier per decade in western North America in correlation with increases in winter and spring temperatures (Schwartz, Ahas, and Aasa 2006; Ault et al. 2011). In the Southwestern United States, changes in the phenology of bird species corresponding to climate change have been documented for over a decade. These include earlier egg-laying by Mexican jays (*Aphelocoma ultramarina*) (Brown, Li, and Bhagabati 1999), earlier appearance of American robins (*Turdus migratorius*) at a given elevation (Inouye et al. 2000), and earlier arrival of migratory birds to their breeding range (MacMynowski et al. 2007). Earlier emergence of adult butterflies in some areas of the Southwest also has been attributed to climate change (Forister and Shapiro 2003).

Data from a high-elevation research station in the Rocky Mountains,[ii] where air temperatures are increasing, demonstrated that from 1976 through 2008, yellow-bellied marmots (*Marmota flaviventris*) weaned their young approximately 0.17 days earlier each year (Ozgul et al. 2010). Earlier emergence from hibernation (Inouye et al. 2000), giving birth earlier in the season, changes in weaning time, and extended duration of growing seasons were associated with larger animals at the start of hibernation and increases in abundance of the animals (Ozgul et al. 2010). These apparent responses to higher temperatures may be short-term, especially if long, dry summers become more frequent, and may decrease growth rates and increase mortality rates. In the same geographic area, higher temperatures and less precipitation have been associated with a change in flowering phenology across meadows. Blooming of some forbs is occurring earlier, which increases the probability of mortality from a late frost (Inouye 2008). Abundance of flowers in the middle of the growing season has decreased, which may reduce probabilities of persistence of insects that feed on and pollinate the flowers throughout the summer (Aldridge et al. 2011).

There is less evidence of changes in phenology in apparent response to climatic changes in the arid lowlands of the Southwest than in moister, higher-elevation regions such as the Rocky Mountains. Nevertheless, examination of twenty-six years of data on flowering phenology along an elevational gradient in the Catalina Mountains of south-central Arizona suggests the onset of summer flowering is strongly associated with the amount and timing of July precipitation (Crimmins, Crimmins, and Bertelsen 2011). In deserts, soil moisture can have a greater effect on phenology than does temperature (Kimball et al. 2009), and plants at higher elevations, which typically receive more precipitation than lower elevations, may have a greater probability of becoming moisture-stressed than those at lower elevations (Bradley and Mustard 2008; Crimmins, Crimmins, and Bertelsen 2011).

Interactions in freshwater systems

Documented changes in hydrology associated with increases in air temperature in the Southwest and throughout the western United States include earlier spring runoff and peak flows, increases in evapotranspiration, and decreases in summer flows (Stewart, Cayan, and Dettinger 2005; Knowles, Dettinger, and Cayan 2006; Painter et al. 2007). However, most research on how freshwater species respond to climate change has focused on physiological responses to temperature and flow rather than on interactions among species. Changes in frequency of flooding or changes in the seasonal pattern of high flows may change the timing of species interactions (Wenger et al. 2011). For example, changes in flooding and flow patterns can affect the timing of fish spawning, increase the probability that eggs will be scoured from gravel nests, wash away newly emerged fry, and change which fish species are present in streams where fall- and spring-spawning salmonids both live or where there may be a high probability of colonization by a given invasive species (Warren, Ernst, and Baldigo 2009).

Higher air temperatures also may lead to changes in food quantity for coldwater fishes. For example, metamorphosis of a mayfly *(Baetis bicaudatus)* that is common in high-elevation streams and is an important prey item is triggered by increased water temperature (Harper and Peckarsky 2006). Mayflies emerge when peak flows subside and protruding rocks become available for egg-laying. Mayflies emerging in years with relatively low streamflow were smaller on average than in years with higher streamflow, when emergence of adults was delayed and the period of feeding by larvae extended (Peckarsky, Encalada, and McIntosh 2011).

8.3 Southwestern Deserts

Changes in the magnitude, frequency, or timing of precipitation and increases in temperature and atmospheric concentrations of carbon dioxide likely will affect soil organisms, vegetation composition, and ecosystem processes in Southwestern deserts, which are defined here as areas with less than 10 inches (around 250 mm) of mean annual precipitation. Frequent but low-volume summer rains increase mortality of organisms in the soil crust that otherwise maintain soil fertility and stability (Belnap, Phillips, and Miller 2004; Reed et al. 2012). In Southwestern deserts—unlike in regions with more precipitation—low concentrations of soil carbon limit the abundance and activity of soil biota and thus their ability to retain nutrients (Kaye et al. 2011). Therefore, nutrients in surface soils are easily absorbed by plants in wet years, especially if the preceding years were dry and nutrient-rich dust accumulated on the soil surface (Hall et al. 2011; Thomey et al. 2011). More plant growth results in higher nutrient retention by plants in wet years, but low retention in dry years, increasing the probability that nutrients will be lost from the ecosystem (Evans et al. 2001; Hall et al. 2011). These phenomena are especially pronounced in areas dominated by invasive non-native annual grasses because in wet years the amount of vegetation in these areas generally is higher than in communities of native perennial plants. Thus, highly variable precipitation can result in large fluctuations of nutrients in soils and plants. In addition, changes in the species of plants that are present in a given location affect soil biota and nutrient cycling (Belnap and Phillips 2001).

Precipitation patterns affect which species of plants are present in a given location. In some desert shrubs, primary production—the amount of energy from the sun that is converted to chemical energy (organic compounds) by an ecosystem's photosynthetic plants during a given time period—is positively correlated with winter or summer precipitation, but not autumn or spring precipitation (Schwinning et al. 2002). For example, long-term primary production in creosote bush (*Larrea tridentata),* a dominant shrub in hot Southwestern deserts (D'Odorico et al. 2010), is thought to increase as the number of years with relatively abundant summer rainfall increases. Shrubs with green stems, such as Mormon tea (*Ephedra*), can photosynthesize in winter and thus take advantage of high soil moisture. In some regions, native grasses require multiple consecutive wet years to persist. The probability of multiple consecutive wet years is projected to decrease as climate changes (Peters et al. 2011). Primary production by annual plants, by contrast, can increase quickly in wet years, but because germination of these plants is limited in dry years, their abundance and distribution is expected to fluctuate widely in the future. High annual biomass can increase the probability of fires. Fires often result in mortality of the perennial plants, further changing which species of plants are present (Brooks and Pyke 2001). The Southwest currently has a pronounced cycle of fire in regions dominated by invasive non-native grasses (D'Antonio and Vitousek 1992; Brooks et al. 2004), and climate change is likely to increase the number and intensity of such fires (Abatzoglou and Kolden 2011).

How plants respond to increasing temperatures and decreasing precipitation is expected to vary among plant species in Southwestern deserts as a function of both direct thermal effects and associated decreases in soil moisture (Munson et al. 2012). Photosynthetic pathway, or type of metabolism, can affect the response of plants to temperature. For example, plants with crassulacean acid metabolism (CAM; plants that store carbon dioxide at night and thus minimize water loss during the day) use water more efficiently than plants with C4 metabolism, which lose little water during the day. Both CAM and C4 plants use water more efficiently than plants with C3 metabolism, which grow and lose water during the day (Collins et al. 2010; Morgan et al. 2011). Thus, increases in temperature and concomitant decreases in soil moisture are expected to increase the competitive advantage of CAM plants relative to C4 plants and of both CAM and C4 plants relative to C3 plants. Changes in biomass of both CAM and C4 plants in response to increases in temperature may be minimal, but there are exceptions (Munson, Belnap, and Okin 2011; Throop et al. 2012). Season of activity also may affect how plants respond to changes in temperature: plants that are dormant in winter (e.g., saltbush [*Atriplex*]) may lose biomass during relatively high-temperature years, whereas those that are active year-round (such as blackbrush [*Coleogyne*] and juniper [*Juniperus*]) may increase in biomass during those years (Munson, Belnap, and Okin 2011).

In contrast to predictions that increases in temperature will negatively affect C3 plants, higher nighttime temperatures increased establishment and survival of creosote bush (*Larrea tridentata*), a perennial C3 shrub (D'Odorico et al. 2010). The presence of the shrub raised ground temperatures, which was associated with increases in the plant's abundance. Thus, as shrubs expand throughout the Southwest, regional temperatures or temperatures in microhabitats of some species may increase to a greater extent than projected by climate models.

There is no clear evidence that non-native invasive plants will be more likely to survive and reproduce than native plants as climate changes, given that responses to climate change will vary by species. However, if changes in climate increase the probability of non-native plant invasion, then their generally high reproductive capacity and dispersal rates, rapid growth, and ability to adapt to short-term environmental variability may increase the probability they will become established and persist, in some cases quite rapidly (Pysek and Richardson 2007; Willis et al. 2010).

Increasing concentrations of atmospheric carbon dioxide may offset the effects of changes in other climatic variables, increasing the difficulty of accurately projecting responses to environmental change. For instance, although increases in temperature and decreases in soil moisture likely will benefit C4 plants more than C3 plants, increases in carbon dioxide likely will benefit C3 plants more than C4 plants (Morgan et al. 2011).

Increases in carbon dioxide also may increase the biomass of annual non-native grasses (Ziska, Reeves, and Blank 2005) and generally benefit invasive plants more than native plants (Bradley et al. 2010).

Changes in climate will affect how much carbon is contained in the vegetation and soils of deserts of the Southwest. The amount of above-ground plant biomass decreased as temperature increased and precipitation decreased in central New Mexico (Anderson-Teixeira et al. 2011). On the Colorado Plateau, drought was associated with a substantial decrease in photosynthetic production of organic compounds, with summer rains rarely resulting in net increase in biomass (Bowling et al. 2010). Spring uptake of carbon was associated with deep soil moisture, which required relatively high precipitation in the prior autumn and winter; projections suggest such precipitation is less likely to occur in the future. In more-arid grasslands of the warm deserts, establishment of non-native annual grasses can increase soil carbon due to increases in primary productivity relative to that in communities where non-natives are absent (Ziska, Reeves, and Blank 2005). Nevertheless, increases in soil carbon often are transient, and the conversion of sagebrush (*Artemisia* spp.) steppe to cheatgrass (*Bromus tectorum*) can result in long-term depletion of soil organic matter (Norton et al. 2004) and reduction of above-ground carbon sequestration (Bradley et al. 2006). In addition, the presence of non-native species generally increases fire frequency, leading to substantial declines in soil carbon and nutrients (Brooks and Pyke 2001).

8.4 Southwestern Forests

Temperature, precipitation, and pests and pathogens

Geographically widespread and rapid increases in rates of mortality of coniferous trees believed by scientists to result from drought and higher temperatures have been documented for old forests throughout the western United States (van Mantgem et al. 2009). Annual mortality throughout the region has at least doubled since 1995, with mortality rates increasing over time (van Mantgem et al. 2009). Mortality rates of all major genera of trees have increased, suggesting that relatively predictable changes in the proportion of species with different characteristics, such as life history traits (e.g., shade intolerance), size, forest stand density or forest fragmentation, are unlikely to be the primary cause of the mortality (van Mantgem et al. 2009).

Tree mortality in forests and woodlands from outbreaks of bark beetles and fire has been attributed to changes in climate, particularly higher temperatures and lower precipitation (Swetnam and Betancourt 1998; Breshears et al. 2005; Westerling et al. 2006; Allen et al. 2010). Williams and colleagues (2010) estimated that since 1980, levels of tree mortality have been higher and more spatially extensive than during the 90-year record, including those during a period of drought in the 1950s (Breshears et al. 2005).

At a number of sites across the Southwest, rapid and nearly complete mortality of pinyon pine (*Pinus edulis*), a dominant, widespread species, was attributed to drought accompanied by unusually high temperatures from 2000 to 2003. Mortality approaching 90% was documented for trees at high-elevation sites in Colorado and Arizona that are near the upper elevational limit of pinyon pine and where precipitation and water availability are relatively high compared to other locations where the species occurs (Breshears et al. 2005). Most of the mortality occurred in response to outbreaks of bark beetle (*Ips confusus*), which have been correlated with shifts in temperature and precipitation. For example, higher temperatures lead to water stress that can greatly increase the probability that pinyon pine will die in response to bark beetles (Bentz et al. 2010). Even droughts of relatively short duration may be sufficient to cause widespread die-off of pinyon pine if temperatures increase (Adams et al. 2009). Extensive tree mortality caused by bark beetles was estimated to have occurred across at least 12% of Southwestern forests and woodlands between 1997 and 2008 (Breshears et al. 2005; Williams et al. 2010). As of 2010, bark beetles were estimated to have affected more than twice the forest area burned by wildfires in Arizona and New Mexico in recent decades (USFS 2007; Williams et al. 2010).

As both summer and winter temperatures increase, beetles have erupted in high-elevation stands of white pine (*Pinus albicaulis*) in the Rocky Mountains where only intermittent attacks occurred during the past century (Raffa et al. 2008). Population sizes of two bark beetle species that have caused extensive mortality in Southwestern forests—the mountain pine beetle (*Dendroctonus ponderosae*) and spruce beetle (*Dendroctonus rufipennis*)—are expected to increase as temperature and the incidence of drought increases, albeit with considerable variability over time and geographic area (Bentz et al. 2010).

Rapid mortality of mature aspen (*Populus tremuloides*), known as sudden aspen decline, also has been reported throughout the Southwest and other regions within the United States and Canada (Frey et al. 2004; Fairweather, Geils, and Manthei 2007; Worrall et al. 2008). The decline is characterized by dieback within two to six years in apparently healthy stands of mature aspen and poor generation of suckers. Drought was identified as a major cause of recent diebacks (Hogg, Brandt, and Kochtubajda 2005; Fairweather, Geils, and Manthei 2007; Hogg, Brandt, and Michaelian 2008; Worrall et al. 2008; Rehfeldt, Ferguson, and Crookston 2009). Mortality resulted from various combinations of insects and pathogens, including *Cytospora* canker, usually caused by poplar borers (*Valsa sordida*) and bark beetles (Worrall et al. 2008). In documented cases of sudden aspen decline in both Colorado and Arizona, mortality generally decreased as elevation increased. Average mortality of aspen in dry sites below around 7,500 feet (2,300 meters) was greater than 95% from 2000 to 2007 (Fairweather, Geils, and Manthei 2007). The area with climate currently suitable for aspen growth and survival (that is, not accounting for potential evolutionary adaptation to climate change) is projected to decrease by 10% to 40% by 2030 (Rehfeldt, Ferguson, and Crookston 2009).

Fire

Climate affects both fuel availability and flammability, and the relative role of each in causing wildfires varies across ecosystem types (Littell et al. 2009; Westerling 2010). In dense forests that typically have infrequent but severe fires, fuel flammability is closely related to climate during the peak fire season. In comparison, moisture availability affects the amount of fine surface fuels in forests with more frequent, but lower-severity fires (Westerling et al. 2003; Swetnam and Betancourt 1998; Littell et al. 2009; Westerling 2010). However, regional incidence of forest wildfires is generally associated with drought—and higher temperatures and an earlier spring are expected to exacerbate drought and its effects on the extent of forest wildfires (Brown et al. 2008; Littell et al 2009; Westerling 2010; Schoennagel, Sherriff, and Veblen 2011).

The area of forest and woodland burned in the western United States by wildfires that actively were suppressed was more than five times larger during the period 1987–2003 than during 1970–1986, and was associated with increases in temperature and earlier spring snowmelt (Figure 8.1) (Westerling et al. 2006). This increase primarily was due to lightning-ignited wildfires. Forests and woodlands in the six Southwestern states accounted for a third of the increase in fires that exceeded 494 acres (200 hectares) in the western United States. The area burned in the Southwest increased more than 300% relative to the area burned during the 1970s and early 1980s (Figure 8.2, data updated from Westerling et al. 2006).

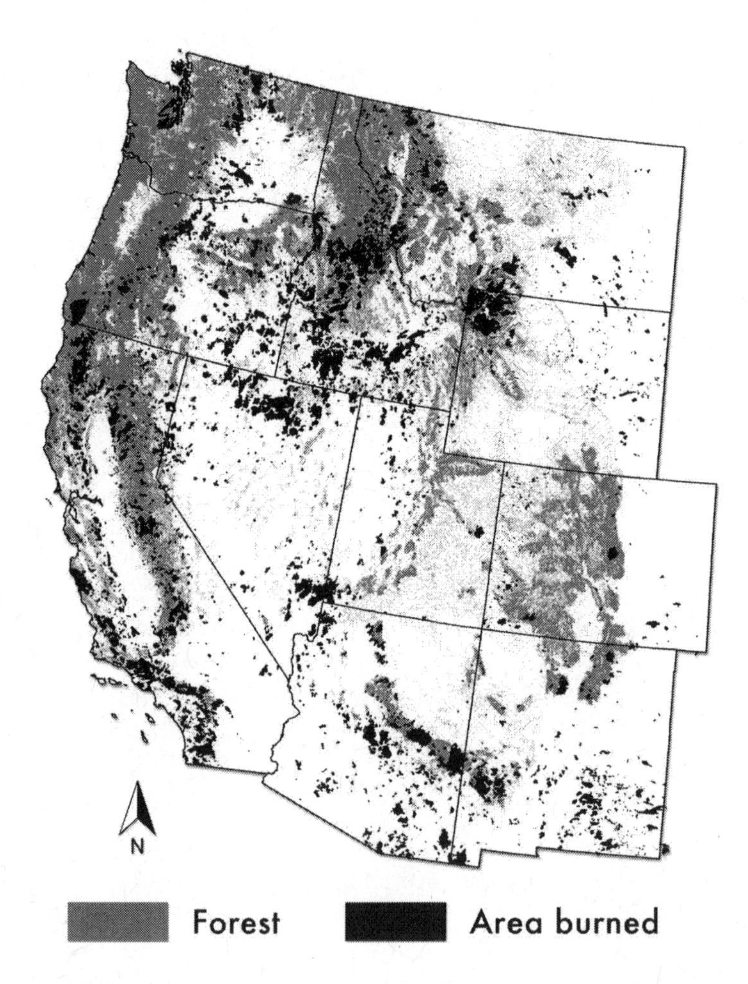

Figure 8.1 Areas of the western United States burned in large (> 1000 acres [400 ha]) fires, 1984–2011. Dark shading shows fires in areas classified as forest or woodland at 98-feet (30-meter) resolution by the LANDFIRE project (http://www.landfire.gov/). Fire data from 1984–2007 are from the Monitoring Trends in Burn Severity project (http://www.mtbs.gov/) and fire data from 2008–2011 are from the Geospatial Multi-Agency Coordination Group (http://www.geomac.gov/).

If fuels are available, the area of forest burned may increase substantially as temperature and evapotranspiration increase. The National Research Council (2011) projected that if temperature increases by 1.8°F (1°C), there will be a 312% increase in area burned in the Sierra Nevada, southern Cascades, and Coast Ranges of California; a 380%

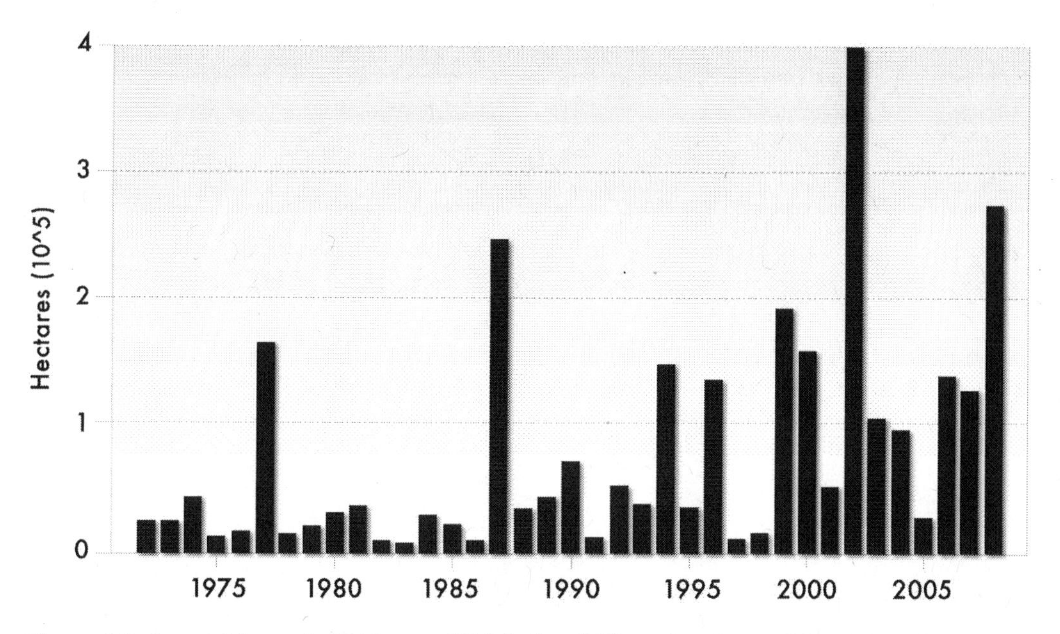

Figure 8.2 Area of large (>1,000 acre [400 ha]) wildfires that burned lands dominated by forest and woodland and managed by the U.S. Bureau of Indian Affairs, U.S. National Park Service, and U.S. Forest Service in Arizona, California, Nevada, New Mexico, and Utah. Data from Westerling, Turner, Smithwick et al. (2011 online supplement); U.S. Department of the Interior (2008 fire data); and U.S. Department of Agriculture (https://fam.nwcg.gov/fam-web/kcfast/mnmenu.htm).

increase in the mountains of Arizona and New Mexico; a 470% increase on the Colorado Plateau; and a 656% increase in the southern Rocky Mountains. Using finer spatial and temporal resolutions, allowing for nonlinear relations between variables, and examining a broad range of climate scenarios, Westerling and Bryant (2008) and Westerling, Bryant, and others (2011) similarly projected increases in the probability of large fires (100% to 400%) and burned area (100% to more than 300%) for much of Northern California's forests across a range of scenarios of climate, population growth, and development footprints. The greatest increases in burned area (at least 300%) were projected in models that were based on an emissions scenario associated with relatively dry conditions in which increases in temperature were greater than 5.4°F (3°C). Spracklen and others (2009) projected increases in burned area by 2050 ranging from 43% in Arizona and New Mexico to 78% in Northern California and 175% in the Rocky Mountains, given temperature increases of 2.7° to 3.6°F (1.5°C to 2°C). Mid-twenty-first-century increases in area burned in Northern California projected by Spracklen and others (2009) were comparable to those projected by Westerling, Bryant, and others (2011).

All of the studies cited in this section employed statistical models that assume interactions among climate, vegetation, and wildfire are similar to those in currently managed ecosystems and incorporate scenarios of future climate. As fuel characteristics are altered by the cumulative effects of climate and disturbance, however, these interactions may change. Also, the range of climate variability in recent decades for which comprehensive fire histories exist is small compared to what is projected under many scenarios that assume current rates of increase in greenhouse gas emissions. Accordingly, the accuracy of projections from the statistical models may decrease as changes in climate exceed the historical record. Furthermore, the resources and strategies applied to managing fire and other ecosystem processes may change in the future, with unknown effects. For example, Stephens, Martin, and Clinton (2007) estimated that fire suppression played a role in reducing the annual area burned in California during the 1900s to a tenth of prehistoric levels. Such reductions in area burned are widely thought to have contributed to increases in fuel densities and fire severity in forests throughout the Southwest that had frequent, low-severity surface fires in prehistoric times (e.g., Fulé, Covington, and Moore 1997; Miller et al. 2009; Allen et al. 2010).

Species in the Southwest are known or hypothesized to be responding directly or indirectly to changes in climate via changes in geographic distributions, phenology, and interspecific interactions. In some cases, responses at the level of individual plants and animals, populations, or species lead to changes in ecosystem structure and function, including disturbances such as fire. If past and current relations between species and environmental variables are well understood and can be described mathematically, then the responses of ecosystems and their component plants and animals can be projected given different scenarios of future climate. However, the accuracy of the projections depends in part on the accuracy with which climate variables can be projected, the similarity of future to past and current values of climate variables, and the extent to which species adapt to environmental change through evolution or short-term changes in physiology and behavior.

References

Abatzoglou, J. T., and C. A. Kolden. 2011. Climate change in western US deserts: Potential for increased wildfire and invasive annual grasses. *Rangeland Ecology and Management* 64:471–478, doi:10.2111/REM-D-09-00151.1

Adams, H. D., M. Guardiola-Claramonte, G. A. Barron-Gafford, J. Camilo Villegas, D. D. Breshears, C. B. Zou, P. A. Troch, and T. E. Huxman. 2009. Temperature sensitivity of drought-induced tree mortality portends increased regional die-off under global-change-type drought. *Proceedings of the National Academy of Sciences* 106:7063–7066.

Aldridge, G., D. W. Inouye, J. R. K. Forrest, W. A. Barr, and A. J. Miller-Rushing. 2011. Emergence of a mid-season period of low floral resources in a montane meadow ecosystem associated with climate change. *Journal of Ecology* 99:905–913.

Allen, C. D, A. K. Macalady, H. Chenchouni, D. Bachelet, N. McDowell, M. Vennetier, T. Kitzberger, et al. 2010. A global overview of drought and heat-induced tree mortality reveals emerging climate change risks for forests. *Forest Ecology and Management* 259:660–684.

Anderson-Teixeira, K. J., J. P. DeLong, A. M. Fox, D. A. Brese, and M. E. Litvak. 2011. Differential responses of production and respiration to temperature and moisture drive the carbon balance across a climatic gradient in New Mexico. *Global Change Biology* 17:410–424.

Archer, S. R., and K. I. Predick. 2008. Climate change and ecosystems of the southwestern United States. *Rangelands* 30:23–28.

Ault, T. R., A. K. Macalady, G. T. Pederson, J. L. Betancourt, and M.D. Schwartz. 2011. Northern hemisphere modes of variability and the timing of spring in western North America. *Journal of Climate* 24:4003–4014, doi:10.1175/2011JCLI4069.1.

Belnap, J., and S. L. Phillips. 2001. Soil biota in an ungrazed grassland: Response to annual grass (*Bromus tectorum*) invasion. *Ecological Applications* 11:1261–1275.

Belnap, J., S. L. Phillips, and M. E. Miller. 2004. Response of desert biological soil crusts to alterations in precipitation frequency. *Oecologia* 141:306–316.

Bentz, B. 2008. Western U.S. bark beetles and climate change. N.p.: U.S. Forest Service, Climate Change Resource Center. http://www.fs.fed.us/ccrc/topics/bark-beetles.shtml.

Bentz, B. J., J. Régnière, C. J. Fettig, E. M. Hansen, J. L. Hayes, J. A. Hicke, R. G. Kelsey, J. F. Negrón, and S. J. Seybold. 2010. Climate change and bark beetles of the western United States and Canada: Direct and indirect effects. *BioScience* 60:602–613.

Bonfils, C., B. D. Santer, D. W. Pierce, H. G. Hidalgo, G. Bala, T. Das, T. P. Barnett, et al. 2008. Detection and attribution of temperature changes in the mountainous western United States. *Journal of Climate* 21:6404–6424.

Both, C., M. van Asch, R. B. Bijlsma, A. B. van den Burg, and M. E. Visser. 2009. Climate change and unequal phenological changes across four trophic levels: Constraints or adaptations? *Journal of Animal Ecology* 78:73–83.

Bowling, D. R., S. Bethers-Marchetti, C. K. Lunch, E. E. Grote, and J. Belnap. 2010. Carbon, water, and energy fluxes in a semiarid cold desert grassland during and following multiyear drought. *Journal of Geophysical Research* 115: G04026, doi:10.1029/2010JG001322.

Bradley, B. A., D. M. Blumenthal, D. S. Wilcove, and L. H. Ziska. 2010. Predicting plant invasions in an era of global change. *Trends in Ecology and Evolution* 25:310–318.

Bradley, B. A., R. A. Houghton, J. F. Mustard, and S. P. Hamburg. 2006. Invasive grass reduces aboveground carbon stocks in shrublands of the western U.S. *Global Change Biology* 12:1815–1822.

Bradley, B. A. and J. F. Mustard. 2008. Comparison of phenology trends by land cover class: A case study in the Great Basin, USA. *Global Change Biology* 14:334–346.

Breshears, D. D., N. S. Cobb, P. M. Rich, K. P. Price, C. D. Allen, R. G. Balice, W. H. Romme, et al. 2005. Regional vegetation die-off in response to global-change-type drought. *Proceedings of the National Academy of Sciences* 102:15144–15148.

Brooks, M. L., C. M. D'Antonio, D. M. Richardson, J. B. Grace, J. E. Keeley, J. M. DiTomaso, R. J. Hobbs, M. Pellant, and D. Pyke. 2004. Effects of invasive alien plants on fire regimes. *Bioscience* 54:677–688.

Brooks, M. L., and D. A. Pyke. 2001. Invasive plants and fire in the deserts of North America. In *Proceedings of the invasive species workshop: The role of fire in the control and spread of invasive species*, ed. K. E. M. Galley and T. P. Wilson, 1-14. Misc. Pub. No. 11. Tallahassee, FL: Tall Timbers Research Station.

Brown, J. L., S. H. Li, and N. Bhagabati. 1999. Long-term trend toward earlier breeding in an American bird: A response to global warming? *Proceedings of the National Academy of Sciences* 96:5565–5569.

Brown, P. M., E. K. Heyerdahl, S. G. Kitchen, and M. H. Weber. 2008. Climate effects on historical fires (1630–1900) in Utah. *International Journal of Wildland Fire* 17:28–39.

Cayan, D. R., T. Das, D. W. Pierce, T. P. Barnett, M. Tyree, and A. Gershunov. 2010. Future dryness in the southwest US and the hydrology of the early 21st century drought. *Proceedings of the National Academy of Sciences* 107:21271–21276.

Chevin, L.-M., R. Lande, and G. M. Mace. 2010. Adaptation, plasticity, and extinction in a changing environment: Towards a predictive theory. *PLoS Biology* 8 (4): e1000357, doi:10.1371/journal.pbio.1000357.

Cleland, E. E., I. Chuine, A. Menzel, H. A. Mooney, and M. D. Schwartz. 2007. Shifting plant phenology in response to global change. *Trends in Ecology and Evolution* 22:357–365.

Collins, S. L., J. E. Fargione, C. L. Crenshaw, E. Nonaka, J. T. Elliott, Y. Xia, and W. T. Pockman. 2010. Rapid plant community responses during the summer monsoon to nighttime warming in a northern Chihuahuan Desert grassland. *Journal of Arid Environments* 74:611–617.

Comer, P., D. Faber-Langendoen, R. Evans, S. Gawler, C. Josse, G. Kittel, S. Menard, et al. 2003. Ecological systems of the United States: A working classification of U.S. terrestrial systems. Arlington, VA: NatureServe.

Crimmins, T. M., M. A. Crimmins, and C. D. Bertelsen. 2011. Onset of summer flowering in a "Sky Island" is driven by monsoon moisture. *New Phytologist* 191:468–479.

D'Antonio, C., and P. Vitousek. 1992. Biological invasions by exotic grasses, the grass-fire cycle and global change. *Annual Review of Ecology and Systematics 23:63–88.*

D'Odorico, P., J. D. Fuentes, W. T. Pockman, S. L. Collins, Y. He, J. S. Medeiros, S. DeWekker, and M.E. Litvak. 2010. Positive feedback between microclimate and shrub encroachment in the northern Chihuahuan Desert. *Ecosphere* 1: Article 17, doi:10.1890/ES10-00073.1.

Evans, R. D., R. Rimer, L. Sperry, and J. Belnap. 2001. Exotic plant invasion alters nitrogen dynamics in an arid grassland. *Ecological Applications* 11:1301–1310.

Fairweather, M. L., B. W. Geils, and M. Manthei. 2007. Aspen decline on the Coconino National Forest. In *Proceedings of the 55th Annual Western International Forest Disease Work Conference,* comp. N. McWilliams and P. Palacios, 53–62. Salem: Oregon Department of Forestry.

Fleishman, E., and D. D. Murphy. 2012. Minimizing uncertainty in interpreting responses of butterflies to climate change. In *Ecological Consequences of Climate Change: Mechanisms, Conservation, and Management*, ed. J. Belant and E. Beever, 55–66. London: Taylor and Francis.

Forister, M. L., A. C. McCall, N. J. Sanders, J. A. Fordyce, J. H. Thorne, J. O'Brien, D. P. Waetjen, and A. M. Shapiro. 2010. Compounded effects of climate change and habitat shift patterns of butterfly diversity. *Proceedings of the National Academy of Sciences* 107:2088–2092.

Forister, M. L., and A. M. Shapiro. 2003. Climatic trends and advancing spring flight of butterflies in lowland California. *Global Change Biology* 9:1130–1135.

Franklin, J., K. E. Wejnert, S. A. Hathaway, C. J. Rochester, and R. N. Fisher. 2009. Effect of species rarity on the accuracy of species distribution models for reptiles and amphibians in Southern California. *Diversity and Distributions* 15:167–177.

Frey, B. R., V. J. Lieffers, E. H. Hogg, and S. M. Landhäusser. 2004. Predicting landscape patterns of aspen dieback: Mechanisms and knowledge gaps. *Canadian Journal of Forest Research* 34:1379–1390.

Fulé, P. Z., W. W. Covington, and M. M. Moore. 1997. Determining reference conditions for ecosystem management of southwestern ponderosa pine forests. *Ecological Applications* 7:895–908.

Hall, S. J., R. A. Sponseller, N. B. Grimm, D. Huber, J. P. Kaye, C. Clark, and S. Collins. 2011. Ecosystem response to nutrient enrichment across an urban airshed in the Sonoran Desert. *Ecological Applications* 21:640–660.

Harper, M. P., and B. L. Peckarsky. 2006. Emergence cues of a mayfly in high-altitude stream systems: Potential response to climate change. *Ecological Applications* 16:612–621.

Hellmann, J. J., J. E. Byers, B. G. Bierwagen, and J. S. Dukes. 2008. Five potential consequences of climate change for invasive species. *Conservation Biology* 22:534–543.

Hellmann, J. J., and M. E. Pfrender. 2011. Future human intervention in ecosystems and the critical role for evolutionary biology. *Conservation Biology* 25:1143–1147.

Hogg, E. H., J. P. Brandt, and B. Kochtubajda. 2005. Factors affecting interannual variation in growth of western Canadian aspen forests during 1951–2000. *Canadian Journal of Forest Research* 35:610–622.

Hogg, E. H., J. P. Brandt, and M. Michaelian. 2008. Impacts of a regional drought on the productivity, dieback, and biomass of western Canadian aspen forests. *Canadian Journal of Forest Research* 38:1373–1384.

Inouye, D. W. 2008. Effects of climate change on phenology, frost damage, and floral abundance of montane wildflowers. *Ecology* 89:353–362.

Inouye, D. W., B. Barr, K. B. Armitage, and B. D. Inouye. 2000. Climate change is affecting altitudinal migrants and hibernating species. *Proceedings of the National Academy of Sciences* 97:1630–1633.

Intergovernmental Panel on Climate Change (IPCC). 2007. *Climate change 2007: Impacts, adaptation and vulnerability. Contribution of Working Group II to the Fourth Assessment Report of the Intergovernmental Panel on Climate Change*, ed. M. L. Parry, O. F. Canziani, J. P. Palutikof, P. J. van der Linden, and C. E. Hanson. Cambridge: Cambridge University Press.

Kaye, J. P., S. E. Eckert, D. A. Gonzalez, J. O. Allen, S. J. Hall, R. A. Sponseller, and N. B. Grimm. 2011. Decomposition of urban atmospheric carbon in Sonoran Desert soils. *Urban Ecosystems* 14:737–754, doi:10.1007/s11252-011-0173-8.

Kellermann, V., B. V. Heerwaarden, C. M. Sgro, and A. A. Hoffmann. 2009. Fundamental evolutionary limits in ecological traits drive *Drosophila* species distributions. *Science* 325:1244–1246.

Kelly, A. E., and M. L. Goulden. 2008. Rapid shifts in plant distribution with recent climate change. *Proceedings of the National Academy of Sciences* 105:11823–11826.

Kier, G., H. Kreft, T. M. Lee, W. Jetz, P. L. Ibisch, C. Nowicki, J. Mutke, and W. Barthlott. 2009. A global assessment of endemism and species richness across island and mainland regions. *Proceedings of the National Academy of Sciences* 106:9322–9327.

Kimball, S., A. L. Angert, T. E. Huxman, and D. L. Venable. 2009. Contemporary climate change in the Sonoran Desert favors cold-adapted species. *Global Change Biology* 16:1555–1565.

Knowles, N., M. D. Dettinger, and D. R. Cayan. 2006. Trends in snowfall versus rainfall in the western United States. *Journal of Climate* 19:4545-4559.

Lenihan, J. M., R. Drapek, D. Bachelet, and R. P. Neilson. 2003. Climate change effect on vegetation distribution, carbon, and fire in California. *Ecological Applications* 13:1667–1681.

Littell, J. S., D. McKenzie, D. L. Peterson, and A. L. Westerling. 2009. Climate and wildfire area burned in western U.S. ecoprovinces, 1916–2003. *Ecological Applications* 19:1003–1021.

Loarie, S. R., B. E. Carter, K. Hayhoe, S. McMahon, R. Moe, C. A. Knight, and D. D. Ackerly. 2008. Climate change and the future of California's endemic flora. *PLoS One* 3: e2502.

MacDonald, G. M. 2010. Climate change and water in southwestern North America. *Proceedings of the National Academy of Sciences* 107:21256–21262.

MacMynowski, D., T. Root, G. Ballard, and G. Geupel. 2007. Changes in spring arrival of Nearctic-Neotropical migrants attributed to multiscalar climate. *Global Change Biology* 13:2239–2251.

Marushia, R. G., M. W. Cadotte, and J. S. Holt. 2010. Phenology as a basis of management of exotic annual plants in desert invasions. *Journal of Applied Ecology* 47:1290–1299.

Miller, J. D., H. D. Safford, M. Crimmins, and A. E. Thode. 2009. Quantitative evidence for increasing forest fire severity in the Sierra Nevada and southern Cascade Mountains, California and Nevada, USA. *Ecosystems* 12:16–32.

Miller-Rushing, A. J., T. T. Høye, D. W. Inouye, and E. Post. 2010. The effects of phenological mismatches on demography. *Philosophical Transactions of the Royal Society B* 365:3177–3186.

Miller-Rushing, A. J., and D. W. Inouye. 2009. Variation in the impact of climate change on flowering phenology and abundance: An examination of two pairs of closely related wildflower species. *American Journal of Botany* 96:1821–1829.

Møller, A. P., D. Rubolini, and E. Lehikoinen. 2008. Populations of migratory bird species that did not show a phenological response to climate change are declining. *Proceedings of the National Academy of Sciences* 105:16195–16200.

Morgan, J. A., D. R. LeCain, E. Pendall, D. M. Blumenthal, B. A. Kimball, Y. Carrillo, D. G. Williams, J. Heisler-White, F. A. Dijkstra, and M. West. 2011. C4 grasses prosper as carbon dioxide eliminates desiccation in warmed semi-arid grassland. *Nature* 476:202–206.

Moritz, C., J. L. Patton, C. J. Conroy, J. L. Parra, G. C. White, and S. R. Beissinger. 2008. Impact of a century of climate change on small mammal communities in Yosemite National Park, USA. *Science* 322:261–264.

Munson, S. M., J. Belnap, and G. S. Okin. 2011. Responses of wind erosion to climate-induced vegetation changes on the Colorado Plateau. *Proceedings of the National Academy of Sciences* 108:3854–3859.

Munson, S. M., R. H. Webb, J. Belnap, J. A. Hubbard, D. E. Swan, and S. Rutman. 2012. Forecasting climate change impacts to plant community composition in the Sonoran Desert region. *Global Change Biology* 18:1083–1095.

National Research Council. 2011. *Climate stabilization targets: Emissions, concentrations, and impacts over decades to millennia.* Washington, DC: National Academies Press.

Nicotra, A. B., and A. M. Davidson. 2010. Adaptive phenotypic plasticity and plant water use. *Functional Plant Biology* 37:117–127.

Norton, J. B., T. A. Monaco, J. M. Norton, D. A. Johnson, and T. A. Jones. 2004. Soil morphology and organic matter dynamics under cheatgrass and sagebrush-steppe plant communities. *Journal of Arid Environments* 57:445–466.

Ozgul, A., D. Z. Childs, M. K. Oli, K. B. Armitage, D. T. Blumstein, L. E. Olson, S. Tuljapurkar, and T. Coulson. 2010. Coupled dynamics of body mass and population growth in response to environmental change. *Nature* 466:482–485.

Painter, T. H., A. P. Barrett, C. C. Landry, J. C. Neff, M. P. Cassidy, C. R. Lawrence, K. E. McBride, and G. L. Farmer. 2007. Impact of disturbed desert soils on duration of mountain snow cover. *Geophysical Research Letters* 34: L12502, doi:10.1029/2007GL030284.

Pan, L., S. Chen, D. Cayan, M. Lin, Q. Hart, M. Zhang, Y. Liu, and J. Wang. 2010. Influences of climate change on California and Nevada regions revealed by a high-resolution dynamical downscaling study. *Climate Dynamics* 37:2005-2020, doi 10.1007/s00382-010-0961-5.

Parmesan, C. 1996. Climate effects on species range. *Nature* 382:765–766.

—. 2006. Ecological and evolutionary responses to recent climate change. *Annual Review of Ecology and Systematics* 2006:637–669.

—. 2007. Influences of species, latitudes and methodologies on estimates of phenological response to global warming. *Global Change Biology* 13:1860–1872.

Parmesan, C., and G. Yohe. 2003. A globally coherent fingerprint of climate change impacts across natural systems. *Nature* 421:37–42.

Parra, J. L., and W. B. Monahan. 2008. Variability in 20th century climate change reconstructions and its consequences for predicting geographic responses of California mammals. *Global Change Biology* 14:2215–2231.

Peckarsky, B. L., A. C. Encalada, and A. R. McIntosh. 2011. Why do vulnerable mayflies thrive in trout streams? *American Entomologist* 57:152–164.

Peters, D. P. C., J. Yao, O. E. Sala, and J. Anderson. 2011. Directional climate change and potential reversal of desertification in arid and semiarid ecosystems. *Global Change Biology* 18:151-163, doi:10.1111/j.1365-2486.2011.02498.x.

Pysek, P. and D. M. Richardson. 2007. Traits associated with invasiveness in alien plants: Where do we stand? In *Biological invasions*, ed. P. Pysek and D. M. Richardson, 97-126. Berlin: Springer.

Raffa, K. F., B. H. Aukema, B. J. Bentz, A. L. Carroll, J. A. Hicke, M. G. Turner, and W. H. Romme. 2008. Cross-scale drivers of natural disturbances prone to anthropogenic amplifications: The dynamics of bark beetle eruptions. *BioScience* 58:501–517.

Reed, S. C., K. K. Coe, J. P. Sparks, D. C. Housman, T. J. Zelikova, and J. Belnap. 2012. Changes to dryland rainfall results in rapid moss mortality and altered soil fertility. *Nature Climate Change* 2:752-755, doi:10.1038/nclimate1596.

Reed, T. E., D. E. Schindler, and R. S. Waples. 2011. Interacting effects of phenotypic plasticity and evolution on population persistence in a changing climate. *Conservation Biology* 25:56–63.

Rehfeldt, G. E., N. L. Crookston, M. V. Warwell, and J. S. Evans. 2006. Empirical analyses of plant-climate relationships for the western United States. *International Journal of Plant Sciences* 167:1123–1150.

Rehfeldt, G. E., D. S. Ferguson, and N. L. Crookston. 2009. Aspen, climate, and sudden decline in western USA. *Forest Ecology and Management* 258:2352–2364.

Schoennagel, T., R. L. Sherriff, and T. T. Veblen. 2011. Fire history and tree recruitment in the upper montane zone of the Colorado Front Range: Implications for forest restoration. *Ecological Applications* 21:2210–2222.

Schwartz, M. D., R. Ahas, and A. Aasa. 2006. Onset of spring starting earlier across the Northern Hemisphere. *Global Change Biology* 12:343–351.

Schwinning, S., K. Davis, L. Richardson, and J. R. Ehleringer. 2002. Deuterium enriched irrigation indicates different forms of rain use in shrub/grass species of the Colorado Plateau. *Oecologia* 130:345–355.

Seager, R., and G. A. Vecchi. 2010. Greenhouse warming and the 21st century hydroclimate of southwestern North America. *Proceedings of the National Academy of Sciences* 107:21277–21282.

Spracklen, D. V., L. J. Mickley, J. A. Logan, R. C. Hudman, R. Yevich, M. D. Flannigan, and A. L. Westerling. 2009. Impacts of climate change from 2000 to 2050 on wildfire activity and carbonaceous aerosol concentrations in the western United States. *Journal of Geophysical Research* 114: D20301.

Stephens, S. L., R. E. Martin, and N. E. Clinton. 2007. Prehistoric fire area and emissions from California's forests, woodlands, shrublands, and grasslands. *Forest Ecology and Management* 251:205–216.

Stewart, I. T., D. R. Cayan, and M. D. Dettinger. 2005. Changes toward earlier streamflow timing across western North America. *Journal of Climate* 18:1136–1155.

Swetnam, T. W., and J. L. Betancourt. 1998. Mesoscale disturbance and ecological response to decadal climatic variability in the American Southwest. *Journal of Climate* 11:3128–3147.

Thackeray, S. J., T. H. Sparks, M. Frederiksen, S. Burthe, P. J. Bacon, J. R. Bell, M. S. Botham, et al. 2010. Trophic level asynchrony in rates of phenological change for marine, freshwater and terrestrial environments. *Global Change Biology* 16:3304–3313.

Thomey, M. L., S. L. Collins, R. Vargas, J. E. Johnson, R. F. Brown, D. O. Natvig, and M. T. Friggens. 2011. Effect of precipitation variability on net primary production and soil respiration in a Chihuahuan Desert grassland. *Global Change Biology* 17:1505–1515.

Thorne, J. H., B. J. Morgan, and J. A. Kennedy. 2008. Vegetation change over sixty years in the central Sierra Nevada, California, USA. *Madroño* 55:223–237.

Throop, H. L., L. G. Reichman, O. E. Sala, and S. R. Archer. 2012. Response of dominant grass and shrub species to water manipulation: An ecophysiological basis for shrub invasion in a Chihuahuan Desert grassland. *Oecologia* 169:373–383, doi:10.1007/s00442-011-2217-4.

Thuiller, W., D. M. Richardson, and G. F. Midgley. 2007. Will climate change promote alien plant invasions? *Ecological Studies* 193:197–211.

U.S. Forest Service (USFS). 2007. Forest insect and disease conditions in the Southwestern Region, 2006. USFS Publication PR-R3-16-2. Albuquerque: U.S. Forest Service, Southwestern Region, Forestry and Forest Health.

U.S. Geological Survey (USGS). National Gap Analysis Program. 2004. Provisional digital land cover map for the southwestern United States. Version 1.0. Logan: Utah State University, College of Natural Resources, RS/GIS Laboratory.

van Mantgem, P. J., N. J. Stephenson, J. C. Byrne, L. D. Daniels, J. F. Franklin, P. Z. Fulé, M. E. Harmon, et al. 2009. Widespread increase of tree mortality rates in western United States. *Science* 323:521–524.

Visser, M. E. 2008. Keeping up with a warming world: Assessing the rate of adaptation to climate change. *Proceedings of the Royal Society of London Series B* 275:649–659.

Walther, G., E. Post, P. Convey, A. Menzel, C. Parmesan, T. J. C. Beebee, J.-M. Fromentin, O. Hoegh-Guldberg, and F. Bairlein. 2002. Ecological responses to recent climate change. *Nature* 416:389–395.

Warren, D. R., A. G. Ernst, and B. P. Baldigo. 2009. Influence of spring floods on year-class strength of fall- and spring-spawning salmonids in Catskill mountain streams. *Transactions of the American Fisheries Society* 138:200–210.

Wenger, S. J., D. J. Isaak, C. H. Luce, H. M. Neville, K. D. Fausch, J. B. Dunham, D. C. Dauwalter, et al. 2011. Flow regime, temperature, and biotic interactions drive differential declines of trout species under climate change. *Proceedings of the National Academy of Sciences* 34:14175–14180.

Westerling, A. L. 2010. Wildfires. In *Climate change science and policy*, ed. S. Schneider, A. Rosencranz, M. Mastrandrea, and K. Kuntz-Duriseti, 92–103. Washington, DC: Island Press.

Westerling, A. L., T. J. Brown, A. Gershunov, D. R. Cayan, and M. D. Dettinger. 2003. Climate and wildfire in the western United States. *Bulletin of the American Meteorological Society* 84:595–604.

Westerling, A. L., and B. P. Bryant. 2008. Climate change and wildfire in California. *Climatic Change* 87 (Suppl. 1): S231–S249.

Westerling, A. L., B. P. Bryant, H. K. Preisler, T. P. Holmes, H. G. Hidalgo, T. Das, and S. Shrestha. 2011. Climate change and growth scenarios for California wildfire. *Climatic Change* 109 (Suppl. 1): S445–S463.

Westerling, A. L., H. G. Hidalgo, D. R. Cayan, and T. W. Swetnam. 2006. Warming and earlier spring increases western U.S. forest wildfire activity. *Science* 313:940–943.

Westerling, A. L., M. G. Turner, E. H. Smithwick, W. H. Romme, and M. G. Ryan. 2011. Continued warming could transform Greater Yellowstone fire regimes by mid-21st century. *Proceedings of the National Academy of Sciences* 108:13165–13170.

Williams, A. P., C. D. Allen, C. I. Millar, T. W. Swetnam, J. Michaelsen, C. J. Still, and S. W. Leavitt. 2010. Forest responses to increasing aridity and warmth in the southwestern United States. *Proceedings of the National Academy of Sciences* 107:21289–21294.

Williams, J. W., and S. T. Jackson. 2007. Novel climates, no-analog communities, and ecological surprises. *Frontiers in Ecology and the Environment* 5:475–482.

Willis, C. G., B. Ruhfel, R. B. Primack, A. J. Miller-Rushing, and C. C. Davis. 2008. Phylogenetic patterns of species loss in Thoreau's woods are driven by climate change. *Proceedings of the National Academy of Sciences* 105:17029–17033.

Willis, C. G., B. R. Ruhfel, R. B. Primack, A. J. Miller-Rushing, J. B. Losos, and C. C. Davis. 2010. Favorable climate change response explains non-native species' success in Thoreau's woods. *PLoS One* 5: e8878, doi:10.1371/journal.pone.0008878.

Willis, K. J., and G. M. MacDonald. 2011. Long-term ecological records and their relevance to climate change predictions for a warmer world. *Annual Review of Ecology, Evolution, and Systematics* 42:267-287, doi:10.1146/annurev-ecolsys-102209-144704.

Wolkovich, E. M., and E. E. Cleland. 2010. The phenology of plant invasions: A community ecology perspective. *Frontiers of Ecology and the Environment* 9:287–294.

Worrall, J. J., L. Egeland, T. Eager, R. A. Mask, E. W. Johnson, P. A. Kemp, and W. D. Shepherd. 2008. Rapid mortality of *Populus tremuloides* in southwestern Colorado, USA. *Forest Ecology and Management* 255:686–696.

Yang, L. H., and V. H. W. Rudolf. 2010. Phenology, ontogeny and the effects of climate change on the timing of species interactions. *Ecology Letters* 13:1–10.

Ziska, L. H., J. B. Reeves III, and B. Blank. 2005. The impact of recent increases in atmospheric CO_2 on biomass production and vegetative retention of cheatgrass (*Bromus tectorum*): Implications for fire disturbance. *Global Change Biology* 11:1325–1332.

Endnotes

i The ecological systems were grouped into nine land-cover classes (alpine tundra; wetland, riparian, or playa; shrubland; sparsely vegetated or barren; forest; grassland or prairie; open water; developed or agriculture; and disturbed land).

ii Located near Gothic, Colorado, at an elevation of around 9,500 feet (2,900 meters).

Chapter 9

Coastal Issues

COORDINATING LEAD AUTHORS

Margaret R. Caldwell (Stanford Woods Institute for the Environment, Stanford Law School), Eric H. Hartge (Stanford Woods Institute for the Environment)

LEAD AUTHORS

Lesley C. Ewing (California Coastal Commission), Gary Griggs (University of California, Santa Cruz), Ryan P. Kelly (Stanford Woods Institute for the Environment), Susanne C. Moser (Susanne Moser Research and Consulting, Stanford University), Sarah G. Newkirk (The Nature Conservancy, California), Rebecca A. Smyth (NOAA, Coastal Services Center), C. Brock Woodson (Stanford Woods Institute for the Environment))

EXPERT REVIEW EDITOR

Rebecca Lunde (NOAA)

Executive Summary

The California coast is constantly changing due to human development and physical forces. With the increase in climate impacts—including sea-level rise, ocean warming, ocean acidification, and increased storm events—effects of these physical forces will be more significant and will present substantial risks to coastal areas in the future. Natural ecosystems, coastal development, economic interests, and even cultural attachment to the coast will be at risk. Given the high concentration of coastal development, population, infrastructure, and economic activity in coastal counties, continued and growing pressure to protect these assets and activities from rising sea levels is expected.

Chapter citation: Caldwell, M. R., E. H. Hartge, L. C. Ewing, G. Griggs, R. P. Kelly, S. C. Moser, S. G. Newkirk, R. A. Smyth, and C. B. Woodson. 2013. "Coastal Issues." In *Assessment of Climate Change in the Southwest United States: A Report Prepared for the National Climate Assessment,* edited by G. Garfin, A. Jardine, R. Merideth, M. Black, and S. LeRoy, 168–196. A report by the Southwest Climate Alliance. Washington, DC: Island Press.

We have identified the following seven key messages that highlight major climate issues facing the California coast:

- The future severity of coastal erosion, flooding, inundation, and other coastal hazards will increase due to sea-level rise and continued coastal development. (high confidence). Any increased intensity and/or increased frequency of storm events will further aggravate the expected impacts. (medium confidence)
- The implications of global sea-level rise for coastal areas cannot be understood in isolation from other, shorter-term sea-level variability related to El Niño-Southern Oscillation (ENSO) events, storms, or tides. The highest probability and most damaging events through the year 2050 will be large ENSO events when elevated sea levels occur simultaneously with high tides and large waves. Between 2050 and 2100, or when sea levels approach ~14–16 inches above the 2000 baseline, the effects of sea-level rise (flooding and inundation) and combined effects of sea-level rise and large waves will result in property damage, erosion, and flood losses far greater than experienced now or in the past. (high confidence)
- Ocean warming affects a range of ecosystem processes, from changes in species distribution to reduced oxygen content and sea-level rise. (medium-high confidence) However, there is considerable uncertainty about how changes in species distributions and lower oxygen content of ocean waters will impact marine ecosystems, fisheries, and coastal communities.
- Ocean acidification is a significant threat to calcium-carbonate-dependent species and marine ecosystems. (high confidence) There is substantial uncertainty about acidification's precise impacts on coastal fisheries and marine food webs along the West Coast.
- Coastal development and other land uses create impediments to the natural migration of coastal wetlands through "hardening" of the coastline (e.g., seawalls, revetments, bulkheads) and by the occupation and protection of space into which wetlands might otherwise migrate. (high confidence) In developing their land-use and other plans, communities need to take into account that an increase in coastal development and other hardening may result in medium- to long-term loss of coastal wetlands and the numerous benefits these habitats provide. (high confidence)
- Critical infrastructure, such as highways and railroads (see Chapter 14), power plants and transmission lines (see Chapter 12), wastewater treatment plants, and pumping stations, have been located along the coast where they are already exposed to damage from erosion or flooding. With rising sea level, risks to vital public infrastructure will increase and more infrastructure will be exposed to future damage from erosion and flooding. (high confidence) Much of the U.S. infrastructure is in need of repair or replacement and the California coast is no exception; impacts from climate change will add to the stress on communities to maintain functionality. (high confidence)
- Coastal communities have a variety of options and tools at hand to prepare for climate change impacts and to minimize the severity of now-unavoidable

consequences of climate warming and disruption. While many coastal communities are increasingly interested in and have begun planning for adaptation, the use of these tools as well as development and implementation of adaptive policies are still insufficient compared with the magnitude of the expected harm. (high confidence)

9.1 Coastal Assets

People are drawn to the coast for its moderate climate, scenic beauty, cultural and ecological richness, rural expanses, abundant recreational opportunities, vibrant economic activity, and diverse urban communities.[i] More than 70% of California residents live and work in coastal counties (U.S. Census Bureau n.d.). Over the last thirty-eight years, the California coastal county population has grown 64%, from about 16.8 million in 1970 to 27.6 million in 2008 (NOEP 2012). Almost 86% of California's total gross domestic product comes from coastal counties (NOEP 2010).

Population density, along with the presence of critical infrastructure and valuable real estate along the coast, accentuates the importance of the coast to the region's economy. California has the nation's largest ocean-based economy, valued at approximately $46 billion annually, with over 90% of this value coming from (1) tourism and recreation, and (2) ports and harbors (Kildow and Colgan 2005).

In addition, the state's natural coastal systems perform a variety of economically valuable functions, including water quality protection, commercial and recreational fish production, plant and wildlife habitat, flood mitigation, recreation, carbon storage, sediment and nutrient transport, and storm buffering. The non-market value of coastal recreation in California alone exceeds $30 billion annually (Pendleton 2009). These benefits, provided at almost no cost, would be impossible to replicate with human-engineered solutions.

9.2 Observed Threats

Overview

Human development and physical forces are constantly changing the coast. Just as growth in coastal populations and economic development have reshaped the coastline with new homes, roads, and infrastructure, so too have physical forces and processes—including waves, tides, currents, wind, storms, rain, and runoff—combined to accrete (build up), erode, and continually reshape the coastline and modify coastal ecosystems. With the increase in the rate of sea-level rise and warmer ocean temperatures related to global climate change, the effects of these physical forces will grow more significant and harmful to coastal areas over time.

Threats to the physical environment

The physical forces and processes that take place in the coastal environment occur across different spatial and temporal scales. The Pacific Basin, including the ocean off California, oscillates between warm and cool phases of the Pacific Decadal Oscillation (PDO), which is associated with differences in atmospheric pressure over the Pacific

Ocean. Ultimately, wind patterns and storm tracks are affected. El Niño-Southern Oscillation (ENSO) events tend to have stronger effects during warm phases of the PDO and are typified by warmer ocean water and higher sea levels, more rainfall and flooding, and more frequent and vigorous coastal storms, which result in greater beach and bluff erosion (Storlazzi and Griggs 2000). These conditions also affect relative abundance of important coastal forage fisheries, such as sardines and anchovies (Chavez et al. 2003).

Sea level along the coast of California has risen gradually over the past century (by about 8 inches [20 cm]), a rate that will accelerate in the future (see Figure 9.1). Sea-level rise alone, however, will have far less impact on the shoreline, infrastructure, or habitat over the next 30 or 40 years than will the combination of elevated sea level, high tides, and storm waves associated with large ENSO events. Moreover, the effects of less severe ENSO events will be magnified by progressively higher sea levels; as a result, coastal communities can expect more severe losses from these events than they have experienced in the past (see Figure 9.2).

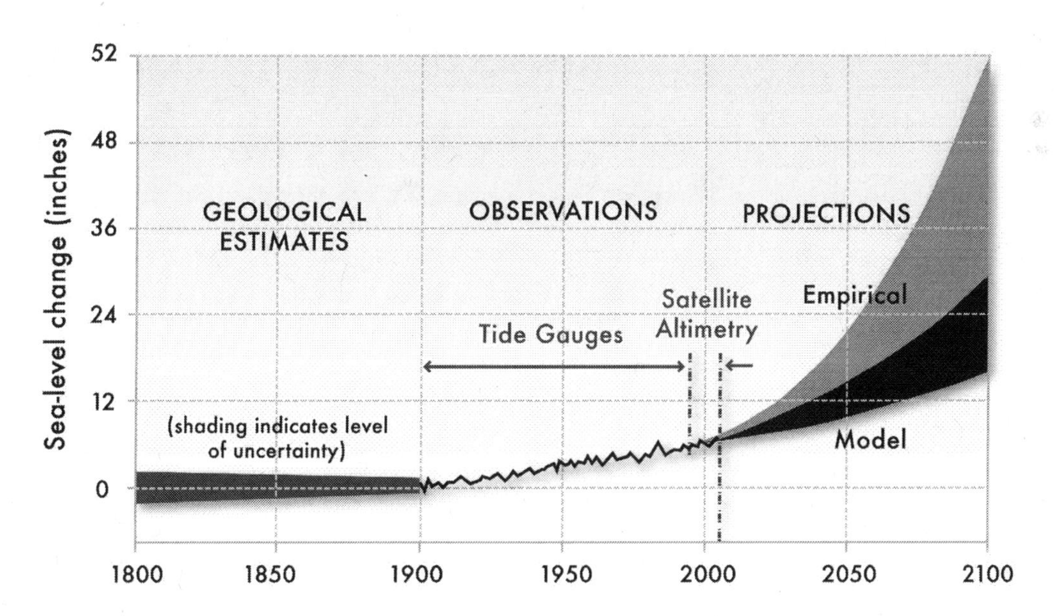

Figure 9.1 Past, present, and future sea-level rise. Geologic and recent sea-level histories (from tide gauges and satellite altimetry) are combined with projections to 2100 based on climate models and empirical data. Modified with permission from Russell and Griggs (2012, Figure 2.1).

Furthermore, changes in global climate cycles, such as the PDO, may soon become an imminent and significant factor in accelerating regional sea-level rise. While over the past century there has been a gradual increase in global sea levels, since about 1993, California tide gauges have recorded very little long-term change in sea level. This "flat" sea level condition had been out of sync with the prevailing global rise in sea level and the historic trends in sea-level rise along the West Coast. The PDO causes differences in sea-surface elevation across the Pacific. Sea levels have been higher in the Western Pacific and lower along the California coast over the past two decades, coinciding with a warm phase of the PDO (see Box 9.1). This recent warm phase appears to have been related to a dramatic change in wind stress (the dragging force of air moving over a surface) (Bromirski et al. 2011). The predominant wind stress regime along the U.S. West Coast served to mitigate the trend of rising sea level, suppressing regional sea-level rise below the global rate. A change in wind stress patterns over the entire North Pacific may result in a resumption of sea-level rise along the West Coast approaching or exceeding the global mean sea-level rise rate (Bromirski et al. 2011).

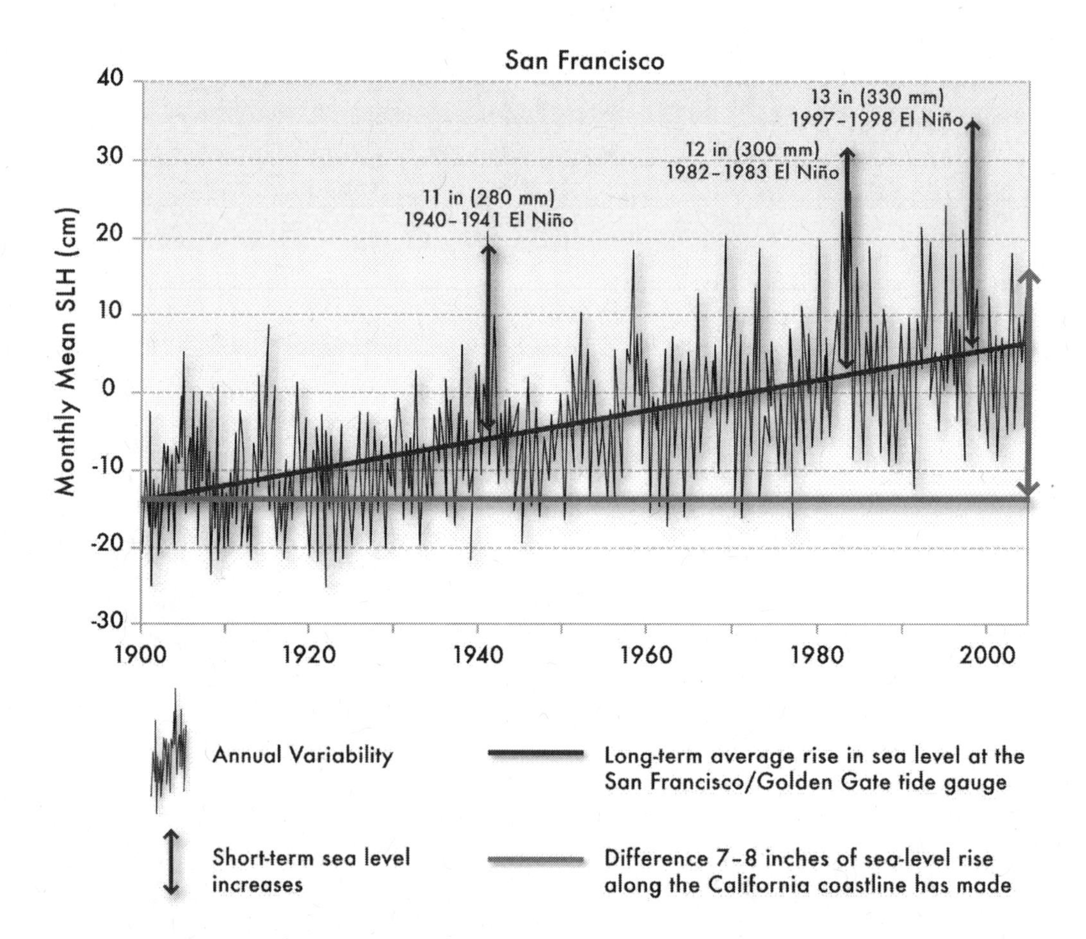

Figure 9.2 Sea-level rise and El Niño events. The implications of sea-level rise for coastal California cannot be understood in isolation from other, shorter-term sea-level variability related to El Niño events, storms, or extreme tides that affect the coast. As historical experience has shown, the greatest damage to coastal areas has occurred during large El Niño events (for example in 1940–41, 1982–83, and 1997–98) when short-term sea-level increases occurred simultaneously with high tides and large waves. If sea level were still at the same elevation in 2005 as it was in 1900, a major El Niño event like that in 1997–98 would fall within the "noise" of today's interannual variability. As sea level is continuing to rise, the impacts of future large ENSO events will be greater than those historic events of similar magnitude, exposing coastal areas to the combined effects of sea-level rise, elevated sea levels from El Niño events, and large waves. Source: Pacific Decadal Oscillation monthly values index (http://jisao.washington.edu/pdo/), NOAA Earth System Research Laboratory Multivariate ENSO Index (http://www.esrl.noaa.gov/psd/enso/mei/#ref_wt3), Wolter and Timlin (2011).

Box 9.1

Coastal Development During Cool PDO Phase

Comparison of periods of coastal development. In the graph below (a), red corresponds to periods with positive or warm PDO conditions and blue corresponds to negative or cool PDO conditions. The vertical axis is a dimensionless PDO index based on North Pacific sea surface temperature variability.

The maps show the increase in housing density [difference in housing units per km^2] along the Southern California Bight that occurred (b) during the extended cool PDO period from about 1950 to 1980 and (c) during the extended warm PDO period from about 1980 to 2010.

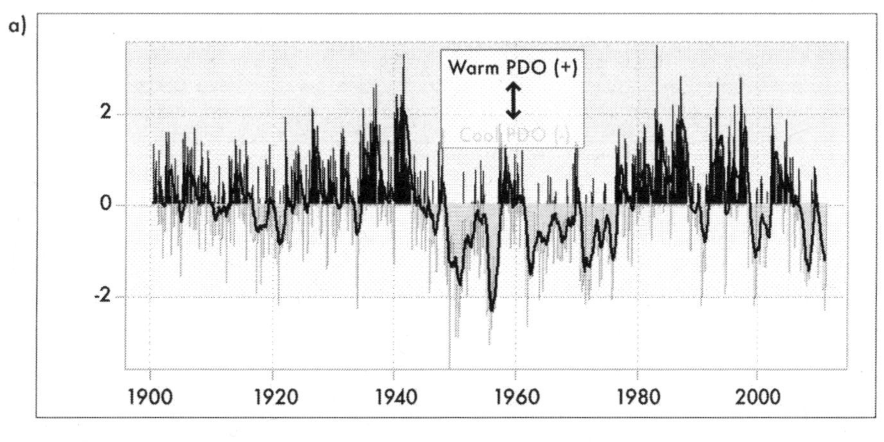

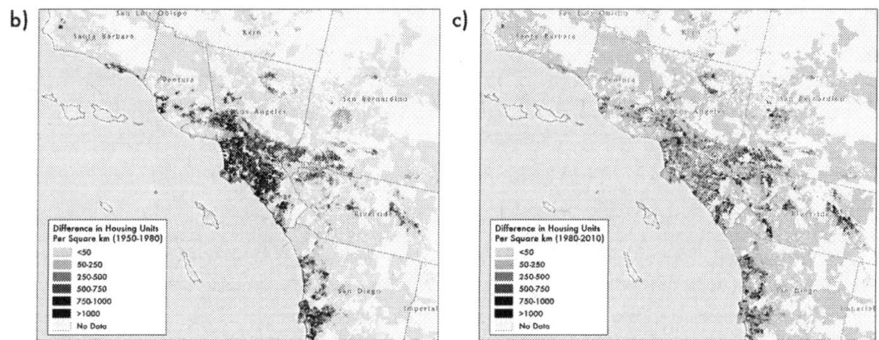

Figure 9.3 Monthly value for the Pacific Decadal Oscillation. The period from about 1945 to 1978 was a cool PDO period marked by an overall calm or benign coastal climate, but also was a period of intensive growth and development along the California coast. The vertical axis is a dimensionless PDO index based on North Pacific sea-surface temperature variability. 9.3(a) adapted from Pacific Decadal Oscillation monthly values index, http://jisao.washington.edu/pdo/; data in 9.3(b) and (c) from http://silvis.forest.wisc.edu/old/Library/HousingDataDownload.php?state=California&abrev=CA; see also Hammer et al. (2004)..

Threats to the built environment

The nature of most human development is increasingly in conflict with the physical and climatic forces that occur along the coast. Efforts to protect development through shoreline armoring and beach nourishment are very costly and often negatively impact coastal ecosystems (Caldwell and Segal 2007). Armoring the coast with hard structures may inhibit natural sediment movement, and thus prevent accretion to and landward migration of beach and other coastal ecosystems. Armoring can also increase vulnerability by encouraging development in erosion or flood-prone areas and giving people who live behind coastal armoring installations a false sense of security (Dugan et al. 2008).

Increasing demand for freshwater resources in coastal areas for domestic, agricultural, and industrial uses adds stress to the provision of surface water and ground water supplies. Increased withdrawals from rivers and streams damage the habitats of anadromous fish (species that spend most of their lives in the ocean but hatch and spawn in freshwater). The overdraft of coastal aquifers increases seawater intrusion, which requires water wells in these areas to be either deepened or abandoned, or water supplies to be imported (Hanson, Martin, and Koczot 2003). Terrestrial runoff and wastewater discharges can be harmful to coastal areas. Their effects are exacerbated when heavy rainfall washes large amounts of fertilizers and other pollutants from the land or causes wastewater systems to overflow and send untreated or inadequately treated wastes into streams, estuaries, and the ocean (Ho Ahn et al. 2005). Finally, the loss of wetlands due to increasing urbanization and development will reduce the resiliency of these coastal ecosystems (CNRA 2010).

9.3 Ocean and Coastal Impacts to Ecosystems

Overview

The global ocean—in particular the Pacific Ocean for the U.S. West Coast—plays a significant role in shaping coastal ecosystem processes. As climate and ocean chemistry continue to change, significant alterations in the composition, structure, and function of coastal ecosystems are anticipated. These changes will manifest most clearly as a result of rising sea levels, changing ocean temperatures, and increasing acidity of coastal waters—each of which is discussed below. The relationship between humanity and the coastal environment will inevitably shift in response to dynamic ocean and coastal ecosystems, and each of the changes enumerated above is likely to intensify threats to human development in coastal regions.

Sea-level rise

As sea levels rise, tidal wetlands and beaches will accrete vertically to keep up, become inundated, or "migrate" landward. Their fate depends on whether there is adequate sediment from nearby watersheds to increase wetland elevation as the sea rises and on the availability of space into which wetlands can migrate (CNRA 2010). Coastal development affects this by altering sediment availability (through, for instance, reduction of sand discharge from streams through the construction of dams and debris basins, and by eliminating bluff erosion through coastal armoring) and by occupying or protecting space into which wetlands might otherwise migrate. The loss of coastal wetlands causes the loss of the numerous benefits they provide, including flood protection, water treatment, recreation, carbon sequestration, biodiversity, and wildlife habitat (see Box 9.2) (King, McGregor, and Whittet 2011). Specifically, with a rise in sea level projected to be as high as 4.6 feet (1.4 meters) by 2100, approximately 97,000 acres of coastal wetlands in California will potentially be inundated. Nearly 55% of these wetland areas may be able to migrate inland successfully with no loss of function; however, about 45% could lose either their habitat functions or their ability to migrate (see Figure 9.4) (Heberger et al. 2009).

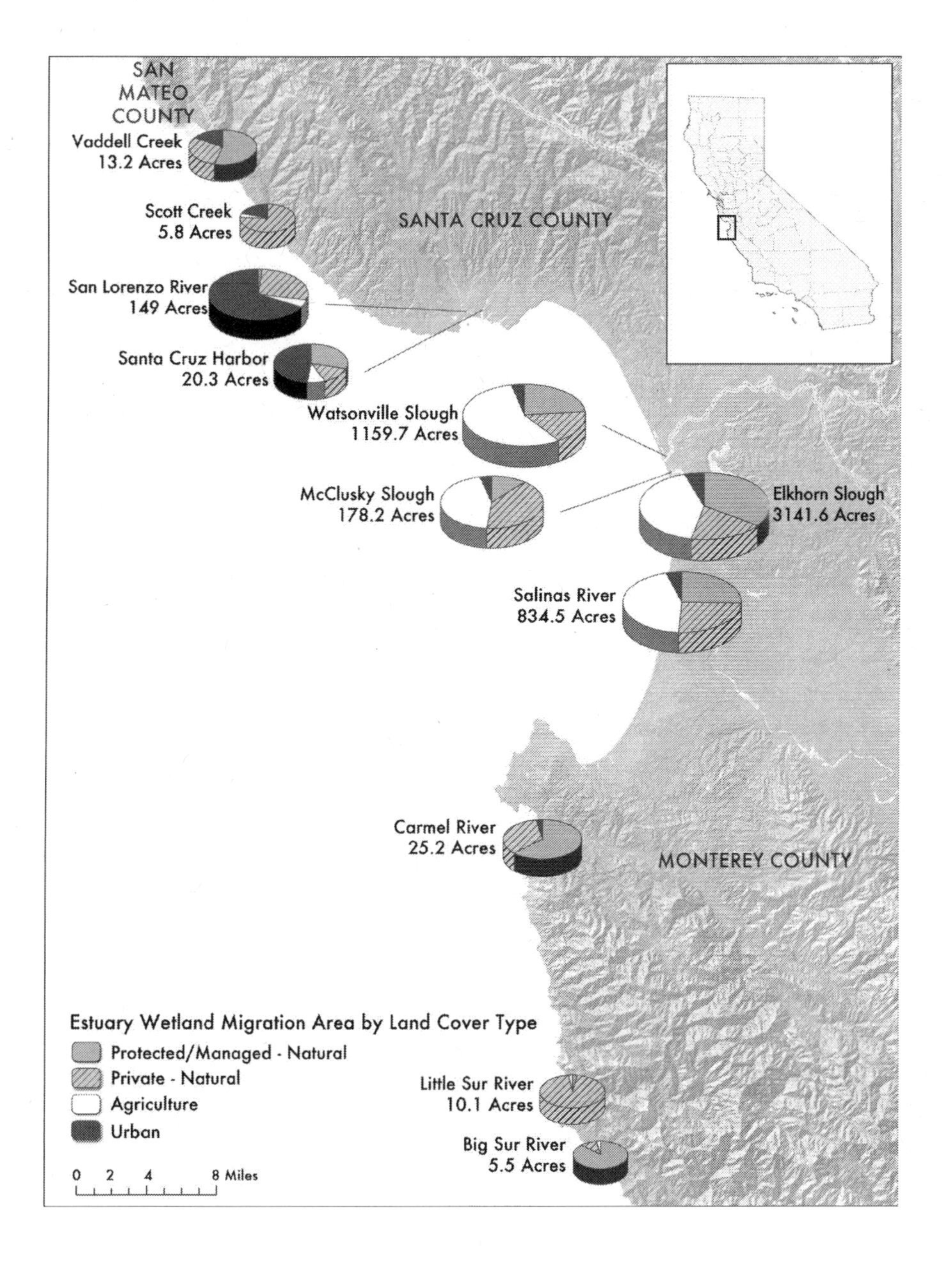

Figure 9.4 Estuary wetland migration area by land-cover type in the Monterey Bay region. Different land-use types will have different capacities to accommodate wetland migration, ranging from urban areas (which are unlikely to accommodate migration at all) to public natural areas (which will likely accommodate migration completely). Between these extremes are agricultural areas and privately owned natural areas, both of which could accommodate migration if landowners choose not to prevent it, such as by not fortifying or armoring their lands. Adapted from Gleason et al. (2011, 23); Heberger et al. (2009).

Box 9.2

Adaptation to Climate Change in the Marine Environment: The Gulf of Farallones and Cordell Bank National Marine Sanctuaries

In 2010, a joint advisory committee for the Gulf of Farallones and Cordell Bank National Marine Sanctuaries, located off the central California coast, published a report on climate change impacts (Largier, Cheng, and Higgason 2010). The study determined that climate change will affect the region's marine waters and ecosystems through a combination of physical changes—including sea-level rise, coastal erosion and flooding, changes in precipitation and runoff, ocean-atmosphere circulation, and ocean water properties (such as acidification due to absorption of atmospheric CO_2)—and biological changes, including changes in species' physiology, phenology, and population connectivity, as well as species range shifts. With this foundational document in hand, sanctuary managers held a series of workshops aimed at developing an adaptation framework that involved both the sanctuaries and their partners onshore and in the marine environment. From those efforts and underlying studies, they determined that the success of adaptation strategies for the marine environment will depend not only on the magnitude and nature of climatic changes, but also on the pressures that already exist in marine environments, including the watershed drainage to the sanctuaries. For example, an adaptation strategy for estuaries and near-shore waters that addresses the changing timing and amount of water from spring snowmelt or more frequent winter storms will also require knowledge about whether the watershed is urbanized, agricultural, or relatively undeveloped. Efforts to foster marine ecosystem adaptation to climate change will require both stringent measures that reduce the global concentration of CO_2 in the atmosphere and the reduction of additional pressures on the regional marine environment (e.g. air pollution, runoff from land into the ocean, waste disposal, and the loss of the filtering and land stabilization services of coastal wetlands) (Kelly et al. 2011).

In 2010, the California Ocean Protection Council issued interim guidance for state and local agencies to use for project planning and development in response to projected sea-level rise (see Table 9.1; CCAT 2010). The same year, the state governments of California, Oregon, and Washington, along with federal agencies—the National Oceanic and Atmospheric Administration, the U.S. Geological Survey, and the Army Corps of Engineers—initiated a study with the National Research Council (NRC) to develop regional West Coast estimates of future sea-level rise to better inform state and local planning and agency decisions (Schwarzenegger 2008). The NRC report was released in June 2012.

Changes or trends for other coastal environmental conditions, such as atmospheric temperatures, precipitation patterns, river runoff and flooding, wave heights and run-up (waves reaching landward), storm frequency and intensity, and fog persistence, are less well understood, often to the point of uncertainty about the direction of change, much less its extent for a specific region. In addition, there will be other changes from rising sea level. For example, "extreme events"—such as the contemporary understanding of 100-year floods—will occur more frequently as a result of both higher coastal

Table 9.1 Static sea-level rise projections (without considering storm events) using the year 2000 as the baseline sea level (California Climate Action Team Sea-Level Rise Interim Guidance Document)

Year	Scenario	Average of Models	Range of Models
2030		7 in (18cm)	5–8 in (13–21 cm)
2050		14 in (36 cm)	10–17 in (26–43 cm)
2070	Low	23 in (59 cm)	17–27 in (43–70 cm)
	Medium	24 in (62 cm)	18–29 in (46–74 cm)
	High	27 in (69 cm)	20–32 in (51–81 cm)
2100	Low	40 in (101 cm)	31–50 in (78–128 cm)
	Medium	47 in (121 cm)	37–60 in (95–152 cm)
	High	55 in (140 cm)	43–69 in (110–176 cm)

Note: For dates after 2050, three different values for sea-level rise are included, based on the IPCC 2007 low, medium, and high GHG emission scenarios as follows: B1 for low projections, A2 for the medium projections, and A1FI for the high projections.

In contrast to the Sea-Level Rise Interim Guidance Document, in this assessment report we refer to the B1 emissions scenario as "low emissions" and the A2 emissions scenario as "high emissions."

Sources: Vermeer and Rahmstorf (2009), IPCC (2007).

storm surges due to sea-level rise and from inland runoff due to extreme rainfall events (see Chapter 7). In addition, tides will extend farther inland in coastal streams and rivers, and saltwater will penetrate farther into coastal aquifers (Loaiciga, Pingel, and Garcia 2012).

Changes in ocean temperature and dynamics

Direct climate change impacts, such as warming sea surface temperatures and ocean acidification, are expected to accelerate or exacerbate the impacts of *present* threats to coastal ecosystems, including pollution, habitat destruction, and over-fishing (Scavia et al. 2002). Warming atmospheric temperatures have already led to an increase in surface-water temperatures and a decrease in the oxygen content of deeper waters (Bograd et al. 2008; Deutsch et al. 2011). Elevated surface temperatures and higher nutrient runoff have led to increased harmful algal blooms and increases in hypoxia in the coastal ocean (Kudela, Seeyave, and Cochlan 2010; Ryan, McManus, and Sullivan 2010). As oceans warm, species adapted to these conditions may be able to expand their native ranges and migrate into ("invade") new regions (see Figure 9.5). For example, Humboldt squid have recently invaded central and Northern California waters, preying on species of commercial importance such as Pacific hake (Zeidberg and Robison 2007). In addition, warmer waters lead to habitat loss for species that are adapted to very specific temperature ranges (Stachowicz et al. 2002). Along with range expansion, the number of invasive species, the rate of invasion, and resulting impacts will increase as coastal ocean waters warm (Stachowicz et al. 2002).

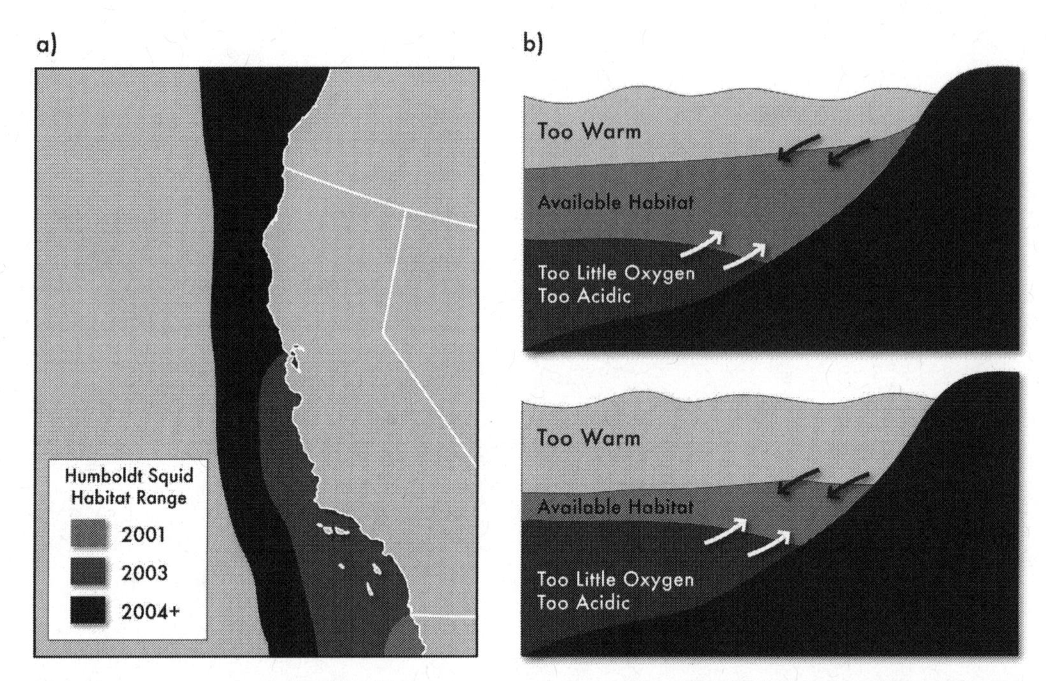

Figure 9.5 Impacts of climate change on marine species distributions and habitat. Many marine species are confined to particular habitats based on water temperature, salinity, or depth. In panel (a) Humboldt squid are confined at their northern edge by temperature. In 2003, the northern edge reached the mouth of the San Francisco Bay, but has recently expanded as far north as Alaska. In panel (b) some fish have limited habitat due to temperature levels of shallower waters above and oxygen or acidity levels of deeper waters below. As surface waters warm and oxygen minimum zones expand or acidity increases, the available habitat for these species is compressed. This leads to lower available resources and potentially increased predation. Source: Bograd et al. (2010), Stramma et al. (2011).

Changes in climate will alter the wind fields that drive coastal upwelling (Bakun 1990; Checkley and Barth 2009; Young, Zieger, and Babanin 2011). It is not clear, however, if changing wind patterns will increase or decrease coastal upwelling or whether each may occur in different locations. Warming surface waters are expected to increase stratification (rate of change in density over depth) and may deepen the thermocline (an abrupt temperature gradient extending from a depth of about 300 feet to 3,000 feet [100m to 1000m]), resulting in a decrease in the amount of nutrients that are delivered to the surface. The timing of seasonal upwelling—during which cold, nutrient-rich water rises to the surface—may shift. Such a mismatch between physical and ecological processes can lead to significant ecosystem consequences (Pierce et al 2006; Barth et al. 2007; Bakun et al. 2010). It is generally accepted that upwelling will be affected by climate change; however, experts differ over what specific changes will occur (Bakun 1990;

Checkley and Barth 2009; Young, Zieger, and Babanin 2011). Hypoxic events—the occurrence of dangerously low oxygen levels that can lead to widespread die-offs of fish or other organisms—will increase as stratification and coastal agricultural runoff increase (Chan et al. 2008). Moreover, as waters warm, they become less able to hold oxygen, resulting in long-term reductions in ocean oxygen content (Bograd et al. 2008; Deutsch et al. 2011). Because warmer coastal waters already are closer to hypoxic thresholds, weaker phytoplankton blooms and smaller nutrient inputs could initiate hypoxia, possibly leading to more frequent, larger, or longer-lasting events, even in regions that have not previously experienced hypoxia.

Ocean acidification

Increased atmospheric CO_2 continues to dissolve in the ocean, making the ocean significantly more acidic than during the preindustrial age (Feely, Doney, and Cooley 2009). Lower pH (more acidic) seas will alter marine ecosystems in ways we do not fully understand, but several predictions are clear: (1) there will be ecological winners and losers as species respond differently to a changing environment (Kleypas et al. 2006; Fabry et al. 2008; Ries et al. 2009; Kroeker et al. 2010); (2) areas of coastal upwelling and increased nutrient runoff will be the most affected (see Figure 9.6) (Kleypas et al. 2006; Feely et al. 2008; Cai et al. 2011; Kelly et al. 2011); and (3) an increase in the variance of pH in nearshore waters may be more biologically important than the changing global average pH, as high frequency peaks in the amount of CO_2 dissolved in water can push marine species beyond their physiological tolerance limits (Thomsen et al. 2010; Hofmann et al. 2011).

Marine food webs are shifting in the already-acidified ocean. Higher CO_2 increases algal growth while hindering the development of shells and other hard parts in mollusks, corals, and other marine animals. These changes are already having direct economic effects: upwelling-intensified acidification has severely harmed several years of hatchery-bred oyster larvae, sending reverberations throughout the industry (Welch 2010; Barton et al. 2012). While the oyster fishery is a relatively small segment of the U.S. seafood industry, about 75% ($3 billion) of the overall industry directly or indirectly depends upon calcium carbonate (the component of shell material that dissolves in lower pH waters). An acidified ocean may change which species the industry targets for cultivation (Cooley and Doney 2009; Langston 2011). Beyond these ecological and economic impacts, ocean acidification is also anticipated to pose direct threats to human health by increasing the number and intensity of harmful algal blooms, which can result in amnesic shellfish poisoning (a disease in humans caused by ingestion of toxins that concentrate in shellfish (Sun et al. 2011; Tatters, Fu, and Hutchins 2012). Existing policy tools to combat the effects of ocean acidification include improved coastal management and more stringent pollution controls under the U.S. Clean Water Act, but addressing the root cause will require reducing atmospheric CO_2 globally (Kelly et al. 2011).

9.4 Coastal Impacts to Communities

Overview

Development along coasts often places residences, coastal tourism development, community resources, and public infrastructure at risk from floods and/or ongoing coastal

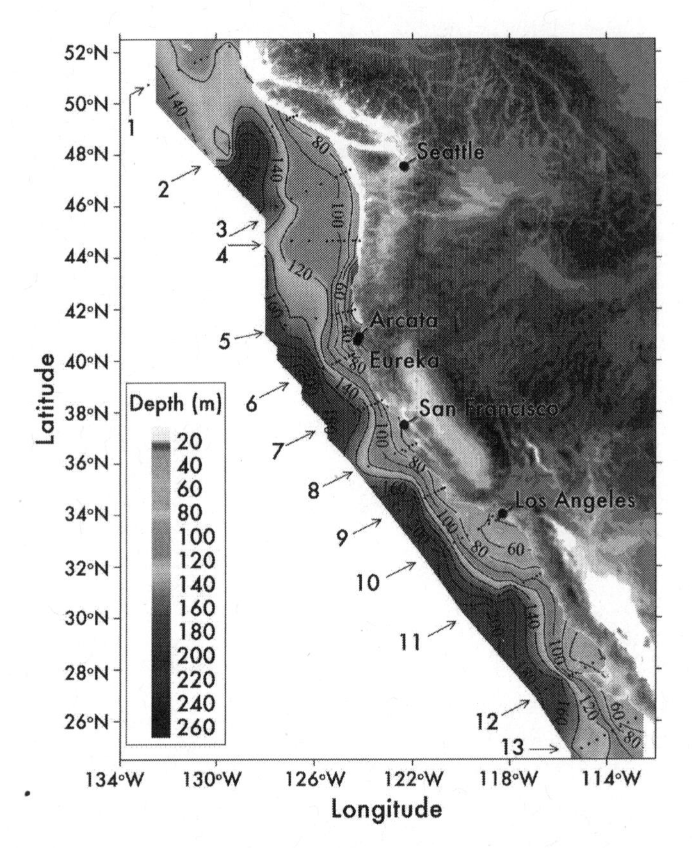

Figure 9.6 Coastal impacts of ocean acidification. This image depicts the aragonite saturation depth on the continental shelf of western North America; lighter shades indicate shallower depths. Aragonite is one of the two most common forms of calcium carbonate, which forms naturally in almost all mollusk shells. Below this depth, it becomes difficult for mollusks and other species to precipitate the calcium carbonate necessary to make shell material. Corrosive waters—those that begin to dissolve calcium carbonate—now occur at shallower depths than in the past because the ocean is absorbing increasing amounts of CO_2 from the atmosphere. Note that in transect 5, corrosive water reaches the ocean surface north of Eureka and Arcata, California. Modified from Feely et al. (2008), reprinted with permission from the American Association for the Advancement of Science.

erosion (see also sections below and Chapter 14 [for more on transportation infrastructure] and Chapter 12 [for more on power plants and energy infrastructure]). Sea-level rise will expand the areas at risk from flooding, accelerate erosion of coastal bluffs and dunes, and, as discussed earlier, permanently inundate large areas of coastal wetlands (Heberger et al. 2009; Revell et al. 2012). Table 9.2 shows some of the current and future vulnerabilities to both flooding and erosion related to increased exposure of the coastal bluff base to expected rise in sea level of 4.6 feet (1.4-meters) by 2100 with no additional development along the coast beyond what existed in 2000 (Revell et al. 2012). In addition, based on a methodology that correlates bluff erosion with increased frequency of exposure to wave attack, erosion could claim as much as nearly 9,000 acres of dunes and 17,000 acres of coastal bluffs from the open ocean coast between the California-Oregon border and Santa Barbara County (Revell et al. 2012).

Table 9.2 Estimated flood and erosion losses for California associated with future sea-level rise on the ocean and bay shoreline

Asset or Concern	Risk [a]	Number or Dollar Amount at Risk in 2000	Number or Dollar Amount at Risk in 2100 (with 55 Inches of Rise in Sea Level)	Dominant Location of Risk [b] (2000/2100)
People	100-year Flood	260,000	410,000	SF Bay/SF Bay
Replacement value of buildings	100-year Flood	$50 billion	$109 billion	Both/SF Bay
People	Erosion [c]		14,000	ND/Ocean
Number of land parcels lost	Erosion [c]		10,000	ND/Ocean
Value of property loss	Erosion [c]		$14 billion	ND/Ocean
Schools	100-year Flood	65	137	Both/SF Bay
Healthcare facilities	100-year Flood	20	55	SF Bay/SF Bay
Police, fire and training areas	100-year Flood	17	34	SF Bay/SF Bay
Hazardous waste sites	100-year Flood	134	332	ND/ND
Highway and (road) miles	100-year Flood	222 (1,660)	430 (3,100)	Ocean/Ocean
Power plants number and (megawatt capacity)	100-year Flood		30 (10,000)	ND/Ocean
Waste treatment plants number and (millions of gallons/day capacity)	100-year Flood		29 (530)	ND/SF Bay

Note:

(a) Flood risks and erosion risks are not mutually exclusive; many of the same assets will be at risk from both flood and erosion.

(b) ND is no data – often since the analysis looked at change from the current conditions.

(c) Erosion impacts were only examined for the open ocean coast from Del Norte County through Santa Barbara. Estimates for erosion losses do not include Ventura, Los Angeles, Orange, or San Diego counties. Nor do estimates include San Francisco Bay, since, "In San Francisco, however, the erosion-related risk is small." (Heberger et al. 2009)

Source: Heberger et al. (2009).

Airport infrastructure

Several of California's key transportation facilities are situated in coastal areas and will be affected by rising sea level. For example, the runways at both the San Francisco (SFO) and Oakland (OAK) International Airports will begin to flood with a 16-inch (40-cm) sea-level rise (within the 2050 sea-level rise scenarios in Table 9.1) (see Figure 9.7). This change would severely impact not only airlines and passengers internationally, but also air cargo, as SFO is the largest air cargo handler in the region and expects to double cargo throughput in the next thirty years. Given the importance of these airports to the local and regional economy, an increase in runway elevation, floodwalls, or the development of some alternative response strategies will be required to avoid these debilitating impacts (Metropolitan Transportation Commission 2004).

Figure 9.7 Impacts to San Francisco and Oakland International Airports. The impacts to San Francisco and Oakland airports will require planning and resources to ensure that these major economic drivers for the San Francisco Bay Area can continue to operate in the future. Modified from Siegel and Bachand (2002), Knowles (2008); see also http://www.bcdc.ca.gov/planning/climate_change/index_map.shtml, Central Bay West Shore map 16 & 55, Central Bay East Shore map 16 & 55.

Vehicular transportation infrastructure

The rise in sea level and increased frequency and intensity of storm events will lead to a combination of increased shoreline inundation and landslides induced by rainfall or wave erosion. With even small sections of roadways disrupted due to these processes, the greater transportation network will be at risk. As a result, the California Department of Transportation prepared guidance on incorporating sea-level rise into project programming and design (Caltrans 2011) (see Box 9.3). (For further discussion of climate impacts on transportation systems, see Chapter 14.)

Box 9.3

The Role of Adaptation in California Ports

California's three major ports—Los Angeles, Long Beach, and Oakland—had a combined throughput of over $350 billion[7] in cargo in 2009, equivalent to 13% of the GDP of the six Southwest states. With such noted economic importance, port authorities are starting to address issues related to sea-level rise. While the greater water depth that will accompany rising sea level will help deeper draft ships, many landside changes will be needed (see Chapter 14). The Port of Long Beach plans to rebuild the Gerald Desmond Bridge because the air gap (the space between the bottom of the bridge and the top of a ship) is restricting some ship transit to times of low tide. The Port of Los Angeles and the Rand Corporation prepared a climate adaptation study to consider the impacts from rising sea levels on the port. The creation and funding of additional protection or response plans for these ports—and their associated costs—is inevitable (Metropolitan Transportation Commission 2004).

Economy, culture, and identity

A large part of California culture and identity is invested in ocean and coastal resources and shoreline access, including beach-going, surfing, kayaking, hiking, and diving, as well as recreational and commercial fishing. Thus, the socio-economic impacts to coastal communities from sea-level rise go well beyond losses to buildings, properties, and infrastructure. For example, estimated losses to the Venice Beach community from a 100-year flood event after a 4.6-foot (1.4-meter) sea-level rise (the high 2100 scenario from Table 9.1) are $51.6 million (an increase of $44.6 million over the present risk), which includes loss of tax revenue, beach-going spending, ecological value, and other societal costs (King, MacGregor, and Whittet 2011). However, such estimates depend on a set of assumptions which—while reasonable—involve significant uncertainties. For example, the 1983 ENSO event caused over $215 million in damage statewide (in 2010 dollars; Griggs and Brown 1998); a similar event in 2100 would be significantly more damaging under conditions of higher sea level, more intensive development, and greater property values (Griggs and Brown 1998). Thus, future losses may be higher than the best available current economic science suggests. In addition, Native American communities, such as the Yurok and Wiyot of Northern California, are also examining traditional uses of coastal areas and the impacts to tribal lands of sea-level rise (including loss of land due to inundation), undertaking coastal restoration projects, assessing the impacts to salmon of overall ecosystem changes. (For further discussion of the impacts of climate change on the lands and resources of Native nations, see Chapter 17.)

9.5 Managing Coastal Climate Risks

Overview

Due to the high concentration of coastal development, population, infrastructure, and economic activity in coastal counties, continued and growing pressure to protect these assets and activities from rising sea levels is expected. Further concentration of wealth, infrastructure, and people along the coast—which historically has resulted in the tendency to protect and harden developed shorelines—is expected to increase the risk of loss of the remaining natural coastal ecosystems in these areas (see Box 9.4) (CNRA 2009; Hanak and Moreno 2011). About 40% of the backshore area (above the high-water line) along the California coast is in public ownership (federal and non-federal)[ii] (USACE 1971), about 107 miles or 10% of the state's coastline had already been hardened as of 2001 (Griggs, Patsch, and Savoy 2005), and more than 90% of the coast's historical wetland areas have been lost or converted due to diking, drainage, and development (Dahl 1990; Van Dyke and Wasson 2005; Gleason et al. 2011).

Box 9.4

The Role of Insurance and Incentives in Coastal Development

Many federally and state-funded actions and programs continue to protect and subsidize high-risk coastal development by shifting the cost of flood protection and storm recovery from property owners and local governments to state and federal taxpayers. For example, the Federal Emergency Management Agency's (FEMA) National Flood Insurance Program (NFIP) offers flood insurance rates that do not reflect the full risk that policyholders face. In addition, the Army Corps of Engineers frequently funds and executes structural shoreline protection projects, while federal and state post-disaster recovery funding and assistance encourages replacing or rebuilding structures with a high level-of-risk exposure (Bagstad, Stapleton, and D'Agostino 2007). These programs work together to distort market forces and favor the movement of people to the coasts. Meanwhile, reinsurance companies and experts studying the insurance market increasingly urge that premiums better reflect actual risks to ensure a reliable insurance system as climate risks increase (Lloyd's of London 2006, 2008; Kunreuther and Michel-Kerjan 2009; Pacific Council on International Policy 2010).

The NFIP is over-exposed and is running a deficit as of 2010 of nearly $19 billion (Williams Brown 2010). To reduce the financial burdens on the flood insurance program and decrease overall vulnerability, FEMA also administers several grant programs designed to mitigate flood hazards prior to disasters occurring (FEMA 2010). These programs are often used for pre-disaster structural flood mitigation measures, but have also been used for structure acquisition, property buy-outs, and demolition or relocation (Multihazard Mitigation Council 2005). The resulting open space is required to be protected in perpetuity, simultaneously providing natural resource benefits and vulnerability-reduction benefits (FEMA 2010).

As mentioned previously, in 2011 the California Ocean Protection Council issued interim sea-level rise guidance for state and local agencies, thus implementing one of the key strategies proposed in California's first statewide climate change adaptation plan (CNRA 2009; California Ocean Protection Council 2011). While not mandatory, this guidance gives state and local government agencies and officials a scientific basis to vet planning and permitting decisions. The guidance will need to be updated regularly as new scientific information becomes available (e.g., through the 2012 NRC study on sea-level rise along the West Coast). Pragmatically, the management and planning mechanisms through which local governments in California are making adaptive changes include general plan updates, climate action or adaptation plans, local coastal program updates, local hazard mitigation plans, implementing regulations (such as tax or building codes), and special or regular infrastructure upgrades (Moser and Ekstrom 2012). A selected list of climate adaptation planning resources can be found in Table A9.1.

Adaptation options

Coastal managers have several adaptation options (USAID 2009; NRC 2010[iii]; Grannis 2011; NOAA 2011; Russell and Griggs 2012) that typically fall into three categories.

First, structural protection measures such as seawalls and revetments (hardened surface built to protect an embankment) as well as beach replenishment have frequently been the preferred option for local governments trying to protect public shorelines and adjacent coastal properties (which are part of the property and commercial tax base) and maintain or enhance opportunities for coastal tourism. Historically, many beach nourishment projects in California have been opportunistic in the sense that they were a means of disposing sand dredged from harbors or produced from coastal construction rather than stand-alone projects for nourishing beaches.

Hardening the shoreline along the coast has resulted in harmful environmental and ecological impacts, both directly in front of and downdrift from the hardened shoreline. Such impacts include passive erosion or beach loss in front of the hardened shoreline, visual impacts (see Figure 9.8), and reduced public access along the shoreline (Griggs 2005). The impacts of shoreline protection within interior waterways—such as bays or estuaries—include loss of tidal prisms (the volume of water leaving an estuary at ebb tide) and loss of coastal wetlands, which contain important bird habitat, fishery nursing grounds, and the capacity of such natural buffers to retain flood waters (Griggs et al. 1997; Runyan and Griggs 2003).

Second, adaptation measures that continue to allow coastal occupancy and yet aim to reduce risks of coastal erosion and flooding are common elements of hazard-mitigation plans and land-use planning (local coastal programs) under the California Coastal Act in coastal communities. These measures include adjustments to building codes (such as requirements for the use of flood-prone basements within flood zone areas) or modifications to standards for development and coastal construction (such as setbacks for building from the shoreline, limits to how much land surface can be made impervious, the amount of freeboard required between the ground and the first inhabited floor, and other flood protection measures, including stormwater retention and treatment on the property).

Generally speaking, measures depend on the environment in which development is situated. For cliff and bluff top construction, zoning or construction policies may contain standards for cliff edge setbacks and requirements to improve onsite water drainage to minimize cliff erosion may be considered (Griggs, Pepper, and Jordan 1992). For low lying areas, coastal plains, beaches, and bayside waterfront areas, development standards requiring construction above base flood elevations and setbacks from high-risk flood and/or erosion areas may be most relevant.

Figure 9.8 Coastal armoring in Southern California. One-third of the shoreline of Southern California (including Ventura, Los Angeles, Orange and San Diego Counties) has now been armored. The photo shows the shoreline in 2010 in Encinitas, in northern San Diego County. Photo courtesy of Kenneth and Gabrielle Adelman of the California Coastal Records Project (http://www.californiacoastline.org/).

Finally, a variety of adaptation measures focus on reducing long-term exposure to the risks associated with climate change and coastal hazards. Such measures might take the form of planned retreat from the shoreline, but might also include the restoration of natural coastal buffers, such as dunes and wetlands. Of particular value are strategies and policies that incorporate natural resource values and management (California Coastal Act 1976; UNCBD 2009). Such ecosystem-based adaptation is an approach that simultaneously builds ecological resilience and reduces the vulnerability of both human and natural communities to climate change.[iv] It is based on the premise that sustainably managed ecosystems can provide social, economic, and environmental benefits, both directly through the preservation of innately valuable biological resources and indirectly through the protection of ecosystem services that these resources provide humans (Coll, Ash, and Ikkala 2009; World Bank 2010).

Level of preparedness and engagement in adaptation planning

A 2005 survey of California coastal counties and communities assessed coastal managers' awareness of the risks associated with climate change and the degree to which they had begun preparing for, planning for, and actively managing these risks in their coastal management activities (Moser 2007; Moser and Tribbia 2007). The vast majority of surveyed coastal managers were of the opinion that climate change is real and is already happening and were significantly concerned about the associated risks. As of 2005, however, very few local governments had taken up the challenge of developing strategies to deal with those risks.

In a follow-up survey conducted in summer 2011, some important shifts could be noted (see Figure 9.9) (Hart et al. 2012). Most remarkable was the increase in the level of activity on adaptation from 2005 to 2011. In 2005, only two of the responding coastal counties and one of the participating cities had climate change plans in place, and four counties and six cities were developing such a plan. By 2011, many more coastal communities in California had begun examining and planning for the impacts of climate change.[v]

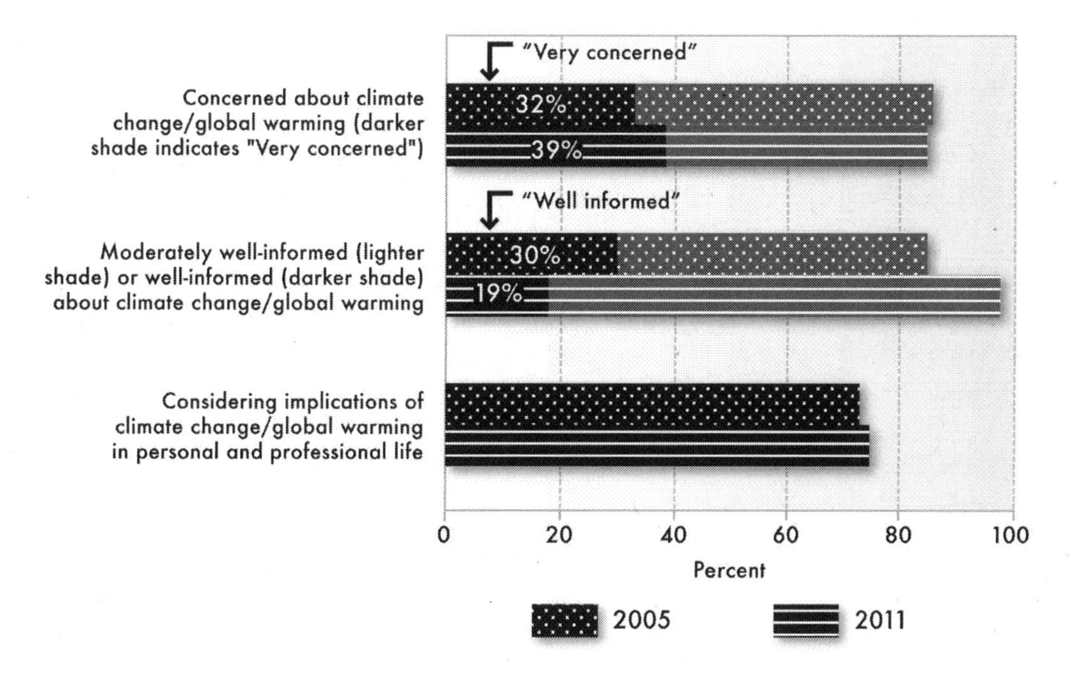

Figure 9.9 California coastal managers' attitudes toward climate change. Well over 80% of California coastal managers are concerned with climate change. The proportion saying they are "very concerned" increased significantly over the past six years. Meanwhile, local managers feel only moderately well informed, indicating a significant need for education. One indication of their readiness to advance adaptation planning is the high proportion of respondents (75%) who report that they already consider the implications of climate change in their personal and professional lives. Source: Moser (2007), Moser and Tribbia (2007), Hart et al. (2012).

A separate analysis highlighted six municipalities in California that have developed local climate adaptation plans or components thereof (Georgetown Climate Center 2012). An Ocean Protection Council resolution passed in June 2007 encouraged Local Coastal Plan (LCP) amendments to address sea-level rise, yet few local governments have even begun the process of considering such LCP amendments. Of particular regional significance is also the overall slow response of the region's major ports and marine facilities, albeit not a unique response among North American or international ports where adaptation planning is just beginning (California Ocean Protection Council 2007; California State Lands Commission 2009; Becker et al. 2012).

Thus, while the state of California has been fairly progressive in adaptation planning (for example, with a Climate Adaptation Strategy, CalAdapt website, and the 2012 sea-level rise study completed by the National Research Council), most adaptation actions will be implemented locally or regionally (often with state and federal support and permits). However, few local governments have begun taking steps to implement either their own plans or the State's existing recommendations.

Barriers to adaptation

Several studies have examined impediments or barriers to adaptation for individuals, communities, organizations, and entire nations.[vi] Increasing empirical evidence from California strongly confirms the presence of barriers to adaptation in coastal communities (Hanak and Moreno 2011). In the above-mentioned 2005 survey of local jurisdictions, coastal managers considered their top barriers to adaptation management to be local monetary constraints, insufficient staff resources, lack of supportive funding from state and federal sources, the all-consuming nature of currently pressing issues, and the lack of a legal mandate to undertake adaptation planning (Moser and Tribbia 2007). When asked again in 2011, the lack of funding to prepare and implement a plan, lack of staff resources to analyze relevant information, and the all-consuming currently pressing issues were again mentioned as overwhelming hurdles for local coastal professionals, followed (with far less frequency) by issues such as lack of public demand to take adaptation action, lack of technical assistance from state or federal agencies, and lack of coordination among organizations (Moser and Ekstrom 2012). Case study research in two cities and two counties in the San Francisco Bay Area found that institutional barriers dominate, closely followed by attitudinal barriers among decision makers. Funding-related barriers were important, but ranked only third (Storlazzi and Griggs 2000; Griggs, Patsch, and Savoy 2005; Kildow and Colgan 2005; Moser and Ekstrom 2012).

There is additional independent evidence that local jurisdictions vary considerably in their technical expertise and capacity to engage in effective coastal land-use management and that they do not use available management tools to the fullest extent possible to improve coastal land management overall (Tang 2008, 2009). For example, experts assert that the California Coastal Act and the Public Trust Doctrine are considerably underutilized in protecting public trust areas and the public interest (Caldwell and Segall 2007; Peloso and Caldwell 2011). Thus, the persistence of this range of institutional, attitudinal, economic, and other adaptation barriers goes a long way toward accounting for the low level of actual preparedness and lack of active implementation of adaptation strategies in coastal California.

The in-depth case studies conducted in the San Francisco Bay Area, however, also reveal that local communities have many opportunities, assets, and advantages that can help them avoid adaptation barriers in the first place, or which they can leverage in efforts to overcome those barriers they encounter (see Chapter 19, Box 19.4). Among the most important of these advantages and assets are people and existing plans and policies that facilitate and allow integration of adaptation and climate change (Moser and Ekstrom 2012) (see also Chapter 18, Section 18.7).

In conclusion, adaptation in coastal California is an emerging mainstream policy concern wherein institutions and the individuals involved—along with supporting financial and technical resources—pose the greatest barriers and constitute the greatest assets in avoiding and overcoming them. While some barriers originate from outside sources (such as the national economic crisis or federal laws and regulation) and communities require state and federal support to overcome entrenched challenges (such as legal and technical guidance or fiscal support), local communities have the power and control to overcome many of the challenges they face (Moser and Ekstrom 2012; for further discussion of the effects of climate change on urban areas, see Chapter 13, and for a discussion of local adaptation and mitigation choices, see Chapter 18).

References

Bagstad, K. J., K. Stapleton, and J. R. D'Agostino. 2007. Taxes, subsidies, and insurance as drivers of United States coastal development. *Ecological Economics* 63:285–298.

Bakun, A. 1990. Global climate change and intensification of coastal ocean upwelling. *Science* 247:198–201.

Bakun, A., D. B. Field, A. Redondo-Rodriguez, and S. J. Weeks. 2010. Greenhouse gas, upwelling-favorable winds, and the future of coastal ocean upwelling ecosystems. *Global Change Biology* 16:1213–1228.

Barth, J. A., B. A. Menge, J. Lubchenco, F. Chan, J. M. Bane, A. R. Kirincich, M. A. McManus, et al. 2007. Delayed upwelling alters nearshore coastal ocean ecosystems in the Northern California current. *Proceedings of the National Academy of Sciences* 104:3719–3724.

Barton, A., B. Hales, G. Waldbusser, C. Langdon, and R. A. Feely. 2012. The Pacific oyster, *Crassostrea gigas*, shows negative correlation to naturally elevated carbon dioxide levels: Implications for near-term ocean acidification effects. *Limnology and Oceanography* 57:698–710.

Becker, A., S. Inoue, M. Fischer, and B. Schwegler. 2012. Climate change impacts on international seaports: Knowledge, perceptions, and planning efforts among port administrators. *Climatic Change* 110:5–29, doi:10.1007/s10584-011-0043-7.

Bograd, S. J., C. G. Castro, E. Di Lorenzo, D. M. Palacios, H. Bailey, W. Gilly, and F. P. Chavez. 2008. Oxygen declines and the shoaling of the hypoxic boundary in the California Current. *Geophysical Research Letters* 35: L12607, doi:10.1029/2008GL034185.

Bograd, S. J., W. J. Sydeman, J. Barlow, A. Booth, R. D. Brodeur, J. Calambokidis, F. Chavez, et al. 2010. Status and trends of the California Current region, 2003–2008. In *Marine ecosystems of the North Pacific Ocean, 2003–2008*, ed. S.M. McKinnell and M.J. Dagg, 106–141. PICES Special Publication 4. Sidney, BC: North Pacific Marine Science Organization.

Bromirski, P. D., A. J. Miller, R. E. Flick, and G. Auad. 2011. Dynamical suppression of sea level rise along the Pacific Coast of North America: Indications for an imminent acceleration. *Journal of Geophysical Research* 116: C07005, doi:10.1029/2010JC006759.

Cai, M., J. S. Schwartz, B. R. Robinson, S. E. Moore, and M. A. Kupl. 2011. Long-term annual and seasonal patterns of acidic deposition and stream water quality in a Great Smoky Mountains high-elevation watershed. *Water Air Soil Pollution* 219:547–562, doi: 10.1007/s11270-010-0727-z.

Caldwell, M., and C. H. Segall. 2007. No day at the beach: Sea level rise, ecosystem loss, and public access along the California coast. *Ecology Law Quarterly* 34:533–578.

California Climate Action Team (CCAT). 2010. State of California sea-level rise interim guidance document. California: Coastal and Ocean Working Group of the California Action Team and Ocean Protection Council. http://www.slc.ca.gov/Sea_Level_Rise/SLR_Guidance_Document_SAT_Responses.pdf.

California Coastal Act of 1976, Pub. Res. Code § 30000 et. seq. 2010.

California Department of Transportation (Caltrans). 2011. *Guidance on incorporating sea level rise: For use in the planning and development of Project Initiation.* N.p.: Caltrans, Climate Change Workgroup.

California Natural Resources Agency (CNRA). 2009. *The California Climate Adaptation Strategy 2009.* A Report to the Governor of the State of California (Draft). Sacramento: CNRA.

—. 2010. *State of the State's Wetlands Report: 10 Years of challenges and progress.* Sacramento: CNRA.

California Ocean Protection Council. 2007. Resolution of the California Ocean Protection Council on climate change, June 14, 2007. http://www.opc.ca.gov/2007/06/resolution-of-the-california-ocean-protection-council-on-climate-change/.

—. 2011. Resolution of the California Ocean Protection Council on sea-level rise, March 11. http://www.opc.ca.gov/2011/04/resolution-of-the-california-ocean-protection-council-on-sea-level-rise/. California State Lands Commission. 2009. *A report on sea level rise preparedness: Staff report to the California State Lands Commission.* http://www.slc.ca.gov/reports/sea_level_report.pdf.

Chan, F., J. A. Barth, J. Lubchenco, A. Kirincich, H. Weeks, W. T. Peterson, and B. A. Menge. 2008. Emergence of anoxia in the California Current large marine ecosystem. *Science* 319: 920.

Chavez, F. P., J. Ryan, S. E. Lluch-Cota, and M. Ñiquen. 2003. From anchovies to sardines and back: Multidecadal change in the Pacific Ocean. *Science* 299:217–221.

Checkley, D. M., and J. A. Barth. 2009. Patterns and processes in the California Current system. *Progress in Oceanography* 83:49–64. doi:10.1016/j.pocean.2009.07.028.

Coll, A., N. Ash, and N. Ikkala. 2009. Ecosystem-based adaptation: A natural response to climate change. Gland, Switzerland: International Union for the Conservation of Nature (IUCN).

Cooley, S. R., and S. C. Doney. 2009. Anticipating ocean acidification's economic consequences for commercial fisheries. *Environmental Research Letters* 4: 024007, doi:10.1088/1748-9326/4/2/024007.

Cruce, T. L. 2009. Adaptation planning – What U.S. states and localities are doing. Pew Center on Global Climate Change Working Paper. http://www.c2es.org/docUploads/state-adapation-planning-august-2009.pdf.

Dahl, T. E. 1990. Wetlands losses in the United States 1780's to 1980's. Washington, DC: U.S. Fish and Wildlife Service / Jamestown, ND: Northern Prairie Wildlife Research Center.

Deutsch, C., H. Brix, T. Ito, H. Frenzel, and L. Thompson. 2011. Climate-forced variability of ocean hypoxia. *Science* 333:336–339, doi:10.1126/science.1202422.

Dugan, J. E., D. M. Hubbard, I. F. Rodil, D. L. Revell, and S. Schroeter. 2008. Ecological effects of coastal armoring on sandy beaches. *Marine Ecology* 28:160–170.

Ekstrom, J., S. C. Moser, and M. Torn. 2011. *Barriers to climate change adaptation: A diagnostic framework.* Public Interest Energy Research (PIER) Final Report CEC-500-2011-004. Sacramento: California Energy Commission.

Fabry, V. J., B. A. Seibel, R. A. Feely, and J. C. Orr. 2008. Impacts of ocean acidification on marine fauna and ecosystem processes. *ICES Journal of Marine Science* 65:414-432.

Federal Emergency Management Agency (FEMA). 2010. Hazard mitigation assistance unified guidance. http://www.fema.gov/library/viewRecord.do?id=4225.

Feely, R. A., S. C. Doney, and S. R. Cooley. 2009. Ocean acidification: Present conditions and future changes in a high-CO_2 world. *Oceanography* 22 (4): 37–47.

Feely, R. A., C. L. Sabine, J. M. Hernandez-Ayon, D. Ianson, and B. Hales. 2008. Evidence for upwelling of corrosive acidified water onto the continental shelf. *Science* 320:1490–1492.

Georgetown Climate Center. 2012. State and local adaptation plans. http://www.georgetownclimate.org/node/3325.

Gleason M. G., S. Newkirk, M. S. Merrifield, J. Howard, R. Cox, M. Webb, J. Koepcke, et al. 2011. *A conservation assessment of West Coast (USA) estuaries.* Arlington, VA: The Nature Conservancy.

Grannis, J. 2011. *Adaptation tool kit: Sea-level rise and coastal land use; How governments can use land-use practices to adapt to sea-level rise.* Washington, DC: Georgetown Climate Center.

Griggs, G. B. 2005. The impacts of coastal armoring. *Shore and Beach* 73 (1): 13–22.

Griggs, G. B., and K. Brown. 1998. Erosion and shoreline damage along the central California coast: A comparison between the 1997-98 and 1982-83 winters. *Shore and Beach* 66 (3): 18–23.

Griggs, G. B., K. Patsch, and L. Savoy. 2005. *Living with the changing California coast.* Berkeley: University of California Press.

Griggs, G., J. Pepper, and M. Jordan. 1992. *California's coastal hazards: A critical assessment of existing land-use policies and practices.* Santa Cruz: University of California-Santa Cruz, California Policy Seminar.

Griggs, G. B., J. F. Tait, L. J. Moore, K. Scott, W. Corona, and D. Pembrook. 1997. *Interaction of seawalls and beaches: Eight years of field monitoring, Monterey Bay, California.* U.S. Army Corps of Engineers (USACE), Waterways Experiment Station Contract Report CHL-97-1. Washington, DC: USACE.

Hammer, R. B., S. I. Stewart, R. Winkler, V. C. Radeloff, and P. R. Voss. 2004. Characterizing spatial and temporal residential density patterns across the U.S. Midwest, 1940-1990. *Landscape and Urban Planning* 69:183–199.

Hanak, E., and G. Moreno. 2011. California coastal management with a changing climate. *Climatic Change* 111:45–73, doi:10.1007/s10584-011-0243-1.

Hanson, R. T., P. Martin, and K. M. Koczot. 2003. *Simulation of ground-water/surface-water flow in the Santa Clara–Calleguas Ground-Water Basin, Ventura County, California.* U.S. Geological Survey Water–Resources Investigations Report No. 02-4136. Sacramento: U.S. Geological Survey.

Hart, J. F., P. Grifman, S. C. Moser, A. Abeles, M. Meyers, S. Schlosser, and J. A. Ekstrom. 2012. *Rising to the challenge: Results of the 2011 California Adaptation Needs Assessment Survey.* Report No. USCSG-TR-01-2012. Los Angeles, CA: University of Southern California Sea Grant.

Heberger, M., H. Cooley, P. Hererra, P. H. Gleick, and E. Moore. 2009. *The impacts of sea-level rise on the California coast.* Final Paper CEC-500-2009-024-F. Sacramento: California Climate Change Center.

Ho Ahn, J., S. B. Grant, C. Q. Surbeck, P. M. DiGiacomo, N. P. Nezlin, and S. Jiang. 2005. Coastal water quality impact of stormwater runoff from an urban watershed in Southern California. *Environmental Science and Technology* 39:5940–5953.

Hofmann, G. E., J. E. Smith, K. S. Johnson, U. Send, L. A. Levin, F. Micheli, A. Paytan, et al. 2011. High-frequency dynamics of ocean pH: A multi-ecosystem comparison. *PLoS ONE* 6: e28983, doi:10.1371/journal.pone.0028983.

Intergovernmental Panel on Climate Change (IPCC). 2007. *Climate change 2007: Synthesis report. Contribution of Working Groups I, II and III to the Fourth Assessment Report of the Intergovernmental Panel on Climate Change,* eds. R. K. Pachauri and A. Reisinger. Geneva: IPCC. http://www.ipcc.ch/pdf/assessment-report/ar4/syr/ar4_syr.pdf.

Kelly, R. P., M. M. Foley, W. S. Fisher, R. A. Feely, B. S. Halpern, G. G. Waldbusser, and M. R. Caldwell. 2011. Mitigating local causes of ocean acidification with existing laws. *Science* 332:1036–1037.

Kildow, J., and C. S. Colgan. 2005. *California's ocean economy. Report to the Resources Agency, State of California.* N.p.: The National Ocean Economics Program. http://resources.ca.gov/press_documents/CA_Ocean_Econ_Report.pdf.

King, P. G., A. R. MacGregor, and J. D. Whittet. 2011. *The economic costs of sea-level rise to California beach communities.* Sacramento: California Department of Boating and Waterways.

Kleypas, J. A., R. A. Feely, V. J. Fabry, C. Langdon, C. L. Sabine, and L. L. Robbins. 2006. *Impacts of ocean acidification on coral reefs and other marine calcifiers: A guide for future research.* Report of a workshop held 18–20 April 2005, St. Petersburg, FL, sponsored by NSF, NOAA, and the U.S. Geological Survey. Seattle: NOAA, Pacific Marine Environmental Laboratory.

Knowles, N. 2009. *Potential inundation due to rising sea levels in the San Francisco Bay region.* Final Paper CEC-500-2009-023-F. Sacramento: California Climate Change Center.

Kroeker, K. J., R. L. Kordas, R. N. Crim, and G. G. Singh. 2010. Meta-analysis reveals negative yet variable effects of ocean acidification on marine organisms. *Ecology Letters* 13:1419–1434, doi:10.1111/j.1461-0248.2010.01518.x.

Kudela, R. M., S. Seeyave, and W. P. Cochlan. 2010. The role of nutrients in regulations and promotion of harmful algal blooms in upwelling systems. *Progress in Oceanography* 85:122–135.

Kunreuther, H. C., and E. O. Michel-Kerjan. 2009. *At war with the weather: Managing large-scale risks in a new era of catastrophes.* Cambridge: Massachusetts Institute of Technology Press.

Langston, J. 2011. *Northwest ocean acidification: The hidden costs of fossil fuel pollution.* N.p.: Sightline Institute.

Largier, J. L., B. S. Cheng, and K. D. Higgason, eds. 2010. *Climate change impacts: Gulf of the Farallones and Cordell Bank National Marine Sanctuaries.* Report of a Joint Working Group of the Gulf of the Farallones and Cordell Bank National Marine Sanctuaries Advisory Councils. Marine Sanctuaries Conservation Series ONMS-11-04.

Lloyd's of London. 2006. Climate change: Adapt or bust. 360 Risk Project No.1: Catastrophe trends. London: Lloyd's. http://www.lloyds.com/~/media/3be75eab0df24a5184d0814c32161c2d.ashx.

—. 2008. Coastal communities and climate change: Maintaining future insurability. London: Lloyd's. http://www.lloyds.com/~/media/Lloyds/Reports/360%20Climate%20reports/360_Coastalcommunitiesandclimatechange.pdf.

Loaiciga, H. A., T. J. Pingel, and E. S. Garcia. 2012. Sea water intrusion by sea-level rise: Scenarios for the 21st century. *Ground Water* 50:7–47.

Metropolitan Transportation Commission (MTC). 2004. *Regional goods movement study for the San Francisco Bay Area.* Final summary report. Oakland, CA: MTC. http://www.mtc.ca.gov/pdf/rgm.pdf.

Moser, S. C. 2007. Is California preparing for sea-level rise? The answer is disquieting. *California Coast and Ocean* 22 (4): 24-30.

—. 2009. *Good morning, America! The explosive U.S. awakening to the need for adaptation.* Sacramento: California Energy Commission / Charleston, SC: NOAA Coastal Services Center. http://csc.noaa.gov/publications/need-for-adaptation.pdf.

Moser, S. C., and J. A. Ekstrom. 2010. A framework to diagnose barriers to climate change adaptation. *Proceedings of the National Academies of Science* 107:22026–22031.

—. 2012. *Identifying and overcoming barriers to climate change adaptation in San Francisco Bay: Results from case studies.* Public Interest Energy Research White Paper CEC-500-2012-034. Sacramento: California Climate Change Center.

Moser, S. C., and J. Tribbia. 2007. Vulnerability to inundation and climate change impacts in California: Coastal managers' attitudes and perceptions. *Marine Technology Society Journal* 40 (4): 35-44.

Multihazard Mitigation Council. 2005. *Natural hazard mitigation saves: An independent study to assess the future savings from mitigation activities.* Washington, DC: National Institute of Building Sciences (NIBS).

National Ocean Economics Program (NOEP). 2010. NOEP data. http://www.oceaneconomics.org.

—. 2012. Population and housing data: California. http://www.oceaneconomics.org/Demographics/demogResults.asp?selState=6&selCounty=06000&selYears=All&cbCoastal=show&cbCZM=show&selShow=PH&selOut=display&noepID=unknown. Accessed January 20, 2012.

National Oceanic and Atmospheric Administration (NOAA). 2011. *Adapting to climate change: A planning guide for state coastal managers*. Silver Spring, MD: NOAA Office of Ocean and Coastal Resource Management.

National Research Council (NRC). 2010. *America's Climate Choices: Adapting to the Impacts of Climate Change*. Washington, DC: National Academies Press.

—. 2012. *Sea-level rise for the coasts of California, Oregon and Washington: Past, present and future*. Washington, DC: National Academies Press.

Pacific Council on International Policy (PCIP). 2010. *Preparing for the effects of climate change: A strategy for California; A report by the California Adaptation Advisory Council to the State of California*. Los Angeles: PCIP.

Peloso, M.E., and Caldwell, M.R. 2011. Dynamic property rights: The Public Trust Doctrine and takings in a changing climate. *Stanford Environmental Law Journal* 30 (1):51–120.

Pendleton, L. H. 2009. The economic value of coastal and estuary recreation. In *The economic and market value of America's coasts and estuaries: What's at stake*, ed. L. H. Pendleton. N.p.: Coastal Ocean Values Press.

Pierce, S. D., J. A. Barth, R. E. Thomas, and G. W. Fleischer. 2006. Anomalously warm July 2005 in the Northern California Current: Historical context and the significance of cumulative wind stress. *Geophysical Research Letters* 33: L22S04, doi:10.1029/2006GL027149.

Revell, D. L., B. Battalio, B. Spear, P. Ruggiero, and J. Vandever. 2012. A methodology for predicting future coastal hazards due to sea-level rise on the California coast. *Climatic Change* 109 (Suppl. 1): S251–S276.

Ries, J. B., A. L. Cohen, and D. C. McCorkle. 2009. Marine calcifiers exhibit mixed responses to CO_2-induced ocean acidification. *Geology* 37:1131–1134.

Runyan, K. B., and G. B. Griggs. 2003. The effects of armoring sea cliffs on the natural sand supply to the beaches of California. *Journal of Coastal Research* 19:336–347.

Russell, N. L., and G. B. Griggs. 2012. *Adapting to sea-level rise: A guide for California's coastal communities*. Sacramento: California Energy Commission, Public Interest Environmental Research Program.

Ryan, J. P., M. A. McManus, and J. M. Sullivan. 2010. Interacting physical, chemical, and biological forcing of phytoplankton thin-layer variability in Monterey Bay, California. *Continental Shelf Research* 30:7–16.

Scavia, D., J. C. Field, D. F. Boesch, R. W. Buddemeier, V. Burkett, D. R. Cayan, M. Fogarty, et al. 2002. Climate change impacts on U.S. coastal and marine ecosystems. *Estuaries* 25:149–164.

Schwarzenegger, Arnold. 2008. Executive Order S-13-08. November 14, 2009. Section 1. http://gov.ca.gov/news.php?id=11036.

Siegel, S. W., and P. A. M. Bachand. 2002. *Feasibility analysis, South Bay salt pond restoration, San Francisco Estuary, California*. San Rafael, CA: Wetlands and Water Resources.

Stachowicz, J. J., J. R. Terwin, R. B. Whitlatch, and R. W. Osman. 2002. Linking climate change and biological invasions: Ocean warming facilitates nonindigenous species invasions. *Proceedings of the National Academy of Sciences* 99:15497–15500.

Storlazzi, C.D. and G. B. Griggs. 2000. Influence of El Niño-Southern Oscillation (ENSO) events on the evolution of central California's shoreline. *Geological Society of America Bulletin* 112:236–249.

Stramma, L., E. D. Prince, S. Schmidtko, J. Luo, J. P. Hoolihan, M. Visbeck, D. W. R. Wallace, P. Brandt, and A. Körtzinger. 2011. Expansion of oxygen minimum zones may reduce available habitat for tropical pelagic fishes. *Nature Climate Change* 2:33–37, doi:10.1038/NCLIMATE1304.

Sun, J., D. A. Hutchins, Y. Feng, E. L. Seubert, D. A. Caron, and F. X. Fu. 2011. Effects of changing pCO_2 and phosphate availability on domoic acid production and physiology of the marine harmful bloom diatom *Pseudo-nitzschia* multiseries. *Limnology and Oceanography* 56:829–840.

Tang, Z. 2008. Evaluating local coastal zone land use planning capacities in California. *Ocean and Coastal Management* 51:544–555.

—. 2009. How are California local jurisdictions incorporating a strategic environmental assessment in local comprehensive land use plans? *Local Environment* 14:313–328.

Tatters, A. O., F-X. Fu, and D. A. Hutchins. 2012. High CO_2 and silicate limitation synergistically increase the toxicity of *Pseudo-nitzschia fraudulenta*. *PLoS ONE* 7: e32116, doi:10.1371/journal.pone.0032116.

Thomsen, J., M. A. Gutowska, J. Saphörster, A. Heinemann, K. Trübenbach, J. Fietzke, C. Hiebenthal, et al. 2010. Calcifying invertebrates succeed in a naturally CO_2-rich coastal habitat but are threatened by high levels of future acidification. *Biogeosciences* 7:3879–3891, doi:10.5194/bg-7-3879-2010.

U.N. Convention on Biological Diversity (CBD). 2009. *Connecting biodiversity and climate change mitigation and adaptation: Report of the Second Ad Hoc Technical Expert Group on Biodiversity and Climate Change*. CBD Technical Series No. 41. Montreal: Secretariat of the CBD.

U.S. Agency for International Development (USAID). 2009. *Adapting to coastal climate change: A guidebook for development planners*. http://www.crc.uri.edu/download/CoastalAdaptationGuide.pdf.

U.S. Army Corps of Engineers (USACE). 1971. *National shoreline study: California regional inventory*. San Francisco, CA: USACE, Army Engineer Division.

U.S. Census Bureau. n.d. State and county QuickFacts: California. http://quickfacts.census.gov/qfd/states/06000.html (accessed January 20, 2012; last updated September 18, 2012).

Van Dyke, E., and K. Wasson. 2005. Historical ecology of a Central California estuary: 150 years of habitat change. *Estuaries* 28:173–189.

Vermeer, M., and S. Rahmstorf. 2009. Global sea level linked to global temperature. *Proceedings of the National Academy of Sciences* 106:21527–21532.

Welch, C. 2010. Acidification threatens wide swath of sea life. *Seattle Times*, July 31, 2010. http://seattletimes.nwsource.com/html/localnews/2012502655_acidification01.html.

Williams-Brown, O. 2010. *National Flood Insurance Program: Continued actions needed to address financial and operational issues*. Testimony before the Subcommittee on Housing and Community Opportunity, Committee on Financial Services, U.S. House of Representatives. Report GAO-10-631T. Washington, DC: U.S. Government Accountability Office.

Wolter, K., and M. S. Timlin. 2011. El Niño/Southern Oscillation behaviour since 1871 as diagnosed in an extended multivariate ENSO index. *International Journal of Climatology* 31:1074–1087.

World Bank. 2010. *Convenient solutions to an inconvenient truth: Ecosystem-based approaches to climate change*. Washington, DC: World Bank, Environment Department.

Young, I. R., S. Zieger, and A. V. Babanin. 2011. Global trends in wind speed and wave height. *Science* 332:451–455, doi: 10.1126/science.1197219.

Zeidberg, L. D., and B. H. Robison. 2007. Invasive range expansion by the Humboldt Squid, *Dosidicus gigas*, in the eastern North Pacific. *Proceedings of the National Academy of Sciences* 104:12948–12950.

Appendix

See following page.

Table A9.1 Selected Resources in Support of Coastal Adaptation

Name	Website	Description
CalAdapt	http://cal-adapt.org/	Localized, searchable climate change projections for California
California Climate Change Portal	http://www.climatechange.ca.gov/	Research results on climate change, its impacts on California (including coasts), and the state adaptation strategy
California Ocean Protection Council	http://www.opc.ca.gov/	Sea-level rise guidance for state and local agencies, funding opportunities
USGS Coastal Vulnerability Assessment	http://woodshole.er.usgs.gov/ project-pages/cvi/	Historical sea-level rise and erosion hazards (not including future risks)
NOAA Coastal Services Center	http://collaborate.csc.noaa.gov/ climateadaptation/default.aspx	Wide range of information and tools for impacts and vulnerability assessments, adaptation planning, visualization, communication, stakeholder engagement, etc.
Rising Sea Net	http://papers.risingsea.net/	Sea-level rise, impacts (erosion, flooding, wetlands), adaptation options, costs, legal issues (property rights, rolling easements etc.)
Georgetown Law Center - Adaptation	http://www.georgetownclimate. org/adaptation	Searchable database of case studies, adaptation plans, sea-level rise tool kit, and other documents

Endnotes

i The term "coast" refers to the open coast and estuaries.

ii There is no direct link between land ownership and reliance upon shoreline armoring. Public ownership does not guarantee a natural shoreline, and since much of the public backshore may be used for public infrastructure such as roads or parking lots, armoring might also be present. Conversely, private ownership does not necessarily mean there will be development or that coastal armoring will be present. In general however, areas of open space or with low-intensity development are most likely to experience natural shoreline dynamics without human interference.

iii In NRC (2010), see in particular Section 3 and pp. 117–119, which list different coastal adaptation options.

iv Examples of ecosystem-based approaches along the California shoreline include managed retreat (or realignment) projects at Pacifica State Beach and the Surfers Point project at Ventura Beach, both of which improved recreation and habitat values while reducing long-term costs and exposure to risks. Additional case studies illustrating both climate change risks and the efforts made to date toward adapting to them can be found in the state's 2009 *Climate Adaptation Strategy*, in the Pacific Council on International Policy's 2010 advisory report for the state, in Chapter 18 of this report, and in case studies cited throughout this chapter.

v Of the 162 survey responses, which represented 14 coastal counties and 45 coastal municipalities, only 10% had not begun looking at climate change impacts at all, 40% were in the early stage of understanding the potential impacts of climate change and their local vulnerabilities, 41% had entered the more advanced stage of planning for those impacts, and another 9% were implementing one or more identified adaptation options. More detailed survey results have shown that communities are still early in their respective processes, but a clear increase in engagement has been confirmed by several other studies (Hanak and Moreno 2011; Moser 2009; Tang 2009; Cruce 2009).

vi See the extensive literature review in Ekstrom, Moser, and Torn (2011) and Moser and Ekstrom (2010).

vii Based on reportings from the ports; see http://www.portoflosangeles.org/maritime/growth.asp; http://logisticscareers.lbcc.edu/portoflb.htm; http://www.portofoakland.com/maritime/facts_comm_02.asp.Chapter 10

Chapter 10

Water: Impacts, Risks, and Adaptation

COORDINATING LEAD AUTHOR

Bradley Udall (University of Colorado)

EXPERT REVIEW EDITOR

Gregory J. McCabe (U.S. Geological Survey)

Executive Summary

This chapter focuses on societal vulnerabilities to impacts from changes in sources, timing, quantity, and quality of the Southwest's water supply. It addresses both vulnerabilities related to environmental factors (such as wildfire risk and increased stream temperatures) and issues related to water management (such as water and energy demand, and reservoir operation). The chapter describes water management strategies for the coming century, including federal, regional, state, and municipal adaptation initiatives.

- The water cycle is a primary mechanism by which the earth redistributes heat. Climate change has already altered the water cycle and additional changes are expected. A large portion of the Southwest is expected to experience reductions in streamflow and other water stresses in the twenty-first century (Bates et al. 2008; Karl, Melillo and Peterson 2009; Seager and Vecchi 2010; Reclamation 2011d). (high confidence)
- Changes in water supplies lead to a wide range of societal vulnerabilities that impact almost all human and natural systems, including agriculture, energy, industry, domestic, forestry, and recreation (Westerling et al. 2006; Ray et al. 2008; Williams et al. 2010). (high confidence)
- Considerable resources are now being allocated by larger water entities to understand how to adapt to a changing water cycle. A full range of solutions involving both supply and demand are being examined. Most smaller utilities have not begun the process of adapting. To date, adaptation progress has been modest (Reclamation 2011a; WUCA 2010). (high confidence)

Chapter citation: Udall, B. 2013. "Water: Impacts, Risks, and Adaptation." In *Assessment of Climate Change in the Southwest United States: A Report Prepared for the National Climate Assessment,* edited by G. Garfin, A. Jardine, R. Merideth, M. Black, and S. LeRoy, 197–217. A report by the Southwest Climate Alliance. Washington, DC: Island Press.

- There is a mismatch between the temporal and spatial scales at which climate models produce useful outputs and the scales that are useful to water decision makers. Differing temperature and precipitation responses across models, lack of realistic topography, lack of realistic monsoon simulation, and lack of agreement about the future characteristics of the El Niño-Southern Oscillation (ENSO) all provide significant uncertainty. It is not clear if this uncertainty can be reduced (Nature Editorial Board 2010; Kerr 2011a, 2011b; Kiem and Verdon-Kidd 2011). (high confidence)
- Water supplies in the Southwest are already stressed due to many non-climatic factors. Population growth, endangered species, expensive infrastructure, and legal and institutional constraints all impede solutions. Both climate and non-climate stresses and barriers must be addressed to achieve practical solutions (Reclamation 2005; Lund et al. 2010). (high confidence)
- Twentieth-century water management was based in part on the principle that the future would look like the past. Lack of a suitable replacement for this principle, known as *stationarity*, is inhibiting the process of adaptation and the search for solutions (Reclamation 2005; Milly et al. 2008; NRC 2009; Means et al. 2010; Kiem and Verdon-Kidd 2011). (high confidence)
- Data collection, monitoring, and modeling to support both science and management are critical as the water cycle changes (WestFAST 2010). (high confidence)

10.1 Introduction

This chapter breaks with traditional climate change assessments of the water sector by focusing primarily on emerging adaptation activities being pursued by water providers rather than on either the changes to water cycle or impacts and risks to, and vulnerabilities of, human and natural systems. This altered focus occurs because the mandate of this assessment was to identify important new findings since 2009, the date of the last U.S. national assessment on climate change (Karl, Melillo and Peterson 2009). In most cases, the science about water-cycle changes and human and natural system impacts, risks, and vulnerabilities has changed little over the last three years. During this same period, however, numerous adaptation initiatives have been pursued by water managers and providers in the West. These activities are predominantly new, important, and pertinent to this assessment. It is critical to note that these nascent efforts have produced important documents and networks of knowledgeable experts, but few other tangible products or projects.

In the interest of providing a broader context to these adaptation initiatives, this chapter also summarizes some important information from traditional water-sector assessments about water-cycle changes, impacts, risks, and vulnerabilities. Much of this information is also present in other chapters of this assessment but is repeated here for completeness.

This chapter provides a broad historical overview of water development in the Southwest; briefly discusses the physical impacts to the water cycle that occurred prior to the twentieth century (as deduced from paleoclimate proxies), have occurred during the twentieth century, and are projected to occur in the twenty-first century (material covered in more detail in Chapters 4, 5, 6 and 7); provides a survey of the impacts, risks, and associated vulnerabilities to human and natural systems deriving from changes to the water cycle; and then presents in detail adaptation activities being pursued at different levels of government. Boxes within the chapter discuss the SECURE Water Act, and vulnerabilities to the Colorado River and the Sacramento-San Joaquin Bay Delta complex (see also discussion of Rio Grande Basin in Chapter 16, Section 16.5.1).

THE SUPER SECTOR. For more than 100 years, Southwestern water managers at all levels of government have managed to deliver water to homes, industry, and agriculture through periods of excess and of shortage. These deliveries occurred reliably despite population growth in the six Southwestern states from approximately 5 million persons in the early 1900s to about 56 million in 2010. The passage of the federal 1902 Reclamation Act and numerous state, regional, and municipal actions led to the development of substantial water infrastructure in the West. This infrastructure now serves many purposes, including for agricultural and municipal supplies, recreation, flood control, and environmental needs.

Interstate compacts apportioned the flow of rivers among and between states, while throughout most of the West the doctrine of prior appropriation[i] determined how water was allocated within states (Wilkinson 1992; Hundley 2009). As increases in consumptive use (water that is not returned to a water system after use, as for example water lost through evapotranspiration of crops) occurred during the twentieth century, environmental conflicts arose on almost all Western rivers (Reisner 1993). Water demands for endangered species and other environmental purposes in recent years also have altered water management practices (NRC 2004; Adler 2007; NRC 2010). During the twentieth century, water diversions by humans have substantially reduced flows at river mouths (Pitt et al. 2000; Lund et al 2010; Sabo et al. 2010).

In recent years, municipal per capita water demand has been on a downward trend over large portions of the Southwest. Many discussions are occurring throughout the West on how to manage water in the twenty-first century under conditions of multiple stresses (Isenberg et al. 2007; Colorado Interbasin Compact Committee 2010; Blue Ribbon Committee of the Metropolitan Water District 2011; Reclamation 2011a).

Water is a "super sector" that has direct and indirect connections to perhaps all natural and human systems. In many cases water has no substitute. Agriculture relies on water provided by irrigation. Energy production usually needs water for cooling, just as the transport of water often requires substantial energy. Native Americans rely upon water for agriculture and also to fulfill traditional cultural and spiritual needs. Ecosystems depend critically on the quality, timing, and amounts of water. It is difficult to overstate the importance of water, especially in the arid Southwest.

10.2 Physical Changes to the Water Cycle

The water cycle is an important physical process that transports and mixes heat globally and locally. Widespread changes to the water cycle are anticipated as the earth warms and many changes have already been noted that are related to precipitation patterns and intensity; incidence of drought; melting of snow and ice; atmospheric vapor, evaporation, and water temperatures; lake and river ice; and soil moisture and runoff (Karl,

Melillo and Peterson 2009). Global climate models have consistently shown such changes—including the magnitude and direction (increases or decreases) and spatial patterns of these changes—since the earliest days of climate modeling (Manabe and Wetherald 1975).

Widespread changes to the climate of the Western United States have occurred over the last fifty years. These include higher temperatures, earlier snowmelt runoff, more rain, less snow, and shifts in storm tracks. Some of these changes have been directly attributed to human activities, such as greenhouse gas (GHG) emissions (Barnett et al. 2008). During the same period, no changes have been detected in the region's total annual precipitation or in daily extreme precipitation (Chapter 5).

As discussed in Chapter 5, paleoclimate studies indicate that the period since 1950 has been warmer in the Southwest than during any comparable period in at least 600 years. Reconstructions of drought (from tree rings and other "proxy" records) indicate that the most severe and sustained droughts during the period 1901 through 2010 were exceeded in severity and duration by several paleodroughts in the preceding 2,000 years.

Recent research suggests that the deposition of airborne dust on snowpack in the Colorado River Basin has reduced runoff by 5% on average (Painter et al. 2010). Such dust has become more prevalent since European settlement of the American West.

In addition, recent research confirms a long-standing concern that the large spatial scales in the current generation of global climate models (GCMs) poorly represent the effects of topography on precipitation processes, especially in the Intermountain West (Rasmussen et al. 2011). Numerous studies using GCMs have attempted to quantify the effects of increasing temperatures and changes in precipitation on future runoff in the Southwest. In general these studies show declines in the southern Southwest and increases in the northern Southwest (see Chapter 6). Almost all studies show decreasing April 1 snow water equivalent (the amount of water contained in a snowpack), and declines in late summer runoff (Brekke et al. 2007; Ray et al. 2008; Reclamation 2011d).

Sensitivity studies attempt to quantify future changes in runoff without relying on GCM projections that combine changes in temperature and precipitation. Using a hydrology model driven by temperature and precipitation when temperature is varied and precipitation is held constant for every 1°F (0.6°C) increase in temperature, sensitivity studies show there is a decrease in Colorado River streamflow at Lees Ferry of 2.8% to 5.5%. Similarly, holding temperature constant, each 1% change in precipitation (either an increase or decrease) converts into a 1% to 2% change in runoff (Vano, Das, and Lettenmaier 2012).

The state of Colorado recently estimated that in the Upper Colorado River Basin, irrigated-agriculture requirements could increase by 20% and the growing season could lengthen by 18 days in 2040 (AECOM 2010). Demand studies are highly dependent on the method used to calculate actual and potential evapotranspiration[ii] (Kingston et al. 2009).

10.3 Human and Natural Systems Impacts, Risks and Vulnerabilities

Climate change will affect a large number of human and natural sectors that rely on water. Many of these impacts have been well documented, both in this report and elsewhere

(Kundzewicz 2007; Bates et al. 2008; Ray et al. 2008; CDWR 2009a). A short summary of these issues follows.

Water demands for agriculture and urban outdoor watering will increase with elevated temperatures. Higher temperatures will raise evapotranspiration by plants, lower soil moisture, lengthen growing seasons, and thus increase water demand.

Changes in snowpack, the timing of streamflow runoff, and other hydrologic changes may affect reservoir operations such as flood control and storage. For example, reservoirs subject to flood control regulations may need to evaluate their operations to compensate for earlier and larger floods. Reduced inflows to reservoirs may cause insufficient or unreliable water supplies (Rajagopalan et al. 2009). Changes in the timing and magnitude of runoff will affect the operation of water diversion and conveyance structures.

Although other factors such as land-use change generally have a greater impact on water quality, "water quality is sensitive both to increased water temperatures and changes in patterns of precipitation" (Backlund et al. 2008, p.8). For example, changes in the timing and rate of streamflow may affect sediment load and levels of pollutants, potentially affecting human health. Heavy downpours have been associated with beach closings in coastal areas due to the flushing of fecal material through storm drains that end at the ocean (Karl, Melillo and Peterson 2009). Water quality changes are expected to impact both urban and agricultural uses.

Stream temperatures are expected to increase as the climate warms, which could have direct and indirect effects on aquatic ecosystems, including the spread of in-stream, non-native species and aquatic diseases to higher elevations, and the potential for non-native plant species to invade riparian areas (Backlund et al. 2008). Changes in streamflow intensity and timing may also affect riparian ecosystems; see further discussion in Chapter 8.

Changes in long-term precipitation and soil moisture can affect groundwater recharge rates. This may reduce groundwater availability in some areas (Earman and Dettinger 2011). Also, higher sea levels can promote the intrusion of salt water into coastal freshwater aquifers (Sherif and Singh 1999).

Earlier runoff and changes in runoff volumes may complicate the allocation of water in prior-appropriation systems and interstate water compacts, affecting which right-holders receive water and operations plans for reservoirs (Kenney et al. 2008). In one study, the City of Boulder, Colorado, found that its upstream junior reservoir storage rights may allow more storage of water when runoff occurs earlier in the year, because downstream senior agricultural diverters will not be able to use the water during shorter daylight hours (Averyt et al. 2011). Reductions in Colorado River flows could affect the multi-state allocation of water via the Colorado River Compact (Barnett and Pierce 2008).

Water demands and their associated pumping and treatment costs may be affected by a changing climate. Warmer air temperatures may place higher demands on hydropower reservoirs for peak energy periods. Reductions in flows for hydropower or changes in timing may reduce the reliability of hydropower. Reliable, instantaneously available hydropower is currently used in some cases to backup intermittent renewable energy sources. Warmer lake and stream temperatures may mean more water must be used to cool power plants (Carter 2011).

Box 10.1

Colorado River Vulnerabilities

The Colorado River drains approximately 15% of the area of the continental United States and most of the American Southwest. In the United States it serves over 35 million people in seven states and irrigates over 3 million acres. In Mexico it irrigates over 500,000 acres and also meets some limited municipal demand along the international border. The river is subject to a series of interstate compacts including the original 1922 compact, legal rulings, federal legislation, and an international treaty. This "Law of the River" is said to be the most complex legal arrangement over any river in the world. Changes to any of the agreements generally take years of negotiations.

Although the river has been over-allocated for many years, only in recent years have actual demands exceeded supplies. The U.S. Bureau of Reclamation, which has a prominent role in overseeing the river, projects this imbalance to widen in the coming years due to increasing growth and declining flows due to climate change (Reclamation 2011a). For allocation purposes the compact breaks the river into two parts, the Upper Basin (Wyoming, Colorado, Utah and New Mexico) and the Lower Basin (California, Arizona, and Nevada) (Meyers 1967).

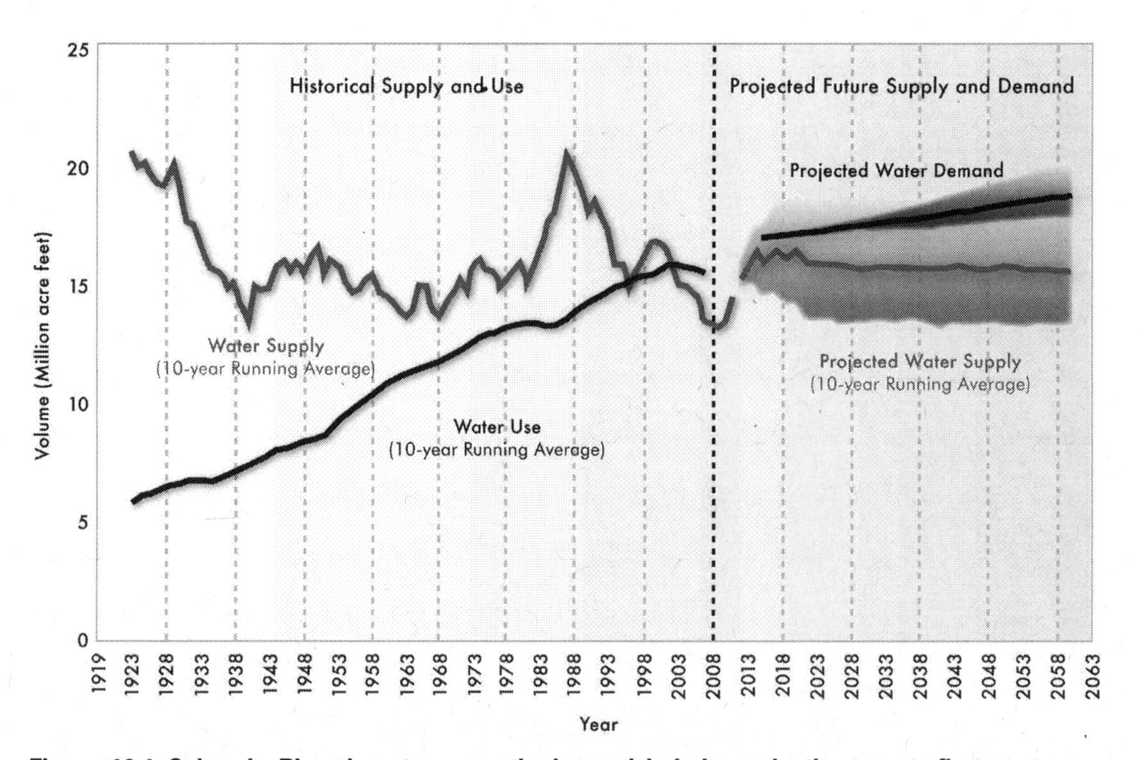

Figure 10.1 Colorado River long-term supply-demand imbalance in the twenty-first century. Reproduced from the U.S. Bureau of Reclamation (Reclamation 2011a).

Box 10.1 (Continued)

Colorado River Vulnerabilities

There are two major social vulnerabilities in the basin, one for the Upper Basin, and one for the Lower Basin.

For the Upper Basin, it is not known how much additional water (if any) exists to develop. This uncertainty is due to both natural climate variability as well as a wide range of projected future declines in flows. These declines are projected to range from 5% to 20% by 2050 (Hoerling et al. 2009). Overuse of water and hence violation of the 1922 Compact by the Upper Basin could lead to the curtailment of water to major Upper Basin water users (including Albuquerque, Salt Lake City, Denver, and most other Front Range municipalities in Colorado), with potentially very large economic impacts. Despite the uncertainty of future water availability and the consequences of overdevelopment, plans to develop additional supplies are being discussed in Colorado and Utah. Colorado is currently investigating how to administer such an unprecedented event (Kuhn 2009).

The Lower Basin is currently relying on unused water from the Upper Basin to which it has no long-term legal right. If this surplus of unused water were to cease to be available either because of climate change or increased Upper Basin use, the Law of the River would force water shortages almost entirely on Arizona (Udall 2009). Arizona has long been unsuccessful at its attempts to procure a larger share of Colorado flows to cover its current overuse. In addition, the current legal arrangements to protect Lake Mead contents by requiring delivery reductions at specified lake elevations fail to indicate what actions will be taken once Lake Mead falls below elevation 1025 feet, approximately 25% of capacity. Several recent studies have suggested that Lakes Mead and Powell, the two largest reservoirs in the United States, could face very large fluctuations or even empty under Upper Basin demand increases and declining flows (Barnett and Pierce 2008; Rajagopalan et al. 2009).

There are also significant environmental vulnerabilities. The Colorado River also has a number of endangered species in both the Upper Basin and Lower Basin. Although an endangered fish recovery program is in place in the Upper Basin and a multi-species conservation plan exists for the Lower Basin (Adler 2007), in recent years no water has reached the ocean in Mexico. Without new international arrangements, environmental flows in this reach are unlikely to occur on a regular basis (Luecke et al. 1999; Pitt et al. 2000; Pitt 2001). The United States and the seven basin states would like Mexico to share in any shortages that may be required to manage the system during extraordinary drought. Although such shortages were anticipated by the 1922 Compact, no agreement has been reached. Transnational negotiations are in progress with Mexico to resolve deliveries to that nation during extraordinary drought.

A study supported by Reclamation and the seven basin states is currently underway to identify and analyze long-term solutions for the supply/demand imbalance.

Changes in air, water, and soil temperatures will affect the relationships among forest ecosystems, surface and ground water, wildfires, and insect pests. Water-stressed trees, for example, are more vulnerable to pests (Williams et al. 2010).

The effects of forest fires alter the timing and amount of runoff and increase the sediment loads in rivers and reservoirs. Denver Water, for example, has expended considerable resources to dredge sediment from reservoirs after recent fires (Yates and Miller 2006).

Changes in reservoir storage will affect lake recreation, just as changes in streamflow timing and amounts affect such activities as rafting and trout fishing. Changes in the character and timing of precipitation and the ratio of snowfall to rainfall will continue to influence winter recreational activities and tourism (Ray et al. 2008).

Box 10.2

Sacramento-San Joaquin Bay Delta Vulnerabilities

The functioning of the Sacramento–San Joaquin Bay Delta is the most critical water issue in California and arguably the most pressing water problem in the United States. This confluence of California's two major river systems—the largest estuary on the West Coast—is used as a natural conveyance facility to move water for 25 million people. Seventy percent of the state's water moves southward from the Sacramento River, through the delta, to canals that supply both Central Valley agriculture and the municipal and industrial demands in the Los Angeles metroplex.

The delta has been substantially modified by humans from its original state and is highly vulnerable to shutdown due to both physical and legal issues (CDWR 2005; NRC 2010, 2011, forthcoming). Within the delta, approximately sixty islands sit below or near sea level and are protected by 1,300 miles of aging levees. These levees are subject to failure from sea-level rise, subsidence, freshwater flooding, earthquakes, and poor levee maintenance (Mount and Twiss 2005) . Failure of the levees from any cause could cause a massive influx of sea water from the San Francisco Bay into the freshwater delta, thus curtailing the movement of freshwater through the delta. Disruption of the flow could cost upwards of $30 billion and require many years to fix (Benjamin and Assoc. 2005). Both the State Water Project and the federal Central Valley Project are at risk (CDWR 2009b; Lund et al. 2010).

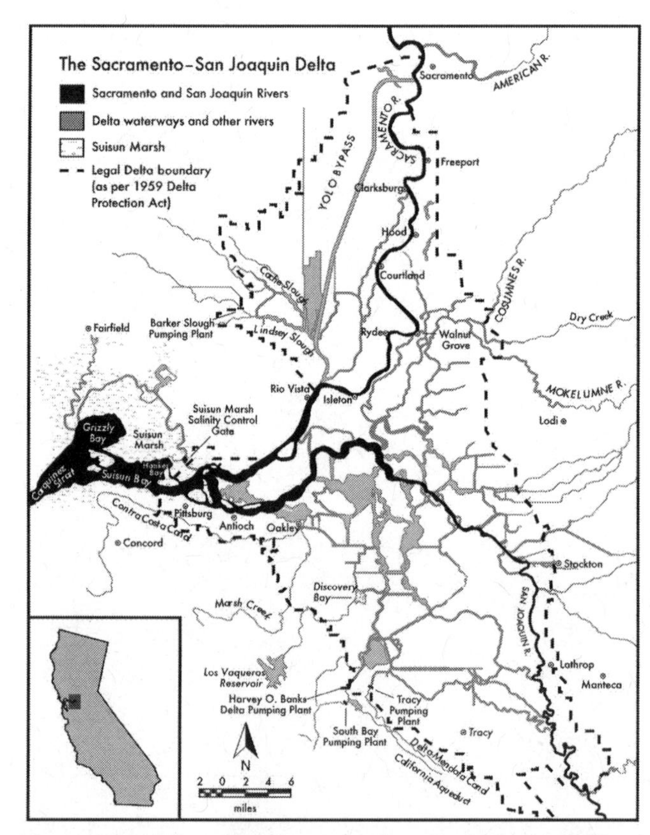

Figure 10.2 Map of Sacramento-San Joaquin Bay Delta. Reproduced with permission from the Public Policy Institute of California (Lund et al. 2007, Figure 1.1).

Box 10.2 (Continued)

Sacramento-San Joaquin Bay Delta Vulnerabilities

In addition to its physical vulnerabilities, the delta also is home to several threatened and endangered species and many invasive species. To protect endangered species, the cross-delta pumps have been shut down for short periods in recent years by federal court order (NRC 2010).

A $11 billion bond issue has been proposed to build a canal around the periphery of the delta but has not yet been put on the ballot in part due to California's continuing budgetary problems and disputes over the impacts of the canal. In 1982, a similar peripheral canal was heavily rejected by voters (Orlob 1982; Hundley 2001).

Besides its vulnerable water infrastructure, the delta is traversed by other key infrastructure including major north-south and east-west highways, electrical power lines, gas lines, and rail lines, all of which are threatened by flooding from the two rivers and by sea-level rise (Lund et al. 2010).

All of these factors have created a contentious situation. Over the last ten years, federal, state, municipal, agricultural, and environmental interests have engaged in a variety of complex and expensive stakeholder initiatives in an attempt to create solutions acceptable to all parties (Owen 2007; Isenberg et al. 2007; Isenberg et al. 2008).

10.4 Water Sector Adaptation Activities

Federal, state, regional, and municipal water management entities over the last five years or so have made substantial investments to understand the physical impacts to water supplies under a changing climate. Additional but more limited work has focused on societal vulnerabilities to these impacts. Many supply-side and demand-side adaptation strategies and solutions are now being considered. The principal challenges and barriers to climate-change adaptation include (1) uncertain, rapidly moving, and, in some cases, contentious scientific studies, and (2) physical, legal, and institutional constraints on strategies and solutions. Adaptation strategies and solutions are generally very specific to a region, limiting widespread application. Twentieth-century water planning was based in part on the idea that climatic conditions of the past would be representative of those in the future; but this model is much less useful in the twenty-first century. Reservoir size, flood control operations, and system yield calculations were all predicated on this important concept, known as stationarity. Replacing this fundamental planning model, or paradigm, is proving to be extremely difficult (Milly et al. 2008; Barsugli et al. 2009; CDWR 2009a; Brown 2010). The unreliability of regional projections has hindered planning efforts; water managers cannot simply replace historical flow sequences in their planning models with projected flows (Kerr 2011b). The rest of this section describes the various adaptation activities being pursued by water managers in the Southwest.

10.5 Planning Techniques and Stationarity

In the late twentieth century, water planning was aided by simulation models driven by historic flow sequences. Flow sequences derived from paleoclimatic evidence were later added to these simulations to test systems under further or more extreme climate variability. Scientists who conducted early climate-change studies used the same simulation models driven by crude GCM-derived future-flow sequences. As statistical downscaling became prevalent (see Chapter 6), hydrology models were used to construct streamflows using highly resolved spatial and temporal inputs of temperature, precipitation, and sometimes other variables. Ultimately, a number of concerns surfaced after deeper analysis of these projections occurred.

GCM-related concerns include widely varying future GHG emissions pathways, differing climate-model responses to GHGs, poorly resolved topography, varying responses to the North American monsoon, and wide ranges of projected precipitation. Concerns about statistical downscaling arose from its use of historical climate data (with the implicit acceptance of stationarity) to build statistical models and from the substantially different results obtained using equally valid statistical techniques. Collectively, these issues caused debate about the suitability of adaption actions relying on GCM projections (Kerr 2011a, 2011b). (See the section on model uncertainties in Chapter 19 for further exploration of this topic.)

Some scientists have cautioned about overreliance on climate change science that is regionally focused (Nature Editorial Board 2010). Water managers have now begun to investigate other methods for decision support, including decision analysis,[iii] scenario planning,[iv] robust decision making,[v] real options,[vi] and portfolio planning[vii] (Means et al. 2010).

In the absence of an alternative to assuming stationarity in management and planning, the National Research Council suggests that "Government agencies at all levels and other organizations, including in the scientific community, should organize their decision support efforts around six principles of effective decision support: (1) begin with users' needs; (2) give priority to process over products; (3) link information producers and users; (4) build connections across disciplines and organizations; (5) seek institutional stability; and (6) design processes for learning" (NRC 2009, p. 2).

10.6 Potential Supply and Demand Strategies and Solutions

Water strategies and solutions to meet the needs of Southwestern population growth range from increasing supplies to decreasing demands. Many of these could also be employed as climate-change adaptation strategies. Examples of these strategies include new dams (in California and Colorado), desalination (San Diego), basin imports via pipeline (in St. George, Utah, and the Front Range of Colorado), municipal conservation, permanent transfers from agriculture (Colorado Springs), water markets, land fallowing (Los Angeles), canal lining (San Diego), retirement of grass lawns through financial incentives (Las Vegas), groundwater banking (Arizona), water re-use (Orange County, California. and Aurora, Colorado), new water rate structures, consumer education, indoor fixture rebates (Denver), new landscape and xeriscape design, water-loss management from leaky mains, and aquifer storage and recovery (Arizona) (Western Resource Advocates 2005). Per-capita demand in recent years has been reduced in many Southwestern cities through active demand-management programs (Gleick 2010; Cohen 2011) (see also Chapter 13, Figure 13.10).

10.7 Barriers to Climate Change Adaptation

Effective climate-change adaptation will require advancements in climate science. As mentioned above, climate models and downscaling are not yet creating projections that can adequately and accurately inform adaptation efforts. Climate variability—both in nature and in climate model projections—can also confound analysis and adaptation planning. Among others, the Water Utility Climate Alliance suggests that climate model outputs suitable for water resource decision making may be a decade or more away (Barsugli et al. 2009). Model improvements in precipitation projections are unlikely to occur in the near-term to medium-term (Hawkins and Sutton 2011).

Adaptation is also constrained by numerous non-climate factors. Western water management in particular is limited by a variety of federal and state laws, interstate compacts, court cases, infrastructure capacities, hydropower considerations, and regulations pertaining to flood control, endangered species, and environmental needs. Infrastructure is also expensive to build and maintain. Many solutions improve one area's welfare at the expense of another and numerous stakeholder groups desire input into the process. Solutions can take years to discover and implement (Coe-Juell 2005; Jenkins 2008). All of these factors must be considered when designing responses.

10.8 Federal Adaptation Initiatives

The federal government has twenty or more agencies with an interest in water management (Udall and Averyt 2009). Historically, coordination of these agencies has been limited, but the last five years has seen the birth of many interagency adaptation activities related to water. A 2009 federal law, the SECURE ("Science and Engineering to Comprehensively Understand and Responsibly Enhance") Water Act (Public Law 111-11), provided the impetus for some of the coordination. New interagency coordinating groups include Climate Change and Water Working Group (CCAWWG), the Western Federal Agency Support Team (WestFAST), and the Water Resources Working Group of the Council on Environmental Quality's Interagency Climate Change Adaptation Task Force. Other federal collaborative efforts include the National Integrated Drought Information System (NIDIS), NOAA's Regional Integrated Sciences and Assessments (RISA), EPA's Climate Ready Water Utilities Working Group, and the DOI Landscape Conservation Cooperatives and Climate Science Centers. Federal climate-change adaptation efforts are in an early formative stage, but they can be expected to grow and evolve in the coming years (WestFAST 2010; Interagency Climate Change Adaptation Task Force 2011). (Relatedly, see Chapter 2, Table 2.1 for a selected list of federal-agency climate assessments that also, directly or indirectly, address issues of water resources.)

10.9 SECURE Water Act Overview

The SECURE Water Act directed the U.S. Bureau of Reclamation to establish a climate-change adaptation program in coordination with the U.S. Geological Survey (USGS), National Oceanic and Atmospheric Administration (NOAA), state water agencies, and NOAA's university-based RISA program. Other sections of the act authorized grants to improve water management, required assessment of hydropower risks, created an intra-governmental climate-change and water panel, promoted enhanced water data collection, and called for periodic water-availability and water-use assessments. (See Box 10.3 for specific Department of the Interior implementation actions since its passage.)

Box 10.3

Department of the Interior SECURE Implementation Actions

Congress passed the SECURE Water Act to promote climate-change adaptation activities in the federal government, especially within the Department of the Interior. The department established the WaterSMART program to assist with the implementation of the Act in 2010. Among other activities, WaterSMART has funded twelve "basin studies" in the West, six of which are in the Southwest. Basin studies investigate basins where supply and demand imbalances exist or are projected, and define options for meeting future demands. Each basin study will provide projections of future supply and demand, analyze how existing infrastructure will perform in the face of changing water supplies, develop options to improve operations, and make recommendations for optimizing future operations and infrastructure. The Colorado River was one of the first basin studies announced and an interim report for this study was released in early 2011 (Reclamation 2011a)

Table 10.1 Selected projections for natural flows in major southwest rivers in 2020, 2050, and 2070

Gauge Location	2020s Median Flow	2050s Median Flow	2070s Median Flow
Colorado River above Imperial Dam	-2%	-7%	-8%
Colorado River at Lees Ferry	-3%	-9%	-7%
Rio Grande at Elephant Butte Dam	-4%	-13%	-16%
Sacramento River at Freeport	3%	3%	-4%
Sacramento-San Joaquin Rivers at Delta	3%	1%	-4%
San Joaquin River at Friant Dam	1%	-9%	-11%

Note: Changes are relative to simulated 1990-1999.

Source: Reclamation (2011d).

Box 10.3 (Continued)

Department of the Interior SECURE Implementation Actions

with completion anticipated in 2012. Other Southwest basin studies underway include the Truckee River in California, the Klamath River in California and Oregon, and the Santa Fe River in New Mexico. Preliminary studies were also begun in 2011 for the Greater Los Angeles area and the Sacramento–San Joaquin Basin.

SECURE requires regular reports to Congress beginning in 2012 and every five years thereafter. In April 2011, Reclamation released its first report which quantified the risks from climate change to the quantity of water resources in seven Reclamation basins, defined the impacts of climate change on Reclamation operations, provided a mitigation and adaptation strategy to address each climate change impact, and outlined its coordination activities with respect to the USGS, NOAA, USDA and appropriate state water resource agencies (Reclamation 2011c).

Reclamation has also issued other SECURE documents. In March 2011, Reclamation released bias-corrected and spatially downscaled surface-water projections for several large Reclamation basins as part of a "West-wide Climate Risk Assessment" (Reclamation 2011d). Reclamation acknowledges that the projections suffer from a lack of model calibration and that this problem must be addressed in the next iteration of projections. Projections for 2050 showed anticipated declines of around 10% in annual runoff in the southern portion of the Southwest with a distinct north to south gradient of declining flows (see Table 10.1 and Figure 10.3).

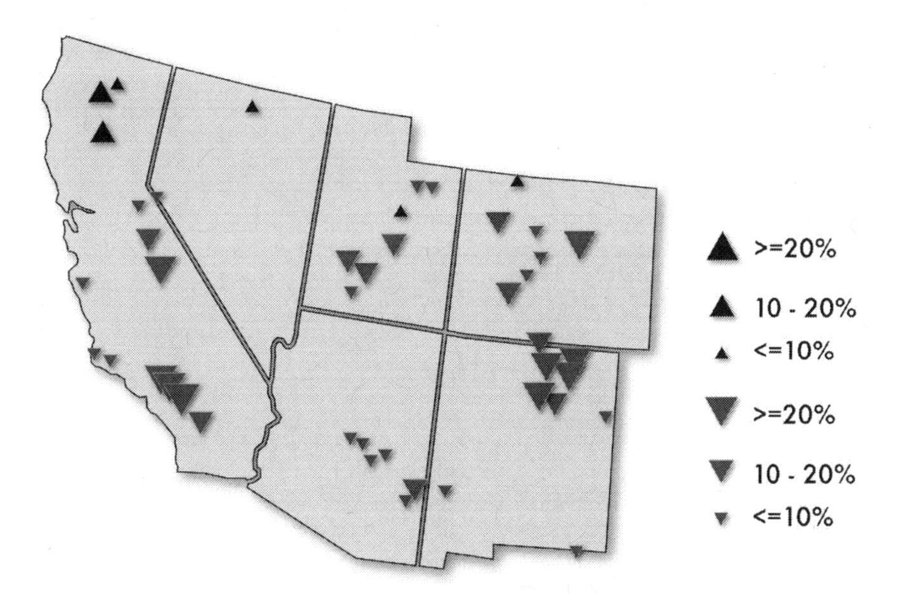

Figure 10.3 Ensemble median percentage change in annual runoff (2050s vs. 1990s) in the Southwest region. Reproduced from the U.S. Bureau of Reclamation (Reclamation 2011d, Figure 65).

10.10 Western States Federal Agency Support Team (WestFAST)

WestFAST is a collaboration among eleven federal agencies with water management responsibilities in the West (WestFAST 2011). The effort began in 2008 to coordinate federal water resource management goals with the needs of the Western States Water Council and its parent, the Western Governors' Association. WestFAST works on (1) climate change, (2) water availability, water use and re-use, and (3) water quality. In 2010, WestFAST produced an inventory of its agency efforts on water and climate change, and supported NIDIS and the newly created Landscape Conservation Cooperatives (WestFAST 2010).

10.11 Climate Change and Water Working Group (CCAWWG)

In 2007, NOAA, Reclamation, and USGS jointly created CCAWWG. The group was later expanded to include the Environmental Protection Agency, the U.S. Army Corps of Engineers, and the Federal Emergency Management Agency, and the name was changed slightly to reflect its now national scope. The purpose of this ongoing effort is to work with water managers to understand their needs, and to foster collaborative efforts across the federal and non-federal scientific community to address these needs in a way that capitalizes on interdisciplinary expertise, shares information, and avoids duplication.[viii]

CCAWWG produced a document in 2009 describing the challenges of adapting to climate change (Brekke et al. 2009). In addition, CCAWWG plans to produce four related documents, two on user needs and on two on science strategies, one each for short-term and long-term problems. The long-term user needs assessment was released in 2011 (Brekke et al. 2011). In 2010, CCAWWG published a literature synthesis of climate change studies for use in planning documents such as environmental impact statements and biological assessments under the Endangered Species Act. This document was updated in 2011 (Reclamation 2011b). The geographic focus of these literature syntheses is the Upper and Lower Colorado and the Sacramento-San Joaquin basins. CCAWWG, along with the RISAs, recently started an authoritative training program to facilitate the translation and application of emerging science and technical capabilities into water-resource planning and technical studies.

10.12 State Adaptation Efforts

Most Southwestern states have begun to categorize the impacts of climate change on water supplies. New Mexico, Utah, Colorado, and California have produced documents describing climate impacts on water resources and, in some cases, societal vulnerabilities to water resources under a changing climate (D'Antonio 2006; Steenburgh et al. 2007; Ray et al. 2008; CDWR 2009a).

The state of California has invested heavily in climate-change studies relating to water resources (Vicuna and Dracup 2007). In 2006, California released *Progress on Incorporating Climate Change into Management of Water* (CDWR 2006). Its 2009 state water plan contains substantial analysis of the impacts of climate change and the strategies necessary to adapt to it (CDWR 2009a). The California Energy Commission, with independent funding, has solicited numerous reports on the impacts of climate change on water, energy, agriculture, and many other topics and has worked closely with other state agencies.[ix]

In 2008, Colorado Front Range water utilities, in partnership with the Colorado Water Conservation Board (CWCB), the Water Research Foundation, and the Western Water Assessment RISA, investigated the impacts of climate change with the Joint Front Range Climate Variability Study (Woodbury et al. 2012). In 2010, the CWCB also funded the Colorado River Water Availability Study to assess changes in the timing and volume of runoff in the Colorado River Basin under several climate change scenarios for 2040 and 2070 (AECOM 2010). Colorado produced a directory of state adaptation activities related to climate variability and climate change in 2011 (Averyt et al. 2011).

Despite all of this adaptation-focused information-gathering activity in the Southwest, few if any water-related decisions have been made due to these actions. This is in part due to the wide range of projections for both temperature increases and precipitation changes from climate models. Decision makers everywhere are struggling to obtain *actionable science,* defined as "data, analysis, forecasts that are sufficiently predictive, accepted and understandable to support decision making" (Kerr 2011a, 1052). A related issue is modification of decision making and planning processes to incorporate non-stationarity.

10.13 Regional and Municipal Adaptation Efforts

The Water Utility Climate Alliance (WUCA), a consortium of ten large water utilities serving 43 million persons across the United States, was created in 2007 to (1) improve and expand climate-change research, (2) promote and collaborate in the development of adaptation strategies, and (3) identify and minimize greenhouse gas emissions. WUCA and participating scientists have published two documents, one on how to improve climate models and the other on useful techniques for decision making under uncertainty (Barsugli et al. 2009; Means et al. 2010). Six of the ten WUCA utilities are located in the Southwest. WUCA members serve on climate-related research review panels, have provided keynote addresses at major conferences, and are on the Federal Advisory Committee for the 2013 National Climate Assessment. Several RISAs recently joined with WUCA utilities to identify how climate models can be used in impact assessments in the *Piloting Utility Model Applications for Climate Change* project.[x]

Major municipal utilities in the Southwest (in San Francisco, the greater Los Angeles area, Las Vegas, Denver, and Salt Lake City) now have personnel dedicated to studying the impacts of climate change on their systems (see also Chapter 13, Section 13.1.4). Reclamation and other federal agencies in the Southwest also now have scientific staff whose primary mission is to research, understand, and communicate climate-change impacts.

The Western States Water Council, an affiliate of the Western Governors' Association (WGA), has convened multiple meetings over the last few years on the topics of drought and climate. WGA was instrumental in the creation of NIDIS by Congress in 2006. In 2009, WGA convened a climate adaptation working group designed to determine appropriate uses of climate-adaptation modeling, and identify and fill existing gaps in climate adaptation efforts at WGA. Since 2006, WGA has released several reports that cover water and climate (WGA 2006, 2008, 2010).

References

Adler, R. W. 2007. *Restoring Colorado River ecosystems: A troubled sense of immensity*. Washington, DC: Island Press.

AECOM. 2010. *Colorado River Water Availability Study Phase I report draft*. Colorado Water Conservation Board. http://cwcb.state.co.us/technical-resources/colorado-river-water-availability-study/Documents/CRWAS1Task10Phase1ReportDraft.pdf.

Averyt, K., K. Cody, E. Gordon, R. Klein, J. Lukas, J. Smith, W. Travis, B. Udall, et al. 2011. *Colorado Climate Preparedness Project final report*. N.p.: Western Water Assessment.

Backlund, P., A. Janetos, and D. S. Schimel. 2008. *The effects of climate change on agriculture, land resources, water resources, and biodiversity in the United States*. U.S. Climate Change Science Program Synthesis and Assessment Product 4.3 (CCSP-SAP 4.3). http://www.sap43.ucar.edu/.

Barnett, T. P., and D. W. Pierce. 2008. When will Lake Mead go dry? *Water Resources Research* 44: W0301.

Barnett, T. P., D. W. Pierce, H. G. Hidalgo, C. Bonfils, B. D. Santer, T. Das, G. Bala, et al. 2008. Human-induced changes in the hydrology of the western United States. *Science* 319:1080–1083.

Barsugli, J., C. Anderson, J. Smith, and J. Vogel. 2009. *Options for improving climate modeling to assist water utility planning for climate change*. N.p.: Water Utility Climate Alliance.

Bates, B. C., Z. W. Kundzewicz, S. Wu, and J. P. Palutikof, eds. 2008. *Climate change and water*. Technical Paper of the Intergovernmental Panel on Climate Change. Geneva: IPCC. http://www.ipcc.ch/pdf/technical-papers/climate-change-water-en.pdf.

Benjamin, Jack R. and Assoc. 2005. *Preliminary seismic risk analysis associated with levee failures in the Sacramento–San Joaquin Delta*. http://www.water.ca.gov/floodmgmt/dsmo/sab/drmsp/docs/Delta_Seismic_Risk_Report.pdf.

Blue Ribbon Committee of the Metropolitan Water District. 2011. *Report of the Blue Ribbon Committee. Metropolitan Water District of Southern California*. http://www.mwdh2o.com/BlueRibbon/pdfs/BRCreport4-12-2011.pdf.

Brekke, L., B. Harding, T. Piechota, B. Udall, C. Woodhouse, and D. Yates. 2007. Review of science and methods for incorporating climate change information into Bureau of Reclamation's Colorado River Basin planning studies. In *Colorado River interim guidelines for Lower Basin shortages and coordinated operations for Lake Powell and Lake Mead: Final environmental impact statement*, Appendix U. Boulder City, NV: U.S. Bureau of Reclamation, Lower Colorado Region.

Brekke, L., J. E. Kiang, R. Olsen, R. Pulwarty, D. Raff, D. P. Turnipseed, R. S. Webb and K. D. White. 2009. *Climate change and water resources management: A federal perspective*. U.S. Geological Survey Circular 1331. Reston, VA: USGS.

Brekke, L., K. White, R. Olsen, E. Townsley, D. Williams, F. Hanbali, C. Hennig, C. Brown, D. Raff, and R. Wittler. 2011. *Addressing climate change in long-term water resources planning and management: User needs for improving tools and information*. N.p.: U.S. Bureau of Reclamation / U.S. Army Corps of Engineers. http://www.usbr.gov/climate/userneeds/.

Brown, C. 2010. The end of reliability. *Journal of Water Resources Planning and Management* 136:143–145.

California Department of Water Resources (CDWR). 2005. *Sacramento–San Joaquin Delta overview*.

—. 2006. *Progress on incorporating climate change into planning and management of California's water resources*. Technical Memorandum Report. Sacramento: CDWR. http://www.water.ca.gov/climatechange/docs/DWRClimateChangeJuly06.pdf.

—. 2009a. Managing for an uncertain future. In *California Water Plan update 2009: Integrated Water*

Management; Volume 1: The strategic plan, Chapter 5. CDWR Bulletin 160-09. http://www.waterplan.water.ca.gov/cwpu2009/index.cfm.

—. 2009b. Sacramento-San Joaquin River Delta. In *California Water Plan update 2009: Integrated Water Management; Volume 3,* Delta. CDWR Bulletin 160-09. http://www.waterplan.water.ca.gov/cwpu2009/index.cfm.

Carter, N. T. 2011. *Energy's water demand: Trends, vulnerabilities, and management.* CRS Report for Congress R41507. Washington, DC: Congressional Research Service.

Coe-Juell, L. 2005. The 15-mile reach: Let the fish tell us. In *Adaptive governance: Integrating science, policy, and decision making.* New York: Columbia University Press.

Cohen, M. J. 2011. Municipal deliveries of Colorado River Basin water. Oakland, CA: Pacific Institute. http://www.pacinst.org/reports/co_river_municipal_deliveries/.

Colorado Interbasin Compact Committee. 2010. Report to Governor Ritter and Governor-Elect Hickenlooper.

D'Antonio, J. 2006. *The impact of climate change on New Mexico's water supply and ability to manage water resources.* N.p.: New Mexico Office of the State Engineer/Interstate Stream Commission. http://www.nmdrought.state.nm.us/ClimateChangeImpact/completeREPORTfinal.pdf.

Earman, S., and M. Dettinger. 2011. Potential impacts of climate change on groundwater resources – A global review. *Journal of Water and Climate Change* 2:213–229.

Gleick, P. H. 2010. Roadmap for sustainable water resources in southwestern North America. *Proceedings of the National Academy of Sciences* 107:21300–21305.

Hawkins, E., and R. Sutton. 2011. The potential to narrow uncertainty in projections of regional precipitation change. *Climate Dynamics* 37:407–418.

Hoerling, M., D. Lettenmaier, D. Cayan, and B. Udall. 2009. Reconciling projections of Colorado River streamflow. *Southwest Hydrology* 8 (3): 20-21, 31.

Hundley, N. 2001. *The great thirst: Californians and water—A history.* Berkeley: University of California Press.

—. 2009. *Water and the West: the Colorado River Compact and the politics of water in the American West.* Berkeley: University of California Press.

Interagency Climate Change Adaptation Task Force. 2011. *National action plan: Priorities for managing freshwater resources in a changing climate.* Washington, DC: Executive Office of the President of the United States. http://www.whitehouse.gov/sites/default/files/microsites/ceq/2011_national_action_plan.pdf.

Isenberg, P., M. Florian, R. M. Frank, T. McKernan, S. W. McPeak, W. K. Reilly, and R. Seed. 2007. *Our vision for the California Delta.* N.p.: State of California Resources Agency, Governor's Delta Vision Blue Ribbon Task Force. http://deltavision.ca.gov/BlueRibbonTaskForce/FinalVision/Delta_Vision_Final.pdf.

—. 2008. *Delta Vision strategic plan.* N.p.: State of California Resources Agency, Governor's Delta Vision Blue Ribbon Task Force. http://deltavision.ca.gov/strategicplanningprocess/staffdraft/delta_vision_strategic_plan_standard_resolution.pdf.

Jenkins, M. 2008. Peace on the Klamath. *High Country News,* June 23, 2008.

Karl, T. R., J. M. Melillo, and T. C. Peterson. 2009. *Global climate change impacts in the United States.* Cambridge: Cambridge University Press. http://downloads.globalchange.gov/usimpacts/pdfs/climate-impacts-report.pdf.

Kenney, D. S., R. Klein, C. Goemans, C. Alvord, and J. Shapiro. 2008. *The impact of earlier spring snowmelt on water rights and administration: A preliminary overview of issues and circumstances in the western states.* N.p.: Western Water Assessment. http://wwa.colorado.edu/western_water_law/docs/WRCC_Complete_Draft_090308.pdf.

Kerr, R. A. 2011a. Time to adapt to a warming world, but where's the science? *Science* 334:1052–1053.

—. 2011b. Vital details of global warming are eluding forecasters. *Science* 334:173–174.

Kiem, A. S., and D. C. Verdon-Kidd. 2011. Steps toward "useful" hydroclimatic scenarios for water resource management in the Murray-Darling Basin. *Water Resources Research* 47: W00G06.

Kingston, D. G., M. C. Todd, R. G. Taylor, J. R. Thompson, and N. W. Arnell. 2009. Uncertainty in the estimation of potential evapotranspiration under climate change. *Geophysical Research Letters* 36: L20403.

Kuhn, E. 2009. Managing the uncertainties of the Colorado River system. In *How the West was warmed: Responding to climate change in the Rockies,* ed. B. Conover, 100–110. Boulder, CO: Fulcrum.

Kundzewicz, Z. W., L. J. Mata, N. W. Arnell, P. Döll, P. Kabat, B. Jiménez, K. A. Miller, T. Oki, Z. Sen, and I. A. Shiklomanov. 2007. Freshwater resources and their management. In *Climate change 2007: Impacts, adaptation and vulnerability. Contribution of Working Group II to the Fourth Assessment Report of the Intergovernmental Panel on Climate Change*, ed. M. L. Parry, O. F. Canziani, J. P. Palutikof, P. J. van der Linden, and C. E. Hanson, Chapter 3. Cambridge: Cambridge University Press.

Luecke, D. F., J. Pitt, C. Congdon, E. P. Glenn, C. Valdés-Casillas, and M. Briggs. 1999. *A delta once more: Restoring riparian and wetland habitat in the Colorado River Delta*. Washington, DC: Environmental Defense Fund. http://www.edf.org/sites/default/files/425_delta.pdf.

Lund, J. R., E. Hanak, W. Fleenor, R. E. Howitt, J. Mount, and P. Moyle. 2007. *Envisioning futures for the Sacramento–San Joaquin Delta*. San Francisco: Public Policy Institute of California. http://www.ppic.org/main/publication.asp?i=671.

—. 2010. *Comparing futures for the Sacramento-San Joaquin Delta*. Freshwater Ecology Series 3. Berkeley: University of California Press.

Manabe, S. and R. Wetherald. 1975. The effects of doubling the CO_2 concentration on the climate of a General Circulation Model. *Journal of the Atmospheric Sciences* 32:3–15.

Means, M. L. E. III, M. Laugier, J. Daw, L. Kaatz, and M. Waage. 2010. *Decision support planning methods: Incorporating climate change uncertainties into water planning*. Water Utility Climate Alliance (WUCA) White Paper. San Francisco: WUCA.

Meyers, C. 1967. The Colorado River. *Stanford Law Review* 19:11–75.

Milly, P. C. D., J. Betancourt, M. Falkenmark, R. M. Hirsch, Z. W. Kundzewicz, D. P. Lettenmaier, and R. J. Stouffer. 2008. Stationarity is dead: Whither water management? *Science* 319:573–574.

Mount, J., and R. Twiss. 2005. Subsidence, sea level rise, and seismicity in the Sacramento-San Joaquin Delta. *San Francisco Estuary and Watershed Science* 3 (1): Article 5.

National Research Council (NRC). 2004. *Endangered and threatened fishes in the Klamath River Basin: Causes of decline and strategies for recovery*. Washington, DC: National Academies Press.

—. 2009. *Informing decisions in a changing climate*. Washington, DC: National Academies Press.

—. 2010. *A scientific assessment of alternatives for reducing water management effects on threatened and endangered fishes in California's Bay Delta*. Washington, DC: National Academies Press.

—. 2011. *A review of the use of science and adaptive management in California's Draft Bay Delta conservation plan*. Washington, DC: National Academies Press.

—. Forthcoming. *Sustainable water and environmental management in the California Bay-Delta*. Washington, DC: National Academies Press.

Nature Editorial Board. 2010. Validation required. *Nature* 463:849.

Orlob, G. T. 1982. An alternative to the Peripheral Canal. *Journal of Water Resources Planning and Management* 108:123–141.

Owen, D. 2007. Law, environmental dynamism, reliability: The rise and fall of CALFED. *Enviromental Law Journal* 37:1145.

Painter, T. H., J. S. Deems, J. Belnap, A. F. Hamlet, C. C. Landry, and B. Udall. 2010. Response of Colorado River runoff to dust radiative forcing in snow. *Proceedings of the National Academy of Sciences* 107:17125–17130.

Pitt, J. 2001. Can we restore the Colorado River delta? *Journal of Arid Environments* 49:211–220.

Pitt, J., D. F. Luecke, M. J. Cohen, E. P. Glenn, and C. Valdés-Casillas. 2000. Two nations, one river: Managing ecosystem conservation in the Colorado River Delta. *Natural Resources Journal* 819:819–864.

Rajagopalan, B., K. Nowak, J. Prairie, M. Hoerling, B. Harding, J. Barsugli, A. Ray, and B. Udall. 2009. Water supply risk on the Colorado River: Can management mitigate? *Water Resources Research* 45: W08201.

Rasmussen, R., C. H. Liu, K. Ikeda, D. Gochis, D. Yates, F. Chen, M. Tewari, et al. 2011. High-resolution coupled climate runoff simulations of seasonal snowfall over Colorado: A process study of current and warmer climate. *Journal of Climate* 24:3015–3048.

Ray, A., J. Barsugli, K. Averyt, K. Wolter, M. Hoerling, N. Doesken, B. Udall and R. S. Webb. 2008. Climate change in Colorado: A synthesis to support water resources management and adaptation. N.p.: Western Water Assessment. http://wwa.colorado.edu/CO_Climate_Report/index.html

Reisner, M. 1993. *Cadillac desert: The American West and its disappearing water*. New York: Penguin.

Sabo, J. L., T. Sinha, L. C. Bowling, G. H. Schoups, W. W. Wallender, M. E. Campana, K. A. Cherkauer, et al. 2010. Reclaiming freshwater sustainability in the Cadillac Desert. *Proceedings of the National Academy of Sciences* 107:21263–21270.

Seager, R., and G. A. Vecchi. 2010. Greenhouse warming and the 21st century hydroclimate of southwestern North America. *Proceedings of the National Academy of Sciences* 107:21277–21282.

Sherif, M. M., and V. P. Singh. 1999. Effect of climate change on sea water intrusion in coastal aquifers. *Hydrological Processes* 13:1277–1287.

Steenburgh, J., D. Bowling, T. Garrett, R. Gillies, J. Horel, R. Julander, D. Long, and T. Reichler. 2007. *Climate change and Utah: The scientific consensus*. N.p.: Utah Governor's Blue Ribbon Advisory Council on Climate Change.

Udall, B. 2009. Water in the Rockies: A twenty-first-century zero-sum game. In *How the West was warmed: Responding to climate change in the Rockies,* ed. B. Conover, 111-121. Boulder, CO: Fulcrum.

Udall, B., and K. Averyt. 2009. A critical need: A national interagency water plan. *Southwest Hydrology* 8(1): 18-19, 30.

U.S. Bureau of Reclamation (Reclamation). 2005. *Water 2025: Preventing crises and conflict in the West.* Water 2025 status report. http://permanent.access.gpo.gov/lps77383/Water%202025-08-05.pdf

—. 2011a. *Interim Report No. 1, Colorado River Basin Water Supply and Demand Study*. Boulder City, NV: U.S. Bureau of Reclamation, Lower Colorado Region. http://www.usbr.gov/lc/region/programs/crbstudy/report1.html.

—. 2011b. *Literature synthesis on climate change implications for water and environmental resources.* 2nd ed. Technical Memorandum 86-68210-2010-03. Denver: U.S. Bureau of Reclamation, Research and Development Office. http://www.usbr.gov/research/docs/climatechangelitsynthesis.pdf

—. 2011c. *SECURE Water Act Section 9503(c) – Reclamation, Climate Change and Water, Report to Congress, 2011.*

—. 2011d. *West-wide climate risk assessments: Bias-corrected and spatially downscaled surface water projections.* Technical Memorandum 86-68210-2011-01. Denver: U.S. Bureau of Reclamation, Technical Service Center.

Vano, J. A., T. Das, and D. Lettenmaier. 2012. Hydrologic sensitivities of Colorado River runoff to changes in precipitation and temperature. *Journal of Hydrometeorology* 13:932–949.

Vicuna, S., and J. A. Dracup. 2007. The evolution of climate change impact studies on hydrology and water resources in California. *Climatic Change* 82:327–350.

Westerling, A. L., H. G. Hidalgo, D. R. Cayan, and T. W. Swetnam. 2006. Warming and earlier spring increase western U.S. forest wildfire activity. *Science* 313:940–943.

Western Governors' Association (WGA). 2006. *Water needs and strategies for a sustainable future.* Denver: WGA. http://www.westgov.org/wga/publicat/Water06.pdf.

—. 2008. *Water needs and strategies for a sustainable future: Next steps.* Denver: WGA. http://www.westgov.org/wga/publicat/water08.pdf.

—. 2010. *Climate adaptation priorities for the western states: Scoping report.* Denver: WGA.

Western Resource Advocates. 2005. *Smart water: A comparative study of urban water use efficiency across the Southwest.* Boulder, CO: Western Resource Advocates.

WestFAST. 2010. *Water-climate change program inventory.* Murray, UT: Western States Federal Agency Support Team (WestFAST) Agencies. http://www.westgov.org/wswc/westfast/reports/climateinventory.pdf.

—. 2011. WestFAST agencies calendar year 2011 work plan. Murray, UT: Western States Federal Agency Support Team (WestFAST) Agencies. http://www.westgov.org/wswc/westfast/workplan/westfast_2011_workplan.pdf.

Wilkinson, C. F. 1992. *Crossing the next meridian: Land, water, and the future of the West.* Washington, DC: Island Press.

Williams, A. P., C. D. Allen, C. I. Millar, T. W. Swetnam, J. Michaelsen, C. J. Still, and S. W. Leavitt. 2010. Forest responses to increasing aridity and warmth in the southwestern United States. *Proceedings of the National Academy of Sciences* 107:21289–21294.

Woodbury, M., M. Baldo, D. Yates, and L. Kaatz. 2012. *Joint Front Range climate change vulnerability study.* Denver: Water Research Foundation.

Miller, K., and D. Yates. 2006. *Climate change and water resources: A primer for municipal water providers.* 1P-5C-91120-05/06-NH. N.p.: Awwa Research Foundation / American Water Works Association / IWA Publishing. http://waterinstitute.ufl.edu/WorkingGroups/downloads/WRF%20Climate%20Change%20DocumentsSHARE/Project%202973%20-%20Climate%20Change%20and%20Water%20Resources.pdf.

Endnotes

i Under prior appropriation, the first person or entity to establish a water right by putting it to "beneficial" use has a right to the full amount from available supplies before a junior appropriator (one who came later) can use his.

ii Evapotranspiration is composed of evaporation from water surfaces and the soil and transpiration of water by plants. Transpiration is the process by which plants take up and use water for cooling and for the production of biomass. Evapotranspiration is frequently measured in two ways: (1) the amount that occurred and (2) the potential amount that would have occurred if enough water had been present to meet all evaporation and transpiration needs. In arid areas the actual amount is frequently less than the potential amount.

iii Decision analysis, according to the Water Utility Climate Alliance (WUCA), is where uncertainties can be well described and decision trees can be used to find optimal solutions.

iv Scenario planning in this context is a tool in which key uncertainties are identified and future scenarios are constructed around these uncertainties. The hope is that different scenarios will identify common, robust approaches for managing the range of uncertainties.

v Robust decision making is a technique that combines classic decision analysis with scenario planning to identify coping strategies that are robust over a variety of futures.

vi Real options is a type of financial based planning method for uncertainty. WUCA describes it as a type of cash flow analysis that includes flexible implementation. It uses classical decision analysis with hedging concepts from financial planning.

vii Portfolio planning is a financial tool where a portfolio is selected to minimize risk and to hedge against future uncertainty.

viii See http://www.esrl.noaa.gov/psd/ccawwg/.

ix See http://www.energy.ca.gov/.

x See http://www.wucaonline.org/html/actions_puma.html.

Chapter 11

Agriculture and Ranching

COORDINATING LEAD AUTHOR

George B. Frisvold (University of Arizona)

LEAD AUTHORS

Louise E. Jackson (University of California, Davis), James G. Pritchett (Colorado State University), John P. Ritten (University of Wyoming)

REVIEW EDITOR

Mark Svoboda (University of Nebraska, Lincoln)

Executive Summary

This chapter reviews the climate factors that influence crop production and agricultural water use. It discusses (a) modeling studies that use climate-change model projections to examine effects on agricultural water allocation and (b) scenario studies that investigate economic impacts and the potential for using adaptation strategies to accommodate changing water supplies, crop yields, and pricing. The chapter concludes with sections on ranching and drought and on disaster-relief programs.

- Under warmer winter temperatures, some existing agricultural pests can persist year-round, while new pests and diseases may become established. While crops grown in some areas might not be viable economically under future climate conditions, other crops could replace them. (high confidence).
- Many important costs of climate change to agriculture will be adjustment costs. The suitability of production in an area depends not only on climate, but also on the presence of complementary infrastructure such as irrigation conveyance systems and specialized agricultural processing and handling facilities, as well

Chapter citation: Frisvold, G. B., L. E. Jackson, J. G. Pritchett, and J. P. Ritten. 2013. "Agriculture and Ranching." In *Assessment of Climate Change in the Southwest United States: A Report Prepared for the National Climate Assessment*, edited by G. Garfin, A. Jardine, R. Merideth, M. Black, and S. LeRoy, 218–239. A report by the Southwest Climate Alliance. Washington, DC: Island Press.

as transportation and energy supply networks. Relocating this complementary infrastructure may be costly, especially if climate change occurs quickly. Moreover, growers in a region may be unfamiliar or inexperienced with crops suitable for the new climate. Adjustment costs can be substantial in tree-crop production, which requires large up-front capital investments and with many years between the time trees are planted and when they produce sellable output. (medium-high confidence).

- Because agriculture accounts for 79% of Southwest water withdrawals, water management and reduction of agricultural water demand are important means to adapt to climate change. Conservation strategies implemented by water managers and agricultural users tend to be more economical than developing new supplies. Options for managing demand may include addressing water pricing and markets, providing incentives to adopt water-saving irrigation technology, reusing tailwater, or shifting to less water-intensive crops. (medium-high confidence).
- The evidence supporting the widely held belief that simply improving on-farm irrigation efficiency conserves water is weak, however. Claims of water conservation are often made at the farm level. Improved application efficiency means that crops take up a higher percentage of applied water. However, this means that less water is available to recharge aquifers or serve as return flows for downstream uses. At the basin- or watershed-scale, increased application efficiency can reduce water available for these other uses. (high confidence)
- Diverse studies using mathematical programming modeling to combine economic and hydrological models have generated some consistent lessons. First, agriculture-to-urban water transfers could significantly reduce the costs of adjusting to regional water shortages. Agriculture would be the sector that alters water use the most, protecting municipal and industrial uses. Second, growers have numerous lower-cost alternatives to fallowing land as a response to drought, such as shifting crop mix, input substitution (e.g. substituting land for water), deficit irrigation, and investments in improved irrigation technologies. To facilitate transfers, additional investments in infrastructure to store and convey water would likely be required. Third, the costs of compliance with environmental regulations, especially those that protect endangered aquatic species, will represent significant adaptation costs. (high confidence)
- Irrigators also could adapt better to climate variability by increased use of water management information that is already available. The California Irrigation Management Information System (CIMIS), a weather information network for irrigation management developed and operated by the California Department of Water Resources, benefits growers via higher yields, lower water costs, and higher crop quality. CIMIS has been estimated to generate $64.7 million in benefits per year at an annual cost to the state of less than $1 million. (medium-high confidence)
- Public and private entities can more effectively deliver web-based information and decision-making tools for climate-change adaptation if they consider

constraints faced by the intended users. As of 2007, there were 29 Southwestern counties where fewer than 30% of agricultural producers had access to high-speed Internet service. Access is particularly low in the Four Corners region, which has a relatively large population of Native American farmers and ranchers. (medium-high confidence)

11.1 Distinctive Features of Southwestern Agriculture

Agriculture in the Southwest has distinctive features that influence how the sector responds to climate variability and change. First, the region accounts for more than half of the nation's production of high-value specialty crops (fruits, vegetables, and nuts). California has the most counties where specialty crops (including melons and potatoes) account for a large share of total agricultural sales (Figures 11.1). Other areas that are important in terms of specialty crops include southwestern Arizona, the San Luis Valley of Colorado, and chili- and pecan-growing areas of New Mexico (along the Rio Grande Valley).

Irrigation plays a critical role in the region. Excluding Colorado, which has significant dryland wheat production, more than 92% of the region's cropland is irrigated. Irrigated crops account for an even larger share of sales revenues. Agricultural uses of

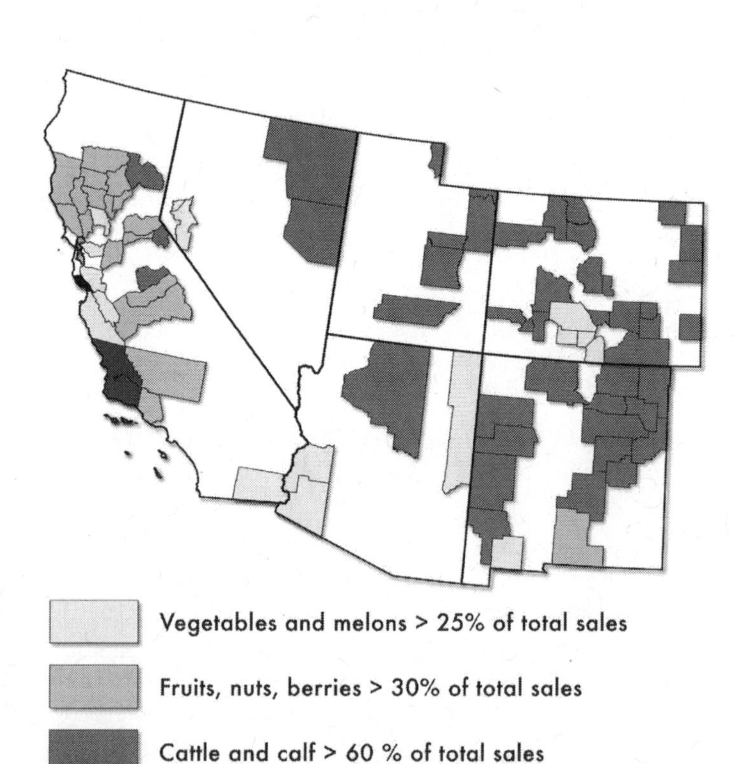

Figure 11.1 Agricultural sales by county.
Source: USDA (2009, 2012).

water (for irrigation and livestock watering) account for 79% of all water withdrawals in the region. As a result, small changes in agricultural water use can have relatively large effects on the water that is available for households, industrial use, and riparian ecosystems.

The region is characterized by extensive surface water infrastructure—including dams, reservoirs, canals, pipelines, and pumping stations—managed by the U.S. Bureau of Reclamation, state water agencies, and local irrigation districts. These systems not only capture and store vast quantities of water; they also transport it over large distances, geographically "decoupling," in terms of climate-change feedbacks, many of the region's water users from its water sources. Not all agricultural areas within the region have access to this extensive surface-water network. Many locations, therefore, are highly dependent on groundwater for irrigation (Figure 11.2). Depletion of groundwater resources in these areas, as measured by increases in the average depth-to-water of wells, presents problems for irrigators, including increased costs for the energy needed to pump the water higher to reach the surface. If groundwater levels fall sufficiently, irrigators may incur additional costs to lower pumps within the well, deepen wells, or dig replacement wells. From 1994 to 2008, according to the USDA Farm and Ranch Irrigation Survey (2010), depth-to-groundwater for irrigation wells increased in all states but Nevada (Figure 11.3).

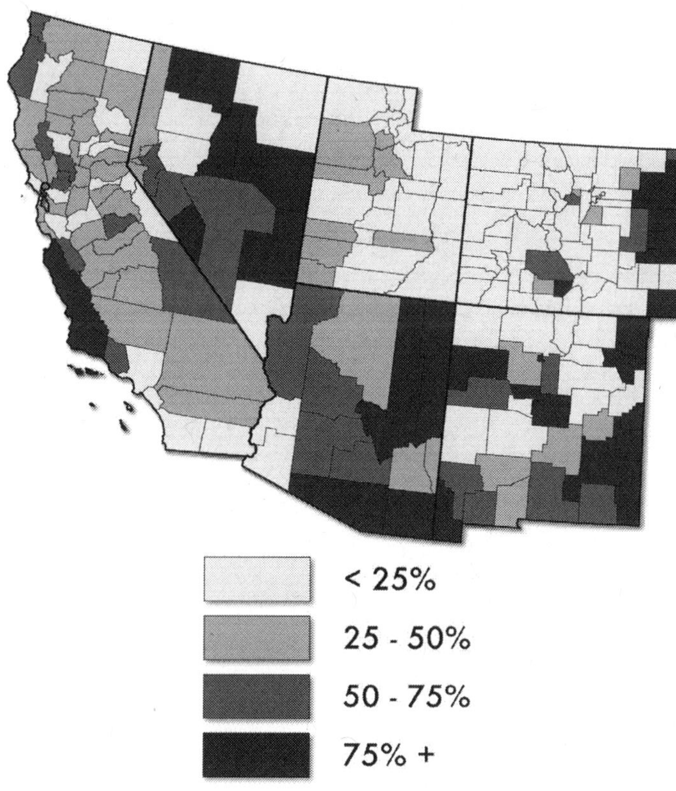

Figure 11.2 Groundwater irrigation withdrawals as a share of total irrigation withdrawals. Source: USGS (2005), Kenny et al. (2009).

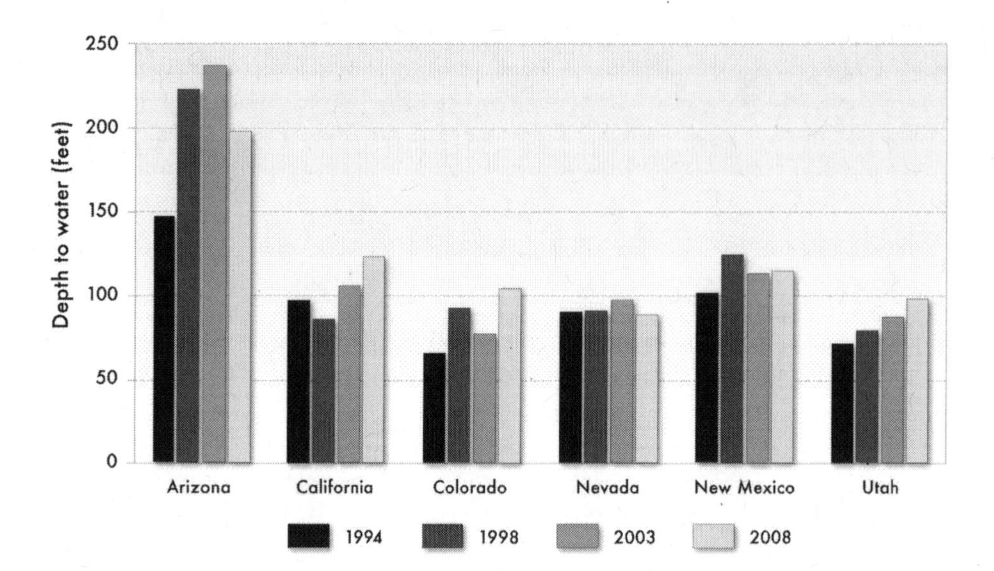

Figure 11.3 Depth to water of irrigation wells. Source: USDA (2010).

The livestock sector, especially cattle ranching and dairies, is also economically important in the region. Cattle account for most of the agricultural sales in many New Mexico and Colorado counties (Figure 11.1). Cattle ranches rely on rain-fed forage on grazing lands, making these enterprises sensitive to changes in climate. Much acreage and irrigation water is devoted to alfalfa and other hay, which provide important forage for the region's dairies as well as supplemental cattle feed.

Other major field crops include cotton in California, Arizona, and New Mexico; durum wheat in Southern California and Arizona; winter wheat in Colorado and California; and corn in eastern Colorado. An emerging challenge to crop production is the rise of glyphosate-resistant (i.e., herbicide-resistant) weeds (Price et al. 2011; CAST 2012; Norsworthy et al. 2012).

11.2 Implications for Specialty Crops

The future presents special challenges and opportunities for producers of high-value crops such as fruits, vegetables, and nuts. Demand for these crops is projected to increase over the next forty years, correlated with expected population and income growth in the United States and throughout the Pacific Rim (Howitt, Medellín-Azuara, and MacEwan 2009, 2010). Compared to field crops, demand for these high-value crops is price inelastic, meaning that demand falls little with price increases. This also means that small reductions in output lead to relatively large increases in price. Thus, price increases that accompany climate-induced losses in output can partially offset the reduced volume sold and thereby buffer producers of these crops from the effects of climate change (though there are obvious increased costs for consumers).

Climate change implies that locations best suited for production of high-value crops will change over time. Fewer frosts may make production of certain vegetables and tree crops more viable in some regions. Yet, for some stone fruits and nuts that require a minimum amount of chill time,[i] reductions in chill hours from a warming climate may reduce the profitability of production in areas where they are currently grown. In addition, many crops have threshold tolerances to high temperatures during key stages of crop development, such as pollination, while unseasonal precipitation or adverse temperatures might harm product quality during fruit development. Climate change and extreme weather are more likely to affect horticultural crops (fruits, vegetables, and ornamental plants) because they have high water content and because sales depend on good visual appearance and flavor (Backlund, Janetos, and Schimel 2008).

Under warmer winter temperatures, existing pests can persist year-round, while new pests and diseases may become established (Gutierrez et al. 2008). A study of Yolo County, California, found that warm-season crops grown there today—tomatoes, cucumbers, sweet corn, and peppers—might not be viable economically under future climate conditions. However, other crops—melon and sweet potatoes in the summer, lettuce or broccoli in the winter—could replace them (Meadows 2009; Jackson et al. 2011).

Quiggin and Horowitz (1999, 2003) note that many important costs of climate change to agriculture will be adjustment costs. The optimal location for producing specific crops will change. Established farms have infrastructure in place—for energy supply, irrigation systems, and grain storage, for example—that will be expensive to relocate, especially if climate change occurs quickly. Moreover, growers in a region may be unfamiliar or inexperienced with crops suitable for the new climate. Adjustment costs can be substantial in tree-crop production: production requires large up-front capital investments, with a stretch of years between the time trees are planted and when they produce sellable output. For example, almonds, apricots, peaches, and plums average four non-bearing years, citrus averages five to six years, while pecans average eight years (Berck and Perloff 1985). If growers reduce the number of trees as a short-term response to drought, it will reduce the region's ability to produce tree crops for many years thereafter. Another strategy—*relocating* where trees are grown—may represent a significant adjustment cost. Adjustment costs also are likely to occur when farmers change irrigation or fertilization practices or other management operations to cope with changes in resource availability, decrease greenhouse gas emissions from nitrous oxide, or to increase carbon storage in soil or the wood of trees (Hatfield et al. 2011).

11.3 On-farm Water Management

Because agriculture accounts for 79% of water withdrawals in the Southwest, methods to manage and reduce agricultural water demand are an important means to adapt to climate change (Levite, Sally, and Cour 2003; Joyce et al. 2006; Joyce et al. 2010). Local conservation strategies implemented by water managers and agricultural users tend also to be more economical than developing new supplies (Kiparsky and Gleick 2003). Options for managing demand may include addressing water pricing and markets, setting allocation limits, improving water-use efficiency, providing public and private incentives to adopt water-saving irrigation technology, reusing tailwater (excess surface water draining from an irrigated field), shifting to less water-intensive crops, and fallowing (Tanaka et al. 2006).

One way to adapt to climate-change-induced water shortages is to shift the mix of crops grown. Table 11.1 shows ranges in water application rates by crop, state, and irrigation technology in acre-feet per acre. An acre-foot is the amount of water required to cover one acre of water one foot deep. Crops in warmer Arizona tend to have higher application rates, while those in Colorado tend to have the lowest. Crops irrigated by sprinkler irrigation systems (sometimes referred to as pressurized systems) have lower application rates than those irrigated by gravity systems, which rely on flooding fields (Kallenbach, Rolston, and Horwath 2010). Sprinkler irrigation includes center-pivot, mechanical-move, hand-move, and non-moving systems (the last used mostly for perennial crops). Rather than using gravity, these systems rely on mechanically generated pressure to pump water to crops.

Table 11.1 Ranges of water application rates (acre-feet of water applied per acre) by state and irrigation technology for different crops grown in Southwestern states

	Minimum	Median[a]	Maximum
Orchards, Vineyards, Nuts	0.3 (Colorado/Drip[b])	2.7 (California/Sprinkler[b])	6.5 (Arizona/Gravity)
Alfalfa	1.6 (Colorado/Sprinkler)	3.1 (Nevada/Sprinkler)	6.4 (Arizona/Gravity)
Sugar Beets[c]	3.7 (Colorado/Sprinkler)		5.3 (Colorado/Gravity)
Cotton	2.2 (New Mexico/Sprinkler)	3.1 (California/Gravity)	4.8 (Arizona/Gravity)
Corn/silage	1.4 (Colorado/Sprinkler)	2.7 (Utah/Sprinkler)	4.7 (Arizona/Gravity)
Corn/grain	1.5 (New Mexico/Gravity)	2.1 (California/Gravity)	4.2 (Arizona/Gravity)
Other Hay	1.3 (Colorado/Sprinkler)	2.1 (New Mexico/Sprinkler)	4.2 (Arizona/Gravity)
Rice	4.1 (California/Gravity)	4.1 (California/Gravity)	4.1 (California/Gravity)
Wheat	1.3 (Colorado/Sprinkler)	2.3 (California/Gravity)	3.6 (Arizona/Gravity)
Barley	1.2 (Utah/Sprinkler)	1.7 (Colorado/Sprinkler)	3.6 (Arizona/Gravity)
Vegetables	1.7 (Colorado/Sprinkler)	2.8 (Nevada/Gravity)	3.5 (Arizona/Sprinkler)
Sorghum	0.6 (Colorado/Sprinkler)	1.7 (California/Sprinkler)	3.5 (Arizona/Gravity)

Note:

a. In cases where the median value was between two actual observations, the value of the observation with the higher application rate is reported.

b. Sprinkler irrigation includes center-pivot, mechanical-move, hand-move, and non-moving systems (the last used mostly for perennial crops). Rather than using gravity, these systems rely on mechanically generated pressure to pump water to crops. Low-flow irrigation methods, which include drip, trickle, and micro-sprinkler methods are not included in this definition of sprinkler, but treated as a separate category by USDA.

c. Only two observations.

Source: USDA (2010).

Irrigators also could adapt better to climate variability by increased use of water-management information that is already available. For example, Parker and others (2000) found the California Irrigation Management Information System (CIMIS), a weather information network for irrigation management developed and operated by the California Department of Water Resources, to be highly valuable to agriculture. The crop evapotranspiration (ET) data provided by CIMIS allows farmers to better match irrigation water applications to crop needs. This reduces risks from climate variability. Growers benefit from higher yields and lower water costs. Improved water management also increases fruit size, reduces mold, and enhances product appearance, all of which can fetch higher crop prices. They estimated that use of CIMIS reduces California's agricultural water applications by 107,300 acre-feet annually. Drought in 1989 appeared to stimulate a large increase in the number of growers and crop consultants who use CIMIS. Parker and others (2000) estimated CIMIS generated $64.7 million in benefits from higher yields and lower water costs, at an annual cost to the state of less than $1 million. CIMIS has also improved pest control and promoted use of integrated pest management techniques, which can reduce costs and improve worker safety by reducing pesticide applications.

Other Southwestern states also provide on-line databases and support tools for water management. For example, the Arizona Meteorological Network (AZMET) provides on-line, downloadable weather data and information for Arizona agriculture. Data include temperature (air and soil), humidity, solar radiation, wind (speed and direction), and precipitation as well as computed variables such as heat units (degree days), chill hours, and crop evapotranspiration. AZMET also provides ready-to-use summaries and special reports that interpret weather data such as Weekly Cotton Advisories. The Lettuce Ice Forecast Program provides temperature forecasts for the vegetable production in Yuma County. The Colorado Agricultural Meteorological Network (CoAgMet) provides daily crop-water use or evapotranspiration reports that can improve irrigation scheduling. In addition to providing raw data, the system allows users to generate customized, location-specific cropwater-use reports.

Public and private entities can more effectively deliver information or develop tools for decision making for climate-change adaptation if they consider constraints faced by the intended users. In 1996, the National Weather Service (NWS) offices discontinued issuing local agricultural weather forecasts in response to budget cuts and to avoid competing with privately supplied forecasts. The expense of privately provided forecasts may pose a barrier to some agricultural information users (Schneider and Wiener 2009). For many Southwestern farmers and ranchers, access to high-speed Internet service remains problematic. As of 2007, there were twenty-nine Southwestern counties where fewer than 30% of agricultural producers have such access (Figure 11.4). Access is particularly low in the Four Corners region of Arizona, Utah, and New Mexico, which has a relatively large population of Native American farmers and ranchers. In rural areas, radio, and television are still widely used for weather information (Schneider and Wiener 2009). Frisvold and Murugesan (2011) found that access to satellite television was a better predictor of weather information use by agricultural producers than was access to the Internet. Emphasis on encouraging commercial weather information providers may be limiting development of applications through these popular media (Schneider and Wiener 2009).

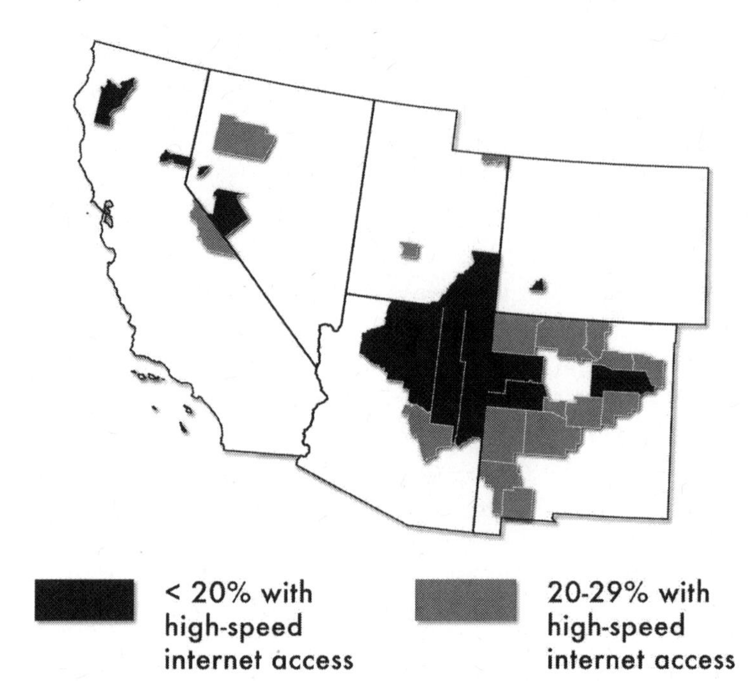

Figure 11.4 Counties in the Southwest in which agricultural producers have limited access to high-speed internet service. Source: USDA (2009, 2012).

Improving irrigation efficiency is frequently cited as a promising response to climate change or water scarcity in general (Parry et al. 1998; Wallace 2000; Ragab and Prudhomme 2002; Mendelson and Dinar 2003; Jury and Vaux 2006; Rockstrom, Lannerstad, and Falkenmark 2007). It can allow individual irrigators to save water costs and improve yields, thus increasing profits. However, improving on-farm application efficiency does not necessarily conserve water (Caswell and Zilberman 1986; Huffaker and Whittlesey 2003; Peterson and Ding 2005; Frisvold and Emerick 2008; Ward and Pulido-Velazquez 2008). Increased on-farm application efficiency means that the crop—rather than its surrounding soil— takes up a greater share of the water that is applied. However, this also means that less water returns to the system as a whole (as groundwater recharge or surface-water return flow). Other downstream irrigators (or other users) often count on this return flow or recharge for their water supplies. Similarly, reducing the water lost through the conveyance system means more of the water diverted reaches a crop, but also results in lower return flows or recharge that is no longer available to other irrigators, urban water users, or ecosystems. While fisheries and aquatic habitat depend on return flows, these "uses" typically do not hold legally recognized water rights that can contest any harm done by water transfers (Chong and Sunding 2006). Many riparian systems now depend on these return flows, which are subject to changes in managed, hydrological systems that have been altered to accommodate human water uses (Wiener et al. 2008). In some cases, minimum flow requirements have been established under the Clean Water Act and Endangered Species Act (see Ward and others [2006] and Howitt, MacEwen and Medellín-Azuara [2009] for analysis of additional costs of maintaining minimum flow requirements). However, litigation and implementation have been contentious, with variable outcomes (Moore, Mulville and Weinberg 1996; Benson 2004). Thus, what may seem a rational response to water scarcity by irrigators at the farm level, may exacerbate water scarcity problems at the basin scale. Policies to increase irrigation efficiency with the hope of freeing up water for other uses may fail to conserve water.

11.4 System-wide Water Management: Lessons from Programming Models of Water Allocation

A number of studies have used mathematical programming models to assess how different areas of the Southwest may respond to different drought, water shortage, or climate-change scenarios—which may include changes in temperature, level and type of precipitation, and the timing of mountain snowmelt and runoff into lower elevation agricultural regions (Cayan et al. 2008). To varying degrees, these studies link physical water supply and associated hydrologic information to economic models. Model solutions find the least-cost response to water shortages given system constraints. A common finding of these studies is that the agricultural sector makes large adjustments in water use, land use, and cropping patterns that allow urban and industrial water uses to remain largely unchanged.

Sustained drought in California

Harou and colleagues (2010) examined the effects of severe, sustained drought in California. Their drought simulations, based on records of ancient (paleo-) climates, assumed streamflows that are 40% to 60% of the current mean flows, with no intervening year wet enough to fully replenish reservoirs. This drought scenario is similar to the effects under "dry forms" of climate warming: those with projected reductions in precipitation. The analysis examined potential impacts to agriculture and the rest of California's economy in 2020. The model simulated allocation and storage of water to minimize costs of water scarcity and system operation. The costs in the drought scenario were borne largely by agriculture, limiting costs to the state's overall economy. The costs of water shortages were greatest in agriculture, except in Southern California where urban costs dominated. Large differences in scarcity costs across sectors and regions created incentives to transfer water from lower-valued agricultural to higher-valued urban uses, where value is determined by user willingness to pay for additional water. The study also calculated costs of maintaining required environmental flows and found these could be quite high, especially for the Sacramento-San Joaquin River Delta. Results also suggested there are large benefits to improving and expanding the conveyance infrastructure to facilitate movement of water.

Water availability and crop yields in California

Howitt, Medellín-Azuara, and MacEwan (2009, 2010) simulated the effects of changes in water availability and crop yields in California in 2050. While statewide agricultural land use and water use were projected to decline by 20% and 21% respectively, total agricultural revenues fell by less: 11%. There were large reductions in acreage of water-intensive crops and small shifts in others. In Southern California agriculture, two factors reduced negative impacts to farmers. Crop price increases accompanied production declines, while farmers also shifted to high-value crops. The greatest reductions in output were among field crops, with relatively less change among fruit and vegetable crops.

In contrast, total urban water use fell by 0.7%. Results assumed that between now and 2050, a more economical means of transferring water from Northern to Southern California will be developed. Absent climate change, growing demand for high-value specialty crops is expected to drive an increase of 40% for California's agricultural revenues by 2050. Under the dry-climate warming scenario, however, revenues are projected to grow 25% by 2050.

Effects of adaptation measures in California

Medellín-Azuara and others (2008) examined the consequences of various adaptation measures in California for 2050 in a dry-warming climate change scenario. They also made assumptions about baseline changes to water demand and land use by 2050. The model allowed water to be allocated to maximize net benefits of the state's water supply, given infrastructure and physical constraints. Water markets implicitly allowed water to flow to higher-valued urban uses. Institutional barriers to water transfers were not modeled. The simulation projected that statewide costs would rise substantially when water markets were geographically restricted. Urban water users in Southern California would purchase water from central and Northern California, while Southern California agriculture would maintain senior water rights to the state's allocation of Colorado River water. Agriculture in the Sacramento, San Joaquin, and Tulare basins would face large economic losses from reduced water availability and lower yields. Grower losses would be only partially compensated by revenues from water sales to urban areas. Rules for water storage and conjunctive management of surface and groundwater would have to change to improve management of the statewide system.

Economic and land-use projections in California

Tanaka and others (2006) combined climate scenarios for 2100 with economic and land-use projections for California. Climate scenarios were based on both wet and dry forms of climate warming. Changes in seasonal water flows ranged from a 4.6-million-acre-foot (maf) increase to a 9.4 maf decrease. Given hydrologic and conveyance constraints, water was allowed to flow from lower-valued to higher-valued uses. Dry warming scenarios presented the greatest challenges to California agriculture. Modeled simulations projected the transfer of water from Southern California agricultural users to urban users. Many of these transfers have already subsequently occurred. In the simulations, Southern California urban users also imported more water from northern agricultural areas. Agricultural water users in the Central Valley were shown to be most vulnerable to dry warming under this simulation; under the driest scenarios, their water use declined by one-third. Although in the simulation agricultural producers received some compensation from agricultural-to-urban water use transfers, transfer income was insufficient to compensate for all the costs of reduced water supplies. Agricultural producers altered irrigation technology in response to water shortages. While statewide agricultural water deliveries fell 24% and irrigated acreage fell 15%, agricultural income was reduced only 6%. Income fell less than water deliveries because farmers adapted by changing both irrigation technologies and crop mix. Farmers reduced production of lower valued crops, while maintaing production of higher valued ones.

Dry warming scenarios substantially increased the costs to agriculture (and other users) of maintaining water supplies for environmental protection. Under the driest warming scenarios, expansion of storage infrastructure yielded few benefits, while expansion of conveyance systems yielded benefits in every year.

Agriculture in Nevada's Great Basin

On a smaller geographic scale, Elbakidze (2006) examined potential impacts of climate change on agriculture in the Truckee Carson Irrigation District of Nevada's Great Basin. He considered scenarios based on two general circulation models (the Canadian and Hadley GCMs) for 2030, which projected warmer temperatures but wetter conditions and increased streamflow. The study also considered scenarios with reduced streamflow. Streamflow scenarios were examined both in isolation and combined with assumed yield increases or yield increases accompanied by price decreases. The crops included alfalfa, other hay, and irrigated pasture. In this study, agricultural returns increased with increased streamflow and decreased with decreased streamflow, but the changes were asymmetric: economic losses under reduced streamflow conditions were much larger than gains realized under increased streamflow conditions. The model assumed that existing infrastructure was sufficient to handle increased streamflow. Benefits of increased streamflow also were dependent on the growers' ability to increase their agricultural acreage.

Water transfers in Rio Grande Basin

Booker, Michelsen, and Ward (2005) examined the role of water transfers in mitigating costs of severe, sustained drought in the Upper Rio Grande Basin, stretching from southern Colorado, through New Mexico, and into West Texas. (The 1938 Rio Grande Compact governs water allocations between the three states.) Their modeling framework was not based on a specific climate-change scenario, but considered droughts that reduced basin inflows to 75% and 50% of the long-term mean. In 2002, inflows actually had fallen to 37% of mean. Under the scenario using existing institutions, surface-water allocations were not transferred between different institutional users, such as cities and irrigation districts. Agriculture accounted for the bulk of water-use reductions and economic losses. The cities of Albuquerque and El Paso did not alter consumption, but shifted to more expensive groundwater sources. Under the intra-compact trading scenarios, transfers were permitted between users *within* states. For example, trades occured between New Mexico agriculture and Albuquerque, and separately between West Texas agriculture and El Paso. Intra-compact trading would reduce economic losses from drought by 20%. Under interstate trading scenario, trades were allowed between all users in New Mexico and Texas. Under this scenario, El Paso and Albuquerque would rent water from the Middle Rio Grande Conservancy District (MRGCD) instead of pumping groundwater. The more MRGCD cut back on water use, the less Elephant Butte Irrigation District did so. Interstate water trading reduced the total economic losses from drought by one-third. The simulation results suggest *potential* gains from expanded water trading. Urban uses in Albuquerque remained unaffected, while those in El Paso fell by 1.1% at most. The researchers pointed out that there would be additional transaction costs associated with establishing and expanding water markets and for designing policy instruments to address third-party damages from transfers. They also note that there do exist institutional and legal impediments to trading water across state lines.

Severe drought in Rio Grande Basin

Ward and others (2006) also modeled impacts of severe drought in the Upper Rio Grande Basin. Increasingly severe drought scenarios were combined with minimum instream flow requirements for endangered fish protection. Agriculture again absorbed most of the shock in response to water shortages and environmental requirements, both in terms of reduced water use and economic losses. The largest absolute losses were in Colorado's San Luis Valley, where relatively high-value crops are grown.

Drought and Arizona's agriculture

The U.S. Bureau of Reclamation's Final Environmental Impact Statement for the *Colorado River Interim Guidelines for Lower Basin Shortages and Coordinated Operations for Lakes Powell and Mead* (Reclamation 2007) considered how Arizona agriculture would be affected by a shortage declaration on the Colorado River.[ii] The Final EIS analysis was not based on any explicit climate change scenarios. Baseline values were based on historical flows, but sensitivity analysis did include some drought scenarios. Other research has suggested that climate change would increase the likelihood of future shortage declarations (Christensen and Lettenmaier 2007; Seager et al. 2007; Rajagopalan et al. 2009). The study assumed the only adaptation mechanism available to agriculture is land fallowing, with crops providing the lowest returns per acre-foot of water fallowed first. Fallowing was possible for alfalfa, durum wheat, and cotton, while it was assumed that high-value specialty crops would continue to be grown. For most shortage scenarios, the bulk of shortage costs were felt in central Arizona and in Mohave County in northwest Arizona.

Reduced water supplies across the Southwest

Frisvold and Konyar (2011) simulated the impacts of reducing agricultural water supplies in Arizona, Colorado, Nevada, New Mexico, and Utah. The model did not include potential barriers to transferring water between uses, regions, or states. Nor did the model include urban sectors, but it accounted for how regional agricultural markets were linked to the broader U.S. and export markets. Possible adaptations included deficit irrigation[iii] (which may apply less water than that needed to maximize output *per acre*), changing the crop mix, and changing input mix. The costs of water shortages to irrigated agriculture using a combination of these strategies was 75% lower than under a scenario where the only adaptation mechanism was land fallowing. Similar to the Reclamation analysis cited above, results suggested that reducing cotton and alfalfa production would be most effective. Similar to the Howitt, Medellín-Azuara, and MacEwan (2009, 2010) California study results, agricultural output declined primarily for commodity crops, with little change to high-value specialty crops. Although the model treated the entire region in aggregate, the largest reductions came from crops grown in central Arizona, which holds junior water rights to Colorado River water. Crops grown in western Arizona were little affected. With high-value crops and senior rights to Colorado

River water, western Arizona would remain a national center of specialty crop production. Model results also suggested there would be relatively large losses to livestock and dairy producers from reduced supplies of alfalfa and feed grains.

Lessons from simulation studies

These mathematical programming model studies varied in many dimensions: period, geographic scope, crop coverage, hydrologic detail, and assumed climate/water shock. Taken together, however, one can draw some general lessons from their results. First, based on these simulations, agriculture would be the sector that alters its water use the most, to adapt to regional water shortages and protect municipal and industrial (M&I) uses. Agriculture would buffer urban users from water shocks, and thus serves an important insurance function. Second, although fallowing irrigated land is one response to drought, growers would have numerous lower-cost options. Third, important factors in adapting to water shortages would be the costs of complying with environmental regulations, especially those that protect endangered aquatic species. Fourth, additional investments in infrastructure to store and convey water would likely be required to reduce negative effects of dry warming or increase the benefits of wet warming. Fifth, and perhaps most importantly, water transfers would have the potential to significantly reduce the costs of adjusting to water shortages under dry warming scenarios. Agriculture-to-urban transfers would increase income for agricultural areas, partially compensating for losses from reduced water use. Currently, however, many institutional restrictions limit the transfer of water across jurisdictions, basins, or state lines. The flexibility provided by water transfers also would depend on future investment in complementary infrastructure.

11.5 Ranching Adaptations to Multi-year Drought

Southwestern cattle ranches depend on rain-fed forage grasses to feed cattle. Only a small portion of pastureland is irrigated. Drought reduces forage production on livestock grazing lands and is a major concern among ranchers (Coles and Scott 2009). In much of the region, rainfall occurs during the winter, and a rise in winter temperatures may increase forage production compared to present conditions. Climate change could offer possibilities for range improvement through the introduction of alternate forage species or of trees and shrubs that increase shade for livestock and soil fertility. Climate change may increase pasture productivity via CO_2 enrichment on plant growth and because warmer temperatures would lengthen the growing season. This should reduce the need of ranchers to store forage to feed animals over the winter. Increased temperatures are expected to increase the variability of precipitation (Izaurralde et al. 2011). Thus, a challenge for ranchers will be to manage this variability.

In the case of severe drought, ranchers can adapt by: (1) purchasing additional feed, (2) reducing herd size through selling of stock, (3) leasing additional grazing land, or (4) temporarily over-grazing lands. These adaptations are not without negative consequences. Cattle sold prematurely at lower weights fetch lower prices and sales prices during droughts can be low because many ranchers are selling simultaneously. Herd liquidation makes restocking herds in future years more expensive. Overgrazing can reduce the long-term productivity of grazing lands. All these factors may reduce both the short-term returns and longer-term debt and borrowing capacity of ranchers. Drought may also affect the price of hay, which ranchers might use for supplemental feed. However, Bastian and colleagues (2009) and Ritten, Frasier, Bastian, Paisley and colleagues (2010) suggest that irrigated hay production and statewide markets for hay reduce the risks of adverse economic impacts from drought. Costs of hay, however, can be high if pervasive drought means that it must be transported over long distances.

Ranchers face two types of risk: price risk and weather risk. They can limit their exposure to price risk through use of futures and options contracts, but to do so means that ranchers must consider price and weather risk jointly. Recent research that focuses on strategies to adapt to multi-year droughts has important implications. Such research does not directly address adaptation to particular climate-change scenarios. However, results are relevant for considering climate scenarios that project continued and prolonged drought. Important considerations for ranchers seeking to adapt to multi-year drought are (a) the length and the severity of the drought and (b) when the drought occurs in the cattle price cycle.

An example of an adaptation strategy for ranchers is to provide supplemental feeding to cattle in addition to pursuing a baseline strategy of herd liquidation (Bastian et al. 2009; Ritten, Frasier, Bastian, Paisley et al. 2010). Supplemental feeding appears to be the better long-term strategy. It allows more animals to be sold after the drought (when prices are higher) and avoids aggressive culling of herds during drought, which would have higher restocking costs. Research findings, however, suggest that there is no single "right" strategy and that the advantages of supplemental feeding depend on where a ranch is in the price cycle.

Another example of an adaptation strategy is a "flexible" rather than "conservative" approach to drought management of livestock operations (Torell, McDaniel and Koren 2011). A conservative approach would maintain low baseline stocking rates, thus requiring little sell-off in response to drought. This approach reduces adjustment costs of destocking and restocking, but does not fully utilize available forage in good years. It thus misses out on opportunities to make high returns in years with abundant forage. The flexible approach would adjust herd size to fit forage productivity and lease additional grazing land during droughts (as opposed to simply destocking). This approach allows ranchers to capitalize on good forage conditions and avoids problems of overgrazing during drought. There are costs to this approach, however. Additional grazing land with suitable forage may be scarce if drought is geographically pervasive and there are added costs of transporting livestock. High transportation costs could make the flexible approach economically unfeasible.

Ritten, Frasier, Bastian, and Gray (2010) also compared a flexible strategy to a fixed cattle-stocking strategy over a multi-year horizon and accounting for uncertainty. Optimal stocking depends on rangeland health, which varies with grazing pressure and growing season precipitation. Compared to the scenario of fixed stocking at levels recommended by the Natural Resource Conservation Service, a scenario of flexible stocking would increase average annual revenues 40%, while reducing profit variability. Over a variety of climate projections, Ritten, Frasier, Bastian and Gray found increased variability of annual precipitation to be a greater threat to ranch profitability than changes in projected average precipitation. For scenarios with more variable precipitation, average stock rates would decline but also would vary more under the flexible system. Variable precipitation could pose problems for cow/calf producers, who maintain a base herd of cows that produces calves for sale after weaning (or until they are yearlings), and so are less flexible than stocker operations. Stocker operations in contrast purchase weaned cattle in the spring to put to pasture before sale. They are more flexible because they do not need to maintain a base herd and stocker purchases can be made based on anticipated forage conditions. The cattle industry could adapt by shifting to more flexible cow-calf-yearling operations that could take better advantage of good years, while selling yearlings early to avoid damaging the range in lean years. Ritten, Bastian, and colleagues (2010) cite this as among the most profitable long-term strategies for cattle producers dealing with prolonged drought.

Torell, McDaniel, and Koren (2011) found the potential gains from this flexible strategy also depend on when drought occurs in the timing of the cattle price cycle. Potential gains from a flexible strategy are greater if forage productivity is more variable from year to year. The approach, however, entails higher costs and financial risks. Further, the approach may be more appropriate in cooler climates. In short-grass prairies, such as in New Mexico, the estimated large gains from the flexible strategy rely on perfect climate forecasts to make management decisions. In actuality, key decisions about livestock purchases depend on past conditions and the well-intended but imperfect 90-day seasonal forecasts of the National Weather Service Climate Prediction Center. Climate forecasts that are more accurate and have a longer lead-time could increase the value of a flexible grazing strategy. At present, the quality of forecasts is not sufficient to make this a preferable strategy in short-grass prairie systems. However, this example illustrates how improved climate forecasts could help ranching adapt to climate change.

11.6 Disaster Relief Programs and Climate Adaptation

Agricultural producers may take a variety of actions to reduce risks from drought, flood, and other weather-related events. They can diversify the mix of the crops they grow, adopt irrigation and pest control practices to protect yields, enter into forward or futures contracts,[iv] or make use of weather or other data to time operations to reduce risk. Increasingly, farmers have diversified their household incomes by relying on both farm and non-farm jobs. Disaster relief programs affect producer incentives for managing risks because they alter the costs and benefits of these and other risk-reducing measures.

Congress has traditionally provided regular disaster payments to growers on an ad hoc basis in response to natural disasters and weather extremes that lowered crop yields or forage production. Ad hoc payments have been criticized because of their expense and because they maintain economic incentives to continue production in areas susceptible to agronomic risks. The Federal Crop Insurance Act of 1980 and subsequent legislation attempted to establish crop insurance, rather than disaster payments, as the main vehicle for managing farm risk. While the number of producers covered under federally subsidized crop insurance has risen, ad hoc disaster payments have continued, averaging about $1 billion annually.

The most recent Farm Bill (2008) established several new disaster relief programs also intended to replace ad hoc payments. The largest program was the Supplemental Revenue Assistance Payments Program (SURE), which pays producers for crop revenue losses from natural disaster or adverse weather. It compensates producers for a portion of their losses not eligible for payments under crop-insurance policies (Shields 2010). A producer can become eligible for payments if a disaster is declared in that producer's county or a contiguous county. Eligible producers need show only a 10% yield loss on one crop to qualify for payments. Outside of designated counties, producers must show a 50% loss of a crop. The SURE program has proved to be complex to administer in part because of how it interacts with crop insurance payments. Payments are often delayed for a year or more after actual losses.

Some researchers have raised concerns that the program encourages more risky behavior by producers (Barnaby 2008; Schnitkey 2010; Shields 2010; Smith and Watts 2010). Small changes in yield, even one bushel per acre, can mean the difference between receiving large payments or no payments. This makes it difficult for producers to determine year to year if they will be eligible or for program payments. It also makes it difficult for administrators to gauge whether producers are actively trying to avoid yield losses. Payments are more likely to be triggered if producers raise a single crop in a county that has high yield risk than if they grow a more diversified mix of crops. In some cases, producers may receive higher revenues by simply allowing their crops to fail (Smith and Watts 2010). Figure 11.5 shows the counties that received two-thirds of SURE payments disbursed in the Southwestern states to date. Fourteen counties, primarily in dryland wheat producing areas, account for most of the payments. Most are counties with payments triggered every year.

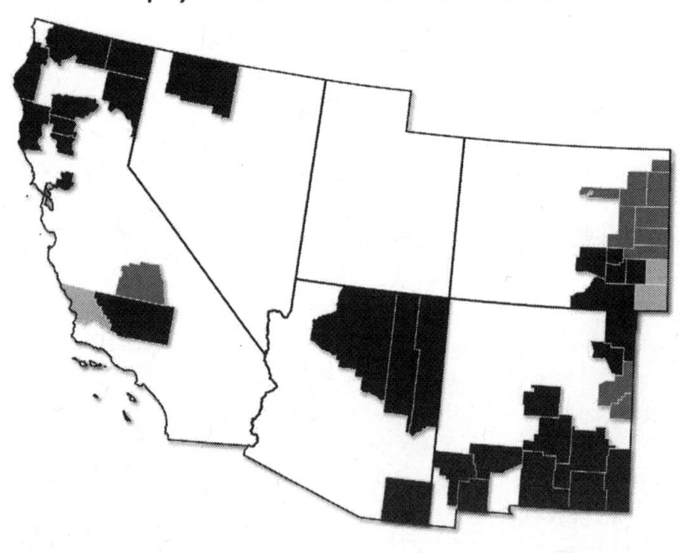

Figure 11.5 Counties that have accounted for two-thirds of all payments disbursed to Southwestern States under the Livestock Forage Disaster Program (LFP) and the Supplemental Revenue Assistance (SURE) Program. Source: USDA (n.d.).

Another disaster relief program established under the 2008 Farm Bill was the Livestock Forage Disaster Program (LFP). The program compensates livestock producers for losses related to drought or fires on grazing lands. For drought compensation, producers must have livestock in counties rated by the U.S. Drought Monitor[v] (Svoboda et al. 2002) as having severe, extreme, or exceptional drought. Payment levels rise with the length and severity of drought. Producers may also qualify if they normally graze livestock on federal lands where federal agencies have banned grazing because of occurrence of fire. LFP has certain advantages over SURE from a risk-management perspective. First, payments are determined by the Drought Monitor rather than disaster designations, which do not necessarily follow clear, severity-related guidelines. Second, because the Drought Monitor releases information weekly, processing and payment of claims is much faster. Third, and perhaps most importantly, payments based on county-level drought or fire conditions mean that payment levels are relatively independent of producer decisions. Thus, there is much less reward for producers failing to limit risk. Figure 11.5 also shows counties that have received two-thirds of all LFP payments disbursed to the Southwestern states to date. It illustrates where drought and fire risks and livestock forage production intersect.

Lobell, Torney, and Field (2011) examined data on federal crop insurance indemnity payments and disaster payments in California from 1993 to 2007. Grapes accounted for the largest number of indemnity claims, followed closely by wheat. Tree crops and grapes accounted for 75% of all indemnity payments. Excess moisture was the most common cause of both insurance and disaster payments, followed by cold spells, then heat waves. The effect of climate change on these payments remains difficult to predict. Less frequent cold extremes would tend to reduce payments, while heat waves would tend to increase them. There remains a high degree of variability in projections of precipitation intensity, flooding risk, and other hydrological risks (Lobell, Torney, and Field 2011). Given the economic significance of damage from wet events, better projections of these extreme events are important.

The Southwest region is characterized by irrigation-dependent production of high-value specialty crops that are vulnerable to excess moisture, followed by cold, then heat. The region also is characterized by ranching and dryland wheat production, both of which are sensitive to fluctuations in precipitation. In both areas, improved projections of precipitation will be crucial for agricultural adaptation. Another key area of uncertainty is knowledge about when improvements in irrigation efficiency actually reduce consumptive use of water on a basin-wide scale and when it actually increases consumptive water use. Finally, many of the costs of climate change to agricultural producers are adjustment costs. Effects of climate change on both tree-crop and livestock production will be long-lived, with short-term shocks having repercussions over several years of tree and animal production cycles.

References

Aron, R. 1983. Availability of chilling temperatures in California. *Agricultural Meteorology* 28:351–363.

Backlund, P., A. Janetos, and D. Schimel. 2008. *The effects of climate change on agriculture, land resources, water resources, and biodiversity*. U.S. Climate Change Science Program, Synthesis and Assessment Product 4.3 (CCSP-SAP 4.3).

Baldocchi, D., and S. Wong. 2008. Accumulated winter chill is decreasing in the fruit growing regions of California. *Climatic Change* 87 (Suppl. 1): S153–S166.

Barnaby Jr., G. A. SURE calculator (new standing disaster aid). *Ag Manager, Kansas State Research and Extension.*

Bastian, C. T., P. Ponnamanneni, S. Mooney, J. P. Ritten, W. M. Frasier, S. I. Paisley, M. A. Smith, and W. J. Umberger. 2009. Range livestock strategies given extended drought and different price cycles. *Journal of the American Society of Farm Managers and Rural Appraisers* 72:153–63.

Benson, R. 2004. So much conflict, yet so much in common: Considering the similarities between western water law and the Endangered Species Act. *Natural Resources Journal* 44:29–76.

Berck, P., and J. M. Perloff. 1985. A dynamic analysis of marketing orders, voting, and welfare. *American Journal of Agricultural Economics* 67:487–496.

Booker, J. F., A. M. Michelsen, and F. A. Ward. 2005. Economic impact of alternative policy responses to prolonged and severe drought in the Rio Grande Basin. *Water Resources Research* 41: W02026, doi:10.1029/2004WR003486.

Caswell, M. F., and Zilberman, D. 1986. The effects of well depth and land quality on the choice of irrigation technology. *American Journal of Agricultural Economics* 68:798–811.

Cayan, D. R., E. P. Maurer, M. D. Dettinger, M. Tyree, and K. Hayhoe. 2008. Climate change scenarios for the California region. *Climatic Change* 87 (Suppl. 1): S21–S42.

Chong, H., and D. Sunding, 2006. Water markets and trading. *The Annual Review of Environment and Resources* 31:239–264.

Christensen, N. S., and D. P. Lettenmaier. 2007. A multimodel ensemble approach to assessment of climate change impacts on the hydrology and water resources of the Colorado River Basin. *Hydrology and Earth System Sciences* 11:1417–1434.

Coles, A. R., and C. A. Scott. 2009. Vulnerability and adaptation to climate change and variability in semi-arid rural southeastern Arizona, USA. *Natural Resources Forum* 33:297–309.

Council for Agricultural Science and Technology (CAST) (2012). *Herbicide-resistant weeds threaten soil conservation gains: Finding a balance for soil and farm sustainability.* CAST Issue Paper 49. Ames, IA: CAST.

Elbakidze, L. 2006. Potential economic impacts of changes in water availability on agriculture in the Truckee and Carson River Basins, Nevada, USA. *Journal of the American Water Resources Association* 42:841–849.

Fereres, E., and M. A. Soriano. 2007. Deficit irrigation for reducing agricultural water use. *Journal of Experimental Botany* 58:147–159.

Frisvold, G., and K. Emerick. 2008. Rural-urban water transfers with applications to the U.S.-Mexico border region. In *Game theory and policy making in natural resources and the environment,* ed. A. Dinar, J. Albiac, and J. Sanchez-Soriano, 155-180. New York: Routledge Press.

Frisvold, G., and K. Konyar. 2012. Less water: How will agriculture in southern Mountain states adapt? *Water Resources Research* 48: W05534, doi:10.1029/2011WR011057.

Frisvold, G., and A. Murugesan. 2011. Use of climate and weather information for agricultural decision making. In *Adaptation and resilience: The economics of climate, water, and energy challenges in the American Southwest*, ed. B. Colby and G. Frisvold. Washington, DC: Resources for the Future.

Gutierrez, A. P., L. Ponti, C. K. Ellis, and T. d'Oultremont. 2008. Climate change effects on poikilotherm tritrophic interactions. *Climatic Change* 87:167–192.

Harou, J. J., J. Medellín-Azuara, T. Zhu, S. K. Tanaka, J. R. Lund, S. Stine, M. A. Olivares, and M. W. Jenkins. 2010. Economic consequences of optimized water management for a prolonged, severe drought in California. *Water Resources Research* 46: W05522, doi:10.1029/2008WR007681.

Hatfield, J. L., K. J. Boote, B. A. Kimball, L. H. Ziska, R. C. Izaurralde, D. Ort, A. M. Thomson, and D. Wolfe. 2011. Climate impacts on agriculture: Implications for crop production. *Agronomy Journal* 103:351–370.

Howitt, R. E., D. MacEwan, and J. Medellín-Azuara 2009. Economic impacts of reductions in Delta exports on Central Valley agriculture. *Agricultural and Resource Economics Update* 12 (3): 1–4.

Howitt, R. E., J. Medellín-Azuara, and D. MacEwan, D. 2009. *Estimating economic impacts of agricultural yield related changes*. Final Report CEC-500-2009-042-F. Sacramento: California Climate Change Center.

—. 2010. Climate change, markets, and technology. *Choices* 3rd Quarter 2010, 25(3).

Huffaker, R., and N. Whittlesey. 2003. A theoretical analysis of economic incentive policies encouraging agricultural water conservation. *Water Resources Development* 19:37–55.

Izaurralde, R. C., A. M. Thomson, J. A. Morgan, P. A. Fay, H. W. Polley, and J. L. Hatfield. 2011. Climate impacts on agriculture: Implications for forage and rangeland production. *Agronomy Journal* 103:371–381.

Jackson, L. E., S. M. Wheeler, A. D. Hollander, A. T. O'Geen, B. S. Orlove, J. W. Six, D. A. Sumner, et al. 2011. Case study on potential agricultural responses to climate change in a California landscape. *Climatic Change* 109 (Suppl. 1): S407–S427.

Joyce, B., D. R. Purkey, D. Yates, D. Groves, and A. Draper. 2010. *Integrated scenario analysis for the 2009 California Water Plan update*. Sacramento: California Department of Water Resources.

Joyce, B., S. Vicuña, L. Dale, J. Dracup, M. Hanemann, D. Purkey, and D. Yates. 2006. *Climate change impacts on water for agriculture in California: A case study in the Sacramento Valley*. White Paper CEC-500-2005-194-SF. Sacramento: California Climate Change Center.

Jury, W. A, and H. J. Vaux. 2006. The role of science in solving the world's emerging water problems. *Proceedings of the National Academy of Sciences* 102:15715–15720.

Kiparsky, M., and P. H. Gleick. 2003. *Climate change and California water resources: A survey and summary of the literature*. Oakland, CA: Pacific Institute.

Kallenbach, C. M., D. E. Rolston, and W. R. Horwath. 2010. Cover cropping affects soil N_2O and CO_2 emissions differently depending on type of irrigation. *Agriculture, Ecosystems & Environment* 137:251–260.

Kenny, J. F., N. L. Barber, S. S. Hutson, K. S. Linsey, J. K. Lovelace, and M. A. Maupin. 2009. *Estimated use of water in the United States in 2005*. U.S. Geological Survey Circular 1344. Reston, VA: USGS.

Lévite, H., H. Sally, and J. Cour. 2003. Testing water demand management scenarios in a water-stressed basin in South Africa: Application of the WEAP model. *Physics and Chemistry of the Earth* 28:779–786.

Lobell, D., A. Torney, and C.B. Field. 2011. Climate extremes in California agriculture. *Climatic Change* 109 (Suppl 1): S355–S363, doi: 10.1007/s10584-011-0304-5

Meadows, R. 2009. UC scientists help California prepare for climate change. *California Agriculture* 63:56–58, doi: 10.3733/ca.v063n02p56.

Medellín-Azuara, J., J. J. Harou, M. A. Olivares, K. Madani, J. R. Lund, R. E. Howitt, S. K. Tanaka, and M. W. Jenkins. 2008. Adaptability and adaptations of California's water supply system to dry climate warming. *Climatic Change* 87 (Suppl. 1): S75–S90.

Mendelsohn, R., and A. Dinar. Climate, water, and agriculture. *Land Economics* 79:328–341.

Moore, M., A. Mulville, and M. Weinberg. 1996. Water allocation in the American West: endangered fish versus irrigated agriculture. *Natural Resources Journal* 36:319–358.

Norsworthy, J. K., S. Ward, D. Shaw, R. Llewellyn, R. Nichols, T. Webster, K. Bradley, et al. 2012. Reducing the risks of herbicide resistance: Best management practices and recommendations. Weed Science 60 (Suppl. 1): 31–62, doi: 10.1614/WS-D-11-00155.1.

Parker, D., D. R. Cohen-Vogel, D. E. Osgood, and D. Zilberman. 2000. Publicly funded weather database benefits users statewide. *California Agriculture* 54 (3): 21–25.

Parry, M., N. W. Arnell, H. Hulme, R. Nicholls, and M. Livermore. 1998. Adapting to the inevitable. *Nature* 395:741.

Peterson, J. M., and Y. Ding. 2005. Economic adjustments to groundwater depletion in the high plains: Do water-saving irrigation systems save water? *American Journal of Agricultural Economics* 87:147–159.

Price, A. J., K. S. Balkcom, S. A. Culpepper, J. A. Kelton, R. L. Nichols, and H. Schomberg. 2011. Glyphosate-resistant Palmer amaranth: A threat to conservation tillage. *Journal of Soil and Water Conservation* 66:265–275.

Quiggin, J., and J. Horowitz. 1999. The impact of global warming on agriculture: A Ricardian analysis; Comment. *American Economic Review* 89:1044–1045.

—. 2003. Costs of adjustment to climate change. *Australian Journal of Agricultural and Resource Economics* 47:429–446.

Ragab, R., and C. Prudhomme. 2002. Climate change and water resources management in arid and semi-arid regions: Prospective and challenges for the 21st century. *Biosystems Engineering* 81:3–34.

Rajagopalan, B., K. Nowak, J. Prairie, M. Hoerling, B. Harding, J. Barsugli, A. Ray, and B. Udall. 2009. Water supply risk on the Colorado River: Can management mitigate? *Water Resources Research* 45: W08201, doi:10.1029/2008WR007652.

Ritten, J., C. T. Bastian, S. I. Paisley, and M. Smith. 2010. Long term comparison of alternative range livestock management strategies across extended droughts and cyclical prices. *Journal of the American Society of Farm Managers and Rural Appraisers* 73:243–252.

Ritten, J. P., W. M. Frasier, C. T. Bastian, and S. T. Gray. 2010. Optimal rangeland stocking decisions under stochastic and climate-impacted weather. *American Journal of Agricultural Economics* 92:1242–1255.

Ritten, J. P., W. M. Frasier, C. T. Bastian, S. I. Paisley, M. A. Smith, and S. Mooney. 2010. A multiperiod analysis of two common livestock management strategies given fluctuating precipitation and variable prices. *Journal of Agricultural and Applied Economics* 42:177–191.

Rockstrom, J., M. Lannerstad, and M. Falkenmark. 2007. Increasing water productivity through deficit irrigation: Evidence from the Indus plains of Pakistan. *Proceedings of the National Academy of Sciences* 104:6253–6260.

Schneider, J. M., and J. Wiener. 2009. Progress toward filling the weather and climate forecast needs of agricultural and natural resource management. *Journal of Soil and Water Conservation* 64:100A–106A.

Schnitkey, G. 2008. SURE window closes September 30. Urbana-Champaign: University of Illinois at Urbana-Champaign, Department of Agricultural and Consumer Economics. http://www.farmdoc.illinois.edu/manage/newsletters/fefo10_16/fefo10_16.html.

Seager, R., M. Ting, I. Held, Y. Kushnir, J. Lu, G. Vecchi, H-P. Huang, et al. 2007. Model projections of an imminent transition to a more arid climate in southwestern North America. *Science* 316:1181–1184, doi:10.1126/science.1139601.

Shields, D. A. 2010. *A whole-farm crop disaster program: Supplemental Revenue assistance payments (SURE).* Congressional Research Service (CRS) Report for Congress R40452. Washington, DC: CRS.

Smith, V., and M. Watts. 2010. The new Standing Disaster Program: A SURE invitation to moral hazard behavior. *Applied Economic Perspectives and Policy* 32:154–69.

Svoboda, M., D. LeComte, M. Hayes, R. Heim, K. Gleason, J. Angel, B. Rippey et al. 2002. The Drought Monitor. *Bulletin of the American Meteorologial Society* 83:1181–1192.

Tanaka, S. K., T. Zhu, J. R. Lund, R. E. Howitt, M. W. Jenkins, M. Pulido-Velazquez, M. Tauber, R. S. Ritzema, and I. C. Ferreira. 2006. Climate warming and water management adaptation for California. *Climatic Change* 76:361–387.

Torell, L. A., K. C. McDaniel, and V. Koren. 2011. Estimating grass yield on blue grama range from seasonal rainfall and soil moisture measurements. *Rangeland Ecology and Management* 1:56–66.

U.S. Bureau of Reclamation (Reclamation). 2007. *Final environmental impact statement, Colorado River interim guidelines for lower basin shortages and coordinated operations for Lakes Powell and Mead*. Boulder City, NV: U. S. Bureau of Reclamation, Lower Colorado Region.

U.S. Department of Agriculture (USDA). 1999. Farm and ranch irrigation survey (1998). In *1997 Census of agriculture,* Vol. 3, *Special studies,* Part 1. AC97-SP-1. Washington, DC: USDA Natural Agricultural Statistics Service (NASS).

—. 2004. Farm and ranch irrigation survey (2003). In *2002 Census of agriculture,* Vol. 3, *Special studies,* Part 1. AC-02-SS-1. Washington, DC: USDA–NASS.

—. 2009. United States summary and state data. In *2007 Census of agriculture,* Vol. 1, *Geographic area series,* Part 51. AC-07-A-51. Washington, DC: USDA–NASS.

—. 2010. Farm and ranch irrigation survey (2008). In *2007 Census of agriculture,* Vol. 3, *Special studies,* Part 1. AC-07-SS-1. Washington, DC: USDA–NASS.

—. 2012. 2007 Census ag atlas maps. http://www.agcensus.usda.gov/Publications/2007/Online_Highlights/Ag_Atlas_Maps/. Washington, DC: USDA–NASS.

—. n.d. Disaster assistance programs: Payments by state for the 5 disaster programs (SURE, LFP, ELAP, LIP, TAP), county data. Washington, DC: USDA Farm Services Agency. http://www.fsa.usda.gov/FSA/webapp?area=home&subject=diap&topic=landing (accessed May 12, 2012).

U.S. Geological Survey (USGS). 2005. Estimated use of water in the United States: County-level data for 2005. http://water.usgs.gov/watuse/data/2005/.

Wallace, J. S. Increasing agricultural water use efficiency to meet future food production. *Agriculture, Ecosystems and Environment* 82:105–119.

Ward, F. A., B. H. Hurd, T. Rahmani, and N. Gollehon. 2006. Economic impacts of federal policy responses to drought in the Rio Grande Basin. *Water Resources Research* 42: W03420, doi:10.1029/2005WR004427.

Ward, F. A., and M. Pulido-Velazquez. 2008. Water conservation in agriculture can increase water use. *Proceedings of the National Academy of Sciences* 105:18215–18220.

Wiener, J., R. Crifasi, K. Dwire, S. Skagen, and D. Yates. 2008. Riparian ecosystem consequences of water redistribution along the Colorado Front Range. *Water Resources Impact* 10 (3): 18–21.

Endnotes

i Chill time is the accumulation of hours between 32°F–45°F (0°C–7°C) during bud dormancy (Aron 1983; Baldocchi and Wong 2008).

ii This would be declared by the Secretary of the Interior in response to specific conditions agreed upon by the seven states participating in the Colorado River Compact: the six Southwestern states considered in this report and Wyoming.

iii Rather than emphasizing maximizing yield (crop output per acre), deficit irrigation focuses more on achieving greater output per unit of water applied (Fereres and Soriano 2007). The strategy can involve some sacrifice of yield, but can use less water.

iv Both forward and futures contracts are agreements to buy and sell an asset at a specified time and price in the future, with both terms agreed upon today. Futures contracts are standardized contracts traded on commodity exchanges, while forward contracts are bilateral agreements between two parties.

v See http://droughtmonitor.unl.edu/.

Chapter 12

Energy: Supply, Demand, and Impacts

COORDINATING LEAD AUTHOR

Vincent C. Tidwell (Sandia National Laboratories)

LEAD AUTHORS

Larry Dale (Lawrence Berkeley National Laboratory), Guido Franco (California Energy Commission)

CONTRIBUTING AUTHORS

Kristen Averyt (University of Colorado), Max Wei (Lawrence Berkeley National Laboratory), Daniel M. Kammen (University of California-Berkeley), James H. Nelson (University of California- Berkeley)

REVIEW EDITOR

Ardeth Barnhart (University of Arizona)

Executive Summary

Energy is important to the Southwest United States, where 12.7% of the nation's energy is produced (extracted or generated) and 12.1% is consumed. The region is in the favorable position of having low per-capita energy consumption (222 million BTUs per person) relative to that of the nation as a whole (302 million BTUs per person); nevertheless, disruption of power has significant economic implications for the region (e.g., LaCommare and Eto 2004; Northwest Power and Conservation Council 2005). Climate change itself, as well as strategies aimed at mitigation and adaptation have the potential to impact the production, demand, and delivery of energy in a number of ways.

Chapter citation: Tidwell, V. C., L. Dale, G. Franco, K. Averyt, M. Wei, D. M. Kammen, and J. H. Nelson. 2013. "Energy: Supply, Demand, and Impacts." In *Assessment of Climate Change in the Southwest United States: A Report Prepared for the National Climate Assessment*, edited by G. Garfin, A. Jardine, R. Merideth, M. Black, and S. LeRoy, 240–266. A report by the Southwest Climate Alliance. Washington, DC: Island Press.

- Delivery of electricity may become more vulnerable to disruption due to climate-induced extreme heat and drought events as a result of:
  - increased demand for home and commercial cooling,
  - reduced thermal power plant efficiencies due to high temperatures,
  - reduced transmission line, substation, and transformer capacities due to elevated temperatures,
  - potential loss of hydropower production,
  - threatened thermoelectric generation due to limited water supply, and
  - the threat of wildfire to transmission infrastructure.

  (medium-high confidence)

- Climate-related policies have the potential to significantly alter the energy sector. A shift from the traditional fossil fuel economy to one rich in renewables has significant implications for related water use, land use, air quality, national security, and the economy. The vulnerability of the energy system in the Southwest to climate change depends on how the energy system evolves over this century. (medium-high confidence)

12.1 Introduction

Energy consumption in the Southwest United States was 12,500 trillion British thermal units (BTUs) in 2009, equal to 222 million BTUs per person (EIA 2010). Any change or disruption to the supply of energy is likely to have significant impacts. For example, a study found that electrical power blackouts and "sags" cost the United States about $80 billion every year in lost services, industrial capacity, and gross domestic product (LaCommare and Eto 2004).

This chapter provides an objective assessment of the vulnerability of the energy infrastructure of the Southwest to the effects of projected climate change. There are a number of ways in which the Southwest's energy infrastructure could be affected by climate change, such as by increased peak electricity demand for cooling, damage to energy infrastructure by extreme events, disruption of hydroelectric and thermoelectric generation due to high temperatures or restricted water availability, and evolution of the energy portfolio both in terms of electricity generation and transportation fuel choices (such as moving from fossil fuel-fired electricity generation to utilization of more renewable sources).

This chapter reviews primary energy production and consumption in the Southwest, both past and present, then focuses on potential climate impacts on energy infrastructure, defining for each the exposure pathways, possible extent of impact, adaptation/mitigation options, and data gaps.

12.2 Energy in the Southwest: Past and Present

Energy is an important resource in the Southwest, both in terms of production and consumption. According to the U.S. Energy Information Administration (EIA 2010), primary energy production in the Southwest was 9,200 trillion BTUs in 2009, 12.7% of total

U.S. production. Since 1960, primary energy production in the Southwest has grown 180%. In 2009, consumption outpaced production by 3,300 billion BTUs. However, the per capita consumption in the Southwest of 222 million BTUs per person is significantly below the national average of 302 million BTUs per person. The 2009 consumption level represents a 255% increase since 1960.

Natural gas represents 43% of the primary energy production in the Southwest (Figure 12.1), followed by crude oil (21%), coal (19%), renewable energy[i] (10%), and nuclear electric power (7%). Since 1960, there has been a strong increase in the production of natural gas, coal, nuclear electric power, and renewable energy, while crude oil production decreased. Significant differences in energy production by state are evident across the Southwest (Figure 12.1). California, Colorado, and New Mexico are among the nation's top ten energy-producing states, while Nevada is ranked 47th. Colorado, New Mexico, and Utah have significant natural gas production, while California leads the region in crude oil and renewable energy production.

In 2009, 36% of the 12,500 trillion BTUs consumed in the region were associated with the transportation sector, with the remaining consumption spread relatively evenly

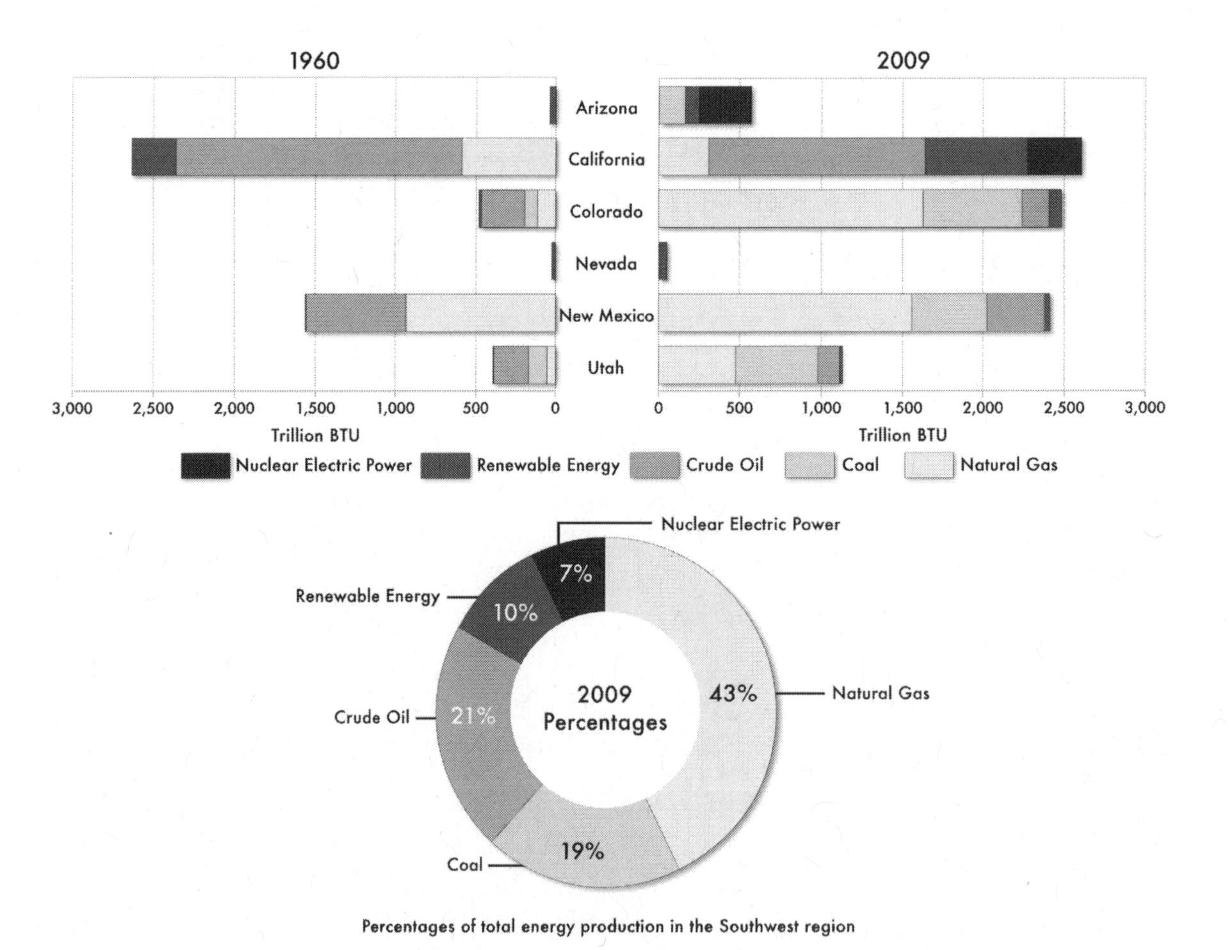

Figure 12.1 Energy production in the Southwest region. Source: EIA (2010).

across the industrial, residential, and commercial sectors (EIA 2010). Forty-two percent of this demand was met with petroleum products, 32% by natural gas, 13% by coal, 8% by renewable sources, and 5% by nuclear electric power (EIA 2010). In total, 87% of total consumption was met with fossil fuels. The most notable change since 1960 is the growing share of demand now met with coal and nuclear electric power. Energy consumption also varies significantly by state (Figure 12.2). Consumption is dominated by California, followed distantly by Arizona and Colorado. However, when considered on a per capita basis a very different picture emerges. California has the lowest per capita consumption at 220 million BTU per person, followed by Arizona (220), Nevada (270), Utah (270), Colorado (290), and New Mexico (330), with all but one below the national average. Unlike the other states, California has a disproportionately high use of petroleum (50% of demand) but low use of coal (only 1%). Arizona has the highest proportion of demand met through non-fossil fuels (24%), while Utah has the lowest (2%).

Specific to electricity, 487 million megawatt hours (MWh) were generated in the Southwest in 2009, a 43% increase over 1990 levels. The associated fuel mix for this generation includes 42% natural gas, 30% coal, 13% nuclear, 8% hydroelectric, and 7% other

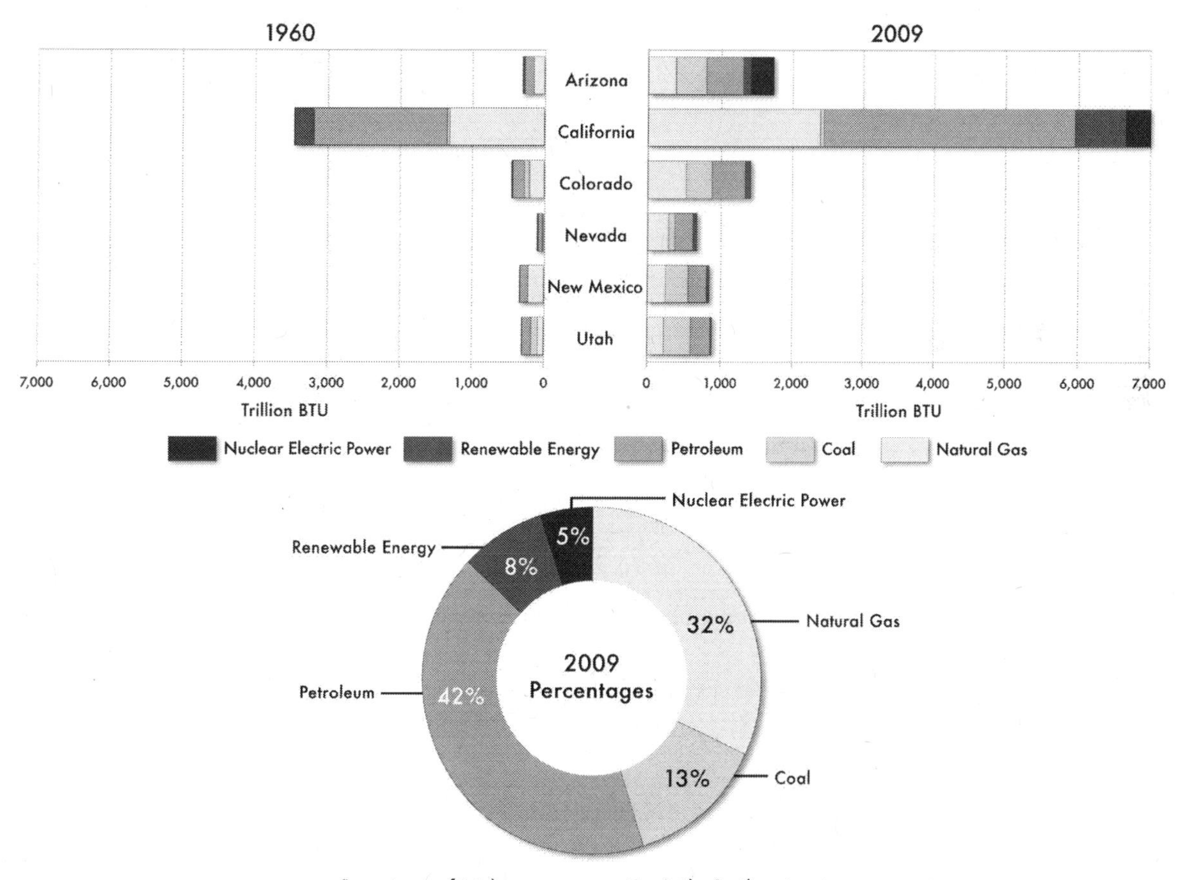

Figure 12.2 Energy consumption in the Southwest region. Source: EIA (2010).

renewable (EIA 2010). Since 1990, significantly less coal has been used to produce electricity, with a commensurate increase in generation by natural gas (Figure 12.3). In 2009 almost 65% of the region's electricity was produced in California and Arizona. Even so, California imported approximately 25% of its electricity, while in contrast Arizona exported 53% of its production. New Mexico, Utah, and Nevada are also net exporters of electricity.

The mix of fuels used to generate electricity differs across the six states (Figure 12.3). Most notable is the sizeable utilization of coal in Colorado, New Mexico, and Utah. California and Nevada make up for their low coal use largely with natural gas. Nuclear electric power is generated only in Arizona and California. In terms of renewables, including large-scale hydropower, California was responsible for 73% of the region's generation (EIA 2010).

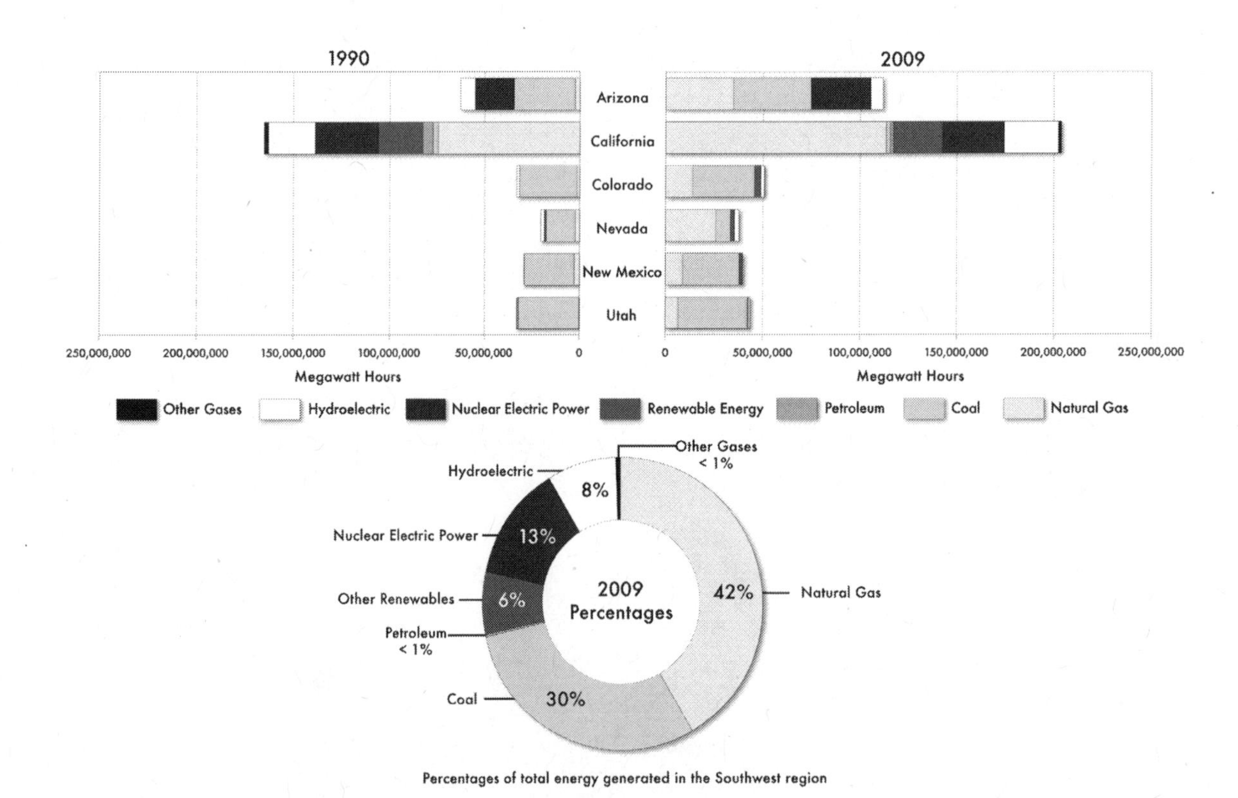

Figure 12.3 Electricity generation in the Southwest region. Source: EIA (2010).

In 2009, total carbon dioxide (CO_2) emission from energy consumption in the region was 728 million metric tons, nearly a third of which was for generation of electricity (EIA 2010). Electricity emissions have grown by 50% over 1990 levels while electricity generation has only increased by 43% over the same period of time. Although California produces 42% of the region's electricity, it is responsible for only 21% of the region's CO_2 emissions from electricity, not counting imports, largely due to generation that favors natural gas and renewables. However, California alone is responsible for 68% of the region's vehicle CO_2 emissions (217 million metric tons).

Figure 12.4 shows the locations of key energy infrastructure in the Southwest (data taken from EIA 2010). Features shown include power plants (distinguished by fuel type), petroleum refineries, coal mines, and transmission lines (of 345kV or more).

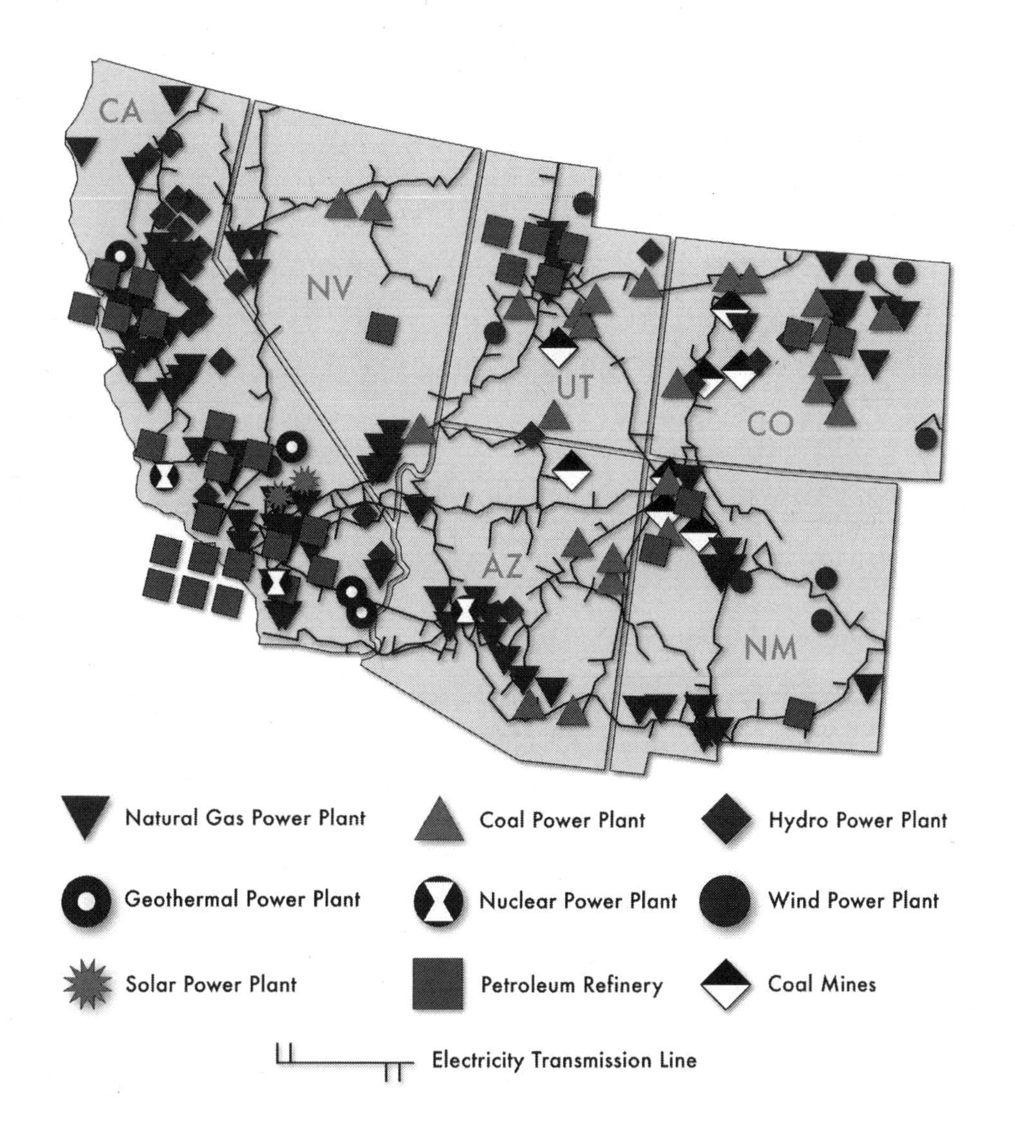

Figure 12.4 Energy distribution in the Southwest region. Source: EIA (2010).

12.3 Potential Climate Impacts on Energy

This section discusses the energy future of the Southwest, particularly the potential effects of climate change on energy demand, supply, and delivery.[ii] The potential impacts are organized and discussed according to eight distinct themes: energy demand, electricity generation, electricity distribution, energy infrastructure, renewable source intensity, evolution of the energy sector, primary energy production, and the cost of climate change. Each theme is explored in terms of linkages to climate change: observed climate-related change; projected range of climate-induced change, mitigation, and adaptation strategies; and related data and knowledge gaps.

Climate-induced impacts to peak and annual electricity demand

A potentially important impact of projected climate change would be on peak electricity demand due to changes in peak summer afternoon temperatures. Climate warming in the region is expected to increase peak period electricity demands for summer cooling and possibly increase peak winter heating electricity demand as well if electricity is used for space heating (Wei et al. 2012. Projected climate-induced changes in mean annual and extreme temperatures, including projected changes in cooling degree days, can be found in Chapters 6 and 7.

Using recent data, Sathaye and others (2011) correlated temperatures above 77ºF (25ºC) with statewide peak loads during the month of August from 2003 to 2009. In California, 90th percentile per-capita peak loads (a measure of extreme energy demands) are projected to increase between 10% and 20% at the end of the century due to the effects of climate change on summer weekday afternoon temperatures.[iii] Others have also analyzed the influence of increasing temperature on energy demand in California (Miller et al. 2007; Franco and Sanstad 2008). No similar studies have been conducted for the Southwest as a whole, but a similar range of impacts on peak demand is expected. One approach would be to use the projections of Franco and Sanstad (2008) who provide an overview and a simple methodology to estimate annual and peak demand using historical and projected temperature data.

Vulnerability of electricity generation to climate impacts

The areas in which electric power plants are vulnerable to direct climate impacts include reductions in power plant efficiency, loss of hydropower generation, and disruption of thermoelectric production.

TEMPERATURE IMPACTS ON NATURAL GAS TURBINES. A warming climate would decrease the capacity and efficiency of a natural gas turbine in several ways. Warmer air is less dense than cooler air, so the air mass of the turbine at higher temperatures is less for a given volume intake. In addition, ambient, or background, temperature influences the air's specific volume, which in turn influences the work of and the power consumed by the compressor. Finally, the pressure ratio within the turbine is reduced at higher temperatures, reducing mass flow (Kehlhofer et al. 2009).

The relationship between temperature and natural gas power plant performance varies by the type of natural gas power plant, the cooling equipment installed at the plant, and the geographic location of the plant. In many applied studies, the basic power output-to-temperature relationship is assumed linear, such that power plant capacity is decreased 0.25% to 0.5% for every °F increase in ambient air temperature (Sathaye et al. 2012). Maulbetsch and DiFilippo (2006) estimated the relationship between ambient temperature and the capacity potential of natural gas power plants. They found that combined-cycle power plant capacity drops 0.15% to 0.25% per °F and air-cooled

Box 12.1

Compounding Impacts of Drought

Delivery of electricity may become more vulnerable to disruption due to climate-induced extreme heat and drought events as a result of:

- increased demand for home and commercial cooling,
- reduced power-plant efficiencies due to high temperatures,
- reduced transmission-line, substation, and transformer capacities due to elevated temperatures,
- potential loss of hydropower production,
- threatened thermoelectric generation due to limited water supply, and
- threat of wildfire to transmission infrastructure.

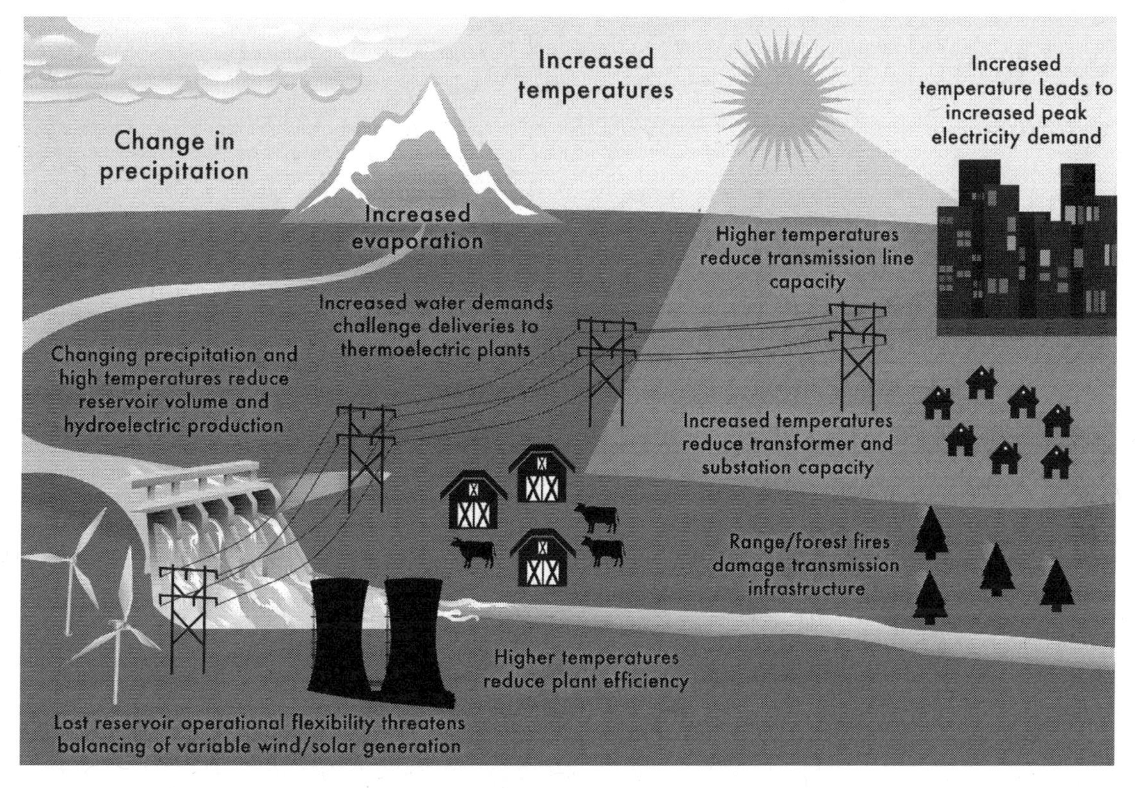

Figure 12.5 Compounding impacts of drought on energy.

combined-cycle power plant (dry cooling) capacity drops 0.35% per °F rise in ambient temperature. Using these assumptions, climate change projections suggest natural gas thermal power plant capacity in California could drop on average between 2% and 5%, and as much as 6% during hot summer afternoons by the end of the century. During peak load periods, reserve margins can be low and natural gas power plants are running near capacity.

Similar studies of climate change impacts on power plants in other Southwestern states have not been undertaken. However, the temperature increases simulated for those states match or exceed temperature simulations for California and a similar impact on power plant performance should be anticipated for those plants as well. Nevertheless, additional research is needed to estimate temperature impacts on different types of electricity generating capacity and on power plant performance outside of California. Types of electric power plants that may be affected by climate warming include coal power, nuclear power, wind, solar, bio-power, and geothermal plants as well as natural gas-fired plants.

A variety of adaptive measures may be taken by utility planners to offset the impact of climate change on thermal power plant performance, such as installing new types of cooling equipment or simply expanding the existing capacity in anticipation of future losses.

CLIMATE IMPACTS ON HYDROPOWER GENERATION. Hydropower generation depends on such factors as total runoff, timing of runoff, reservoir operations, and the profile of electricity demand, each of which is vulnerable to climate change. Projected changes in runoff due to potential decreased precipitation and increased evaporation (Chapter 6) means less water for hydropower production. Earlier snowmelt (Chapter 6) and shifts in the frequency of extreme events (Chapter 7) could lead to important changes to the timing of streamflow and thus reservoir storage. Climate change could also impact water-demand regimes downstream of the reservoir, such as with urban (Chapter 13) and agricultural (Chapter 11) uses. Hydropower production could be compromised as reservoir operations are adjusted to address changing patterns of reservoir storage coupled with changes in competing water uses for in-stream flow requirements, flood control, water supply, and recreation (National Energy Education Development Project 2007). Also, the way in which hydropower is utilized for electricity production may change in the future (for example, with increased use to balance intermittent loads), so that changes in water availability may have different impacts than they do today.

Since 1990, hydroelectric power has satisfied about 11% of the Southwest's total electricity demand (46.9 gigawatt hours [GWh]), of which, as mentioned earlier, California supplied about 34.3 GWh (73%) (EIA 2010). Hydropower production in the region varied widely year to year, ranging from 31.1 GWh (8% of regional demand) to 66.7 GWh (16% of regional demand). Similar variability was noted in the California production record. A unique feature of California hydropower is that much of it (greater than 70%) is generated by a fleet of more than 150 small, single-purpose reservoirs located high in the Sierra Nevada that depend on snowmelt (Madani and Lund 2009). These reservoirs have little storage capacity and thus are uniquely vulnerable to shifts in streamflow that might result from climate change (Vicuña et al. 2008).

Recent droughts provide insight into climate-related impacts on hydropower production. Severe drought in California and the Pacific Northwest significantly reduced hydroelectric power generation in 2001, contributing to tight electricity supplies and high prices (the crisis was enhanced by market manipulation). While significant outages were largely avoided, there were financial impacts due to the large increase in prices (BPA 2002). The Northwest Power and Conservation Council (2005) estimated the total regional economic impact of the drought to be between $2.5 and $6 billion. Impacts of the drought were intensified by energy-water interdependencies linking the California coastal basins, Columbia River Basin, and Colorado River Basin (Cayan et al. 2003). These cross-basin linkages provide the opportunity to compensate hydropower losses if, during dry conditions in one basin, the others experience wetter conditions, or to intensify the effects if two or more basins are simultaneously drier (as was the case in 2001).

Over 70% of hydropower production in the Southwest is associated with the Colorado River and with high-elevation dams in the Sierra Nevada (EIA 2010). Several studies suggest that both systems are vulnerable to future climate change. Specifically, a recent water-budget analysis showed a 50% chance that minimum power pool levels in both Lake Mead and Lake Powell will be reached under current conditions by 2017 if no changes in water allocation from the Colorado River system are made. This would impact over 3,700 MW of generating capacity on the lower Colorado River (EIA 2010). Such a result would be driven by a combination of climate change associated with global warming, the effects of natural climate variability, and the current operating status of the reservoir system (Barnett and Pierce 2008).

Numerous studies have explored potential future climate impacts on Sierra Nevada hydropower production. Harou and others (2010) used geologic evidence suggesting two extreme droughts occurred in California during the last few thousand years, each 120 to 200 years long, with mean annual streamflows 40% to 60% of the historical average. They used an engineering-economic simulation model to evaluate impacts under such conditions to power production, irrigation, and environmental resources, based on projected demands in 2020. Under current operating rules, a 60% reduction in hydroelectric power generation was projected for low-elevation hydropower units that traditionally contribute about 30% of the hydropower generation in California (Phinney et al. 2005). Another study investigated the impact of increased air temperatures of up to 11°F (6°C) on mean annual streamflow, peak runoff timing, and duration of low-flow conditions. Vulnerabilities to these three flow characteristics were found to differ by region and hence their susceptibility to altered hydropower production (Null, Viers, and Mount 2010). Other studies have used climate data from multiple general circulation models (GCMs) under different greenhouse gas (GHG) emissions scenarios to investigate potential impacts on California hydropower production (Madani and Lund 2010) and adaptation options (e.g., K. P. Georgakakos et al. 2011; A. P. Georgakakos et al. 2011). For scenarios resulting in reduced annual streamflow, hydropower production is reduced but to a lesser extent than revenues, reflecting the ability of the system to store water when energy prices are low for use when prices are high. The opposite was true when projected annual flows increased (Vicuña et al. 2008).

Opportunities to mitigate the impact of climate change on hydropower production include expansion of hydropower resources (DOE 2010). According to preliminary estimates from the Electric Power Research Institute (2007), the United States has additional water power resource potential of more than 62,000 MW (or about 79% of current conventional hydroelectric capacity). This includes efficiency upgrades at existing hydroelectric facilities and development of new low-impact facilities. Issues of cost, environmental impacts, and suitability of construction sites are potentially limiting factors. Adaptation opportunities also exist, primarily in the form of altered reservoir operations. Several of the studies noted above for the California Sierra Nevada reservoirs have shown that climate impacts could be mitigated through adaptive management strategies (Harou et al. 2010; A. P. Georgakakos et al. 20011). However, identifying "optimal strategies" that balance multiple objectives (hydropower, water supply, flood control, recreation, and instream flow) is problematic as it would require agreement among stakeholders to change longstanding operational rules.

Two key data gaps are apparent. First, there has been relatively little effort to directly quantify potential climate change impacts on hydropower production in the Colorado River Basin; however, a recent study by the U.S. Bureau of Reclamation (2011) should help to fill this gap. The second gap is in evaluating the broad implications of climate change for the entire, interconnected West, including evaluating concurrent impacts on hydropower and thermoelectric production and on shifts in electric power demand.

CLIMATE AND DROUGHT IMPACTS ON THERMOELECTRIC GENERATION. Thermoelectric generation can be limited when power plant water supplies are threatened, such as when the surface of a water supply drops below intake structures of the generating plant (NETL 2009b), when access to water is limited by priority of power plant water rights (Stillwell, Clayton, and Webber 2011), or when water discharge temperatures exceed environmental limits (Averyt et al. 2011).

There are no examples from the Southwest where drought has threatened thermoelectric power production. However plants have been shut down in other regions due to drought, including the Alabama Browns Ferry facility in 2007, 2010, and 2011 (Nuclear Regulatory Commission [NRC] 2007, 2010, 2011); the Exelon Quad Cities, Illinois, plant in 2006 (NRC 2006); the Minnesota Prairie Island plant in 2006 (NRC 2006); and plants in France in 2003 (De Bono et al. 2004). All these cases involve nuclear plants whose production was limited by effluent temperatures exceeding regulated discharge limits. In contrast, the 2011 drought in Texas threatened to impact a variety of thermoelectric power plants there due to limited water availability (e.g., Galbraith 2012).

There are several studies that have assessed power plants' vulnerability to drought. Each utilized different criteria to assess vulnerability and each focused on different subsets of the electric power industry. A study by NETL (2009a) identified five plant sites in four western states with a total capacity of 3,284 MW as vulnerable on the basis of shallow intake structures (NETL 2009b) and historic severity of local drought measures. In a study focused on coal-fired generation, twenty-six power plants in the Southwest region were identified as vulnerable based on eighteen unique indicators of water demand and supply (NETL 2010). A recent study by the Union of Concerned Scientists (2011) identified twenty power plants in the Southwest that discharge cooling water at a maximum temperature that exceeds 90°F (32°C), most of which is into streams with high biodiversity. Taking a slightly different approach, Sovacool (2009) identified Denver, Las Vegas, and San Francisco as places where water resources are likely to be scarce or declining due to growing and competing demands for thermoelectric and non-thermoelectric water supply. Based on these studies it is difficult to assess exposure of thermoelectric power production to drought/climate change as none of the studies consider the existence and robustness of contingency plans that individual plants may have in place.

Mitigation of drought vulnerability can be achieved through integrated water-energy planning (e.g., Western Governors' Association 2010, 2011) or utilization of non-potable water sources.[iv] Adaptation measures can be achieved through contingency planning and include options such as allowing power plants to lease water from farmers with senior water rights in times of drought or developing backup groundwater sources. Dry cooling (which uses air rather than water to cool a power plant's working fluid) is another option that is already utilized to generate 4,150 MW in the intermountain West (Cooley, Fulton and Gleick 2011). However, dry cooling is more costly, requires additional land, and results in reduced generation efficiencies on hot days. Dry-cooled plant capacity declines on hot days more than water-cooled plant capacity, but hybrid units are also available. In some cases, dry-cooled plants lose 0.5% of their capacity for every 1°F increase in peak temperature—roughly twice the loss of wet cooled plants (Sathaye et al. 2012).

To fully evaluate the vulnerability of thermoelectric power production to drought will require assessing local hydrologic conditions together with plant-level operational characteristics and institutional controls. Plant contingency plans must also be reviewed in terms of their robustness in the context of future climate extremes.

Vulnerability of electricity distribution to climate impacts

This section discusses the vulnerabilities of electricity distribution infrastructure to direct climate impacts, specifically the temperature impacts on electric transmission line capacity (220 KV and higher) and on substation/transformer capacity.

TEMPERATURE IMPACTS ON TRANSMISSION LINE CAPACITY. Transmission lines incur incremental power losses as the temperatures of conductors increase (IEEE 2007). In general, higher temperatures increase the resistance of a conductor, which decreases the carrying capacity of the transmission line and requires additional generation to offset the increased resistance over the lines. Climate change is expected to increase mean and peak period temperatures across the Southwest, thereby increasing demand for electricity while at the same time increasing the resistance of transmission lines and decreasing their carrying capacity. One study of the impact of a 9°F increase in ambient air temperature suggests that while climate warming would cause very little resistance loss, capacity losses might be more significant, amounting to an additional 7% to 8% of peak (Sathaye et al. 2012).

The potential for climate change-induced line capacity losses should be researched further, including identification of operating practices that minimize loss and of new design parameters (e.g., underground transmission lines). Also, utilities generally count on the presence of at least 2 feet per second of wind on hot days. If a "zero-wind" condition should happen during extremely high temperatures, conductor temperatures may rise into the "emergency" range (i.e., above 212°F [100°C]), where continued operation may cause excessive conductor sag, permanent damage, and even lead to wildfires. Further investigation into the effects of climate change on the probability and duration of no-wind conditions on hot days is necessary to more accurately evaluate the impacts on transmission capacity (Sathaye et al. 2012).

One option for coping with the impacts of climate change on the transmission system is decentralized generation: producing a larger fraction of the power at or near the end-use reduces the line capacity requirements. Examples of such generation are low-pressure methane to support local industry (Welsh et al. 2010) or even utilization of traditional windmills to circumvent electric-powered groundwater pumping. Another option for avoiding the impacts of higher temperatures would be to place transmission lines underground.

TEMPERATURE IMPACTS ON SUBSTATION/TRANSFORMER CAPACITY. Major substations contain clusters of transformers that allow alternating current voltage to be "stepped up" or "stepped down" between various components of the power system. A number of studies have looked at the performance of transformers under different operating conditions, including changing ambient temperatures (Lesieutre, Hagman, and Kirtley 1997; Swift et al. 2001; Li and Zielke 2003; Li et al. 2005; Askari et al. 2009). Higher ambient temperatures reduce the peak-load capacity of banks of transformers in substations.[v] High minimum temperatures can affect transformer performance as well as high peak temperatures. For example, in some extreme cases, excessive hot spot conductor temperature within the transformer can lead to catastrophic failure of the transformer, so improved methods to monitor these internal temperatures are proposed (Lesieutre, Hagman, and Kirtley 1997).

Other studies have quantified the general relationships between air temperature and transformer lifespan and capacity (Swift et al. 2001; Li et al. 2005). The basic relationship of power capacity to temperature used in most studies is linear, with some research suggesting that transformer capacity decreases approximately 0.35% for each 1°F of higher ambient temperature (Li et al. 2005).

One study indicates that substations in California could lose, on average, an additional 1.6% to 2.7% capacity by the end of the century (Sathaye et al. 2012). Other parts of the Southwest with similar or higher temperature increases from climate change would face similar substation capacity losses as well. Judging by the spatial distribution of the simulated change in the number of days with maximum temperatures greater than 95°F (35°C), the southern and eastern parts of the Southwest are more at risk to lost substation peak capacity than areas in the region's north and west.

Utility planners may take several adaptive measures to offset future losses to substations and increase their capacities, including proactively installing new types of cooling.

Vulnerability of infrastructure to indirect climate impacts

This section looks at two key indirect vulnerabilities from climate change: risk of wildfire to electric transmission lines and the effects of sea-level rise and coastal inundation on power plants and substations.

WILDFIRE RISK TO ELECTRICITY TRANSMISSION. Fire is a natural component of ecosystems in western North America, and its occurrence is strongly correlated with climate variability. Weather-related effects on fire include behavior (wind conditions), fuels (combustible material), and ignitions (lightning). Wildfires are also greatly affected by moisture availability, as influenced by temperature, precipitation, snowpack, and other meteorological factors, all of which may be impacted by climate change.

Increases in the size and frequency of wildfires in the Southwest will increasingly affect electricity transmission lines (see Box 12.2). Transmission line-related impacts from wildfires are not restricted to the actual destruction of the structures (Aspen Environmental Group 2008; personal communication, Fishman and Hawkins, CAISO [California Independent System Operator Corporation], 2009). In fact, only smaller lines may be directly destroyed in a wildfire, because these types of power lines are typically built with wooden poles. But the transmission capacity of a line can be affected by the heat, smoke, and particulate matter from a fire, even if there is no actual damage to the

Box 12.2

Transmission Line Exposure to Wildfire

The most recent study of fire risk in California suggests that most transmission lines in the state will be exposed to an increased probability of fire risk across a range of different climate models, emission scenarios, and time periods (Sathaye et al. 2012). This figure summarizes changes by the end of the century in the length of lines exposed to areas of either increasing or decreasing burned area, the latter being more common. At the end of the century some key transmission lines crossing between California and Oregon and between northern and Southern California may face a particularly high fire risk. Certain areas will be less at risk. For example, transmission lines passing through the desert areas in southeastern California are projected to see reduced exposure to wildfires in the future, in part due to changes in vegetation projected for the future.

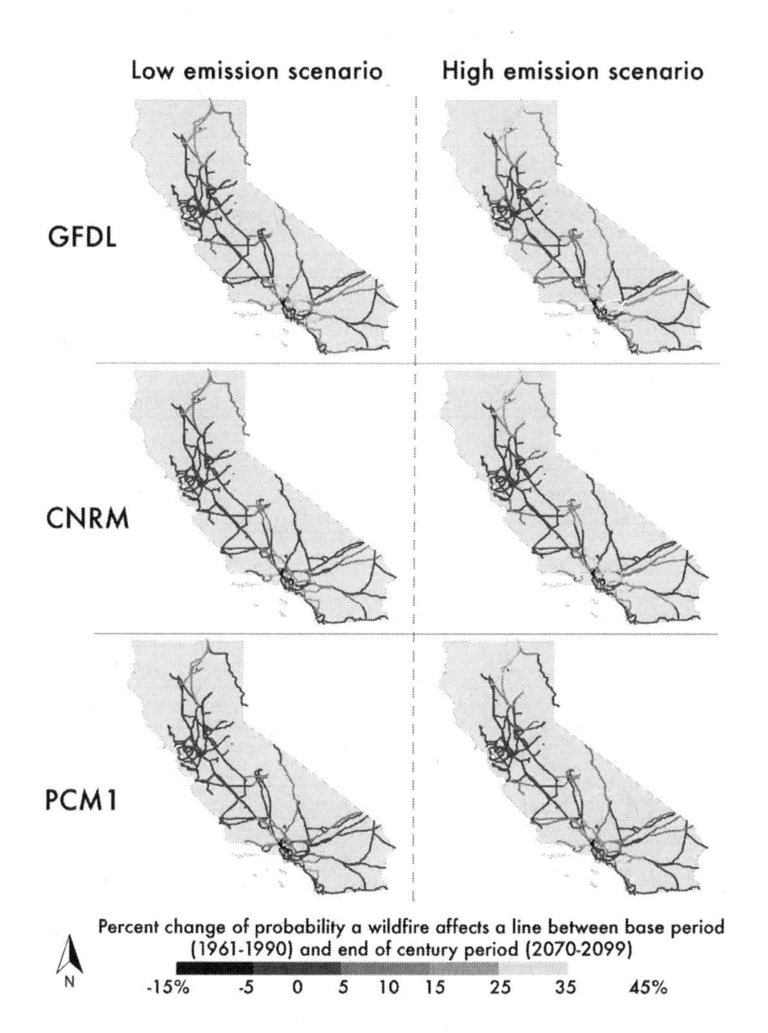

Figure 12.6 Transmission line exposure to wildfire. The probability of damage is taken as the product of the probability fire and the relative length of the line in a given region. Used in this study are the GFDL (Geophysical Fluid Dynamics Laboratory), PCM1 (Parallel Climate Model), and CNRM (Centre National de Recherches Météorologiques) GCM, and the high- and low-emissions scenarios as defined by the Intergovernmental Panel on Climate Change (IPCC) (Nakićenović and Swart 2000). Modified from Sathaye et al. (2012).

physical structure. The insulators that attach the lines to the towers can accumulate soot, creating a conductive path and causing leakage currents that may force the line to be shut down. Ionized air in smoke can act as a conductor, causing arcing either between lines, or between lines and the ground, that results in a line outage. Finally, even if the lines are protected from fire, the effects of firefighting can also negatively affect transmission operation, such as by aircraft dumping loads of fire retardant that can foul the lines or through preventive shutdowns for safety measures.

Several studies have shown that climate change will increase the size and frequency of wildfires in California, which leads the region in wildfire-related economic losses (Flannigan et al. 2000; Lutz et al. 2009; Westerling et al. 2009). Of the ten largest wildfires in California's history, seven have occurred since 2001 (CDFFP 2009). Texas, Arizona, and New Mexico have all experienced record-breaking wildfire seasons in recent years as a result of increasing temperatures and advanced evaporation (Samenow 2011).

Many adverse effects of wildfire risk to the transmission grid can be avoided by careful planning, increasing fire corridors around transmission lines, building excess transmission capacity, and using new transmission line materials that provide better protection against the effects of soot and heating.

There are no studies of fire risk to the transmission grid in the Southwest, outside of California. Since fire frequency and cost due to climate change are likely to be particularly high, studies of this issue across the Southwest are recommended. (For further discussion of the effects of climate change on the occurrence of wildfires in the Southwest, see Chapter 8, Section 8.4.2.)

PROJECTED IMPACTS OF SEA-LEVEL RISE AND COASTAL INUNDATION ON POWER PLANTS AND SUBSTATIONS. California is the only state in the Southwest region facing impacts from sea-level change (see a detailed discussion in Chapter 9). Mean sea level along California's coast has risen at a rate of about 8 inches (20 cm) per century for several decades—a rate that may increase (Cayan et al. 2009). Mean high water, which poses an even more significant threat, is increasing at an even faster rate (Flick et al. 2003). Extreme surge events at high tides, often provoked by winter storms, are also expected to increase (Cayan et al. 2009). The confluence of these three trends puts increasing amounts of coastal energy infrastructure at risk: recent studies of sea level along the California coast indicate that twenty-five to thirty power plants—including thirteen located in the San Francisco Bay Area—are at risk of impact from what would be currently a 100-year flood (which will become more frequent) caused by a 4.6-foot sea-level rise (Figure 12.7) (Sathaye et al. 2012).[vi] Many of the plants shown to be at risk, though, are likely to be retired over the next few decades.

The vulnerability of power plants to flooding is very site-specific and more information should be gathered on a site-by-site basis. It should be noted that in addition to increasing flood risk, sea-level rise may accelerate shoreline erosion and destabilize coastal power plants and related infrastructure. Further studies are needed to evaluate the risk posed by soil erosion to energy infrastructure along the coastline.

Adaptive measures may be taken to protect power plants against flooding and destabilization from sea-level rise. For example, new power plants should be constructed at higher elevations, farther from the ocean, and out of reach of tidal flooding. In addition, existing plants may be protected by building higher levees.

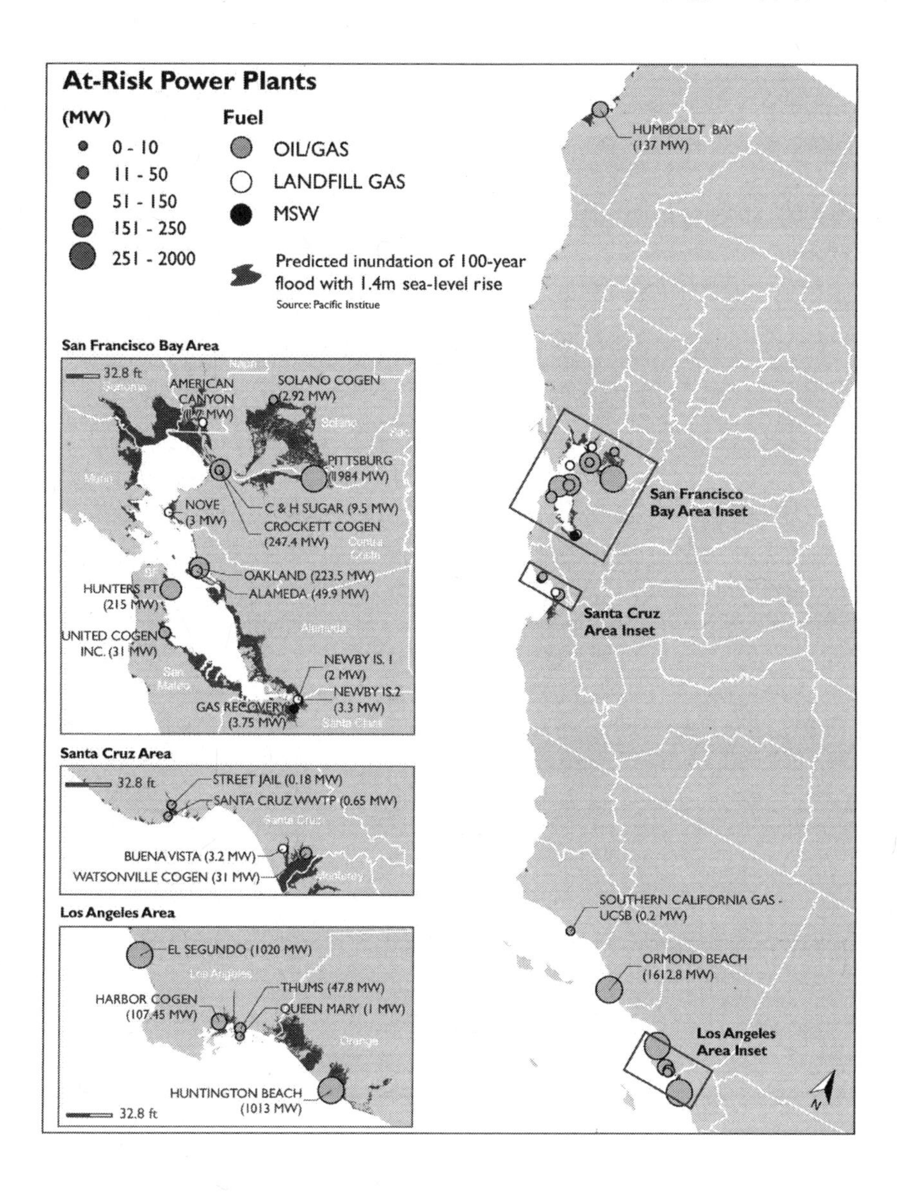

Figure 12.7 Power plants potentially at risk to a 100-year flood with 1.4-m (4.6-ft) sea-level rise. Reproduced from Sathaye et al. (2012).

Climate impacts on renewable source intensity

To estimate the potential impacts of climate change on renewable sources of energy, spatial details are needed on how the electricity system would evolve. For example, if climate change causes shifts in wind regimes, areas now suitable for wind farms may be less suitable in the future and new areas may become an attractive option for wind energy development. Few studies, however, have been published on this topic for the Southwest.

WIND POWER. Some research suggests that wind speeds in the United States as a whole are declining, based on the 50th and the 90th percentile wind speeds from 1973 to 2005 (Pryor et al. 2009; Pryor and Ledolter 2010). However, for the Southwest region no significant changes in wind speeds are reported (Pryor and Ledolter 2010).[vii]

Pryor and Barthelmie (2011) estimated potential changes of wind-energy density (the mean annual energy, in watts, available per square meter of the area swept by a turbine) for the conterminous United States using results from outputs from a handful of regional climate models (with spatial resolutions of about 31 miles [50 km]), driven by the outputs from atmospheric-ocean global climate models. The reported changes in wind-energy density for the next fifty years are not beyond what could be expected from natural variability. This implies that current and planned wind power facilities likely are not in danger of experiencing reductions in their ability to generate electricity, and areas that are presently suitable for wind-power generation would not be affected by climate change at least for several decades.

Rasmussen, Holloway, and Neme (2011) attempted to estimate the potential impacts of climate change on three California sites with significant installed wind capacity: Altamont Pass (562 MW), San Gorgonio Pass (710 MW), and Tehachapi Pass (359 MW). The authors used the same outputs from the regional climate models as did Pryor and Barthelmie (2011). Rasmussen , Holloway, and Neme (2011) reported wide disagreement between the regional climate models on how wind resources would be affected. This may be explained in part by the fact that the 31-mile spatial resolution of the regional climate models are much wider than the relatively narrow passes where wind resources in California are located (i.e., the models cannot resolve the topographic features in these passes that contribute to the acceleration of the wind). Mansbach and Cayan (2012) used a different approach, developing a statistical model that relates large-scale atmospheric features that are supposed to be adequately modeled by the global climate models with local wind speed at three California wind farms. Their results were mixed, precluding any conclusions about the potential impacts of climate change on wind farms.

In summary, no trend in wind-energy density has been observed in the Southwest and it seems likely that wind power will continue to be a viable resource in the Southwest. However, further analyses and modeling studies are needed to resolve the current discrepancy between model results and to address the deficiencies with the existing studies.

SOLAR POWER. Climate change can affect the amount of solar irradiation reaching ground level mainly via its effects on clouds. Unfortunately, the simulation of clouds remains one of the main sources of uncertainty in the projections of future climate regimes (IPCC 2007). Global and regional climate models are implemented at geographical resolutions too coarse to allow for the simulation of clouds directly from physical principles. Instead, they use statistical relationships (parameterizations) developed using historical data. For this reason, caution is in order with the interpretation of estimated ground-level solar radiation information generated by global and regional climate models.

Crook et al. (2011) used the results from two atmospheric-ocean global climate models developed in the United Kingdom to estimate the impacts of climate change on the future of photovoltaic (PV) and concentrated solar power (CSP) units at the global level. For California and Nevada they reported reductions in power output for both PV and CSP units on the order of a few percent at the end of this century, in comparison with historical conditions. But an uncertainty analysis suggests a low level of confidence in their results for this region. In addition, as suggested by Rasmussen, Holloway, and Neme (2011), reliance on just a few models is not advisable because modeling results at the regional level can change, depending on the model used for the study.

High ambient temperatures affect the performance of some PV systems (Kawajiri, Oozeki, and Genchi 2011) such as crystalline silicon PV modules, which are one of the most popular technologies on the market but do not perform well under the dry and hot conditions common to the Southwest. There are, however, other PV technologies that are less sensitive to ambient temperatures (Crook et al. 2011).

In summary, solid scientific information is lacking on how climate change would affect the amount of solar irradiation reaching ground level in the Southwest and, therefore, how it would affect PV and CSP systems. But higher ambient temperatures will have some detrimental effects on some PV system technologies.

Evolution of the energy sector

Climate-related policies have the potential to significantly alter the energy sector. A shift from the traditional fossil fuel economy to one rich in renewable energy sources has significant implications for related water use, land use, air quality, national security, and the economy. This section explores potential evolutionary paths for both electricity generation and transportation fuels.

EVOLUTION OF THE ELECTRICITY GENERATION FUEL MIX. Climate change and the policies adopted to address climate change are likely to cause a shift in the fuel mix used to generate electricity. There are multiple sources of information on potential global and U.S. energy scenarios (e.g., Wilbanks et al. 2007), but they lack regional detail and do not allow for an estimation of how the energy system in the Southwest could be transformed. Nevertheless, two potential illustrative scenarios are explored that, broadly speaking, are consistent with a business-as-usual path and with a future in which strong mandates to reduce greenhouse gas emissions materialize. These two scenarios could be seen as compatible with the spirit of the IPCC's high- and low-emissions scenarios, respectively, as discussed before in this report (see Chapter 2), though they do not specifically represent these two global emission scenarios.

The business-as-usual scenario comes from the "reference case" presented in the U.S. Energy Information Administration's *Annual Energy Outlook 2011* report (EIA 2010). Results for the EIA's Southwest region are used; however, it should be noted that the regional boundaries used by EIA do not fit perfectly the boundaries for the Southwest used in this report (the EIA boundary captures roughly 80% of the electricity generation within the Southwest region delimited for this report). The reference case assumes that current laws and regulations remain in place, such as the mandates in California, Arizona, Colorado, and Nevada to increase the amount of electricity generated from renewable sources of energy. This is reflected in the contribution of renewable sources of energy going from 16% in 2009 to about 27% in 2035, the end year of the EIA simulations. This new electricity generation would come from a mix of wind, solar, and geothermal with limited additions of hydroelectric. The electricity generation from nuclear units and coal-burning power plants would remain at about the same levels as in the recent past but minor increases in generation from natural-gas-burning power plants is observed in the EIA scenario. We speculate that under this business-as-usual scenario the same general trend would continue past 2035.

There are multiple options with regard to the evolution of the electricity system in the Southwest if strong measures are taken to reduce greenhouse gas emissions. A new study, for example, uses a long-term capacity expansion model for the electricity system for the Western United States (Wei et al. 2012), taking into account hourly electricity demand and the hourly availability of generating resources, especially intermittent sources such as solar and wind power plants. This modeling system uses geographical information about electricity generation and demand to estimate the regional evolution of the electricity system. As shown in Figure 12.8, by 2050, solar, wind, and geothermal resources would become major sources of electricity in the Southwest. Electricity generation from coal-burning power plants without carbon capture and sequestration disappears in this potential scenario.

The evolution of the electricity system will impact water resources. Thermal electricity generation withdraws copious amounts of water for cooling. While only a very small fraction of the water withdrawn is actually consumed (i.e., evaporated to the atmosphere) the water that is returned to its source, such as rivers and the ocean, is not pristine, containing at least some levels of thermal pollution (Averyt et al. 2011; Cooley, Fulton and Gleick 2011). Different power plants and energy systems can have substantially different levels of impacts on water resources: energy crops for biofuels can require from very little to a substantial amount of water; wind resources do not withdraw water; and solar power generation requires from very little to large amounts of water, depending on the conversion technology (Cooley, Fulton and Gleick 2011; Kenney and Wilkinson 2012).

EVOLUTION OF THE TRANSPORTATION FUELS MIX. The transportation system in the Southwest will likely mirror the transportation system in the rest of the country (see Chapter 14). The EIA estimates an increase of more than 15% in the amount of total liquid fuels consumed for the U.S. transportation sector in 2035 as compared with 2009 consumption for the reference case (EIA 2010). As indicated before, the reference case assumes that only current laws and regulations remain in place. EIA also reports an "extended policies" case for which EIA assumes a 3% annual increase of corporate average fuel economy (CAFE) standards until 2025, then no change until 2035. In this case, total liquid fuel consumption for the U.S. transportation sector would increase only by about 4% from 2009 to 2035. Hybrid electric cars and/or cars fueled by alternative fuels play an important role in achieving the more stringent CAFE standards.

If drastic reductions in greenhouse gas emissions are required, several studies have found that for the United States it is more practical and less costly to reduce emissions in the electricity-generating sector and to electrify the rest of the sectors as much as possible (Clarke et al. 2007; Fawcett et al. 2009; National Research Council 2010); the situation is not different for the Southwest (Wei et al. 2012). Biofuels could play an important role in the availability of liquid fuels for the transportation sector (Parker et al. 2010) but could only replace a small percent of current levels of primary energy consumption (Field et al. 2008). At the same time, liquid biofuels could play a key role in replacing transportation services that cannot be easily electrified, such as air transport. Several studies suggest that biofuels must be produced in a way that does not increase net greenhouse-gas emissions, does not hinder food security, and does not result in negative ecological impacts (e.g., Tilman et al. 2009). In addition, new studies suggest that it is also important to consider the biogeophysical consequences of bioenergy crops, such as local cooling or warming, or changes in water demand (e.g., Georgescu, Lobell, and Field 2009, 2011).

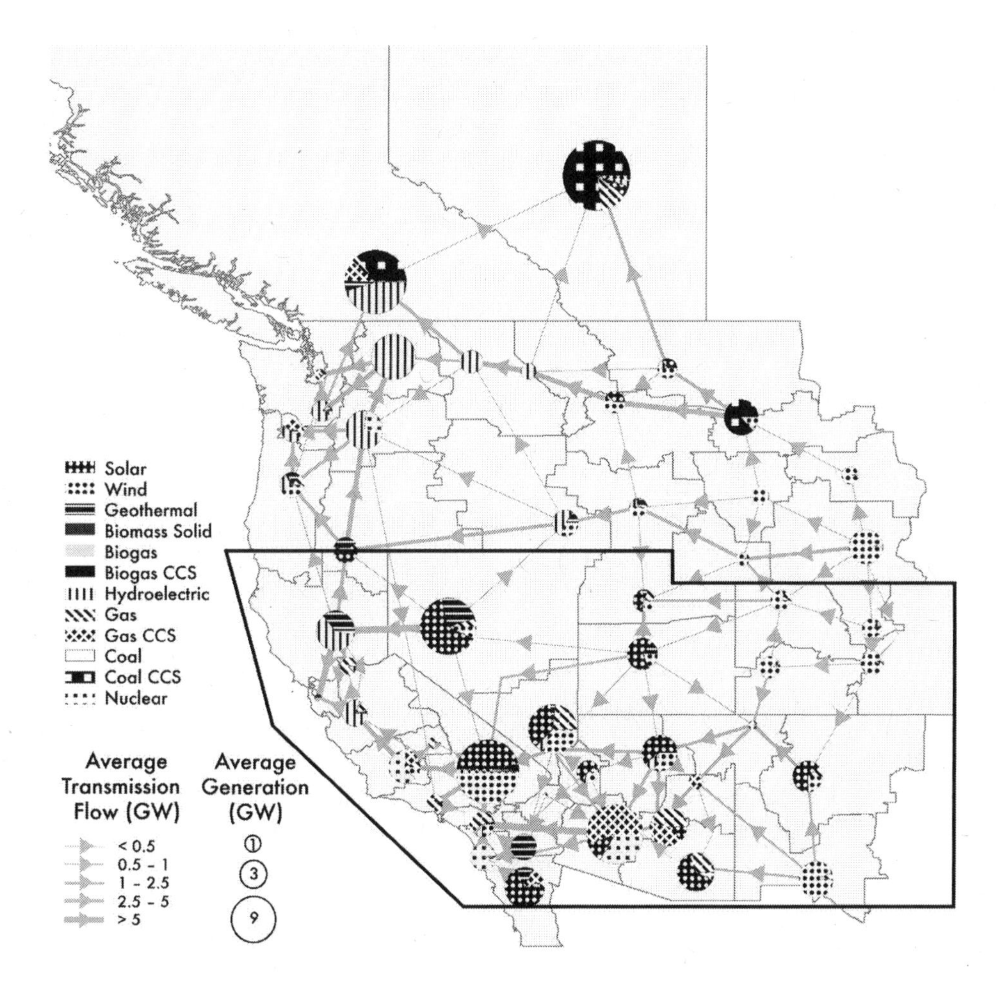

Figure 12.8 Projected generation and transmission flow in 2050. The map shows projected average generation and average transmission flow between load areas (electricity demand zones) under high regulation of green-house gas emissions. The size of each pie represents the amount of generation in the load area in which the pie resides. The flows of electricity are depicted as straight lines for clarity. Reproduced from Wei et al. (2012).

Climate impacts on primary energy production

As noted above, the Southwest accounts for 12.7% of our nation's primary energy production. Climate change has the potential to influence this production in at least three ways. First, energy policies aimed at mitigating the impacts of climate change could significantly alter the mix of primary fuels used to generate electricity and fuel transportation (see above), thus influencing the demand for the primary fuels. Specifically, emission standards could cause a move to renewables and away from coal and petroleum. However, demand for natural gas and uranium would likely increase, potentially offsetting losses in coal production. Potential losses of petroleum production will be more difficult to offset as water availability will challenge biofuel production in the Southwest (DOE 2011). Because these primary energy sources are traded internationally, emission policies would need to be matched in much of the rest of the world before significant impacts would be felt locally.

Second, climate change could increase the demand for electricity: higher temperatures are likely to be met with higher demands for cooling (Section 12.3.1). Alternatively, higher cooling needs could spawn improvement in cooling/energy efficiencies or higher energy prices, which would reduce electricity demand.

Third, climate change could impact the availability of water needed for primary energy extraction and processing (DOE 2006). Reduced water supplies (see Chapters 6 and 10) could affect current production and especially the development of new resources. A particularly good example is the ongoing struggles over water and oil shale development in the Piceance Basin (Western Resource Advocates 2009).

Cost of climate change

In 2009, total expenditures for end-use energy (such as retail electricity and motor gasoline) in the Southwest was about $160 billion. California contributed about 66% to these expenditures (EIA 2010). Any changes in energy prices due to climate change or climate change policies will have a direct economic impact on consumers. For example, since retail electricity expenditure in the Southwest was about $52 billion in 2009 (EIA 2010), the postulated increases in electricity demand due to higher temperatures discussed previously would have represented a few billion dollars per year for the current electricity generation and demand system. The estimation of changes of energy expenditures in the rest of this century due to physical changes in our climate is extremely difficult because it will depend on multiple uncertain factors such as the price of energy, population growth, technology evolution, and human behavior.

Climate policies directly limiting greenhouse gas emissions or indirectly affecting emissions via requirements such as renewable portfolio standards will also affect the cost of energy. At the same time, the potential benefits of policies such as increasing energy security, limiting climate change, reducing air pollution, and perhaps creating local jobs must be considered. The authors are not aware of a comprehensive study along these lines for the Southwest.

Climate policies at the national and international levels designed to reduce greenhouse gas emissions will directly impact the energy sector, given that currently fossil-fuel combustion is by far the dominant source of these emissions. Conventional modeling studies at the national level suggest that drastic reductions of greenhouse gases would reduce gross domestic product (GDP) by at most a few percent, delaying a given GDP level only for a few years (National Research Council 2010). However, as indicated before, this represents at best only a partial economic analysis because it does not take into account the benefits of limiting the impacts of climate change. In addition, the economic models used in these studies assume perfectly functioning markets and any restrictions on emissions, by design, result in economic penalties (DeCanio 2003). Ironically, these models do not take into account the effect that climate change would have on economic activity and energy demand, for example, diverting economic resources to mitigate climate damages (Hallegatte and Hourcade 2007). More realistic economic models are sorely needed.

References

Askari, M., M. Kadir, W. Ahmad, and M. Izadi. 2009. Investigate the effect of variations of ambient temperature on HST of transformer. In *2009 IEEE Student Conference on Research and Development, 16–18 November 2009*, 363–376. N.p.: Institute of Electrical and Electronics Engineers.

Aspen Environmental Group. 2008. Effects of wildfires on transmission line reliability. In *Draft environmental impact report/ environmental impact statement and proposed land use amendment, Sunrise Powerlink Project*, Attachment 1–A. http://www.cpuc.ca.gov/environment/info/aspen/sunrise/deir/apps/a01/App%201%20ASR%20z_Attm%201A-Fire%20Report.pdf.

Averyt, K., J. Fisher, A. Huber-Lee, A. Lewis, J. Macknick, N. Madden, J. Rogers, and S. Tellinghuisen. 2011. *Freshwater use by U.S. power plants: Electricity's thirst for a precious resource.* A report of the Energy and Water in a Warming World Initiative. Cambridge, MA: Union of Concerned Scientists.

Barnett, T. P., and D. W. Pierce. 2008. When will Lake Mead go dry? *Water Resources Research* 44: W03201, doi:10.1029/2007WR006704.

Beard, L. M., J. B. Cardell, I. Dobson, F. Galvan, D. Hawkins, W. Jewell, M. Kezunovic, T. J. Overbye, P. K. Sen, and D. J. Tylavsky. 2010. Key technical challenges for the electric power industry and climate change. *IEEE Transactions on Energy Conversions* 25:465–473.

Bonneville Power Administration (BPA). 2002. *Guide to tools and principles for a dry year strategy (draft).* http://www.bpa.gov/power/pgp/dryyear/08-2002_Draft_Guide.pdf.

California Department of Forestry and Fire Protection (CDFFP). 2009. 20 largest California wildland fires (by acreage burned). N.p.: CDFFP. http://www.fire.ca.gov/about/downloads/20LACRES.pdf.

Cayan, D., M. Tyree, M. Dettinger, H. Hidalgo, T. Das, E. Maurer, P. Bromirski, N. Graham, and R. Flick. 2009. *Climate change scenarios and sea level rise estimates for the California 2008 Climate Change Scenario Assessment.* Draft Report CEC-500-2009-014D. Sacramento: California Climate Change Center.

Cayan, D. R., M. D. Dettinger, R. T. Redmond, G. J. McCabe, N. Knowles, and D. H. Peterson. 2003. The transboundary setting of California's water and hydropower systems, linkages between the Sierra Nevada, Columbia, and Colorado hydroclimates. In *Climate and water: Transboundary challenges in the Americas*, ed. H. Diaz and B. Morehouse, Chapter 11. Dordrecht: Kluwer.

Clarke, L., J. Edmonds, H. Jacoby, H. Pitcher, J. Reilly, and R. Richels. 2007. *Scenarios of greenhouse gas emissions and atmospheric concentrations*. Synthesis and Assessment Product 2.1a; Report by the U.S. Climate Change Science Program and the subcommittee on Global Change Research. Washington, DC: U.S. Department of Energy, Office of Biological and Environmental Research.

Cooley, H., J. Fulton, and P. H. Gleick. 2011. *Water for energy: Future water needs for electricity in the intermountain West.* Oakland, CA: Pacific Institute.

Crook, J. A., L. A. Jones, P. M. Forster, and R. Crook. 2011. Climate change impacts on future photovoltaic and concentrated solar power energy output. *Energy & Environmental Science* 4:3101–3109, doi:10.1039/C1EE01495A.

De Bono, A., P. Peduzzi, S. Kluser, and G. Giuliani. 2004. *Impacts of summer 2003 heat wave in Europe*. Environmental Alert Bulletin 2. Nairobi: United Nations Environment Programme.

Decanio, S. J. 2003. *Economic models of climate change: A critique.* Basingstoke, UK: Palgrave Macmillan.

Ebinger, J. and W. Vergara. 2011. Climate impacts on energy systems: Key issues for energy sector adaption. Washington, DC: World Bank.

Electric Power Research Institute (EPRI). 2007. Assessment of waterpower potential and development needs. Palo Alto, CA: EPRI.

Energy Information Administration (EIA), 2010. *Annual energy outlook 2011 with projections to 2035*. Report No. DOE/EIA-0383(2011). Release date December 16, 2010. Washington, DC: DOE. http://www.eia.gov/oiaf/aeo/gas.html.

Fawcett, A. A., K. V. Calvin, F. C. de la Chesnaye, J. M. Reilly, and J. P. Weyant. 2009. Overview of EMF 22 U.S. transition scenarios. *Energy Economics* 31: S198–S211.

Field, C. B., J. E. Campbell, and D. B. Lobell. 2008. Biomass energy: The scale of the potential resource. *Trends in Ecology & Evolution* 23:65–72.

Flannigan, M. D., B. J. Stocks, and B. M. Wotton. 2000. Climate change and forest fires. *Science of the Total Environment* 262:221–229.

Franco, G., and A. Sanstad. 2008. Climate change and electricity demand in California. *Climatic Change* 87:139–151.

Galbraith, K. 2012. Texas senate hears warning on drought and electricity. *The Texas Tribune,* January 10, 2012.

Georgakakos, A. P., H. Yao, M. Kistenmacher, K. P. Georgakakos, N.E. Graham, F.-Y. Cheng, C. Spencer, and E. Shamir. 2011. Value of adaptive water resources management in Northern California under climatic variability and change: Reservoir management. *Journal of Hydrology* 412/413:34–46.

Georgakakos, K. P., N. E. Graham, F.-Y. Cheng, C. Spencer, E. Shamir, A. P. Georgakakos, H. Yao, and M. Kistenmacher. 2011. Value of adaptive water resources management in northern California under climatic variability and change: Dynamic hydroclimatology. *Journal of Hydrology* 412/413:47–65.

Georgescu, M., D. B. Lobell, and C. B. Field. 2009. Potential impact of U.S. biofuels on regional climate. *Geophysical Research Letters* 36: L21806.

—. 2011. Direct climate effects of perennial bioenergy crops in the United States. *Proceedings of the National Academy of Sciences,* published online, doi: 10.1073/pnas.1008779108.

Hallegatte, S., and J. Hourcade. 2007. Why economic growth dynamics matter in assessing climate change damages: Illustration on extreme events. *Ecological Economics* 62:330–340.

Harou, J. J., J. Medellín-Azuara, T. Zhu, S. K. Tanaka, J. R. Lund, S. Stine, M. A. Olivares, and M. W. Jenkins. 2010. Economic consequences of optimized water management for a prolonged, severe drought in California. *Water Resources Research* 46: W05522, doi:10.1029/2008WR007681.

Institute of Electrical and Electronics Engineering (IEEE). 2007. IEEE 738-2006 – Standard for calculating the current-temperature of bare overhead conductors. Piscataway, NJ: IEEE Standards Association.

Intergovernmental Panel on Climate Change (IPCC). 2007. *Climate change 2007: The physical science basis. Contribution of Working Group I to the Fourth Assessment Report of the Intergovernmental Panel on Climate Change,* ed. S. Solomon, D. Qin, M. Manning, Z. Chen, M. Marquis, K. B. Averyt, M. Tignor and H. L. Miller. Cambridge and New York: Cambridge University Press. http://www.ipcc.ch/ipccreports/ar4-wg1.htm.

Karl, T. R., J. M. Mellilo, and T. C. Peterson, eds. 2009. *Global climate change impacts in the United States*. Cambridge: Cambridge University Press. http://downloads.globalchange.gov/usimpacts/pdfs/climate-impacts-report.pdf.

Kawajiri, K., T. Oozeki, and Y. Genchi. 2011. Effect of temperature on PV potential in the world. *Environmental Science & Technology* 45:9030–9035.

Kehlhofer, R., F. Hanemann, F. Stirnimann, and B. Rukes. 2009. *Combined-cycle gas and steam turbine power plants*. 3rd ed. Tulsa, OK: PennWell Publishing.

Kenney, D. S., and R. Wilkinson, eds. 2012. *The water-energy nexus in the American West*. Cheltenham, UK: Edward Elgar Publishing.

LaCommare, K. H, and J. H. Eto. 2004. *Understanding the cost of power interruptions to U.S. electricity consumers*. Report No. LBNL-55718. Berkeley, CA: Ernest Orlando Lawrence Berkeley National Laboratory.

Lesieutre, B. C., W. H. Hagman, and L. Kirtley. 1997. An improved transformer top oil temperature model for use in an on-line monitoring and diagnostics system. *IEEE Transactions on Power Delivery* 12:249–256.

Li, X., and G. Zielke. 2003. A study on transformer loading in Manitoba: Peak load ambient temperature. *IEEE Transactions on Power Delivery* 18:1249–1256.

Li, X., R. Mazur, D. Allen, and D. Swatek. 2005. Specifying transformer winter and summer peak-load limits. *IEEE Transactions on Power Delivery* 20:185–190.

Lutz, J.A., J.W. van Wagtendonk, A.E. Thode, D.D. Miller and J.F. Franklin. 2009. Climate, lightning ignitions, and fire severity in Yosemite National Park, California, USA. *International Journal of Wildland Fire* 18:765–774.

Madani, K., and J. R. Lund. 2009. Modeling California's high-elevation hydropower systems in energy units. *Water Resources Research* 45: W09413, doi:10.1029/2008WR007206.

—. 2010. Estimated impacts of climate warming on California's high-elevation hydropower. *Climatic Change* 102:521–538. doi: 10.1007/s10584-009-9750-8.

Mansbach, D., and D. Cayan. 2012. Formulation of a statistical downscaling model for California site winds, with application to 21st century climate scenarios. Draft final report. Sacramento: California Energy Commission, Public Interest Energy Research (PIER) Program.

Maulbetsch, J. S., and M. N. DiFilippo. 2006. *Cost and value of water use at combined-cycle power plants*. PIER Final Project Report CEC-500-2006-034. Sacramento: California Energy Commission, Public Interest Energy Research (PIER) Program.

Miller, N. L., J. Jin, K. Hayhoe, and M. Auffhammer. 2007. *Climate change, extreme heat, and electricity demand in California*. PIER Project Report CEC-500-2007-023. Sacramento: California Energy Commission, Public Interest Energy Research (PIER) Program.

Nakićenović, N., and R. Swart, eds. 2000. *Special report on emissions scenarios: A special report of Working Group III of the Intergovernmental Panel on Climate Change*. Cambridge: Cambridge University Press.

National Energy Education Development (NEED) Project, 2007. Hydropower. In *Secondary Energy Infobook*, 24–27. Manassas, VA: NEED Project.

National Energy Technology Laboratory (NETL), 2009a. *An analysis of the effects of drought conditions on electric power generation in the western United States*. Report no. DOE/NETL-2009/1365. Pittsburgh, PA: NETL.

—. 2009b. *Impact of drought on U.S. steam electric power plant cooling water intakes and related water resource management issues*. Report no. DOE/NETL-2009/1364. Pittsburgh, PA: NETL.

—. 2010. *Water vulnerabilities for existing coal-fired power plants*. Report no. DOE/NETL-2010/1429. Pittsburgh, PA: NETL.

Northwest Power and Conservation Council. 2005. *The fifth Northwest electric power and conservation plan*. http://www.nwcouncil.org/energy/powerplan/5/Volume1_screen.pdf.

National Research Council. 2010. *Limiting the magnitude of future climate change*. Washington, DC: National Academies Press.

Nuclear Regulatory Commission (NRC). 2006. Power reactor status reports, July–August. Washington, DC. http://www.nrc.gov/reading-rm/doc-collections/event-status/reactor-status.

—. 2007. Power reactor status reports, July. http://www.nrc.gov/reading-rm/doc-collections/event-status/reactor-status.

—. 2010. Power reactor status reports, July–August. http://www.nrc.gov/reading-rm/doc-collections/event-status/reactor-status.

—. 2011. Power reactor status reports, July. http://www.nrc.gov/reading-rm/doc-collections/event-status/reactor-status.

Null, S. E., J. H. Viers, and J. F. Mount. 2010. Hydrologic response and watershed sensitivity to climate warming in California's Sierra Nevada. *PLoS ONE* 5 (4): e9932, doi:10.1371/journal.pone.0009932.

Parker, N., P. Tittmann, Q. Hart, R. Nelson, K. Skog, A. Schmid, E. Gray, and B. Jenkins. 2010. Development of a biorefinery optimized biofuel supply curve for the western United States. *Biomass and Bioenergy* 34:1597–1607.

Phinney, S., R. McCann, and G. Franco. 2005. *Potential changes in hydropower production from global climate change in California and the western United States*. Consultant report CEC-700-2005-010. Sacramento: California Energy Commission.

Pryor, S. C., and R. J. Barthelmie. 2011. Assessing climate change impacts on the near-term stability of the wind energy resource over the United States. *Proceedings of the National Academy of Sciences*, published online, doi: 10.1073/pnas.1019388108.

Pryor, S. C., R. J. Barthelmie, D. T. Young, E. S. Takle, R. W. Arritt, D. Flory, W. J. Gutowski Jr., A. Nunes, and J. Road. 2009. Wind speed trends over the contiguous United States. *Journal of Geophysical Research* 114: D14105, doi:10.1029/2008JD011416.

Pryor, S. C., and J. Ledolter. 2010. Addendum to "Wind speed trends over the contiguous United States". *Journal of Geophysical Research* 115: D10103, doi:10.1029/2009JD013281.

Rasmussen, D. J., T. Holloway, and G. F. Neme. 2011. Opportunities and challenges in assessing climate change impacts on wind energy: A critical comparison of wind speed projections in California. *Environmental Research Letters* 6:024008, doi:10.1088/1748-9326/6/2/024008.

Samenow, J. 2011. Las Conchas fire near Los Alamos largest in New Mexico history. Posted to *Capitol Weather Gang* blog July 1, 2011. http://www.washingtonpost.com/blogs/capital-weather-gang/post/las-conchas-fire-near-los-alamos-largest-in-new-mexico-history/2011/07/01/AGcNXptH_blog.html.

Sathaye, J., L. Dale, P. Larsen, G. Fitts, K. Koy, S. Lewis, and A. Lucena. 2012. *Estimating risk to California energy infrastructure from projected climate change.* Report No. CEC-500-2012-057. Sacramento: California Energy Commission.

Sovacool, B. K. 2009. Running on empty: The electricity-water nexus and the U.S. electric utility sector. *Energy Law Journal* 30 (11): 11-51.

Stillwell, A. S., M. E. Clayton, and M.E. Webber. 2011. Technical analysis of a river basin-based model of advanced power plant cooling technologies for mitigating water management challenges. *Environmental Research Letters* 6:034015.

Swift, G. W., E. S. Zocholl, M. Bajpai, J. F. Burger, C. H. Castro, S. R. Chano, F. Cobelo, et al. 2001. Adaptive transformer thermal overload protection. *IEEE Transactions on Power Delivery* 16:516–521.

Tilman D., R. Socolow, J. A. Foley, J. Hill, E. Larson, L. Lynd, S. Pacala, et al. 2009. Beneficial biofuels—The food, energy, and environmental trilemma. *Science* 325:270–271.

Union of Concerned Scientists (UCS). 2007. Rising temperatures undermine nuclear power's promise. Union of Concern Scientists backgrounder. Washington, DC: UCS. http://www.nirs.org/climate/background/ucsrisingtemps82307.pdf.

U.S. Bureau of Reclamation. 2011. *Interim report No. 1, Colorado River Basin water supply and demand study*. Boulder City, NV: U.S. Bureau of Reclamation, Lower Colorado Region. http://www.usbr.gov/lc/region/programs/crbstudy/report1.html.

U.S. Department of Energy (DOE). 2006. *Energy demands on water resources: Report to Congress on the interdependency of energy and water*. Albuquerque: Sandia National Laboratories. http://www.sandia.gov/energy-water/docs/121-RptToCongress-EWwEIAcomments-FINAL.pdf.

—. 2010. *Water power for a clean energy future.* Report No. DOE/GO-102010-3066. N.p.: DOE, Office of Energy Efficiency and Renewable Energy. http://www1.eere.energy.gov/water/pdfs/48104.pdf.

—. 2011. *U.S. billion-ton update: Biomass supply for a bioenergy and bioproducts industry.* ORNL/TM-2011/224. Oak Ridge, TN: Oak Ridge National Laboratory.

Vicuña, S., R. Leonardson, M. W. Hanemann, L. L. Dale, and J. A. Dracup. 2008. Climate change impacts on high elevation hydropower generation in California's Sierra Nevada: A case study in the upper American River. *Climatic Change* 87: S123–S137.

Wei, M., J. H. Nelson, M. Ting, C. Yang, D. Kammen, C. Jones, A. Mileva, J. Johnston and R. Bharvirkar. 2012. *California's carbon challenge: Scenarios for achieving 80% emissions reductions in 2050.* Berkeley, CA: Lawrence Berkeley National Laboratory.

Welsh, R., S. Grimberg, G. W. Gillespie, and M. Swindal. 2010. Technoscience, anaerobic digester technology and the dairy industry: Factors influencing North country New York dairy farmer views on alternative energy technology. *Renewable Agriculture and Food Systems* 25:1701–1780.

Westerling, A. L., B. P. Bryant, H. K. Preisler, H. G. Hidalgo, T. Das, and S. R. Shrestha. 2009. *Climate change, growth and California wildfire.* Draft Paper CEC-500-2009-046D. Sacramento: California Climate Change Center.

Western Governors' Association. 2010. Policy resolution 10-15: Transmission and the electric power system. http://www.westgov.org/energy.

—. 2011. Policy resolution 11-7: Water resource management in the West. http://www.westgov.org/initiatives/water.

Western Resource Advocates. 2009. *Water on the rocks: Oil shale water rights in Colorado.* Boulder, CO: Western Resource Advocates. http://www.westernresourceadvocates.org/land/wotrreport/index.php.

Wilbanks, T. J., V. Bhatt, D. E. Bilello, S. R. Bull, J. Ekmann, W. C. Horak, Y. J. Huang, et al. 2007. *Effects of climate change on energy production and use in the United States.* Synthesis and Assessment Product 4.5; Report by the U.S. Climate Change Science Program and the subcommittee on Global Change Research. Washington, DC: U.S. Department of Energy, Office of Biological and Environmental Research.

Endnotes

i Includes biofuel-based transportation fuels and electricity generated with renewable energy (e.g., solar, wind, geothermal, biomass, hydroelectric).

ii Several reports have considered potential climate impacts on the energy industry at the national and international levels (e.g., Karl, Melillo, and Peterson 2009; Beard et al. 2010; Ebinger and Vergara 2011). Here we direct attention to the Southwest region of the United States.

iii These projections are similar to estimates presented in another recent study of California peak loads and climate change (Miller et al. 2007), which projects 90th percentile peak demand increases of 6.2% to 19.2 % under the IPCC's (2007) high-emissions scenario.

iv Non-potable sources are now required in California for new permitting of thermoelectric water use (California Water Code, Section 13552).

v An 86°F (30°C) ambient temperature approximately corresponds to a 248°F (120°C) hot spot conductor temperature at a typical transformer (Swift et al. 2001).

vi Caution needs to be taken, however, as the analysis by Sathaye and colleagues (2011) was conducted at a scoping level and site-specific analyses are necessary to determine actual risks.

vii The authors caution about the reliability of these reported trends, however, given the potential problems with the wind measurements due to such factors as changes in the location of the monitoring stations, degradation in the performance of the instruments used to measure wind speed, and changes in land use close to the monitoring stations.

Chapter 13

Urban Areas

COORDINATING LEAD AUTHOR

Stephanie Pincetl (Institute of the Environment and Sustainability, University of California, Los Angeles)

LEAD AUTHORS

Guido Franco (California State Energy Commission), Nancy B. Grimm (Arizona State University), Terri S. Hogue (Civil and Environmental Engineering, UCLA), Sara Hughes (National Center for Atmospheric Research), Eric Pardyjak (University of Utah, Department of Mechanical Engineering, Environmental Fluid Dynamics Laboratory)

CONTRIBUTING AUTHORS

Alicia M. Kinoshita (Civil and Environmental Engineering, UCLA), Patrick Jantz (Woods Hole Research Center)

REVIEW EDITOR

Monica Gilchrist (Local Governments for Sustainability)

Executive Summary

The unique characteristics of Southwest cities will shape both the ways they will be impacted by climate change and the ways the urban areas will adapt to the change. The Southwest represents a good portion of the arid and semi-arid region of North America and many of its cities rely on large-scale, federally built water storage and conveyance

---

Chapter citation: Pincetl, S., G. Franco, N. B. Grimm, T. S. Hogue, S. Hughes, E. Pardyjak, A. M. Kinoshita, and P. Jantz. 2013. "Urban Areas." In *Assessment of Climate Change in the Southwest United States: A Report Prepared for the National Climate Assessment*, edited by G. Garfin, A. Jardine, R. Merideth, M. Black, and S. LeRoy, 267–296. A report by the Southwest Climate Alliance. Washington, DC: Island Press.

structures. Water regimes in this part of the country are expected to be significantly impacted by climate change because of higher temperatures, reduced snowpack, and other factors, including possibly reduced or more unpredictable patterns of precipitation, which will affect cities and their water supplies. Further, the cities are likely to experience greater numbers of high-temperature days, creating vulnerabilities among populations who lack air conditioning or access to cooling shelters. Myriad and overlapping governmental organizations are responsible for public goods and services in the region, as in other parts of the country. Their jurisdictions generally do not correspond to ecosystem or watershed boundaries, creating mismatches for climate adaptation programs and policies and significant barriers to cooperation and collaboration. Finally, many local governments are facing budget constraints, making it difficult to plan and implement new programs to anticipate the potential impacts of climate change.

In summary:

- The water supplies of Southwest cities, which are located in arid and semi-arid regions and rely on large-scale, federally built water storage and conveyance structures, will be less reliable due to higher temperatures, reduced snowpack, and other factors, including possibly reduced precipitation. (high confidence)
- Some Southwest cities are likely to experience greater numbers of extreme high-temperature degree days; residents who lack air conditioning or access to cooling shelters will be especially vulnerable to these changes. (high confidence)
- The large metropolitan areas that concentrate most of the population in the Southwest are governed by counties, cities, and hundreds or thousands of special districts, which makes coordination complex and therefore decreases the capacity for cities to adapt to climate change. (high confidence)
- Within metropolitan regions, substantial differences in fiscal capacity and in political and decision making capacities to plan and implement new programs to anticipate the potential impacts of climate change reduce the capacity for cities to adapt to climate change. (high confidence)
- Options for decreasing urban vulnerability to climate change include making data available to improve targeted programs for energy conservation (high confidence). For example, utility data (such as electric and gas bills) have heretofore been considered confidential, so understanding energy or gas use in a city, by land use, building type, or sociodemographic profile must be derived from surveys or national models. This makes it impossible to target specific energy use for reduction or as a model for conservation. (high confidence)
- Data availability will also improve the ability to develop new approaches to understanding urban energy flows (including wastes such as GHGs). Urban metabolism, for example, quantifies inputs and outputs to cities (Pincetl, Bunje and Holmes 2012) and can include the life-cycle analysis of supply chains that supply cities (Chester, Pincetl and Allenby 2012). Supply chains (which span all movement and storage of raw materials, manufacturing, and finished goods from point of origin to point of consumption) are poorly understood, and therefore their greenhouse gas (GHG) emissions are difficult to identify and reduce.[i] (high confidence)

- Monitoring for climate-related indices is weak throughout the region, and the data that are collected are usually not synthesized in ways that are useful to understand the urban causes and potential impacts of climate change. Most climate studies have been done at macro-scales or pertain to specific issues such as water supply. More local data, such as on water use in urban areas, would be useful in managing for climate change adaptation and planning for future impacts. (high confidence)

13.1 Cities in the Southwest

The importance of urban lands in climate change has been articulated in a previous assessment, which discussed the potential of urban planning, urban land management systems, and urban land regulation in addressing climate change challenges (Blanco et al. 2011). This chapter focuses on U.S. urban development impacts on climate and the potential effects of climate change on cities in the Southwest. Southwest cities grew throughout the twentieth century—a period of resource and land abundance—circumstances that shaped their land use, their residents' dependency on automobiles for transportation, the choice of building types, and patterns of resource use. Southwest cities have continued to grow at a tremendous pace, particularly the arid cities of Phoenix and Las Vegas (Figure 13.1).

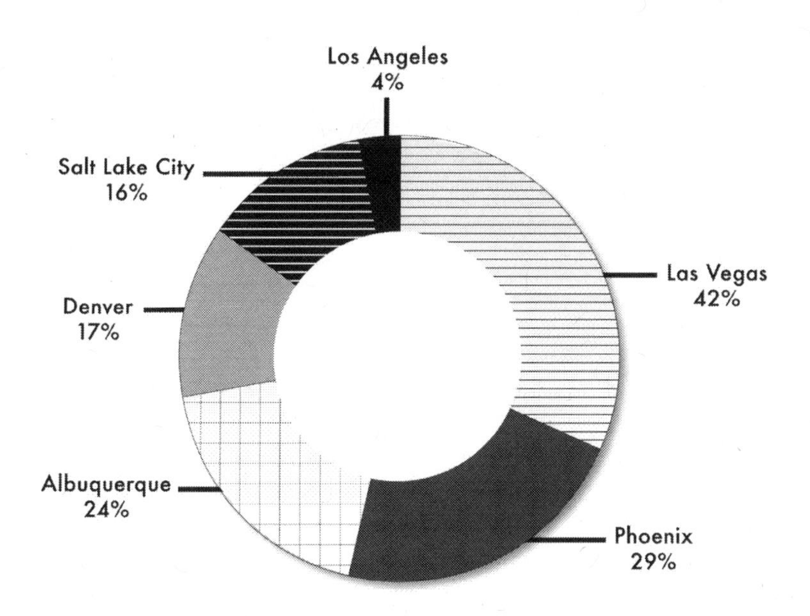

Figure 13.1 Population change in Southwest cities, 2000–2010. Source: U.S. Census Bureau, 2010 data (http://2010.census.gov/2010census/).

Continued growth of suburban and urban areas in the Southwest will affect their vulnerability to climate change, depending on factors such as the geographical distribution of this growth (see Chapter 3). Figure 13.2 shows one potential growth pattern that is compatible with the IPCC high-emissions (A2) scenario (Nakićenović and Swart 2000; Bierwagen et al. 2010), a scenario in which there is continued population growth. Two immediate conclusions can be reached about potential impacts in urban areas taking into account the projected potential changes in climate discussed in Chapters 6 and 7. First, cities would grow in areas that are projected to experience more rapid warming for the rest of this century. Second, urbanization would occur in areas that are projected to experience less predictable precipitation.

Areas of new urbanization in (shown in Figure 13.2), such as the Central Valley of California, also tend to coincide with the areas that are susceptible to large floods, as shown in Figure 13.3. Somewhat paradoxically, the probability of large flooding events rises with climate change in the Central Valley (Das et al. 2011), even if total precipitation levels go down.

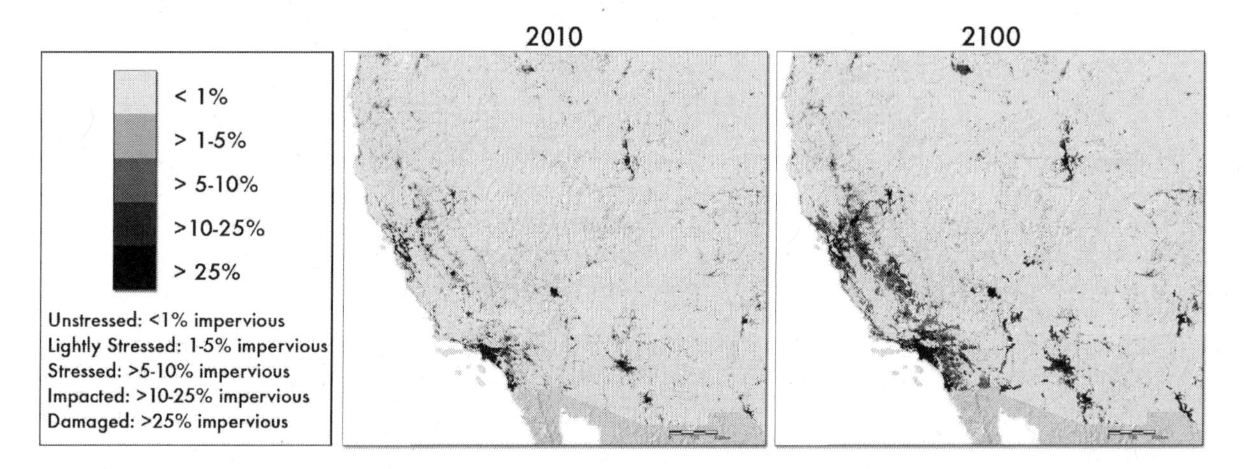

Figure 13.2 Human settlements (impervious surfaces) in the Southwest, 2000 and 2100. The 2100 scenario is compatible with the IPCC A2 high-emissions scenario. Reproduced from the ICLUS web viewer (http://134.67.99.51/ICLUSonline/, accessed on 2/12/2012).

Observed changes in climatic trends in major cities in the Southwest

Major cities in the Southwest and other parts of the United States are already experiencing changes in temperature. It is unclear, however, if these changes are mostly due to changes in land cover (as from the urban heat island effect)[ii] or if they are a manifestation of regional changes in climate. A recent study by Mishra and Lettenmaier (2011) tackled this issue by developing time series (data) of temperature and precipitation for 100 major cities and their surrounding rural areas in the United States and comparing their trends. Where the trend for an urban area is somewhat different than its surrounding rural area, changes in land cover (i.e., urbanization) may play a role. Therefore these changes may be reversible by making changes in urban morphology aimed at reducing the urban heat island. Mishra and Lettenmaier concluded that for the Southwest, changes in nighttime minimum temperatures and heating and cooling degree days are due to regional changes in climate, and urbanization is not the main reason for the already observed warming in cities (see Figure 13.4). The implications of this finding are that: (1) cities should prepare for increases in energy demand for space cooling and a reduction in energy demand for space heating, and (2) reducing the heat island effect should help but may not completely eliminate overall warming in the long term.

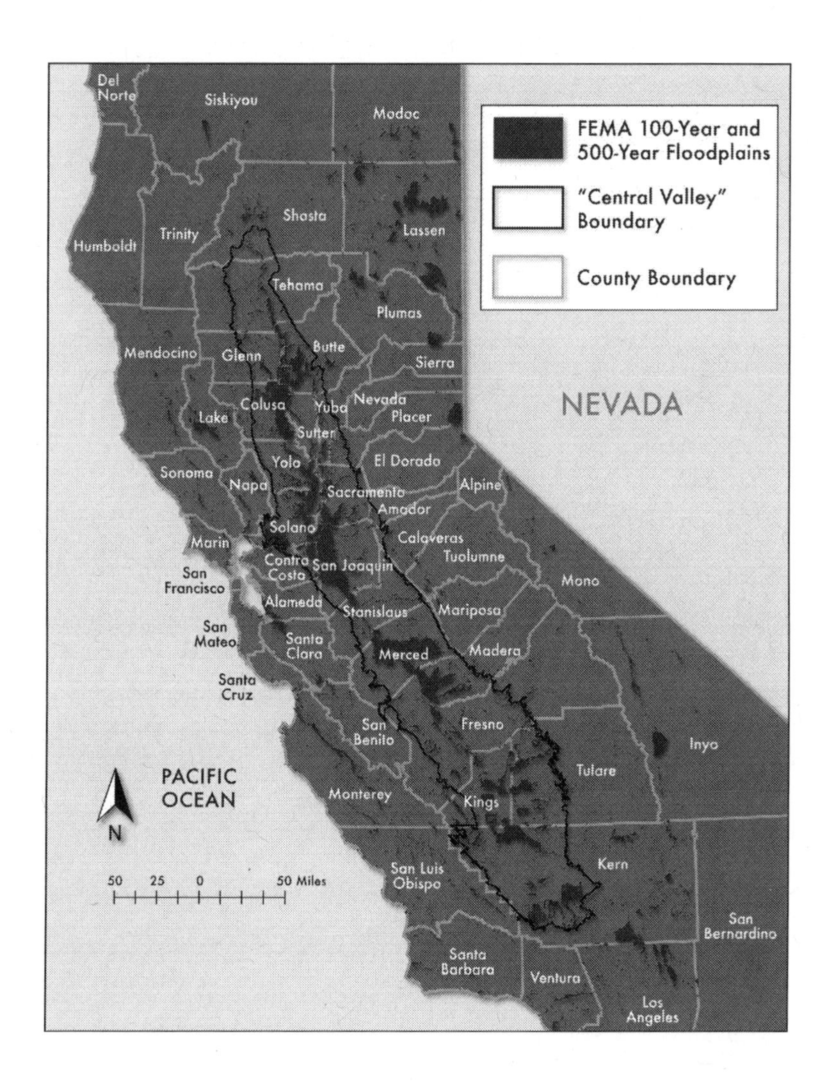

Figure 13.3 FEMA 100-year and 500-year floodplains in California. Adapted from Galloway et al. (2007).

The potential for extreme precipitation events is important for urban managers to consider because the amount of rain and duration of these events determine the needed design capacity of the stormwater infrastructure. Substantial increases in extreme precipitation events may result in the failure of stormwater systems if new extreme precipitation levels are outside their design envelope. So far, the historical trends for extreme precipitation for cities in the Southwest are less clear, with no uniform regional trends as shown in Figure 13.5, suggesting that there is not a clear imminent risk to stormwater systems from flooding of this type. As reported in Chapter 7, climate projections for the Southwest and throughout the country suggest an increase in extreme precipitation events but these projections are highly uncertain. It is possible that the climate-change signal is still emerging from the "noise" created by climate variability. In any event, flood control managers in California are preparing for potentially unusually high incidences of precipitation and consequent flooding (see, for example the projects listed on the California Dept. of Water Resources FloodSAFE website).[iii]

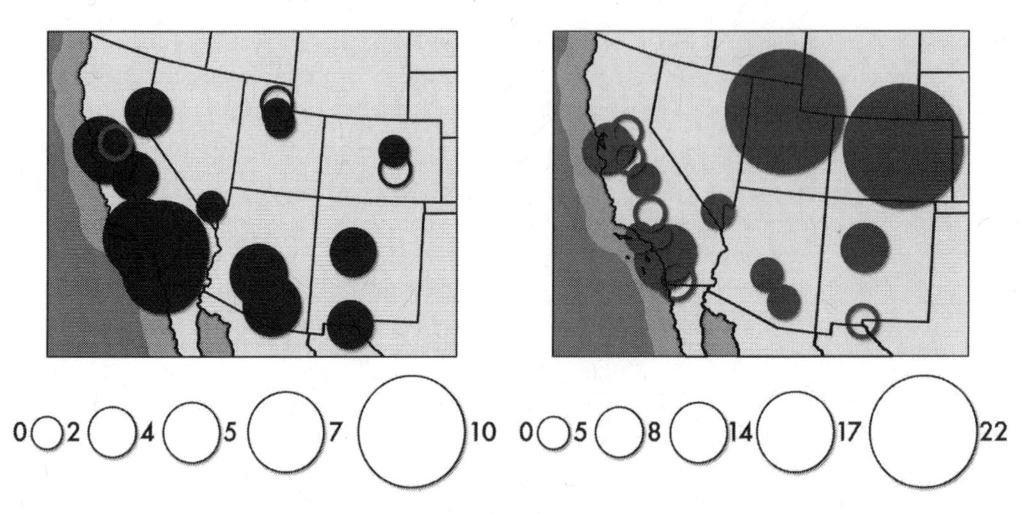

Figure 13.4 Trends in heating and cooling degree days in major urban areas of the Southwest. Trends in heating degree-days (left) and cooling degree-days (right) are based on 65°F (18.3°C) and 75°F (23.9°C) bases, respectively, for the period 1950 to 2009 in seventeen urban areas. Dark grey circles represent decreasing trends and light grey increasing trends. The trends are proportional to the diameter of the circles. Filled circles represent statistically significant trends. See glossary for definitions of heating and cooling degree-days. Adapted from Mishra and Lettenmaier (2011) with permission from the American Geophysical Union.

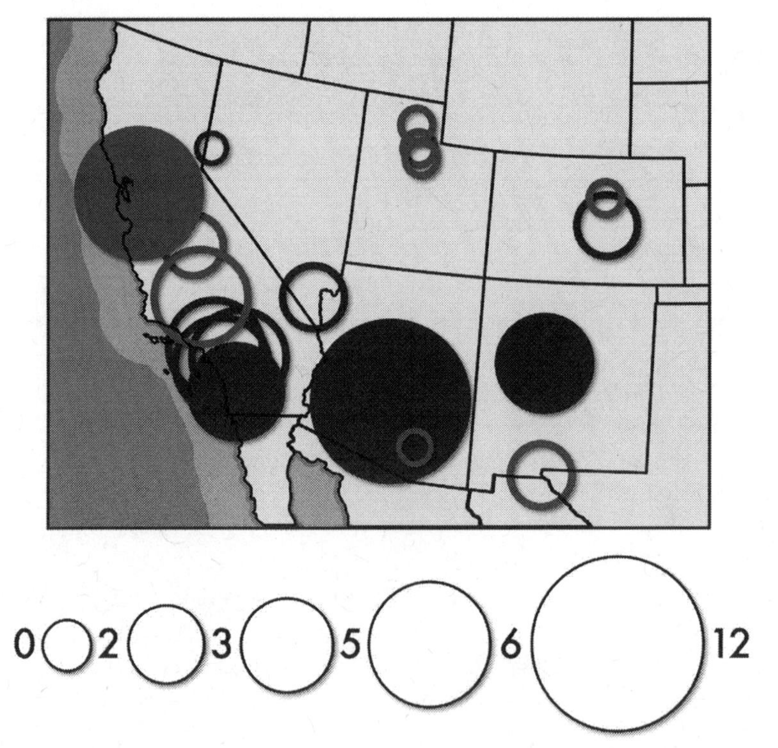

Figure 13.5 Trends in the annual 1-day maximum daily precipitation amounts in major urban areas of the Southwest. Dark grey circles represent decreasing trends and light grey increasing trends. The trends are proportional to the diameter of the circles. Filled circles represent statistically significant trends. Adapted from Mishra and Lettenmaier (2011) with permission from the American Geophysical Union.

Urban processes that contribute to climate change

The contribution of urban areas in the Southwest to climate change is a function of a variety of features: urban form (dense or sprawling); the allocation of land to commercial, industrial, and residential uses, and their spatial disposition; infrastructure (including building technologies); impermeable surfaces; surface albedo; water supply and disposal systems; and transportation systems. Yet the body of research and amount of available data about the impact of such features on emissions and climate are relatively small. For example, the size of buildings, their construction materials, and building standards for energy efficiency all have greenhouse gas (GHG) emissions implications, but the exact relationships are not well understood. Levels of affluence and consumption in cities; whether cities depend on long or short supply chains for their goods and services; the energy used in the manufacturing and distribution of goods: all these play significant but poorly accounted-for roles in contributing to climate change. These energy-related flows are components of a city's *urban metabolism* (Kennedy, Cuddihy, and Engel-Yan 2007; Chester, Martin, and Sathaye 2008; Kennedy, Pincetl, and Bunje 2010). Comprehensive accounting that includes life-cycle analyses and cradle-to-grave GHG accounting of cities' metabolisms simply does not exist. Nor are the redirection of flows (such as water reuse and recycling or advanced nutrient recycling) well documented and quantified. Thus, both a city's contributions to climate change and the effects of its efforts to reduce those impacts remain unquantified due to lack of observational data.

Specifically, the dynamics of an increase in urban heat, or *heat flux,* that results from the transfer of heat energy to the atmosphere from pavement and other hard surfaces, heat generation from vehicles, and the consumption of electricity and heating fuel, are poorly understood. Sailor and Lu (2004) estimated these fluxes for a number of U.S. cities including Los Angeles and Salt Lake City, and Grossman-Clarke and colleagues (2005) estimated them for different land covers in Phoenix during the summer (see Figure 13.6). They found most cities have peak values during the day of approximately 30 to 60 watts per square meter (W/m^2, the power per unit area radiated by a surface). Salt Lake City and Los Angeles had relatively low fluxes compared to other cities, with peak values less than 15 W/m^2 and 35 W/m^2, respectively. Salt Lake City's flux level was particularly low due to low population density. For all of the cities (across the United States) analyzed in the Sailor and Lu study, heating generated from vehicles was the dominant cause of heat flux in the summer, accounting for 47% to 62% of it. Wintertime heating was also a very important cause, but less so in Southwest cities where winters are not as cold as in other parts of the United States. Recently, Allen and colleagues (2010) developed a global model for human-caused fluxes and found that globally the average daily urban heat flux due to human causes has a range of 0.7 W/m^2 to 3.6 W/m^2. Globally, they found heat release from buildings to be the most important contributor.

Several studies have included the effects of anthropogenic heat fluxes in global climate models (GCMs) (Flanner 2009; McCarthy, Best, and Betts 2010; McCarthy et al. 2012). While these models are at the global scale and cannot effectively quantify specific urban areas, they provide some insight into the importance of these fluxes. For example, McCarthy et al. (2011) ran simulations with both CO_2 doubling and anthropogenic urban heat fluxes. They found that by 2050 in the Los Angeles area, the number of hot days experienced in urban areas would be similar to the number of hot days experienced in

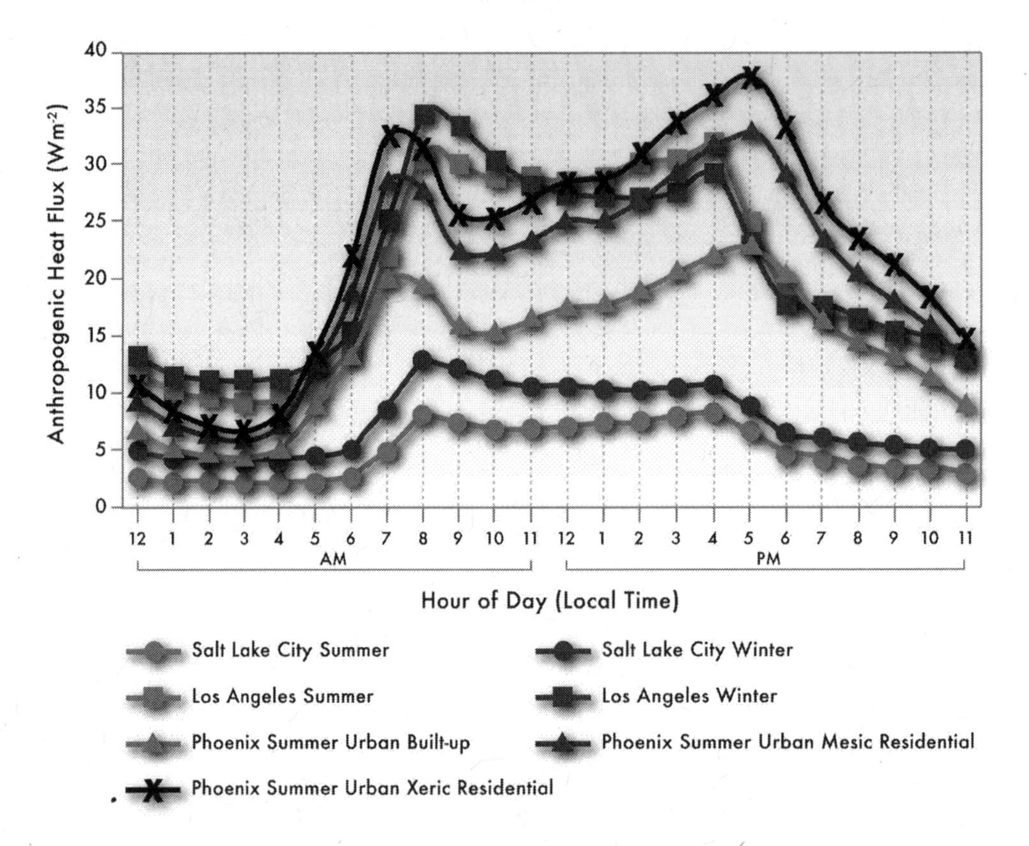

Figure 13.6 Anthropogenic heat flux estimates for three Southwest cities, shown in watts per square meter. Urban built-up (no vegetation), urban mesic residential (well-watered flood or overhead irrigated), and urban xeric residential (drought-adapted vegetation with drip irrigation) are distinguished by their type of vegetation and irrigation, listed in parentheses. Adapted from Sailor and Lu (2004) and Grossman-Clarke et al. (2005).

rural areas. However, the annual frequency of hot *nights* would increase more in the city than in the rural areas for the CO_2 doubling scenario. In the city of Los Angeles the number of hot nights would increase by two days with 20 W/m^2 of anthropogenic heating and by ten days with 60 W/m^2.

About 60 million people live in the Southwest, the majority of whom reside in major metropolitan urban centers and consume the majority of goods and services. As indicated before, good information is lacking about the direct and indirect GHG emissions from urban populations. However, new studies have shown that the net emissions associated with imports and exports of goods and services in the United States are substantial (Peters and Hertwich 2008; Davis and Caldeira 2010), as shown in Figure 13.7.

Estimates of consumption-based emissions (which take into account net imports and exports of goods and services) for the United States are about 12% higher than production-based inventories (i.e., conventional inventories) (Davis and Caldeira 2010). Preparing consumption-based inventories for cities should be a priority to identify potential unrecognized sources of indirect emissions. By implementing life-cycle analysis for goods and services, such information could be developed.

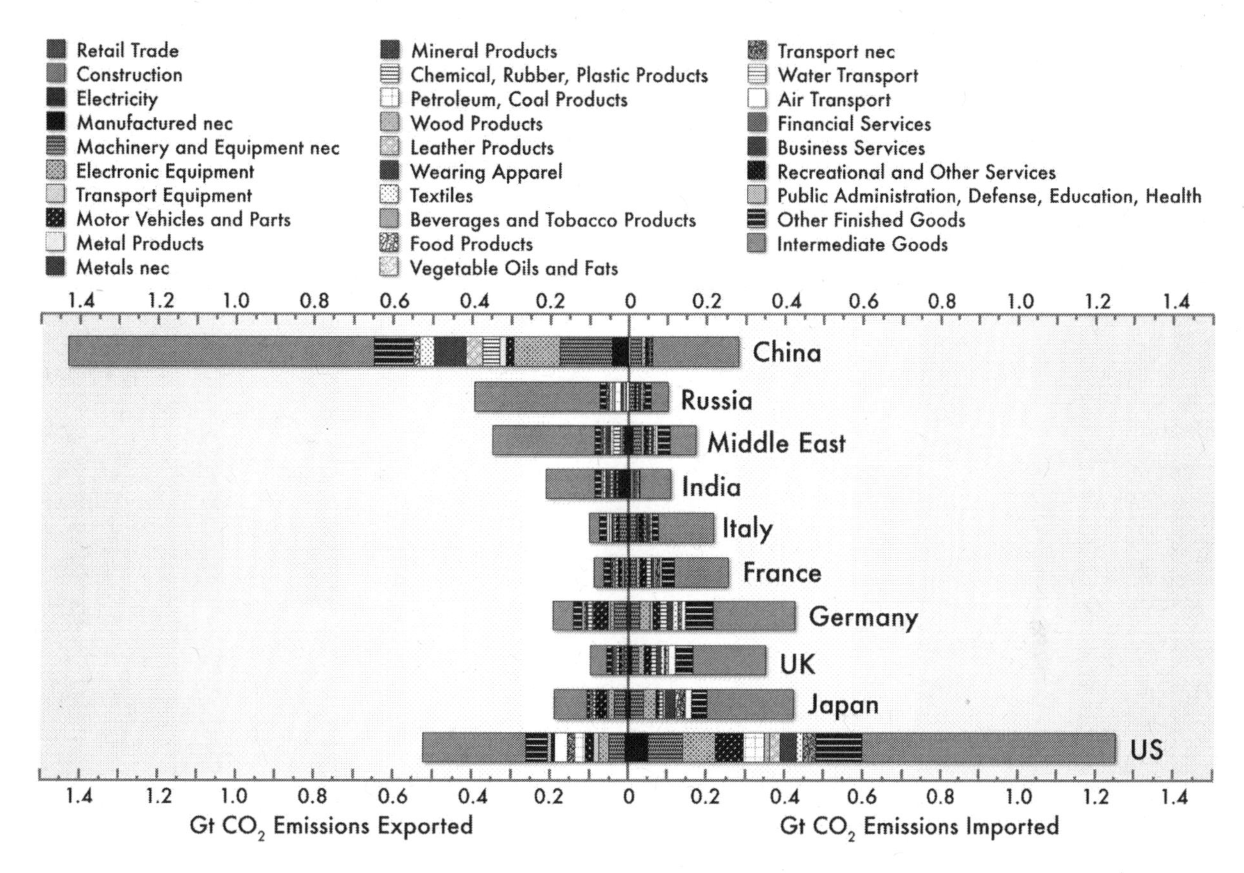

Figure 13.7 Embodied CO_2 emissions associated with goods and services exported (left) and imported (right), for selected countries. The U.S. imports a substantial volume of CO2 emissions embodied in machinery, electronics, motor vehicles and parts, chemical/rubber/plastic products, and intermediate goods (the long light gray segment accounting for roughly 50% of imported emissions). Reproduced from Davis and Caldeira (2010).

While direct-emission measurements of GHGs in cities are rare, a number of urban studies do exist (e.g., Velasco and Roth 2010; Crawford et al. 2011 and references therein; Ramamurthy and Pardyjak 2011). These researchers have begun to quantify CO_2 emissions from different types of urban surfaces using the eddy covariance (EC) method[iv] that has been employed by researchers of non-urbanized ecosystem sites around the world for many years (Aubinet et al. 2000; Baldochhi 2003). The dearth of urban data is partly a result of a number of difficulties in making these measurements (Velasco and Roth 2010).[v] The measurements are, however, very important as they provide spatial and temporal information about the sources and sinks[vi] of emissions that cannot be obtained from conventional inventory methods. A framework for a global Urban Flux Network[vii] now exists to help identify those cites that currently measure fluxes or have recently made such measurements around the world. Only two sites are currently listed in the Southwestern United States: Salt Lake City (Ramamurthy and Pardyjak 2011) and a USGS-operated site in Denver.

While only a limited number of urban EC studies exist, a general understanding of important mechanisms related to the emissions process is starting to form. For example, as shown in Figure 13.8, for a wide range of urban areas around the world there is a surprisingly strong correlation between the proportion of vegetated area and net CO_2 fluxes during the summer. Because of their very low urban density and the presence of substantial vegetation (e.g., urban forests), suburban areas such as in Salt Lake City are relatively small net producers of CO_2. These human-planted urban forests in semi-arid climates provide benefits such as CO_2 sequestration and microclimate mediation during the summer (e.g., temperatures are reduced from shading and evapotranspiration-related cooling), but require irrigation. Increased water demands from urban vegetation, while potentially mitigating some aspects of climate change, could increase water use in urban areas already challenged by scarce water resources. This trade-off is still not well-quantified.

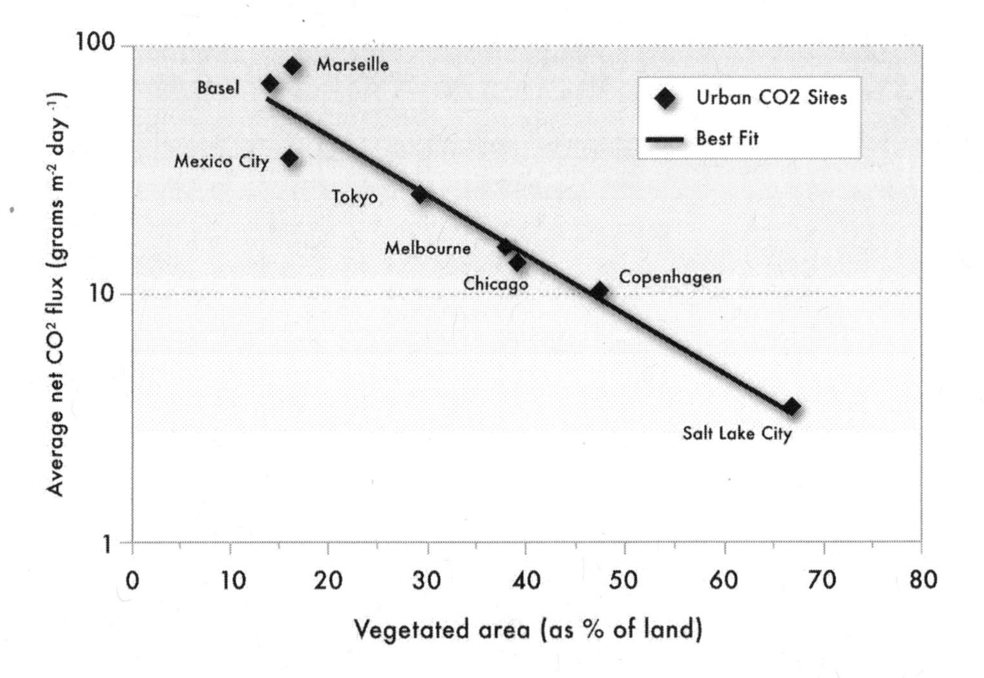

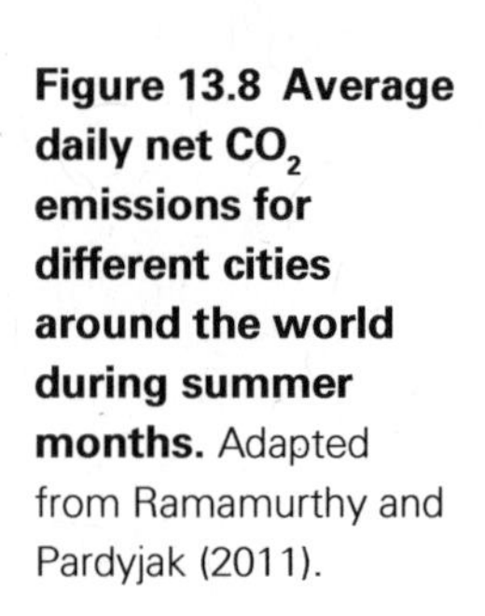

Figure 13.8 Average daily net CO_2 emissions for different cities around the world during summer months. Adapted from Ramamurthy and Pardyjak (2011).

Yet, even the most densely vegetated suburban and urban areas have been found to be net sources of CO_2 (Velasco and Roth 2010). Some well-vegetated suburban areas such as Baltimore take in more CO_2 than they release in the summer through uptake by abundant foliage, yet they are still annual net sources of CO_2 (Crawford et al. 2011). Salt Lake City has a very large urban forest (Pataki et al. 2009); the suburban area monitored there showed significant periods of CO_2 uptake during the daytime in summer, but daytime fluxes still were a net source of CO_2. More research is needed to correlate urban ecosystem parameters with gas exchanges to better understand the mechanisms of transfer. In addition, it is important to recognize that the quantity of CO_2 sequestered by vegetation is dwarfed by urban emissions.

Atmospheric CO_2 concentrations in urban areas are usually much higher than annual global averages because of the local sources of emissions. For example, measurements in Phoenix showed peak CO_2 ambient concentrations that were up to 75% higher than in its rural areas (Idso, Idso, and Balling 2001). In Salt Lake City, Pataki, Bowling and Ehleringer (2003) found peak CO_2 ambient concentrations during wintertime atmospheric inversions that were around 60% higher than in its rural areas, while summertime afternoon values were very close to background levels. Jacobson's 2010(a) Los Angeles study of the potential implications of these "urban domes" of CO_2, found that higher ambient CO_2 concentrations result in small but important increases in air pollution (ozone and particulate matter; see Figure 13.9).[viii] This is in addition to potential increases in air pollution from a global increase of GHGs in the atmosphere reported by others (e.g., EPA 2009; Jacobson and Street 2009; Zhao et al. 2011). Local control of CO_2 emissions would provide a means of improving urban air quality conditions in cities if the urban "dome effect" is confirmed.

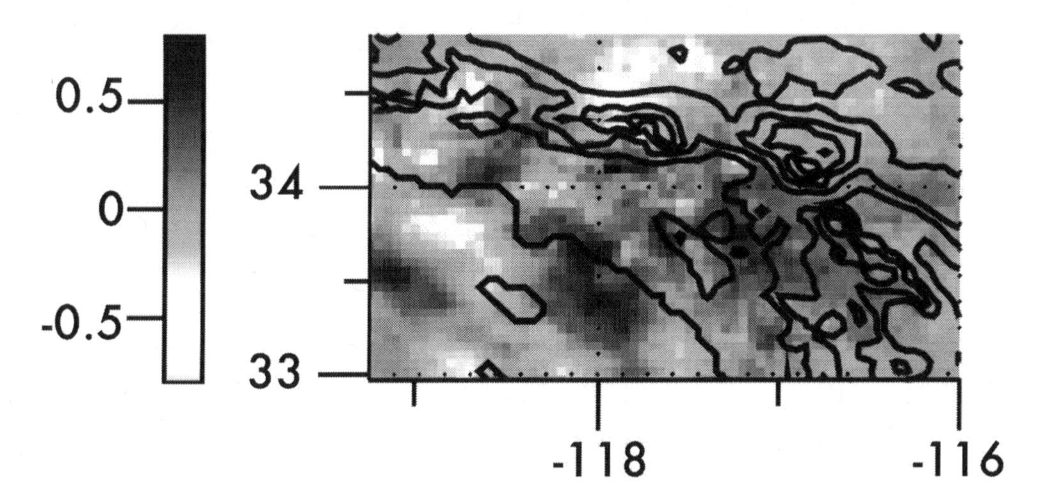

Figure 13.9 Changes of ozone (O_3) concentrations due to the "CO_2 dome" effect in Los Angeles. Modeled differences in ozone concentration in parts per billion from two simulations (with and without CO_2 emissions in Los Angeles); August-October. Contour lines indicate topography. The darker areas inland, to the south and west of the San Gabriel and Santa Ana Mountains (e.g., at approximately 33.7°N, 117.2°W), show increases in surface ozone, due to increased CO_2 aloft. Increased ozone is implicated in air pollution-related deaths. Adapted with permission from the American Chemical Society (Jacobson 2010a, 2501).

Climate change will affect Southwest cities differently, due to their unique geographical locations, settlement histories, population growth rates, shapes and infrastructure, economies, and socio-demographic characteristics. Impacts to residents will in turn depend on where they live and their own capacities and incomes. Southwestern cities—especially the largest metropolitan areas in each of the Southwestern states—have shared characteristics that may cause climate to impact them differently than cities in other regions of the country. Historic development paths of cities continue to influence development patterns and constrain the potential for mitigation and adaptation to climate change. Cities are the products of a particular historical period; most in the Southwest were shaped by the concerns and aspirations of the early twentieth century. These cities were built in a time when there seemed to be no resource constraints, and so their location and form may be less appropriate and functional than in generations past, as climate changes over the next century and beyond. What were hot summers in Phoenix, for example, may become extremely hot summers by the second half of the century, affecting generations that are not yet born.

Government characteristics of large metropolitan regions in the Southwest

The largest metropolitan regions in each state are Albuquerque (New Mexico), Denver (Colorado), Las Vegas (Nevada), Los Angeles (California), Phoenix (Arizona), and Salt Lake City (Utah). One characteristic of Southwest cities is that they are often part of much larger urbanized regions. For example, the city of Los Angeles is one of eighty-eight cities in Los Angeles County, a fully urbanized political jurisdiction. To distinguish Los Angeles from the other cities within the county in terms of its climate contributions or impacts is difficult, as all are intertwined through shared infrastructure and airsheds (shared paths of airflow and pollutants). Thus, one of the important obstacles for cities relative to potential impacts of climate change and adaptation is coordinating governance in complex, fragmented metropolitan regions.

Jurisdictional boundaries in these metropolitan areas are particularly important for the management of environmental resources. Political jurisdictions—such as cities, counties, and special district governments—are superimposed upon ecosystems, watersheds, groundwater resources, and climate zones in ways that do not conform to their physical processes and properties, making it challenging to manage them in a coherent or integrated fashion. There are few requirements for coordinated management or integrated approaches across jurisdictions with regard to infrastructure, natural resources, or any of the daily tasks of local government. This jurisdictional fragmentation of the built environment and infrastructure is complex and place-specific, making it complicated to manage emissions from these large urban areas or to plan and implement mitigation and adaptation measures to address climate impacts. For example, coordinated watershed management for greater water recapture and reuse is difficult because of the number of jurisdictions that have to be integrated (Green 2007). Jurisdictional complexity, differences in scale, and differences in the way data are gathered and made available are significant problems to overcome in order to understand regional contributions to climate change and how those regions may respond.

Southwest cities as distinctive federal creations

To understand the distinctiveness of Southwest cities, it is useful to put their development in historical context. As the nation developed, lands west of the 100th meridian were a source of interest to the federal government (because of their potential) and of special concern (due to their aridity). For the region to become populous and develop a viable economy, providing water was essential (see also Chapter 10). Localities, territories, and states lacked sufficient resources to develop the size and scale of water projects necessary to move water long distances and store it for when it was needed, or the infrastructure to harness major rivers such as the Colorado River (Hundley 2001). Federal water infrastructure was the major factor (driver) for growth of Southwest cities. Federal investments in water development were accompanied by expansion of the electric grid and provision of electricity (Lowitt 1984), which also contributed to urban growth (Table 13.1). At the turn of the twentieth century, the federal government invested the financial resources of the nation to build Southwest water projects, enabling both large-scale agriculture and urban development to occur. Federal water development projects harnessed the Colorado River and other rivers to provide essential water for the growth of the Los Angeles region, and subsequently Phoenix and Las Vegas. Denver benefited from the Colorado-Big Thompson Project. Salt Lake City too complemented its water resources with the Central Utah Project that included water storage (Lowitt 1984). Though Albuquerque relies on groundwater and some surface water, federal funds were provided to relieve flooding and drainage issues in the area, and to bring Rio Grande water in for irrigators and urban use. There are also many small water providers in these cities and region that may rely on local water sources as well. Understanding the full water supply system in these metropolitan regions is very difficult as there are many retail water suppliers, created over time as city regions grew, that buy water from large water wholesale agencies like the Southern California Metropolitan Water District, and/or have small local water resources they sell directly or blend with water from large-scale suppliers. Federally subsidized water projects made water abundant and inexpensive in the Southwest. Water agencies' mission became one of meeting demand from what seemed to be limitless water availability (Hundley 2001).

With increasing uncertainty about snowpack and rainfall due to potential climate change impacts, the historic allocation of Colorado River water distributed by federal infrastructure is once again becoming an increasingly contentious issue among Southwestern states and between competing urban and agricultural demands (see also

Table 13.1 Major water supply projects for Southwestern cities

City	Major Water Supply Projects
Albuquerque	Rio Grande Project (1906–1952), Elephant Butte Dam (1916), Middle Rio Grande Conservancy District (1928– present), Rio Grande Compact (1938–present), San Juan-Chama Project (1962–present)
Denver	Colorado-Big Thompson Project (1938–1957)
Las Vegas	Colorado River dams (1932–1961), Southern Nevada Water Authority (~1947)
Los Angeles	Colorado River dams (1932–1961), Central Valley Project (1937–1979)
Phoenix	Central Arizona Project (1946–1968), Horseshoe Dam (1944–1946)
Salt Lake City	Central Utah Project (1956)

Chapter 10, Box 10.1). Natural areas and ecosystems have been the last to receive rights to water, due to the relatively recent acknowledgement of the importance of ecosystem services (which recognize the value of services provided by an ecosystem, such as recreation, flood control, and reduction of nitrates and other contaminants) and the protection of endangered species. Water rights are hierarchical in time, the oldest users having the first rights under prior appropriation water law. The new ecological concerns have added yet one more water client. Water rights among the different water constituents in the West raise delicate policy questions about the best use of water: for irrigation, for municipal and industrial use, or for ecosystems. Colorado in 1973 recognized the importance of protecting streamflows for the preservation of the natural environment and has a program of water rights acquisitions through the Colorado Water Conservation Board. In New Mexico, in-stream water transfers are left largely to the state engineer, but it is unclear who is eligible to transfer the rights or to hold them. California's in-stream protections derive from the federal Central Valley Project Improvement Act, passed in 1992, which mandated changes in the management of the Central Valley Project to protect and restore habitat for fish and wildlife and has influenced water management throughout the state. Nevertheless, during drought years there is contention about water allocation, and in a future of restricted or unpredictable water supply, determining priorities for water will become more pressing. There are currently no institutions or frameworks to resolve these trade-offs, either within the states themselves or among them. Historically, abundant and inexpensive water (supplied by the federal water systems and more recently by state systems) fueled expectations of an infinite water supply and enabled profligate water use in the Southwest, including extensive outdoor water use in residential areas. Fortunately this is beginning to change (Cohen 2011), but expectations of an infinite water supply were directly linked to federal investments in the West. Urban water use is now consistently declining in every Southwest city, which may reflect increasing awareness of scarcity and the implementation of water conservation policies (Figure 13.10).

Concerns about demands for water and other resources for the Southwest and its emerging cities also led to the setting aside of lands in forests and mountains that still remained in the public domain at the end of the nineteenth century. At that time, no large-scale water transfer systems had been put in place to bring water to growing cities from far-flung places, and the growing understanding that poor forest practices led to floods provided a strong justification for watershed protection. President Benjamin Harrison, for example, designated the lands surrounding the Los Angeles Basin as national forests in 1892 (Pincetl 1999). This federal policy shift resulted in most of the contemporary metropolitan regions in the Southwest being surrounded by public lands that provide important ecological services, including flood control and recreation. This shift also created an extensive wildland-urban interface. Despite their size, the largest Southwestern cities are relatively isolated from other metropolitan areas, are often surrounded by public lands, and rely heavily on imported water (see the satellite images of Los Angeles and Phoenix in Figure 13.11).

Federal spending in the Southwest region during and after the Second World War also fueled growth and created multiplier effects throughout the economy. Airfields and military bases were built, Los Alamos National Laboratory was created in New Mexico, and other investments in the aerospace and ancillary industries provided the

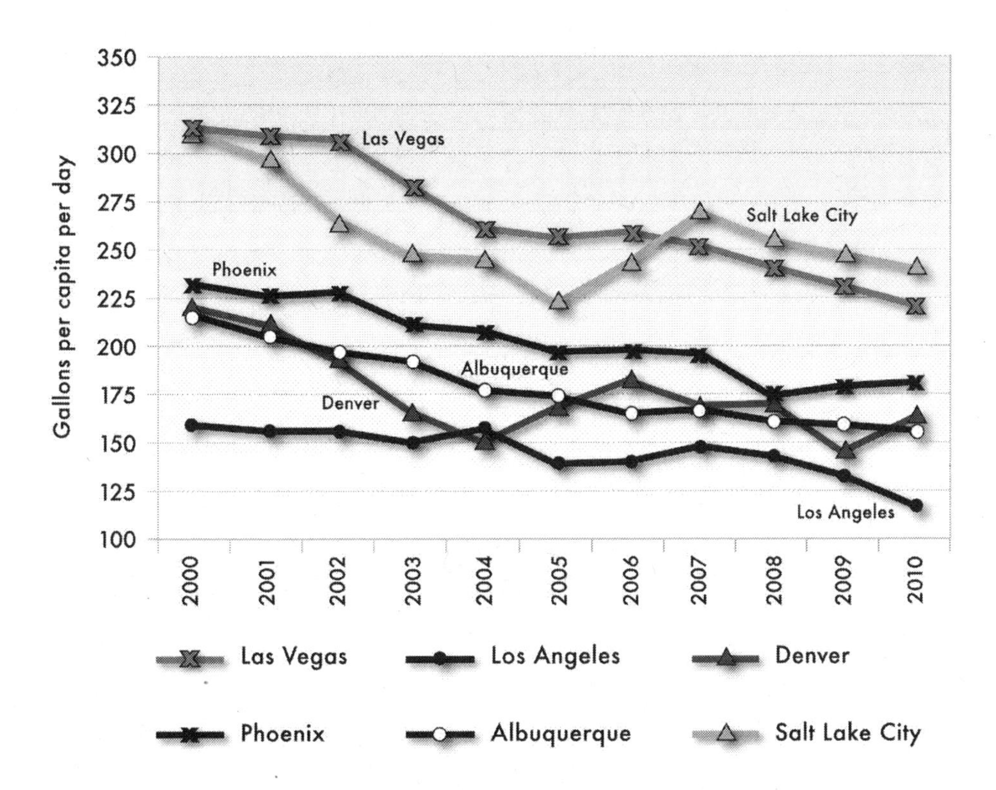

Figure 13.10 Per capita water use in Southwest cities (2000–2010). Source: Great Western Institute (2010), Salt Lake City Department of Public Utilities (SLCDPU 2009), Los Angeles Department of Water and Power (LADWP 2011), City of Albuquerque (http://www.cabq.gov/), City and County of Denver (http://www.denvergov.org/), Southern Nevada Water Authority (http://snwa.com/), Salt Lake City Public Utilities (http://www.slcclassic.com/utilities/), City of Phoenix (http://phoenix.gov/), Los Angeles Department of Water and Power (http://www.ladwp.com/).

employment base for urban development. The Southwest became the home of new techniques for mass home building (also made possible by the expanding water supplies) and populations in the metropolitan areas grew rapidly (Nash 1985; Kupel 2003). The Kaiser Company, for example, built worker housing in Los Angeles near its manufacturing facilities, pioneering the development of planned, dense, automobile-dependent, single-family tracts with nodal shopping malls and other services. Home building was modeled on assembly-line aircraft construction, making it possible to considerably accelerate the pace of construction (Hise 1997). These factors have made cities of the Southwest both expansive and relatively denser than other cities in the country: ten of the fifteen densest metropolitan areas in the United States are located in California, Nevada, and Arizona (Eidlin 2010). Growth of the Southwest cities coincided with both automobile-dominant transportation and federal investment in it, including the Federal Highway Act of 1956, a Cold War-related national system built for defense purposes. As a result, the morphology of Southwest cities is densely suburban. Nationally, the automobile-dependent urban form is a product of post-war suburban growth, with highways and home mortgages subsidized by the federal government.

Figure 13.11 Satellite image of urban Los Angeles (top) and Phoenix (bottom). The two cities are surrounded by undeveloped mountainous areas and public lands. Image by Geology.com using Landsat data from NASA.

13.2. Pathways Through which Climate Change Will Affect Cities in the Southwest

Fire hazards

The extensive public lands surrounding these major metropolitan regions and the corresponding urban-wildland interface make them susceptible to increased wildfires driven by a drier climate, extensive and scattered urbanization in the public lands, a history of fire suppression, and changing vegetation in the natural lands themselves.

The Southwest cities are not equally prone to wildfires (Figure 13.12) nor are they equally likely to suffer increased fire impacts due to differences in the types of ecosystems in the surrounding natural lands. But for the cities at risk, the cost of fire protection is significant. Increased fire incidence will cause property damage and impose related costs, some of which are only beginning to be understood. Issues such as who should pay for fire protection can be contentious, as can be the development of new building regulations for greater fire resistance and land-use regulations to prevent construction in high-fire-zone areas (Pincetl et al. 2008). Less evident impacts are also likely. For example, in the Los Angeles National Forest Station Fire, vegetation burned that had not burned since before the introduction of air pollution controls. Stormwater samples taken after the Station Fire showed high levels of heavy metals that had been deposited before Clean Air Act requirements were imposed (Burke et al. 2011). Water from the front range of the Los Angeles National Forest is a key source of groundwater recharge, and infiltration basins have been inundated with these post-fire pollutants. Except for the study of the L.A. Station Fire, little or no monitoring of such impacts has been done. Increased incidents of urban-fringe fires will require improved post-fire monitoring and management and treatment of stormwater runoff to reduce impacts to city water supplies and downstream ecosystems (see also Chapter 3, Section 3.2.1, and Chapter 8, Section 8.4.2).

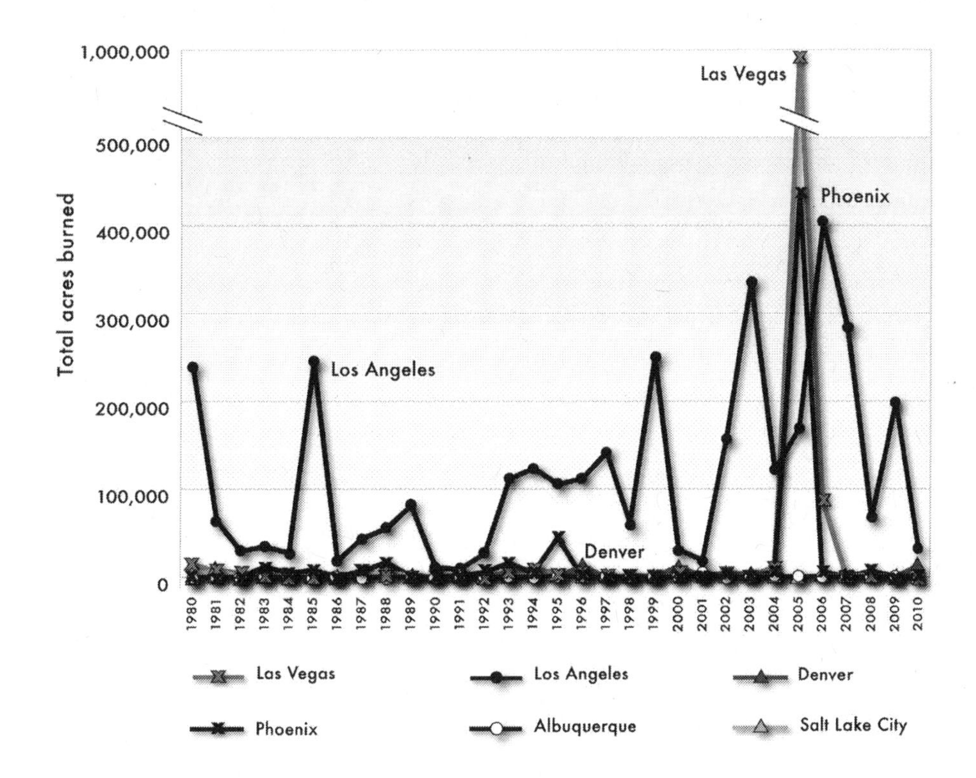

Figure 13.12 Total acres burned in wilderness/urban interface zones of six Southwestern cities. Source: U.S. Department of the Interior's Geospatial Multi-Agency Coordination Group (GeoMAC) Wildland Fire Support (http://www.geomac.gov), State of California Fire and Resource Assessment Program (http://frap.cdf.ca.gov).

The built environment

The built environment itself can be a conduit for climate impacts. High percentages of impermeable surfaces like asphalt—which is commonly used in cities—increase surface temperatures, amplify heat waves, and reduce stormwater infiltration, contributing to potential flooding. Lack of energy-conservation standards increases the vulnerability of residents to heat waves. For example, Arizona, Colorado, and Utah have no statewide energy codes for building construction, deferring to the localities to develop their own. In contrast, California that has energy conservation regulations at the state level, continues to lead the nation in building-energy efficiency. State-level regulations create an even playing field, relieving localities from having to develop their own codes, and provide the technical expertise that can be required. Instead, currently there is a patchwork of different energy-conservation standards, and some states have none at all. New Mexico, Nevada, and California have statewide mandatory requirements to which specific cities have imposed additional requirements.[ix] Building-energy standards are a method to reduce the impact of extreme heat incidences.

In each of the cities, energy providers have instituted financial incentives for conservation and in some cases the use of alternative energies; city and county governments have also developed various types of regulatory frameworks to encourage efficiency and "green" building (Table 13.2) (see also Chapter 12, Section 12.3.5).

Cities in the Southwest also have very different patterns of infrastructure use and regulation. The percentage of each city's population using public transit, for example, ranges from 1.6% in Albuquerque to 6.2% in Los Angeles (Figure 13.13). According to the American Association of State Highway and Transportation Officials, light-duty trucks and automobiles contribute 16.5% of U.S. GHG emissions. Cities have done inventories on their own vehicle use and GHG emissions, but city-wide and county-wide GHG emissions are not available across the Southwest though under Senate Bill 375 in

Table 13.2 Energy-efficiency incentives and regulations in Southwestern cities

City	Financial Incentives	Rules, Regulations and Policies
Albuquerque	Green Building Incentive program	Energy conservation code
Denver	Energy-efficiency rebates from service provider	"Green building" requirement for city-owned buildings
Las Vegas	Energy-efficiency rebates from service provider	County-wide energy conservation code
Los Angeles	Renewable energy and energy-efficiency support and rebate programs from service provider	County-wide green building programs and LEED certification for public buildings
Phoenix	Renewable energy and energy-efficiency support and rebate programs from service provider	Design standards for city buildings; renewable energy portfolio goals
Salt Lake City	Renewable energy and energy-efficiency support and rebate programs from service provider	Green power purchasing by city; high-performance buildings requirement

Sources: City of Albuquerque (http://www.cabq.gov/); City and County of Denver (http:// www.denvergov.org/); City of Las Vegas (http://www.lasvegasnevada.gov/); City of Los Angeles (http://www.lacity.org/); City of Phoenix (http://phoenix.gov/); Salt Lake City (http://www.slgov.com/).

California, such inventories are being conducted. Still, fossil-fuel combustion generates GHGs and in regions where there is a greater reliance on fossil fuels and on single-occupancy vehicles, there will be more production of GHGs. While public transportation also emits GHGs, reducing overall vehicle miles traveled (as by single occupancy vehicles rather than multiple passenger public transportation) will reduce regional GHG emissions. (More information on passenger travel and emissions can be found in Chapter 14, Section 14.2.)

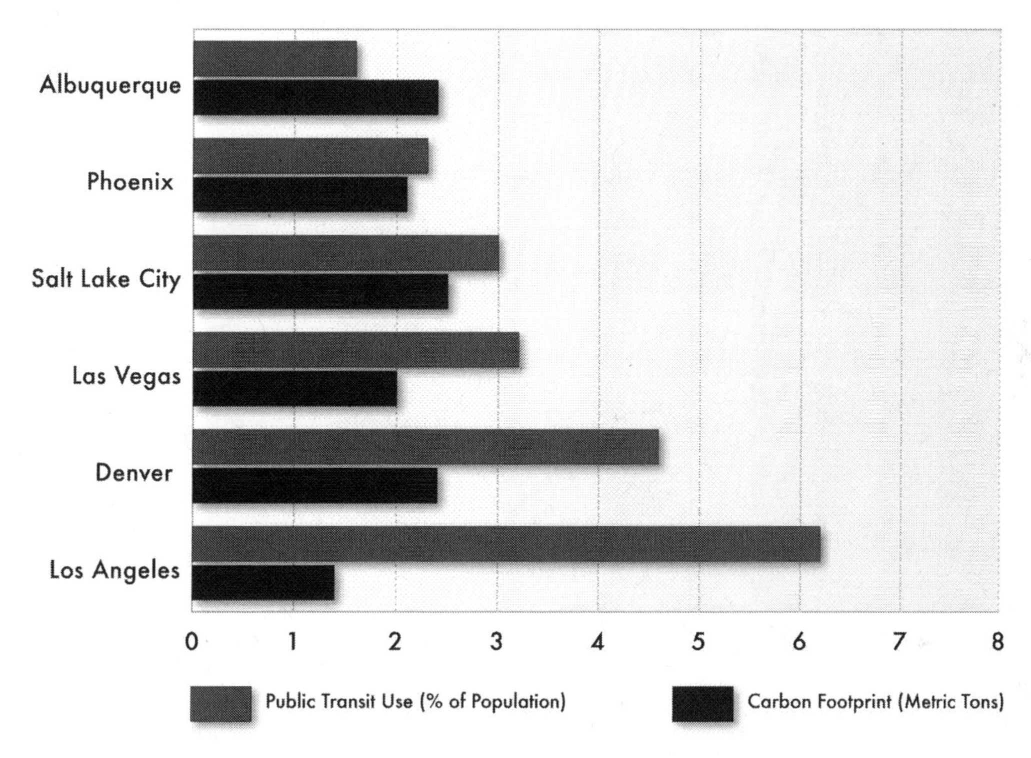

Figure 13.13 Public transit use and carbon footprints of Southwest cities. Source: U.S. Census Bureau American Community Survey 2009 (http://www.census.gov/acs/www/data_documentation/2009_release/), and Brown, Sarzinski and Southworth (2008).

Smart growth and new urbanism initiatives have been influential in the Southwest. The concept of smart growth is to place new development near existing urban infrastructure, especially transportation, and to make urban areas more compact. New urbanism promotes the creation and restoration of diverse, walkable, compact, and mixed-use communities based on specific design principles (Haas 2012). These ideas for new ways to build cities were inspired by concerns about health, walkability, and livability, as well as cost. Living near transit lines and in walkable neighborhoods has multiple benefits; it so happens that these benefits are lined up with the mitigation of climate change as well. One major initiative has been the California and Utah Regional Blueprint Planning Program, with the federal Environmental Protection Agency acting as a partner. In the Salt Lake City region, this effort, along with the infrastructure development for the 2002

Olympics, created the infrastructure and incentives for more compact development and development adjacent to already developed areas. For the major cities in California, the process has engaged thousands of residents in articulating a vision for the long-term future of their region and understanding the implications of different types of growth relative to impacts on land use, transportation, energy, and (increasingly) climate. This has also led to yearly *California Regional Progress* reports, available online at the California Department of Transportation website. The Phoenix, Mesa, and Valley Metro Rail systems are also receiving assistance from EPA to develop a regional strategy that will encourage compact, mixed-use, and transit-oriented development (see its Region 9 SmartGrowth web page[x]).

Climate change and urban water

In the Southwestern United States, potential impacts of climate change on water resources have been summarized by numerous researchers (e.g., Knowles and Cayan 2002; Miller, Bashford, and Strem 2003; Hayhoe et al. 2004; Mote et al. 2005; Cayan et al. 2010; MacDonald 2010; and Chapters 6 and 10 of this report). Most climate models indicate that the Southwest will become drier in the twenty-first century, and that there will be increased frequencies of extreme weather events, including drought, flooding, and heat waves (IPCC 2007). Increasing temperatures are expected to alter precipitation patterns (i.e. volume, frequency, and intensity) and correspondingly alter regional streamflow patterns. Increasing temperatures will impact urban populations in the Southwest; their impacts may be already felt in communities such as Phoenix, whose annual number of misery days (days where people feel strongly impacted by temperature and there can be adverse health impacts) has been increasing[xi] (Figure 13.14) (Ruddell et al. forthcoming). Annual minimum temperatures in all six of the Southwest cities considered here are also increasing (Figure 13.15). For further discussion on the effects of climate change on human health in urban areas, see Chapter 15.

Precipitation patterns in the Southwest are typically highly variable (Figure 13.16). Climate change is anticipated to make variable and extreme precipitation even more common and to result in changes in flood frequency and extreme runoff events (Lopez, Hogue, and Stein 2011). Highly structured and in-filled cities (cities that have high proportions of impermeable surfaces due to roads and buildings, little open space, and existing infrastructure such as for stormwater) have little capacity to adapt to increasing flows—for example by devoting existing open spaces to stormwater infiltration—and so may be especially vulnerable to extreme flooding. Enhanced, intensified water flows will increase the wash-off of suspended sediments and other pollutants, degrading water quality, as was the case in the Station Fire in Los Angeles mentioned above (Benitez-Gilabert, Alvarez-Cobelas, and Angeler 2010; Lopez, Hogue, and Stein in review). Altered flow regimes and degraded water quality also have significant implications for downstream ecosystems that receive polluted urban stormwater. There have been some initial efforts to restore such ecosystems. For example, since 1997, the Wetlands Recovery Project (WRP) in Southern California has invested over $500 million in the acquisition and restoration of coastal wetlands. Unfortunately, they are now at accelerated risk of degradation due to increased potential for increased high-precipitation events and fires. Wetlands in other parts of the Southwest are at similar risk of fire impacts on water quality.

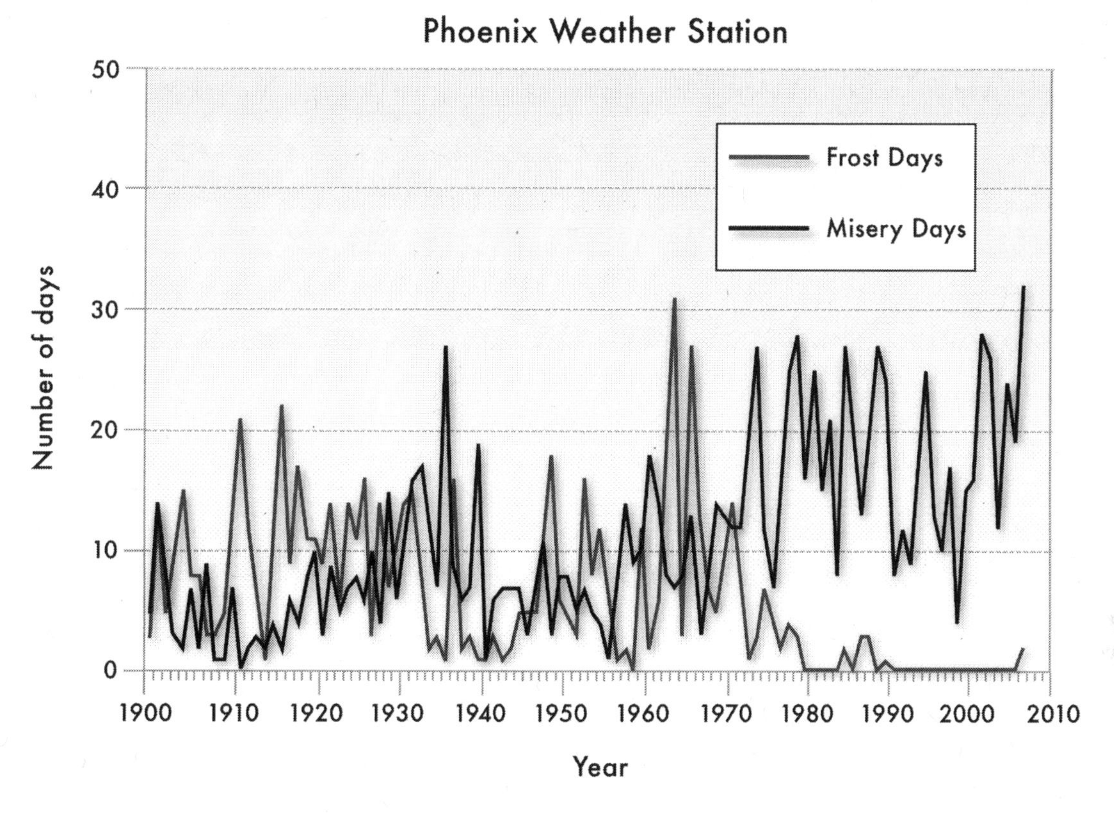

Figure 13.14 Number of misery and frost days in Phoenix, 1900–2010. Note the increase in misery days, and decrease in frost days during the last 40 years. Source: Ruddell et al. (2012).

Temperature increases will also impact vegetation across the Southwest, increasing evapotranspiration rates. If this holds true, water demand in urban ecosystems will increase. This is especially true in cities with extensive vegetation that is not climate-appropriate and has heavy water demand. For instance, the city of Los Angeles likely uses 40% to 60% of its residential water for outdoor and landscaping application (LADWP 2011), much of this going to non-native and non-climate-appropriate species such as turf grass. The trend toward urban tree planting to reduce the urban heat island effect and provide other benefits, as exemplified in the Los Angeles Million Tree Planting program, may also lead to unintended increased water demand (Pataki et al. 2011).[xii] However, landscaping and greenspace (protected and reserved areas of undeveloped land) are unequally distributed in many of the Southwest urban centers. Many residents will be limited in their capacity to plant and maintain greenspace relative to local resources and incomes, compromising their ability to mitigate increased temperatures and their associated energy needs. There are complex environmental justice implications in the distribution of greenspace as many low income neighborhoods suffer from lack of tree canopy and other vegetation and so are hotter (Grossman-Clarke et al. 2010; Chow, Chuang, and Gober 2012; Pincetl et al. 2012).

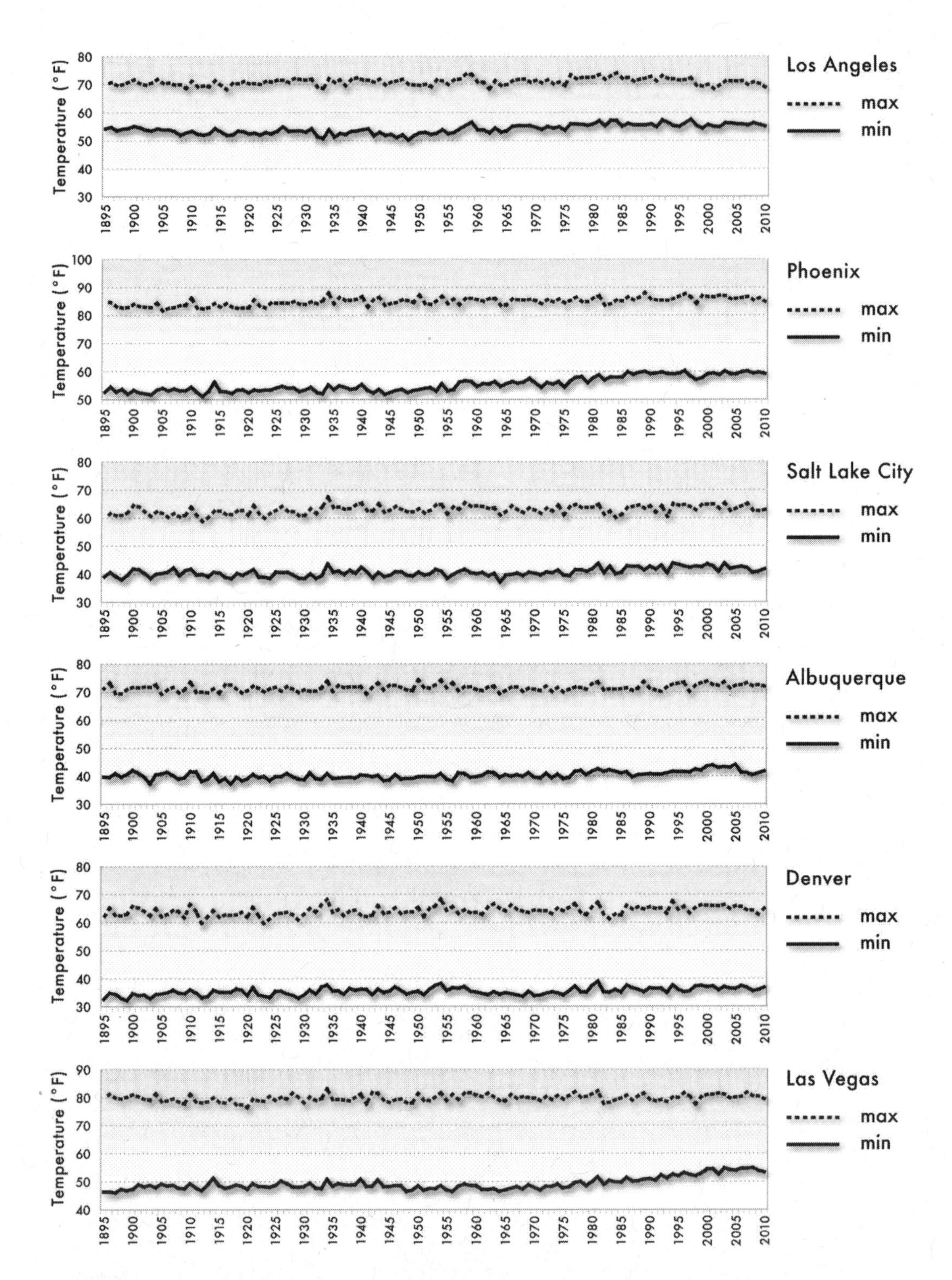

Figure 13.15 Average annual minimum temperatures for six Southwestern cities (in °F). Source: Prism Climate Group (2011), Oregon State University, http://prism.oregonstate.edu.

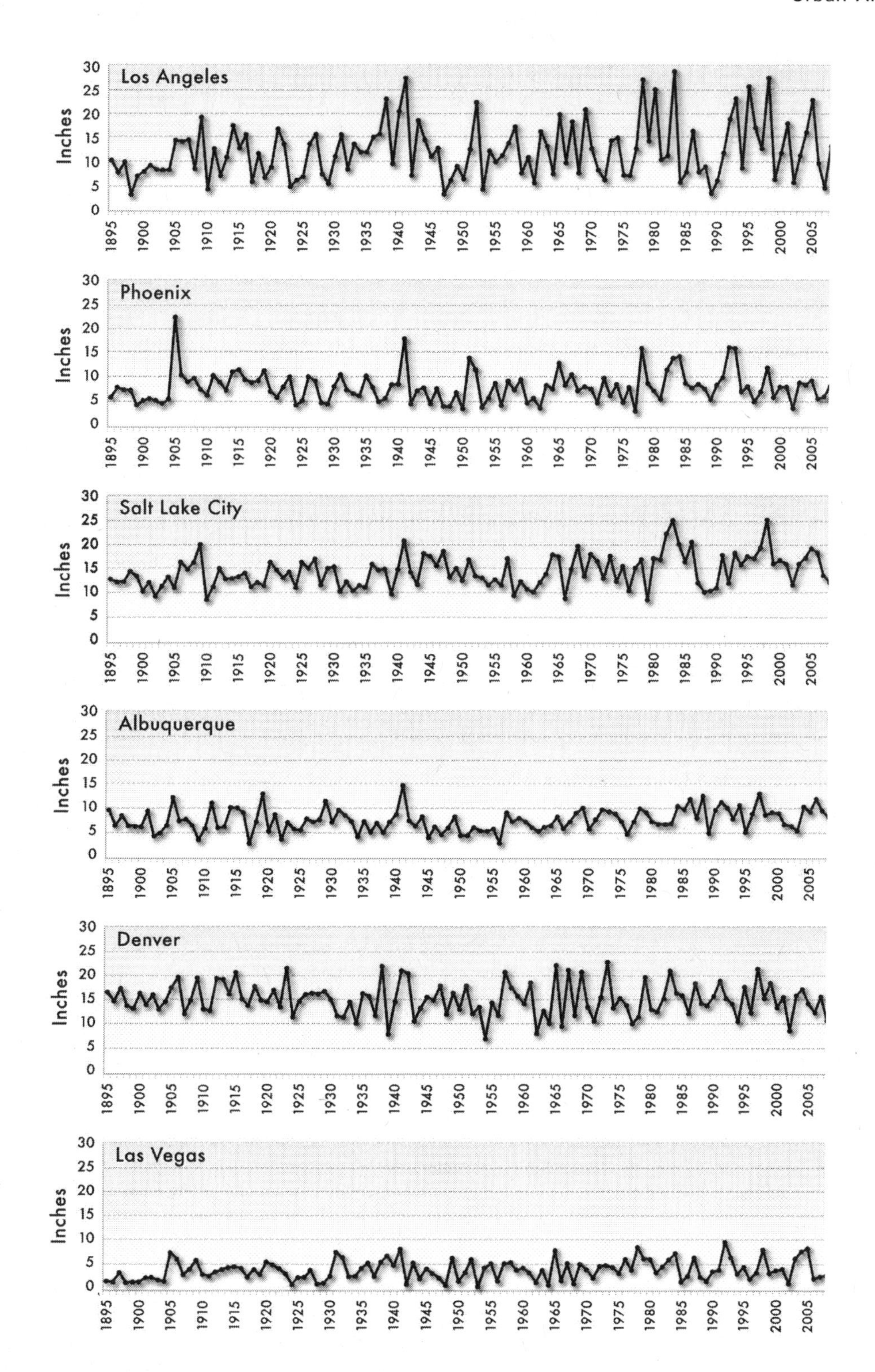

Figure 13.16 Average annual rainfall (in inches) in six Southwestern cities, 1895–2010.
Source: Prism Climate Group (2011), Oregon State University, http://prism.oregonstate.edu.

Sea-level rise in Southwest cities is a third consequence of rising temperatures. Its potential impacts are described further in Chapter 9. Observations over the last century show sea-level rise of about 8 inches (about 203 mm) along California's coast (see Chapter 9, section 9.2.2). Sea-level rise has a range of associated consequences, but a key concern for Southwest coastal cities, such as Los Angeles, is the salinization of groundwater and estuaries, which reduces freshwater availability (Bloetscher et al. 2010; Quevauviller 2011). Sea-level rise is expected to shift the fresh water/saline interface inland due to salt water intrusion, potentially contaminating groundwater supplies used in urban coastal cities. Ultimately, this may result in increased costs associated with infrastructure to segregate sea water from fresh water (these include barrier well injections[13] and/or the implementation of additional barrier systems) (Webb and Howard 2011).

Wastewater and stormwater infrastructure, as well as reclamation and recharge projects are also vulnerable to sea-level rise, especially in low-lying coastal regions (Bloetscher et al. 2010). Model simulations by Webb and Howard (2011) indicate that aquifers can take several centuries to gain equilibrium following a cessation in sea-level rise, largely dependent on the aquifer's properties.

Sea-level rise will likely affect private property and also infrastructure such as sewage treatment plants located along the coast, as well as roads and rail lines. A number of coastal cities and regions are beginning to prepare for the potential of this impact, integrating costs and plans in their capital improvement programs. The Port of Los Angeles is planning for adaptation to sea-level rise by seeking expert advice, contracting with Rand Corporation to identify key vulnerabilities and develop a set of general approaches based on alternative models. (See additional discussion in Chapter 9, Box 9.4). Such planning raises questions of jurisdictional responsibility, as infrastructures such as sewage treatment plants often serve regional cities. Land-use impacts may affect only some jurisdictions and not others, and funding can be complicated as well.

13.3 Critical Missing Data and Monitoring in Cities

For complex urban regions, where the majority of the Southwest population lives, greater inter-jurisdictional collaboration of data collection and planning will be required to allow cities to adapt to climate change and become more resilient. One such approach is the Los Angeles Regional Collaborative for Climate Action and Sustainability,[xiv] which operates at the level of Los Angeles County (the county encompasses eighty-eight cities, 10 million inhabitants and over 1,000 special districts). The goal of the collaborative is to develop a plan for climate action and sustainability that draws on the strengths of each member to build an integrated and coherent response to potential impacts of climate change. Best management practices across the region will be shared, as well as funding for programs and projects. Although few examples of this type of initiative currently exist, it represents a feasible strategy for Southwest cities to move adaptation forward. (See additional examples in Chapter 18.)

Quantification of energy flows into urban regions and pollution sinks is also required so that carbon-mitigation strategies can be based on rigorous data analyses that incorporate a life-cycle-based understanding of the generation of GHGs. Tracking of pollution sinks—such as methane from landfills or the deposition of air pollutants on soils that may then become captured in runoff—will help determine the impacts of urban systems. Such urban energy flow studies—known as urban metabolism—can improve our understanding of a host of climate-change vulnerabilities. A key need in conducting such analyses are geographically specific data, such as household, commercial, and industrial energy-use data, that are currently not available due to privacy concerns. Coupling household-level energy use with land-use data will reveal important aspects of urban activities and help identify mechanisms by which they may be improved.

Observations on land-atmosphere interactions in Southwest urban centers are also lacking. For example, while local temperature data are generally available from government agencies, the scattered distribution of temperature gauges often do not accurately reflect the spectrum of microclimates in an urban area. There is also a need for more measurements of atmospheric CO_2 and urban energy fluxes, as discussed, and for distributed runoff data. Other key data that would be useful in evaluating urban water consumption patterns and trade-offs are vegetation type and species used in urban areas and separate metering for indoor and outdoor water use. Air-quality monitoring is equally sparse and often limited to criteria pollutants. GHG monitoring is nonexistent, as are comprehensive data on the health impacts of heat waves and air-quality deterioration.

Cities in the Southwest of the United States have unique but regionally shared characteristics. They are artifacts of federal policy, including protection of the public lands that surround many of them. Federal land, transportation, and water policies have shaped the urban form, creating a dense urban sprawl characterized by thirsty, climate-inappropriate vegetation in urban areas that are dependent on inexpensive and abundant fuels and water. The impacts of climate change on these arid cities are largely centered on probable water scarcities over the course of the twenty-first century, punctuated by extreme weather events that will bring flooding, fires, and extreme heat events. Fortunately, water consumption per capita is beginning to decrease in most western states (Cohen 2011). In some states there is increased investment in and ridership of public transportation, offering the hope that transportation-related GHG emissions will decrease. Measures such as capturing landfill GHGs are advancing as well, and numerous cities and towns are planning, directly or indirectly, for climate change impacts, including requiring buildings to become more energy-efficient.

The fragmented governance of these cities makes coordinated and integrated programs and responses difficult, and budgetary constraints are significant. Governance experiments such as the Los Angeles Regional Collaborative for Climate Action and Sustainability provide a vision of a process that could effectively coordinate climate responses in fragmented Southwest cities. Governance and fiscal capacity will play an important role in the ability of regions to adapt.

References

Allen, L., F. Lindberg, and C. S. B. Grimmond. 2010. Global to city scale urban anthropogenic heat flux: Model and variability. *International Journal of Climatology* 31:1990–2005.

Aubinet, M., A. Grelle, A. Ibrom, U. Rannik, J. Moncrieff, T. Foken, A. S. Kowalski, et al. 2000. Estimates of the annual net carbon and water exchange of forests: The EUROFLUX methodology. *Advances in Ecological Research* 30:113–175.

Aubinet, M.,Vesala, T., and D. Papale, eds. 2012. *Eddy covariance: A practical guide to measurement and data analysis*. Dordrecht: Springer.

Baldocchi, D. 2003. Assessing the eddy covariance technique for evaluating carbon dioxide exchange rates of ecosystems: Past, present and future. *Global Change Biology* 9:479–492.

Benitez-Gilabert, M., M. Alvarez-Cobelas, and D. G. Angeler. 2010. Effects of climatic change on stream water quality in Spain. *Climatic Change* 103:339–352.

Bierwagen, B. G., D. M. Theobald, C. R. Pyke, A. Choate, P. Groth, J. V. Thomas, and P. Morefield. 2010. National housing and impervious surface scenarios for integrated climate impact assessments. *Proceedings of the National Academy of Sciences* 107:20887–20892.

Blanco, H., P. McCarney, S. Parnell, M. Schmidt, and K. C. Seto. 2011. The role of urban land in climate change. In *Climate change and cities: First assessment report of the Urban Climate Change Research Network*, ed. C. Rosenzweig, W. D. Solecki, S. A. Hammer, and S. Mehrotra, 217–248. Cambridge: Cambridge University Press.

Bloetscher, F., D. E. Meeroff, B. N. Heimlich, A. R. Brown, D. Bayler, and M. Loucraft. 2010. Improving resilience against the effects of climate change. *Journal of the American Water Works Association* 102:36–46.

Brown, M. A., A. Sarzinski, and F. Southworth. 2008. *Shrinking the carbon footprint of metropolitan America*. Washington, DC: Brookings Institution.

Burke, M., T. S. Hogue, J. Barco, C. Wessel, A. Kinoshita, and E. Stein. 2011. Dynamics of pre- and post-fire pollutant loads in an urban fringe watershed. In *Southern California Coastal Water Research Project 2011 annual report*, ed. K. Schiff and K. Miller, 61–70. Costa Mesa, CA: Southern California Coastal Water Research Project. ftp://ftp.sccwrp.org/pub/download/DOCUMENTS/AnnualReports/2011AnnualReport/ar11_061_070.pdf.

Cayan, D. R., T. Das, D. W. Pierce, T. P. Barnett, M. Tyree, and A. Gershunov. 2010. Future dryness in the southwest U.S. and the hydrology of the early 21st century drought. *Proceedings of the National Academy of Sciences* 107:21271–21276.

Chester, M., E. Martin, and N. Sathaye. 2008. Energy, greenhouse gas, and cost reductions for municipal recycling systems. *Environmental Science and Technology* 42:2142–2149, doi:10.1021/es0713330.

Chester, M., S. Pincetl, and B. Allenby. 2012. Avoiding unintended tradeoffs by integrating life-cycle impact assessment with urban metabolism. *Current Opinion in Environmental Sustainability* 4:451–457.

Chow, W. T., W. Chuang, and P. Gober. 2012. Vulnerability to extreme heat in metropolitan Phoenix: Spatial, temporal and demographic dimensions. *The Professional Geographer* 64:286–302.

Cohen, M. J. 2011. *Municipal deliveries of Colorado River Basin water*. Oakland, CA: Pacific Institute.

Crawford, B., C. S. B. Grimmond, and A. Christen. 2011. Five years of carbon dioxide fluxes measurements in a highly vegetated suburban area. *Atmospheric Environment* 45:896–905.

Davis, S. J., and K. Caldeira, 2010. Consumption-based accounting of CO_2 emissions. *Proceedings of the National Academy of Sciences* 107:5687–5692.

Das, T., M. D. Dettinger, D. R. Cayan, and H. G. Hidalgo. 2011. Potential increase in floods in California's Sierra Nevada under future climate projections. *Climatic Change* 109 (Suppl. 1): S71–S94, doi:10.1007/s10584-011-0298-z.

Feigenwinter, C., R. Vogt, and A. Christen. 2012. Eddy covariance measurements over urban areas. In *Eddy covariance: A practical guide to measurement and data analysis,* ed. M. Aubinet, T. Vesala, and D. Papale. Dordrecht: Springer.

Flanner, M. G. 2009. Integrating anthropogenic heat flux with global climate models. *Geophysical Research Letters* 36: L02801.

Eidlin, E. 2010. What density does not tell us about sprawl. *Access* 37 (Fall 2010): 2–9.

Galloway, G. E., J. J. Boland, R. J. Burby, C. B. Groves, S. L. Longville, L. E. Link, Jr., J. F. Mount, et al. 2007. *A California challenge: Flooding in the Central Valley; A report from an independent review panel to the Department of Water Resources, State of California.* http://www.water.ca.gov/news/newsreleases/2008/101507challenge.pdf.

Great Western Institute. 2010. *SWSI conservation levels analysis final report*. Prepared for the Colorado Water Conservation Board.

Green, D. 2007. *Managing water: Avoiding crisis in California*. Berkeley: University of California Press.

Grossman-Clarke, S., J. A. Zehnder, W. L. Stefanov, Y. Liu, and M. A. Zoldak. 2005. Urban modifications in a mesoscale meteorological model and the effects on near-surface variables in an arid metropolitan region. *Journal of Applied Meteorology* 44:1281–1297.

Grossman-Clarke, S., J. A. Zehnder, T. Loridan and S. C. Grimmond. 2010. Contribution of land use changes to near surface air temperatures during recent summer extreme heat events in the Phoenix metropolitan area. *Journal of Applied Meteorology and Climatology* 49:1649–1664.

Gutzler, D., and T. Robbins. 2011. Climate variability and projected change in the western United States: Regional downscaling and drought statistics. *Climate Dynamics* 37:835–839.

Haas, T., ed. 2012. *Sustainable urbanism and beyond*. New York: Rizzoli.

Hayhoe, K., D. Cayan, C. B. Field, P. C. Frumhoff, E. P. Maurer, N. L. Miller, S. C. Moser, et al. 2004. Emissions pathways, climate change, and impacts on California. *Proceedings of the National Academy of Sciences* 101:12422–12427, doi:10.1073/pnas.0404500101.

Hise, G. 1997. *Magnetic Los Angeles: Planning the twentieth-century metropolis*. Baltimore, MD: Johns Hopkins University Press.

Hundley, N. 2001. *The great thirst: Californians and water; A history*. Berkeley: University of California Press.

Idso, C. D., S. B. Idso, and R. C. Balling, Jr. 2001. An intensive two-week study of an urban CO_2 dome in Phoenix, Arizona, USA. *Atmospheric Environment* 35:995–1000.

Intergovernmental Panel on Climate Change (IPCC). 2007. *Climate change 2007: The physical science basis. Contribution of Working Group I to the Fourth Assessment Report of the Intergovernmental Panel on Climate Change,* ed. S. Solomon, D. Qin, M. Manning, Z. Chen, M. Marquis, K.B. Averyt, M. Tignor and H.L. Miller. Cambridge: Cambridge University Press.

Jacobson, M. Z. 2010a. Enhancement of local air pollution by urban CO_2 domes. *Environmental Science & Technology* 44:2497–2502.

—. 2010b. Short-term effects of controlling fossil-fuel soot, biofuel soot and gases, and methane on climate, Arctic ice and air pollution health. *Journal of Geophysical Research* 115: D14209, doi:10.1029/2009JD013795.

Jacobson, M. Z., and D. G. Street. 2009. Influence of future anthropogenic emissions on climate, natural emissions, and air quality. *Journal of Geophysical Research* 114: D08118, doi:10.1029/2008JD011476.

Knowles, N., and D. R. Cayan. 2002. Potential effects of global warming on the Sacramento/San Joaquin watershed and the San Francisco estuary. *Geophysical Research Letters* 29:1891, doi:10.1029/2001GL014339.

Kennedy C., J. Cuddihy, and J. Engel-Yan. 2007. The changing metabolism of cities. *Journal of Industrial Ecology* 11:43–59, doi:10.1162/jie.2007.1107.

Kennedy, C., S. Pincetl, and P. Bunje. 2010. The study of urban metabolism and its applications to urban planning and design. *Environmental Pollution* 159:1965–1973, doi:10.1016/j.envpol.2010.10.022.

Kupel, D. E. 2003. *Fuel for growth: Water and Arizona's urban environment*. Tucson: University of Arizona Press.

Los Angeles Department of Water and Power (LADWP). 2011. Draft 2010 Urban Water Management Plan. Los Angeles: LADWP. http://www.ladwp.com.

Lopez, S. R, T. S. Hogue, and E. Stein. In review. A framework for evaluating regional hydrologic sensitivity to climate change using archetypal watershed modeling. *Hydrology and Earth System Science*.

Lowitt, R. 1984. *The New Deal and the West*. Norman: University of Oklahoma Press.

MacDonald, G. M. 2010. Water, climate change, and sustainability in the Southwest. *Proceedings of the National Academy of Sciences* 107:21256–21262.

McCarthy, M. P., M. J. Best, and R. A. Betts. 2010. Climate change in cities due to global warming and urban effects. *Geophysical Research Letters* 37: L09705, doi:1029/2010GL042845.

McCarthy, M. P., C. Harpham, C. M. Goodess, and P. D. Jones. 2012. Simulating climate change in UK cities using a regional climate model, HadRM3. *International Journal of Climatology* 32:1875–1888, doi:10.1002/joc.2402.

Matonse, A. H., D. C. Pierson, A. Frei, M. S. Zion, E. M. Schneiderman, R. Mukundan, and S. M. Pradhanang. 2011. Effects of changes in snow pattern and the timing of runoff on NYC water supply system. *Hydrological Processes* 25:3278–3288.

Miller, N. L., K. E. Bashford, and E. Strem. 2003. Potential impacts of climate change on California hydrology. *Journal of the American Water Resources Association* 39:771–784, doi:10.1111/j.1752-1688.2003.tb04404.x.

Mishra, V., and D. P. Lettenmaier. 2011. Climatic trends in major U.S. urban areas, 1950–2009. *Geophysical Research Letters* 38: L16401, doi:10.1029/2011GL048255.

Mote, P. W., A. F. Hamlet, M. P. Clark, and D. P. Lettenmaier. 2005. Declining mountain snowpack in western North America. *Bulletin of the American Meteorological Society* 86:39–49, doi:10.1175/BAMS-86-1-39.

Nakićenović, N., and R. Swart, eds. 2000. *Special report on emissions scenarios: A special report of Working Group III of the Intergovernmental Panel on Climate Change*. Cambridge: Cambridge University Press.

Nash, G. D. 1985. *The American West transformed: The impact of the Second World War*. Lincoln: University of Nebraska Press.

Pataki, D. E., D. R. Bowling, and J. R. Ehleringer. 2003. Seasonal cycle of carbon dioxide and its isotopic composition in an urban atmosphere: Anthropogenic and biogenic effects. *Journal of Geophysical Research* 108:4735, doi:10.1029/2003JD003865.

Pataki, D. E., P. C. Emmi, C. B. Forster, J. I. Mills, E. R. Pardyjak, T. R. Peterson, J. D. Thompson, and E. Dudley-Murphy. 2009. An integrated approach to improving fossil fuel emissions scenarios with urban ecosystem studies. *Ecological Complexity* 6:1–14.

Pataki, D., H. McCarthy, E. Litvak, and S. Pincetl. 2011. Transpiration of urban forests in the Los Angeles metropolitan area. *Ecological Applications* 21:661–677.

Peters, G. P., and E. G. Hertwich. 2008. CO_2 embodied in international trade with implications for global climate policy. *Environmental Science & Technology* 42:1401–1407.

Pincetl, S. 1999. *Transforming California: The political history of land use in the state*. Baltimore, MD: Johns Hopkins University Press.

Pincetl, S., P. E. Bunje, and T. Holmes. 2012. An expanded urban metabolism method: Towards a systems approach for assessing urban energy processes and causes. *Land Use and Urban Planning* 107:193–202, doi:10.1016/j.landurbplan.2012.06.006.

Pincetl, S., T. Gillespie, D. E. Pataki, S. Saatchi, and J. D. Saphores, 2012. Urban tree planting programs, function or fashion? Los Angeles and urban tree planting campaigns. *GeoJournal*, published online, doi:10.1007/s10708-012-9446-x.

Pincetl S., P. W. Rundel, J. Clark De Blasio, D. Silver, T. Scott, and R. Halsey. 2008. It's the land use, not the fuels: Fires and land development in Southern California. *Real Estate Review* 37:25–42.

Quevauviller, P. 2011. Adapting to climate change: Reducing water-related risks in Europe—EU policy and research considerations. *Environmental Science and Policy* 14:722–729.

Ramamurthy, P., and E. R. Pardyjak. 2011. Toward understanding the behavior of carbon dioxide and surface energy fluxes in the urbanized semi-arid Salt Lake Valley, Utah, USA. *Atmospheric Environment* 45:73–84.

Ruddell, D., D. Hoffman, O. Ahmad, and A. Brazel. Forthcoming. An analysis of historical threshold temperatures for Phoenix (urban) and Gila Bend (desert). *Climate Research* 54.

Sailor D. J., and L. Lu. 2004. A top-down methodology for developing diurnal and seasonal anthropogenic heating profiles for urban areas. *Atmospheric Environment* 38:2737–2748.

Salt Lake City Department of Public Utilities (SLCDPU). 2010. *2009 Water master conservation plan.* Salt Lake City: SLCDPU.

U.S. Environmental Protection Agency (EPA). 2009. *Assessment of the impacts of global change on regional U.S. air quality: A synthesis of climate change impacts on ground-level ozone; An interim report of the U.S. EPA Global Change Research Program.* EPA/600/R-07/094F.

Velasco E., and M. Roth. 2010. Cities as net sources of CO_2: Review of atmospheric CO_2 exchange in urban environments measured by eddy covariance technique. *Geography Compass* 4:1238–1259.

Webb, M. D., and K. W. F. Howard. 2011. Modeling the transient response of saline intrusion to rising sea-levels. Ground Water 49:560–569.

Zhao, Z., S. Chen, M. J. Kleeman, and A. Mahmud. 2011. The impact of climate change on air quality related meteorological conditions in California – Part II: Present versus future time simulation analysis. *Journal of Climate* 13:3362–3376.

Endnotes

i An urban metabolism refers to the total urban systems flows of materials, energy and inputs, and outputs in the form of waste. Supply chains are components of the urban metabolism.

ii Urban heat island effect was defined as "the relative warmth of a city compared with surrounding rural areas, associated with changes in runoff, the concrete jungle effects on heat retention, changes in surface albedo, changes in pollution and aerosols, and so on" by the IPCC (2007).

iii See http://www.water.ca.gov/floodsafe/.

iv The EC method is a widely used micrometeorological technique designed to measure turbulent exchanges of mass, momentum, and heat between an underlying surface and the atmosphere (see Aubinet, Vesala and Papale 2012 and references within). For CO_2 exchange, rapid measurements of vertical velocity fluctuations and CO_2 mixing ratio are made on a tower well above the buildings and trees of an urban surface in the so-called constant flux layer. From these quantities, a covariance is computed (Baldocchi 2003). If appropriate assumptions are satisfied, the covariance is a measure of the net differences between the uptake of CO_2 by photosynthesis and the emission of CO_2 by anthropogenic and biological processes.

v Difficulties are both practical and technical. Practical difficulties include funding for such equipment as flux towers, their siting in urban areas, and funds to conduct the monitoring and data analysis. Additional technical difficulties exist related to quantifying important contributions to fluxes, such as those related to complex distributions of sources and sinks and their relationship to advection and non-homogeneous surfaces that are common in urban areas (Feigenwinter, Vogt, and Christen 2012).

vi A *source* is a process or activity through which a greenhouse gas is released into the atmosphere. A *sink* is something that acts as a reservoir to absorb it on a short- or long-term basis.

vii See http://www.urban-climate.org.

viii CO_2 concentrations can cause higher levels of PM 2.5 by increasing vapor pressures in some locations (Jacobson 2010b).

ix See http://www.energycodes.gov/states/state (U.S. Department of Energy's Building Energy Codes Program).

x See http://www.epa.gov/region9/climatechange/smart-growth.html.

xi Misery days are days when the temperature maximum is greater than or equal to 110°F or when the temperature minimum is less than 32°F.

xii Stomata, the microscopic pores on the leaves and stems of plants, are the means by which plants transpire, or lose water vapor to the atmosphere. Although there is some debate on plant stomatal response to increasing temperatures, a significant body of research indicates that evapotranspiration rates (the combination of evaporation and transpiration) may increase (Gutzler and Robbins 2011; Matonse et al. 2011; Lopez, Hogue, and Stein in review).

xiii A barrier well intrusion barrier is a well used to inject water into a fresh water aquifer to prevent the intrusion of salt water.

xiv See http://www.environment.ucla.edu/larc/.

Chapter 14

Transportation

COORDINATING LEAD AUTHOR

Deb A. Niemeier (University of California, Davis)

LEAD AUTHORS

Anne V. Goodchild (University of Washington), Maura Rowell (University of Washington), Joan L. Walker (University of California, Berkeley), Jane Lin (University of Illinois, Chicago), Lisa Schweitzer (University of Southern California)

REVIEW EDITOR

Joseph L. Schofer (Northwestern University)

Executive Summary

The Southwest transportation network includes major freeways, rail corridors of national importance, and major port- and border-crossing facilities. Recent passenger-travel trends suggest that vehicle ownership and per capita vehicle miles traveled (VMT) may have stabilized across the Southwest, which may be partly attributed to the economic recession as well as transportation planning strategies such as pricing, transit service improvements, managed lanes, and changes in land-use configurations. However, the Southwest appears poised to show gains in rail-freight traffic due to imports of foreign products, often in containerized cargo or bulk materials.

The following key messages highlight major climate issues facing the Southwest transportation sector:

- Many transportation infrastructure projects, currently in planning, design, or construction, do not necessarily address the potential effects of climate change. As climate change effects begin to manifest, design and operational vulnerabilities of these transportation system elements will appear. (high confidence)

Chapter citation: Niemeier, D. A., A. V. Goodchild, M. Rowell, J. L. Walker, J. Lin, and L. Schweitzer. 2013. "Transportation." In *Assessment of Climate Change in the Southwest United States: A Report Prepared for the National Climate Assessment*, edited by G. Garfin, A. Jardine, R. Merideth, M. Black, and S. LeRoy, 297–311. A report by the Southwest Climate Alliance. Washington, DC: Island Press.

- Alternative-fuel vehicle sales steadily increased throughout the Southwest until 2008. Yet, hybrid and alternative-fuel vehicles constitute less than 5% of the total passenger vehicle fleet in the Southwest. Increased heat events, which are confidently projected for the region, may increase vehicle air-conditioner usage and emissions and decrease fuel economy. (high confidence)
- The seaports of Los Angeles and Long Beach comprise the largest port complex in the United States and handle 45% to 50% of the containers shipped into the United States. Direct impacts of projected climate changes (such as sea-level rise and flooding) to California ports, include more frequent dredging of harbors and channels, realignments of port infrastructure—such as, jetties, docks, and berths—relative to rising waterline. (medium-high confidence)
- Extreme heat events, projected to increase during the course of the next 100 years, can shorten the life of pavements. Roadway deterioration will have an impact on all trade—including local trade circulation—that occurs between the Southwest and other U.S. regions, and trade between the Southwest and Mexico. (medium-low confidence)
- Increased precipitation intensity, which some studies project for the Southwest region, is associated with reductions in traffic safety, decreases in traffic efficiency—such as speed and roadway capacity—and increases in traffic accidents. (medium-low confidence)

14.1 Introduction

The transportation system in the Southwest comprises a number of major freeways, more than 514,000 lane-miles of rural roads, and more than 350,000 lane-miles of urban roads (FHWA 2011). Rail corridors of national importance and major port and border-crossing facilities also serve the region. Recent national statistics show about 484,000 million vehicle miles traveled (VMT) in the Southwest in 2008, roughly 16% of the national total (BTS 2008). After a number of years in which per capita VMT increased rapidly throughout the United States, per capita passenger VMT in the Southwest tended to be relatively stable or even declined during the late 1990s. Yet, in certain parts of the Southwest total VMT continued to increase.

Increased transportation activity combined with an expanding economy until about 2007 and increased electricity generation significantly contributed to the long-term rise in total CO_2 emissions generated by fossil-fuel combustion. In 2009, transportation uses accounted for about one-third of the total CO_2 emissions generated by fossil fuels (EPA 2012). California's transportation-related CO_2 emissions, which are higher on average than most states, were close to 40% of the state's total CO_2 emissions (California Air Resources Board [CARB] 2008), while Colorado's transportation-related emissions account for about 24% of the state's greenhouse gas (GHG) emissions (Climate Action Panel 2007). Despite increased numbers of "clean" vehicles and reduced tailpipe emissions of traditionally regulated pollutants, the proportion of total GHG from transportation increased slightly from 29.1% in 1990 to 31.2% in 2009 (EPA 2012). This may be attributable to increased VMT.

This chapter begins by describing current trends in passenger and freight transportation in the Southwest. The chapter then reviews the potential effects that climate change may have on transportation infrastructure, on the movement of passengers and goods, and on the risks to infrastructure integrity. A concluding overview examines the uncertainties associated with estimating future climate impacts and how these uncertainties, coupled with the timescales upon which infrastructure decisions normally are made, complicate adaptation planning and management.

14.2 Passenger Transportation Trends in the Southwest

While the Southwest states vary in their approaches to reducing GHG, all rely on a similar suite of options that include increased use of cleaner and more efficient vehicle technologies, new incentives to encourage people to change their travel behavior, and cleaner burning fuels. The U.S. Environmental Protection Agency (EPA) sets emissions standards for motorized vehicles nationally; however, the state of California has passed its own legislation regulating vehicle GHG emissions. The California standards are stricter than the national standards and were subsequently adopted by Arizona and New Mexico.[i] There also have been changes in vehicle fleet composition over time.

The success of hybrid-electric and alternative-fuel vehicles has been notable in the last decade. Not unexpectedly, California has led the way in terms of sales: one in four hybrid vehicles sold nationwide between 2003 and 2007 were purchased in California (Figure 14.1). Alternative-fuel vehicle sales steadily increased throughout the Southwest until 2008. While electric vehicles comprise a quarter of the total alternative-fuel vehicles registered in California, their share remains negligible in other Southwest states, where cars using an ethanol-fuel blend tend to dominate the alternative-fuel vehicle market. Although these figures are encouraging, hybrid and alternative-fuel vehicles constitute less than 5% of the total passenger vehicle fleet in the Southwest.

As fuel efficiency rises, the cost of driving declines, which historically has increased travel. Recent trends, however, suggest that vehicle ownership and per capita VMT may have stabilized across the Southwest (Figure 14.2), likely aided by the economic recession but also helped by transportation planning strategies such as pricing, transit service improvements, managed lanes, and changes in land-use configurations. Drops seen in the late 2000s in registered new hybrid vehicles, vehicles owned per capita, and vehicle miles traveled per capita are likely to be largely due to the effects of economic recession.

14.3 Freight Movement in the Southwest

Freight transportation includes both pick-up and delivery services and the movement of goods into and out of a region. Pick-up and delivery services include package-delivery services, such as UPS and Federal Express, as well as waste and recycling pick-up. Over the past thirty years, increased use of lean supply chains, "just-in-time" manufacturing, and Internet shopping has increased the demand for this sector. Broadly speaking, truck delivery is generally more efficient with respect to VMT and CO_2 emissions than having shoppers make individual trips to commercial centers. Nationally, pickup and delivery freight is expected to grow with increased use of delivery services (Golob and Regan 2001).

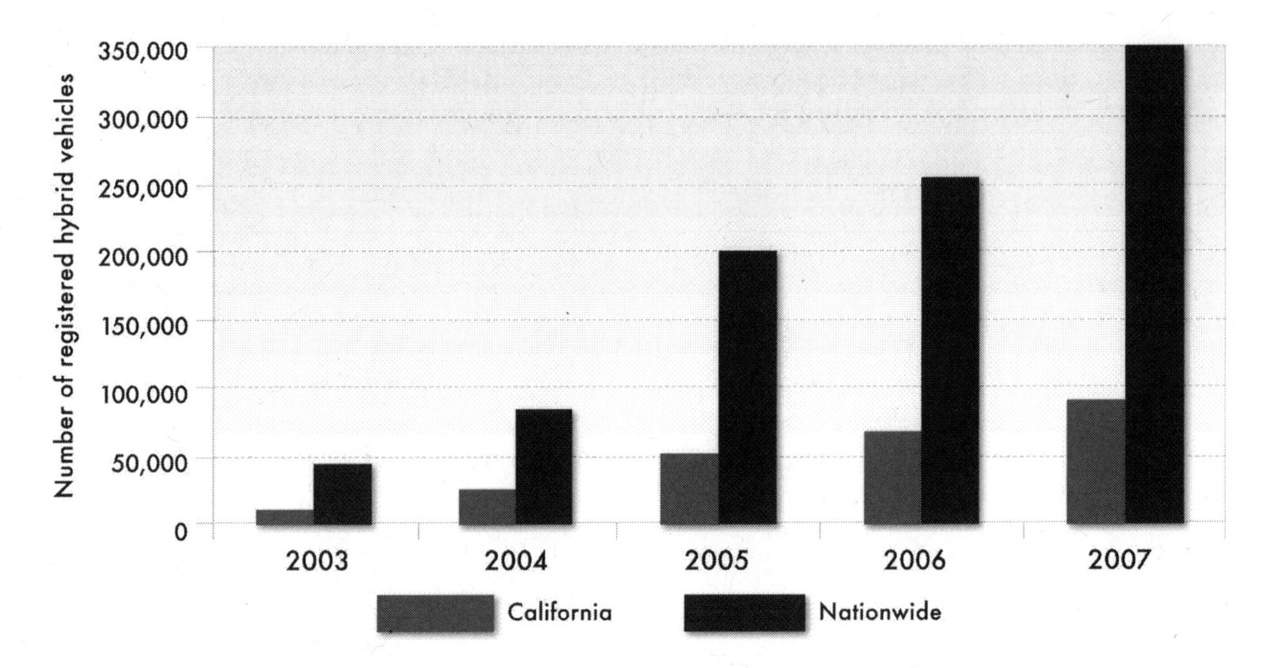

Figure 14.1 Number of new registered hybrid vehicles in California and throughout the United States. Source: RITA (2008); state transportation statistics.

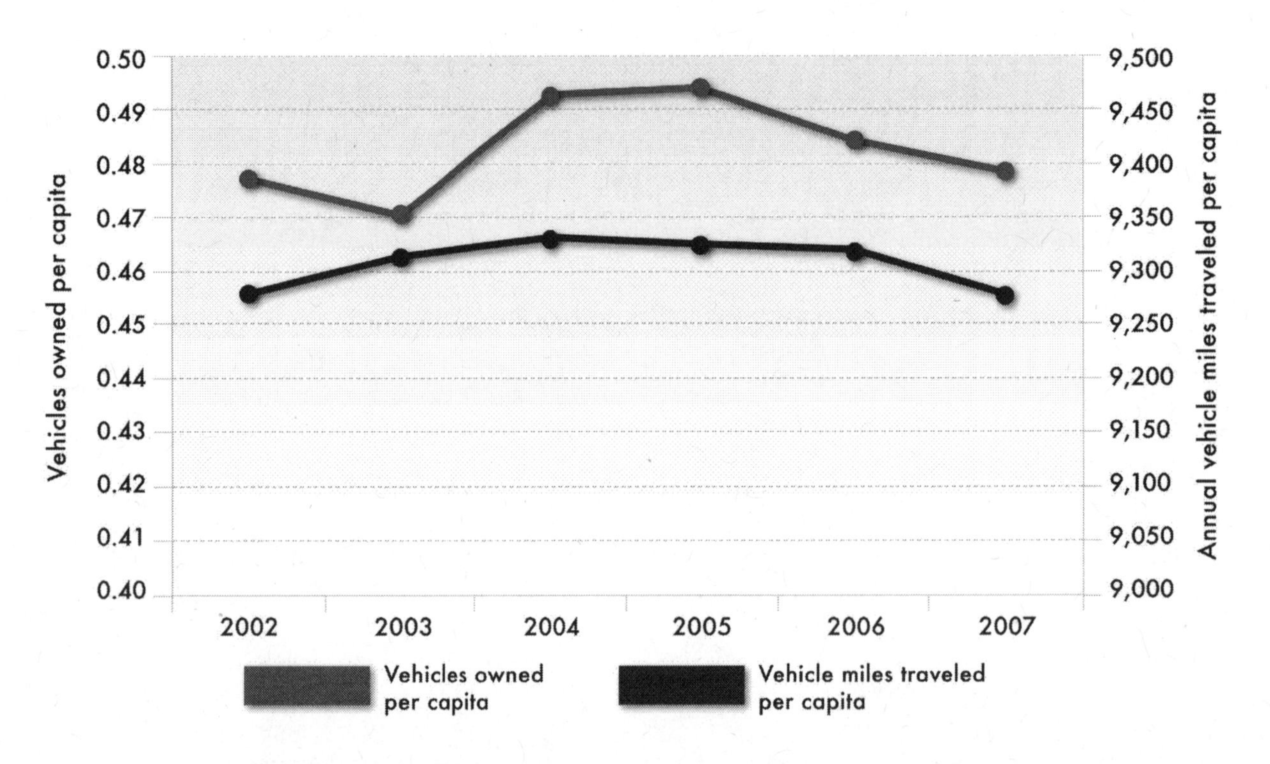

Figure 14.2 Per capita vehicle ownership and annual vehicle miles traveled in the Southwest. Source: RITA (2008); state transportation statistics.

Transportation services distribute Southwest-produced agricultural, commodity, and manufactured goods across and outside the region. Freight export volumes moved by the trucking sector have stayed reasonably constant over the last three decades. Economic conditions currently suggest that export cargo volumes will increase (WTO 2011), which may in turn increase use of rail for cargo, particularly for non-time-sensitive items such as empty containers, waste, and recyclable materials.

The Southwest is poised to show gains in rail-freight traffic due to containerized and bulk foreign imports. The volume of this cargo grew dramatically between 1990 and 2008 (WTO 2011). In 2010, the value of goods imported by the Port of Los Angeles was estimated at $293.1 billion, compared with $32.7 billion in goods imported by land into California from Mexico the same year.

Foreign imports are typically transported from a seaport or across a land border to an intermodal terminal, handling facility, or distribution center from which the goods are then distributed throughout the United States. The seaports of Los Angeles and Long Beach comprise the largest port complex in the United States and handle 45% to 50% of the containers shipped into the United States. Their regional and national importance is illustrated by the 2002 lockout at the Port of Los Angeles, which is estimated to have cost the U.S. economy $1 billion per day (Cohen 2002). Of the containers unloaded at the Port of Los Angeles, 77% leave California; roughly half of those leave by rail and half by truck transport (Heberger et al. 2009). As both fuel prices and the cost of CO_2 emissions rise, a propensity for using rail is likely to emerge (Siikavirta et. al. 2008; TEMS 2008). However, diversion of large amounts of cargo from trucks to rail is not likely to happen in the immediate future due to railway congestion and the mature state of freight movements via truck.

The Southwest also trades goods within the United States. Domestic freight uses the same transportation network as international freight and is subject to the same surface transportation rates and policies. Domestic freight is also intertwined with foreign trade in that many of the raw materials and equipment needed in domestic production are imported from other countries.

14.4 Impacts of Climate Change

Climate effects will vary by location within the Southwest. Sea-level rise is expected to be a significant issue for California, for example, while potential changes in temperature and precipitation would pose significant challenges for Arizona and Nevada. The force of these effects will be highly variable, but nonetheless will result in significant costs to infrastructure (Cambridge Systematics 2009). This section reviews the types of direct and indirect impacts to transportation services that are likely to emerge as a result of sea-level rise, extreme heat events, and increased precipitation intensity.

Direct impacts

FLOODING. Flooding of coastal infrastructure, coupled with increased intensity of storm events and land subsidence, poses the greatest potential threat to surface transportation systems in California (NRC 2008). Without the adoption of adaptive measures, a sea-level rise as great as 4.6 feet (1.4 meters, as projected in the high-emissions scenario; see also Chapter 9) would expose California's transportation infrastructure to the flooding of nearly 3,500 miles of roadways and 280 miles of rail lines (Heberger et al. 2009). The rate at which sea-level rise is projected to increase represents one of the "most troublesome aspects of projected climate change" (Knowles et al. 2009, 1).

Coastal regions of California bear the majority of this risk, with vulnerability split roughly equally between the San Francisco Bay Area and the Pacific Coast (see Figure 14.3). Among the areas affected, communities of color, low-income populations, and critical safety, energy, and public health infrastructure would be disproportionately affected. While coastal erosion has also been identified as a significant problem in California, the statewide flooding risk exceeds that of erosion (Heberger et al. 2009).

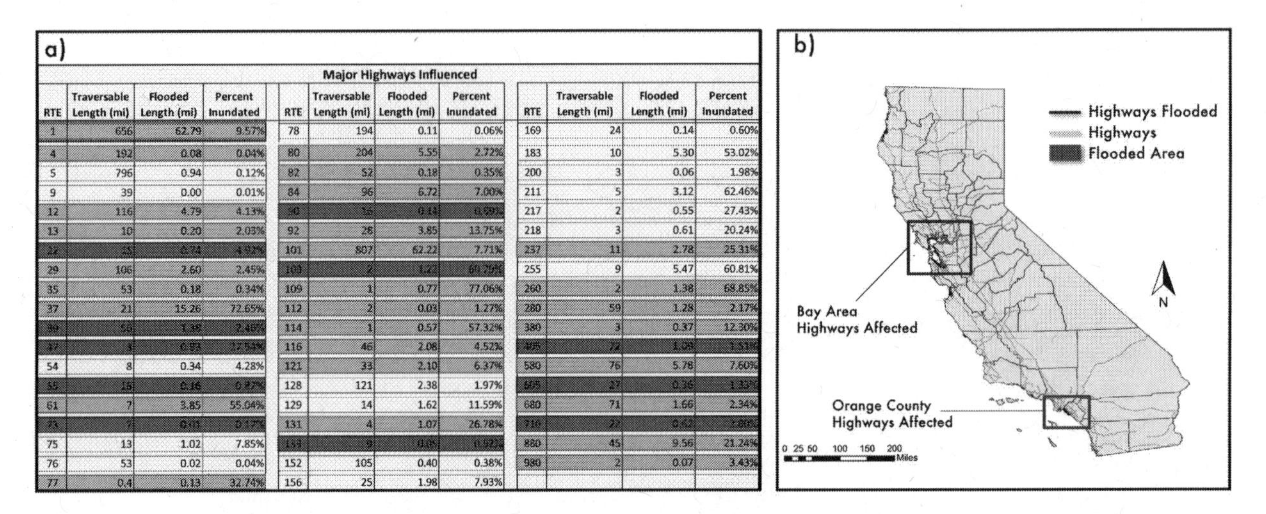

a)

Major Highways Influenced

RTE	Traversable Length (mi)	Flooded Length (mi)	Percent Inundated
1	656	62.79	9.57%
4	192	0.08	0.04%
5	796	0.94	0.12%
9	39	0.00	0.01%
12	116	4.79	4.13%
13	10	0.20	2.03%
22	15	0.74	4.92%
29	106	2.60	2.45%
35	53	0.18	0.34%
37	21	15.26	72.65%
39	56	1.38	[illegible]
47	3	0.53	17.54%
54	8	0.34	4.28%
55	15	0.16	0.87%
61	7	3.85	55.04%
73	7	0.01	0.17%
75	13	1.02	7.85%
76	53	0.02	0.04%
77	0.4	0.13	32.74%

RTE	Traversable Length (mi)	Flooded Length (mi)	Percent Inundated
78	194	0.11	0.06%
80	204	5.55	2.72%
82	52	0.18	0.35%
84	96	6.72	7.00%
90	16	0.14	[illegible]
92	28	3.85	13.75%
101	807	62.22	7.71%
103	2	1.22	[illegible]
109	1	0.77	77.06%
112	2	0.03	1.27%
114	1	0.57	57.32%
116	46	2.08	4.52%
121	33	2.10	6.37%
128	121	2.38	1.97%
129	14	1.62	11.59%
131	4	1.07	26.78%
[illegible]	9	0.05	[illegible]
152	105	0.40	0.38%
156	25	1.98	7.93%

RTE	Traversable Length (mi)	Flooded Length (mi)	Percent Inundated
169	24	0.14	0.60%
183	10	5.30	53.02%
200	3	0.06	1.98%
211	5	3.12	62.46%
217	2	0.55	27.43%
218	3	0.61	20.24%
237	11	2.78	25.31%
255	9	5.47	60.81%
260	2	1.38	68.85%
280	59	1.28	2.17%
380	3	0.37	12.30%
405	72	1.09	1.51%
580	76	5.78	7.60%
605	27	0.36	1.33%
680	71	1.66	2.34%
710	22	0.62	[illegible]
880	45	9.56	21.24%
980	2	0.07	3.43%

Figure 14.3 California highways affected by 140cm of sea-level rise. Source: Heberger (2009), Knowles (2009), Pacific Institute GIS data downloads (http://www.pacinst.org/reports/sea_level_rise/data/index.htm).

Flooding on the region's roadways will damage the physical infrastructure and require increased maintenance (Heberger et al. 2009). Inundated roadways will obstruct freight by delaying deliveries and forcing changes in route (CCCEF 2002) and disrupt international and domestic supply chains that depend on reliable delivery of goods.

Both flooding and rising sea levels can change coastal ports by creating deeper water. Deeper water allows vessels with deeper hulls to safely navigate a channel. While deeper water also leaves less clearance under bridges, most bridges over shipping lanes are already set high in order to accommodate large ships (Titus 2002; Heberger et al. 2009). However, the Golden Gate Bridge could block large vessels if sea level were to rise by four to five feet (Perez 2009). In addition, increases in storm surges would increase siltation and require more frequent dredging of harbors and channels; storm surges would require bridges to be built stronger and possibly higher to accommodate higher tides (Titus 2002); and bridges and port infrastructure would need additional protection from corrosion as the salt concentration and water levels change (PIANC EnviCom Task Group 3, 2006).

Other needed changes include port infrastructure realignments relative to the waterline, such as to docks, jetties, dry/wet/cargo docks, berths, and other port facilities, and modification of roll-on/roll-off operations to correct for new deck heights (Caldwell et al. 2000). Advancing saltwater in upstream channels may also change sediment location and create sandbars that can obstruct safe navigation (Titus 2002). To maintain safe channels, dredging will have to increase and pilots will need access to updated seafloor mapping.

Summer melting of Arctic ice may allow for a longer Arctic shipping season. The usability of the Northwest Passage for commercial marine shipping is highly uncertain and at best is predicted to vary year to year. But Canada's International Policy Statement predicted in 2005 that the Northwest Passage would be sufficiently ice-free for regular use during summer as early as 2015. Arctic shipping lanes would provide a route that is 5,000 nautical miles shorter for Asia-to-Europe trade than would a route passing through the Panama Canal. Vessels too large for the Panama Canal may be attracted to the Northwest Passage as an alternative to truck or rail transport across the United States. The use of the Arctic shipping lane rather than unloading in California and trucking across the United States would reduce cargo volumes in California ports (Pharand 2007). This change in demand could lessen the vulnerability of California ports, but it would also reduce economic activity in the region's transportation sector.

EXTREME HEAT EVENTS. Extreme heat events affect the duration of roadways and infrastructure. Extended periods of heat can shorten the life of and deteriorate pavements, force thermal expansion of bridges (thus delaying bridge operations and impacting their attendant maritime commerce), and deform the alignment of rail lines. Roadway deterioration will have an impact on all trade—including local trade circulation—that occurs between the Southwest and the remainder of the United States as well as trade between the Southwest and Mexico.[ii]

High temperatures can force rail lines out of alignment in what are called "sun kinks." Such a condition was responsible for injuring 100 people in a passenger train derailment near Washington, D.C., in 2002. The CSX Corporation, a freight transportation provider and owner of the rail line, initiated temporary speed restrictions after the incident, which slowed supply chains. These rail slowdowns could become more problematic as the frequency of extreme events due to climate change occur (Caldwell et al. 2000).

CHANGES IN PRECIPITATION. Changes in precipitation—specifically changes in intensity, frequency, and seasonality—also represent a significant threat to the Southwest transportation infrastructure. Compared to temperature, precipitation changes are more difficult to predict because precipitation is highly variable and localized. However, published studies tend to agree that while most of the Southwest is unlikely to see increases in total annual precipitation (Seager et al. 2007), increased precipitation intensity is likely (Alpert et al. 2002; Groisman et al. 2004; Groisman et al. 2005).

Increased precipitation intensity likely will result in one of more of the following: decreases in traffic demand; reductions in traffic safety; and decreases in the efficiency of operational features, such as speed, capacity, or travel-time variability (Table 14.1). Not surprisingly, severe weather events both decrease traffic demand and increase traffic accidents. Studies show traffic demand (measured by traffic volume) can change by anywhere from 5% to 80% due to severe weather events (e.g., Hanbali and Kuemmel 1993; Maze, Agarwal, and Burchett 2006).

Table 14.1 Potential impacts of precipitation events on transportation operations in the Southwest

Change in Precipitation	Impacts on Land Transportation Operations	Impacts on Marine Transportation Operations	Impacts on Air Transportation Operations
Increase in precipitation intensity and stormwater runoff	• Increased delay • Increased traffic disruption • Reduced safety and maintenance	Increased delay	• Increased delay • Increased stormwater runoff, causing flooding, delays, and airport closings • Impact on emergency evacuation planning, facility maintenance, and safety management
Increase in drought conditions	• Increased susceptibility to wildfires, causing road closures and reduced visibility	Impacts on river transportation routes and seasons	• Increased susceptibility to wildfires causing reduced visibility
More frequent strong hurricanes	• Interrupted travel and shipping • More frequent and more extensive emergency evacuations	Increased need for emergency evacuation planning, facility maintenance, and safety management	• More frequent interruptions in air service

Source: NRC (2008)

Depending on the level of planning and preparation undertaken by transportation providers, climate change may substantially and directly impact transportation operations as well as transportation infrastructure. For example, although a submerged jetty can be replaced or reconfigured, until this work is completed, it can no longer support the mobility of goods. Failing infrastructure cannot fulfill the role for which it was designed. Without advance planning to address and adapt to weather conditions that could reduce or limit infrastructure capacity, key infrastructure is at risk of being substantially less available. Table 14.2 summarizes the range of expected direct impacts to transportation infrastructure of climate change.

Indirect impacts

VEHICLE EMISSIONS. Heat events in the Southwest may increase air-conditioner usage in vehicles, which may bump up the total emissions. The U.S. EPA's Supplemental Federal Test Procedure (SFTP) for air conditioning (SC03) shows that total vehicular emissions increase 37% when air conditioning is turned on while driving, while fuel economy drops as much as 43% in a high-fuel-economy vehicles and 13% in conventional vehicles (Farrington and Rugh 2000).

Table 14.2 Potential impacts of climate on transportation infrastructure in the Southwest

Climate Change Factor	Impacts on Land Transportation Infrastructure	Impacts on Marine Transportation Infrastructure	Impacts on Air Transportation Operations
Sea-level rise and more frequent heavy flooding	• Inundation of roads and rail lines in coastal areas • More frequent or severe flooding of underground tunnels and low-lying infrastructure • Erosion of road base and bridge supports • Bridge scour • Loss of coastal wetlands and barrier shoreline • Land subsidence	• Reduced effectiveness of harbor and port facilities to accommodate higher tides and storm surges • Reduced clearance under waterway bridges • Changes in navigability of channels	• Inundation of airport runways located in coastal areas
Rising temperature and increase in heat waves	• Thermal expansion on bridge expansion joints and paved surfaces • Concerns regarding pavement integrity (e.g., softening), traffic-related rutting, migration of liquid asphalt • Rail-track deformities	• Low water levels • Extensive dredging to keep shipping channels open	• Heat-related weathering and buckling of airport and runway pavements and concrete facilities • Heat-related weathering of vehicle stock
Increase in precipitation intensity	• Increased flooding of roadways, railroads, and tunnels • Overloaded drainage systems • Increased road washout • Increased soil-moisture levels affecting structural integrity	• Changes in underwater surface and buildup of silt and debris	• Impacts on structural integrity of airport facilities • Destruction or disabling of navigation aid instruments • Damage to runway, pavement drainage systems, and other infrastructure
Increase in drought conditions	• Increased susceptibility to wildfires that threaten transportation infrastructure directly • Increased susceptibility to mudslides	• Reduced river flow and shipping capacity	• Increased susceptibility to wildfires that threaten airport facilities directly
More frequent strong hurricanes	• Increased threat to stability of bridge decks • Increased damage to signs, lighting fixtures, and supports • Decreased expected lifetime of highways exposed to storm surge	• Damage to harbor infrastructure from waves and storm surges • Damage to cranes and other dock and terminal facilities	• Damage to terminals, navigation aids, fencing around perimeters, and signs, etc.

Source: NRC (2008) and Karl, Melillo and Peterson (2009).

ECONOMY. The indirect economic effects of climate change on transportation infrastructure might include the shifting of production centers for agriculture, forestry, and fisheries. While some predictions show that climate change would increase U.S. agricultural production overall, some parts of the country likely would benefit more than others, such as areas at higher latitudes (see Chapter 11). Geographic shifts in the agricultural, forestry, and fishery industries would necessitate shifts in transportation routing patterns as well, prompting the need for new infrastructure. Southwest agricultural exports may decrease and imports may increase (Reilly et al. 2003). There may also be downward cost effects on the food chains of local and regional agriculture (NRC 2010).

On a larger scale, changes in international imports and exports of agricultural products may shift seaport traffic (Caldwell et al. 2000; NRC 2008; Koetse and Rietveld 2009). Major storm events may also require evacuations of coastal areas, which could disrupt normal trade flow. In the short term, product shortages and supply-chain disruptions could increase costs for shippers, carriers, retailers, manufacturers, and others reliant on the normal flow of goods (Ivanov et al. 2008).

If the major ports in California cannot handle their usual volume due to climate-caused damage, delay, or obstruction, ports in Oregon and Washington (or elsewhere) may need to be used instead. Such a diversion could tax smaller ports and their transportation network, and add travel time to the movement of goods throughout the United States. Diverting cargo to ports in Canada or Mexico—also an option—would hurt the economies of the Southwest and the United States (MARAD 2009).

HEALTH. Indirect health effects associated with added transportation infrastructure stress have been less emphasized in the literature, yet are critically important. Transportation serves as an essential component that both defines and responds to housing and settlement patterns. This relationship determines access to goods and services. Thus, when climate change alters the environmental context of human populations and settlements, the transportation system is also altered. Forecasting these health effects hinges on predicting both the type and magnitude of environmental change and their associated impacts on human populations, settlements and the transportation system. Disadvantaged and elderly populations, who are traditionally under-served by their transportation systems, are likely to be hardest hit by climate-change effects (see detailed discussion in Chapter 15 about the effects of climate change on the health of human populations in the Southwest).

14.5 Major Vulnerabilities and Uncertainties

There are many uncertainties associated with estimating future climate impacts. These uncertainties, coupled with the timescales on which infrastructure decisions are normally made, complicate responses. For example, new infrastructure construction can take as long as twenty years, with much of that time in the planning and engineering phases. Many transportation infrastructure projects already underway (in planning, design, or construction) were developed under priorities different than those of today and did not necessarily consider climate change. As the effects of climate change begin to manifest, the design and operational vulnerabilities of the transportation system will appear.

Disruptions to the transportation system

Disruptions to the transportation system, whether caused by climate change or other factors, have major economic effects on transportation system users. Climate change has the ability to impact all modes of passenger and freight transportation, including roads, bridges, tunnels, rail, public transportation, air transport, the vehicles that use these facilities, and the energy sources (gas, electric, etc.) that fuel them. Higher fuel and power consumption and the potential disruption of fuel and electric supplies cause prices to rise. Disruptions may prevent some trips from being made altogether. Most critically, when individuals can't get to work, they lose productivity and wages. For example, the system-wide transportation damage caused by the 1996 flash floods in Chicago prevented some commuters from reaching Chicago for up to three days (NRC 2008). Disruptions within a single link (for example, the collapse of the I-35W Mississippi River Bridge) can have ramifications on congestion levels throughout an urban area. Other, longer-term ramifications include relocation costs for households that need to follow the jobs, which further stress transport networks designed for lower demand and lead to increased congestion.

Studies of economic impacts of climate change have mainly focused on the costs of rebuilding infrastructure and costs related to freight movement. The economic impacts to passenger travel, even basic estimates of time lost, have not been actively researched. The key variables necessary to quantify economic impacts are loss of human life, economic productivity, and relocation costs. Other damages are difficult to quantify or estimate, such as breaks in social networks and families, anxiety, and stress. All such social and economic changes can have health implications. While short-term effects may be relatively easy to quantify, long-term effects are more important. The magnitude of the economic consequences will depend on the links within the disrupted network, the properties of both the transport network (including levels of redundancy), transport demand (the amount and location of desired travel), and the duration of the event (including recovery and rebuilding time).

The international goods movement system relies on goods supply and demand, international collaboration, physical and natural infrastructure, and favorable economic conditions. For example, the shipping community foresees climate change to be an issue but does not have the capacity to adequately predict and proactively combat its effects (see also Chapter 9).

Ports are a major intermodal connection, transferring containers and bulk goods from ships to trucks and railways. Ports comprise the harbor, berths, terminals, cranes, and surface transportation connections. Harbors must allow safe passage of ships. Larger ships require that the harbor allow deep draft vessels. Height of bridges is also a factor in safe ship passage. Once moored in the berth, another limiting factor is the size of the port cranes. Cranes are located on the port terminals and extend over ships to pick up and lower containers. The cranes must be large enough to reach across an entire ship. Increasing ship size therefore drives port infrastructure development. A consequence of larger container ships is the transformation of international shipping from a linear system to a hub and spoke system (Notteboom 2004; RITA 2011): cargo ships increasingly service only a few ports (called *load centers*) per region, with smaller vessels then distributing goods regionally. To increase their competiveness and likelihood of becoming a regional load center, ports are making large investments such as dredging to increase their water depth, buying larger cranes, constructing new terminals, and raising bridges to add to ships' height clearance.

Part of the competitive strategy of American ports is to promote the use of green technology and environmentally sustainable practices at their facilities. Planning for climate change, however, has not been a primary concern during port infrastructure development (IFC International 2008). As climate effects are felt, other routings and ports may become more competitive, for example, the Panama Canal (as well as the Northwest Passage, discussed previously). Opened in 1914, the Panama Canal connects the Pacific and Atlantic Oceans and now facilitates the passage of forty vessels a day (Autoridad del Canal de Panamá 2011). Even though only 25% of the world's fleet can fit through the canal locks, 4% of global trade and a much higher percentage of all U.S.-destined trade pass through the Panama Canal (Rosales 2007). The canal is currently undergoing an expansion that will allow the larger cargo ships to traverse the canal and unload cargo on the U.S. East Coast rather than on the West Coast. In anticipation of this diversion of cargo, East Coast ports are investing in their facilities (e.g., dredging, raising bridges, building new terminals) but with little planning for climate change (IFC International 2008). One study found that the cost of investment in Arctic-capable ships that could use the Northwest Passage to move between Japan and Newfoundland, Canada, would be recovered, but the investment would be uneconomical if the trip were extended to New York (Somanathan, Flynn, and Szymanski 2007). There is high uncertainty involved with potential use of the Northwest Passage for shipping, including under what jurisdiction the passage would fall. The United States and Canada are contesting whether the passage falls within Canadian territorial waters or should be considered an international strait. Other uncertainties are whether and to what extent the Arctic will be ice-free, whether navigational aids of the largely uncharted Arctic will be sufficient for safe passage, and whether use of the Northwest Passage will prove economical (Griffiths 2004; Birchall 2006; Somanathan, Flynn, and Szymanski 2007).

A key uncertainty to any assessment of potential vulnerabilities is the demand for the movement of goods within the United States and internationally. Historically, domestic vehicle miles travelled (by both passengers and goods), has tracked very closely with GDP. While this has changed somewhat in the last decade, the two are still highly correlated. Growth in world trade volume has outpaced world gross domestic production over the last decade. Global and domestic economic activity is a key driver of the demand for the movement of goods, as evidenced by the drop in demand after the 2008 economic collapse. The pattern of production and consumption is also uncertain, but determines the demand for goods movement both around the globe, within the United States, and within metropolitan regions. Of course local and regional effects depend on many things, including land-use policies, property values, and government incentives both in the United States and around the globe. Finally, given the volume of CO_2 produced in the distribution of freight, another key uncertainty is the price shippers and carriers will be expected to pay for CO_2 emissions. Any pricing of emissions (not just those from mobile sources) will directly affect global and local trade and the cost of transportation.

References

Alpert, P., T. Ben-Gai, A. Baharad, Y. Benjamini, D. Yekutieli, M. Colacino, L. Diodato, et al. 2002. The paradoxical increase of Mediterranean extreme daily rainfall in spite of decrease in total values. *Geophysical Research Letters* 29:1536.

Autoridad del Canal de Panamá, Operations Department. 2011. Monthly canal operations summary—February 2011. OP's Advisory to Shipping No. A-04-2011. http://www.pancanal.com/common/maritime/advisories/2011/a-04-2011.pdf.

Birchall, S. J. 2006. Canadian sovereignty: Climate change and politics in the Arctic. *Arctic* 59 (2): iii–iv.

Caldwell, H., K. H. Quinn, J. Meunier, J. Suhrbier, and L. Grenzeback. 2000. *Potential impacts of climate change on freight transportation.* Washington, DC: U.S. Department of Transportation, Center for Climate Change and Environmental Forecasting.

California Air Resources Board (CARB). 2008. *Climate change scoping plan: A framework for change.* Sacramento: CARB.

Cambridge Systematics, Inc. 2009. *Transportation adaptation to global climate change.* Washington, DC: Bipartisan Policy Center.

Center for Climate Change and Environmental Forecasting (CCCEF). 2002. *The potential impacts of climate change on transportation: Federal Research Partnership Workshop, October 1–2, 2002; Summary and discussion papers.* Washington, DC: U.S. Dept. of Transportation, CCCEF. http://climate.dot.gov/documents/workshop1002/workshop.pdf.

Climate Action Panel. 2007. *Final report of the Climate Action Panel, Colorado Climate Project of the Rocky Mountain Climate Organization.* Louisville, CO: Rocky Mountain Climate Organization.

Cohen, S. S. 2002. Economic impact of a West Coast dock shutdown. Berkeley, CA: Berkeley Roundtable on the International Economy. http://brie.berkeley.edu/publications/ships 2002 final.pdf.

Commission for Environmental Cooperation (CEC). 2011. *Destination sustainability: Reducing greenhouse gas emissions from freight transportation in North America.* Montreal: CEC. http://www.cec.org/Storage/99/9783_CEC-FreightTransport-finalweb_en.pdf.

Farrington, R., and M. Rugh. 2000. *Impact of vehicle air-conditioning on fuel economy, tailpipe emissions, and electric vehicle range.* Preprint. NREL/CP-540-28960. Golden, CO: National Renewable Energy Laboratory (NREL).

Federal Highway Administration (FHWA). 2011. Highway statistics 2009. Washington, DC: U.S. Department of Transportation, FHWA.

Golob, T. F., and A. C. Regan. 2001. Impacts of information technology on personal travel and commercial vehicle operations: Research challenges and opportunities. *Transportation Research Part C: Emerging Technologies* 92:87–121.

Griffiths, F. 2004. Pathetic fallacy: That Canada's Arctic sovereignty is on thinning ice. *Canadian Foreign Policy* 11 (3): 1–16.

Groisman, P. Y., R. W. Knight, T. R. Karl, D. R. Easterling, B. Sun, and J. H. Lawrimore. 2004. Contemporary changes of the hydrological cycle over the contiguous United States: Trends derived from in situ observations. *Journal of Hydrometeorology* 5:64–85.

Groisman, P. Y., R. W. Knight, D. R. Easterling, T. R. Karl, G. C. Hegerl, and V. N. Razuvaev. 2005. Trends in intense precipitation in the climate record. *Journal of Climate* 18:1326–1350.

Hanbali, R. M., and D. A. Kuemmel. 1993. Traffic volume reductions due to winter storm conditions. In *Transportation Research Record No. 1387, Snow removal and ice control technology: Papers presented at the 3rd International Symposium on Snow Removal and Ice Control Technology, September 14–18, 1992, Minneapolis, Minnesota,* 159–164. Washington, DC: Transportation Research Board.

Heberger, M., H. Cooley, P. Herrera, P. H. Gleick, and E. Moore. 2009. *The impacts of sea-level rise on the California coast.* Final Paper CEC-500-2009-024-F. Sacramento: California Climate Change Center.

IFC International. 2008. *Planning for climate change at U.S. ports.* White paper. N.p.: U.S. Environmental Protection Agency. http://www.epa.gov/sectors/pdf/ports-planing-for-cci-white-paper.pdf.

Ivanov, B., G. Xu, T. Buell, D. Moore, B. Austin, and Y. Wang. 2008. *Storm-related closures of I-5 and I-90: Freight transportation economic impact assessment report, winter 2007-2008.* Final Research Report WA-RD 708.1. Olympia: Washington State Department of Transportation.

Karl, T. R. , J. M. Mellilo, and T. C. Peterson, eds. 2009. *Global climate change impacts in the United States.* Cambridge: Cambridge University Press. http://downloads.globalchange.gov/usimpacts/pdfs/climate-impacts-report.pdf.

Knowles, N. 2009. Potential inundation due to rising sea levels in the San Francisco Bay Region. Final Paper CEC-500-2009-023F. Sacramento: California Climate Change Center.

Koetse, M. J., and P. Rietveld. 2009. The impact of climate change and weather on transport: An overview of empirical findings. *Transportation Research Part D: Transport and Environment* 14:205–221, doi:10.1016/j.trd.2008.12.004.

Maritime Administration (MARAD). 2009. *America's ports and intermodal transportation system.* Washington, DC: U.S. Department of Transportation, MARAD. http://www.glmri.org/downloads/Ports&IntermodalTransport.pdf.

Maze, T. H., M. Agarwal, and G. Burchett. 2006. Whether weather matters to traffic demand, traffic safety, and traffic operations and flow. In *TRB 85th Annual Meeting compendium of papers CD-ROM.* Washington, DC: Transportation Research Board.

Notteboom, T. E. 2004. Container shipping and ports: An overview. *Review of Network Economics* 3 (2), published online, doi:10.2202/1446-9022.1045.

National Research Council (NRC). Committee on Climate Change and U.S. Transportation. 2008. *Potential impacts of climate change on U.S. transportation.* Transportation Research Board Special Report 290. Washington, DC: National Research Council.

—. Committee on Twenty-First Century Systems Agriculture. 2010. *Toward sustainable agriculture in the 21st century.* Washington, DC: National Academies Press.

Perez, P. 2009. *Potential impacts of climate change on California's energy infrastructure and identification of adaptation measures.* Staff Paper CEC 150-2009-001. Sacramento: California Energy Commission. http://www.energy.ca.gov/2009publications/CEC-150-2009-001/CEC-150-2009-001.PDF.

Pharand, D. 2007. The arctic waters and the Northwest Passage: A final revisit. *Ocean Development & International Law* 38:3–69, doi:10.1080/00908320601071314.

PIANC EnviCom Task Group 3. 2006. *Waterborne transport, ports, and waterways: A review of climate change drivers, impacts, responses, and mitigation.* Brussels: PIANC. http://www.pianc.org/downloads/envicom/envicom-free-tg3.pdf.

Reilly, J., F. Tubiello, B. McCarl, D. Abler, R. Darwin, K. Fuglie, S. Hollinger, et al. 2003. U.S. agriculture and climate change: New results. *Climatic Change* 57:43–67.

Research and Innovation Technology Administration (RITA). 2008. State transportation statistics 2008. Washington, DC: RITA. http://www.bts.gov/publications/state_transportation_statistics/state_transportation_statistics_2008/index.html.

—. 2011. *America's container ports: Linking markets at home and abroad.* Washington, DC: RITA. http://www.bts.gov/publications/americas_container_ports/2011/.

Rosales, M. 2007. The Panama Canal Expansion Project: Transit maritime mega project development, reactions, and alternatives from affected people. Ph.D. diss., University of Florida.

Seager, R., M. Ting, I. Held, Y. Kushnir, J. Lu, G. Vecchi, H. Huang, et al. 2007. Model projections of an imminent transition to a more arid climate in southwestern North America. *Science* 316:1181–1184.

Siikavirta, H., M. Punakivi, M. Karkkainen, and L. Linnanen. 2008. Effects of E-commerce on greenhouse gas emissions: A case study of grocery home delivery in Finland. *Journal of Industrial Ecology* 6:83–97.

Somanathan, S., P. C. Flynn, and J. K. Szymanski. 2007. Feasibility of a sea route through the Canadian Arctic. *Maritime Economics and Logistics* 9:324–334.

Titus, J. 2002. Does sea level rise matter to transportation along the Atlantic coast? In *The potential impacts of climate change on transportation, Federal Research Partnership Workshop, October 1-2, 2002: Summary and discussion papers*, 135-150. N.p.: U.S. Department of Transportation, Center for Climate Change and Environmental Forecasting. http://climate.dot.gov/documents/workshop1002/titus.pdf.

Transportation Economics and Management Systems, Inc. (TEMS). 2008. *Impact of high oil prices on freight transportation: Modal shift potential in five corridors; Executive summary*. Washington, DC: U.S. Maritime Administration (MARAD). http://www.marad.dot.gov/documents/Modal_Shift_Study_-_Executive_Summary.pdf.

U.S. Environmental Protection Agency (EPA). 2012. *Inventory of U.S. greenhouse gas emissions and sinks: 1990-2010*. EPA 430-R-12-001. Washington, DC: EPA. http://www.epa.gov/climatechange/ghgemissions/usinventoryreport.html.

World Trade Organization (WTO). 2011. International trade statistics. Geneva: WTO. http://www.wto.org/english/res_e/statis_e/statis_e.htm.

Endnotes

i States adopting the California standards can be found at: http://www.c2es.org/what_s_being_done/in_the_states/vehicle_ghg_standard.cfm.

ii The Southwest states contain major land ports between the two countries such as San Ysidro/Tijuana and Calexico/Mexicali (CEC 2011).

Chapter 15

Human Health

COORDINATING LEAD AUTHORS

Heidi E. Brown (University of Arizona), Andrew C. Comrie (University of Arizona), Deborah M. Drechsler (California Air Resources Board)

LEAD AUTHORS

Christopher M. Barker (University of California, Davis), Rupa Basu (California Office of Environmental Health Hazard Assessment), Timothy Brown (Desert Research Institute), Alexander Gershunov (Scripps Institution of Oceanography), A. Marm Kilpatrick (University of California, Santa Cruz), William K. Reisen (University of California, Davis), Darren M. Ruddell (University of Southern California)

REVIEW EDITOR

Paul B. English (California Department of Public Health)

Executive Summary

Global climate models project changes in precipitation patterns, drought, flooding, and sea-level rise, and an increase in the frequency, duration, and intensity of extreme heat events throughout the Southwest. The challenge for the protection of public health is to characterize how these climate events may influence health and to establish plans for mitigating and responding to the health impacts. However, the effects of climate change on health vary across the region, by population, and by disease system, making it difficult to establish broad yet concise health promotion messages that are useful for developing adaptation and mitigation plans.

Techniques are increasingly available to quantify the health effects resulting from climate change and to move forward into predictions that are of sufficient resolution to

---

Chapter citation: Brown, H. E., A. C. Comrie, D. M. Drechsler, C. M. Barker, R. Basu, T. Brown, A. Gershunov, A. M. Kilpatrick, W. K. Reisen, and D. M. Ruddell. 2013. "Human Health." In *Assessment of Climate Change in the Southwest United States: A Report Prepared for the National Climate Assessment*, edited by G. Garfin, A. Jardine, R. Merideth, M. Black, and S. LeRoy, 312–339. A report by the Southwest Climate Alliance. Washington, DC: Island Press.

establish policy guidelines. Strides are being made in assigning cost to both the positive and negative effects on health of proposed climate-mitigation strategies or the lack thereof. As a result, more tools are available for cities and states to develop mitigation and adaptation plans that are specifically tailored to their populations.

For this assessment, we identify six key messages that relate to climate change and health in the Southwest:

- Climate change will exacerbate heat-related morbidity and mortality. (high confidence)
- Climate change will increase particulate matter levels from wildfires with subsequent effects on respiratory health. (medium-high confidence)
- Climate change will influence vector-borne disease prevalence, but the direction of the effects (increased or decreased incidence) will be location- and disease-specific. (medium-high confidence)
- Disadvantaged populations are expected to bear a greater burden from climate change as a result of their current reduced access to medical care and limited resources for adaptation strategies. (high confidence)
- Certain climate-change mitigation strategies have costs and benefits relevant to public health. Considering health costs (positive and negative) will more accurately represent the costs and benefits of the mitigation strategies. (medium-high confidence)
- Mitigation and adaptation plans tailored to the specific vulnerabilities of cities and states will lessen the impacts of climate change. (medium-high confidence)

15.1 Introduction

Summer season average temperatures in the Southwest United States are projected to be up to 9°F (approximately 5°C) higher than the present by the end of the twenty-first century (see Chapters 6 and 7 for details on climate change predictions in the Southwest). Global climate models also forecast changes in precipitation patterns, drought, flooding, and sea-level rise, and an increase in the frequency, duration, and intensity of extreme heat events throughout the Southwest. These climate changes will vary across the region, however, they are sufficient to threaten human health and well-being (Kunkel, Pielke, and Changnon 1999; Parmesan, Root, and Willig 2000; Baker et al. 2002; Christensen et al. 2004; Meehl and Tebaldi 2004; Harlan et al. 2006; Ruddell et al. 2010).

Unaddressed, there is a reasonable probability that climate change will have a negative impact on health in some Southwest human populations. There is uncertainty as to the timing, magnitude, and locations of these negative impacts. Population demographics, geographical differences, and socioeconomic factors that influence vulnerability also contribute to the uncertainty (Ebi et al. 2009). The complexity of interactions between these factors will require analysis and planning from multiple perspectives, including federal, state, tribal, and local governments, academia, the private sector and nongovernmental organizations (Frumkin et al. 2008).

This chapter focuses on those health effects related to climate change that will likely disproportionately affect the Southwest. It begins with a discussion of climate-related health issues of current concern in the Southwest followed by a brief review of the mechanisms in which climate change influences health. While the health impacts expected from climate change can be estimated at a qualitative level, it is difficult to quantify the effects. Thus, the qualitative discussion as to how climate-change will influence health outcomes and the growing body of quantitative literature that has reported observed or predicted climate-change-related outcomes are discussed separately. Since few studies relevant to this assessment have been performed in Southwest states other than California, in some cases the conclusions presented are based on extrapolation of those findings to the rest of the Southwest. The chapter closes with a discussion of key uncertainties and highlights several key points for public health planning for climate change.

15.2 Current Climate-Related Health Concerns in the Southwest

Climate, even without considering climate change, influences the health of residents in the Southwest in several ways. First, the topographical and climate variability of the Southwest, with its extreme geographical and climatic conditions, is greater than that in any other region of the United States. In addition, several health concerns exist only or primarily in the Southwest. Finally, there is variation in the vulnerability (i.e., the sensitivity, resiliency, and adaptive capacity) of individuals and groups of people within the region (Patz et al. 2005; Bell 2011).

Climate-related exposures can be the direct cause of morbidity (illness) or mortality (death), such as death from hyperthermia. Climate-related exposures can also be a contributing cause of health problems by exacerbating an already existing medical condition—such as heart disease—or can exert indirect effects, as by inducing changes in the ranges of vectors (organisms such as mosquitoes that transmit disease from one host to another) that can introduce health effects to populations who have no previous history of infection. In this section, we discuss health issues related to air quality, heat extremes, wildfires, and the ecology that disproportionately affect the Southwest. These illnesses connote a considerable health burden in this region.

Air quality

Air-pollution exposure is associated with mortality and morbidity (EPA 2006, 2009). The U.S. Environmental Protection Agency (EPA) sets health-based National Ambient Air Quality Standards (NAAQS) for ozone, for particulate matter smaller than 2.5 microns in diameter (PM2.5)[i] and smaller than 10 microns in diameter (PM10), and for four other environmental pollutants (Figure 15.1). Ozone and PM2.5 are considered the greatest threats to human health. Climate is not a factor in setting these standards, since they are based solely on health effects attributable to air-pollutant exposure (Figure 15.2).

The Clean Air Act requires that all states attain the NAAQS. If a state is not in attainment, it must develop state implementation plans (SIPs) that outline how attainment will be reached by a specified date. Currently, all or parts of forty-eight counties in the Southwest do not attain the 8-hour ozone standard (EPA 2011a). Thirty-six of these counties are in California (which has fifty-eight counties total), eight are in Colorado (with sixty-four counties), two are in Arizona (sixteen counties), and two are in Utah (twenty-nine counties). All or part of thirty-eight counties in the Southwest do not attain

Figure 15.1 Hazy view of Los Angeles. The visible smog translates into poor air quality with negative consequences for cardiorespiratory health. Photo courtesy of David Iliff. License: CC_BY_SA 3.0.

the PM2.5 NAAQS (EPA 2011b) including twenty-nine in California, two in Arizona, and seven in Utah. These counties encompass the major metropolitan areas of each state and consequently are home to significant fractions of each state's population.

OZONE. Ozone is a form of oxygen that forms naturally in the stratospheric portion of the atmosphere, where it absorbs most of the sun's UV radiation. In the lower atmosphere ozone is considered an ambient air pollutant, produced through chemical reactions between nitrogen oxides and hydrocarbons typically emitted by the burning of fossil fuels. The health effects of ozone were most recently assessed as part of the 2008 review of the ozone NAAQS (EPA 2006). The EPA concluded that short-term ozone exposure is associated with acute reductions in lung function, increased respiratory symptoms (such as shortness of breath, pain on deep breath and coughing, airway inflammation, and hyperresponsiveness), and increased respiratory hospital admissions and emergency department visits. Some literature suggests an association between ozone and cardiovascular morbidity, as well as mortality in people who have chronic cardiopulmonary disease. Long-term ozone exposure has not been clearly linked with health outcomes, except for structural changes in the airways of chronically exposed animals. People who are physically active outdoors, such as children, outdoor workers, and recreational and professional athletes, are at greatest risk of adverse health effects from ozone exposure.

PARTICULATE MATTER. The most recent assessment of the health effects of PM2.5 is part of the ongoing review of the PM NAAQS (EPA 2009). The EPA concluded that both daily and long-term exposures to PM2.5 are associated with mortality for cardiovascular causes, particularly in the elderly who have pre-existing cardiovascular disease. In comparison, the relationship of long-term and short-term PM2.5 exposure to illness and death (other than that from cardiovascular causes) has not been as consistently demonstrated. PM2.5 exposure is associated with hospitalizations and emergency department visits for exacerbation of pre-existing cardiopulmonary disease, mainly in the elderly. Reduced lung function growth,[ii] increased respiratory symptoms, and asthma exacerbation have been noted in children. Several studies report an increased risk of mortality and respiratory infections in infants exposed to elevated PM2.5 concentrations.

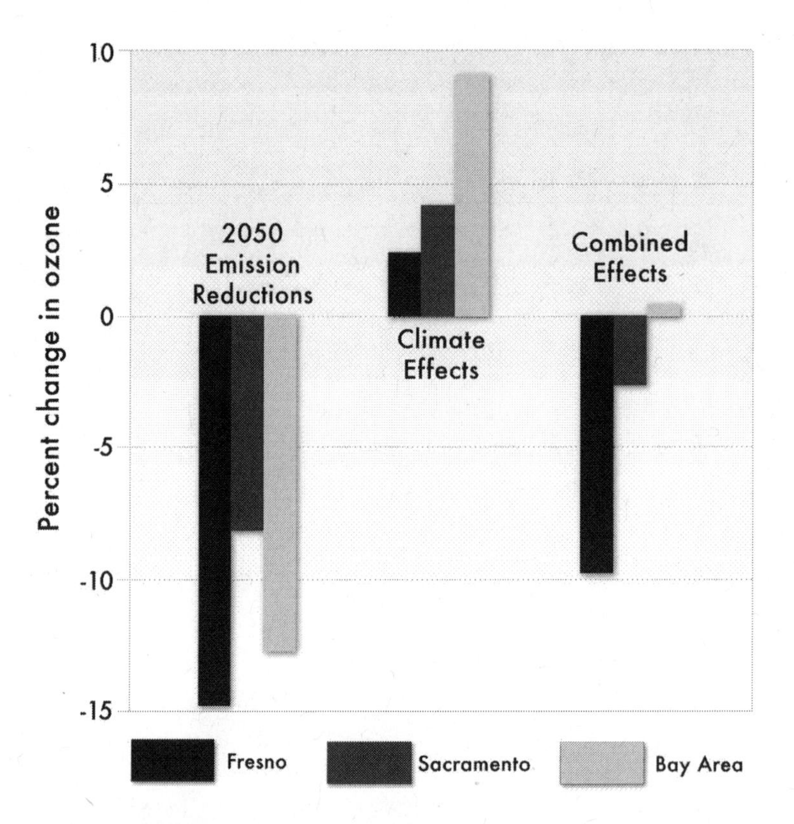

Figure 15.2 Projected ozone response to climate in three urban areas of California, 2050. A recent EPA-funded study at the University of California, Berkeley evaluated the effects of variables associated with anticipated changes in climate and ozone precursor emissions for cities in California's Central Valley (Fresno and Sacramento) and San Francisco Bay Area. In the set of bars at left, expected reductions (due to existing and projected control measures) in ozone precursor emissions in 2050 would reduce ozone levels, assuming no change in climate. In the middle set of bars, changes in climate variables (such as higher temperatures) and expected increases in nitrogen oxide and hydrocarbon emissions lead to higher regional ozone concentrations. When these two effects are combined on the right, the benefits of the emissions reductions are partially or completely offset by climate-related increases in ozone, especially in the San Francisco Bay Area. Thus, climate change will make ozone standards more difficult to attain and maintain, and will increase control costs. Adapted with permission from American Geophysics Union (Steiner et al. 2006; California Air Resources Board, http://www.arb.ca.gov/).

Heat extremes

Heat stress is the leading weather-related cause of death in the United States (CDC 2006; Kalkstein and Sheridan 2007; Sherwood and Huber 2010). Based on death certificates, an estimated 400 deaths each year are directly attributed to heat-related causes (CDC 2006), with the largest number occurring in Arizona (CDC 2005). However, both heat mortality and morbidity are believed to be significantly underreported (CDC 2006). Moreover, heat exposure can cause morbidity directly and also through exacerbation of preexisting chronic disease, particularly of the circulatory system (reviewed in Drechsler 2009).

Heat waves (periods of abnormally elevated temperature) can considerably increase the number of cases of direct and indirect heat-related mortality and morbidity (Semenza et al. 1996; Smoyer 1998; Naughton et al. 2002; Weisskopf et al. 2002; Knowlton et al. 2009; Ostro et al. 2009). The specific temperature associated with a heat wave varies by location because it is defined in relation to local normal conditions (see also Chapters 5 and 7 for more detailed discussions about heat waves in the Southwest).

Wildfires

The high frequency of wildfires in the Southwest presents several health concerns. Wildfire smoke can lead to PM2.5 levels that greatly exceed national standards (Phuleria et al. 2005; Wu, Winer, and Delfino, 2006), contributing to adverse health effects. Other health concerns related to wildfires include death and burn injuries through direct contact with the fire or from indirect effects such as evacuation and dislocation, physical loss of home or other property, and increased risk of mudslides during subsequent rainstorms. Whether or not health effects derive from a specific wildfire depends on many factors, including the proximity of the fire to a population; the size, intensity, and duration of the fire; and whether the smoke plume moves across a populated area.

Permissive ecology

Climate affects the seasonality, geographic distribution, and transmission frequency of infectious illnesses through the regulation of permissive habitat for pathogen establishment. Many of the climate-related infectious illnesses that occur in the United States (e.g., West Nile virus, influenza, food- and water-borne pathogens) also affect the Southwest. These illnesses contribute considerably to U.S. morbidity and mortality. For example, food-borne illnesses affect 25% of the U.S. population and cause some 76 million cases annually (Mead et al. 1999).

Certain infectious diseases are more prevalent or are almost exclusively found in the Southwest. The Centers for Disease Control and Prevention (CDC) reports that from 2005 to 2009, most cases of valley fever (99.1% of 44,029 cases nationwide), plague (93% of 43 cases), and Hanta pulmonary syndrome (61% of 136 cases) occurred in the six Southwestern states (Figure 15.3) (Hall-Baker et al. 2009; Hall-Baker et al. 2010; Hall-Baker et al. 2011; McNabb et al. 2007; McNabb et al. 2008).

15.3 Climate Change and Potential Health Implications

Climate change is expected to increase injury and death related to extreme events, to alter the distribution of infectious diseases, and to exacerbate current climate-related health issues (Frumkin et al. 2008). These changes in climate have varying effects depending on location, event, population susceptibility, and disease. Here we discuss the changes in climate and the way in which they influence human health outcomes. The focus is on health effects that affect the Southwest disproportionately compared to other parts of the United States. The 2006 California heat wave is presented as a "case study" because it had the characteristics of future heat waves predicted for the Southwest.

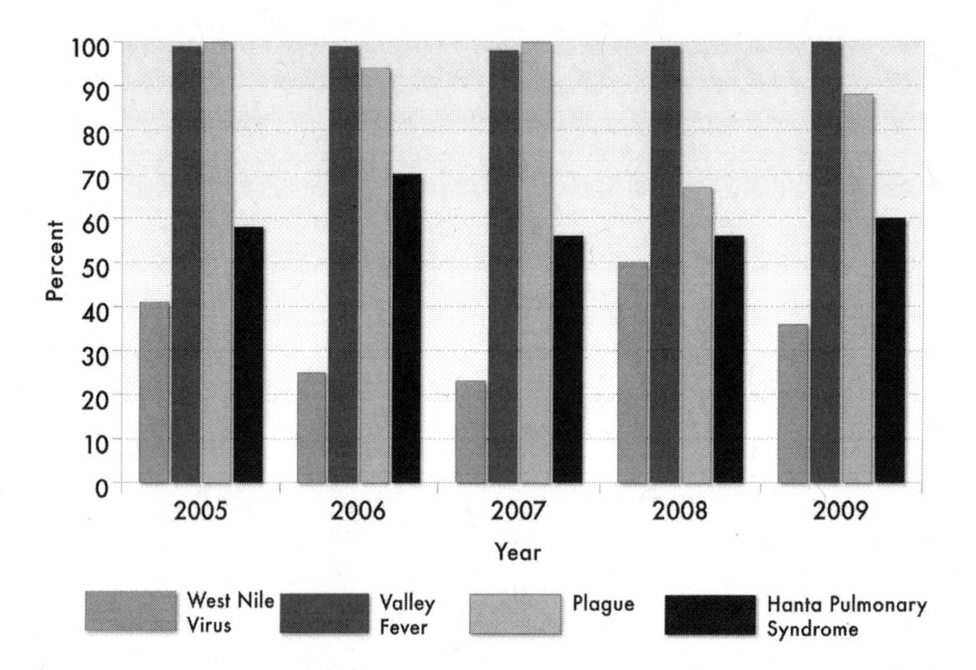

Figure 15.3 Incidence of selected diseases in the Southwest as a percent of total for the United States. The proportion of the total cases in the United States of West Nile virus, valley fever, plague, and Hanta pulmonary syndrome reported annually from the Southwest to the CDC shows consistency from year to year. Data from cases reported to the CDC.

Emissions and air pollution

Rising temperature will accelerate atmospheric chemical reactions, tending to increase concentrations of ozone and possibly PM2.5 (Steiner et al. 2006; Jacobson 2008; Kleeman 2008; Mahmud et al. 2008; Millstein and Harley 2009). However, other meteorological characteristics, such as relative humidity, wind speed, and mixing height[3] also interact with temperature and influence pollutant concentrations (Steiner et al. 2006; Jacobson 2008; Kleeman 2008; Mahmud et al. 2008; Millstein and Harley 2009). Changes in ozone and PM2.5 levels are unlikely to be uniform across an air basin. There remain many uncertainties in our current understanding of the influence of meteorological parameters on air quality. In addition, current strategies for modeling ozone cannot yet factor in the reduction of air pollution that is expected from regulations and control strategies that will be adopted to meet NAAQS attainment requirements. Thus, current knowledge is inadequate to project future health impacts. However, without implementation of new air-pollution-control strategies, the concentrations of these pollutants could increase with climate change, and consequently contribute to increased air pollution-related health effects (Knowlton et al. 2004; Peng et al. 2011).

Increases in extreme events

Increased climate extremes (heat waves and winter cold) and their direct effects, along with indirect effects related to vector populations, are expected to have an overall negative effect on human health (McMichael 2001). In particular, future heat waves are expected to be more humid (Gershunov, Cayan and Iacobellis 2009), with higher overnight low temperatures. Heat waves are expected to increase more in coastal areas compared to inland (Guirguis and Gershunov forthcoming; see also Chapter 7). In addition to rising temperatures in the Southwest, climate models predict that the frequency, intensity, spatial extent, and duration of heat waves will continue increasing through the remainder of this century (Gershunov, Cayan, and Iacobellis 2009; Climate Action Team 2010; see also Chapters 6 and 7).

HEAT-RELATED MORTALITY AND MORBIDITY. Heat stress (the physiological response to excessive heat) can lead to morbidity and mortality and is the primary health-related threat to human health and well-being related to climate change both nationally and within the Southwest (Sherwood and Huber 2010). Heat stress is greater when elevated temperatures continue for several days (Kalkstein et al. 1996; Ruddell et al. 2011) or when conditions are hot and more humid. Humid heat poses a greater physiological stress than dry heat because it reduces the body's ability to cool itself through evaporation (Gagge 1981; Horvath 1981). Increases in daily mortality have been observed to vary by community and the intensity, duration, and timing of the heat event (Anderson and Bell 2011).

Increased physiological stress due to global temperature rise; more frequent, humid, intense, and longer lasting heat waves; and intensification of heat stress by urban heat islands will likely increase heat-related morbidity and mortality in the Southwest (Oke 1982; Brazel et al. 2000; Meehl and Tebaldi 2004; Rosenzweig et al. 2005; IPCC 2007; see also discussion in Chapter 7). This trend will be exacerbated by a projected demographic shift toward an older population (Figure 15.4) (Basu, Dominici and Samet 2005; Basu and Ostro 2008; Sheridan et al. 2011). Basu and Malig (2011) found that higher temperatures were associated with significant reductions in life expectancy and did not only affect extremely frail individuals.

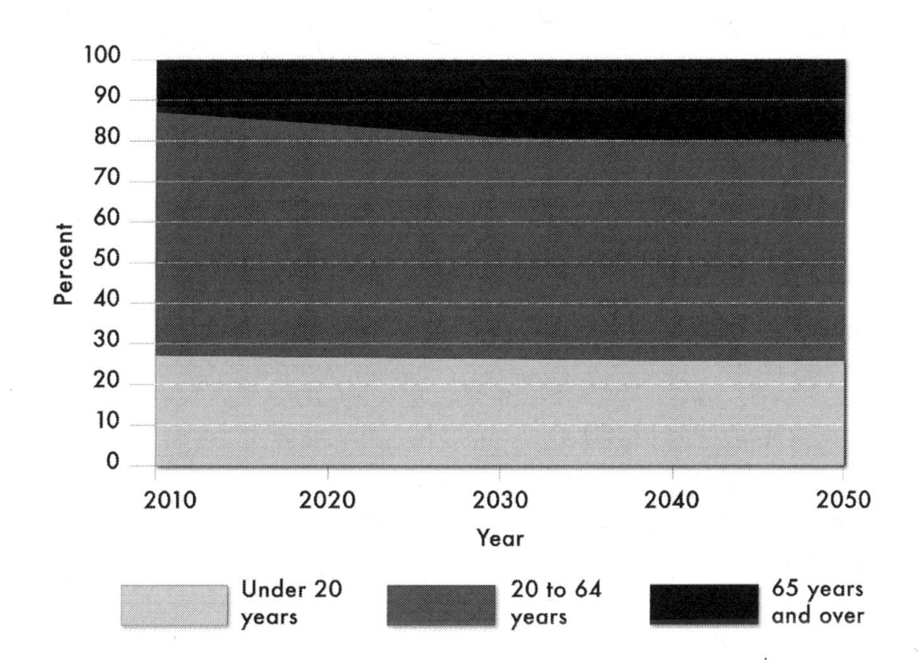

Figure 15.4 U.S. census population predictions by age. Predictions that are stratified by age show an increase over the coming decades in the proportion of the 65-and-older population. The decadal estimates were generated using 2000 U.S. Census data. Source: Vincent and Velkoff (2010).

Outcomes of an unprecedented ten-day humid heat wave in California during late July 2006 are instructive for considering future climate-change-related mortality and emergency room (ER) visits. Studies showed that excess mortality related to the heat wave was at least three times greater than what was reported by coroners and that a greater proportion of deaths occurred in the elderly (Ostro et al. 2009). Knowlton et al. (2009) and Gershunov et al. (2011) confirmed that there were proportionally more ER visits in the cooler coastal areas than in the hotter inland regions, likely because populations living in the cooler parts of the state are less physiologically adapted to heat exposure, have less air conditioning, and are less knowledgeable about protective behaviors. Ownership and usage of air conditioners has been shown to reduce the adverse effects of increased temperature on some chronic health outcomes (Ostro et al. 2010).

Collectively, these studies suggest that: (1) mortality to which heat is a contributing cause will continue to exceed that in which heat is the underlying cause; (2) absolute risks of heat-related mortality will continue to be high in hotter areas, although coastal areas may see greater increases in risk over time with climate change; (3) relatedly, changes in risk of heat-related mortality and morbidity will likely be greater in areas that currently have relatively low peak temperatures (for example coastal or high-altitude areas), and consequently are less adapted to heat than areas that currently have frequent periods of high temperature; and (4) an aging population will increase the size of the at-risk population.

WILDFIRES. Climate change is expected to increase wildfire frequency and size (Westerling et al. 2006, 2009; Flannigan et al. 2009), which in turn will increase the contribution of wildfire smoke to ambient PM2.5 levels. Forest flammability will increase because of projected changes in the patterns of precipitation and drought, and their attendant effects on vegetation stressors such as insects, disease, soil-moisture loss, and tree mortality (Figure 15.5). These changes, along with human factors and land management practices (such as livestock grazing, wildfire suppression, and increased human presence) are expected to increase the number of wildfires, their contribution to ambient air pollution levels, and related health effects (see Chapter 8, Section 8.4.2).

Studies in the Southwest (Shusterman, Kaplan, and Canabarro 1993; Lipsett et al. 1994; Vedal and Dutton 2006) have found no significant relationship between wildfire smoke (measured as PM2.5) and mortality during fire periods. Fire-related deaths were principally related to burns, even in individuals who also had smoke-inhalation injury (Shusterman, Kaplan, and Canabarro 1993). In the Southwest, fire smoke exposure has been associated with respiratory and eye symptoms (Sutherland et al. 2005; Künzli et al. 2006). ER visits and unscheduled physician's visits increase during wildfire periods (Duclos, Sanderson, and Lipsett 1990; Shusterman, Kaplan, and Canabarro 1993; Lipsett et al. 1994; Vedal 2003; Künzli et al. 2006). Künzli and others (2006) also found that non-asthmatic children were more affected than asthmatic children, probably because asthmatic children were more likely to take preventive actions, including remaining indoors, reducing physical activity, using air conditioning, and wearing masks when outdoors. This is the first published study that showed a benefit from adopting these actions. Delfino and colleagues (2009) reported a 34% increase in asthma hospital admissions during the 2003 Southern California wildfires that increased PM2.5 by an average of 70 micrograms (one millionth of a gram) per cubic meter. The greatest increase in

Figure 15.5 2010 Wildfire in Great Sand Dunes National Park, Colorado. © University Corporation for Atmospheric Research.

risk was for people over 65 years of age and for children up to four years of age. Risk of asthma admissions for school-age children was not statistically significant. The greatest increases in risk for hospital admissions were for acute bronchitis and pneumonia, particularly among the elderly, with no significant change in risk for cardiovascular effects or disease. Together these studies suggest that people with acute or chronic respiratory disease at the time of wildfire smoke exposure are the individuals most at-risk.

VALLEY FEVER. The majority of valley fever (coccidioidomycosis) cases in the United States occur in the Southwest (99.1% or 43,634 cases from 2005 to 2009), almost exclusively in Arizona and California (Hall-Baker et al. 2009; Hall-Baker et al. 2010; Hall-Baker et al. 2011; McNabb et al. 2007; McNabb et al. 2008), and mainly affecting people over age 65 (CDC 2003). Potential infection occurs when a dry spell desiccates the soil-dwelling fungus and subsequent soil disruption releases the spores, which are then inhaled. It is hypothesized that moisture in the soil preceding a dry spell promotes fungal growth (Kolivras et al. 2001; Comrie 2005; Comrie and Glueck 2007; Tamerius and Comrie 2011). Changes in extreme climatic events are expected to influence the growth and airborne release of this fungus.

Investigations into the relationship of weather patterns and valley fever found a weak correlation (4% of variance explained) between disease incidence and preceding precipitation for California (Zender and Talamantes 2006) while a substantially stronger correlation was found in two Arizona counties (69% of variance in Maricopa and 54% in Pima Counties explained; Tamerius and Comrie 2011). These disparities are likely related to the complexity of the relationships between climate factors and the fungus, as well as human susceptibility, habitat availability, model and variable selection, and data quality. This complexity impedes precise predictions of how climate change will influence the future incidence of valley fever.

Long-term warming trend

The warming climate is expected to increase the length of the freeze-free season (the time between the last frost of spring and first autumn frost). These changes are in turn expected to influence many vector-borne diseases. The life cycles of vectors and their hosts are influenced by temperature and other climate factors (Gage et al. 2008), which in turn influence the time between vector infection to disease transmission (Reisen et al. 1993; Reisen, Fang and Martinez 2006). Warm temperatures increase the rate at which vector populations grow, speed vector reproductive cycles (Reisen 1995), shorten the time between exposure and infectivity (Reisen et al. 1993; Gubler et al. 2001; Gage et al. 2008; Reisen, Fang and Martinez 2006), and increase vector-host contact rates (Patz et al. 2003).

Vector ranges may change over time, depending on whether or not future local climate is suitable for a given vector (Box 15.1). Thus, previously unexposed populations may become exposed, while some currently exposed populations may no longer be exposed (Lafferty 2009). Understanding the effect of climate on vectors, hosts, and pathogens can help us estimate the geographic extent and intensity of disease risk. Research is continuing to improve our understanding of the association between disease incidence and climate, which brings us closer to predicting the future disease risk.

Box 15.1

Entomologic Risk

Entomologic risk is a term used in describing the distribution of insects with respect to insect-borne disease. It is a helpful topic here as it highlights how a vector can exist in the absence of the occurrence of disease. For example, the dengue and yellow fever vector, Aedes aegypti, occurs in Tucson, Arizona (first noted in 1946, but more recently since 1994 [Merrill, Ramberg, and Hagedorn 2005]), but neither dengue nor yellow fever are known to occur. This notion can be expanded to other health concerns: behavioral adaptation to heat can reduce the possibility of heat- related illness in spite of increasing heat events.

MOSQUITO-BORNE DISEASES. Mosquitos are a vector for the transmission of pathogens worldwide. Their abundance varies over time and space due to variations in temperature and the water available for their larvae (Barker, Eldridge, and Reisen 2010; Morin and Comrie 2010). Abundance of vertebrate hosts and the extent of suitable habitat can also be influenced by climate. Certain peridomestic birds (species that live around human habitation) are effective hosts for viruses such as West Nile virus (Reisen, Fang, and Martinez 2005; Kilpatrick 2011). Models that incorporate the interaction of vectors, hosts, and pathogens in a manner to also be explicit with respect to time and spaces are exceedingly complex. Quantifying these interactions, which are influenced by climate in multiple ways, creates a challenge for predicting with precision the future health consequences of these diseases.

Long-term warming trends are expected to have several effects on mosquito-borne diseases. Longer-living vectors will increase the likelihood that a mosquito will obtain a potentially infectious blood meal, survive the pathogen's incubation period, and become infectious (Gubler et al. 2001; Cook, McMeniman, and O'Neill 2008). More days with temperatures exceeding minimum thresholds will increase vector-host contacts and decrease the period of time it takes for a mosquito to become able to transmit infection after ingesting an infecting blood meal (Kilpatrick et al. 2008; Hartley et al. 2012). Above certain maximum temperature thresholds, however, the mortality of adult female mosquitos increases by around 1% per day for each 1.8°F (1°C) increase in temperature (Reeves et al. 1994). This reduced survival may be compensated for by increased mosquito biting rate and viral replication (Delatte et al. 2009; Hartley et al. 2012). These thresholds vary by vector species and their location, making uniform predictions, such as country-wide predictions about changes in generic mosquito-borne disease risk, inappropriate.

PLAGUE. Plague is a flea-borne bacterial disease maintained in rodents, with occasional spill-over to humans and companion animals. Ninety-three percent (40 cases) of all U.S. plague cases reported between 2005 and 2009 were from the six Southwestern states (Hall-Baker et al. 2009; Hall-Baker et al. 2010; Hall-Baker et al. 2011; McNabb et al. 2007; McNabb et al. 2008).

Most plague outbreaks occur when temperatures are between 75°F and 80°F (24°C and 27°C) and cease at higher temperatures (Brooks 1917; Davis 1953; Cavanaugh and Marshall 1972; Cavanaugh and Williams 1980; Gage et al. 2008; Brown et al. 2010). Global climate cycles, such as the Pacific Decadal Oscillation, as well as local meteorology, influence year-to-year differences in human plague cases (Parmenter et al. 1999; Enscore et al. 2002; Ben Ari et al. 2008). To date, these findings rely on retrospective analyses of climate and disease incidence without predicting future risk associated with climate change.

15.4 Observed and Predicted Effects on Health from Climate Change

Though the publication of research investigating the association between climate and health is growing, statistical models capable of predicting future health impacts are limited (Ebi et al. 2009). The World Health Organization estimated that global warming caused 140,000 excess deaths in 2004 compared to 1970 (WHO 2010). Climate change alters the distribution of physical exposures, which in turn changes the distribution of vulnerable populations and increases the likelihood of adverse impacts to those populations. As discussed in the previous section, climate change is expected to exacerbate several current health concerns, and alter the distribution of vulnerabilities by age, geographical, and socioeconomic factors on a local level (Ebi et al. 2009). This section summarizes recent findings that show already observed climate-change-related health effects or provide quantitative predictions of impacts.

Air quality

The health impacts of PM2.5 and ozone exposure are proportional to their concentrations in the ambient air, which is influenced by emissions, atmospheric chemistry, and emissions-reduction regulations driven by the National Ambient Air Quality Standards (NAAQS). NAAQS are reevaluated approximately every five years, and are revised if new information suggests that the existing NAAQS are inadequate. Overall, concentrations of these two pollutants and the number of air basins in the Southwest that do not attain the NAAQS have declined due to emissions-reduction regulations. However, several parts of the Southwest, particularly portions of California, do not attain the ozone and/or PM2.5 NAAQS (EPA 20011a, 20011b).

Some greenhouse-gas-reduction regulations provide co-benefits—multiple and ancillary health benefits of a program, policy, or intervention—by reducing emissions of other chemicals, such as hydrocarbons and nitrogen oxides that combine to form ozone[iv] and PM2.5 (CARB 2008). Energy demand is projected to increase in the future due to a growing population and a warmer climate, which could increase emissions of ozone precursors and PM2.5 from some sectors (Climate Action Team 2010). Land use planning and policy changes will also affect emissions, particularly those from the transportation sector. Increases and decreases in emissions from various sectors, along with new emissions-reduction regulations and control technologies, will determine future attainment of the NAAQS. Overall, climate change is likely to make it harder to achieve and maintain attainment with the NAAQS.

ASTHMA AND ALLERGIES. As climate warms, data show earlier and longer spring bloom for many plant species, which has led to a general increase in plant biomass and pollen generation, triggering allergies and asthma cases (Weber 2012). A recent EPA review (EPA 2008) of the likely influence of climate change on bioallergens (pollens and molds) concluded that: pollen production is likely to increase in most parts of the United States; earlier flowering is likely to occur for numerous species of plants; changes in the distribution of pollen-producing species are likely, including the possibility of extinction of some species; intercontinental dispersal of bioallergens is possible (Figure 15.6), facilitating the introduction of new aeroallergens into the United States and increases in allergen content, and thus, potency of some aeroallergens are possible. Concomitant exposure to ozone and allergens may also lead to greater allergic responses than exposure to allergens alone (Molfino et al. 1991; Holz et al. 2002; Chen et al. 2004).

Heat-related mortality and morbidity

U.S. heat-related deaths declined between 1964 and 1998 (Davis et al. 2003), likely due to more air conditioning, improved medical care, and better public awareness programs, as well as other infrastructural and biophysical adaptations. However, heat-related mortality and morbidity still occur throughout the Southwest region, particularly associated with intense heat waves. Based on historical data, without additional adaptations, mortality and morbidity will increase as the climate warms (Karl, Melillo and Peterson 2009).

A few studies, all focused on California, have quantitatively estimated future heat-related mortality (Hayhoe et al. 2004; Drechsler et al. 2006; Ostro, Rauch, and Green 2011;

Figure 15.6 Dust storm on Interstate 10 near Phoenix, July 8, 2007. Photo courtesy of Los Cuatro Ojos blog (http://loscuatroojos.com/2007/07/28/dust-storms-over-phoenix-looks-worse-than-it-is/).

Sheridan et al. 2011). Sheridan and colleagues (2011) suggest that by the 2090s, heat waves lasting two weeks or longer will occur about once per year throughout the state and that ten-day or longer heat waves could increase nearly ten times under a higher emissions scenario. Without new adaptations, most of California's urban areas could see significantly higher heat-related mortality by the 2090s, with a significant portion of the increase attributable to California's aging population (Table 15.1). Projected heat-related mortality is not uniform statewide, ranging from a 1.9-fold increase in San Francisco to a 7.5-fold increase in San Diego, compared to present. Incorporation of adaptations into the modeling only partially mitigated these increases. Ostro, Rauch, and Green (2011) estimated that heat-related mortality in California could increase up to three times by the mid-twenty-first century, with about a third of the increase offset by a 20% increase in air conditioning prevalence.

Vector-borne disease

The IPCC's Fourth Assessment Report (2007) states that physical changes in the climate system will alter the "spatial distribution of some infectious diseases," such as dengue fever, malaria, and West Nile virus (WNV). In 2007, California declared a State of Emergency "due to the increasing risk of West Nile virus transmission" (CDPH 2007). The complexity of these systems makes forecasting the effects of climate change on disease outcomes difficult. However, researchers are increasingly able to generate future climate predictions.

Table 15.1 Estimated future heat-related mortality in nine metropolitan statistical areas of California

	Mean Annual Heat-Related Mortality (Age 65 and older)		
Urban Area	**20th Century**	**2090s – Medium Growth**	**2090s – No growth**
Fresno	15	192 – 266	26 – 36
Los Angeles	165	1,501 – 2,997	368 – 732
Oakland	49	413 – 726	85 – 149
Orange County	44	395 – 742	105 – 194
Riverside	60	741 – 1,063	113 – 162
Sacramento	27	275 – 440	55 – 88
San Diego	68	750 – 1,865	207 – 511
San Francisco	53	161 – 247	71 – 110
San Jose	27	256 – 411	44 – 69
TOTAL	**508**	**4,684 – 8,757**	**1,074 – 2,051**

Source: Sheridan et al. (2011).

WEST NILE VIRUS. Recent models that focus on the physiological responses of mosquitos to changes in climate allow us to estimate WNV transmission using future temperature predictions (Figure 15.7). Comparisons of the Southern San Joaquin Valley to the otherwise comparable (in terms of vegetation, land use, rainfall, and human population) but 5.4°F–9°F (3°C–5°C) warmer Coachella Valley of California illustrates the likely influence of a warming climate on arboviruses (viruses spread by arthropod vectors) such as WNV (Reisen, Fang, and Martinez 2006). The transmission season in Coachella starts almost a month earlier than in the San Joaquin Valley and may persist through winter. However, above high temperature thresholds, the maximal intensity of pathogen transmission from mosquito to host may not differ markedly. WNV may be transmitted throughout the winter at very southern latitudes (Tesh et al. 2004; Reisen et al. 2006), whereas at more northern latitudes or high elevations temperatures drop below virus developmental thresholds for extended periods and vector populations enter diapause (a period in which growth or development is suspended), which interrupts virus transmission (Reisen, Smith, and Lothrop 1995; Reisen, Meyer and Milby 1986).

Warming trends apparently allow non-diapausing (not becoming dormant) portions of *Culex* mosquito population to persist through the winter (Reisen et al. 2010), especially in urban environments (Andreadis, Armstrong, and Bajwa 2010). Overall, it is likely that shorter winters will allow transmission of some arboviruses like WNV to continue throughout the year. Highly efficient transmission of WNV is already evident during the summer, and little change is expected with warming trends. However, warming in the currently cooler and densely populated areas along the California coast and in the foothills of the Sierra Nevada and Rocky Mountains will increase risk of transmission of WNV and other mosquito-borne pathogens into new areas.

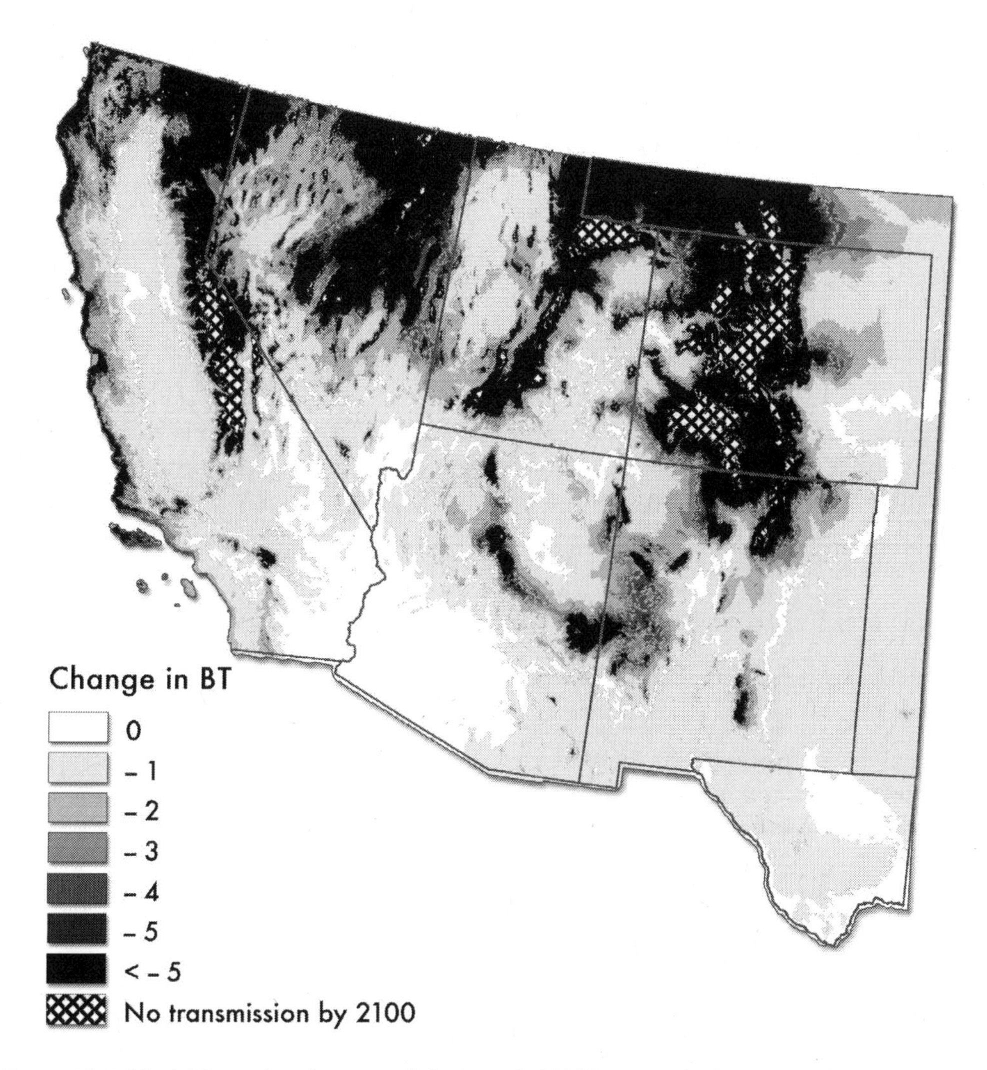

Figure 15.7 Model-based estimates of changes in WNV transmission potential during June, based on expected shortening of the incubation period in mosquitoes. Shown is the decrease in the average number of mosquito bites necessary to get from infection to transmission (BT) for WNV by 2100, based on future temperatures derived for mean scenarios used by Dettinger (2005). Parts of the region with cooler climates show a progressive decrease in BT during June by 2100, increasing the risk of transmission, whereas areas with extremely warm or extremely dry desert climates show no change in BT from current estimates. Source: PRISM Climate Group, Oregon State University, http://www.prism.oregonstate.edu.

15.5 Uncertainties

Quantitative estimates of future health impacts of climate change are difficult and uncertain for several reasons. These include: a dearth of adequate health data and sufficiently downscaled climate data; incomplete understanding of how non-climate-related factors modify risk; incomplete understanding of the relationships between climate, health, and disease processes; and the influence of physiological, behavioral, and societal adaptations. In addition, range shifts for infectious diseases or new introductions are difficult to predict. The Southwest has many large population centers located near earthquake faults and in coastal areas that can be struck by tsunamis. Although not the product of climate change, the Southwest has experienced major earthquakes that inflicted crippling damage to infrastructure in its major population centers, and it continues to be at risk of further earthquakes. A major seismic event preceding or overlapping extreme climate events would further complicate planning and execution of adaptation plans. These issues increase uncertainty for planning adaptive capacity and interventions.

Availability of high-quality health data

Deficiencies in health data quality limit our ability to characterize the relationship of climate change and health and to develop predictive models for climate-related health impacts. Multi-year data sets of consistent quality and from multiple locations, with high spatial and temporal resolution, are needed to assess how risk changes over time and to estimate future impacts on a regional basis (Frumkin et al. 2008; Ebi et al. 2009; English et al. 2009). Data on the spatial distribution of vector-borne and fungal diseases, the organisms that transmit them, and their seasonal abundance are needed to estimate future infectious disease impacts (Bush et al. 2011).

Climate data

Uncertainties in climate modeling are discussed in Chapters 6 and 7 of this report. Adequately downscaled climate projections are needed to quantify health effects at the local level (Ebi et al. 2009).

Disease complexity

Diseases are physiologically complex, and many factors interact with or modify disease (Box 15.2). For example, humans have extensive physiological and behavioral capabilities that allow them to adapt to the usual temperature conditions where they live. The use of physiological, societal, and behavioral data in predictive modeling is limited by data availability and by our understanding of the interactions. These limitations are further confounded by our ability to interpret how physiology, society, and behavior will change with climate change. Failure to incorporate these factors increases uncertainty in predicting health outcomes (Randolph 2009) and may lead to spurious attribution of risk (Campbell-Lendrum and Woodruff 2006). A recent review on climate and human health over millennia suggests that modern societies may be less flexible and thus vulnerable (McMichael 2012). The complexities of human behavior and adequate characterization of the entities involved in disease transmission create uncertainty in estimating future climate-related health outcomes.

Box 15.2

Linking Data with Models: Lessons from Mosquito-borne Disease

A key challenge in predicting the impacts of climate change on disease is determining the net impact of a multitude of individual effects, including some that offset others. The difficulty is to predict the net impact of all these climate influences acting simultaneously. Mathematical models provide a powerful method for this integration.

For example, a strong empirical test of the impacts of global warming on mosquito-borne disease transmission would require data (preferably weekly or monthly) on temperature, rainfall, mosquito abundance, infection prevalence in mosquitoes, and infection prevalence in humans (ideally age-structured), in addition to human density and vector control efforts over the time period in question. Since substantial year-to-year variation in climate is ubiquitous, a minimum time series of a decade is likely needed to successfully disentangle the influence of the myriad influences that control incidence of mosquito-borne disease in humans. Unfortunately, such datasets are rare or non-existent for even the most important human infectious diseases.

New disease introduction

Predicting where new pathogens will emerge is difficult. Certain socio-economic, environmental, and ecological factors are common in areas where infectious diseases tend to emerge, which may help identify risk areas (Jones et al. 2008). Global travel and trade have already contributed to emergence of pathogens in new areas (Gubler 2002; Tatem, Hay, and Rogers 2006; Randolph and Rogers 2010). Climate- and habitat-based models can identify habitat suitable for the species of interest and identify theoretical geographic distributions. However, the emergence of new pathogens in an area is more complicated, requiring suitable hosts and environmental conditions to facilitate arrival, establishment, and spread (Randolph and Rogers 2010).

15.6 Public Health Planning for Climate Change

Current climate-related human health effects, combined with forecasts for a warming climate and an increase in the vulnerable population, demand development and implementation of mitigation and adaptation strategies (Bollen et al. 2009; Jack and Kinney 2010). Immediate improvements to health would result from strengthening public health infrastructure to respond to climate-induced threats (Costello et al. 2011). The initial challenge is to assess linkages between climate and human health at city, state, and regional levels, and develop mitigation and adaptation plans for each spatial scale (Frumkin et al. 2008). Only certain states and cities in the Southwest currently have action plans (e.g., Boulder, Colorado; Phoenix, Arizona; California; Colorado; and New Mexico—for examples see ACCAG 2006, City of Boulder 2002, and NMCCAG 2006), with most focusing on reduction in anthropogenic waste heat (heat produced by human activities for which there is no useful application, such as the heat generated to cool a structure), air quality improvement, promotion of active lifestyles, and efficient use of natural resources.

Barriers to developing climate action plans include reduced tax revenue due to the 2007 financial downturn, along with competing budget priorities that have reduced the public funding available for long-term investment in mitigations to reduce human vulnerability to climate change. Several of the Southwest states (Arizona, California, Nevada) have been particularly affected by the ongoing recession (Heinberg 2011), including large reductions in the workforce of many local health departments (NACCHO 2010). Institutional barriers can limit development of partnerships between public, private, and non-profit groups to address climate change concerns (Harlan and Ruddell 2011), leading to lack of coordination, inefficient use of resources, or continued neglect of the most vulnerable and disenfranchised. Moreover, little is known about the cost-effectiveness of many climate change mitigation actions (Kalkstein et al. 1996; Kalkstein and Sheridan 2007). Together these factors provide challenges to action plan development. A recent publication provides a framework for identifying and organizing barriers to managing the risk and impacts of climate change (Ekstrom, Moser, and Torn 2011).

Adaptation

The health sector is more involved in adapting to climate change than mitigating it (Frumkin et al. 2008). Surveillance (for temperature-related morbidity and mortality, adverse effects related to air pollution and wildfire smoke exposure, and vector-borne diseases) is a key adaptation strategy that must be coordinated with first-alert systems and emergency services. For example, Los Angeles County has established an automated, near-real-time surveillance system to detect health changes in incidence of increased morbidity and mortality related to environmental stresses (LACDPH 2006). Public communications to both policy makers and the public should clearly emphasize that (a) climate change is already upon us, (b) it will be bad for human health, (c) the consequences will be worse if no adaptation or mitigation measures are taken, and (d) there are actions we can take individually and as a society that will simultaneously reduce the consequences of climate change and improve health (Maibach et al. 2011). To be effective, the framing of climate-change communication should be sensitive to concerns of various segments of the population. Public education campaigns, in multiple languages, should emphasize protective behaviors to reduce risk and provide care for vulnerable individuals and groups (Naughton et al. 2002; Weisskopf et al. 2002; Drechsler 2009). Access to cooling centers and other services to prevent climate-related morbidity and mortality are needed, particularly for the elderly, infirm, and economically disadvantaged. Vector-control programs and occupational safety standards for outdoor workers should be reviewed and strengthened. The federal block grant LIHEAP program[5] provides reduced energy rates during the cooling season for low-income residents of Arizona, Nevada, and New Mexico. Reduced energy rates are available only during the heating season in California, Utah, and Colorado. The influence of climate change on emissions and atmospheric chemistry will need to be included in future planning efforts to reduce emissions and attain the health-based NAAQS.

Co-benefits

Though the health sector usually focuses on adaptation to climate change, evidence is increasing that certain mitigation policies also provide ancillary health benefits. Quantifying the economic benefits to health may provide additional support for the implementation of these mitigation policies and help reduce the future public health impacts of climate change. For example, many actions that reduce greenhouse gas emissions also reduce emissions of PM2.5 and ozone precursors. Community designs that promote walking and bicycling to reduce emissions from vehicles also can help improve the health of individuals (Frumkin et al. 2008). An emerging body of research demonstrates a large potential source of health co-benefits from different mitigation strategies is the physical activity component of active transport (Woodcock et al. 2009; Grabow et al. 2011; Maizlish et al. 2011; Rabl and de Nazelle 2012). Health consequences of the mitigation activities themselves should be incorporated into cost-benefit analyses of mitigation strategies (Haines et al. 2009; Costello et al. 2011).

References

Anderson, G. B., and M. L. Bell. 2011. Heat waves in the United States: Mortality risk during heat waves and effect modification by heat wave characteristics in 43 U.S. communities. *Environmental Health Perspectives* 119:210–218.

Arizona Climate Change Advisory Group (ACCAG). 2006. *Climate change action plan, August 2006*. Phoenix: ACCAG.

Andreadis, T. G., P. M. Armstrong, and W. I. Bajwa. 2010. Studies on hibernating populations of *Culex pipiens* from a West Nile virus endemic focus in New York City: Parity rates and isolation of West Nile virus. *Journal of the American Mosquito Control Association* 26:257–264.

Baker, L. C., A. J. Brazel, N. Selover, C. Martin, N. McIntyre, F. R. Steiner, and A. Nelson. 2002. Urbanization and warming of Phoenix (Arizona, USA): Impacts, feedback, and mitigation. *Urban Ecosystems* 6:183–203.

Barker, C. M., B. F. Eldridge, and W. K. Reisen. 2010. Seasonal abundance of *Culex tarsalis* and *Culex pipiens* complex mosquitoes (Diptera: Culicidae) in California. *Journal of Medical Entomology* 47:759–768.

Basu, R., F. Dominici, and J. M. Samet. 2005. Temperature and mortality among the elderly in the United States: A comparison of epidemiologic methods. *Epidemiology* 16:58–66.

Basu, R., and B. Malig. 2011. High ambient temperature and mortality in California: Exploring the roles of age, disease, and mortality displacement. *Environmental Research* 111:1286–1292.

Basu, R., and B. D. Ostro. 2008. A multicounty analysis identifying the populations vulnerable to mortality associated with high ambient temperature in California. *American Journal of Epidemiology* 168:632–637.

Bell, E. 2011. Readying health services of climate change: A policy framework for regional development. *American Journal of Public Health* 101:804–813.

Ben Ari, T., A. Gershunov, K. L. Gage, T. Snall, P. Ettestad, K. L. Kausrud, and N. C. Stenseth. 2008. Human plague in the USA: The importance of regional and local climate. *Biological Letters* 4:737–740.

Bollen, J., B. Guay, S. Jamet, and J. Corfee-Morlot. 2009. *Co-benefits of climate change mitigation policies: Literature review and new results*. Working Paper 693. Paris: Organisation for Economic Cooperation and Development (OECD), Economics Department.

Brazel, A. J., N. Selover, R. Vose, and G. Heisler. 2000. The tale of two climates: Baltimore and Phoenix urban LTER sites. *Climate Research* 15:123–135.

Brooks, R. J. 1917. The influence of saturation deficiency and of temperature on the course of epidemic plague. *Journal of Hygiene* 5:881–899.

Brown, H. E., P. Ettestad, P. J. Reynolds, T. L. Brown, E. S. Hatton, J. L. Holmes, G. E. Glass, K. L. Gage, and R. J. Eisen. 2010. Climatic predictors of the intra- and inter-annual distributions of plague cases in New Mexico based on 29 years of animal-based surveillance data. *American Journal of Tropical Medicine and Hygiene* 82:95–102.

Bush, K. F., G. Luber, S. R. Kotha, R. Dhaliwal, V. Kapil, M. Pascual, D. G. Brown, et al. 2011. Impacts of climate change on public health in India: Future research directions. *Environmental Health Perspectives* 119:765–770, doi:10.1289/ehp.1003000.

California Air Resources Board (CARB). 2008. *Climate change scoping plan: A framework for change.* Sacramento: CARB. http://www.arb.ca.gov/cc/scopingplan/document/adopted_scoping_plan.pdf.

California Climate Action Team (CCAT). 2010. *Climate Action Team Biennial Report, April 2010.* N.p.: CCAT. http://www.energy.ca.gov/2010publications/CAT-1000-2010-004/CAT-1000-2010-004.PDF.

California Department of Public Health (CDPH). 2007. West Nile virus state of emergency: 2007 summary report. Sacramento: CDPH.

Campbell-Lendrum, D., and R. Woodruff. 2006. Comparative risk assessment of the burden of disease from climate change. *Environmental Health Perspectives* 114:1935–1941.

Cavanaugh, D. C., and J. D. Marshall. 1972. The influence of climate on the seasonal prevalence of plague in the Republic of Vietnam. *Journal of Wildlife Diseases* 8:85–94.

Cavanaugh, D. C. and J. E. Williams. 1977. Plague: Some ecological interrelationships. In *Proceedings of the International Conference on Fleas, Ashton Wold, Peterborough, UK, 21–25 June 1977*, ed. R. Traub and H. Starcke. Rotterdam: A. A. Balkema.

Centers for Disease Control (CDC). 2003. Increase in coccidioidomycosis: Arizona, 1998–2001. *MMWR Morbidity and Mortality Weekly Report* 52:109–12.

Centers for Disease Control and Prevention (CDC). 2005. Heat-related mortality: Arizona, 1993–2002 and United States, 1979–2002. *Morbidity and Mortality Weekly Report* 54:628–630.

—. 2006. Heat-related deaths: United States, 1999-2003. *Morbidity and Mortality Weekly Report* 55 (29): 796–798.

Chen, L. L., I. B. Tager, D. B. Peden, D. L. Christian, R. E. Ferrando, B. S. Welch, and J. R. Balmes. 2004. Effect of ozone exposure on airway responses to inhaled allergen in asthmatic subjects. *Chest* 125:2328–2335.

Christensen, N. K., A. W. Wood, N. Voisin, D. P. Lettermaier, and R. N. Palmer. 2004. The effects of climate change on the hydrology and water resources of the Colorado River Basin. *Climatic Change* 62:337–363.

City of Boulder. 2002. *City of Boulder: Climate Action Plan*. Boulder, CO: City of Boulder.

Comrie, A. C. 2005. Climate factors influencing coccidioidomycosis seasonality and outbreaks. *Environmental Health Perspectives* 113:688–692.

Comrie, A. C., and M. E. Glueck. 2007. Assessment of climate-coccidioidomycosis model - Model sensitivity for assessing climatologic effects on the risk of acquiring coccidioidomycosis. *Annals of the New York Academy of Science* 1111:83–95.

Cook, P. E., C. J. McMeniman, and S. L. O'Neill. 2008. Modifying insect population age structure to control vector-borne disease. *Advances in Experimental Medicine and Biology* 627:126–140.

Costello, A., M. Maslin, H. Montgomery, A. M. Johnson, and P. Ekins. 2011. Global health and climate change: Moving from denial and catastrophic fatalism to positive action. *Philosophical Transactions of the Royal Society A* 369:1866–1882.

Davis, D. H. S. 1953. Plague in Africa from 1935 to 1949: A survey of wild rodents in African territories. *Bulletin of the World Health Organization* 9:655–700.

Davis, R. E., P. C. Knappenberger, P. J. Michaels, and W. M. Novicoff. 2003. Changing heat-related mortality in the United States. *Environmental Health Perspectives* 111:1712–1718.

Delatte, H., G. Gimonneau, A. Triboire, and D. Fontenille. 2009. Influence of temperature on immature development, survival, fecundity, and gonotrophic cycles of *Aedes albopictus*, vector of chikungunya and dengue in the Indian Ocean. *Journal of Medical Entomology* 46:33–41.

Delfino, R. J., S. Brummel, J. Wu, H. Stern, B. Ostro, M. Lipsett, A. Winer, et al. 2009. The relationship of respiratory and cardiovascular hospital admissions to the Southern California wildfires of 2003. *Occupational and Environmental Medicine* 66:189–197.

Dettinger, M. D. 2005. From climate-change spaghetti to climate-change distributions for 21st century California. *San Francisco Estuary and Watershed Science* 3 (1): Article 4. http://repositories.cdlib.org/jmie/sfews/vol3/iss1/art4.

Drechsler, D. M. 2009. *Climate change and public health in California*. Final Report CEC-500-2009-034-F. Sacramento: California Climate Change Center. http://www.energy.ca.gov/2009publications/CEC-500-2009-034/CEC-500-2009-034-F.PDF.

Drechsler, D. M., N. Motallebi, M. Kleeman, D. Cayan, K. Hayhoe, L. S. Kalkstein, N. M. Miller, S. Sheridan, J. Jin, and R. A. VanCuren. 2006. *Public health-related impacts of climate change in California*. White Paper CEC-500-2005-197-SF. Sacramento: California Climate Change Center. http://www.energy.ca.gov/2005publications/CEC-500-2005-197/CEC-500-2005-197-SF.PDF.

Duclos, P., L. M. Sanderson, and M. Lipsett. 1990. The 1987 forest fire disaster in California: Assessment of emergency room visits. *Archives of Environmental Health* 45:53–58.

Ebi, K. L., J. Balbus, P. L. Kinney, E. Lipp, D. Mills, M. S. O'Neill, and M. L. Wilson. 2009. U.S. funding is insufficient to address the human health impacts of and public health responses to climate variability and change. *Environmental Health Perspectives* 117:857–862.

Ekstrom, J. A., S. C. Moser, and M. Torn. 2011. *Barriers to climate change adaptation: A diagnostic framework*. Public Interest Energy Research (PIER) Program final project report CEC-500-2011-004. Sacramento: California Energy Commission.

English, P. B., A. H. Sinclair, Z. Ross, H. Anderson, V. Boothe, C. Davis, K. Ebi, et al. 2009. Environmental health indicators of climate change for the United States: Findings from the State Environmental Health Indicator Collaborative. *Environmental Health Perspectives* 117:1673–1681.

Enscore, R. E., B. J. Biggerstaff, T. L. Brown, R. F. Fulgham, P. J. Reynolds, D. M. Engelthaler, C. E. Levy, et al. 2002. Modeling relationships between climate and the frequency of human plague cases in the southwestern United States, 1960–1997. *American Journal of Tropical Medicine and Hygiene* 66:186–196.

Flannigan, M. D., M. A. Krawchuk, W. J. de Groot, B. M. Wotton, and L. M. Gowman. 2009. Implications of global wildland fire. *International Journal of Wildland Fire* 18:483507.

Frumkin, H., J. Hess, G. Luber, J. Malilay, and M. McGeehin. 2008. Climate change: The public health response. *American Journal of Public Health* 98:435–445.

Gage, K. L., T. R. Burkot, R. J. Eisen, and E. B. Hayes. 2008. Climate and vectorborne diseases. *American Journal of Preventive Medicine* 35:436–450.

Gagge, A. P. 1981. The new effective temperature (ET*) – an index of human adaptation to warm environments. In *Environmental physiology: Aging, heat and altitude: Proceedings of Life, Heat and Altitude Conference, May 15–17, 1979, Las Vegas, NV*, ed. S. M. Horvath and M. K. Yousef, 59-77. New York: Elsevier/North-Holland.

Gershunov, A., D. Cayan and S. Iacobellis. 2009. The great 2006 heat wave over California and Nevada: Signal of an increasing trend. *Journal of Climate* 22:6181–6203.

Gershunov, A, Z. Johnston, H. G. Margolis, and K. Guirguis. 2011. The California heat wave 2006 with impacts on statewide medical emergency: A space-time analysis. *Geography Research Forum* 31:6–31.

Grabow, M. L., S. N. Spak, T. Holloway, B. Stone, A. C. Mednick, and J. A. Patz. 2011. Air quality and exercise-related health benefits from reduced car travel in the midwestern United States. *Environmental Health Perspectives* 120:68–76, doi:10.1289/ehp.1103440.

Gubler, D. J. 2002. Epidemic dengue/dengue hemorrhagic fever as a public health, social and economic problem in the 21st century. *TRENDS in Microbiology* 10:100–103.

Gubler, D. J., P. Reiter, K. L. Ebi, W. Yap, R. Nasci, and J. A. Patz. 2001. Climate variability and change in the United States: Potential impacts on vector- and rodent-borne diseases. *Environmental Health Perspectives* 109:223–233.

Guirguis, K. and A. Gershunov. Forthcoming. California heat waves in the present and future. *Geophysical Research Letters* 39.

Haines, A., A. J. McMichael, K. R. Smith, I. Roberts, J. Woodcock, A. Markandya, B. G. Armstrong, et al. 2009. Public health benefits of strategies to reduce greenhouse-gas emissions: overview and implications for policy makers. *The Lancet* 374:2104–2114.

Hall-Baker, P. A., S. L. Groseclose, R. A. Jajosky, D. A. Adams, P. Sharp, W. J. Anderson, J. P. Abellera, et al. 2011. Summary of notifiable diseases: United States, 2009. *Morbidity and Mortality Weekly Report* 58 (53): 1–100.

Hall-Baker, P. A., E. Nieves, R. A. Jajosky, D. A. Adams, P. Sharp, W. J. Anderson, J. J. Aponte, G. F. Jones, A. E. Aranas, S. B. Katz, et al. 2010. Summary of notifiable diseases: United States, 2008. *Morbidity and Mortality Weekly Report* 57 (54): 1–94.

Hall-Baker, P. A., E. Nieves, R. A. Jajosky, D. A. Adams, P. Sharp, W. J. Anderson, J. J. Aponte, G. F. Jones, A. E. Aranas, A. Rey, et al. 2009. Summary of notifiable diseases: United States, 2007, *Morbidity and Mortality Weekly Report* 56 (53): 1–94.

Harlan, S. L., A. J. Brazel, L. Prashad, W. L. Stefanov, and L. Larsen. 2006. Neighborhood microclimates and vulnerability to heat stress. *Social Science and Medicine* 63:2847–2863.

Harlan, S. L., and D. Ruddell. 2011. Climate change and health in cities: Impacts of heat and air pollution and potential co-benefits from mitigation and adaptation. *Current Opinion in Environmental Sustainability* 3:126–134.

Hartley, D. M., C. M. Barker, A. Le Menach, T. Niu, H. D. Gaff, and W. K. Reisen. 2012. The effects of temperature on the emergence and seasonality of West Nile virus in California. *American Journal of Tropical Medicine and Hygiene* 86:884–894.

Hayhoe, K. D., D. Cayan, C. B. Field, P. C. Frumhoff, E. P. Maurer, N. L. Miller, S. C. Moser, et al. 2004. Emissions pathways, climate change, and impacts on California. *Proceedings of the National Academy of Sciences* 101:12422–12427.

Heinberg, R. 2011. *The end of growth: Adapting to our new economic reality*. Gabriola Island, BC: New Society Publishers.

Holz, O., M. Mücke, K. Paasch, S. Böhme, P. Timm, K. Richter, H. Magnussen, and R. A. Jörres. 2002. Repeated ozone exposures enhance bronchial allergen responses in subjects with rhinitis or asthma. *Clinical and Experimental Allergy* 32:681–689.

Horvath, S. M. 1981. Historical perspectives of adaptation to heat. In *Environmental physiology: Aging, heat and altitude: Proceedings of Life, Heat and Altitude Conference, May 15–17, 1979, Las Vegas, NV*, ed. S. M. Horvath and M. K. Yousef. New York: Elsevier/North-Holland.

Intergovernmental Panel on Climate Change (IPCC) 2007. *Climate change 2007: Synthesis Report. Contribution of Working Groups I, II and III to the Fourth Assessment Report of the Intergovernmental Panel on Climate Change*, ed. R. K. Pachauri and A. Reisinger. Geneva: IPCC. http://www.ipcc.ch/pdf/assessment-report/ar4/syr/ar4_syr.pdf.

Jack, D. W., and P. L. Kinney. 2010. Health co-benefits of climate mitigation in urban areas. *Current Opinion in Environmental Sustainability* 2:172–177.

Jacobson, M. Z. 2008. On the causal link between carbon dioxide and air pollution mortality. *Geophysical Research Letters* 35: L03809, doi:10.1029/2007GL031101.

Jones, K. E., N. G. Patel, M. A. Levy, A. Storeygard, D. Balk, J. L. Gittleman, and P. Daszak. 2008. Global trends in emerging infectious diseases. *Nature* 451:990–994.

Kalkstein, A., and S. Sheridan. 2007. The social impacts of the heat–health watch/warning system in Phoenix, Arizona: Assessing the perceived risk and response of the public. *International Journal of Biometeorology* 52:43–55.

Kalkstein, L. S., P. P. Jamason, J. S. Greene, J. Libby, and L. Robinson. 1996. The Philadelphia hot weather-health watch warning system: Development and application, summer 1995. *Bulletin of the American Meteorological Society* 77:1519–1528.

Karl, T. R., J. M. Melillo, and T. C. Peterson, eds. 2009. *Global climate change impacts in the United States*. Cambridge: Cambridge University Press. http://downloads.globalchange.gov/usimpacts/pdfs/climate-impacts-report.pdf.

Kilpatrick, A. M. 2011. West Nile virus: Globalization, land use, and the emergence of an infectious disease. *Science* 334:323–327.

Kilpatrick, A. M., M. A. Meola, R. M. Moudy, and L. D. Kramer. 2008. Temperature, viral genetics, and the transmission of West Nile virus by *Culex pipiens* mosquitoes. *PLoS Pathogens* 4: e1000092.

Kleeman, M. J. 2008. A preliminary assessment of the sensitivity of air quality in California to global change. *Climatic Change* 87 (Suppl. 1): 273–292.

Knowlton, K., J. E. Rosenthal, C. Hogrefe, C., B. Lynn, S. Gaffin, R. Goldberg, C. Rosenzweig, K. Civerolo, J. Y. Ku, and P. L. Kinney. 2004. Assessing ozone-related health impacts under a changing climate. *Environmental Health Perspectives* 112:1557–1563.

Knowlton, K., M. Rotkin-Ellman, G. King, H. G. Margolis, D. Smith, G. Solomon, R. Trent, and P. English. 2009. The 2006 California heat wave: Impacts on hospitalizations and emergency department visits. *Environmental Health Perspectives* 117:61–67.

Kolivras, K. N., P. S. Johnson, A. C. Comrie, and S. R. Yool. 2001. Environmental variability and coccidioidomycosis (valley fever). *Aerobiologia* 17:31–42.

Kunkel, K. E., R. A. Pielke Jr., and S. A. Changnon. 1999. Temporal fluctuations in weather and climate extremes that cause economic and human health impacts: A review. *Bulletin of the American Meteorological Society* 80:1077–1098.

Künzli, N., E. Avol, J. Wu, W. J. Gauderman, E. Rappaport, J. Millstein, J. Bennion, et al. 2006. Health effects of the 2003 Southern California wildfires on children. *American Journal Respiratory and Critical Care Medicine* 174:1221–1228.

Lafferty, K. D. 2009. The ecology of climate change and infectious diseases. *Ecology* 90:888–900.

Lipsett, M., K. Waller, D. Shusterman, S. Thollaug, and W. Brunner. 1994. The respiratory health impact of a large urban fire. *American Journal of Public Health* 84:434–438.

Los Angeles Department of Public Health (LACDPH). 2006. Emergency Department syndromic surveillance and population-based health monitoring in Los Angeles. Acute Communicable Disease Control Program Special Studies Report. Los Angeles: LACDPH.

Mahmud, A., M. Tyree, D. Cayan, N. Motallebi, and M. J. Kleeman. 2008. Statistical downscaling of climate change impacts on ozone concentrations in California. *Journal of Geophysical Research* 113: D21104, doi:10.1029/2007JD009534.

Maibach, E., A. Leiserowitz, C. Roser-Renouf, C. K. Mertz. 2011. Identifying like-minded audiences for global warming public engagement campaigns: An audience segmentation analysis and tool development. *PLoS ONE* 6: e17571.

Maizlish, N. A., J. D. Woodcock, S. Co, B. Ostro, D. Fairley, and A. Fanai. 2011. *Health co-benefits and transportation-related reductions in greenhouse gas emissions in the Bay Area*. Technical report. Sacramento: California Department of Public Health. http://www.cdph.ca.gov/programs/CCDPHP/Documents/ITHIM_Technical_Report11-21-11.pdf.

McMichael, A. J. 2001. Health consequences of global climate change. *Journal of the Royal Society of Medicine* 94:111–114.

—. 2012. Insights from past millennia into climatic impacts on human health and survival. *Proceedings of the National Academy of Sciences*, published online, doi:10.1073/pnas.1120177109.

McNabb, S. J. N., R. A. Jajosky, P. A. Hall-Baker, D. A. Adams, P. Sharp, W. J. Anderson, J. J. Aponte, et al. 2007. Summary of notifiable diseases: United States, 2005. *Morbidity and Mortality Weekly Report* 54 (53): 292.

McNabb, S. J. N., R. A. Jajosky, P. A. Hall-Baker, D. A. Adams, P. Sharp, C. Worsham, W. J. Anderson, et al. 2008. Summary of notifiable diseases: United States, 2006. *Morbidity and Mortality Weekly Report* 55 (53): 1–94.

Mead, P. S., L. Slutsker, V. Dietz, L. F. McCaig, J. S. Bresee, C. Shapiro, P. M. Griffin, and R. V. Tauxe. 1999. Food-related illness and death in the United States. *Emerging Infectious Diseases* 5:607–625.

Meehl, G. A., and C. Tebaldi. 2004. More intense, more frequent, and longer lasting heat waves in the 21st century. *Science* 305:994–997.

Merrill, S. A., F. B. Ramberg, and H. H. Hagedorn. 2005. Phylogeography and population structure of *Aedes aegypti* in Arizona. *American Journal of Tropical Medicine and Hygiene* 72:304–310.

Millstein, D. E., and R. A. Harley. 2009. *Impact of climate change on photochemical air pollution in Southern California.* Final Paper CEC-500-2009-021-D. Sacramento: California Climate Change Center. http://www.energy.ca.gov/2009publications/CEC-500-2009-021/CEC-500-2009-021-F.PDF.

Molfino, N. A., S. C. Wright, I. Katz, S. Tarlo, F. Silverman, P. A. McClean, J. P. Szalai, M. Raizenne, A. S. Slutsky, and N. Zamel. 1991. Effect of low concentrations of ozone on inhaled allergen responses in asthmatic subjects. *Lancet* 338:199–203.

Morin, C. W., and A. C. Comrie. 2010. Modeled response of the West Nile virus vector *Culex quinquefasciatus* to changing climate using the dynamic mosquito simulation model. *International Journal of Biometeorology* 54:517–529.

National Association of County and City Health Officials (NACCHO). 2010. *Local health department job losses and program cuts: Findings from January/February 2010 survey*. Research Brief. Washington, DC: NACCHO.

Naughton, M., A. Henderson, M. Mirabelli, R. Kaiser, J. Wilhelm, S. Kieszak, C. Rubin, and M. McGeehin. 2002. Heat-related mortality during a 1999 heat wave in Chicago. *American Journal of Preventative Medicine* 22:221–227.

New Mexico Climate Change Advisory Group (NMCCAG). 2006. *Final Report, December 2006*. Albuquerque: New Mexico Environment Department. http://www.nmclimatechange.us/ewebeditpro/items/O117F10150.pdf.

Oke, T. R. 1982. The energetic basis of the urban heat island. *Quarterly Journal of the Royal Meteorological Society* 108:1–24.

Ostro, B. D., S. Rauch, and S. Green. 2011. Quantifying the health impacts of future changes in temperature in California. *Environmental Research* 111:1258–1264, doi:10.1016/j.envres.2011.08.013.

Ostro, B. D., L. A. Roth, R. S. Green, and R. Basu. 2009. Estimating the mortality effect of the July 2006 California heat wave. *Environmental Research* 109:614–619.

Ostro, B. D., L. A. Roth, R. S. Green, B. Malig, and R. Basu. 2010. The effects of temperature and use of air conditioning on hospitalizations. *American Journal of Epidemiology* 172:1053–1061.

Parmenter, R. R., E. P. Yadav, C. A. Parmenter, P. Ettestad, and K. L. Gage. 1999. Incidence of plague associated with increased winter-spring precipitation in New Mexico. *American Journal of Tropical Medicine and Hygiene* 61:814–821.

Parmesan, C., T. L. Root, and M.R. Willig. 2000. Impacts of extreme weather and climate on terrestrial biota. *Bulletin of American Meteorological Society* 81:443–450.

Patz, J. A., D. Campbell-Lendrum, T. Holloway, and J. A. Foley. 2005. Impact of regional climate change on human health. *Nature* 438:310–317.

Patz, J. A., A. Githeko, J. P. McCarty, S. Hussain, U. Confalorieri, and N. de Wet. 2003. Climate change and infectious diseases. In *Climate change and human health: Risks and responses,* ed. A. J. McMichael, D. H. Campbell-Lendrum, C. F. Corvalán, K. L. Ebi, A. Githeko, J. D. Scheraga and A. Woodward, 103–133. Geneva: World Health Organization.

Peng, R. D., J. F. Bobb, C. Tebaldi, L. McDaniel, M. L. Bell, and F. Dominici. 2011. Toward a quantitative estimate of future heat wave mortality under global climate change. *Environmental Health Perspectives* 119:701–706.

Phuleria, H., P. M. Fine, Y. Zhu, and C. Sioutas. 2005. Air quality impacts of the October 2003 Southern California wildfires. *Journal Geophysical Research* 110: D07S20.

Rabl, A., and A. de Nazelle. 2012. Benefits of shift from car to active transport. *Transportation Policy* 19:121–131.

Randolph, S. E. 2009. Perspectives on climate change impacts on infectious diseases. *Ecology* 90:927–931.

Randolph, S. E. and D. J. Rogers. 2010. The arrival, establishment and spread of exotic diseases: Patterns and predictions. *Nature Reviews – Microbiology* 8:361–371.

Reeves, W. C., J. L. Hardy, W. K. Reisen, and M. M. Milby. 1994. Potential effect of global warming on mosquito-borne arboviruses. *Journal of Medical Entomology* 31:323–332.

Reisen, W. K. 1995. Effect of temperature on *Culex tarsalis* (Diptera: Culicidae) from the Coachella and San Joaquin Valleys of California. *Journal of Medical Entomology* 32:636–645.

Reisen, W. K., Y. Fang, H. D. Lothrop, V. M. Martinez, J. Wilson, P. O'Connor, R. Carney, B. Cahoon-Young, M. Shafii, and A. C. Brault. 2006. Overwintering of West Nile virus in Southern California. *Journal of Medical Entomology* 43:344–355.

Reisen, W. K., Y. Fang, and V. M. Martinez. 2005. Avian host and mosquito (Diptera: Culicidae) vector competence determine the efficiency of West Nile and St. Louis encephalitis virus transmission. *Journal of Medical Entomology* 42:367–375.

—. 2006. Effects of temperature on the transmission of West Nile virus by *Culex tarsalis* (Diptera: Culicidae). *Journal of Medical Entomology* 43:309–317.

Reisen, W. K., R. P. Meyer, and M. M. Milby. 1986. Overwintering studies on *Culex tarsalis* (Diptera: Culicidae) in Kern County, California: Temporal changes in abundance and reproductive status with comparative observations on *C. quinquefasciatus* (Diptera: Culicidae). *Annals of the Entomology Society of America* 79:677–685.

Reisen, W. K., R. P. Meyer, S. B. Presser, and J. L. Hardy. 1993. Effect of temperature on the transmission of western equine encephalomyelitis and St. Louis encephalitis viruses by *Culex tarsalis* (Diptera: Culicidae). *Journal of Medical Entomology* 30:151–160.

Reisen, W. K., P. T. Smith, and H. D. Lothrop. 1995. Short term reproductive diapause by *Culex tarsalis* (Diptera: Culicidae) in the Coachella Valley of California. *Journal of Medical Entomology* 32:654–662.

Reisen W. K., T. Thiemann, C. M. Barker, H. Lu, B. Carroll, Y. Fang, and H. D. Lothrop. 2010. Effects of warm winter temperature on the abundance and gonotrophic activity of *Culex* (Diptera: Culicidae) in California. *Journal of Medical Entomology* 47:230-237.

Rosenzweig, C., W. D. Solecki, L. Parshall, M. Chopping, G. Pope, and R. Goldberg. 2005. Characterizing the urban heat island in current and future climates in New Jersey. *Global Environmental Change Part B: Environmental Hazards* 6:51–62.

Ruddell, D. M., S. L. Harlan, S. Grossman-Clarke, and A. Buyanteyev. 2010. Risk and exposure to extreme heat in microclimates of Phoenix, AZ. In *Geospatial techniques in urban hazard and disaster analysis*, ed. P. Showalter and Y. Lu, 179–202. New York: Springer.

Ruddell, D., S. L. Harlan, S. Grossman-Clarke, and G. Chowell. 2011. Scales of perception: Public awareness of regional and neighborhood climates. *Climatic Change*, published online, doi:10.1007/s10584-011-0165-y.

Semenza, J., C. Rubic, K. Falter, J. Selanikio, W. Flanders, H. Howe, and J. Wilhelm. 1996. Heat-related deaths during the July 1995 heat wave in Chicago. *The New England Journal of Medicine* 335:84–90.

Sheridan, S., C. Lee, M. Allen, and L. Kalkstein. 2011. *A spatial synoptic classification approach to projected heat vulnerability in California under future climate change scenarios*. Final report. Sacramento: California Air Resources Board.

Sherwood, S. C., and M. Huber. 2010. An adaptability limit to climate change due to heat stress. *Proceedings of the National Academy of Sciences* 107:9552–9555.

Shusterman, D., J. Z. Kaplan, and C. Canabarro. 1993. Immediate health-effects of an urban wildfire. *Western Journal of Medicine* 158:133–138.

Smoyer, K.E. 1998. A comparative analysis of heat waves and associated mortality in St. Louis, Missouri: 1980 and 1995. *International Journal of Biometeorology* 42:44–50.

Steiner, A. L., S. Tonse, R. C. Cohen, A. H. Goldstein, and R. A. Harley. 2006. Influence of future climate and emissions on regional air quality in California. *Journal of Geophysical Research* 111: D18303, doi:10.1029/2005JD006935.

Sutherland, E. R., B. J. Make, S. Vedal, L. N. Zhang, S. J. Dutton, J. R. Murphy, and P. E. Silkoff. 2005. Wildfire smoke and respiratory symptoms in patients with chronic obstructive pulmonary disease. *Journal of Allergy and Clinical Immunology* 115:420–422.

Tamerius, J. D., and A. C. Comrie. 2011. Coccidioidomycosis incidence in Arizona predicted by seasonal precipitation. *PLoS One* 6: e21009.

Tatem, A. J., S. I. Hay, and D. J. Rogers. 2006. Global traffic and disease vector dispersal. *Proceedings of the National Academy of Sciences* 103:6242–6247.

Tesh, R. B., R. Parsons, M. Siirin, Y. Randle, C. Sargent, H. Guzman, T. Wuithiranyagool, et al. 2004. Year-round West Nile virus activity, Gulf Coast region, Texas and Louisiana. *Emerging Infectious Diseases* 10:1649–1652.

U.S. Environmental Protection Agency (EPA). 2006. *Air quality criteria for ozone and related photochemical oxidants (2006 final)*, vol. I. EPA 600/R-05/004aF. Washington, DC: EPA. http://cfpub.epa.gov/ncea/cfm/recordisplay.cfm?deid=149923.

—. 2008. *Review of the impacts of climate variability and change on aeroallergens and their associated effects (final report)*. EPA/600/R-06/164F. Washington, DC: EPA. http://cfpub.epa.gov/ncea/cfm/recordisplay.cfm?deid=190306.

—. 2009. *Integrated science assessment for particulate matter (final report)*. EPA/600/R-08/139F. Washington, DC: EPA. http://cfpub.epa.gov/ncea/cfm/recordisplay.cfm?deid=216546.

—. 2011a. Area designations for 2008 ground-level ozone standards. Washington, DC: EPA. http://www.epa.gov/ozonedesignations/2008standards/.

—. 2011b. Particulate matter (PM-2.5) 2006 standard nonattainment area state map. Washington, DC: EPA. http://www.epa.gov/airquality/greenbk/rnmapa.html.

Vedal, S. 2003. Wildfire air pollution and respiratory emergency visits: A natural experiment. *American Journal of Respiratory and Critical Care Medicine* 167: A974.

Vedal, S., and S. J. Dutton. 2006. Wildfire air pollution and daily mortality in a large urban area. *Environmental Research* 102:29–35.

Vincent, G. K., and V. A. Velkoff. 2010. *The next four decades, the older population in the United States: 2010 to 2050.* Current Population Reports P25-1138. Washington, DC: U.S. Census Bureau.

Weber, R. W. 2012. Impact of climate change on allergens. *Annals of Allergy, Asthma and Immunology* 108:294–299.

Weisskopf, M. G., H. A. Anderson, S. Foldy, L. P. Hanrahan, K. Blair, T. J. Török, and P. D. Rumm. 2002. Heat wave morbidity and mortality, Milwaukee, Wis, 1999 vs 1995: An improved response? *American Journal of Public Health* 92:830–833.

Westerling, A. L., B. P. Bryant, H. K. Preisler, T. P. Holmes, H. G. Hidalgo, T. Das, and S.R. Shrestha. 2009. *Climate change, growth, and California wildfire*. Final Paper CEC-500-2009-046-F. Sacramento: California Climate Change Center. http://www.energy.ca.gov/2009publications/CEC-500-2009-046/CEC-500-2009-046-F.PDF.

Westerling, A. L., H. G. Hidalgo, D. Cayan, and T. W. Swetnam. 2006. Warming and earlier spring increases western U.S. forest wildfire activity. *Science* 313:940–943.

Woodcock, J., P. Edwards, C. Tonne, B. G. Armstrong, O. Ashiru, D. Banister, S. Beevers, et al. 2009. Public health benefits of strategies to reduce greenhouse-gas emissions: Urban land transport. *The Lancet* 374:1930–1943.

World Health Organization (WHO). 2010. *Climate change and health*. Fact Sheet 266, January 2010. Geneva: WHO.

Wu, J., A. Winer, and R. Delfino. 2006. Exposure assessment of particulate matter air pollution before, during, and after the 2003 southern California wildfires. *Atmospheric Environment* 40:3333–3338.

Zender, C. S., and J. Talamantes. 2006. Climate controls on valley fever incidence in Kern County, California. *International Journal of Biometeorology* 50:174–182.

Endnotes

i The smaller size of PM2.5 particulates allows them to lodge deeply in the lungs.

ii This is a smaller amount of growth in lung function during the child's growth period. Both the growth rate and the attained lung function at adulthood seem to be smaller in children growing up in high PM2.5 areas.

iii Mixing height is the level of the inversion layer. It is like an atmospheric ceiling that limits the volume of air into which air pollution can mix. High air pollution is associated with a low mixing height/inversion layer, while a high mixing height is associated with better air quality.

iv Ozone forms in the atmosphere as the result of reactions involving sunlight and two classes of directly emitted precursors. One group of precursors includes various oxides of nitrogen, such as nitric oxide and nitrogen dioxide, and the other group includes volatile organic compounds (also called reactive organic gases), such as hydrocarbons.

v Low Income Home Energy Assistance Program (LIHEAP), http://www.liheap.ncat.org.

Chapter 16

Climate Change and U.S.-Mexico Border Communities

COORDINATING LEAD AUTHOR

Margaret Wilder (University of Arizona)

LEAD AUTHORS

Gregg Garfin (University of Arizona), Paul Ganster (Institute for Regional Studies of the Californias, San Diego State University), Hallie Eakin (Arizona State University), Patricia Romero-Lankao (NCAR), Francisco Lara-Valencia (Arizona State University), Alfonso A. Cortez-Lara (Colegio de la Frontera Norte), Stephen Mumme (Colorado State University), Carolina Neri (National Autonomous University of Mexico), Francisco Muñoz-Arriola (Scripps Institution of Oceanography)

REVIEW EDITOR

Robert G. Varady (University of Arizona)

Executive Summary

This chapter examines climate-related vulnerability in the western portion of the U.S.-Mexico border region from the Pacific coast of California–Baja California to El Paso–Ciudad Juárez, focusing primarily on border counties in the United States and municipalities in Mexico. Beginning with a brief overview of projected climate changes for the region, the chapter analyzes the demographic, socioeconomic, institutional, and other drivers of climate-related vulnerability, and the potential impacts of climate change across multiple sectors (e.g., water, agriculture and ranching, and biodiverse ecosystems). The border region has higher poverty, water insecurity, substandard housing,

---

Chapter citation: Wilder, M., G. Garfin, P. Ganster, H. Eakin, P. Romero-Lankao, F. Lara-Valencia, A. A. Cortez-Lara, S. Mumme, C. Neri, and F. Muñoz-Arriola. 2013. "Climate Change and U.S.-Mexico Border Communities." In *Assessment of Climate Change in the Southwest United States: A Report Prepared for the National Climate Assessment*, edited by G. Garfin, A. Jardine, R. Merideth, M. Black, and S. LeRoy, 340–384. A report by the Southwest Climate Alliance. Washington, DC: Island Press.

and lack of urban planning relative to the rest of the United States, and multiple socioeconomic asymmetries exist between the U.S. and Mexico sides of the border. These asymmetries create challenges for governance, planning, effective communication of climate-related risks, and design of adaptation strategies. Although they represent an important part of the picture, a comprehensive assessment of regional adaptation strategies was not within the scope of the chapter.

The chapter highlights the following key findings relating to climate change and socioeconomic and cultural diversity, water, wetlands ecosystems, and institutions and governance.

- Climate change exposes the populations in the border region to uneven impacts, due to their cultural and institutional diversity and uneven economic development. (high confidence)
- Climate change exposes sensitive wetland ecosystems, which are hotspots of border region biodiversity, to impacts such as reduced precipitation and extended drought. (high confidence)
- Projected climate changes will put additional pressure on severely stressed water systems and may exacerbate existing vulnerabilities relating to water supply and water quality. Cascading effects of additional stress on water systems include: challenges to energy infrastructure, agriculture, food security, and traditional farming and ranching cultures prevalent in the border region. (medium-high confidence)
- Building adaptive capacity to climate change generally benefits from efforts to cooperate and collaborate to resolve trans-border environmental problems, yet asymmetries in information collection, the definition and scope of problems, and language create challenges to effective cooperation and collaboration. (medium-high confidence)
- Institutional asymmetries, including distinctions in governance approaches—centralized (Mexico) versus decentralized (United States)—and institutional fragmentation and complexity, make the task of collaboration daunting, and reduce the potential adaptive capacity in the region. (medium-high confidence)

16.1 Introduction

While the U.S.-Mexico border has been called a "third country" and has been identified as a distinct region (Anzaldúa 1987), the challenges it faces are due in large measure to its high degree of *integration* into global processes of economic and environmental change. The border region is characterized by a so-called "double exposure" (Leichenko and O'Brien 2008)—meaning that environmental change in the region is driven by accelerated processes of global economic integration (such as foreign-owned industries and international migration) coupled with intensive climate change. It is critical to understand the drivers of climate-related vulnerability and capacities for adaptation in the region in the context of the region's distinct history and contemporary challenges, shared climate regime, transboundary watersheds and airsheds, and interdependent economies and cultures.

This chapter defines the border region and how observed climate trends since 1961 and projected climate change conditions have affected or are likely to affect the region. Next, the chapter provides a framework to understand climate-related vulnerability, adaptive capacity, and adaptation, and examines the major drivers (forces) that lead to vulnerability and the evidence of sectoral impacts of climate change resulting in vulnerability in the border region.

There are three important caveats as to the scope and analysis of this study. First, consistent with the risk-based vulnerability framework suggested by the National Climate Assessment, this discussion assesses the sensitivity, exposure, and capacity for response for a given population or sector. The evidence for this analysis is based on qualitative and (to a lesser extent) quantitative studies in specific contexts within the border region. With regard to the climatology, vulnerability, consequences, and impacts, there are no comprehensive studies that encompass the western portion of the U.S.-Mexico border. The highlighted vulnerabilities presented in the executive summary are those that are sustained by strong evidence from multiple contexts within the region. In most cases (especially drawing on qualitative studies), while the evidence of a vulnerability, impact, or consequence may be strong, it is often not sufficiently calibrated to assess degrees of exposure or sensitivity (of populations or sectors) to a climate risk. Thus, to a large extent, there is an imperfect fit between the kind of evidence available and the requirement to assess precisely the relative exposure and sensitivity. Second, for the purposes of this report, this chapter analyzes and highlights those areas of vulnerability (e.g., urban, agriculture, socioeconomic) judged to be of paramount importance and for which there is robust evidence, while excluding other important regional vulnerabilities (such as health and livelihoods). Third, this chapter focuses on assessing *key vulnerabilities*. It is not within the scope of this chapter to fully represent the adaptation activities—ongoing or planned—that may aid in reducing these vulnerabilities, but some are discussed here. (See also Chapter 18 for a general overview of solutions and choices for responding to climate change in ways that reduce risks and support sustainable development.) Throughout this chapter, four callout boxes present evidence of successful trans-border cooperation or collaboration. Collaboration is a significant component for strengthening the region's adaptive capacity and for building resilience.

16.2 Definition of the Border Region

The analysis here focuses on the U.S.-Mexico border—delimited for the purposes of this chapter to the western portion of the border from San Diego-Tijuana on the Pacific coast to the Paso del Norte area[i] of the Rio Grande—which has been identified as a distinct region that serves as the interface between Mexico and the western United States (Ganster and Lorey 2008) (Figure 16.1). The western portion of the border region corresponds to the definition used in the National Climate Assessment (NCA) for the Southwest region, which includes among the U.S. border states only California, Arizona, and New Mexico (and excludes Texas, which is part of the NCA's Middle West region). Nevertheless, this analysis incorporates the Paso del Norte corridor due to its key importance in the New Mexico portion of the border region.

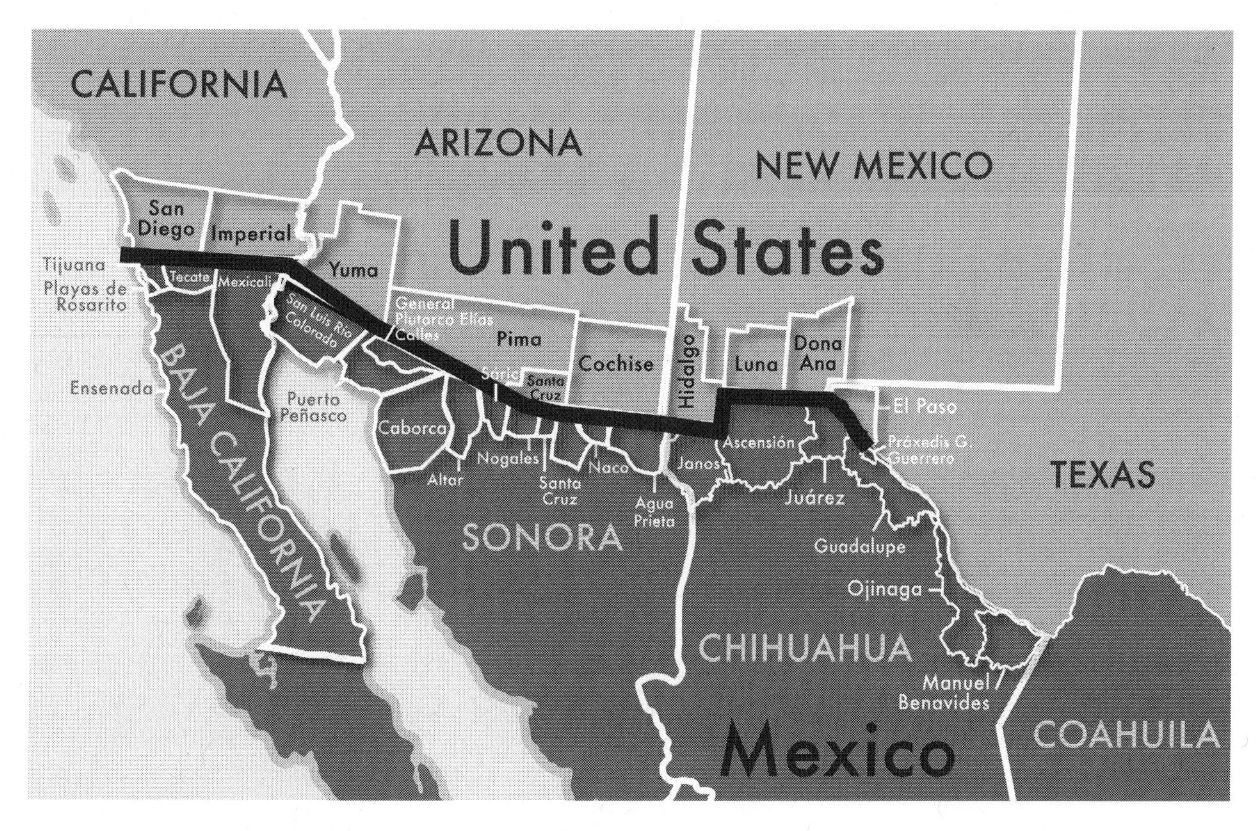

Figure 16.1 Western portion of the U.S.-Mexico border region. Source: EPA (2011).

Most of the border's population is concentrated along the international boundary in fourteen city pairs (eight of them in the western portion)[ii] that constitute binational urban systems. Rural population is scarce except for the irrigated areas of the Colorado River and the Imperial-Mexicali valleys.

The border region can be defined in a number of ways (Ganster and Lorey 2008; Varady and Ward 2009). These include the six Mexican and four U.S. border states, the region of shared culture and language bisected by the border, the watersheds and subbasins along the boundary, the 62-mile zone (100 kilometers) on each side of international line as defined by the La Paz Agreement between Mexico and the United States, or by the administrative boundaries of the U.S. counties and the Mexican municipalities (*municipos*) that abut the international boundary. This chapter covers three U.S. states (California, Arizona, and New Mexico), the El Paso corridor, and three Mexican states (Baja California, Sonora, and Chihuahua). For present purposes, the latter category—border counties and municipios—is most important in terms of societal vulnerability to climate change, given the border population concentration in major urban areas. While the focus is on the region that includes the counties and municipalities along the border, data from these local administrative units are supplemented with state-level data.

16.3 Border Region Climate Variability, Climate Change, and Impacts

The border region considered here is characterized by high aridity and high temperatures. Typically, about half of the eastern part of the region's precipitation falls in the summer months, associated with the North American monsoon, while the majority of annual precipitation in the Californias falls between November and March. The region is subject to both significant inter-annual and multi-decadal variability in precipitation.[iii] This variability, associated with ENSO, has driven droughts and floods and challenged hydrological planning in the region.[iv] Further challenging this understanding is a paucity of data, particularly on the high-altitude mountainous regions in northern Mexico. Differences in the availability of high-quality and continuous meteorological and hydrological records spanning long periods of time, and relatively poor data sharing complicate understanding of the border region's climate. The scarcity of such data makes it difficult to verify climate model projections at fine spatial scales.

Also, reconciling differences in projected changes in temperature, based on global climate model (GCM) studies conducted separately by U.S. and Mexican scientists (Table 16.1),[v] is complicated by the fact that (1) they use different sets of models from the IPCC Fourth Assessment archive; (2) they use different methods of downscaling output from coarse spatial scale models to finer regional spatial scales;[vi] (3) in some cases they do not use the same greenhouse gas (GHG) emissions scenarios; (4) they average future projections for different spans of years; and (5) they use different spans of years for providing a measure of average historical climate. High quality data are essential for statistically downscaling GCM output. Thus, issues with meteorological observations add to several other sources of uncertainty (see discussion in Chapters 2 and 19).

Table 16.1 Summary of projected changes in selected climate parameters

Projected Change	Direction of Change	Border Subregion Affected	Confidence
Average annual temperature	Increasing	Throughout the border region; lowest magnitude of increase is near the coast; greatest is Arizona-Sonora border or New Mexico-Chihuahua border	High
Average summer temperature	Increasing	Throughout the border region; greatest increases in the Sonoran Desert border region	High
Average winter temperature	Increasing	Throughout the border region; greatest increases in the Sonoran Desert border region	High
Average annual maximum temperature	Increasing	Throughout the border region; greatest increases in the eastern Chihuahuan Desert; only estimated south of the border	Medium-High
Average annual minimum temperature	Increasing	Throughout the border region; greatest increases in the Sonoran Desert; only estimated south of the border	Medium-High

Table 16.1 Summary of projected changes in selected climate parameters (Continued)

Projected Change	Direction of Change	Border Subregion Affected	Confidence
Length of freeze-free season	Increasing	Throughout the border region; only estimated north of the border	Medium-High
Annual number of days with maximum temperatures > 100°F	Increasing	Throughout the border region; greatest increases in the central Sonoran Desert border and in northwest Chihuahua	Medium-High
Heat wave duration	Increasing	Throughout the border region	High
Cooling degree days	Increasing	Throughout the border region	High
Cold episodes	Decreasing	Throughout the border region	Medium-High
Annual precipitation	Decreasing	Greatest decreases along the coast and parts of the Arizona-Sonora border	Medium-High
Winter precipitation	Decreasing	Greatest and most consistent decreases (over time) are projected for the Arizona-Sonora border, into the western Chihuahuan Desert	Medium-Low
Spring precipitation	Decreasing	Occurs along the length of the border, from CA coast to NM-TX border (based on studies that only examine the U.S. side of the border)	Medium-High
Summer precipitation	Decreasing	Mid-century decreases are greatest for the Sonoran Desert border region	Medium-Low
Drought	Increasing	Throughout the border region; increasing markedly during the second half of the twenty-first century	High
Colorado River streamflows	Decreasing	Measured at Lees Ferry, AZ	High

Note: See Chapters 2 and 19 for a discussion of how confidence levels are assessed.

Temperature

Overall, climate models show trends of increasing temperatures for the border region; this result is robust throughout the course of the twenty-first century, regardless of which combinations of models, downscaling method, and emissions scenario were used (Tables 16.1 and 16.2 and IPCC 2007b).[vii] For the border region, average annual temperatures are projected to increase on the order of 2°F to 6°F (1°C to 3.5°C) during the mid-century time frame (around 2041–2070, according to the high-emissions scenario), with the greatest increases inland (see Chapter 6, Figure 6.1 and Magaña, Zermeño, and Neri 2012). The magnitude of temperature increases is greatest during the summer, as high as 6°F to 7°F (3°C -4°C) during the mid-century, with areas of especially high increases concentrated in the western Sonoran Desert (Montero and Pérez-López 2010) and the northern Chihuahuan Desert (see Chapter 6, Figures 6.1 and 6.8).[viii]

Associated with the maximum and minimum temperature projections are an array of projections for derived parameters and temperature extremes (Table 16.1). Some key derived and extreme temperature projections include large projected increases in cooling degree days in Southern California and Arizona (up to 100 degree days, using a 65°F (18°C) baseline; see Chapter 6), large increases in the annual number of days with maximum temperatures greater than 100°F (38°C), including increases of more than 30 to 35 days in the central Arizona and northwest Chihuahua border regions (see Chapter 6 and Figure 16.2), increased heat wave magnitude (but with more humidity, therefore having a larger impact on nighttime minimum temperatures; see Chapter 7),[ix] and diminished[x] frequency of cold episodes (see Chapter 7).

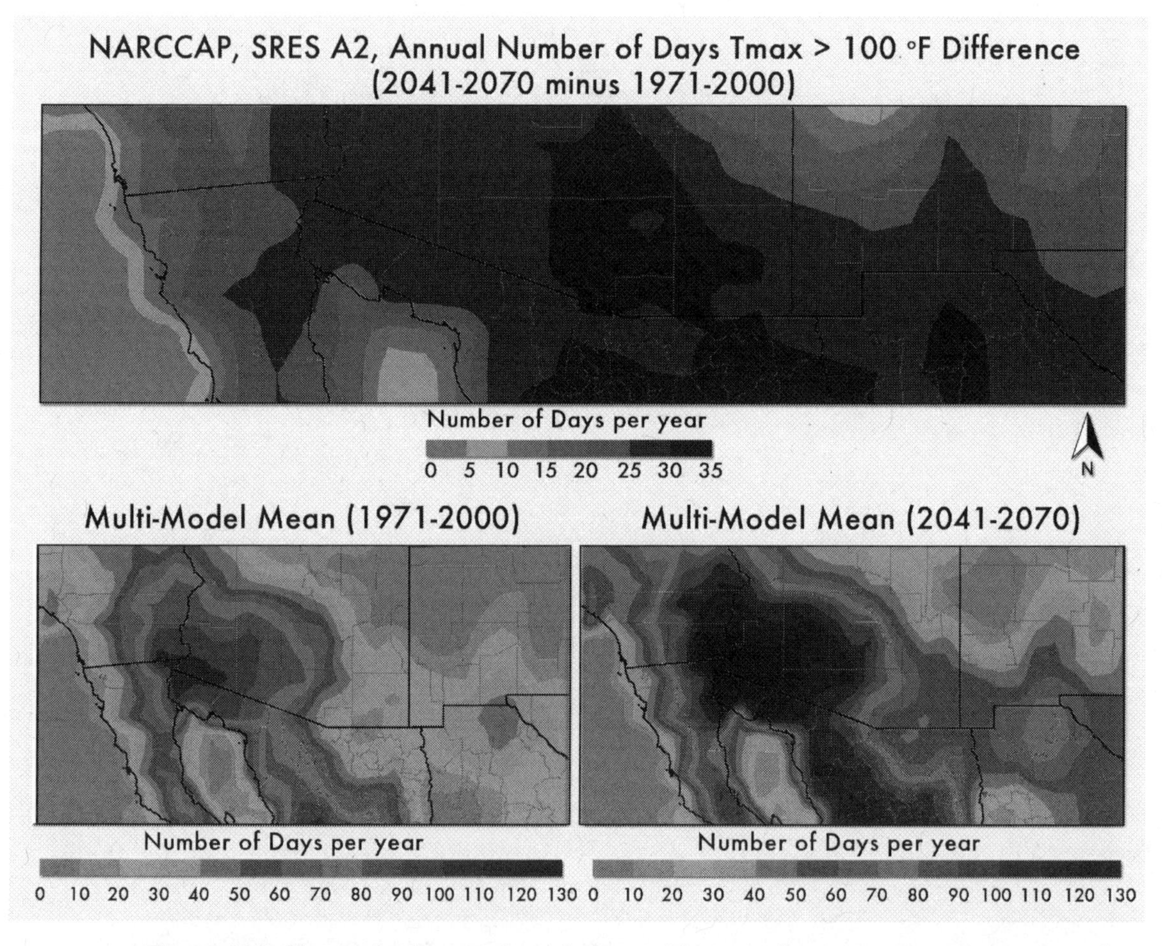

Figure 16.2 Change in the number of days with a maximum temperature greater than 100°F (38°C). The top map shows the change between the NARCCAP (Mearns et al. 2009) multi-model average for 1971–2000 (lower left) and the average for 2041–2070 (lower right). Map generated by Laura Stevens.

Table 16.2 Projected mean annual temperature increases (in °F) in comparison to 1971–2000 along the U.S.-Mexico border, from the California coast to the New Mexico-Texas border

Period	Higher Emissions (SRES A2) Projected Temperature	Lower Emissions (SRES B1) Projected Temperature
2021–2050	2–3°	2–3°
2041–2070	4–5°	3–4°
2070–2099	7–8°	4–5°

Note: Based on studies that only examine the U.S. side of the border
Source: Kunkel (2011).

Table 16.3 Mean temperatures for 1961–1990 (in °F) and projected changes under the high-emissions scenario for 2061–2090, averaged for the Mexican border states

State	1961–1990			2061–2090		
	Winter	Summer	Annual	Winter	Summer	Annual
Baja California	56.2°	83°	69.3°	+5.2°	+5°	+5.2°
Sonora	54.6°	82°	68.2°	+5.9°	+6.5°	+6.4°
Chihuahua	48.1°	76.8°	62.8°	+5°	+4.9°	+5.1°

Source: Montero and Pérez-López (2010).

Precipitation

Future precipitation in the border region, as projected by climate models, is dominated by a continued high degree of annual precipitation variability, indicating that the region will remain susceptible to anomalously wet spells and also remain vulnerable to drought (see Chapters 6 and 7). Precipitation projections have generally low to medium-low confidence, due to variability over shorter periods and the lack of firm consensus among GCM simulations. Nevertheless, spring precipitation is projected to decrease in all but one of sixteen models, exacerbating dryness in the border region's driest season and probably aggravating the dryness that initiates the summer period. Areas that are already prone to little precipitation are expected to see longer runs of days with little or no precipitation.

There is greater confidence in projections of decreased annual precipitation as one moves south. Statistically downscaled studies by Mexican scientists (under the high-emissions scenario) confidently project border region annual precipitation decreases of more than 20% by mid-century, with the largest seasonal decreases projected for winter in the Arizona-Sonora border region (Montero and Pérez-López 2010; Magaña, Zermeño, and Neri 2012).[xi]

Drought

Seager and colleagues (2007) project (under high-emissions SRES A1b) increased drought for a region that encompasses the border region.[xii] Their projections have been confirmed in subsequent studies (e.g., Seager et al., 2009), and independently by Magaña, Zermeño, and Neri (2012).[xiii] Cayan and others (2010) describe a tendency for intensified dryness in hydrological measures in the Southwest from downscaled climate model projections. In the first half of the twenty-first century, Magaña, Zermeño, and Neri's (2012) drought projections exhibit high interannual and multidecadal variability, characteristic of the region.[xiv] Dominguez, Cañon, and Valdes (2010) note that La Niña episodes, which are associated with drought in the border region, may become warmer and drier in the future.

We find that the results of these studies, in conjunction with the temperature and precipitation projections of Montero and Pérez-López (2010) and projections for the U.S. side of the border (summarized in Chapter 6), provide a compelling case for an increased likelihood of drought, with ramifications for northern Mexico water supplies (Magaña, Zermeño, and Neri 2012) and probably for groundwater recharge (e.g., Serrat-Capdevila et al. 2007; Earman and Dettinger 2011; Scott et al. 2012).[xv] Moreover, these assessments are consistent with projections of streamflow for trans-border rivers, such as the Colorado River and Rio Grande (known as the Río Bravo in Mexico, and hereafter Rio Grande), which show decreasing streamflow, lower flow extremes during drought, and potential water resource deficits greater than those previously observed (see Chapters 6 and 7; Hurd and Coonrod 2007; Reclamation 2011).

16.4 Understanding Vulnerability, Risk, and Adaptive Capacity in the Border Region

Definitions and concepts

Vulnerability to climate variability and climate change is the experience (by an individual, household, ecosystem, community, state, country, or other entity) of negative outcomes due to climate stresses and shocks (Leichenko and O'Brien 2008). Experts have approached vulnerability in two distinct but related ways. One approach centers on the underlying political and socioeconomic structures, institutions, and conditions that affect vulnerability, including asymmetries in power and resource distribution (e.g., Adger 2006; Eakin and Luers 2006; Lahsen et al. 2010; Ribot 2010; Sánchez-Rodriguez and Mumme 2010). Vulnerability may be reduced through poverty alleviation and development strategies in developing countries with persistent inequalities (Seto, Sánchez-Rodriguez, and Fragkias 2010). A second approach centers on developing systematic measures of climate-related *risk* in a system, calibrated by exposure and sensitivity (of actors at multiple scales, including households and neighborhoods to cities, states or countries) and the coping (or adaptive) capacity to deal with it (Yohe and Tol 2002; NRC 2010; Moss 2011). Adaptive capacity is the ability (of a household, community, or other unit of organization) to reduce its vulnerability to climate-related risks through coping strategies such as application of social, technical, or financial resources (Yohe and Tol 2002; NRC 2010). Consistent with the NCA framework, the analysis presented here uses the second, risk-based approach, but draws on both types of approaches to provide evidence for its conclusions.

This analysis uses the IPCC definition that "vulnerability is a function of character, magnitude and rate of climate change to which a system is exposed, as well as the system's sensitivity and adaptive capacity" (IPCC 2007a, 6). Risk embodies the likelihood of harm plus the consequences; thus the consequences of a harm occurring are embedded within the concept of vulnerability. Vulnerability to a climate-related risk is mediated by *sensitivity* (for example, by the degree of dependence on resources and activities that are impacted by climate change and non-climate parameters), by *exposure* (for example, the probability of experiencing change in non-climatic and climatic factors), and by *capacity* to cope or to adapt (for example, the demographic, socioeconomic, institutional and technological characteristics that enable response to stress). In general, where resources for coping are relatively abundant, vulnerability is relatively low; but where resources for coping effectively are lacking, vulnerability is typically high. This definition characterizes vulnerability both in terms of stressors and the stressed. Stressors here are regarded as the interactions of economic and cultural globalization, demographic change, and climate change. The vulnerability of the stressed border region includes both the specific attributes of the place and population that transform those stressors into specific risks that threaten the quality of life and the capacities to effectively cope with such stressors. Capacity can be considered a function of assets—financial, material, natural, human, political and social—as well as knowledge, perception of risk, and willingness to act (Grothman and Patt 2005; Moser and Sattherwaite 2010).

The primary determinants and outcomes of vulnerability can vary across scales (from local to international). Vulnerability is ultimately a nested phenomenon in which the impacts and adaptive actions taken at one scale can have ramifications for the whole system (Adger et al. 2009; Eakin and Wehbe 2009). Institutions (the rules, norms and regulations that govern the distribution of resources and their management) are instrumental in mediating risks. Effective cross-scalar governance is thus a critical element of addressing vulnerability (Adger, Arnell and Tompkins 2005), particularly in the border region where trans-border collaboration at multiple governance scales is critical. Trans-border collaboration among formal government agencies and informal governance stakeholders may help reduce regional vulnerability through a shared understanding of the priority vulnerabilities, shared data, and cooperative means of reducing these vulnerabilities (Wilder et al. 2010).[xvi] While non-collaboration has led to less than optimal outcomes in the past (see Box 16.1 for a case study on collaboration and non-collaboration), collaboration may help to reduce regional climate-related vulnerability and to promote appropriate adaptive strategies for the border region (see Box 16.2 for an example on the Colorado River Joint Cooperative Process in the Colorado River delta).

Collaboration takes place at multiple scales in the border region, ranging from formal intergovernmental collaborative agreements (such as the U.S.-Mexico Transboundary Aquifer Assessment Program) to informal networks of local water managers working with the climate research community to develop regional adaptive strategies (such as the Climate Assessment for the Southwest Program in Arizona and New Mexico).

Box 16.1

Case Study 1: Why Is Trans-border Collaboration Important?

Transboundary cooperation to address the impacts of climate variability and climate change is essential to promoting the best outcomes and to building regional adaptive capacity on both sides of the border. Despite formal agreements between the United States and Mexico to cooperate to resolve key transboundary environmental problems (e.g., La Paz Agreement; Minute 306), there are recent important examples where lack of cooperation has led to suboptimal (e.g., win-lose rather than win-win) outcomes:

- In 2002, Mexico invoked its privilege to declare conditions of "extraordinary drought" on the Rio Grande and withheld delivery of irrigation water to Texas farmers, causing millions of dollars in losses.
- When the United States extended the security fence at the border between Nogales, Arizona, and Nogales, Sonora, it was done without reference to local hydrological conditions and without input from officials on the Mexico side. Floodwaters in 2008 became impounded behind the fence on the Nogales, Sonora side of the border, causing millions of pesos worth of damage in Sonora.
- The lining of the All-American Canal (AAC) was completed in 2008, under formal protest and after legal challenges by Mexican and U.S. groups. The change resulted in increased water for households in San Diego County and decreased water for farmers in Baja California. Farmers in the irrigation district of Mexicali had used groundwater recharged by seepage flows from the earthen-lined canals for over sixty years, and concrete-lining of the AAC stopped groundwater recharge and therefore reduced groundwater availability.

Despite these examples of non-collaboration, the trend toward transboundary collaboration has been strong over the last twenty-five years and examples of successful collaboration to reduce environmental vulnerability abound:

- **Emergency Response.** The Border Area Fire Council (BAFC) provides collaborative emergency fire services on both sides of the California-Baja California border. The BAFC was formed during the 1996 fire season to facilitate cross-border assistance for wildfire suppression (GNEB 2008). Operating under a mutual assistance agreement that is updated periodically, BAFC has improved communications across the border, held many joint training exercises, implemented fire safety campaigns on both sides of the border, coordinated development and maintenance of fire breaks along the border, and jointly conducted prescribed burns along the border. BAFC operates in a number of natural protected areas in the region and has improved awareness and protection of biodiversity. It includes more than thirty federal, state, and local organizations representing fire protection, law enforcement, elected officials, the health sector, natural resource managers, and others from both sides of the border. Examples of BAFC's efforts include assistance in the fall of 2007 when sixty Baja California firefighters crossed the border

Box 16.1 (Continued)

Case Study 1: Why Is Trans-border Collaboration Important?

to help with the San Diego County firestorm. Previously in June 2006, ten engines and crews from the California Department of Forestry and Fire Protection had crossed into Baja California to support Mexican fire authorities for six days with a fire that burned 5,200 acres.

- **Scientist-Stakeholder Research.** The Climate Assessment for the Southwest (CLIMAS) at the University of Arizona is a NOAA Regional Integrated Science Assessment program, focused on Arizona and New Mexico. The program brings together scientists and researchers from many disciplines in the natural and social sciences with citizen groups and decision makers to develop a better fit between climate science products (such as forecasts and projections) and the resource managers (such as water or forest managers) and decision makers who use the data. Since 2005, CLIMAS has actively worked with partners at the Colegio de Sonora, Universidad de Sonora, and other Mexican institutions of higher learning to build regional adaptive capacity in the border region via the bilingual Border Climate Summary, workshops with stakeholders and researchers, webinars, and fieldwork focused on identifying common understandings of regional vulnerability and appropriate adaptive strategies. Other funding partners who have collaborated in these projects include the Inter-American Institute for Global Change Research and NOAA's Sectoral Applications Research Program.
- **Trans-border Data Sharing.** The U.S.-Mexico Transboundary Aquifer Assessment Program (TAAP), authorized by U.S. federal law and supported institutionally and financially by both the U.S. and Mexico, is a successful binational program focused on the assessment of shared aquifers. Although the United States did not appropriate funds for TAAP in fiscal year 2011/2012, during this period the Mexican government began funding assessment activities on its side of the border. TAAP is implemented by the U.S. Geological Survey and the state water resources research institutes of Arizona, New Mexico, and Texas, with collaboration from Mexican federal, state, and local counterparts, as well as IBWC and CILA. Two central aims of TAAP include the scientific assessment of shared groundwater resources; and development of dual adaptive-management strategies through expanded binational information flows and data exchange (Wilder et al. 2010; Megdal and Scott 2011). Mutually defined priorities for Arizona's and Sonora's common Santa Cruz and San Pedro aquifers, for example, are meeting human and ecosystem water requirements in the context of growth and climate change (Scott et al. 2012). TAAP is a model of successful trans-border cooperation in data sharing and assessment that supports water-management decision-making in both countries and enhances the adaptive capacity of the region in the face of climate change.

Key drivers of border vulnerability

Growth trends, urban development patterns, socioeconomic factors, and institutions and governance mechanisms can be drivers of border region vulnerability.

CONTEXT-SHAPING VULNERABILITY. Today, rapid growth and uneven economic development are two major contributors to climate-related vulnerability. Institutional asymmetry and governance fragmentation on both sides of the transboundary region create challenges for reducing vulnerability and for trans-border cooperation. Multiple characteristics define the border region, including high rates of poverty in a landscape of uneven economic development; diverse ethnic identities; environmental, social, economic, and cultural interdependency; and rapid growth and urbanization relative to both U.S. and Mexico averages.

Box 16.2

Case Study 2: Colorado River Joint Cooperative Process

Trans-border collaboration is playing a significant role in addressing environmental challenges in the Colorado River delta. The Colorado River Joint Cooperative Process (CRJCP) formed under the auspices of the International Boundary and Waters Commission (IBWC) and its Mexican counterpart (Comisión Internacional de Límites y Agua, CILA) in 2008 to develop "binational processes for meeting municipal, agricultural, and environmental needs" in the delta (Zamora-Arroyo and Flessa 2009). The CRJCP includes government agencies, NGOs, and water stakeholders from both countries. The CRJCP has a difficult task ahead, given that excess flows from the United States are likely to be eliminated in the near future, operational losses are likely to decrease, groundwater supplies will be reduced, and agricultural return flows are likely to decrease as water moves from agriculture to the cities (Zamora-Arroyo and Flessa 2009). The supply of municipal effluent is likely to increase, however, although it may be captured for urban use rather than for ecological flows. A new treaty Minute (Minute 319) adopted on November 20, 2012 establishes a new commitment by the U.S. and Mexico to cooperate around water and ecological needs in the region. The CRJCP is a collaborative model for other trans-border areas and issues. Although it ultimately received formal federal approval in both the United States and Mexico, the CRJCP originated from an informal coalition of local stakeholders that led ultimately to the formal collaborative process.

HISTORY. Tribal peoples occupied the border region for many thousands of years before the arrival of Spanish, Mexicans, and then Americans to the area (see also Chapter 17).[xvii] Today there are twenty-three Native nations on the U.S. side in the border region, and about eight indigenous groups on the Mexican side (Starks, McCormack, and Cornell 2011); some of these peoples (such as the Kumeyaay, the Cocopah/Cúcapa, the Yaquis, and the Tohono O'odham) continue to have strong trans-border ties. They manage diverse lands and water and economic resources in the border region. Spanish colonizers in the sixteenth century expropriated significant land and resources, and mestizos, whose land use practices combined Indian and Hispanic traditions, settled in the border and created many of the current border towns. English-speaking colonizers of the United States introduced their beliefs of commercial capitalism and the frontier vision to land use and resource practices in the nineteenth century. The successive arrival of farmers, workers, investors, migrants, and bureaucrats has continually transformed the border over the last 100 years and created one of the most dynamic and diverse sociocultural landscapes in the world. The diversity of the border region's population—including differences in languages used at home and access to technology—challenges effective communication about climate-related risk.

CONTEMPORARY TRENDS. Currently, most of the U.S.-Mexico border population is concentrated in fourteen fast-growing, paired, adjacent cities with a common history, strong interactions, and shared problems (CDWR 2009). Eight of these binational pairs are on the western end of the border in the area included in this chapter (see Figure 16.3). In 2002, there were approximately 1 million (legal) border crossings daily by residents in the border's twin cities to work, shop, attend classes, visit family, and participate in other activities (GAO 2003); the number of crossings declined to half a million by 2010. Mexican border towns are part of a very centralized national political system, suffer from limited fiscal resources, and lack a tradition of urban planning. Across the border, U.S. towns have had greater political autonomy, are part of a strong and stable national economy, and have broadly applied land use planning and supplied basic infrastructure and services to their residents. Thus, the "twin cities" along the U.S.-Mexico border are places of encounter but also of intense political, social, and physical contrasts.

The per capita income within the U.S. border counties is only about 85% of the U.S. per capita income. If wealthy San Diego County is excluded, the GDP per capita of the border region is only about 64% of the national level (2007 data). In 2006, if the twenty-four U.S. counties along the border were aggregated as the fifty-first state, they would rank 40th in per capita income, 5th in unemployment, 2nd in tuberculosis, 7th in adult diabetes, 50th in insurance coverage for children and adults, and 50th in high school completion—all characteristic of regions of poverty (Soden 2006; and www.bordercounties.org). In the 2010 U.S. Census, Arizona and New Mexico were tied for the fourth-highest poverty level in the United States.

DEMOGRAPHIC DRIVERS. The binational border region from San Diego-Tijuana to Paso del Norte is demographically dynamic, growing much faster than the average of either nation (Figure 16.4).

Since 1970, the U.S. side of the border region has attracted huge flows of domestic migrants—mostly non-Hispanics, seeking a Southwestern "sunbelt" lifestyle (and climate), retirement, or job opportunities—and international immigrants—mostly from Mexico and Central America, seeking jobs and economic opportunities. Between 1983 and 2005, the population almost doubled (from 6.9 million people to over 13 million). Since the economic recession began in 2007–2008, however, growth rates have declined in border states, except Texas (Cave 2009; Frey 2011).[xviii] Growth rates have also declined in many border counties. For example, growth slowed in the counties bordering Sonora, from a rate of 5.3% between 2000 and 2005 to 1.6% between 2007 and 2008 (Mwaniki-Lyman, Pavlakovich-Kochi, and Christopherson, n.d.).

Population projections for the region (reported in USEPA's 2006 *State of the Border Region*) estimate that the region's population will grow to between 16 million and 25 million people by 2030, an increase of 46% (based on the medium scenario analyzed). Declining growth rates since the onset of recession in 2008 may represent slower than projected regional growth.

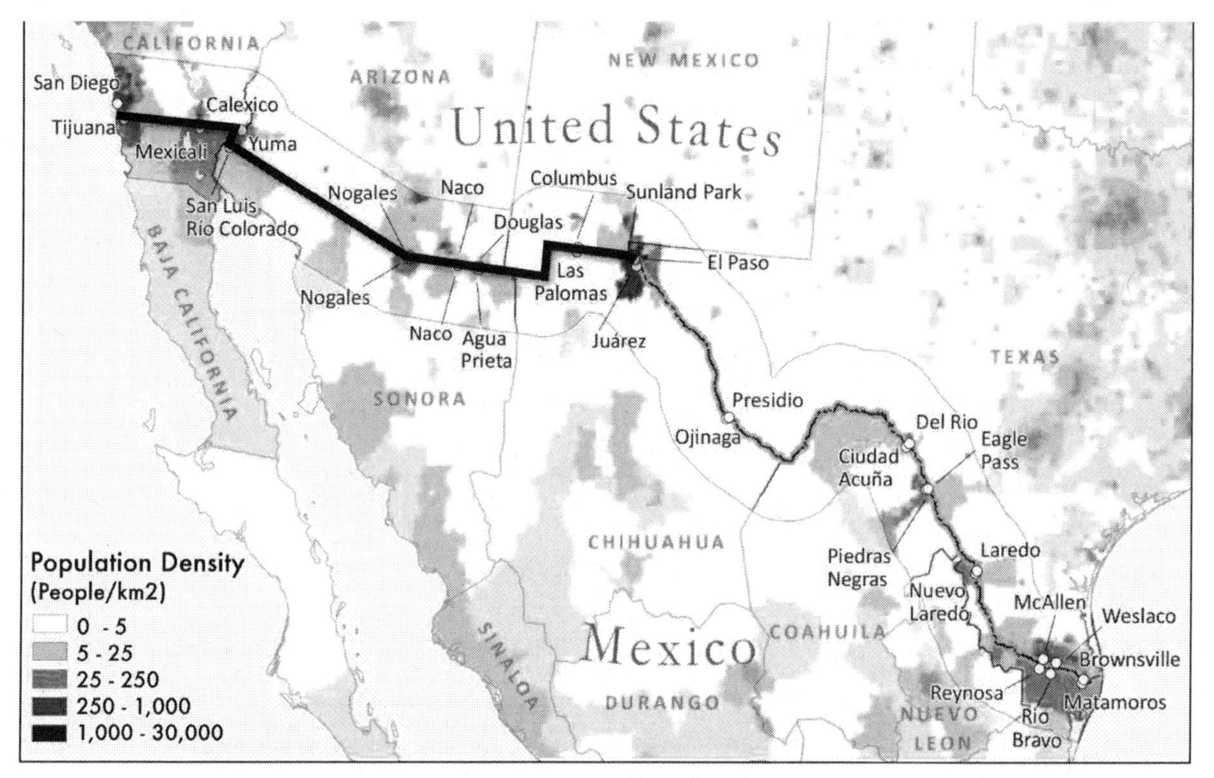

Figure 16.3 Population density in U.S.-Mexico border region. Population density shading refers to the areas on either side of the border, and not the borderline. Reproduced from EPA (2011).

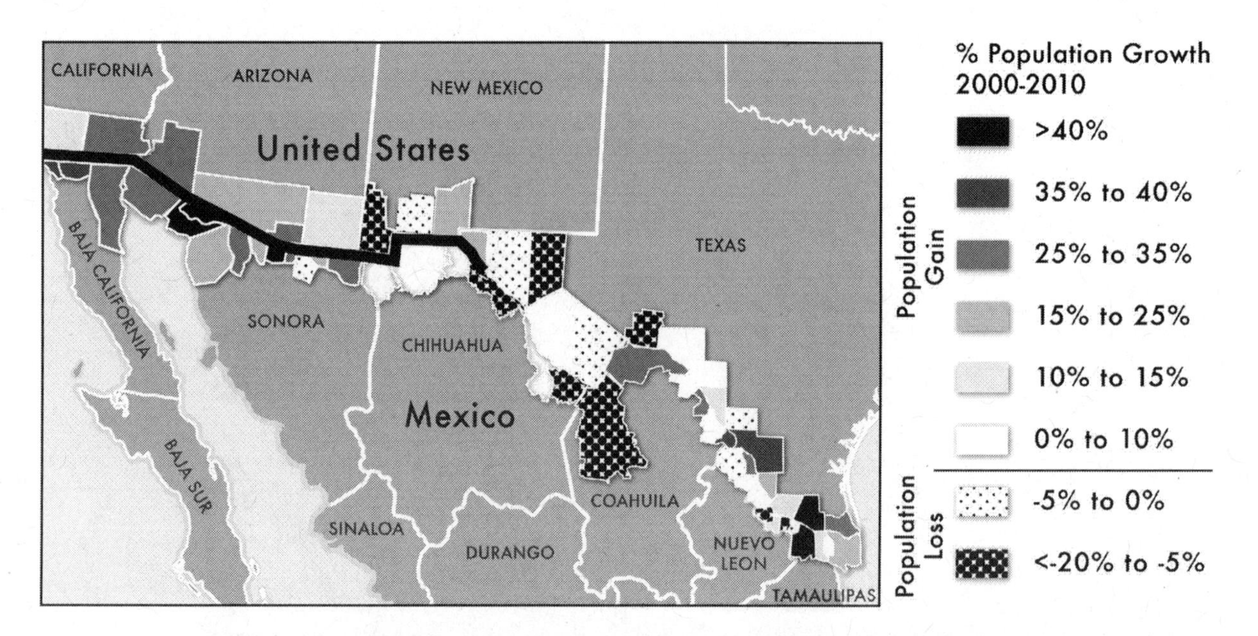

Figure 16.4 Population growth in the western portion of the U.S.-Mexico border region (2000–2010). Adapted from Good Neighbor Environmental Board 14th Report (GNEB 2011). Source: U.S. Census Bureau and Instituto Nacional de Estadistica y Geografia (INEGI).

Ninety percent of the border population resides in cities and the remaining 10% live in smaller tribal and indigenous communities or in rural areas. Over 40% of the region's population resides in California and Baja California, which are home to the major border cities of San Diego, Tijuana, and Mexicali (EPA 2011). Most population growth in the next few decades will occur in mid-size and large urban centers, intensifying border urbanization and metropolization (Lara et al. 2012). Especially on the Mexican side, the pace of urban growth will be highest in the large border cities and municipalities: the proportion of population living in urban Mexican centers with more than 500,000 people is predicted to rise to 58.1% in 2030, from about 44.6% in 2005 (CONAPO 2007). Already, cities like Ciudad Juárez, Chihuahua, and El Paso, Texas, are practically "fused" or continuous across the border and are merging with adjacent cities and towns forming transborder metropolitan corridors.

This analysis indicates that in the future, exposure to climatic stress will not only increase with population growth, particularly in urban areas, but sensitivity may also increase through water-food-energy dependent growth trajectories that will be sensitive to climatic disturbance. Sustainable growth will be critical for adaptation.

Socioeconomic drivers

Multiple studies have identified the border as a region of high social vulnerability due to intersecting processes of rapid growth, domestic and international migration, economic intensification and globalization, and intensive climate change (Liverman and Merideth 2002; Austin et al. 2004; Varady and Morehouse, 2004; Hurd et al. 2006; Ray et al. 2007; Collins 2010; Jepson 2012; Wilder et al. 2010; Wilder et al. 2012). Climate impacts are not uniformly distributed across populations and space but instead affect specific vulnerable populations and places (Romero-Lankao et al. 2012).

Ethnicity is a significant factor in sensitivity and exposure to climate-related risk (Verchick 2008; Morello-Frosch et al. 2009). Hispanics are the largest ethnic group in the border region and in 2008 were 42.2% of the population of the U.S. border counties in the study area; if San Diego with its large non-Hispanic population is excluded, then the U.S. border region is 55.7% Hispanic.[ixx] In California, for example, Latino and African-American communities were found to be more vulnerable to heat exposure and heat stress than the state population as a whole (Morello-Frosch et al. 2009). Research in the South and in the Southwest United States documents a higher climate vulnerability among Latino and African-American populations due to relatively low incomes, substandard housing, structure of employment (e,g, outdoor laborers in landscaping and construction), lack of affordability of utility costs, and lack of transportation (Vásquez-León, West, and Finan 2003; Verchick 2008; Morello-Frosch et al. 2009). Minority communities have a greater exposure to the urban heat island effect and suffer more health problems due to poorer air quality and concentration of industrial uses in the areas where they live (Harlan et al. 2008; Ruddell et al. 2010).

The diverse cultural meanings and practices associated with resource allocation and management traditions (Sayre 2002; Sheridan 2010) are also likely to affect adaptation. For example, the water resources in the Rio Grande Valley, which bisects New Mexico, are challenged by multiple sector claims and increasing demand associated with population growth. These water resources must also serve the traditions and economic needs of Native American tribes and pueblos, and flow through traditional *acequias*—canals—the lifeblood of four-hundred-year-old Hispanic communities (Hurd and Coonrod 2007; Perramond 2012).

In general, climate change research has paid limited attention to socioeconomic vulnerability and adaptation in human communities. This analysis indicates that in the future an increasingly diverse population will be exposed to climatic stress (e.g., floods, storms, hurricanes [in coastal areas], heat waves, and drought) with implications for the languages and technologies used to communicate about climate risks and hazards that affect the region (Vásquez-León, West, and Finan 2003; Morello-Frosch et al. 2009; Wilder et al. 2012). In addition, development initiatives (such as infrastructure, sewerage networks, and improved housing) to address uneven development are critical to future adaptation.

Urbanization, infrastructure, and economy

Regional impacts associated with the climate changes described in Table 16.1 increase the stresses on urban infrastructure (such as energy for cooling) and water (to meet both consumptive and nonconsumptive demands for energy generation), exacerbate air pollution, create public health challenges associated with heat waves, and cause increased demand for urban green spaces. These concepts are explored further in Chapter 13.

Urban vulnerability is structured not only by demographic change, but rather occurs in multiple sectors, including the built environment and urban economy (Romero-Lankao and Qin 2011), especially with the increasing urbanization of poverty (Sánchez-Rodriguez 2008). Three processes of urban change at the city level have relevance for understanding and managing risks from climate variability and climate change generally and in the border region. First, cities have expanded into areas that are prone to droughts, heat waves, wildfires, and floods (Collins, Grineski, and Romo Aguilar 2009; Moser and Sattherwaite 2010; Seto, Sánchez-Rodriguez, and Fragkias 2010). Second, large sections of the urban population along the U.S.-Mexico border live in unplanned communities in "informal" housing, lacking the health and safety standards needed to respond to hazards, and with no insurance (Collins, Grineski, and Romo Aguilar 2009; Wilder et al. 2012). Third, characteristics of the built environment (such as the heat island effect, high levels of atmospheric pollutants, impervious surfaces, and inadequate drainage systems) can amplify the impacts of high temperatures, storms and other hazards associated with climate change (Wilbanks et al. 2007; Romero-Lankao and Qin 2011). While the urban infrastructure of many urban areas on the U.S. side of the border needs major upgrades to prepare for likely climate change impacts (Field et al. 2007), many Mexican cities have the additional burden of overcoming development deficits. Among these deficits are inadequate all-weather roads, lack of paved roads, poor water treatment (lack of water treatment plants or treatment plants with insufficient capacity for drinking water and sewage); decaying water infrastructure, and institutional constraints such as lack of financing from taxes, uncoordinated planning, and competition among agencies for agendas and resources.

Even at the neighborhood scale, certain characteristics of the built environment can amplify risks. For instance, studies showed variations in vegetation and land-use patterns across Phoenix produce an uneven temperature distribution that was correlated with neighborhood socioeconomic characteristics. In other words, affluent areas were less densely settled, had lower mean temperatures, and thus had lower vulnerability to heat stress, while low-income areas had more rental housing, greater prevalence of multi-generational families sharing a household, and a higher prevalence of non-English language speakers (Harlan et al. 2008). These findings point to the need for climate hazards and risks to be communicated to the public in a way that respects the diversity of media, technology, and languages in used in the region. In the long run, these problems could be reduced through improved urban development and investment in marginalized areas.

INFRASTRUCTURE. Dense urban areas in the border region contain substantial populations who are vulnerable to natural disasters linked to climate change because they live in substandard housing in floodplains or on steep slopes or in housing located in areas on the urban periphery that are susceptible to wildfires (GNEB 2008). Tijuana and Nogales, two border cities experiencing rapid population growth, have received an influx of immigrants seeking employment in the maquiladora industry. Many settle in informal (unplanned) colonias (border-region residential communities that are economically distressed and usually underserved by infrastructure) with unsuitable topography, characterized by steep slopes and canyons. With few measures to control erosion, extreme rain events and the prevalent topography lead to runoff and floods during extreme conditions (Cavazos and Rivas 2004; Lara and Díaz-Montemayor 2010).

Border cities are also underserved by water and wastewater infrastructure as well as other urban infrastructure such as paved streets and lighting (Lemos et al. 2002; Jepson 2012). In 2007, for example, the Border Environment Cooperation Commission estimated that there was nearly $1 billion in unmet investment in water and wastewater infrastructure in the border region (BECC 2007). An estimated 98,600 households in the United States and Mexico border region lacked safe drinking water, and an estimated 690,700 homes lacked adequate wastewater collection and treatment services (EPA 2011). Thus, on both sides of the border, large numbers of residents do not have safe potable water piped into their homes and lack proper sewage collection and treatment services (GNEB 2008).

Informal colonias from Tijuana to Nogales to Juárez often are off-the-grid for water, sanitation, and electricity and rely on purchased water from trucks at relatively higher cost than municipal tap water (Cavazos and Rivas 2004; Collins 2010; Wilder et al. 2012). Water-scarce states like Sonora have water rationing in major cities—including the capital, Hermosillo, and its largest border city, Nogales—based on a system known in Spanish as *tandeo*. Basic infrastructure is limited in the informal colonias, and construction on unsafe hillsides in floodplains leads to increased risk to human residents from severe flooding when it rains. Flooding of unpaved roads may disrupt water-truck deliveries for households not on the municipal grid.

ECONOMY. The border is a region of dynamic growth in both industry and employment. The region is of critical value to the global economy and both countries' national economies due to its production of agriculture and manufactured goods. Its economic significance therefore enhances its exposure to climatic stress. Its integration into the global economy means that climate stresses have potential impacts beyond local borders because of the potential of disrupted trade. The economy of the border is highly integrated through manufactured and agricultural trade, export-oriented production and labor, and markets that include cross-border manufacturing clusters in aerospace, electronics, medical devices, automotive products, and other sectors.

Mexico's maquiladora industry experienced declines due to the 2001–2002 recession and the period that followed. Maquiladoras are duty-free, foreign-owned assembly plants responsible for nearly half of Mexico's exports in 2006 (GAO 2003; Robertson 2009). At their peak in 2000, they employed over 1 million people, of which 78% (839,200) were from the five major border cities of Tijuana, Mexicali, and Juárez (and Matamoros and Reynosa in the eastern border region) (GAO 2003).[xx] After 2006, Mexico no longer tracked maquiladora exports separately from its other exports.[xxi]

Cities on the U.S. side of the border have benefited from the substantial flow of trade created by maquiladoras, with more than 500,000 jobs added to the U.S. border region between 1990 and 2006, in services, retail trade, finance, and transportation. While maquiladoras drive higher employment in Mexican border cities, Cañas et al. (2011) found that Texas border cities experienced the highest maquiladora-related employment increases, with El Paso providing the third-most maquiladora-related jobs of all border cities (after McAllen and Reynosa). By comparison, California and Arizona border cities experienced a smaller benefit. Asian production inputs have displaced U.S. suppliers, whose share dropped from 90% in 2000 to 50% in 2006, notably affecting Tijuana maquiladoras and San Diego suppliers. Maquila employment declined as a result of the 2001–2002 recession and global low-wage competition from southeast Asia. By 2006, maquiladoras employed over 750,000 people in border cities (Cañas and Gilmer 2009). Other forms of integration are trade and capital flows.[xxii]

This analysis indicates that urban areas in the border region are vulnerable based on exposure to climate stressors. Urban infrastructure is sensitive to flooding (and related erosion) and drought, and urban-based economic activities of both regional and global consequence may be sensitive to impacts caused by climate stressors (such as water scarcity or water shortage). Urban areas could be set on a more sustainable development path through urban and economic development strategies such as extending water and sanitation networks and improving their efficiency; improving flood and erosion control; promoting water conservation at the household (e.g., rainwater harvesting) and municipal (e.g., expanded water treatment and reuse) levels; improving substandard or inappropriately-sited housing; and extending urban green spaces in low-income areas.

Institutional and governance drivers

Institutional asymmetry and fragmentation—meaning differences in governance frameworks and lack of cohesion and coordination among multiple government agencies and actors on the two sides of the border—create potential vulnerabilities in managing transborder environmental resources. Water management is used here as a lens into institutions and environmental governance in the region. Governance refers to "the set of regulatory processes, mechanisms, and organizations through which political actors influence environmental actions and outcomes" (Lemos and Agrawal 2006, 298). The term encompasses both government and non-government actors, including communities, businesses, and non-governmental organizations. On the U.S. side of the border, water governance is decentralized; on the Mexican side, despite decentralization initiatives codified into national and state laws since 1992, it remains highly centralized (Pineda Pablos 2006; Mumme 2008; Scott and Banister 2008; Wilder 2010; Varady, Salmón Castillo, and Eden forthcoming). U.S. border cities and counties are embedded in systems of water rights and water administration dominated by the four border states—Texas, New Mexico, Arizona, and California—subject to applicable international treaties, interstate river compacts, an assortment of federal laws affecting water development, water quality, and ecological values, and contracts with federal agencies with water-related jurisdictions. Water providers range widely in size, from small local utilities up to giant municipal water providers like the Metropolitan Water District of Southern California and the San Diego Water Authority. Farther east, agencies include El Paso Water Utilities and local municipal water authorities. In irrigation, management ranges from the sprawling Imperial and Coachella irrigation districts, which have Colorado River water entitlements that dwarf those of Nevada and Utah combined, to lesser ones like New Mexico's Mimbres Valley Irrigation Company and Arizona's Upper San Pedro Water District. On the Mexican side of the border, states and municipios as well as irrigation districts are governed by Mexico's National Water Law through the National Water Commission (CONAGUA). The western Mexican border states (Baja California, Sonora, and Chihuahua) each have a state-level water agency that partners with CONAGUA and local water utilities (*organismos operadores*), while irrigation districts remain under the direct oversight of CONAGUA or, in the case of large irrigation districts, are administered by an irrigation district authority with CONAGUA oversight. At the international level, the allocation and management of riparian surface water is governed by several treaties and their amendments and extensions.[xxiii]

The inherent differences between these decentralized and centralized systems of water governance complicate binational cooperation and water planning at the border. Political and administrative decisions on managing scarcities and climatic variation are often achieved more readily in Mexico than in the United States owing to centralized planning in that country.[xxiv] Institutional fragmentation and complexity mark water resource management on the U.S. side, in particular (Mumme 2000; Milman and Scott 2010; Wilder et al. 2012). The treaties and international institutions for coordinating water also have limitations for the management of climate variability. The International Boundary and Water Commission (IBWC) has a limited mandate for coordinating binational activities in times of prolonged drought and lacks basin-wide advisory bodies to assist it as it deals with national, state, and local authorities (Mumme 1986). The IBWC also lacks clear jurisdiction for managing groundwater extraction of groundwater in the border zone (Scott, Dall'erba, and Díaz-Caravantes 2010).

These recognized policy challenges provide a strong rationale for the development of binational watershed partnerships and less formal arrangements aimed at supporting the ecological health of watersheds and water conservation in the border region. Partnerships like the Tijuana Watershed Task Force, the Upper San Pedro River Partnership, and the Santa Cruz River Aquifer Assessment all point in the direction of sustainable initiatives that need be supported and strengthened. Recent IBWC-based efforts have extended the treaty regimes on the Rio Grande and the Colorado River to better address conservation and long-term water supply planning. Programs include an innovative

Water Conservation Investment Fund established at the North American Development Bank (NADB) in 2003 and the 2010 establishment of the binational Consultative Council for the Colorado River to consider shortage challenges of an international nature (Mumme et al. 2009). These arrangements comprise adaptive strategies that help the border region address known shortcomings in current water governance and add to regional resilience (Wilder et al. 2010).

Vulnerability may be reduced and regional resilience increased through flexible and dynamic governance institutions and increased trans-border collaboration at the federal, state, and local scales to share information and data, respond to changing needs and conditions, and resolve transboundary water and other environmental issues via consultative or collaborative processes (see Box 16.3). In addition, a better integration of scientific and technological progress (such as climate variability/climate change monitoring and forecasts or irrigation and water distribution techniques) into planning and operations would help agencies and other governance actors be more responsive to climate change.

Box 16.3

Case Study 3: Reducing Cross-Border Emissions: California-Baja California Cooperation on Greenhouse Gas Emission

As a signatory to the Kyoto Protocol, Mexico initiated greenhouse gas (GHG) inventories, began a voluntary reduction program, and developed GHG management plans as part of a broad national approach characterized by public-private partnerships.[xxxvii] Baja California was one of the first Mexican states to develop a GHG inventory, in March 2010, through cooperation with the Center for Climate Strategies, U.S. Environmental Protection Agency (EPA), and the Border Environment Cooperation Commission (Chacon Anaya et al. 2010). In the absence of national programs, California took the initiative with AB 32, the Global Warming Solutions Act of 2006, which called for reducing by 2020 California's GHG emissions to levels of 1990.

As the two states have moved forward with GHG inventories and the planning process for climate plans, California and Baja California officials have exchanged information and methodologies. This was facilitated by the active involvement of EPA and the Border Environment Cooperation Commission, along with the Environmental Roundtable of the Border Governors Conference. At a local level, the San Diego Association of Governments (SANDAG) has facilitated transborder information exchange on climate change issues and data with counterparts in Baja California. This was accomplished through binational information meetings, including "Binational Seminar: Challenges and Opportunities for Crossborder Climate Change Collaboration" (2009) and "Binational Event: Crossborder Climate Change Strategies" (2010).[xxxviii] The SANDAG efforts have successfully placed GHG and climate change as topics on the planning agenda for local and state authorities in the California-Baja California border region. The Border 2020 binational environmental program will reinforce these regional transboundary efforts by focusing on reducing GHG and on actions to help border communities become more resilient to the effects of climate change. [xxxix]

Drivers of biophysical changes and their impacts

The border region is particularly rich in species and ecosystem diversity. The Good Neighbor Environmental Board (GNEB 2006) reports that the fragile ecosystems of the border region are under threat from drought, invasive species, and urban sprawl. Socioeconomic factors are related to biodiversity loss, in that population growth may drive higher resource use, leading to higher vulnerability to climate change. Biodiversity loss has many potential negative impacts, such as encouraging the encroachment of invasive species, decreasing water-retention capacities, and resulting in fewer locations that can be used as recreational areas or that can sequester carbon dioxide. (See Chapter 8 for more discussion of the benefits of ecosystem processes and biodiversity.)

The ecological features of the border region vary widely. About a dozen transboundary rivers provide water to cities, tribes, and farms in the two countries, including two major rivers, the Colorado River and the Rio Grande, and many smaller sources—such as the Tijuana and New rivers in California and Baja California, the Santa Cruz and San Pedro rivers in southern Arizona and northern Sonora, the Hueco Bolsón and the Mesilla-Conejo-Medanos in the Paso del Norte region, and the Mimbres-Los Muertos aquifer and drainage system in New Mexico. Major desert ecosystems include the Mojave (Imperial Valley, California), Sonoran (southern Arizona and Sonora), and Chihuahuan (eastern Arizona and western New Mexico) Deserts (GNEB 2006; EPA 2011). Features include fertile desert estuaries on the Baja California and Sonora coasts; chaparral-covered coastal plains and oak savannahs in California; deserts of cactus, creosote, mesquite, palo verde, and sagebrush across parts of Arizona and New Mexico, mixed with pine and oak forests in higher mountain elevations; and hilly areas of grasses and mesquite moving eastward into Texas. Coastal zones at the eastern and western ends of the border contain important marine and freshwater habitat (Liverman et al. 1999; Varady et al. 2001; GNEB 2006). As an example, Figure 16.5 indicates the vast ecological resources in protected designations within the Arizona-Sonora portion of the border region.

Within the entire U.S.-Mexico border region (including the eastern portion of the region outside the scope of this chapter), there are over 6,500 animal and plant species (EPA 2011).[xxv] On the Mexican side, 235 species found in the border region are classified in a risk category. Of these, 85 are considered endangered under Mexico law. In the United States, 148 species found in border counties are listed as endangered under the U.S. Endangered Species Act (EPA 2011, 15).

The border fence erected and extended by the United States Department of Homeland Security to prevent undocumented immigration has had extremely negative effects on wildlife, including endangered species, whose habitats and ranges lie in the transboundary region (López-Hoffman, Varady, and Balvanera 2009; Segee and Córdova 2009; Sierra Club 2010).[xxvi] The fence deters virtually all wildlife crossings, cutting animals and reptiles off from sources of water, food, and access to habitat and to potential mates.

Wetlands are a critical source of biodiversity and losses of wetlands may be irreversible, limiting or prohibiting future efforts at restoration (Beibighauser 2007). At-risk estuaries include the Tijuana River and the Rio Grande, including the adjacent Laguna Madre coastal lagoon (Liverman et al. 1999). The Rio Grande is also home to endangered silvery minnows in the last remnant of their historical habitat and to flocks of migrating cranes and geese who gather in vast numbers to rest and refuge in riparian *bosques*

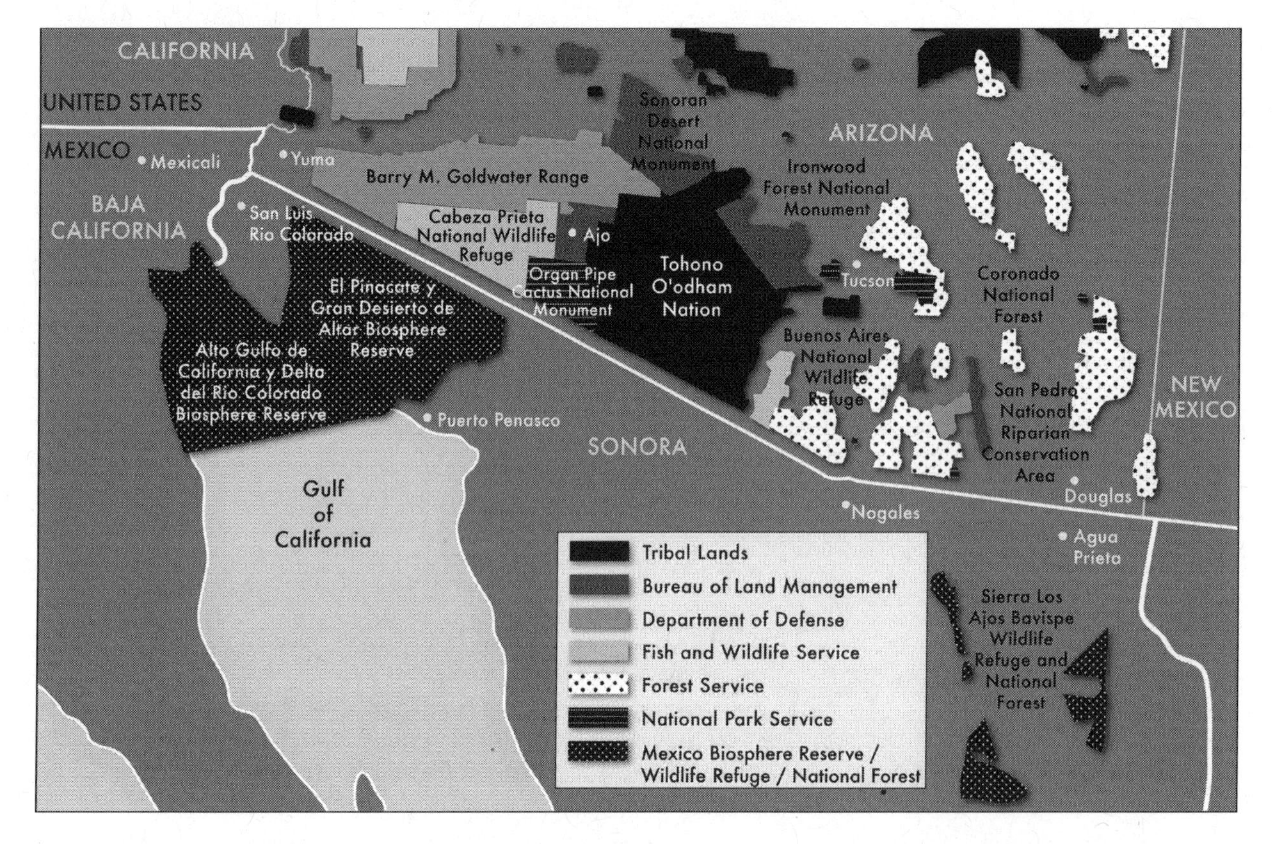

Figure 16.5 Protected areas in western portion of U.S.-Mexico border region. Source: Laird-Benner and Ingram (2011) reprinted with permission from Taylor & Francis.

(woodlands) (Hurd and Coonrod 2008). Native fish, neotropical songbirds, and migratory waterfowl, including threatened and endangered species, have all declined precipitously in recent decades (Lacewell et al. 2010).

The Colorado River delta is a significant border ecosystem that is most at risk from increasing regional water stress. Lacking a dedicated source of water to maintain ecological flows, several wetlands of high resource value are threatened, including the Ciénega de Santa Clara (see case study below) (Glenn et al. 1992; Glenn et al. 1996; Liverman et al. 1999; Pitt and Luecke 2000; Varady et al. 2001; Zamora-Arroyo and Flessa 2009). Two principal vulnerabilities associated with the Lower Colorado River and delta are (1) the lack of dedicated ecological flows to sustain critical wetlands and bird habitat in the delta; and (2) the over-allocation of Colorado River water and over-reliance of the seven U.S. basin states and Sonora and Baja California on its water as a principal source of supply.[xxvii] This latter issue is addressed in the water sector analysis below; the discussion here is on the Colorado River delta ecosystem.

Likely effects of the climate changes described in Table 16.1 are primarily associated with increasing temperatures, declining precipitation and streamflows, and increasing extreme events (i.e., droughts). Expected effects include: constraints on available water supply to major cities reliant on Colorado River water (MacDonald 2010; Woodhouse et al. 2010); increased urban-agriculture competition over water; constraints on meeting increasing regional water-energy demand; and threats to ecosystems of high resource value, including endangered species habitat (Pitt and Luecke 2000; Zamora-Arroyo and Flessa 2009).

The Colorado River delta has been called "one of the most important estuaries in the world" (Zamora-Arroyo and Flessa 2009, 23) and is the largest remaining wetland system in southwestern North America. Although it originally comprised 2 million acres (800,000 hectares) of wetlands habitat, it has shrunk to only 10% of its original size since 99% of the water has been diverted (Zamora-Arroyo and Flessa 2009). These wetland areas are critical stopovers on the Pacific migratory flyway and significant breeding and wintering habitat for 371 bird species (400,000 migratory waterbirds), including endangered species such as the Yuma Clapper Rail (listed in both the United States and Mexico). Both the Andrade Mesa and Ciénega de Santa Clara wetlands in the delta are experiencing water scarcity due to increased demand and changes in water management. The wetlands rely on system inefficiencies (water not used by agriculture or cities), amounting to less than 1% of its original sources (Pitt and Luecke 2000; Zamora-Arroyo and Flessa 2009). These "accidental" sources are now threatened as water managers increase efficiency; for example, the 2008 concrete-lining of the All-American Canal may cut off seepage that has been important in sustaining the Andrade Mesa wetlands.[xxviii]

A Colorado River Joint Cooperative Research Process involving key binational government agencies, non-governmental organizations, and water users has a goal of finding dedicated sources to meet minimum flows required to sustain these critical wetlands (Zamora-Arroyo and Flessa 2009) (see also Box 16.2). The best options to ensure the survival of the delta are agricultural return flows, municipal effluent, and acquisition of new water rights (Zamora-Arroyo and Flessa 2009).

Biodiverse and environmentally significant border ecosystems are exposed to urban encroachment, increasing scarcity of water, and habitat threats, as well as habitat fragmentation and land-use change caused by the U.S. border fence. Endangered species habitat and wetlands systems are sensitive to the increasing scarcity of water to sustain critical habitats. Institutional trans-border collaborations in critical wetlands areas such as the Ciénega de Santa Clara and the Tijuana Estuary (see Box 16.4) are developing adaptive strategies that may add to the sustainability of these areas and will help confront the impacts of future climate change.

16.5 Sectoral Analysis of Border Vulnerability

Water supply and sectoral vulnerability

Climate change in the Southwest will place additional burdens on an already-stressed water system (see Chapter 10). As a general rule across North America, the shift will be from wet to wetter, in wetter areas, and from dry to drier, in arid regions like the border (see Chapter 6). Severely over-drafted aquifers and those aquifers affected by saltwater intrusion are already a challenge for the region (see, for example, Figures 16.6 and 16.7 for northern Mexico). Regional impacts associated with these changes are anticipated to include: a decreased water supply in storage reservoirs for urban use and irrigation, especially in the Colorado system; higher summer temperatures leading to stresses on

Box 16.4

Case Study 4: Collaboration to Protect the Tijuana Estuary

The Tijuana River Estuary is the largest and one of the last remaining large tidal wetlands on the Pacific Coast (Roullard 2005, plates 31-36; Ganster 2010). The 2,500-acre (1,012-hectare) Tijuana River National Estuarine Research Reserve (TRNERR) is situated on the international boundary at the endpoint of the 1,750-square-mile (4,532 square kilometer) binational Tijuana River Watershed. One-third of the watershed is in the United States and the remaining area in Mexico, and includes much of the rapidly urbanizing areas of Tijuana and Tecate. The estuary's diverse contiguous beach, dune, salt marsh, riparian, and upland habitats are home to many rare and endangered species of plants and animals. The estuary is vulnerable to human impacts and the effects of climate change that include sea-level rise, altered precipitation patterns and sedimentation rates, and invasion of exotic species. The likely effects of climate change also pose significant challenges to the viability of past habitat restoration efforts in the estuary (see, for example, Zedler 2001).

In order to make this system more resilient to both watershed and coastal stressors, the Tijuana River Valley Recovery Team was convened in 2008.[xl] This effort brings together over thirty regulatory, funding, and administrative agencies with the scientific community, environmental groups, and other stakeholders. The Recovery Team has produced a "roadmap" that addresses broad ecosystem goals and identifies actions that can facilitate adaptation to climate change, such as controlling cross-border flows of sediment and trash, improving hydrology, changing land use, and restoring habitat (Tijuana River Valley Recovery Team 2012). The plan identifies broad zones of the Tijuana River estuary area that will serve different functions. These include (1) transitional areas designed to accommodate habitat shifts associated with rising sea level, (2) private lands that should be acquired and restored to habitats that can dynamically respond to changing conditions, and (3) lands that will remain in agricultural or recreational use and are protected inundation. The roadmap also specifically calls for the impacts of climate change to be assessed at more precise spatial scales and shorter time scales so that management practices can effectively respond to evolving climate conditions.

energy provision during peak demand; extended and more severe drought periods; and higher evapotranspiration rates (Table 16.1). As Udall (2011, 12) notes, "The past century is no longer a guide to water management" (see also Planning Techniques and Stationarity section, Chapter 10). The principal watersheds in the region are of particular significance to the sustainability of ecosystems and human activities.

The two major transboundary rivers in the border region—the Colorado River and the Rio Grande—are systems where conflicts over water are prevalent (see Chapter 10, Box 10.1). Both the United States and Mexico have aging water infrastructures with a voluminous backlog of needs that are very expensive to fix. As described throughout the present work, water is connected to many other sectors, including energy, transportation, human health, ecosystems, and agriculture. Higher projected temperatures will affect water quality; surface water temperatures are expected to increase, in turn impacting the organisms and species (including humans) that depend on these resources,

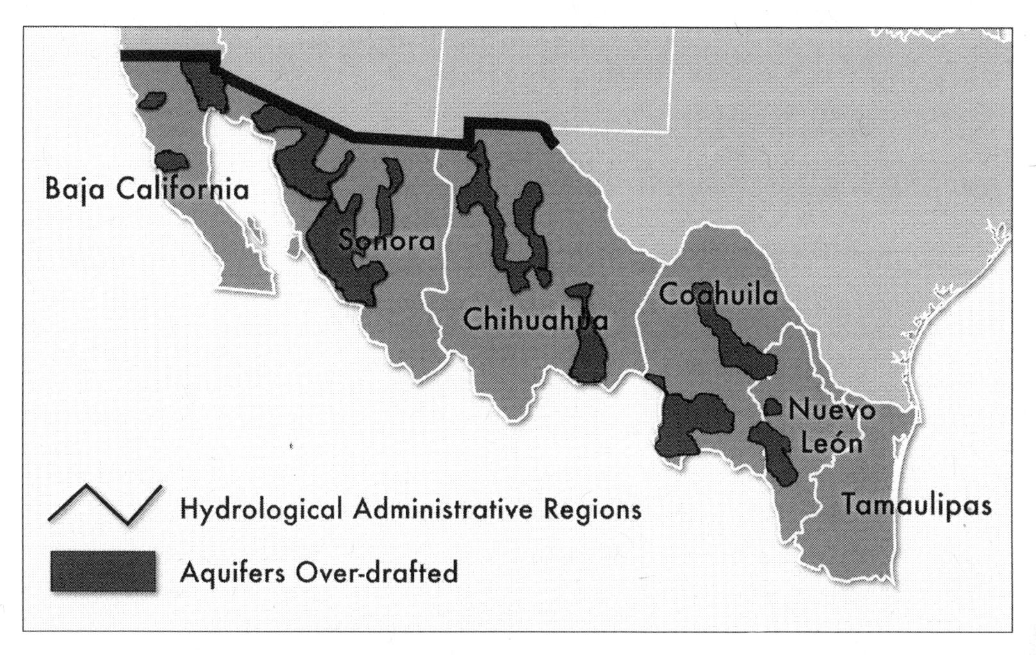

Figure 16.6 Over-drafted aquifers in Mexico. Note the concentration of these in northeast Baja California, along the coast of Sonora, and in the Rio Grande/Río Bravo watershed. Source: CONAGUA (2011, chap. 2, 34).

while groundwater quality in coastal aquifers may be affected by sea-level rise that leads to saltwater intrusion (see Chapter 9, Section 9). Chronic salt accumulation in soils associated with hot and arid climates can produce agricultural losses and places additional restrictions on regional agricultural water management.[ixxx] Scientific research on groundwater is lacking in comparison to knowledge on surface water resources, and the lack is particularly pronounced on the Mexican side (Moreno 2006; Scott, Dall'erba, and Díaz-Caravantes 2010; Granados-Olivas et al. 2012). Also, the effects of climate variability and change on water quality are virtually unexplored territory.

There are almost no natural impoundments of any substantial size in the border region. However, there are a number of man-made reservoirs, most of which are fed by the Colorado River or Rio Grande (examples are the Imperial and Morelos Dams on the Colorado River and the Leasburg and American Dams in the border region on the Rio Grande), and so are replenished by water derived primarily from winter snowpack in distant mountains. Upper Rio Grande flows in particular rely primarily on snowpack (Lacewell et al. 2010). Smaller border-crossing rivers like the Santa Cruz and the San Pedro get their most substantial flows from summer precipitation, and somewhat less from winter storms and local snowpack in high elevation "sky island" mountain ranges. The New River in the Mexicali-Imperial Valley region receives its flow from treated wastewater and agricultural drains. Numerous small reservoirs in the border region capture rainfall and many also store imported water from the major river systems. Of San Diego's twenty-five reservoirs, many import water from the Colorado River and from the California Water Project in Northern California. Coastal Baja California has

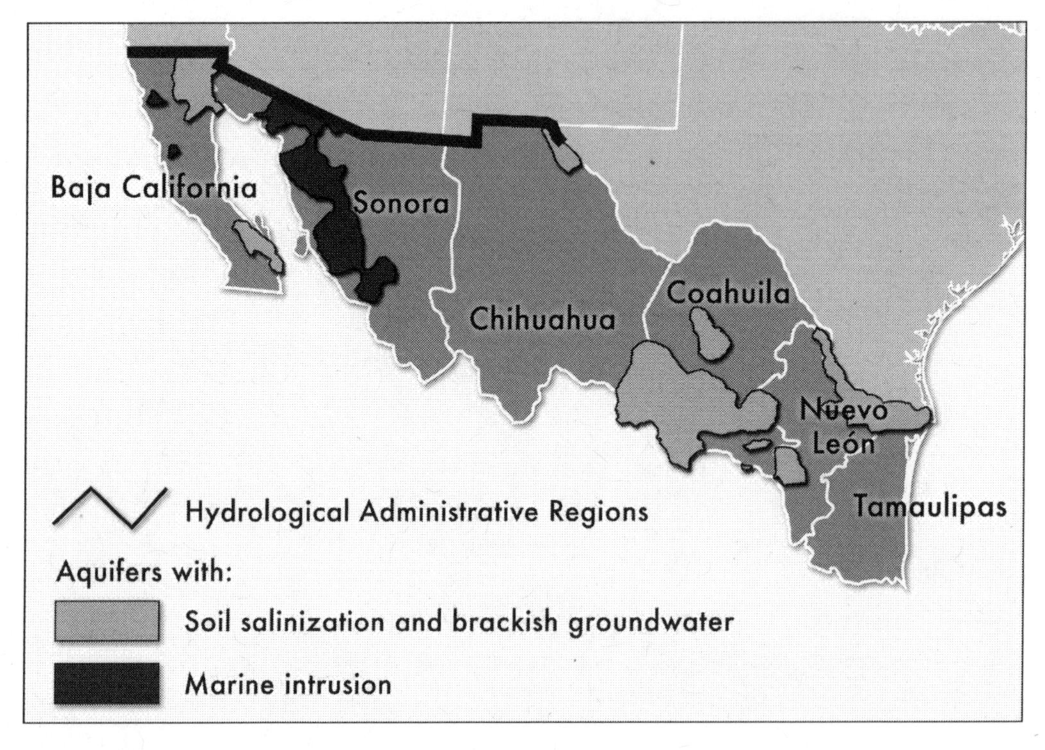

Figure 16.7 Areas in the border region of Mexico affected by saltwater intrusion or saline soil. Note the concentration of these problems in the irrigation districts of northeastern Baja California and along the coast of Sonora, as well as in the Juarez Valley. Source: CONAGUA (2011, chap. 2, 35).

two reservoirs that capture runoff and store water pumped over the mountains from the Colorado River to serve Tijuana, Playas de Rosarito, and Tecate.

Agriculture uses the largest share of water (about 80% of total supply) (McDonald 2010; CONAGUA 2011). The next largest use is municipal/urban, followed by industrial and thermoelectric. Figure 16.8 shows water use in the border states in Mexico.

The eight-year period from 2000 to 2007 was "a period of unprecedented dryness in the Colorado River basin when compared to the roughly 100-year historical record" (CDWR 2009, 21). Modeling by the U.S. Bureau of Reclamation shows that shortages due to drought become "increasingly likely" (CDWR 2009, 21) in the future as water demands increase. Cayan and others (2010, 21271) call the recent drought the "most extreme in over a century." The Colorado system of reservoirs is one of the region's "most important buffers against drought" (MacDonald 2010, 21259). During the early twenty-first century drought, storage levels have "declined precipitously" and could potentially fall below operable levels (MacDonald 2010, 21259). Impacts of the recent drought include: emergency restrictions on outdoor water use (for Tucson and San Diego); reductions in urban water service delivery (e.g., Metropolitan Water District of Southern California in 2009); agricultural revenue losses (documented at $308 million in California statewide); impacts to hydro-generated electricity; and forest loss due to wildfires and spread of bark beetle destruction (MacDonald 2010; see also Chapter 8).

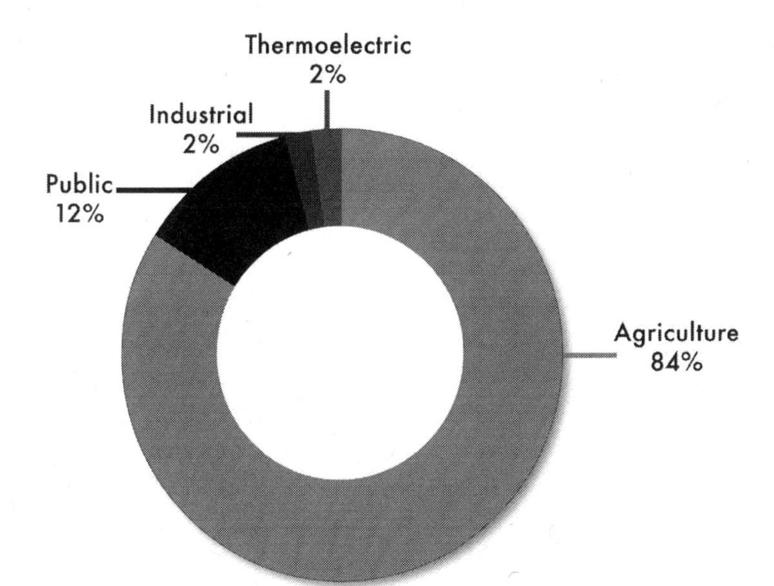

Figure 16.8 Water use in Mexico's border region (includes Region 1, Peninsula Baja California; Region II, Northwest; and Region VI, Rio Bravo). Adapted from CONAGUA (2011, annexes, 128-129, 133).

RIO GRANDE WATERSHED. The Rio Grande has its headwaters in the San Juan Mountains of southern Colorado, flows through New Mexico, forms the international boundary between the United States and Mexico (Figure 16.9), and terminates in the Gulf of Mexico. Its watershed is divided roughly equally between the United States and Mexico. The Upper Rio Grande is defined as the headwaters area in Colorado downstream to Fort Quitman, Texas (about 60 miles downstream from El Paso). The Lower Rio Grande, from Fort Quitman to the Gulf, takes in the river's largest tributaries, including the Pecos River and Devil's River in Texas and the Río Conchos, Río Salado, and Río San Juan in Mexico (CDWR 2009). The Upper Rio Grande system has two large storage reservoirs, Elephant Butte and Caballo Reservoirs, as well as smaller dams.[xxx] Overall, about half of the basin's 19 million acre-feet (MAF) of storage is in Mexico and the other half in the United States (CDWR 2009). A 1938 interstate compact divides the waters of the Upper Rio Grande among Colorado, New Mexico, and Texas. Two treaties between the United States and Mexico govern allocation of water from the river's international reach. Above Fort Quitman, the United States is required annually to deliver 60,000 acre-feet of Rio Grande water at Ciudad Juárez, in accordance with the Convention of 1906.

Significant shared groundwater resources that are critical supply sources for cities in this area include the Hueco Bolson and Mesilla Bolson aquifers in the El Paso–Ciudad Juárez region which are shared among New Mexico, Texas, and Mexico. Overdraft and salinity challenges are major issues for both sides of the border in this region (see Figures 16.6 and 16.7). Groundwater levels and quality have declined precipitously in the most important aquifer, the Hueco Bolson, since 1940 (Granados-Olivas et al. 2012). The water supply for the Upper Rio Grande Basin is fully allocated. Its system of engineered storage and delivery requires precipitation "at the right time, right place, over time, and with adequate quantity" in order to function properly (Lacewell et al. 2010, 105). Changes in the timing and amount of rainfall accompanied by an increase in temperature puts the system in a vulnerable situation (Lacewell et al. 2010).

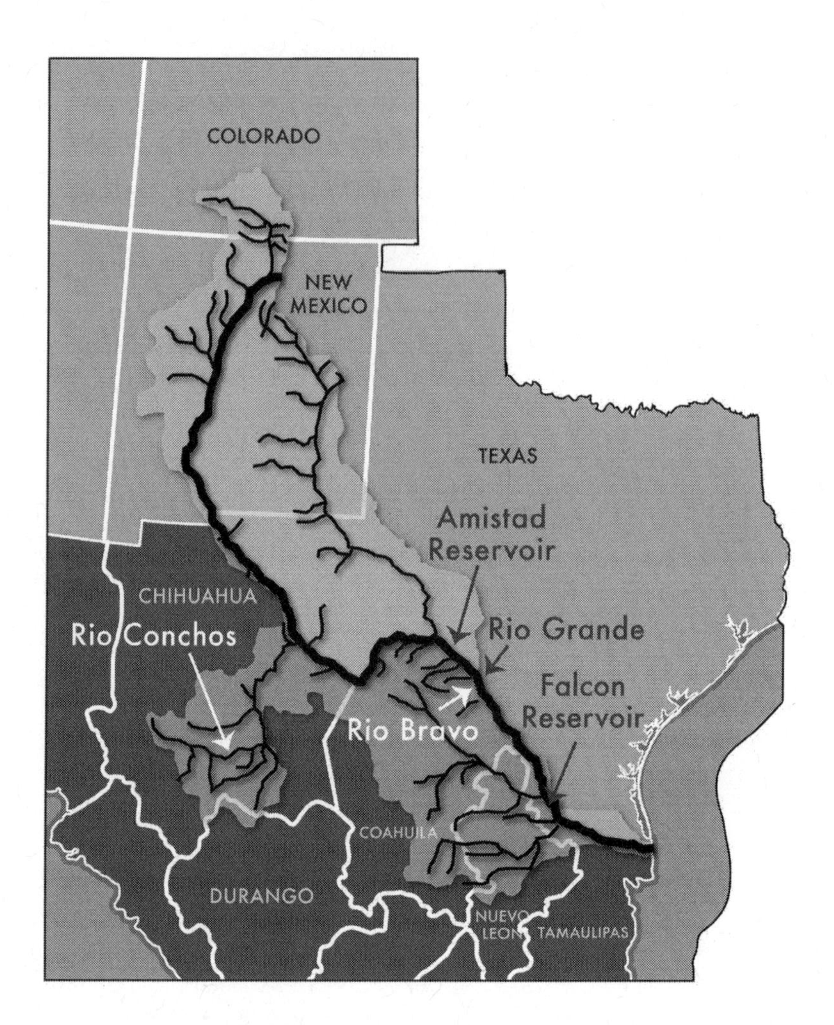

Figure 16.9 Rio Grande Basin.
Source: Lacewell et al. (2010).

The Rio Grande and its associated aquifers are the principal and often only water source for cities and farms from southern Colorado through New Mexico and into far west Texas (Hurd and Coonrod 2007). The vulnerability that these water users face in light of potential climatic and hydrologic changes is indicated not only by their dependence on a sole source of supply but by the oversubscribed claims to and exhaustive use of this source (Hurd et al. 2006; Hurd and Coonrod 2007). Using a hydro-economic model developed for the Upper Rio Grande, Hurd and Coonrod (2007, 2008) identified the following significant vulnerabilities for New Mexico based on a "middle severity" future climate change scenario:

- A reduction in long-run average water supply leading to a 2% reduction by 2030 and 18% by 2080, with the reduction affecting agriculture in 2030 and both agriculture and urban areas in 2080;
- Increases in water prices, as demand exceeds supply due to population growth and projected climate change;
- A concomitant shift in consumptive use by urban areas at the expense of agriculture;

- Secondary economic effects resulting from reduced consumptive use by agriculture, including significant economic losses from reservoir recreation and from job losses in the agriculture sector. For example, the 2030 middle scenario estimates a total economic loss of $8.4 million associated with a 3.5% reduction in agricultural water use, mostly in direct losses to agriculture ($7.1 million). By 2080, total economic losses associated with a 22.5% reduction in water use results in a loss of $61.7 million (based on year 2000 dollars).
- Worrisome impacts on natural ecosystems, in particular on the endangered silvery minnow habitat that lacks dedicated sources with the minimal flows needed to sustain it;
- Increased flooding (this is anticipated but not accounted for by the authors);
- Negative impacts on water quality;
- Negative impacts on native Hispanic communities who are likely to be among the first farmers to experience pressure to transfer water from their acequia systems to cities (see also Perramond 2012).

In the Paso del Norte binational area, key water resource vulnerabilities include:

- The lack of trans-border "data fusion" among governmental agencies;
- Higher evapotranspiration rates associated with increasing irrigation needs under projected climate changes;
- Increasing salinity of groundwater used for agriculture and drinking water (Hurd et al. 2006).

This analysis of water sector vulnerability indicates increasing risk exposure for agriculture, local economies, and ecosystems resulting in potentially serious impacts, including reduced natural water supplies; increased urban-agriculture competition; potentially negative impacts for off-the-grid users, including informal colonias and high-value riparian areas that lack a dedicated source of water. The water-energy infrastructure, especially during summer peak demand and during extended drought periods, will be sensitive to climate change. Traditional farming and ranching cultures may be increasingly exposed to climate-change impacts, resulting in reductions in their production. Finally, fundamental ecosystem changes may ensue, including reductions in soil moisture and increased pest infestations and disease.

Agriculture and ranching

Agriculture and ranching account for a small share of the border region's gross domestic product (GDP). Yet farmers and ranchers are the primary managers of most of the region's water and land resources. In the border region, agriculture accounts for approximately 80% of water consumption. About 74% of Arizona's land and 85% of New Mexico's land is used for farming and ranching (USDA n.d.). Agriculture and ranching in the border region will increasingly have to compete with cities for water. Agriculture and ranching also play an important cultural and political role in the regional identity and traditions, and agricultural ecosystems are significant. Thus climate change-related impacts on the Southwest landscape will most likely have significant impacts on the

Southwest's agricultural sectors. Changes in water availability, vegetation cover, carbon dioxide levels, and frequency of extreme events like floods and drought will impact crop and forage production, increasing costs for both producers and consumers.

The border region contains three major irrigated agriculture areas: Imperial Valley-Coachella (California), Yuma–San Luis Río Colorado–Mexicali (Arizona–Sonora–Baja California), and the Rio Grande Valley (New Mexico/Texas/Chihuahua). Agriculture consumes about 86% of total water resources in the Mexicali Valley (CONAGUA 2008). The Irrigation District 014, Colorado River, encompasses the Mexicali and San Luis Río Colorado valleys and provides water to about 2,500 agricultural operations over an irrigated area of 204,000 hectares (about 455,000 acres). This is one of the most productive agriculture districts in northern Mexico, sustained mainly by water from the Colorado River.[xxxi] Wheat, cotton, and alfalfa are the most important among fifty registered crops. An important secondary water source is a transboundary aquifer (which is recharged by the Colorado River) shared by the United States and Mexico.

Climate change impacts on regional agriculture and livestock in the Mexicali Valley are directly linked to production and productivity reductions. According to the Programa Estatal de Acción ante el Cambio Climático—Baja California (PEAC-BC)[xxxii], during the last three decades, changes in local and regional climate conditions have been and will be impacting agriculture. Preliminary PEAC-BC findings indicate that changing climate conditions will: drastically reduce the quantity and quality of available water; change the distribution and population dynamics of pest infestations and predator species; and cause changes in crop pollinators. A preliminary review suggests that major spring-summer season crops like cotton may be impacted by the more intensive and increased July–August rainfall period by staining the cotton fiber and reducing the quality for international market grades. Fall-winter crops such as wheat may be negatively affected in both yields and quality of grain protein produced because warming will reduce winter chill hours required for optimal results. The expected higher evaporative-transpiration rates will require increased application of irrigation water per acre, resulting in reduced production. Acreage devoted to alfalfa may decline. This, in turn, will affect the regional livestock sector, which will need to obtain more expensive alfalfa from distant suppliers (Cortez-Lara 2011).

In Arizona, agricultural use of irrigated water accounts for about 70% of water use and about 80% of the state's Colorado River allocations. Groundwater aquifers supply roughly half of total agricultural supply in Arizona, and the Colorado River and its tributaries supply the other half (Owen 2008). Agriculture in New Mexico uses almost 78% of the state's water supply (Owen 2008). Farm size and type of farm may be important indicators of the relative vulnerability of agricultural operations to climate or other stresses and factors (Hoppe, Banker, and MacDonald 2010).

In New Mexico, agriculture comprises a $1.7 billion annual industry, around three-quarters of it from livestock. Major crops include forage crops, onion, pecans, and wheat. In Arizona, agriculture is approximately a $2.4 billion annual industry, with over one-third in livestock (Owen 2008). Major crops include forage crops, cotton, lettuce, and wheat. Farmers may opt to alter their crop mix or invest in more water efficient systems as an adaptation strategy.

Cattle ranching in the Southwest relies on rainfed pastures and browse, which are sensitive to precipitation decreases as well as seasonality of precipitation[xxxiii]. Thus, drought is the most significant concern in some areas (Coles and Scott 2009). The yields of cattle (both in numbers of head and weight) will be reduced with declines in average precipitation (Owen 2008). Higher temperatures suppress cattle appetite, but warmer temperatures bode well for winter survival rates (Owen 2008). Small ranching operations are most vulnerable to drought, especially when combined with volatile cattle prices and pressure from urban land markets to subdivide the land (Eakin and Conley 2002).

Farmers with access to groundwater-supplied irrigation prefer dryness so they can control levels of water applied to crops (Vásquez-León, West, and Finan 2003; Coles and Scott 2009). The high cost of electricity for groundwater pumping is a major factor for irrigators (Wilder and Whiteford 2006; Coles and Scott 2009) because both energy and water needs increase with temperature (Garfin, Crimmins, and Jacobs 2007; Scott and Pasqualetti 2010). However, a study in southeast Arizona found that farmers and ranchers made limited use of climate information, preferring to continue customary practices and lacking confidence in linking their livelihoods to seasonal climate forecasts (Coles and Scott 2009). This study, like that of Eakin and Conley (2002), found small operations had less adaptive capacity than larger ones, indicating that scale of operation is a key factor shaping how vulnerability is experienced in the agricultural sector.

On the eastern edge of the area considered in this chapter, Rio Grande waters are impounded in Elephant Butte and Caballo Reservoirs for irrigation of 135,000 acres of land along approximately 200 miles of valley, including land in El Paso County, Texas. Crops include cotton, pecans, dairy, vegetables, and grapes in southern New Mexico and El Paso (Lacewell et al. 2010). Cattle and livestock are also a significant part of the economy.

This analysis indicates that farming and ranching in the border region are exposed to risks from climate change, and are especially sensitive to changes in seasonality and timing of these changes. Adaptation has a vital role in promoting sustainability of these livelihoods and traditional ways of life. Increasing the adaptive capacity of the agriculture and ranching sector to reduce livelihood risks will enhance the sustainability of these sectors. The current and future challenges in this field in relation to water availability must be addressed though a perspective that includes the conjunctive use of surface and groundwater (i.e., coordinated management of these resources to improve efficiency), technological improvement in irrigation, and sustainable crops. Additional discussion on potential impacts from climate change and climate variability on the agricultural sector can be found in Chapter 11.

Wildfire

Wildfires pose a considerable risk to border communities and to communities throughout the Southwest alike. Wildfires in the contiguous Western states in the United States increased by more than 300% from the 1970s to 2005 (Corringham, Westerling, and Morehouse 2008) and are extremely costly in terms of human life, loss of structures, forest mortality, habitat destruction, and direct fire suppression costs.[xxxiv] The years 2006 and 2008 were the worst on record for wildfire activity in the United States (Grissino-Mayer 2010). In California, the two largest wildfires on record and eleven of the twenty largest recorded fires occurred in the past decade (MacDonald 2010). The fire season of 2011 was "record-setting" in Arizona and New Mexico, and the southern border region (southeastern New Mexico and southeastern Arizona) was hardest-hit in each state.[xxxv]

Climatic factors including higher average temperatures since the 1970s and extensive droughts have contributed to conditions for increased wildfire, as have land-use changes and fire-suppression strategies (Williams et al. 2010). The seasonality of temperature and precipitation changes is especially critical; higher temperatures, earlier spring warming, and decreased surface water contribute to an increase in wildfires (MacDonald 2010). Drought-related bark beetle damage has had devastating effects on Southwest forests. Overall, Williams and colleagues (2010) estimate that approximately 2.7% of Southwestern forest and woodland area experienced substantial mortality due to wildfires from 1984 to 2006, and approximately 7.6% experienced mortality due to bark beetles or wildfire during this period.

Wildfire and land-use management play a large role in controlling the outbreak of wildfires, and climate information should be an important aspect of the planning process. Expected climatic changes will alter future forest productivity, disturbance regimes, and species ranges throughout the Southwest (Williams et al. 2010). Peak fire-suppression periods vary from region to region, with important implications for decision making around wildfire (Corringham, Westerling, and Morehouse 2008; Westerling et al. 2011).

While fire managers in the Southwest United States are integrating short-term weather and climate information into their planning, long-term forecasts are less utilized due to a perceived lack of reliability (Corringham, Westerling, and Morehouse 2008). Trans-border emergency response to wildfires is another critical element of effective management. Events such as wildfires "do not respect administrative boundaries" (GNEB 2008, 2). Trans-border communication-sharing and response systems (as appropriate) can add to regional resilience and improve forest sustainability.

References

Adger, W. N. 2006. Vulnerability. *Global Environmental Change* 16:268–281.

Adger, W. N., N. W. Arnell, and E. L. Tompkins. 2005. Successful adaptation across scales. *Global Environmental Change* 15:77–86.

Adger, W. N., S. Dessai, M. Goulden, M. Hulme, I. Lorenzoni, D. R. Nelson, L. O. Naess, J. Wolf, and A. Wreford. 2009. Are there social limits to adaptation to climate change? *Climatic Change* 93:335–354, doi:10.1007/s10584-008-9520-z.

Anzaldúa, G. 1987. *Borderlands/La Frontera*. San Francisco: Aunt Lute Press.

Austin, D. E., E. Mendoza, M. Guzmán, and A. Jaramillo. 2004. Partnering for a new approach: Maquiladoras, government agencies, educational institutions, non-profit organizations, and residents in Ambos Nogales. In *Social costs of industrial growth in northern Mexico*, ed. K. Kopinak, 251–281. San Diego: University of California–San Diego, Center for U.S.–Mexican Studies.

Beibighauser, T. R. 2007. *Wetland drainage, restoration, and repair*. Lexington: The University Press of Kentucky.

Border Environment Cooperation Commission (BECC). 2007. *An analysis of program impacts and pending needs*. http://www.cocef.org/english/VLibrary/Publications/SpecialReports/White%20Paper%20Analysis%20of%20US-MEX%20Border%20Program.pdf.

Brito-Castillo, L., A. Leyva-Contreras, A. V. Douglas, and D. Lluch-Belda. 2002. Pacific Decadal Oscillation and the filled capacity of dams on the rivers of the Gulf of California continental watershed. *Atmósfera* 15:121–136.

Brito-Castillo, L., A. V. Douglas, A. Leyva-Contreras, and D. Lluch-Belda. 2003. The effect of large-scale circulation on precipitation and streamflow in the Gulf of California continental watershed. *International Journal of Climatology* 23:751–768.

Cañas, J., R. A. Coronado, R. W. Gilmer, and E. Saucedo. 2011. *The impact of the maquiladora industry on U.S. border cities.* Research Department Working Paper 1107. Dallas, TX: Federal Reserve Bank of Dallas. http://www.dallasfed.org/assets/documents/research/papers/2011/wp1107.pdf.

Cañas, J, and R.W. Gilmer. 2009. The maquiladora's changing geography. *Southwest Economy* (Second Quarter 2009): 10–14. Dallas, TX: Federal Reserve Bank of Dallas. http://www.dallasfed.org/assets/documents/research/swe/2009/swe0902c.pdf.

California Department of Water Resources (CDWR). 2009. *Water and border area climate change: An introduction; Special report for the XXVI Border Governors Conference.* Sacramento: CDWR.

Cash, D. W., W. C. Clark, F. Alcock, N. M. Dickson, N. Eckley, D. H. Guston, J. Jager, and R. B. Mitchell. 2003. Knowledge systems for sustainable development. *Proceedings of the National Academy of Sciences* 100:8086–8091.

Cavazos, T., and D. Rivas. 2004. Variability of extreme precipitation events in Tijuana, Mexico. *Climate Research* 25:229–243.

Cave, D. 2009. Recession slows growth of population in Sunbelt. New York Times December 24, 2009.

Cayan, D. R., T. Das, D. W. Pierce, T. P. Barnett, M. Tyree, and A. Gershunov. 2010. Future dryness in the southwest US and the hydrology of the early 21st century drought. *Proceedings of the National Academy of Sciences* 107:21271–21276.

Chacon Anaya, D., M. E. Giner, M. Vazquez Valles, S. M. Roe, J. A. Maldonado, H. Lindquist, B. Strode, R. Anderson, C. Quiroz, and J. Schreiber. 2010. *Greenhouse gas emissions in Baja California and reference case projections 1990–2025.* Ciudad Juárez, Chihuahua: Border Environment Cooperation Commission.

Coles, A. R., and C. A. Scott. 2009. Vulnerability and adaptation to climate change and variability in semi-arid rural southern Arizona, USA. *Natural Resources Forum* 33:297–309.

Collins, T. W. 2010. Marginalization, facilitation, and the production of unequal risks: The 2006 Paso del Norte floods. *Antipode* 42:258–289.

Collins, T. W., S. E. Grineski, and M. L. Romo Aguilar. 2009. Vulnerability to environmental hazards in the Ciudad Juárez (Mexico)-El Paso (USA) metropolis: A model for spatial risk assessment in transnational context. *Applied Geography* 29:448–461.

Comisión Nacional del Agua (CONAGUA). 2008. *Estadísticas del agua en México.* México, D.F.: CONAGUA.

—. 2011. *Estadísticas del agua en México, Edición 2011.* México, D.F.: CONAGUA.

Consejo Nacional de Población (CONAPO). 2007. *Delimitación de las zonas metropolitanas de México 2005.* México, D.F.: CONAPO.

Corringham, T. W., A. J. Westerling, and B. J. Morehouse. 2008. Exploring use of climate information in wildland fire management: A decision calendar study. *Journal of Forestry* 106:71–77.

Cortez-Lara, A. 2011. Gestión y manejo del agua: El papel de los usuarios agrícolas del Valle de Mexicali. *Problemas del Desarrollo* 167 (42): 71–96.

Díaz-Castro, S., M. D. Therrell, D. W. Stahle, and M. K Cleaveland. 2002. Chihuahua winter-spring precipitation reconstructed from tree-rings, 1647–1992. *Climate Research* 22:237–244.

Dominguez, F., J. Cañon, and J. Valdes. 2010. IPCC-AR4 climate simulations for the southwestern US: The importance of future ENSO projections. *Climatic Change* 99:499–514.

Eakin, H., and J. Conley. 2002. Climate variability and the vulnerability of ranching in southeastern Arizona: A pilot study. *Climate Research* 21:271–281.

Eakin, H., and A. Luers. 2006. Assessing the vulnerability of social-environmental systems. *Annual Review of Environment and Resources* 31:365–394.

Eakin, H., and M. Wehbe. 2009. Linking local vulnerability to system sustainability in a vulnerability framework: Cases from Latin America. *Climatic Change* 93:355–377.

Earman, S., and M. D. Dettinger. 2011. Potential impacts of climate change on groundwater resources: A global review. *Journal of Water and Climate Change* 2:213–229.

Favre, A., and A. Gershunov. 2009. North Pacific cyclonic and anticyclonic transients in a global warming context: Possible consequences for western North American daily precipitation and temperature extremes. *Climate Dynamics* 32:969–987.

Field, C. B., L. D. Mortsch, M. Brklacich, D. L. Forbes, P. Kovacs, J. A. Patz, S. W. Running, and M. J. Scott. 2007. North America. In *Climate Change 2007: Impacts, adaptation and vulnerability. Contribution of Working Group II to the Fourth Assessment Report of the Intergovernmental Panel on Climate Change,* ed. M. L. Parry, O. F. Canziani, J. P. Palutikof, P. J. van der Linden and C. E. Hanson, 617–652. Cambridge: Cambridge University Press.

Frey, W. H. 2011. 2011 puts the brakes on U.S. population growth. *Up Front,* December 28, 2011. Washington, DC: Brookings Institution. http://www.brookings.edu/opinions/2011/1228_census_population_frey.aspx.

Ganster. P. 2010. La cuenca binacional del Río Tijuana. In *Diagnóstico y priorización socio-ambiental de las cuencas hidrográficas de México,* ed. H. Cotler Avalos. México, D.F.: Instituto Nacional de Ecología.

Ganster, P., and D. E. Lorey. 2008. *The U.S.-Mexican border into the twenty-first century,* 2nd ed. Lanham, MD: Rowman and Littlefield.

Garcia-Cueto, R. O., A. Tejeda-Martínez, and E. Jáuregui-Ostos. 2010. Heat waves and heat days in an arid city in the northwest of Mexico: Current trends and in climate change scenarios. *International Journal of Biometeorology* 54:335–345.

Garfin, G., M.A. Crimmins, and K. L. Jacobs. 2007. Drought, climate variability, and implications for water supply and management. In *Arizona water policy,* ed. B. G. Colby and K. L. Jacobs, Chapter 5. Washington, DC: Resources for the Future.

Glenn, E. P., R. S. Felger, A. Búrquez, and D. S. Turner. 1992. Cienega de Santa Clara: Endangered wetland in the Colorado River Delta, Sonora, Mexico. *Natural Resources Journal* 32:817–824.

Glenn, E. P., C. Lee, R. Felger, and S. Zengel. 1996. Effects of water management on the wetlands of the Colorado River Delta, Mexico. *Conservation Biology* 10:1175–1186.

Gochis, D. J., L. Brito-Castillo, and W. J. Shuttleworth. 2007. Correlations between sea-surface temperatures and warm season streamflow in northwest Mexico. *International Journal of Climatology* 27:883–901.

Good Neighbor Environment Board (GNEB). 2006. *Air quality and transportation and cultural and natural resources on the U.S.-Mexico border: Ninth report of the GNEB to the President and Congress of the U.S.* EPA 130-R-06-002. Washington, DC: EPA.

—. 2008. *Natural disasters and the environment along the U.S.-Mexican border: Eleventh report of the GNEB to the President and Congress of the U.S.* EPA 130-R-08-001. Washington, DC: EPA.

—. 2011. *The potential environmental and economic benefits of renewable energy development in the U.S.-Mexico Border Region: Fourteenth report of the Good Neighborhood Environmental Board to the President and Congress of the United States.* Washington, DC: EPA.

Granados-Olivas, A., B. Creel, E. Sánchez-Flores, J. Chavez, and J. Hawley. 2012. Thirty years of groundwater evolution: Challenges and opportunities for binational planning and sustainable management of the transboundary Paso del Norte watersheds. In The US-Mexican border environment: Progress and challenges for sustainability, ed. E. Lee and P. Ganster, 201–217. SCERP Monograph Series no. 16. San Diego: San Diego State University Press.

Grissino-Mayer, H. D. 2010. Wildfire hazard and the role of tree-ring research. In *Tree rings and natural hazards: A state-of-the-art*, ed. M. Stoffel, M. Bollschweiler, D. R. Butler, and B. H. Luckman, 323–328. Advances in Global Change Research 41. New York: Springer.

Grothman, T., and A. Patt. 2005. Adaptive capacity and human cognition: The process of individual adaptation to climate change. *Global Environmental Change* 15:199–213.

Gutzler, D. S., and T. O. Robbins. 2011. Climate variability and projected change in the western United States: Regional downscaling and drought statistics. *Climate Dynamics* 37:835–839, doi:10.1007/s00382-010-0838-7.

Harlan, S. L., A. J. Brazel, G. D. Jenerette, N. S. Jones, L. Larsen, L. Prashad, and W. L. Stefanov. 2008. In the shade of affluence: The inequitable distribution of the urban heat island. In *Equity and the Environment*, ed. R. C. Wilkinson and W. R. Freudenburg, 173–202. Research in Social Problems and Public Policy 15. Bingley, UK: Emerald Group Publishing.

Higgins, R. W., Y. Chen, and A.V. Douglas. 1999. Interannual variability of the North American warm season precipitation regime. *Journal of Climate* 12:653–680.

Higgins, R. W., and W. Shi. 2001. Intercomparison of the principal modes of interannual and intraseasonal variability of the North American Monsoon System. *Journal of Climate* 14:403–417.

Hoppe, R. A., D. E. Banker, and J. M. MacDonald. 2010. *America's diverse family farms, 2010 edition.* Economic Information Bulletin 96653. Washington, DC: USDA.

Hurd, B., C. Brown, J. Greenlee, A. Granados, and M. Hendrie. 2006. Assessing water resource vulnerability for arid watersheds: GIS-based research in the Paso del Norte region. *New Mexico Journal of Science* 44 (Aug. 2006): 39–61.

Hurd, B. H., and J. Coonrod. 2007. *Climate change and its implications for New Mexico's water resources and economic opportunities.* Las Cruces: New Mexico State University. http://portal.azoah.com/oedf/documents/08A-AWS001-DWR/Omnia/20070700%20Hurd%20and%20Coonrod%20Climate%20Change%20NM%20Water.pdf.

—. 2008. Climate change risks New Mexico's waterways: Its byways and its flyways. *Water Resources Impact* 10 (4): 1–5.

Intergovernmental Panel on Climate Change (IPCC). 2007a. *Climate change 2007: Impacts, adaptation and vulnerability. Contribution of Working Group II to the Fourth Assessment Report of the Intergovernmental Panel on Climate Change*, ed. M. L. Parry, O. F. Canziani, J. P. Palutikof, P. J. van der Linden and C. E. Hanson. Cambridge: Cambridge University Press.

—. 2007b. *Climate change 2007: The physical science basis. Contribution of Working Group I to the Fourth Assessment Report of the Intergovernmental Panel on Climate Change*, ed. S. Solomon, D. Qin, M. Manning, Z. Chen, M. Marquis, K. B. Averyt, M. Tignor, and H. L. Miller. Cambridge: Cambridge University Press.

Jepson, W. 2012. Claiming water, claiming space: Contested legal geographies of water in south Texas. *Annals of the Association of American Geographers* 102:614–631.

Kolef, P., A. Lira-Noriega, T. Urquiza, and E. Morales. 2007. Priorities for biodiversity conservation in Mexico's northern border. In *A barrier to our shared environment: The border fence between the United States and Mexico*, ed. A. Cordova and C. de la Parra. México, D.F.: Instituto Nacional de Ecología.

Kunkel, K. 2011. *Southwest region climate outlooks.* Draft report. Prepared for National Climate Assessment Southwest Region Team.

Lacewell, R. D., A. M. Michelsen, M. E. Rister, and A. W. Sturdivant, 2010. Transboundary water crises: Learning from our neighbors in the Rio Grande (Bravo) and Jordan River watersheds. *Journal of Transboundary Water Resources* 1:95–123.

Lahsen, M., R. Sánchez-Rodriguez, P. Romero-Lankao, P. Dube, R. Leemans, O. Gaffney, M. Mirza, P. Pinho, B. Osman-Elasha, and M. S. Smith. 2010. Impacts, adaptation and vulnerability to global environmental change: Challenges and pathways for an action-oriented research agenda for middle- and low-income countries. *Current Opinion in Environmental Sustainability* 2:364–374.

Laird-Benner, W., and H. Ingram. 2011. Sonoran Desert network weavers: Surprising environmental successes on the U.S./Mexico Border. *Environment* 53:6–16.

Lara, F., A. J. Brazel, M. E. Giner, E. Mahoney, R. Raja, and M. Quintero. 2012. The response of U.S.-Mexico border cities to climate change: Current practices and urgent needs. In *The US-Mexican border environment: Progress and challenges for sustainability*, ed. E. Lee and P. Ganster, 267–288. SCERP Monograph Series no. 16. San Diego: San Diego State University Press.

Lara, F., and G. Diaz-Montemayor. 2010. *City of green creeks: Sustainable flood management alternatives for Nogales, Sonora.* Ciudad Juárez, Chihuahua: Border Environment Cooperation Commission.

Leichenko, R., and K. L. O'Brien. 2008. *Environmental change and globalization: Double exposures.* New York: Oxford University Press.

Lemos, M. C., and A. Agrawal. 2006. Environmental governance. *Annual Review of Environment and Resources* 31:297–325.

Lemos, M. C., D. Austin, R. Merideth, and R. G. Varady. 2002. Public–private partnerships as catalysts for community-based water infrastructure development: The Border WaterWorks program in Texas and New Mexico colonias. *Environment and Planning C: Government and Policy* 20:281–295.

Lemos, M. C., and B. J. Morehouse. 2005. The co-production of science and policy in integrated climate assessments. *Global Environmental Change* 15:57–68.

Liverman, D. M., and R. Merideth. 2002. Climate and society in the U.S. Southwest: The context for a regional assessment. *Climate Research* 21:199–218.

Liverman, D., R. Varady, O. Chavez, O., and R. Sanchez. 1999. Environmental issues along the U.S.-Mexico border: Drivers of change and responses of citizens to institutions. *Annual Review of Energy and the Environment* 24:607–643.

López-Hoffman, L., R. G. Varady, and P. Balvanera. 2009. Finding mutual interest in shared ecosystem services: New approaches to transboundary conservation. In *Conservation of shared environments: Learning from the United States and Mexico*, ed. L. Lopez-Hoffman, E. McGovern, R. G. Varady, and K. W. Flessa, 137–153. Tucson: University of Arizona Press.

MacDonald, G. M. 2010. Water, climate change, and sustainability in the Southwest. *Proceedings of the National Academy of Sciences* 107:21256–21262.

Magaña, V., and C. Conde. 2000. Climate and freshwater resources in northern Mexico: Sonora, a case study. *Environmental Monitoring and Assessment* 61:167–185.

Magaña, V., D. Zermeño, and C. Neri. 2012. Climate change scenarios and potential impacts on water availability in northern Mexico. *Climatic Research* 51:171–184, doi: 10.3354/cr01080.

Mearns, L. O., W. Gutowski, R. Jones, R. Leung, S. McGinnis, A. Nunes, and Y. Qian. 2009. A regional climate change assessment program for North America. *Eos Transactions AGU 90*:311.

Megdal, S. B., and C.A. Scott. 2011. The importance of institutional asymmetries to the development of binational aquifer assessment programs: The Arizona-Sonora experience. *Water* 3:949–963, doi:10.3390/w3030949.

Méndez, M., and V. Magaña. 2010. Regional aspects of prolonged meteorological droughts over Mexico. *Journal of Climate* 23:1175–1188.

Milman, A., and C. A. Scott. 2010. Beneath the surface: Intranational institutions and management of the United States-Mexico transboundary Santa Cruz aquifer. *Environment and Planning* 28:528–551.

Montero Martínez, M., J. Martínez Jiménez, N. I. Castillo Pérez, and B. E. Espinoza Tamarindo. 2010. Escenarios climáticos en México proyectados para el siglo XXI: Precipitación, temperaturas máxima y mínima. In *Atlás de vulnerabilidad hídrica en México ante el cambio climático*, ed. P. F. Martinez Austria and C. Patiño-Gómez, 39–64. Efectos de los Cambios Climáticos en los Recursos Hídricos de México, III. Jiutepec, Morelos: Instituto Mexicano de la Tecnología del Agua.

Morello-Frosch, R., M. Pastor, J. Saad, and S. B. Shonkoff. 2009. The climate gap: Inequalities in how climate change hurts Americans and how to close the gap. *Solutions* 460 (7253): 1–32.

Moreno, J. L. 2006. *Por abajo del agua: Sobreexplotación y agotamiento del acuífero de la Costa de Hermosillo, 1945–2005*. Hermosillo: El Colegio de Sonora.

Moser, C., and D. Sattherwaite. 2010. Towards pro-poor adaptation to climate change in the urban centers of low- and middle-income countries. In *Social dimensions of climate change: Equity and vulnerability in a warming world*, ed. R. Mearns and A. Norton, 231. Washington, DC: World Bank.

Moss, R. 2011. Preliminary guidance on priority topics for the 2013 National Climate Assessment. Presentation to the National Climate Assessment Southwest Region Technical Report—Lead Authors Workshop, University of Colorado-Boulder, August 1–4, 2011.

Mumme, S. P. 1986. Engineering diplomacy: The evolving role of the International Boundary and Water Commission in U.S.-Mexico water management. *Journal of Borderlands Studies* 1:73–108.

—. 2000. Minute 242 and beyond: Challenges and opportunities for managing transboundary groundwater on the U.S.-Mexico border. *Natural Resources Journal* 40:341–379.

—. 2008. From equitable utilization to sustainable development: Advancing equity in water management on the U.S.-Mexico border. In *Water, place and equity*, ed. J. Whiteley, H. Ingram, and R. Perry, 117–146. Cambridge: MIT University Press.

Mumme, S. P., D. Lybecker, O. Gaona, and C. Manterola. 2009. The Commission on Envronmental Cooperation and transboundary cooperation across the U.S.-Mexico border. In *Conservation of shared environments: Learning from the United States and Mexico*, ed. L. Lopez-Hoffman, E. McGovern, R. G. Varady, and K. W. Flessa, 261–278. Tucson: University of Arizona Press.

Mwaniki-Lyman, L., V. Pavlakovich-Kochi, and R. Christopherson. n.d. Arizona-Sonora region population. Tucson: University of Arizona, Eller College of Management Economic and Business Research Center. http://ebr.eller.arizona.edu/arizona_border_region/AZ_demographic.asp.

National Research Council (NRC). 2010. *Adapting to the impacts of climate change*. Washington, DC: National Academies Press.

Owen, G. 2008. Impacts: Agriculture. Tucson: University of Arizona, Southwest Climate Change Network. http://www.southwestclimatechange.org/impacts/people/agriculture.

Pavia, E. G., F. Graef, and J. Reyes. 2006. PDO–ENSO effects in the climate of Mexico. *Journal of Climate* 19:6433–6438.

Pavlakovich-Kochi, V., and A. Charney. 2008. *Mexican visitors to Arizona: Visitor characteristics and economic impacts, 2007-8*. Phoenix: Arizona Office of Tourism.

Pelling, M., C. High, J. Dearing, and D. Smith. 2008. Shadow spaces for social learning: A relational understanding of adaptive capacity to climate change within organizations. *Environment and Planning A* 40: 867–884.

Perramond, E. P. 2012. The politics of scaling water governance and adjudication in New Mexico. *Water Alternatives* 5:62–82.

Pierce, D. W., T. Das, D. R. Cayan, E. P. Maurer, N. L. Miller, Y. Bao, M. Kanamitsu, et al. 2012. Probabilistic estimates of future changes in California temperature and precipitation using statistical and dynamical downscaling. *Climate Dynamics*, published online, doi:10.1007/s00382-012-1337-9.

Pineda Pablos, N., ed. 2006. *La búsqueda de la tarifa justa: El cobro de los servicios de agua potable y alcantarillado en México*. Hermosillo: Colegio de Sonora.

Pitt, J., and D. F. Luecke. 2000. Two nations, one river: Managing ecosystem conservation in the Colorado River delta. *Natural Resources Journal* 49:819–864.

Ray, A. J., G. M. Garfin, M. Wilder, M. Vásquez-León, M. Lenart, and A. C. Comrie. 2007. Applications of monsoon research: Opportunities to inform decisionmaking and reduce regional vulnerability. *Journal of Climate* 20:1608–1627.

Ribot, J. C. 2010. Vulnerability does not fall from the sky: Toward multiscale, pro-poor climate policy. In *Social Dimensions of climate change: Equity and vulnerability in a warming world*, ed. R. Mearns and A. Norton, 47-74. Washington, DC: The World Bank.

Robertson, R. 2009. Mexico and the great trade collapse. N.p.: Centre for Economic Policy Research, VoxEU.org. http://www.voxeu.org/index.php?q=node/4286.

Roullard, P. 2005. Tijuana River estuary: Tijuana River National Estuarine Research Reserve. In *Tijuana River Watershed Atlas*, ed. R. D. Wright and R. Vela, plates 31–36. San Diego: San Diego State University Press.

Romero-Lankao, P., M. Borbor-Córdova, R. Abrutsky, G. Günther, E. Behrenz, and L. Dawidowsky. 2012. ADAPTE: A tale of diverse teams coming together to do issue-driven interdisciplinary research. *Environmental Science & Policy*, published online, doi:10.1016/j.envsci.2011.12.003.

Romero-Lankao, P., and H. Qin, 2011. Conceptualizing urban vulnerability to global climate and environmental change. *Current Opinion in Environmental Sustainability* 3:142–149.

Ruddell, D. M., S. L. Harlan, S. Grossman-Clarke, and A. Buyantuyev. 2010. Risk and exposure to extreme heat in microclimates of Phoenix, AZ. In *Geospatial technologies in urban hazard and disaster analysis*, ed. P. Showalter and Y. Lu, 179–202. New York: Springer.

Sánchez-Rodriguez, R. 2008. Urban sustainability and global environmental change: Reflections for an urban agenda. In *The new global frontier: Urbanization, poverty, and environment in the 21st century*, ed. G. Martine, G. McGranahan, M. Montgomery, and R. Fernández-Castilla, 149–163. London: Earthscan.

Sánchez-Rodriguez, R., and S. P. Mumme. 2010. Environmental protection and natural resources. USMX Working Paper 1-10-01. San Diego: University of California-San Diego, Center for U.S.-Mexican Studies.

Sayre, N. 2002. *Ranching, endangered species, and urbanization in the Southwest: Species of capital*. Tucson: University of Arizona Press.

Scott, C. A., and J. M. Banister. 2008. The dilemma of water management "regionalization" in Mexico under centralized resource allocation. *International Journal of Water Resources Development* 24:61–74.

Scott, C. A., S. Dall'erba, and R. Díaz-Caravantes. 2010. Groundwater rights in Mexican agriculture: Spatial distribution and demographic determinants. *The Professional Geographer* 62:1–15.

Scott, C. A., S. Megdal, L. Antonio Oroz, J. Callegary, and P. Vandervoet. 2012. Effects of climate change and population growth on the transboundary Santa Cruz aquifer. *Climate Research* 51:159–170.

Scott, C. A., and M. J. Pasqualetti. 2010. Energy and water resources scarcity: Critical infrastructure for growth and economic development in Arizona and Sonora. *Natural Resources Journal* 50:645–682.

Seager, R., M. Ting, M. Davis, M. Cane, N. Naik, J. Nakamura, C. Li, E. Cook, and D. W. Stahle. 2009. Mexican drought: An observational modeling and tree ring study of variability and climate change. *Atmósfera* 22:1–31.

Seager, R., M. Ting, I. Held, Y. Kushnir, J. Lu, G. Vecchi, H-P. Huang, et al. 2007. Model projections of an imminent transition to a more arid climate in southwestern North America. *Science* 316:1181–1184.

Secretaría de Agricultura, Ganadería, Desarrollo Rural, Pesca y Alimentación (SAGARPA). Delegación Estatal en Baja California. 2011. *Programa de información y estudios agropecuarios.* Mexicai, B.C.: SAGARPA.

Segee, B., and A. Córdova. 2009. A fence runs through it: Conservation implications of recent U.S. border security legislation. In *Conservation of Shared Environments: Learning from the United States and Mexico,* ed. L. Lopez-Hoffman, E. McGovern, R. G. Varady, and K. W. Flessa, 241–256. Tucson: University of Arizona Press.

Serrat-Capdevila, A., J. B. Valdes, J. Gonzalez-Pérez, K. Baird, L. J. Mata, and T. Maddock III. 2007. Modeling climate change impacts – and uncertainty – on the hydrology of a riparian system: The San Pedro basin (Arizona/Sonora). *Journal of Hydrology* 347:48–66.

Seto, K. C., R. Sanchez-Rodriguez, and M. Fragkias. 2010. The new geography of contemporary urbanization and the environment. *Annual Review of Environment and Resources* 35:167–194.

Sheridan, T. E. 2010. Embattled ranchers, endangered species, and urban sprawl: The political ecology of the new American West. *Annual Review of Anthropology* 36:121–138.

Sierra Club. 2010. *Wild versus wall.* Video. Tucson: Sierra Club Grand Canyon Chapter, Border Campaign. http://arizona.sierraclub.org/conservation/border/borderfilm.asp.

Soden, D. L. 2006. *At the cross roads: US / Mexico border counties in transition.* IPED Technical Reports, Paper 27. El Paso: University of Texas at El Paso, Institute for Policy and Economic Development (IPED). http://digitalcommons.utep.edu/iped_techrep/27.

Stahle, D. W., E. R. Cook, J. Villanueva-Diaz, F. K. Fye, D. J. Burnette, R. D. Griffin, R. Acuna-Soto, R. Seager, and R. Heim, Jr. 2009. Early 21st-century drought in Mexico. *EOS Transactions AGU* 90:89–90.

Starks, R. R., J. McCormack, and S. Cornell. 2011. South: The U.S.-Mexico border region. In *Native nations and U.S. Borders: Challenges to indigenous culture, citizenship, and security,* 33–49. Tucson, AZ: Native Nations Institute for Leadership, Management, and Policy.

Tijuana River Valley Recovery Team (TRVRT). 2012. *Recovery strategy: Living with the water.* San Diego: TRVRT. http://www.waterboards.ca.gov/sandiego/board_info/agendas/2012/Feb/item13/DRAFT_RecoveryStrategy11-22-11_Rev1.pdf.

Udall, B. 2011. Climate change impacts on water in the southwestern U.S. Paper presented at National Climate Assessment Southwest Technical Report Authors' Meeting, Boulder, CO, Aug. 3-5, 2011.

U.S. Bureau of Reclamation. 2011. *SECURE Water Act Section 9503(c) - Reclamation Climate Change and Water, Report to Congress.* Denver: U.S. Bureau of Reclamation. http://www.usbr.gov/climate/SECURE/docs/SECUREWaterReport.pdf.

U.S. Department of Agriculture (USDA). n.d. Major land uses. Data Files. Washington D.C.: U.S. Department of Agriculture, Economic Research Service. http://www.ers.usda.gov/data/MajorLandUses/ (accessed March 1, 2012).

U.S. Environmental Protection Agency (EPA). 2006. *State of the border region, indicators report, 2005.* EPA-160-R-60-001. Washington, D.C.: EPA, Border 2012: U.S.-Mexico Environmental Program.

—. 2011. *State of the border region, indicators report, 2010.* Washington D.C.: EPA, Border 2012: U.S.-Mexico Environmental Program.

U.S. General Accounting Office (GAO). 2003. *International trade: Mexico's maquiladora decline affects U.S.-Mexico border communities and trade; Recovery depends in part upon Mexico's actions.* GAO 03-891. Washington, DC: GAO. http://www.gao.gov/cgi-bin/getrpt?GAO-03-891.

Varady, R., K. Hankins, A. Kaus, E. Young, and R. Merideth. 2001. Nature, water, culture, and livelihood in the lower Colorado River Basin and delta: An overview of issues, policies, and approaches to environmental restoration. In *The lower Colorado River Basin and Delta,* ed. E. P. Glenn, D. Radtke, B. Shaw, and A. Huete, special issue, *Journal of Arid Environments* 49:195–209.

Varady, R., and B. Morehouse. 2004. Cuánto cuesta? Development and water in Ambos Nogales and the upper San Pedro basin. In *The social costs of industrial growth in Northern Mexico,* ed. K. Kopinak, 205–248. San Diego: University of California-San Diego, Center for U.S.-Mexican Studies.

Varady, R., and E. Ward. 2009. Transboundary conservation in the borderlands: What drives environmental change? In *Conservation of Shared Environments: Learning from the United States and Mexico,* ed. L. Lopez-Hoffman, E. McGovern, R. G. Varady, and K. W. Flessa, 9–22. Tucson: University of Arizona Press.

Varady, R., R. Salmón Castillo, and S. Eden. Forthcoming. Key issues, institutions, and strategies for managing transboundary water resources in the Arizona-Mexico border region. In *Shared borders, shared waters: Israeli-Palestinian and Colorado River Basin water challenges*. ed. S. B. Megdal, R. G. Varady, and S. Eden. Boca Raton, FL: CRC Press.

Vásquez-León, M., C. T. West, and T. J. Finan, 2003. A comparative assessment of climate vulnerability: Agriculture on both sides of the U.S.-Mexico border. *Global Environmental Change* 13:159–173.

Verchick, R. M. 2008. Katrina, feminism, and environmental justice. *Cardozo Journal of Law and Gender* 14:791–800.

Westerling, A. L., B. P. Bryant, H. K. Preisler, T. P. Holmes, H. G. Hidalgo, T. Das, and S. R. Shrestha, 2011. Climate change and growth scenarios for California wildfire. *Climatic Change* 109 (Suppl. 1): S445–S463.

Wilbanks, T. J., P. Romero Lankao, M. Bao, F. Berkhout, S. Cairncross, J.-P. Ceron, M. Kapshe, R. Muir-Wood and R. Zapata-Marti. 2007. Industry, settlement and society. In *Climate change 2007: Impacts, adaptation and vulnerability. Contribution of Working Group II to the Fourth Assessment Report of the Intergovernmental Panel on Climate Change,* ed. M.L. Parry, O.F. Canziani, J.P. Palutikof, P.J. van der Linden, and C.E. Hanson, 357–390. Cambridge: Cambridge University Press.

Wilder, M. 2010. Water governance in Mexico: Political and economic apertures and a shifting state-citizen relationship. *Ecology and Society* 15:22.

Wilder, M., C. A. Scott, N. Pineda, R. Varady, and G. M. Garfin, eds. 2012. *Moving forward from vulnerability to adaptation: Climate change, drought, and water demand in the urbanizing southwestern United States and northwest Mexico.* Tucson: University of Arizona, Udall Center for Studies in Public Policy.

Wilder, M., C. A. Scott, N. Pineda Pablos, R. G. Varady, G. M. Garfin, and J. McEvoy. 2010. Adapting across boundaries: Climate change, social learning, and resilience in the U.S.-Mexico border region. *Annals of the Association of American Geographers* 100:917–928.

Wilder, M., and S. Whiteford. 2006. Flowing uphill toward money: Groundwater management and ejidal producers in Mexico's free trade environment. In *Changing structure of Mexico: Political, social and economic prospects,* ed. Laura Randall, 341-358. New York: M. E. Sharpe.

Williams, A. P., C. D. Allen, C. I. Millar, T. W. Swetnam, J. Michaelsen, C. J. Still, and S. W. Leavitt. 2010. Forest responses to increasing aridity and warmth in the southwestern United States. *Proceedings of the National Academy of Sciences* 107:21289–21294.

Woodhouse, C., D. M. Meko, G. M. MacDonald, D. W. Stahle, and E. R. Cook. 2010. 1,200-year perspective of 21st century drought in southwestern North America. *Proceedings of the National Academies of Science* 107:21283–21288.

Yohe, G., and R. S. J. Tol. 2002. Indicators for social and economic coping capacity for moving toward a working definition of adaptive capacity. *Global Environmental Change* 12:25–40.

Zamora-Arroyo, F., and K. Flessa, 2009. Nature's fair share: Finding and allocating water for the Colorado River Delta. In *Conservation of shared environments: Learning from the United States and Mexico*, ed. L. López-Hoffman, E. McGovern, R. G. Varady, and K. W. Flessa, 23–28. Tucson: University of Arizona Press.

Zedler, J. B. 2001. *Handbook for restoring tidal wetlands*. Boca Raton, FL: CRC Press.

Endnotes

i The Paso del Norte area includes the Ciudad Juárez municipality in Chihuahua, El Paso County in Texas, and Doña Ana County in New Mexico.

ii Among the paired cities in the western portion of the border region are: San Diego, California-Tijuana, Baja California Norte; Calexico, California-Mexicali, Baja California Norte; Yuma, Arizona-San Luis Río Colorado, Sonora; Nogales, Arizona-Nogales, Sonora; Naco, Arizona-Naco, Sonora; Douglas, Arizona-Agua Prieta, Sonora; Columbus, Texas-Las Palomas, Chihuahua; El Paso, Texas-Ciudad Juárez, Chihuahua.

iii For further discussion of interannual and multidecadal precipitation variability on the U.S. side of the border, see Chapter 4; for Mexico, see Diaz-Castro et al. 2002; Higgins and Shi 2001; Higgins, Chen, and Douglas 1999; Méndez and Magaña 2010; and Seager et al. 2009.

iv For further discussion of ENSO (El Niño-Southern Oscillation), droughts, floods and hydrological planning on the U.S. side of the border, see Chapters 4 and 5 and Garfin, Crimmins, and Jacobs 2007; for Mexico, see Magaña and Conde 2000; Brito-Castillo et al. 2002; Brito-Castillo et al. 2003; Pavia, Graef, and Reyes 2006; Gochis, Brito-Castillo, and Shuttleworth 2007; Ray et al. 2007; Seager et al. 2009; Stahle et al. 2009; and Méndez and Magaña 2010.

v These studies include Hurd and Coonrod 2007; Dominguez, Cañon, and Valdes 2010; Montero Martinez et al. 2010; Gutzler and Robbins 2011; Kunkel 2011; Reclamation 2011; Magaña, Zermeño, and Neri 2012; Scott et al. 2012; and Chapter 6 of this report.

vi For a discussion of downscaling methods, see Chapter 6, Section 6 of this document.

vii One notable aspect of mean temperature projections for the border region and for western North America more generally is that temperatures are projected to increase over the course of the century, regardless of the emissions scenario.

viii Consistent with these estimates are statistically downscaled projections of increased maximum and minimum temperatures in summer and winter, with the highest minimum temperature increases in the western Sonoran Desert and the highest maximum temperature increases in the northern Chihuahuan Desert. One set of statistically downscaled estimates of temperature changes for the north of the border region (high- and low-emission scenario models) are summarized in Table 16.2 and another set of statistically downscaled estimates for Mexican border states in the region (SRES A2) in are summarized in Table 16.3.

ix Garcia-Cueto, Tejeda-Martinez, and Jáuregui-Ostos (2010) note that, in the historic record for the border city of Mexicali, Baja California Norte, the duration and intensity of heat waves have increased for all summer months, there are 2.3 times more heat waves now than in the decade of the 1970s, and that the high-emissions SRES A2 projections show that for the 2020s, 2050s, and 2080s, heat waves could increase (relative to 1961–1990), by 2.1, 3.6, and 5.1 times, respectively.

x A special consideration for the western part of the border region in the wintertime could be that the circulation may change so that there will be fewer cyclones and more anticyclones (Favre and Gershunov 2009) resulting in (a) less frequent precipitation (this is well corroborated by several studies and in many models) and also in (b) more frequent cold spells. The second result is less certain, studied only in one model—CNRM-CM3 by Favre and Gershunov (2009), in which Mexican data were explicitly considered. Some recent results from Pierce et al. (2012), based on several models, suggest that the magnitude of cold outbreaks in January (see Pierce et al. 2012, Figure 6) will not likely diminish in California. This signal should probably extend south of the border some into Baja California (see Chapter 7).

xi Although across the border region winter precipitation is perhaps less substantial than summer precipitation, it is during this season that major dams store water that is used in the onset of the agricultural activities in the spring and summer months. Baja California, with its winter-dominated Mediterranean annual cycle of precipitation, is projected to have the highest percent of precipitation decreases among the Mexican states in the U.S.-Mexico border region (Montero Martinez et al. 2010).

xii The projections of Seager and colleagues (2007) are based on GCM analyses of precipitation minus evaporation, from an ensemble of GCMs used in the IPCC Fourth Assessment Report. They note that projected changes in atmospheric circulation, which promote atmospheric stability and poleward expansion of the Hadley Cell, are factors that contribute to projected temperature-driven increases in evaporation and greater aridity.

xiii Magaña, Zermeño, and Neri (2012), using statistically downscaled data from an ensemble of GCMs that use the high-emissions scenario, show large decreases in 24-month Standardized Precipitation Index (a measure of drought) and soil moisture during the second half of the twenty-first century in northwestern Mexico. Similarly, Gutzler and Robbins (2011), using statistically downscaled data from an ensemble of GCMs (SRES A1b) show large increases in the Palmer Drought Severity Index in the northern part of the border region; they note that "the projected trend toward warmer temperatures inhibits recovery from droughts caused by decade-scale precipitation deficits."

xiv Seager et al. (2009) note that this strong natural variability may obscure the development of increasing aridity that is occurring as the result of increasing temperatures and evaporation.

xv See the Executive Summary above for confidence statements pertaining to this summary.

xvi When effective, collaborative networks may become "communities of practice" that pursue new "adaptive pathways"—intentionally adaptive operations or strategies responsive to climatic change—in their respective institutions (Wilder et al. 2010). For a general discussion of the integral role of collaboration (e.g., trust, social learning, iterative interactions, common definitions of challenges)—not related to the border region, see Cash et al. 2003 and Pelling et al. 2008. Relating these aspects of collaboration to scientist-decision maker networks with the goal of co-production of science and policy, see Lemos and Morehouse 2005.

xvii For a concise history of the border region, see Ganster and Lorey 2008.

xviii The Wall Street Journal Online reported that 37% of net new jobs created in the U.S. since the economic recovery began were created in Texas (http://online.wsj.com/article/SB10001424052702304259304576375480710070472.html). Texas leads the nation in minimum-wage jobs (at 9.5 % of total workforce) (CNNMoney, http://money.cnn.com/2011/08/12/news/economy/perry_texas_jobs/index.htm).

ixx Source: U.S. Census Bureau FactFinder, http://factfinder.census.gov/servlet/DTGeoSearchByListServlet?ds_name=PEP_2008_EST&_lang=en&_ts=286892460001.

xx For example, Tijuana, with a population of about 1.2 million, is heavily dependent on maquiladoras with over 600 plants (2002 data, GAO 2003) and is closely tied to the U.S. market.

xxi Robertson (2009) notes that November 1, 2006, the Mexican government formally integrated the firms in the maquiladora industry into the PITEX program (Programas de Importación Temporal para Producir Artículos de Exportación), thus ending the practice of separating maquiladora trade from other manufacturing trade statistics. Beyond this date, statistics specific to maquiladora export are unavailable.

xxii Much of U.S.-Mexico trade occurs between border states. For example, 62% of U.S. exports to Mexico originated in Texas, California, Arizona, and New Mexico; of this, 70% was destined for Mexican border states (GAO 2003). The total actual value of merchandise trade (exports and imports to and from the U.S. and Mexico) in 2008 was $367 billion—a 266% increase since 1994 (EPA 2011). Official data show that the four U.S. border states originated 58.8% of U.S. exports to Mexico (88.8 billion dollars), which is more than twice their 24% share of U.S. GDP (Bureau of Economic Analysis, Trade Stats Express). Retail sales contribute to GDP and economic interdependence at the border. Residents from Tijuana make 1.5 million trips per month into the San Diego area, mainly to shop. In El Paso, Juárez residents account for more than 20% of retail sales (GAO 2003). Cross-border tourism creates positive economic impacts in Arizona-Sonora (Pavlakovich-Kochi and Charney 2008) including jobs, retail sales, and tourism. Tijuana, El Paso, and Nogales, Arizona are all significant ports-of-entry for Mexican agricultural produce.

xxiii Three treaties are of particular importance: the 1906 Water Convention on the Rio Grande River, the 1944 Water Treaty allocating water on the Colorado and Rio Grande Rivers, and the 1970 Boundary Treaty. The International Boundary and Water Commission (IBWC), established in its modern form by the 1944 Water Treaty, oversees implementation of these treaties and is charged with settling all disputes related to these agreements.

xxiv Mexico has begun to decentralize and delegate some authority for water resources to regional watershed councils and the Mexican states. See Ley de Aguas Nacionales y su Reglamento. 1992 rev. Mexico, D.F.: Comisión Nacional de Aguas. Available at: http://www.conagua.gob.mx/CONAGUA07/Publicaciones/Publicaciones/Ley_de_Aguas_Nacionales_baja.pdf; OECD. 2003. Environmental Performance Reviews: Mexico. Paris: Organization for Economic Cooperation and Development, p. 20.

xxv On the Mexico side of the region, inventories have documented the presence of 4,052 plant species; 454 species of invertebrates; 44 species of amphibians (mostly crustaceans); 184 species of reptiles; 1,467 species of birds; and 175 species of mammals (EPA 2011, based on Kolef et al. 2007).

xxvi The Sierra Club's "Wild Versus Wall" video (http://arizona.sierraclub.org/conservation/border/borderfilm.asp) illustrates the negative impacts on wildlife of the border fence.

xxvii The Colorado River has its headwaters in the Rocky Mountains and passes through nine states in two countries, and through the tribal homelands of the Cocopah tribe in the U.S. and the Cúcapa in Sonora. Waters of the Colorado River were allocated in the 1944 Treaty, based on a high-flow year (1922). Under the treaty, the water is shared among seven U.S. basin states (California, Arizona, Nevada, Colorado, Wyoming, and New Mexico) and Mexico is guaranteed 1.5 million acre-feet annually. From its distribution point at the Imperial Dam in Yuma, Arizona, the Colorado River winds to the west and empties into the delta before a trickle (in some years) reaches the Gulf of California. The total watershed of the Colorado is 244,000 square miles. The Colorado River system supports nearly 30 M people along its 1,400 mile (2,250 kilometer) length, 120 miles of which are in Mexico. It irrigates 3.7 million acres of farmland, including 500,000 in Mexico. Major cities in the border region drawing on the Colorado for urban uses include San Diego, San Luis Río Colorado, and Mexicali. Major agricultural areas reliant on surface water from the Colorado include Imperial and Coachella Valleys, and San Luis Río Colorado and Mexicali irrigation districts. All told, more than twenty U.S. Native American tribes have rights to Colorado River water.

xxviii No data are yet available on the impacts of AAC concrete-lining; however, experts have visually observed decreased flows (personal communication, 1/2012, A. Cortez-Lara).

ixxx MacDonald (2010) notes in the U.S. West today these losses are already on the order of $2.5 billon/year.

xxx In addition, the upper Rio Grande receives a trans-basin diversion from Reclamation's San Juan-Chama project (on the Upper Colorado River) of about 94,000 acre-feet annually.

xxxi The 2011-2012 agricultural programs for the Mexicali Valley and San Luis Río Colorado, after the reduced area due to the 2010 earthquake, are authorized to grow 72, 697 hectares of wheat, 32,064 hectares of cotton, and 27,251 hectares of alfalfa (SAGARPA, Delegación Estatal en Baja California, 2011).

xxxii In late 2008 the Secretary of the Environment of the State Government of Baja California formed the PEAC-BC, an interdisciplinary research team that includes research institutes and universities of the region such as the UABC, CICESE, and COLEF. Their aims were to assess current and potential impacts of climate change in Baja California as well as to propose mitigation actions. For more information see http://peac-bc.cicese.mx.

xxxiii The quantity of summer rain can be a major determinant of the number of head produced, but rain that is too heavy can waterlog pastures and wash out roads used to transport cattle to market

(Coles and Scott 2009). Other weather and climate-related sources of vulnerability identified include heavy rains, winds, hail, lightning, and frosts (Coles and Scott 2009).

xxxiv The annual cost of wildland fire suppression in California alone now typically exceeds $200 million (MacDonald 2010). Three simultaneous wildfires in San Diego County in October 2003 and another in October 2007 resulted in 25 deaths, destroyed a total of 3,700 homes, and scorched over 1,850 square miles (3,000 square kilometers) (Grissino-Mayer 2010).

xxxv Approximately 1.1 million acres burned in New Mexico in 2011, more than 4.5 times the state's average of around 242,000 acres. In Arizona, slightly more than 1 million acres burned, more than 5.5 times the state average of about 182,000 acres). Dry conditions desiccated soils and live fuel sources (e.g., grasses, shrubs, and trees) by the spring and a hard February freeze killed many plants and contributed to the fuel build-up (Southwest Climate Outlook, Oct. 25, 2011).

xxxvi The Colorado River Water Delta Trust has identified a minimum base flow need of 63 mcm (51,000 acre-feet). The Trust has acquired 1.7 mcm (1,367 acre-feet), based on a successful collaboration between NGOs and the state of Baja California in securing treated effluent from Mexicali for environmental flows to the Rio Hardy (Zamora-Arroyo and Flessa 2009).

xxxvii http://www.geimexico.org/english.html provides an overview of Mexican efforts.

xxxviii See http://www.sandag.org/index.asp?projectid=235&fuseaction=projects.detail.

ixl U.S. EPA, "Draft Border 2020 Document – for public comment – September 5, 2011," lines 126-153.

xl See http://www.tjriverteam.org.

Chapter 17

Unique Challenges Facing Southwestern Tribes

COORDINATING LEAD AUTHOR

Margaret Hiza Redsteer (U.S. Geological Survey)

LEAD AUTHORS

Kirk Bemis (Zuni Tribe Water Resources Program), Karletta Chief (University of Arizona), Mahesh Gautam (Desert Research Institute), Beth Rose Middleton (University of California, Davis), Rebecca Tsosie (Arizona State University)

REVIEW EDITOR

Daniel B. Ferguson (University of Arizona)

Executive Summary

When considering climate change, risks to Native American lands, people, and cultures are noteworthy. Impacts on Native lands and communities are anticipated to be both early and severe due to their location in marginal environments. Because Native American societies are socially, culturally, and politically unique, conventional climate change adaptation planning and related policies could result in unintended consequences or conflicts with Native American governments, or could prove to be inadequate if tribal consultation is not considered. Therefore, it is important to understand the distinct historical, legal, and economic contexts of the vulnerability and adaptive capacity of Southwestern Native American communities. The key messages presented in this chapter are:

- Vulnerability of Southwestern tribes is higher than that for most groups because it is closely linked to endangered cultural practices, history, water rights, and

Chapter citation: Redsteer, M. H., K. Bemis, K. Chief, M. Gautam, B. R. Middleton, and R. Tsosie. 2013. "Unique Challenges Facing Southwestern Tribes." In *Assessment of Climate Change in the Southwest United States: A Report Prepared for the National Climate Assessment*, edited by G. Garfin, A. Jardine, R. Merideth, M. Black, and S. LeRoy, 385–404. A report by the Southwest Climate Alliance. Washington, DC: Island Press.

socio-economic and political marginalization, characteristics that most Indigenous people share. (high confidence)

- Very little data are available that quantify the changes that are occurring or that establish baseline conditions for many tribal communities. Additional data are crucial for understanding impacts on tribal lands for resource monitoring and scientific studies. (high confidence)
- The scant data available indicate that at least some tribes may already be experiencing climate change impacts. (medium confidence)
- Tribes are taking action to address climate change by instituting climate-change mitigation initiatives, including utility-scale, alternative-energy projects, and energy-conservation projects. Tribes are also evaluating their existing capacity to engage in effective adaptation planning, even though financial and social capital is limited.

17.1 Introduction

The Southwestern United States is home to 182 federally recognized tribes (Federal Register 2010, Figure 17.1). California has the largest number of tribes (109), and the largest Native American population in the country (Table 17.1). Arizona, New Mexico, Colorado, and Utah are also home to seven of the most populous tribes, with populations ranging from 10,000 to over 300,000 (U.S. Census 2010). Nine tribes in the Southwest are considered "large land-holding tribes," five of which are among the ten largest reservations in the United States, ranging in size from 600,000 to 15 million acres (Federal Register 2010). More than one-third of the land in Arizona is tribal land.

Southwestern tribes are situated within all of the region's ecosystems and climatic zones, and the challenges these Native nations face from climate change may be just as varied. For example, tribes with large land holdings, those near the coast or in areas of scarce water, and those with large populations could face challenges different from the challenges faced by smaller tribes or those in or near urban areas. However, special issues confronting most if not all tribes include cultural and religious impacts, impacts to sustainable livelihoods, population emigration, and threats to the feasibility of living conditions. Tribal resources, already stretched to the limit, will have to be improved for tribes to cope adequately with a changing climate. Tribes' unique histories and legal status often results in political marginalization that must be addressed in order for tribes to face these challenges on equal footing with other governments.

Native nations predate the formation of the U.S. government; they entered into treaties with Great Britain and other European countries within their own territories. The United States continued the treaty relationship until 1871, but the nature of the political relationship changed over time. In a famous trilogy of nineteenth-century U.S. Supreme Court cases, Chief Justice John Marshall designated tribal governments as domestic, dependent nations that govern themselves under the protection of federal law. The federal government holds reservation lands in trust for the benefit of Indian nations; U.S. state governments generally may not exercise jurisdiction over reservation lands except when authorized to do so by the federal government (Cohen 2005).

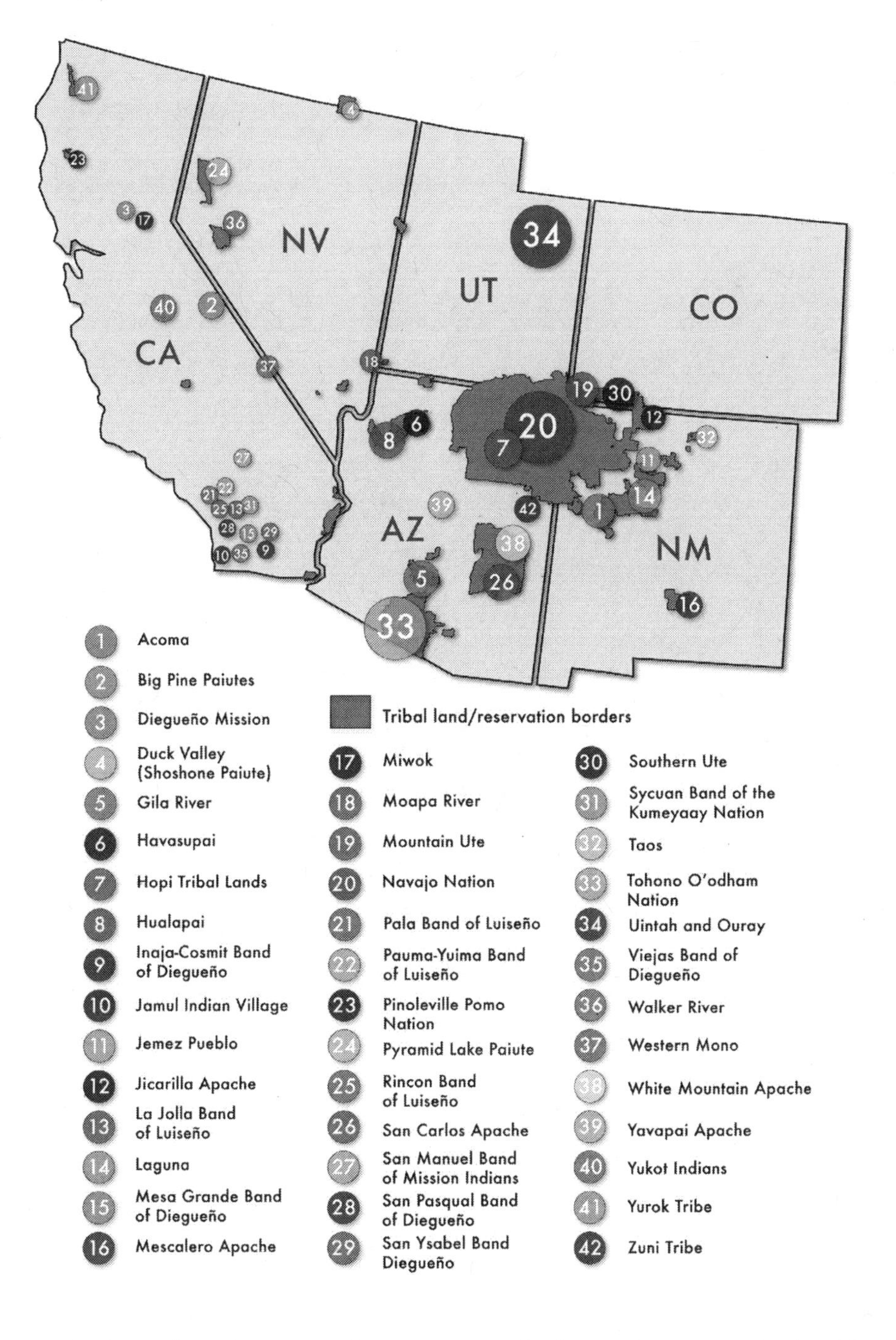

Figure 17.1 Map of southwestern United States showing tribal lands and the location of tribes discussed in text.

The federal government's duty to protect Indian nations, as articulated by Chief Justice John Marshall, is now understood as the federal trust responsibility. The Bureau of Indian Affairs is the agency that directly administers the trust responsibility. However, other agencies that control federal land and other natural resources must protect any applicable tribal rights, including rights to water, fish and wildlife, and cultural resources, such as traditional cultural properties. (Pevar 2012).

Table 17.1 Tribal lands and populations in the Southwestern United States

State	No. of Tribes	Total State Population	Tribal Population	% Tribal Population	Total State (acres)	Approx. Tribal land (acres)	% Tribal Land
AZ	21	6,392,017	294,033	4.6 %	72,982,074	26,273,547	36 %
NM	23	2,059,179	193,562	9.4 %	77,841,869	4,467,287	5.7 %
UT	8	2,763,885	33,166	1.2 %	54,352,753	5,150,817	9.5 %
CO	2	5,029,196	55,321	1.1 %	66,641,485	921,214	1.4 %
NV	19	2,700,551	12,600	1.2 %	70,782,330	1,253,812	1.8 %
CA	109	37,253,956	372,529	1.0 %	104,798,976	407,932	0.4 %
Total	**182**	**56,198,784**	**961,211**	**1.7 %**	**447,399,488**	**48,474,609**	**10.8 %**

Source: Federal Register (2010), U.S. Census (2010).

The federal government's "plenary power" over Indian affairs has resulted in a complex web of statutes that promote a policy of self-determination for tribal governments. Tribes have legal authority to make and enforce their own laws, and to regulate their lands, resources, and members. As U.S. senior district Judge Bruce S. Jenkins has noted (after Duthu 2008, p 4):

> *Modern tribal governments routinely exercise civil governmental authority over a range of day to day activities, much like comparable state and local government entities.... [T]ribal departments and agencies administer and deliver an expanding array of community services—from police, fire, and other emergency services, to education, health, housing, justice, employment assistance, environmental protection, cultural preservation, land use planning, natural resource conservation and management, road maintenance, water and public utilities. Indian tribes fit squarely within the ranks of American civic bodies, sharing the common duty and responsibility to provide essential services to the people of the communities they serve.*

As separate sovereign governments, tribes have the authority to address climate change as an important issue that affects their lands, resources, and traditional practices. Because climate change operates across jurisdictional boundaries, an awareness of tribal rights to water and cultural resources, located both on and off the reservation, are important to understand and evaluate when examining how climate change will affect tribes. This is particularly true for California, where tribes have smaller land holdings and must rely heavily on public lands for resources used in their cultural and religious practices (Anderson 2005).

17.2 The Effects of Marginal Living Conditions and Extreme Climatic Environments

In some cases, Native people and their cultural resources have already been affected by climate change. Reservations were often established in regions that typically have extreme environments and where sustainability of acceptable living conditions is already a challenge. In more arid parts of the Southwest, tribes sometimes have land that is drier—and has more limited access to water—than do their non-Indian neighbors. Large land-holding tribes, particularly in Arizona and Utah, are situated in regions with limited rainfall and water sources of poor quality that non-Native pioneers settling in the West found to be undesirable. For example, Navajo reservation boundaries were established within the driest third of the Navajo traditional homeland (Redsteer, Kelley et al. 2010), and fierce competition among Anglo and Hispanic populations for the best rangelands precluded retention of the more verdant traditional lands for Navajo use (Bailey and Bailey 1986). Helen H. Jackson (1883, 459) in describing changes of land occupation wrote:

> *From tract after tract of [ancestral] lands they have been driven out year by year by the white settlers of the country until they can retreat no further, some of their villages being literally on the last tillable spot on the deserts edge or in mountains far recesses... In southern California today are many fertile valleys which were thirty years ago the garden spots of these same Indians.*

Despite these historical land tenure changes and all the challenges facing Native people today, they continue to practice a lifestyle deeply connected to their natural surroundings. Cultural ties to the land include gathering herbal medicines and native plant foods, subsistence hunting and fishing, and traditional agricultural practices, such as farming and raising sheep. These practices continue to play a role in tribal life, and may also provide significant portions of many tribal economies.

17.3 Current Impacts on Native Lands

Native American cultures are closely linked to local resources in specific ecological niches that are likely to be altered in a changing climate (Kuhnlein and Receveur 1996; Smith et al. 2008; Green and Raygorodesky 2010). Many publications have generally described how tribes could be affected by climate change (see Hanna 2007; National Wildlife Federation 2011). However, few scientific studies address and quantify current climate-change impacts on Native lands and peoples of the United States, except in Alaska (e.g. Cruikshank 2001; Krupnik and Jolly 2002; Parkinson and Berner 2008; Davis 2010; Kofinas et al. 2010; Alexander et al. 2011). The high vulnerability of tribes to climate change and the information available (although limited) suggest that some tribes could be experiencing impacts, even though they lack specific documentation.

Many factors can lead to ecological and environmental change, and clear links of cause and effect need to be established in order to assess the effects that climate has had and might have for Native peoples in the region. In one documented example on the Navajo Nation, long-term trends of increasing temperatures, decreasing snowfall, declining streamflow, and water availability have magnified the impacts of drought that began in 1996 and continues today (Redsteer, Kelley et al. 2010). Streamflow data and historic information on surface-water features (such as springs, lakes and streams) show significant changes over the past century (Redsteer, Kelley et al. 2010). These changes have not coincided spatially or temporally with water development. Many surface-water features are now dry year-round or ephemeral, and began to disappear in the early to mid-1900s. Moreover, significant reductions in the number and length of stream reaches with perennial flow have occurred since 1920, and for some historic ephemeral streams, no flow during spring run-off and summer rains occurs today (Figure 17.2).

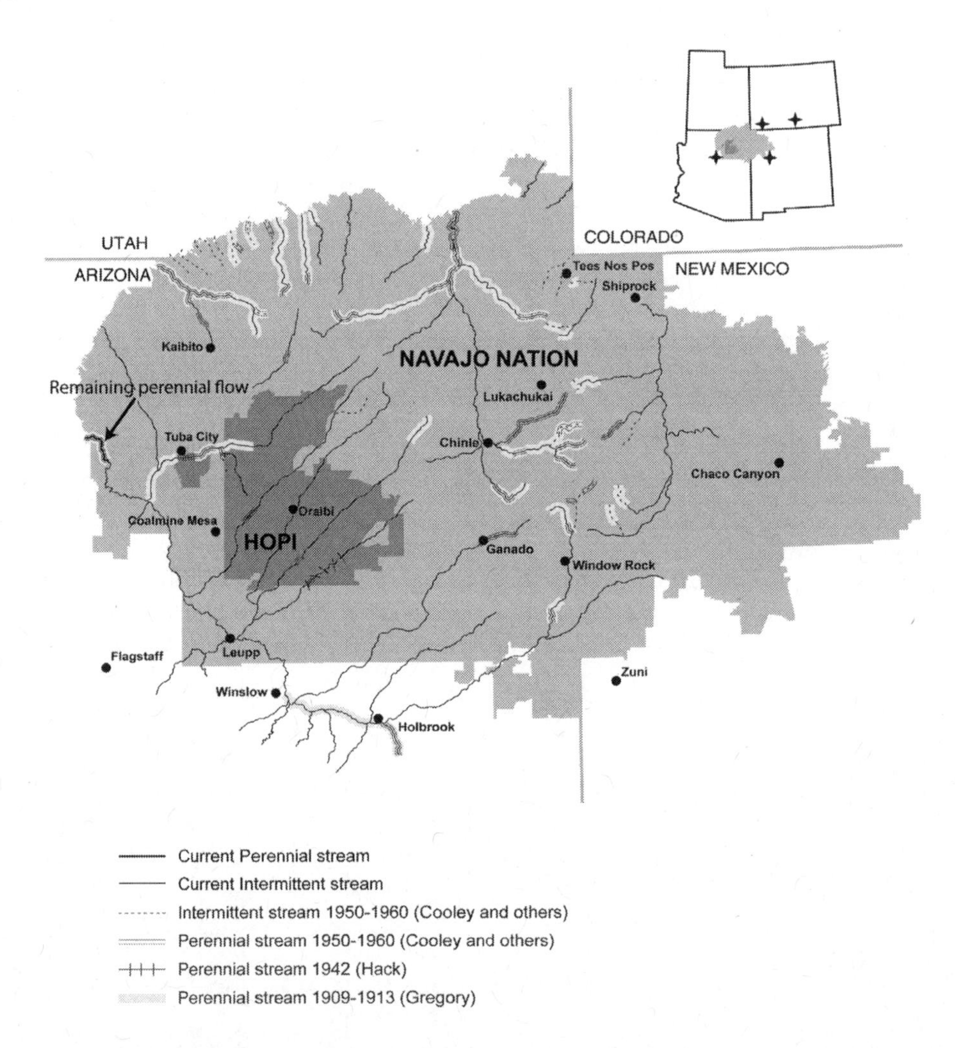

Figure 17.2 Map of the Navajo Nation (and lands of the Hopi Tribe) showing historic changes in perennial flow. The bold black line (shown with arrow) indicates where perennial stream flow exists today. Inset map shows location of Navajo lands, black diamonds specify locations of sacred mountains on the perimeter of Navajo traditional homelands. Data from reports by Herbert Gregory (1917), Hack (1942), Cooley et al. (1969) and USGS stream gauge data (after Redsteer, Kelley et al. 2010).

Interviews with 73 traditional Navajo elders about their observations of weather patterns and their impacts on traditional practices provided detailed accounts describing declines in snowfall, surface water features, and water availability (Redsteer, Kelley and Francis 2011). Other noticeable changes reported in these accounts include the disappearance of springs and the plants and animals found near water sources or in high elevations, such as certain medicinal plants, cottonwood trees, beavers, and eagles. The elders observed changes in the frequency of wind, sand, and dust storms. Navajo traditionalists also mentioned the lack of available water and changing socio-economic conditions as leading causes for the decline in the ability to grow corn and other crops (Redsteer, Kelley et al. 2010). Corn has been central to many Native cultural practices and traditions, including all Puebloan people in the Southwest. The use of corn pollen is also central to every Navajo ceremony.

Although the studies of climate-change impacts to Native people are limited, significant recent climate-related impacts to ecosystems on Native lands have occurred. In 2002, the Southwest experienced one of its most active fire seasons as a result of drought conditions and high winds (Feltz et al. 2002). The Rodeo-Chediski fire in Arizona burned 467,000 acres, setting a record for its immense size. Approximately 25% of the area burned was timber and grazing land belonging to the White Mountain Apache Tribe (Strom 2005; Kuenzi 2006). The fire resulted in areas that were severely burned, with 50% of the area showing no signs of ponderosa pine regeneration, and 16% with no surviving ponderosa pine. These areas are projected to undergo a shift to oak-manzanita shrubland (Strom 2005). The White Mountain Apache land, however, fared better than adjacent Forest Service lands because of the tribe's forest management policy of prescribed burns (Kuenzi 2006).

A continuation of dry, windy conditions in following years also led to record-breaking wildfire conditions that affected tribes in California (FEMA 2004). In October 2003, three simultaneous wildfires, the largest and most deadly in the history of California, destroyed 2,400 homes, killed sixteen people, and charred 376,000 acres in San Diego County. Again in October 2007, nine simultaneous fires of varying sizes burned throughout the county (including the Poomacha fire). These fires required the evacuation of 300,000 people and resulted in the loss of more than 1,800 homes and many other structures, 369,600 acres of land, and nine fire-related deaths. Local firefighting costs in 2007 topped $80 million (City of San Diego 2007).

In the 2007 Poomacha fire, the La Jolla Band of Luiseño Indians and the Rincon Band of Luiseño Indians, who had escaped major damage from the fires in 2003, suffered severe damage to homes and businesses (BIA 2007). Closing the reservations because of the fire caused food shortages, but the damage to tribal communities is more severe than these statistics would suggest. As one tribal member told a reporter, nearby municipalities "are newer places and people can leave and go elsewhere. ... This has been our home for generations. We have ties to the land. We won't go rebuild somewhere else" (Kelly 2007). The Poomacha fire burned 94% of the La Jolla reservation, destroying thick forests of live oak that once shaded homes and provided acorns for generations of Native Americans. "We were already at the bottom of the barrel and now this takes us down even further," said tribal Chairman Tracy Lee Nelson, whose house was destroyed in the fire (Kelly 2007).

Other tribal communities impacted by California wildfires include the Barona Band of Mission Indians (Cedar fire of 2003; Witch fire of 2007), the Inaja-Cosmit Band Indians (Witch fire of 2007), the Mesa Grande Band of Mission Indians (Witch fire of 2007), the Pala Band of Mission Indians (Poomacha fire of 2007), the Pauma Band of Luiseño Indians (Poomacha fire of 2007), the San Pasqual Band of Diegueno Mission Indians (Poomacha fire of 2007), the Iipay Nation of Santa Ysabel (Witch fire of 2007) and the Viejas Band of Kumeyaay Indians (Witch fire of 2007). The Jamul Indian Village and the Sycuan Band of the Kumeyaay Nation were also threatened by the Harris fire of 2007 (BIA 2007).

It is highly likely that increasing fire severity and other climate-related ecosystem impacts are affecting traditional Native foods and resources. Another climate-related impact, "sudden oak death," is a growing concern in California coastal areas and is spread by a pathogen that is sensitive to changes in humidity and temperature (Guo, Kelly, and Graham 2005; Liu et al. 2007). It may have been rare until changes in the environment (related to climate change) and increasing fire frequency led to its increasing prevalence (Rizzo and Garbelotto 2003; Pautasso et al. 2012). Tribes that have used oaks and acorns are numerous, and include Miwok, Western Mono, Yukots, Yurok, Paiute, and various Apache tribes, among many others (Anderson 2005). Acorns are a recognized staple food source of Native Americans in coastal California and the surrounding region, including Paiutes that traversed the Sierra Nevada in historic times to obtain them (Muir 1911). In addition to being a source of traditional foods, oaks are a valued source for traditional medicine and dyes for basketry (Ortiz 2008).

17.4 Potential Rangeland Impacts

Many tribes are dependent on livestock as a significant part of their economy, including the Hopi Tribe, Hualapai Tribe, Jicarilla Apache Nation, Navajo Nation, Pyramid Lake Paiute Tribe, San Carlos Apache Tribe, Southern Ute Indian Tribe, Tohono O'odham Nation, and White Mountain Apache Tribe. Tribal communities dependent on livestock tend to have limited alternative livelihoods, and additional climate-related stresses to the rangeland will further reduce economic resources. Livestock, especially cattle, are a significant source of economic and food security for large numbers of families on the Navajo Nation (Redsteer, Kelley et al. 2010). Stock-raising by large numbers of Navajo families is also important to preserve aspects of traditional culture.

Sand and dust storms

Climate-driven impacts to rangeland include increased mobility of sand dunes and potentially an increase in regional dust storms (Painter et al. 2010; Redsteer, Bogle and Vogel 2011). Sand dunes cover approximately one-third of the Navajo Nation as well as significant areas of Hopi tribal land (Redsteer 2002; Redsteer and Block 2003). Dune fields are susceptible to changes in precipitation, temperature, and wind speed and circulation patterns. In areas of Navajo and Hopi land that have wetter and cooler conditions, vegetation grows on sand dunes and stabilizes them. However, with drought conditions, these dune fields and sheet sands now exist under meteorological conditions where dunes may not have enough moisture to support the plant life necessary to make dunes resistant to wind erosion (Redsteer 2002). Increasing aridity in arid and semi-arid regions is often concurrent with the deterioration of surface vegetation and increasing dune mobility, jeopardizing rangeland productivity (Redsteer and Block 2004). An additional complication is that during floods, new sediment delivered in ephemeral rivers and washes (i.e., drainages that flow only temporarily after precipitation or snowmelt) provides a sand supply for new dune fields (Redsteer, Bogle et al. 2010). The risk of wide-scale movement of sand dunes is high, because the dry spells already make sand dunes more active. With projected warmer and drier conditions, deposits of sand dunes that have been stabilized by vegetation are highly likely to become mobile (Figure 17.3). Once sand dunes are mobile, it is difficult to reverse the process so that stabilization can occur, because vegetation must establish itself on a moving landform (Yizhaq, Ashkenazy, and Tsoar 2009). Very few plants are adapted to surviving abrasion by sand and sand burial (Downes et al. 1977). Currently, dunes are inundating housing, causing transportation problems, and contributing to a loss of rare and endangered native plants and grazing land (Redsteer, Bogle and Vogel 2011).

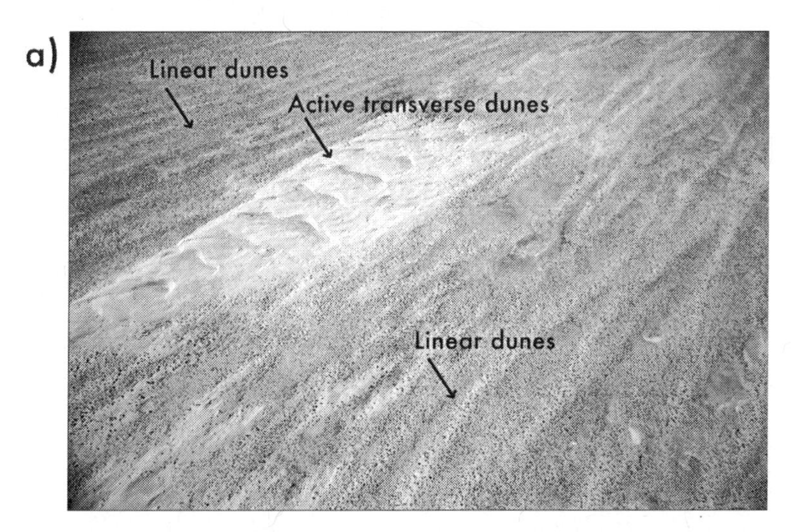

Figure 17.3 Photos of sand dune deposits on Navajo and Hopi land. a) Stabilized linear dunes, with local reactivation forming mobile transverse dunes; and b) Active sand dunes forming downwind of a dry streambed sediment source. Photo courtesy of Margaret Hiza Redsteer.

17.5 Adaptation Strategies and Adaptation Planning

In the past, Native peoples in the Southwest adapted to natural hazards through unique strategies guided by their cultural beliefs and practices (Tsosie 2007). Although many such Native traditions continue today, modern circumstances now make tribes especially vulnerable to climate extremes (National Wildlife Federation 2011). In some cases, modern land-use policies circumvent the ability of Native people to practice traditional adaptation strategies (James, Hall, and Redsteer 2008; Redsteer, Kelley et al. 2010). Tribal environmental and natural-resources management programs are working to address local impacts and tribes have lobbied for adaptation funding from the federal government. According to California Indian Water Commission President Atta Stevenson (a member of the Cahto tribe), "There are numerous climate change conditions we have witnessed and try to adapt to, but climate change is a global crisis without funding resources or commitment by government leadership to address Tribal suffrage and ecological demise of our traditional cultures. We cannot combat this ... alone."[i] In 2011, the Bureau of Indian Affairs (BIA) began to offer grants under a Tribal Climate Change Program, with a total allocation of $319,000.[ii] However, of the fifteen grants awarded, only three were for vulnerability assessments, and none were for climate adaptation planning.[iii] In 2009, the Department of the Interior (DOI) began a Climate Change Adaptation Initiative, setting aside funding for lands under federal jurisdiction, but it has not provided adequate funding for the BIA to assist tribes. Lack of adequate funding is not merely an impediment to adaptation planning, but it also further increases vulnerability to climate change impacts. Jerry Pardilla (2011), the National Tribal Environmental Council Executive Director, describes the situation in these terms:

> *Tribal lands comprise 95 million acres of the 587 million acres or 16 percent of federal land in the Initiative. The BIA has 11 million acres more than the National Park Service, yet the Administration has proposed nearly 50 times more funding for the NPS in FY 2012.*

Despite having few resources, adaptation planning workshops for tribes in the Southwest for both climate change and drought have occurred frequently in the past few years. There has also been a concerted effort by many tribes to forge ahead with adaptation plans (Wotkyns 2011). One example from the Yurok Tribe in the coastal Klamath Basin, California, demonstrates what tribes can implement if the resources are available. Adaptation planning by the Yurok Tribe Environmental Program entails monitoring water, air, and fisheries to understand local effects of climate change. Kathleen Sloan, the Tribe's Environmental Program Director, noted in 2009 that in many areas, the Yurok Tribe is the only entity collecting the data critical for creating climate models for the Lower Klamath. The tribe is training staff to monitor impacts, developing educational materials to encourage participation in adaptation planning, gathering oral histories from tribal elders, and creating a comprehensive prioritization plan to guide future tribal assessments. The Yurok Tribe's plan includes developing regional models to provide information for the Klamath Basin.

Floods and disaster planning

Severe weather events are occurring on tribal lands frequently, resulting in emergency declarations. Severe flooding in Havasu Canyon in 2008 struck a blow to the Havasupai tribal economy from lost tourism revenue. Since then, the Havasupai have experienced repeated flood events, the latest in October 2010, making recovery difficult. Funding from the San Manuel Band of Mission Indians ($1 million), federal and state agencies, and non-profit organizations has helped in recovery efforts. Currently, Havasu Canyon is closed until rehabilitation work and flood mitigation measures are completed (Wotkyns 2011).

The Federal Emergency Management Agency (FEMA) is a major source of flood and disaster assistance that requires ongoing commitments by communities for eligibility. The National Flood Insurance Program is a cornerstone of floodplain management, but only four Southwestern tribes participate in it (FEMA 2011). Many tribes lack zoning laws or floodplain delineation despite retaining jurisdiction over federal lands. Lack of funding, difficult jurisdictional challenges posed by the presence of non-Indian lands, and in some cases the need for approval by the Department of the Interior comprise some of the causes for the lack of regulation. Today, limited resources and abundant low-income housing challenge administration of floodplain ordinances and mandatory insurance for federally financed homes located in floodplains (Bemis 2003). FEMA-approved hazard mitigation plans address other disasters and are a prerequisite to receive certain federal funding. Forty-eight Southwestern tribes participate in these management plans, either through their own plans or by adopting plans developed by other local jurisdictions (FEMA 2009). The requirement for renewal every five years presents a continual challenge to resource-limited tribes, and some plans have expired.

Droughts and drought mitigation planning

Many tribal governments are unprepared to cope with climate extremes because of the poor economic conditions in tribal communities. Tribal water resources on arid reservations are typically marginal and highly susceptible to frequent water shortages. While every Southwestern state has a drought plan, only four Southwestern tribes have completed plans through the Bureau of Reclamation States Emergency Drought Program (National Drought Mitigation Center 2010; Reclamation 2010). Tribes have limited resources to develop and implement these plans. Despite being the first tribe to submit its plan to Congress, the Hualapai Tribe lacks the personnel and funding to perform monitoring and actions triggered pursuant to its plan (Knudson, Hayes, and Svoboda 2007). The Hopi Tribe and Navajo Nation have submitted plans to Congress but also have struggled with monitoring (Ferguson and Crimmins 2009). The large region encompassed by these two tribes (about 30,000 square miles in total) presents challenges for adequate monitoring. With a smaller land area (about 700 square miles), the Zuni Tribe has been able to issue monthly monitoring reports for its plan, but some federal stations it uses are at risk from insufficient federal cost-sharing and the inability to replace cooperators (Bemis 2010).

Recent reports from workshops offered through the National Integrated Drought Information Systems (NIDIS) documented the challenges facing tribes and identified opportunities for assistance (Collins et al. 2010; Ferguson et al. 2011). Chronic underfunding and short-term funding cycles for programs within tribal government leave tribes without the financial and human resources needed to make climate assessments or plan for natural hazards (Ferguson et al. 2011). Planning must also address tribal cultural needs and sovereignty. Modern monitoring, forecasting, and adaptation techniques can sometimes ignore or be inconsistent with traditional Native values, knowledge, and practices. The sovereign status of tribes can create complications or obstacles for collecting data and managing resources. The NIDIS Four Corners Tribal Lands Regional Drought Early Warning System is a pilot project attempting to address some of these issues and provide better information and resources for drought planning (Alvord 2011). By fully involving tribes throughout its course, this pilot project can provide a model for other tribal regions.

17.6 Challenges for Adaptation Planning

In planning to face the effects of climate change, tribal rights to water, cultural resources, and sacred sites located both on and off reservations are likely to be issues connected to adaptation planning. If sacred sites are not recognized, there is a substantial chance of increased conflict, which would constrain or even derail efforts to maintain resilient cultural and natural resources. The challenges of climate change and adaptation planning for federal land managers and for tribes may be difficult because of potential conflicts between the trust responsibility to Indian nations and the mandate of federal agencies to engage in a multiple-use policy. There have been effective partnerships initiated by tribes to address climate-related issues that affect resources and traditional practices, but there are also examples of ineffective communication leading to conflict.

Within the Department of the Interior (DOI), Landscape Conservation Cooperatives (LCCs) were established to develop science capacity to support resolution of resource management issues. The Native American Land Conservancy (NALC) is land conservancy with representatives from the Chemehuevi, Cahuilla, Wyandotte, Seneca, and other tribes. The NALC is providing a strategy for the Desert LCC, with members on the LCC steering committee and science working group, incorporating observations of eastern Mojave Desert traditional ecological knowledge. To this end, the NALC has drafted a white paper that includes historical information, cultural resource concerns, and ways to evaluate and lessen impacts of climate change based on indigenous understanding of the region's sacred sites, areas and landscapes.

17.7 Vulnerability from Economic, Political, and Legal Stresses

Limited resources and poor economic conditions reduce the resilience of tribes to climate change. More than one-quarter of the American Indian and Alaska Native population lives in poverty—a rate more than double the general U.S. population (Sarche and Spicer 2008). Moreover, approximately 13.3% of Native Americans lack accessibility to safe drinking water (Indian Health Service 2007). Income levels and human development indicators such as health and education are significantly lower than those of the rest of the population (U.S. Commission on Civil Rights 2004).

Vulnerability and adaptive capacity

Nevada's largest tribe, the Pyramid Lake Paiute Tribe (PLPT), is deeply connected—culturally, physically, and spiritually—to Pyramid Lake and its ecosystem. Pyramid Lake, at the terminal end of the Truckee-Carson River, is considered "the most beautiful of North America's desert lakes" (Wagner and Lebo 1996, 108). It is home to an endangered fish called cui-ui (*Chasmistes cujus*)—a primary cultural resource—and the threatened Lahontan cutthroat trout. The Paiute tribe's original name is *Kuyuidokado,* or cui-ui eaters. Traditionally, they traveled to the lake for annual cui-ui spawning to gather and dry fish (Wagner and Lebo 1996). The Paiute origin story is based upon the lake and a tufa-rock formation called the Stone Mother, resembling a woman whose tears created the lake (Wheeler 1987). Fishing and recreational activities are central to the PLPT economy. Wetlands also provide reeds for basketry, a symbol of Native identity. Although some cultural practices have been lost due to impacts from non-tribal settlement and exploitation, the PLPT continues to hold steadfast to their cultural connection to the lake. The tribe protects the lake via water rights negotiations for endangered species protection, by creating and enforcing policies on water quality, maintaining minimum in-stream flows for spawning, and by funding fisheries management activities.

The case of the Pyramid Lake Paiute Tribe exemplifies the vulnerabilities that tribes face from climate change (Tsosie 2007; Shonkoff et al. 2011). The tribe's vulnerability is related to cultural dependence on the lake, but external socio-economic factors also influence its adaptive capacity, amplifying potential impacts (Gautam and Chief forthcoming). Risk factors include upstream water use by municipal, industrial, and agricultural entities. Spawning and sustenance of endangered cui-ui fish are dependent on both water quantity and quality (Sigler, Vigg, and Bres 1985; USFWS 1992). Water supplies in nearby Carson Basin largely determine how much water reaches Pyramid Lake, particularly in the dry years, as irrigation requirements have senior water rights. Devastating impacts have already occurred from water diversion for agricultural use. Blocked access to upriver spawning grounds during a drought left dying fish for two miles downstream of the Derby Dam. Cattle encroachment upon wetlands occurs during droughts that reduce available forage. Limited economic opportunities and dwindling federal support constrain the tribe's adaptation capacity. In a survey, 73% of PLPT respondents said they believe climate change is occurring and that humans play a role, whereas 63% of rural Nevadans believe climate change is occurring, with only 29% attributing a human role (personal communication from Z. Liu, 2012). Factors such as a remarkable public awareness of climate change, sustainability-based values, the technical capacity for natural resource management, proactive initiatives for invasive-species control, and external scientific networks contribute to PLPT's adaptive capacity.

Water rights

Water rights are closely linked to the vulnerability and adaptive capacity of tribes. The legal basis for tribal water rights is the federal "reserved rights" doctrine, which holds that Indian nations have reserved rights to land and resources in treaties they signed with the United States. In the famous 1908 case of *Winters v. United States,* the Supreme Court held that when the U.S. government establishes a reservation, it also implicitly reserves water rights sufficient to meet the current and future needs of the tribe and the purpose for which the reservation was set aside (including fisheries, where applicable). Thus, the priority date for tribal water rights under the *Winters* doctrine is the date the reservation was established, making many tribal governments in the Southwest senior water resource users with significant adjudication rights (Cohen 2005).

In the 1963 case of *Arizona v. California*, the Supreme Court determined that the only feasible way to quantify tribal rights was by "practicably irrigable acreage" (PIA) on reservations. Difficulties with the PIA quantification of water rights include the differences in the amount of tillable land available from one reservation to another, as well as the water-rights standard being based on the amount of land a tribe has, rather than tribal population. Western water law doctrine may have worked well in early nineteenth and twentieth centuries, but fails to take into account valuable in-stream uses of water such as fish and wildlife habitat (Royster and Blumm 2008). The Arizona Supreme Court extended the PIA standard in 2001, finding that agriculture is not the only means to determine tribal water allocations and that water is also needed by tribes for other purposes.[4] The Arizona Supreme Court found that the purpose of a federal Indian reservation is to serve as a "permanent home and abiding place" to the Native American people living there, and that water allocations must satisfy both present and future needs of reservations as "livable homelands."

Litigation for a determination of water rights on paper is an expensive and lengthy process. Some tribal governments have negotiated settlement agreements, foregoing a significant percentage of their legal claims to water in exchange for a secure allocation and for funding for the infrastructure necessary to gain the actual value of water resources (Clinton et al. 2010). Congressional action is needed to approve settlements and allocate the funding necessary to build water-delivery infrastructure. Between 1986 and 2006 Congress enacted twenty settlements into law (Royster and Blumm 2008). In spite of the cost, some tribes have preferred litigation, because with settlements, tribes invariably give up some measure of their legal rights to water, leaving them in a weaker bargaining position. Many examples of current tribal vulnerabilities are linked to water allocations. In a warmer and drier Southwest, conflicts over water appear imminent.

In the arid Owens Valley of California, spring and summer snowmelt are crucial to water supplies. The Big Pine Paiute Tribe of Owens Valley channeled this runoff to irrigate important food plants, and have observed changes to runoff in the watershed. Repeat photography of upstream Palisade Glacier shows notable shrinkage in recent decades. The tribe shares water with the Los Angeles Department of Water and Power (DWP), which owns nearly all the water rights in the valley. DWP allocates the tribe 1,116.32 acre-feet per year based on a 1939 land exchange between the federal government and the city (Gorin and Pisor 2007). Water shortages are likely to increase DWP's export of water from the Valley, leaving the Big Pine Paiute with an uncertain water supply.

17.8 Climate Change Mitigation Strategies

Despite needing additional resources, tribes are forging ahead to address climate change. Many see climate change mitigation and energy conservation as great financial opportunities that may help address current economic woes and the challenges of a limited resource base. The Pueblo of Jemez has begun constructing a utility-scale solar project in New Mexico. Tribes with mitigation plans include the Gila River Indian Community, Hopi Tribe, Navajo Nation, and Yavapai-Apache Nation. Examples outlined below depict some of the current activities.

The Pinoleville Pomo Nation, in partnership with the University of California, Berkeley, launched a sustainable housing program. Drought conditions within and around the Pinoleville Pomo Nation were taxing residents and the local government resources. Heating and cooling inefficient standard houses funded by the U.S. Department of Housing and Urban Development (HUD) also placed an increased burden on residents (Shelby et al. 2010). A self-sufficient, sustainably focused community model for housing, energy, and water conservation now addresses these issues through the use of solar photovoltaic systems, wind turbine systems, passive and active solar water heaters, grey water systems, and passive building design strategies such as passive solar gain and sun shading.

The Rincon Band of Luiseño Indians, in San Diego County, owns Harrah's Rincon Casino and Resort. When the economy declined in 2008 and 2009, Rincon still invested $13.5 million in energy-efficient retrofits and a one-megawatt solar plant to power the casino (Wolfe 2010). The tribe commissioned the solar plant in conjunction with a casino-wide retrofit of rooftop air-conditioning. A modified chiller plant captures waste heat for hot water in the casino's 662 hotel rooms. According to EPA, this saves 3.3 million kWh/year. The 3,986-panel solar plant provides 90% of the required power for heating, ventilation, and air conditioning, generating enough energy to power 2,200 homes. Through offsets, the solar array also saves 3.5 million KW hours per year, providing enough electricity for 583 individuals (based on average individual use of 6,000 KW hours/year).

Among the tribe's many additional mitigation and environmental sustainability practices are the use of solar induction to heat Harrah's pool and the composting of green waste for the property's organic gardens (diverting 6,000 pounds of waste per month).

LOOKING FORWARD. With continuing climate change effects, Native American lands, communities, cultures, and traditions are at risk. Vulnerability is closely linked to external land use policies, political marginalization, water rights, and poor socio-economic conditions. Tribes will be important parties to any future proceedings that deal with water shortage allocations or coordinated reservoir operations because of their reserved water rights. These issues are likely to intensify in an era of climate change. However, there have been few climate change studies on tribal lands and little documentation of the impacts. Studies that are available show that impacts to tribal resources are already underway in at least in some areas of the Southwest. Additional transformation of ecosystems by fire, pests, and disease, exacerbated by altered climatic conditions, are certain to affect traditional foods and medicines.

Many reservations, particularly those with large land holdings, have insufficient capacity to adequately monitor climatic conditions (Ferguson et al. 2011). Without monitoring, tribal decision makers lack necessary data to quantify and evaluate the changes taking place and to plan and manage resources accordingly. In addition, lack of information from tribal lands that typically have more extreme environments leaves climate scientists without crucial information from areas that are likely to see early impacts from climate change. Most reservations lack the data necessary to contribute to more accurate downscaled climate models, because meteorological monitoring is sparse over areas of significant size. The latest U.S. Census (2010) shows that some reservations are losing the younger segment of their population to emigration; this trend is cause for concern among those in tribal governments who interpret the changing demographics as a sign of untenable living conditions due to dwindling water resources and increasingly desertified rangeland.[v]

Despite all of the challenges, Native communities also have much to offer the climate science community. Native communities have persisted and adapted during periods of wide-ranging natural climate variability. The role of indigenous environmental knowledge has received increasing attention, and studies of local environmental knowledge show that it contributes greatly to our understanding of ecosystem change (e.g. Newton, Paci and Ogden 2005; Green and Raygorodetsky 2010; Pearce et al. 2010; Sanchez-Cortes and Chavero 2010; Alexander et al. 2011; Harris and Harper 2011; Singh, Bhowmik and Pandey 2011). In spite of fewer economic resources, or perhaps because of them, many Southwestern tribal communities are exemplary in their efforts to mitigate climate change, and are actively seeking resources to assist with adaptation.

References

Alexander, C., N. Bynum, E. Johnson, U. King, T. Mustonen, P. Neofotis, N. Oettlé , et al. 2011. Linking indigenous and scientific knowledge of climate change. *BioScience* 477–484.

Alvord, C. 2011. Overview of the NIDIS Four Corners pilot activities. *National Integrated Drought Information System Newsletter* 2 (1): 9–10. http://www.drought.gov/imageserver/NIDIS/newsletter/NIDIS_Newsletter_Winter_2011.pdf

Anderson, K. M. 2005. *Tending the wild: Native American knowledge and the management of California's natural resources*. Berkeley: University of California Press.

Bailey, G., and R. G. Bailey, eds. 1986. *A history of the Navajos: The reservation years*. Santa Fe, NM: School of American Research Press.

Bemis, K. 2003. Zuni Pueblo tribe and the National Flood Insurance Program. Presentation at the Association of State Floodplain Managers Annual Conference, May 10–16, 2003, St. Louis, Missouri.

—. 2010. Zuni Drought Contingency Plan. Presentation at Drought, Water and Climate workshop, 14-15 December, 2010, Washington, D.C. http://www.westgov.org/component/joomdoc/doc_download/1322-bemis-presentation-2010.

Bureau of Indian Affairs (BIA). 2007. Artman to inspect fire damage on La Jolla and Rincon Reservations; will meet with tribal, federal and state officials on relief efforts. News release, October 29, 2007. Washington, DC: U.S. Department of the Interior.

City of San Diego. 2007. After action report: October 2007 wildfires; City of San Diego response. http://www.sandiego.gov/mayor/pdf/fireafteraction.pdf.

Clinton, R., C. Goldberg, R. Tsosie, K. Washburn, and E. R. Washburn. 2010. *American Indian law: Native nations and the federal system*, 6th ed. N.p.: LexisNexis.

Cohen, F., ed. 2005. *Felix Cohen's handbook of federal Indian law*, 2005 ed. N.p.: LexisNexis.

Collins, G., M. H. Redsteer, M. Hayes, M. Svoboda, M. D. Ferguson, R. Pulwarty, D. Kluck, and C. Alvord. 2010. *Climate Change, Drought and Early Warning on Western Native Lands Workshop report: 9-11 June, 2009, Jackson Lodge, Grand Teton National Park*, WY. N.p.: National Integrated Drought Information System.

Cooley, M. E., J. W. Harshbarger, J.P. Akers, and W. F. Hardt. 1969. *Regional hydrogeology of the Navajo and Hopi Indian reservations, Arizona, New Mexico, and Utah, with a section on vegetation*. USGS Professional Paper 521-A. Reston, VA: U.S. Geological Survey.

Cruikshank, J. 2001. Glaciers and climate change: Perspectives from oral tradition. *Arctic Journal* 54:377–393.

Davis, S. H. 2010. Indigenous peoples and climate change. *International Indigenous Policy Journal* 1:1–19. http://ir.lib.uwo.ca/iipj/vol1/iss1/2.

Downes, J., D. Fryrear, R. Wilson, and C. Sabota. 1977. Influence of wind erosion on growing plants. *Transactions of the American Society of Agricultural Engineers* 20:885–889.

Duthu, N. B. 2008. *American Indians and the law*. London: Penguin Books.

Federal Emergency Management Agency (FEMA). 2004. *The California Fires Coordination Group: A report to the Secretary of Homeland Security*. Washington, DC: FEMA.

—. 2009. Hazard mitigation plan status list for Indian tribal governments. http://www.fema.gov/library/viewRecord.do?id=3565.

—. 2011. The National Flood Insurance Program community status book. http://www.fema.gov/fema/csb.shtm.

Federal Register. 2010. Indian entities eligible to receive services from the United States Bureau of Indian Affairs. *Federal Register* 75 (Friday, October 1, 2010): 60810–60814.

Feltz, J. M., M. Moreau, E. Prins, K. McClaid-Cook, and I. F. Brown. 2002. Recent validation studies of the GOES wildfire automated biomass burning algorithm (WF_ABBA) in North and South America.

Ferguson, D., C. Alvord, M. Crimmins, M. H. Redsteer, M. Hayes, C. McNutt, R. Pulwarty, and M. Svoboda. 2011. *Drought preparedness for tribes in the Four Corners region: Workshop report, April 8-9, 2010, Flagstaff, Arizona*. Tucson: Univ. of Arizona, Climate Assessment for the Southwest (CLIMAS).

Ferguson, D., and M. Crimmins. 2009. Who's paying attention to the drought on the Colorado Plateau? *Southwest Climate Outlook*, July 2009: 3–6.

Gautam, M., K. Chief, and W. J. Smith, Jr. Forthcoming. Climate Change in Arid Lands and Native American Socioeconomic Vulnerability: The Case of the Pyramid Lake Paiute Tribe. In "Facing climate change: The experiences of and impacts on U.S. tribal communities, indigenous people, and native lands and resources." *Climatic Change* 115.

Green, D., and G. Raygorodetsky. 2010. Indigenous knowledge of a changing climate. *Climatic Change* 100:239–242.

Gregory, H. E. 1917. *Geology of the Navajo country: A reconnaissance of parts of Arizona, New Mexico, and Utah*. U.S. Geological Survey Professional Paper 93. Reston, VA: USGS.

Gorin, T., and K. Pisor, K., 2007. *California's residential electricity consumption. prices, and bills, 1980-2005*. Staff Paper CEC-200-2007-18. Sacramento: California Energy Commission.

Guo Q., M. Kelly, and C. H. Graham. 2005. Support vector machines for predicting distribution of sudden oak death in California. *Ecological Modeling* 182:75–90.

Hack, J. T. 1942. *The changing physical environment of the Hopi Indians of Arizona*. Papers of the Peabody Museum of Archaeology and Ethnology Volume 35, Issue 1. Cambridge, MA: Harvard University.

Harris, S., and B. Harper. 2011. A method for tribal environmental justice analysis. *Environmental Justice* 4:231–237.

Indian Health Service (IHS). 2007. *Public Law 86-121 Annual Report for 2007*. Rockville, MD: IHS, Sanitation Facilities Construction Program. http://www.ihs.gov/dsfc/documents/SFCAnnualReport2007.pdf.

Jackson, H. H. 1883. *A century of dishonor*. Reproduction. Scituate, MA: Digital Scanning, Inc., 2001.

James, K., D. Hall, and M. H. Redsteer. 2008. Organizational environmental justice with a Navajo (Diné) Nation case example. In *Research in social issues in management*, ed. S. Gilliland, D. Steiner, and D. Skarlicki, 263–290. Greenwich, CT: Information Age Publishing.

Kelly, D. 2007. A struggling tribe faces new hardships. *Los Angeles Times* online, November 22, 2007.

Knudson, C. L., M. J. Hayes, and M. D. Svoboda. 2007. Case study of tribal drought planning: The Hualapai Tribe. *Natural Hazards Review* 8:125–131.

Kofinas, G. P., F. S. Chapin III, S. BurnSilver, J. I. Schmidt, N. L. Fresco, K. Kielland, S. Martin, A. Springsteen, and T. S. Rupp. 2010. Resilience of Athabascan subsistence systems to interior Alaska's changing climate. *Canadian Journal of Forestry Research* 40:1347–1359.

Krupnik, I. and D. Jolly, eds. 2002. *The Earth is faster now: Indigenous observations of Arctic environmental change*. Fairbanks, AK: Arctic Research Consortium of the U.S.

Kuenzi, A.M. 2006. Treatment effects and understory plant community response on the Rodeo-Chediski fire, Arizona. M.S. thesis, Northern Arizona University.

Kuhnlein, H. V., and O. Receveur. 1996. Dietary change and traditional food systems of indigenous peoples. *Journal of Nutrition Annual Review* 16:417–442.

Liu, D., M. Kelly, P. Gong, and Q. Guo. 2007. Characterizing spatial-temporal tree mortality patterns associated with a new forest disease. *Forest Ecology and Management* 253:220–231.

Muir, J. D. 1911. *My first summer in the Sierra*. Reproduction. Lawrence, KS: Digireads.com, 2008.

National Drought Mitigation Center. 2010. The status of state drought plans, December 2010. http://www.drought.unl.edu/Planning/PlanningInfobyState.aspx.

Hanna, J. 2007. *Native communities and climate change: Legal and policy approaches to protect tribal legal rights*. Report pending final review. Boulder: Univ. of Colorado School of Law, Natural Resources Law Center.

National Wildlife Federation (NWF). 2011. *Facing the storm: Indian tribes, climate-induced weather extremes, and the future for Indian country*. Boulder, CO: NWF, Rocky Mountain Research Center.

Newton, J., J. C. D. Paci, and A. Ogden 2005. Climate change and natural hazards in northern Canada: Integrating indigenous perspectives with government policy. *Mitigation and Adaptation Strategies for Global Change* 10:541–571.

Ortiz, B. 2008. Contemporary California Indian uses for food of species affected by *Phytophthora ramorum*. In *Proceedings of the Sudden Oak Death Third Science Symposium, March 5–9, 2007, Santa Rosa, California*, tech coord. S. J. Frankel, J. T. Kliejunas, and K. M. Palmieri. Gen. Tech. Rep. PSW-GTR-214. Albany, CA: U.S. Forest Service, Pacific Southwest Research Station.

Painter, T. H., J. S. Deems, J. Belnap, A. F. Hamlet, C. C. Landry, and B. Udall. 2010. Response of Colorado River to dust radiative forcing in snow. *Proceedings of the National Academy of Sciences*, published online, doi: 10.1073/pnas.0913139107.

Pardilla, J. 2011. *Tribal set-aside sought in DOI Climate Change Adaptation Initiative*. Report for the National Tribal Environmental Council, Albuquerque, NM.

Parkinson, A. J., and J. Berner. 2009. Climate change and impacts on human health in the Arctic: an international workshop on emerging threats and the responses of arctic communities to climate change. International Journal of Circumpolar Health 68:88–95.

Pautasso M., T. F. Döring, M. Garbelotto, L. Pellis and M. J. Jeger. 2012. Impacts of climate change on plant diseases—Opinions and trends. *European Journal of Plant Pathology*. doi: 10.1007/s10658-012-9936-1.

Pearce, T., B. Smit, F. Duerden, J. D. Ford, A. Goose, and F. Kataoyak, F. 2010. Inuit vulnerability and adaptive capacity to climate change in Ulukhatok, Northwest Territories, Canada. *Polar Record* 46:157–177.

Pevar, S. L. 2012. The rights of Indians and tribes, 4th ed. Oxford: Oxford Univ. Press.

Redsteer, M. H. 2002. Factors effecting dune mobility on the Navajo Nation, Arizona, USA. In *Proceedings of the 5th International Conference on Aeolian Research and the Global Change and Terrestrial Ecosystem-Soil Erosion Network, July 22–25, 2002, Lubbock, Texas, USA*, 385. Publication 02-2. Lubbock: Texas Tech University, International Center for Arid and Semiarid Lands Studies.

Redsteer, M. H., and D. Block. 2003. Mapping susceptibility of sand dunes to destabilization on the Navajo Nation, southern Colorado Plateau. *Geological Society of America Abstracts with Programs* 35:170, Paper No. 68-9.

—. 2004. Drought conditions accelerate destabilization of sand dunes on the Navajo Nation, southern Colorado Plateau. *Geological Society of America Abstracts with Programs* 36 (5): 171, Paper No. 66-8.

Redsteer, M. H., K. B. Kelley, H. Francis, and D. Block. 2010. Disaster risk assessment case study: Recent drought on the Navajo Nation, southwestern United States. In *Global Assessment Report on Disaster Risk Reduction 2011, Annexes and Papers, Chapter 3.* http://www.preventionweb.net/english/hyogo/gar/2011/en/what/drought.html.

Redsteer, M. H., R. Bogle, J. Vogel, D. Block, M. Velasco, and B. Middleton. 2010. The history and growth of a recent dune field at Grand Falls, Navajo Nation, NE Arizona. *Geological Society of America Abstracts with Programs* 42 (5): 416, Paper No. 170-5.

Redsteer, M. H., R. C. Bogle, and J. M. Vogel. 2011. *Monitoring and analysis of sand dune movement and growth on the Navajo Nation, southwestern United States.* USGS Survey Fact Sheet 2011-3085. http://pubs.usgs.gov/fs/2011/3085/.

Redsteer, M. H., K. B. Kelley, and H. Francis. 2011. Increasing vulnerability to drought and climate change on the Navajo Nation. Paper GC43B-0928, delivered at American Geophysical Union Annual Meeting, 5-9 December 2011, San Francisco.

Rizzo, D. M., and M. Garbelotto. 2003. Sudden oak death: Endangering California and Oregon forest ecosystems. *Frontiers in Ecology and the Environment* 1:197–204.

Royster, J. V., and M. C. Blumme, eds. 2008. *Native American natural resources law: Cases and materials,* 2nd ed. Durham, NC: Carolina Academic Press.

Sarche, M., and Spicer P., 2008. Poverty and health disparities for American Indian and Alaska Native children: Current knowledge and future prospects. *Annals of the New York Academy of Sciences* 1136:126–136.

Sanchez-Cortes, M. S., and E. L. Chavero. 2010. Indigenous perception of changes in climate variability and its relationship with agriculture in a Zoque community, Chiapas, Mexico. *Climatic Change* 107:363-389, doi:10.1007/s1584-010-9972-9.

Shelby, R., D. Edmunds, A. James, J. A. Perez, Y. Shultz, and T. Angogino. 2010. The co-design of culturally-inspired sustainable housing with the Pinoleville Pomo Nation. Paper presented at Open 2010, the 14th Annual Conference of the National Collegiate Inventers and Innovators Alliance (NCIAA), March 25-27, 2010, San Francisco.

Shonkoff, S. B., R. Morello-Frosch, M. Pastor, and J. Sadd. 2011. The climate gap: Environmental health and equity implications of climate change and mitigation politics in California; A review of the literature. *Climatic Change* 109:485–503, doi:10.1007/s10584-011-0310-7.

Sigler, W. F., S. Vigg, and M. Bres M. 1985. Life history of the cui-ui, *Chasmistes cujus Cope, in Pyramid Lake, Nevada: A review. Western North American Naturalist* 45:571–603.

Singh, R. K., S. N. Bhowmik, and C. B. Pandey. 2011. Biocultural diversity, climate change and livelihood security of the Adi community: Grassroots conservators of the eastern Himalaya Arunachal Pradesh. *Indian Journal of Traditional Knowledge* 10:39–56.

Sloan, K. 2009. Climate change issues and needs for the Yurok Tribe. In *Impacts of climate change on tribes in the United States,* Attachment L. Albuquerque, NM: National Tribal Air Association.

Smith, J. B., S. H. Schneider, M. Oppenheimer, G. W. Yohe, W. Hare, M. Mastrandrea, A. Patwardhan, et al. 2008. Assessing dangerous climate change through an update of the Intergovernmental Panel on Climate Change (IPCC) "reasons for concern." *Proceedings of the National Academy of Science* 106:4133–4137.

Strom, B. A. 2005. Pre-fire treatment effects and post-fire forest dynamics on the Rodeo-Chediski burn area, Arizona. M.S. thesis, Northern Arizona University.

Tsosie, R. 2007. Indigenous people and environmental justice: The impact of climate change. *University of Colorado Law Review* 78:1625–1677.

U.S. Bureau of Reclamation. 2010. Entities eligible to request drought assistance under Title 1 of the Reclamation States Emergency Drought Relief Act of 1991, as amended. http://www.usbr.gov/drought/website-eligible-entities.pdf.

U.S. Census Bureau. 2010. 2010 census demographic profiles. http://2010.census.gov/2010census/data/.

U.S. Commission on Civil Rights. 2004. Broken promises: Evaluating the Native American Health Care System. Washington, DC: U.S. Commission on Civil Rights.

U.S. Fish and Wildlife Service (USFWS). 1992. Cui-ui (*Chasmistes cujus*) recovery plan. Portland, OR: USFWS.

Wagner, P., and M. E. Lebo. 1996. Managing the resources of Pyramid Lake, Nevada, amidst competing interests. *Journal of Soil and Water Conservation* 51:108–117.

Wotkyns, S. 2011. *Tribal climate change efforts in Arizona and New Mexico*. Flagstaff, AZ: Institute for Tribal Environmental Professionals.

Wheeler, S. S., ed. 1987. *The desert lake: The story of Nevada's Pyramid Lake*. Caldwell, ID: Caxton Press.

U.S. Environmental Protection Agency (EPA). N.d.: Rincon band of Luiseno Indians uses solar to save energy. In Clean energy and climate change: Tribes; Tribal renewable energy projects. http://www.epa.gov/region09/climatechange/tribes/index.html, last updated April 19, 2012.

Yizhaq, H., Y. Ashkenazy, and H. Tsoar. 2009. Sand dune dynamics and climate change: A modeling approach. *Journal of Geophysical Research* 114: F01023.

Endnotes

i Stevenson, Atta, (November, 2011)."California Indian Water Commission: Statement on Climate Change." Written communication. The California Indian Water Rights Commission consists of Tribal People dedicated to the protection of sacred sites, tribal water, and other inherent rights. According to Stevenson, "We offer our opinions to give voice to our relatives that cannot speak for themselves, the fish, the trees, winged-ones, etc., and especially the water."

ii U.S. Department of the Interior, Bureau of Indian Affairs (August 2011) Letter to tribal leaders and request for proposals 5p.

iii Jill Sherman-Wayne, (January 2012). Written communication.

iv In re General Adjudication of All Rights to Use Water in the Gila River System and Source, 35 P.3d 68 Ariz. 2001.

v John Leeper (October 2011). Director of Navajo Nation Water Resource Management, written communication.

Chapter 18

Climate Choices for a Sustainable Southwest

COORDINATING LEAD AUTHORS

Diana Liverman (University of Arizona), Susanne C. Moser (Susanne Moser Research and Consulting, Stanford University)

LEAD AUTHORS

Paul S. Weiland (Nossaman Inc.), Lisa Dilling (University of Colorado)

CONTRIBUTING AUTHORS

Maxwell T. Boykoff (University of Colorado), Heidi E. Brown (University of Arizona), Eric S. Gordon (University of Colorado), Christina Greene (University of Arizona), Eric Holthaus (University of Arizona), Deb A. Niemeier (University of California, Davis), Stephanie Pincetl (University of California, Los Angeles), W. James Steenburgh (University of Utah), Vincent C. Tidwell (Sandia National Laboratories)

REVIEW EDITOR

Jennifer Hoffman (EcoAdapt)

Executive Summary

The Southwest faces many stresses from current climate variability and is projected to become a hotspot for climate change. A century of economic and population growth has placed pressures on water resources, energy supplies, and ecosystems. Yet the Southwest

Chapter citation: Liverman, D., S. C. Moser, P. S. Weiland, L. Dilling, M. T. Boykoff, H. E. Brown, E. S. Gordon, C. Greene, E. Holthaus, D. A. Niemeier, S. Pincetl, W. J. Steenburgh, and V. C. Tidwell. 2013. "Climate Choices for a Sustainable Southwest." In *Assessment of Climate Change in the Southwest United States: A Report Prepared for the National Climate Assessment*, edited by G. Garfin, A. Jardine, R. Merideth, M. Black, and S. LeRoy, 405–435. A report by the Southwest Climate Alliance. Washington, DC: Island Press.

also has a long legacy of adaptation to climate variability and of environmental management that has enabled society to live within environmental constraints and to protect large parts of the landscape for multiple uses and conservation. Many different types of organizations and individuals in the Southwest have already taken a variety of steps to respond to climate change; and a wide range of choices are available for those choosing to reduce greenhouse gas (GHG) emissions or implement preparedness and adaptation measures to manage the risks from climate variability and change in the region. Others are pursuing energy and water efficiency, renewable energy, or sustainable agriculture for other reasons but these can also reduce emissions or assist with adaptation.

This chapter features the following key findings:

- The U.S. Southwest is a region with great capacity both to respond to environmental stress and to steward its abundant natural resources. Past efforts to develop its water resources and protect its public lands are indicative of this capacity, and while viewed as successes by many, they also illustrate challenges and trade-offs in policy and actions that can increase resilience for some while increasing vulnerability for others. (medium-high confidence)
- Local and state governments, tribes, private-sector entities, non-profit organizations, as well as individuals are already taking steps to reduce the causes of climate change in the Southwest—though often not solely for climate-mitigation purposes—and there are many lessons to learn from the successes and failures of these early efforts. Few systematic studies have been undertaken to date to evaluate the effectiveness and impacts of the choices made in the Southwest to reduce GHG emissions. (medium-high confidence)
- If the Southwest decides to reduce a proportional share of the emissions recommended (50% to 80% by 2050) by the U.S. National Academy of Sciences and others, the carbon budget for the region between 2012 and 2050 would only be 150–350 million metric tons per year (NRC 2010d). This would be a very challenging but not impossible target to meet. (medium-low confidence)
- There are low-cost, cost-saving, or revenue-generating opportunities for emission reductions in the Southwest, especially in energy efficiency and renewable energy. (medium-high confidence)
- A range of stakeholders are already planning how to prepare for and respond to climate risks in the Southwest, but few have begun implementing adaptation programs due to financial, institutional, informational, political, and attitudinal barriers. Various adaptation options exist in every sector, including many that help society respond to current risks of climate variability and extreme events. (medium-high confidence)
- Many response options simultaneously provide adaptation and mitigation "co-benefits," reducing the causes of climate change while also increasing the preparedness and resilience of different sectors to climate change. Other response options involve trade-offs between increasing emissions or reducing resilience

- More research and monitoring is needed to track and evaluate decision outcomes and to understand the balance and effectiveness of these choices especially under financial constraints. (high confidence)

18.1 Introduction

This chapter provides an integrated overview of solutions and choices for responding to climate change in ways that reduce risks and support sustainable development in the Southwest. The goal is to illustrate the range of choices for responding to climate change, along with some of the relevant trade-offs and opportunities, to inform policy options and decisions. In the context of climate change, risk reduction includes: reducing global GHG emissions to limit global changes; limiting activities locally or regionally (e.g. land use choices) that increase unwanted local or regional climatic changes; taking action now to accommodate and adapt to climate changes to date; and increasing capacity to respond effectively and adapt to future changes.

The chapter begins with a discussion of how the Southwest might choose to secure a sustainable future in the context of climate change. The Southwest has a long history of adapting to environmental stresses and managing resources, which demonstrate the ability of the region to make choices that promote sustainability of ecosystems and natural resources, the economy, and society, but also to minimize some of the risks.

Because some studies have identified the Southwest as a potential hotspot of climate change (see Chapter 5) where changes may start to occur rapidly or unfold particularly severely, this chapter also examines some of the options for transformational adaptation to climate change—rather than make more incremental adjustments to climate risks—in the event it becomes necessary to make significant changes in resource allocation or technology, or to relocate people, ecosystems and infrastructure. The co-benefits and trade-offs in linking mitigation and adaptation are also discussed.

National choices about responding to climate change were recently presented by the National Research Council's *America's Climate Choices* study (NRC 2010a, 2010b, 2010c, 2010d). This study is used as a starting point for identifying some of the options for limiting emissions and adapting to climate change, and analyzing what these options might mean for the Southwest in terms of social, technological, economic, behavioral, and institutional structures and choices.

The chapter also reviews some of the choices and solutions that are already being implemented in the Southwest in response to climate change. These efforts include: regional activities by federal agencies; the plans and activities of states, cities, and communities; key regional collaborations such as those in major river basins; and solutions that have been chosen by businesses, tribes, and civil society organizations.

Finally, we discuss options for integrating mitigation and adaptation activities in ways that mutually support each other, rather than produce difficult trade-offs, and focus on the challenges communities and organizations face in planning and implementing solutions. We also raise the question of what actions may be needed if both global mitigation and regional adaptation fail to minimize climate change and resulting impacts to acceptable levels.

18.2 Defining a Sustainable Approach to Climate Change in the Southwest

For the purposes of this chapter, a "sustainable" Southwest is defined as one where the choices we make in responding to climate change assist in the long-term maintenance of economic, social, and environmental well-being—in other words, in meeting the needs of the present without compromising future generations (Wiek et al. 2012). These choices include reducing the risks of climate change by limiting emissions and making it easier to adapt to the impacts of climate changes that are occurring or will occur. Sustainable solutions endure in the face of continuing climate change and other stresses. The Southwest alone cannot mitigate all global GHG emissions, but the region can choose from many options to reduce its proportional contribution to the global causes of climate change and reduce the region's own vulnerability to climate change.

Climate change is not the only threat to sustainability in the Southwest, so pathways to sustainability involve managing multiple risks to the region. This requires considering not just environmental, economic, and social goals, and addressing climate mitigation and adaptation, but also managing risks and opportunities for the well-being of the region's residents and the Earth system (MacDonald 2010). The best pathways will be those that maximize the benefits for environment, economy, and society while minimizing costs and environmental risks, especially for the most vulnerable. One of the greatest challenges is to be prepared for and able to act in the face of uncertainty while being aware of the possibility of reaching thresholds where conditions deteriorate rapidly (Lempert and Groves 2010; Westley et al. 2011). A sustainable Southwest will need early warning of such risks and plans for responding if and when they occur.

18.3 Making a Sustainable Living in the Southwest: Lessons from History

The history of the Southwest demonstrates a remarkable ability to adapt to the climatic and geographic extremes of the region. Tapping into this ability is key to developing sustainable solutions to future climate change.

Throughout human history, water—in particular the ability to move it across the landscape—has been critical to the growth of societies (Worster 1992). Many of the prehistoric peoples of the Southwest found ways to harvest rainwater and runoff and even developed sophisticated water conveyance systems and other techniques for living in a desert climate. The European settlers who came later established water infrastructure and institutions for the development of cities and agriculture. The development of water resources is one of the most notable stories of settling and living in the Southwest. Key to the rapid population expansion in the Southwest was the construction of massive water projects, especially following the passage of the federal Reclamation Act of 1902 (Hundley 1991) (Figure 18.1). By taming the highly variable flow of rivers such as the Colorado and Rio Grande and creating a vast network of canals and ditches capable of moving water between basins, settlers and the federal government did more than just adapt to the necessities of life in an arid climate—they made it a thriving corner of the nation.

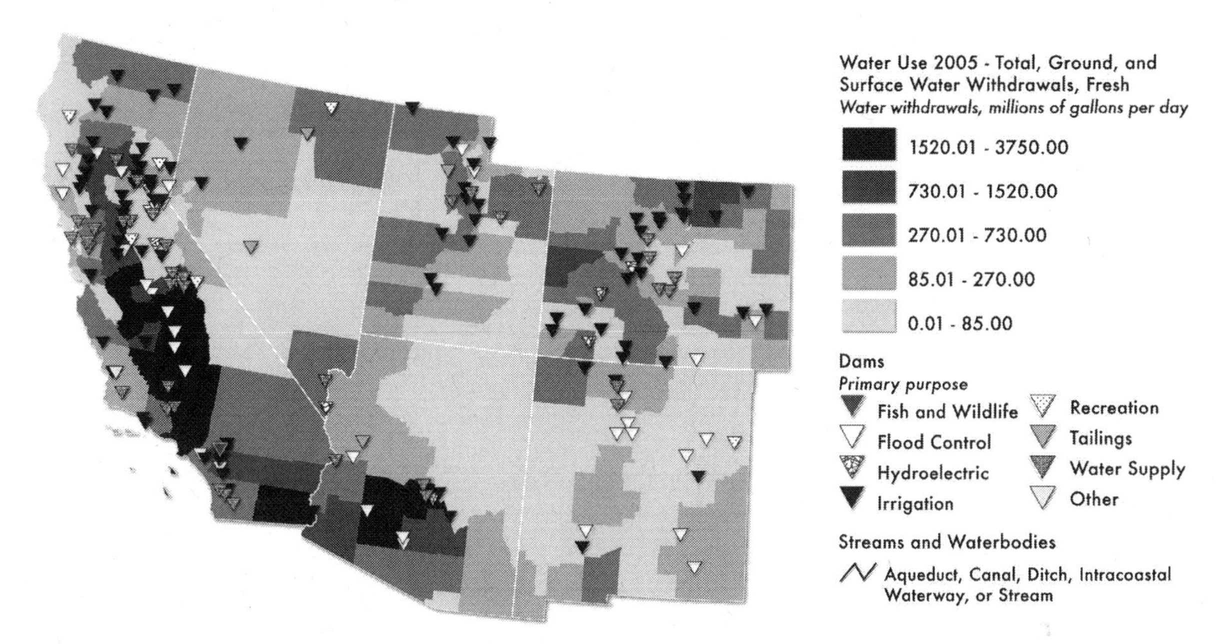

Figure 18.1 Water resource development in the Southwest: 2005 surface-water withdrawals for irrigation and dams. The massive development of water resources stands as one of the grand stories of settling the Southwest. Although this has created its share of environmental and social problems, and there are legitimate questions as to the long-term sustainability of water supply interventions, the systems of dams (shown here), diversions, and management institutions is a testament to the region's ability to invest in managing its environment for economic and social well-being. In light of these incredible efforts to make the Southwest habitable, meeting the new climate challenges of the twenty-first century seems less daunting. Map from *The National Atlas of the United State of America* (http://www.nationalatlas.gov; see also http://www.nationalatlas.gov/mapmaker?AppCmd=CUSTOM&LayerList=wu2005%3B5&visCats=CAT-hydro,CAT-hydro; accessed October 8, 2012).

Water development in the West certainly also created a number of environmental and social problems, and there are legitimate questions as to the long-term ecological, social, and economic sustainability of water demand and use in the region. Fundamental changes to the natural flow of water have had profound consequences for the natural environment. The waters of the Colorado River now rarely reach its mouth at the Gulf of California, drying up a critical and unique wetland (Glenn et al. 1996). The complex plumbing of the Central Valley Project and the State Water Project in California has changed fish communities, water quality, and habitat structure in the Sacramento-San Joaquin Delta (Nobriga et al. 2005). Water availability allowed for massive increases in population throughout the West, in turn increasing the vulnerability of population centers to drought and increased competition between water users (Reclamation 2005).

Despite these consequences, the settlement and watering of the Southwest stands as a reminder of the remarkable effort and funds mustered to transform a dry landscape into one with booming urban centers and extensive and productive agricultural lands.

A second example of choices that created a more sustainable Southwest were the decisions of federal, state, and local governments, as well as private landowners and conservation groups, to set aside vast areas of the West to conserve extractive commodities such as timber and protect scenic beauty, wildlife, habitat, and open space. Twenty-two national parks, nearly 66 million acres of national forests, 74 wildlife refuges, and other protected areas cover more than 165 million acres of the Southwest, conserving natural resources, and providing income to users such as ranchers, loggers, miners, and tourist operators and recreation to millions of residents and tourists (Clawson 1983; Wilkinson 1992) (see also Chapter 3, Section 3.1.3). The Southwest is also home to 120 million acres under the jurisdiction of the Bureau of Land Management (Figure 18.2). While the vast majority of federal public lands were originally created to conserve natural resources for uses in the public interest, such as timber and grazing lands, an environmental protection movement in the 1960s and 1970s led to stronger laws, guiding the management and protection of public lands, and also recognized a number of non-utilitarian uses for the federal domain (Hardt 1994). The Wilderness Act of 1964, Wild and Scenic Rivers Act of 1968, Endangered Species Act of 1973, National Forest Management Act of 1976, Federal Land Policy and Management Act of 1978, and a host of other laws and regulations helped ensure that public lands could be managed and conserved for years to come and that biodiversity would be protected.

This federal land ownership system, which covers nearly 30% of the entire United States (Loomis 2002), helps protect habitat and ecosystem services, facilitates sustainable management of resources, and provides an "insurance policy" for climate adaptation, as land-based resources and economies (such as forestry, tourism, and recreation) as well as species and ecosystems consequently have significant space to migrate to and adjust to the changing conditions. As federal climate-change adaptation response becomes increasingly coordinated, this large area of land can be managed for adaptation and multiple uses in an integrated fashion although multiple jurisdictions can present some barriers.

As with water development, protection of public lands has its challenges. Extractive users, ranchers, recreationalists, and environmentalists struggle with each other and with land-management agencies over appropriate uses of these areas. Yet the wealth of publicly owned land across the United States, especially in the Southwest, is a testament to the willingness of Americans to take proactive steps to prevent the exploitation of resources for the benefit of current and future generations: a sign of a spirit more than capable of tackling the challenges of future climate change.

The Southwest is also leading the economic transformation that has become known as the "green economy," with investments in business ventures that increase energy security, promote sustainability, and reduce environmental impacts (Jones 2009). Colorado and California in particular have supported moves to a green economy where jobs and profits are associated with renewable energy. Colorado has targeted public policy at green energy, attracting venture capital to clean technology of $800 million, and hosting an estimated 17,000 green jobs (see Box 18.5). In Colorado, New Mexico, and Utah, green job growth has outpaced overall job growth (Headwaters Economics 2010). In California, Roland-Holst (2008) estimates that energy efficiency has already generated income savings and created 1.5 million jobs, while redirecting consumption to in-state supply chains. He further estimates that AB 32 (the California Global Warming Solutions Act) will encourage innovation, increase income, and create more than 400,000 new jobs.

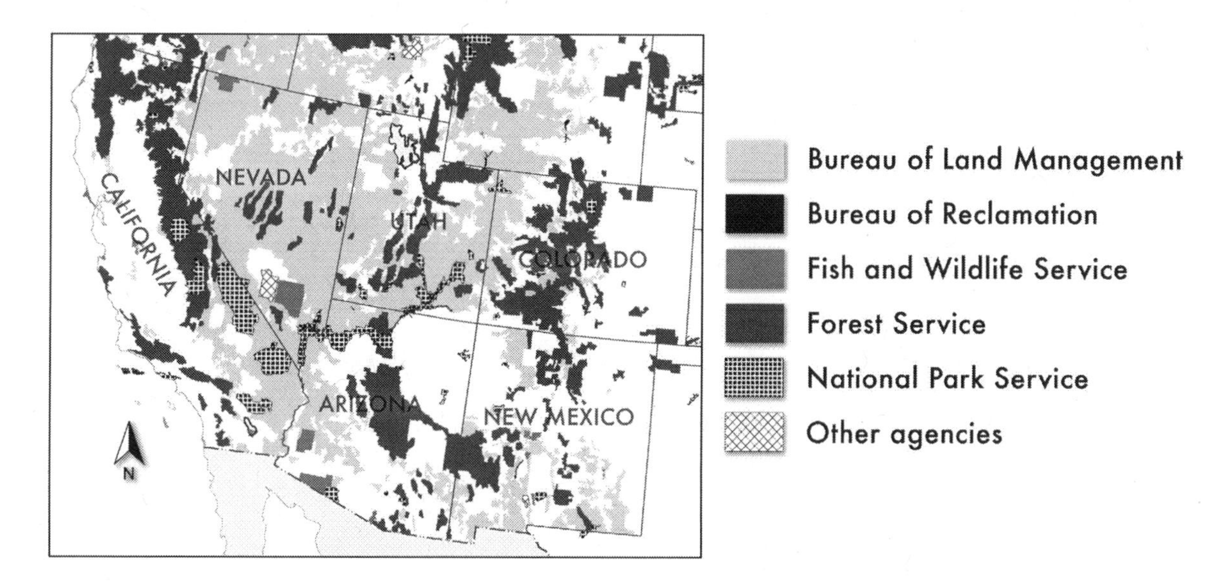

Figure 18.2 Extensive federal lands in the Southwest: A legacy for the future. This map illustrates the legacy of federal land ownership in the Southwest, covering nearly 30 percent of the entire United States. Protected habitat and ecosystem services ensure sustainable management of resources and may be the greatest insurance policy against losses in the future, because natural resource use and biological species can more easily adapt to rapidly changing climatic conditions. Modified from *The National Atlas of the United States of America* (http://www.nationalatlas.gov; see also http://nationalatlas.gov/printable/images/pdf/fedlands/fedlands3.pdf; accessed October 8, 2012).

Other examples of sustainable choices in the Southwest include those cities and communities that have broken with the western model of sprawl, energy-intensive buildings, and dependence on the automobile, to plan more sustainable communities. Sustainable urbanism in the Southwest has included downtown infill, dry landscaping, water reuse, renewable energy development, green-building standards, and public transport to reduce water and energy use, protect green space, and create more livable cities (see, for example, http://www.lgc.org/freepub/healthy_communities/index.html; Garde 2004; Farr 2007; and Chapter 13). Examples of large developments focused on a new sustainable urbanism in the region include Mesa Del Sol, New Mexico; Civano in Tucson, Arizona; Stapleton, Colorado; Mountain House in San Joaquin County, California; and Santa Monica, California.

18.4 Limiting Emissions in the Southwest

To keep human-caused climate change below dangerous levels, the National Research Council (2010d) suggested that the United States and other industrial countries should reduce GHG emissions by 50% to 80% by 2050 compared to 1990 levels. This would give a reasonable chance of keeping atmospheric GHG concentrations below 450 parts per million and limiting overall temperature increases to 3.6°F (2°C) above preindustrial levels. Because annual U.S. carbon dioxide emissions in 1990 were estimated to be 6 gigatons (Gt), a 50% reduction by 2050 would mean reducing emissions to 3 Gt a year, and an 80% cut would be to 1.2 Gt. The NRC estimated that this gives the United States a total carbon budget of 170 Gt to 200 Gt for 2012 to 2050. (The study relied on a wide range of peer reviewed studies and estimated 2008 U.S. emissions to be the equivalent of about 7 Gt of carbon dioxide.) While its recommendation is highly ambitious and challenging, the NRC believes achieving this goal is possible, and more easily so if begun immediately. The study also offers a basket of options for reaching this goal, including choices such as: putting a price on carbon; increasing the energy efficiency of electricity production and transport; moving toward low carbon fuels; increased research and development for carbon capture and storage and new-generation nuclear power generation; and the retirement or retrofit of emission-intensive infrastructure (NRC 2010c, 4–5).

The latest emissions data for CO_2 from fossil fuels in the Southwest shows the region is responsible for 13.4% of the U.S. total in 2009, dominated by emissions from California, which ranks second to Texas in overall emissions (see also Chapter 12). The recently released GHG data reported by large facilities (EPA 2012a) shows that the largest emitters in the Southwest are power-generating plants and oil refineries, with only thirty facilities producing 50% of the emissions from all large facilities (EPA 2012b) (Table 18.1).

Table 18.1 Greenhouse gas emissions by state in the Southwest, shown as CO_2 equivalent emissions (in million metric tons [MMT])

State	CO_2 emissions in MMT (2009)	Percent of Region	Percent of U.S.
Arizona	94	12.95	1.74
California	377	51.93	6.96
Colorado	93	12.81	1.72
Nevada	40	5.51	0.74
New Mexico	58	7.99	1.07
Utah	64	8.82	1.18
Region	726	100	13.4
U.S. total	**5,417**	—	**100**

Source: http://www.epa.gov/statelocalclimate/state/activities/ghg-inventory.html.

Since data for projections of regionally specific carbon emissions scenarios are not available, an estimate of possible regional emission reductions is provided based on the NRC study cited above (NRC 2010d). Assuming global "business-as-usual" emissions were to increase at 3% per year as assumed in several studies (Nakićenović and Swart 2000; Garnaut 2008), the Southwest would have emissions of about 1,000 million metric tons (MMT) in 2020 and 2,400 MMT in 2050.[i] Alternatively, using the lower observed U.S. emissions growth rate of 1.2% per year (1990–2007) (http://epa.gov/climatechange/emissions/index.html), the region would have emissions of around 810 MMT in 2020, and approximately 1,090 MMT in 2050.

For the Southwest to contribute its fair share to reducing emissions by 2050, as NRC recommends, the region would need to reduce emissions to about 150 MMT to 350 MMT per year by 2050. Since this is much lower than projected business-as-usual emissions discussed above—as much as a 90% cut by 2050—we conclude that meeting higher emission reduction goals in the Southwest would be very challenging, but not impossible. Any delay in beginning serious emission reductions would make achieving the region's goal of reducing its proportional share that much harder.

Of the states in the region, only California and Colorado have made commitments to reduce their emissions in line with the 50% to 80% reduction recommended by the NRC. California's goal is to reduce emissions by 80% below 1990 levels by 2050 (State of California, Executive Order S-3-05) and Colorado's is to reduce to 80% below 2005 levels by 2050 (State of Colorado, Executive Order D-004-08). Other states have made more modest or non-binding emission reduction commitments—for example, through the 2007 Western Climate Initiative's target of 15% below 2005 levels by 2020—but some of these commitments have been rescinded or not implemented (http://www.c2es.org/states-regions).

Significant emission reductions can be made at low cost or can save money (see Chapter 12). One estimate for the United States showed that significant emissions reductions could be achieved at a cost of less than $50 per ton of avoided emissions and that almost half of these reductions would actually involve money savings especially from energy efficiency (McKinsey 2007) (Figure 18.3). Many of the money-saving options are relevant to the Southwest and are already being implemented through individual and corporate choices or through government incentives and regulation (see, for example, case studies of local communities at http://www.lgc.org/freepub/energy/index.html). Options include reducing overall energy consumption by driving less or adjusting thermostats, more efficient lighting, more efficient electronic equipment, building insulation, more efficient automobiles, power plant retrofits, and methane management at mines. California has adopted many energy-efficiency strategies over the past several decades, and its economy grew by 80% between 1960 and 2008, with no change in per capita electricity use and a savings of $1,000 per household (Kammen, Kapadia and Fripp 2004; Engel and Kammen 2009; Wei, Patadia, and Kammen 2010).

Some researchers also suggest that the Southwest has a comparative advantage and real opportunities in certain areas of emission reductions, which include solar energy, energy-efficiency savings, and low-carbon electric vehicles (Zweibel, Mason, and Fthenakis 2008; Fthenakis, Mason, and Zweibel 2009). New commercial installations of solar concentrating or solar photovoltaic facilities have been located in the Southwest or are under review in California, Colorado, Arizona, and Nevada. These states lead the country with installed solar photovoltaics and concentrated solar (Gelman 2010). The combination of ample cloud-free days and large areas of land, including abandoned industrial sites, farmland, and public land, represent a regional opportunity for this energy supply. Large solar facilities are not without controversy, however, as they can displace native species, disturb the soil, and may conflict with other human uses of the land.

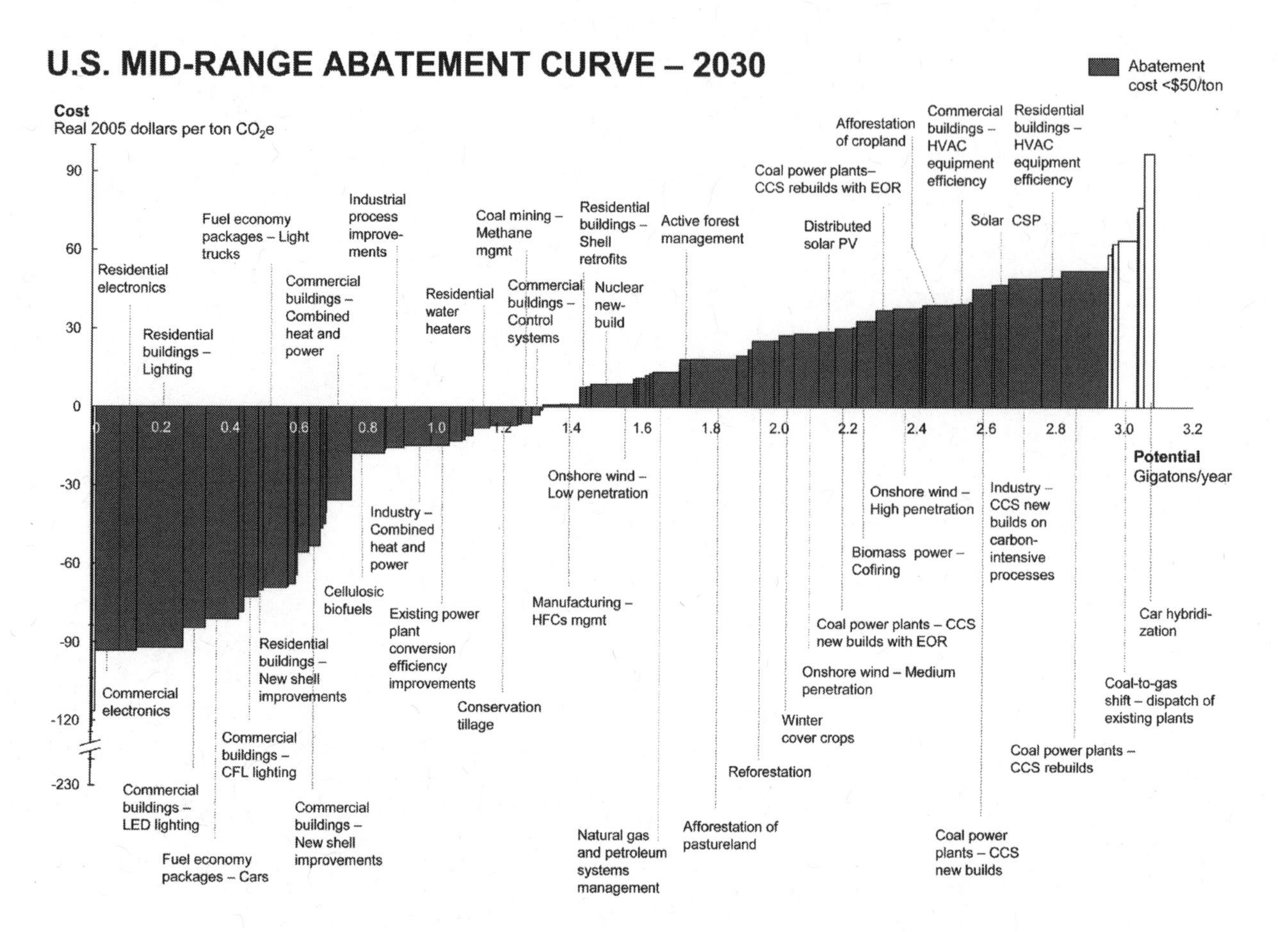

Figure 18.3 McKinsey Mitigation Cost Curve. Governments, for-profit and non-profit organizations, and individuals are already taking steps to reduce the causes of climate change in the Southwest. Many low-cost or negative-cost opportunities for emission reductions (particularly energy efficiency and renewable energy) are available. This well-known graphic shows a wide range of actions that incur cost savings (with "negative costs" shown on the left side of the graphic with bars extending below the horizontal line). Actions to the right of the graphic incur increasingly higher costs. The width of each bar associated with a particular action indicates how much carbon could be abated in 2030 throughout the United States if it were implemented fully (in gigatons of carbon per year). Graph based on McKinsey (2007).

18.5 Adaptation Options in the Southwest

Many of the chapters in this report show that impacts of climate change are not only expected to occur in the future, but are already beginning to manifest across the Southwest. This implies that reducing emissions (i.e., mitigation) cannot be the only response to climate change. Efforts are now also required to prepare for, plan for, and minimize those impacts that cannot be avoided and turn expected climate changes into opportunities wherever possible (i.e., adaptation).[ii]

Box 18.1

Case Studies of Climate Choices for a Sustainable Southwest

Federal Lands and Agency Planning in the Southwest

Federal land and resource management agencies are beginning to incorporate climate change considerations into planning, although efforts are not consistent across agencies (Jantarasami, Lawler, and Thomas 2010). A 2009 Secretarial Order issued at the Department of the Interior spurred individual agencies to begin to incorporate adaptation into individual decisions. The National Park Service's Climate Change Response Program aims to protect park resources from climate change impacts while also using parks to develop knowledge about ecosystem impacts from climate change. A survey of federal land managers in three states in 2011 (Colorado, Wyoming, and Utah) showed that only 6% of their offices were carrying out adaptation plans, but another 25% percent were in the process of developing plans (Archie et al. 2012). A majority were either not currently planning for climate change adaptation (47%) or did not know the status of adaptation planning in their office (24%). Preliminary data indicate that there is some difference in the level of planning among agencies, with the U.S. Fish and Wildlife Service planning at a significantly higher rate than its sister agencies, but it is too early to say why this may be the case. The National Park Service has adopted a range of actions to meet the challenges of climate change in the Southwest region including efforts to reduce energy consumption with a goal of carbon neutrality and through the Climate Friendly Parks program which provides parks with tools to address climate change, including emissions inventories, action plans, and outreach support.

This section focuses on adaptation and basic approaches to it and provides examples of activities already being undertaken. *America's Climate Choices* (NRC 2010a) provides a starting point to lay out a fundamental way of thinking about adaptation to climate change. In it, adaptation is essentially viewed as a challenge in risk management. The Southwest is no stranger to climate-related risks, such as drought, heat extremes, floods, high-wind storms, wildfires, heavy snowfall in the mountains, and cold snaps (Chapters 4, 7, and 8). To reduce the risks from these events in the past, the region's residents, businesses, and planners devised a number of mechanisms, including early warning systems, emergency planning, irrigation systems, building codes, and insurance policies. As the historical patterns of extreme weather events change with a warmer, drier regional climate, the Southwest will need these and additional risk-management tools to prepare for the future so that disruptive events do not become disasters.

Risk management in the face of an uncertain future climate—as defined and discussed in detail in *America's Climate Choices* (NRC 2010a, 2010c)—entails a number of characteristics and iterative, inclusive processes to implement over time. These characteristics and processes are summarized here as generic components that will apply to many if not most adaptation strategies as they are implemented in different sectors:

- **Risk identification, vulnerability assessment, and evaluation.** Scientists and stakeholders jointly identify projected changes in the climate and relevant consequences for particular regions or sectors in light of existing or expected social, economic, and ecological vulnerabilities.
- **Development and assessment of adaptation strategies.** Stakeholders, decision makers, scientists, and engineers assess the costs, benefits, feasibility, and limits of a range of adaptation options.
- **Iterative decision making and deliberate learning.** Many pro-active adaptation decisions will need to be made without "perfect" knowledge of what the future may hold, thus requiring frequent revisiting of decisions and making deliberate efforts at monitoring outcomes and reevaluating them in light of changing knowledge, changing climate, non-climatic stressors, and policy contexts. (This idea and many of those that follow are addressed further in Chapter 19.)
- **Maximizing flexibility.** Whenever decisions with long-term (greater than 30 years) implications can be made incrementally, future risks will be minimized if options for course changes are not foreclosed immediately.
- **Enhancing robustness.** Whenever decisions with long-term implications are being made that can be reversed only at major expense (if at all), future climate risks (and the odds of investing in the wrong option) will be minimized if the considered option(s) will work under a range of plausible future scenarios.
- **Ensuring durability.** To avoid or minimize a perception of economic and social uncertainty, investors, homeowners, and others require some stability to make decisions. Some degree of durability of decisions is needed, with rational adjustments allowed over time.
- **Having a portfolio of approaches.** In a rapidly changing, complex environment, simplistic "fixes," narrow sectoral approaches, or reliance on only a small set of options used in the past are typically insufficient to meet future challenges.
- **Focusing on "no-regrets" options whenever possible.** While any adaptation strategy may involve benefits for some and disadvantages for others, "no-regrets" options are understood as those that would—regardless of the exact unfolding of future climate change—provide the benefit of reducing vulnerability or increasing resilience. For example, improving access for poor, less mobile populations to cooling centers during heat waves would already be beneficial, and will be even more beneficial if and when heat extremes become more common, even if there is some cost involved in providing this service to those currently disadvantaged populations.
- **Focusing on "low-hanging fruit."** Such options are those that are useful for reducing climate risks, are relatively easy to implement, and may not cost much. Examples are avoiding placing more people and assets at risk, improving early-warning or disaster preparedness and response systems, and building climate-change considerations into existing plans for ecosystem restoration or floodplain management.
- **Focusing on building adaptive capacity.** Another very useful strategy already being pursued by a number of institutions and governments in the Southwest—is

to build the capacity to address climate change impacts in the future, including improving understanding of the problem, educating and building awareness among citizens, establishing collaborative ties with others, improving data sharing and communication, setting up stakeholder engagement processes, and developing funding mechanisms.

Box 18.2

Case Studies of Climate Choices for a Sustainable Southwest

Private Sector Responses in the Southwest: Levi Strauss

Levi Strauss & Co. (LS&CO) is a retail company based in California that has started to take steps to mitigate its contribution to climate change. It groups its climate-related goals under three categories: (1) reducing climate-change impacts resulting from production (supply chain focus); (2) reducing impacts from its facilities; and (3) promoting environmentally friendly use and disposal of its products. Since 2007, LS&CO has reduced carbon emissions by 5.84%; this reduction came despite a 6% increase in its real-estate portfolio. The company launched a "Levi's® Water<Less™" jeans product line that reduces both water and energy consumption. LS&CO was also active in supporting The California Global Warming Solutions Act (AB 32) and joined the campaign against Prop 23 that aimed to overturn AB 32 (see also Box 18.4). LS&CO supports a non-profit program focused on teaching irrigation and rainwater-capture techniques in India, Pakistan, Brazil, and Central Africa. Currently 5% of the cotton used in its jeans production is grown using sustainable methods and the company aims to increase this to 20% by 2015.

Source: http://www.levistrauss.com/about/public-policy/environment.

Table 18.2 identifies some of the many options for adapting to climate change in the Southwest, many of which are mentioned in earlier chapters of this report. In addition to adaptation options for specific sectors, any jurisdiction can take steps to develop integrated adaptation plans. This is already being pursued by several entities in the Southwest, including the Western Governors Association, stakeholders in San Diego Bay, the cities of Los Angeles, Salt Lake, San Francisco, and Tucson, the state of California, and a number of regional water utilities.

The actual and potential capacity to adapt to climate variability and change exists at a variety of scales and involves a number of institutions across the Southwest. At a local scale, efforts like watershed protection and restoration conducted by non-governmental organizations and other institutions could minimize potential climate impacts to habitats and ecosystem services (e.g., Carpe Diem West 2011). More formally, a number of municipalities and counties have developed climate adaptation assessments or plans aimed at preparing for future impacts. For example, eight municipalities in the Southwest have formed the Regional Climate Adaptation Planning Alliance to develop a common approach for individual adaptation efforts. Local water providers in Phoenix and Denver have been downscaling climate model data to estimate potential impacts on streamflow, and thus on their long-term water supplies. They are now beginning to explore flexible and incremental actions to respond to such changes if they occur (Quay 2010).

Table 18.2 Adaptation options relevant for the Southwest

Sector	Example Adaptation Strategies
Agriculture	Improved seeds and stock for new and varying climates (and pests, diseases), increase water use efficiency, no-till agriculture for carbon and water conservation, flood management, improved pest and weed management, create cooler livestock environments, adjust stocking densities, insurance, diversify or change production.
Coasts	Plan for sea level rise—infrastructure, planned retreat, natural buffers, land use control. Build resilience to coastal storms—building standards, evacuation plans. Conserve and manage for alterations in coastal ecosystems and fisheries.
Conservation	Information and research to identify risks and vulnerabilities, secure water rights, protect migration corridors and buffer zones, facilitate natural adaptations, manage relocation of species, reduce other stresses (e.g., invasives)
Energy	Increase energy supplies (especially for cooling) through new supplies and efficiency. Use sustainable urban design, including buildings for warmer and variable climate. Reduce water use. Climate-proof or relocate infrastructure.
Fire management	Use improved climate information in planning. Manage urban-wild land interface.
Forestry	Plan for shifts in varieties, altered fire regimes, protection of watersheds and species.
Health and emergencies	Include climate in monitoring and warning systems for air pollution, allergies, heat waves, disease vectors, fires. Improve disaster management. Cooling, insulation for human comfort. Manage landscape to reduce disease vectors (e.g., mosquitos). Public health education and training of professionals.
Transport	Adjust or relocate infrastructure (coastal and flood protection, urban runoff), plan for higher temperatures and extremes.
Urban	Urban redesign and retrofit for shade, energy, and water savings. Adjust infrastructure for extreme events, sea-level rise.
Water management	Enhance supplies through storage, transfers, watershed protection, efficiencies and reuse, incentives or regulation to reduce demand and protect quality, reform or trade water allocations, drought plans, floodplain management. Use climate information and maintain monitoring networks, desalinate. Manage flexibly for new climates not stationarity.

Source: Smith, Horrocks et al. (2011); Smith, Vogel et al. (2011).

Box 18.3

Case Studies of Climate Choices for a Sustainable Southwest

Cities Responding to Climate Change in the Southwest

Cities are emerging as the leaders in setting policies, preparing risk assessments, and setting targets for the reduction of GHG emissions (Rosenzweig 2010; see also Chapter 13). Organizations such as ICLEI, the World Mayors Council on Climate Change, and the C40 Cities Climate Leadership Group have provided successful venues for cities to raise awareness and disseminate best practices (Zimmerman and Faris 2011). Nationwide most of the climate action at the city level is still focused on mitigation (Wheeler 2000; http://www.icleiusa.org/) and recent assessments of these mitigation efforts have been critical of their likelihood to reach stated goals (Willson and Brown 2008). In the Southwest, more than 140 cities are members of ICLEI. Success stories include Fort Collins, Colorado, which has not increased its annual GHG emissions since 2005 despite 5% population growth. Fort Collins is hoping to reduce emissions by 80% below 2005 levels by 2050 (Karlstrom 2010). Salt Lake City, Utah, reduced its GHG emissions by 31% between 2005 and 2009 (Zimmerman and Faris 2011). Cities such as Los Angeles; Boulder City, Nevada; and Pleasanton, California, are also promoting initiatives to expand locally based renewable energy initiatives (Zimmerman and Faris 2011).

Several states have also begun adaptation planning efforts (see Center for Climate and Energy Solutions 2012; Georgetown Law Center 2012). Although California is the only state in the region to have completed a state adaptation plan (see Box 18.4), climate action plans in Arizona, New Mexico, and Colorado call for the development of statewide adaptation activities, and in some sectors—such as water management—adaptation activities are already underway (Chou 2012). Many local governments are also engaging in adaptation planning; to date more than 140 cities in the Southwest are members of ICLEI—Local Governments for Sustainability.[iii] To facilitate such adaptation planning, nine western utilities—together with several from other U.S. regions—have formed the Water Utility Climate Alliance and have been funding research on adaptation strategies for water utilities. This includes a study on advancing climate modeling (Barsugli et al. 2009) and methods for planning adaptation under uncertainty (Means et al. 2010).

18.6 Linking Mitigation and Adaptation

To move toward greater sustainability, both adaptation and mitigation efforts are needed and in some organizations (and households) the same person or group of decision makers are responsible for both activities. While both types of activities have distinct goals, their interaction has four possible outcomes: (1) mitigation positively supports the achievement of adaptation goals; (2) mitigation undermines the achievement of adaptation goals; (3) adaptation supports the achievement of mitigation goals (emission

Box 18.4

Case Studies of Climate Choices for a Sustainable Southwest

California's Climate Policy History and AB 32

The history of climate-change policy making in California is longer than in most other states (Franco et al. 2008). Beginning in 1988, Assembly Bill 4420 (AB 4420) called on the California Energy Commission to lead the preparation of the first scientific assessment of the potential impacts of climate change and of policy options to reduce GHG emissions. It took until 2000 before the first steps were taken to regulate GHG emissions, when Senate Bill 1771 created the non-profit California Climate Action Registry (CA Registry), allowing state organizations to register and track their voluntary emission reductions. Shortly thereafter in 2002, the assembly passed the so-called "Pavley bill" (AB 1493), a ground-breaking law which led to the regulation of GHG emitted from automobiles. After an executive order was signed by Governor Arnold Schwarzenegger in June 2005 (S-3-05), the California state assembly then passed the California Global Warming Solutions Act (AB 32) in 2006, committing the state to reduce GHG emissions statewide by 80% below 1990 levels by mid-century, with an interim goal of capping emissions at 1990 levels by 2020. Several additional laws have been passed since in support of these policy goals, including requirements to generate a growing percentage of electricity from renewable energy and to develop integrated land use and transportation strategies (Franco et al. 2008; NRC 2010c, Box 2.1). Contrary to widespread concerns, the climate-policy initiatives in California appear to have positive economic impacts on the state economy in terms of jobs generated and technological innovation spurred (Roland-Holst 2008; Berck and Xie 2011).

reductions); and (4) adaptation undermines the achievement of mitigation goals. Because funding is often limited and alternatives are not always feasible, in some instances adaptation may have to be chosen even though it increases emissions, or one type of effort must be focused on one rather than the other because of mandates. For example, heat wave response may require extra air conditioning in public buildings or extra groundwater pumping, even when this increases emissions because other options such as desalination are too expensive or simply not available in the near-term. Some renewable energy options may require more water use, thus adding to adaptation challenges. It is important to examine the interaction of mitigation and adaptation in the Southwest because it can help maximize potential co-benefits and reduce potential trade-offs if they cannot entirely be avoided (Scott and Pasqualetti 2010). To the extent trade-offs are perceived by interested stakeholders, they can pose barriers to progress, and thus need careful consideration (Moser 2012). Table 18.3 lists examples of activities particularly relevant in the Southwest region that illustrate these interactions.

While trade-offs should be avoided, stand-alone climate policies that pursue only mitigation or adaptation goals should not be disfavored if they are well indicated and demonstrably useful even if they do not have explicit co-benefits for other policy goals. This may entail difficult political challenges, as it is reasonable to expect that there will be times when true sustainability and successful adaptation require hard choices, including convincing stakeholders that what they perceive as harmful to them could be beneficial to them and the larger community and environment in the long term.

Box 18.5

Case Studies of Climate Choices for a Sustainable Southwest

Colorado's Green Economy

Colorado has a strong focus on the fast-growing clean-energy economy. Between 1998 and 2007, jobs in the U.S. clean-energy sector grew by 9.1%, while those in Colorado's clean-energy sector grew by 18.8% (Pew Charitable Trusts 2009). Colorado has one of the most aggressive Renewable Portfolio Standards (RPS)—a requirement to produce a certain amount of energy from renewable sources—with 30% of energy to be sourced from renewable energy by 2020, according to Headwaters Economics (2010). This RPS was doubled from its previous target when lawmakers observed the ease with which it was being met, together with an influx of jobs in rural areas. Colorado has provided a variety of incentives to promote its clean-energy growth, including direct funding for renewable energy development targeted at residential and commercial buildings. In 2009 Colorado implemented an Energy Efficiency Resource Standard with the goal of achieving 11.5% energy savings by 2020 for investor-owned utilities. Colorado was ranked fifth nationally in terms of total venture capital investment in clean energy between 2006 and 2008, with almost $800 million invested in clean technologies.

18.7 Barriers to Planning for and Implementing Climate Solutions

As adaptation has become a focus of public policy, many states, local governments, tribes, for-profit and non-profit organizations, and individuals have encountered impediments to the development and implementation of mitigation and adaptation efforts. At the same time, researchers have made progress in documenting and examining these impediments, including in the Southwest.

The National Research Council distinguished four basic groups of barriers to climate action: (a) inadequate information and experience, (b) inadequate institutional support, (c) lack of resources and technology, and (d) behavioral impediments (NRC 2010a). These barriers were also found for mitigation (NRC 2010d) and are echoed in other studies (e.g., Post and Altman 1994; Verbruggen et al. 2009; Gifford, Kormos, and McIntyre 2011). More recent studies provide much more detailed insights into the range of impediments that decision makers encounter (e.g., Amundsen, Berglund, and Westskog 2010; Burch 2010; Storbjörk 2010; Ekstrom, Moser, and Torn 2011; Measham et al. 2011; McNeeley 2012; Moser and Ekstrom 2012).

Table 18.3 Examples of synergies and trade-offs between regionally relevant mitigation and adaptation activities and climate-change impacts

Mitigation supports Adaptation	Reforestation increases carbon storage and improves water resources.	Jimenez et al. 2009
	Moving from water-cooled concentrating solar power plants in California and Nevada toward dry cooling helps reduce water needs for the energy sector and leaves resources available for other users.	Schultz, Shelby, and Agogino 2010
	Increased urban tree cover increases carbon storage and shading, resulting in lower cooling-energy demand and fewer heat-related health risks.	Blate et al. 2009
	Installation of renewable energy systems in homes, farms, and tribal land, as well as building retrofits to increase insulation and energy efficiency reduce emissions and produce high-quality jobs, thus increasing income-generating opportunities for communities and lowering their vulnerability to change.	Averyt et al. 2011; Nowak, Crane, and Stevens 2006; Pataki et al. 2006; Chen et al. 2011
Mitigation undermines Adaptation	Carbon capture and storage from coal-burning power plants increases demand on and creates greater competition for regionally scarce water resources.	Averyt et al. 2011
	As hydroelectric power generation declines because of decreased precipitation, water supplies may become insufficient to meet all human and environmental needs, and the power deficit may be made up from CO_2-emitting sources.	Giridharan et al. 2007
	Power generation has occasionally depleted aquifers in the Southwest.	
	Power plants dependent on water cooling will release warmed waters into already warmer rivers and streams, adding further stress on aquatic plants and animals and reducing water quality.	
	The move to renewable energy can be water intensive: U.S. nuclear power plants may require as much as eight times more freshwater than natural gas plants per unit of electricity generated and 11 % more than coal plants. Some concentrating solar power plants consume more water per unit of electricity than the average coal plant.	
	More compact coastal urban design (to reduce transportation-related emissions) may increase the urban heat island effect and could concentrate development in hazardous areas (such as floodplains).	
Adaptation supports Mitigation	Improved forest fuel management (and reduction) decreases the risk of devastating wildfires (and thus large releases of carbon into the atmosphere), and thus maintains watershed health, reduces the risk of landslides, soil erosion, and destruction of infrastructure, and better preserves scarce water resources.	Carpe Diem West 2011

Table 18.3 Examples of synergies and trade-offs between regionally relevant mitigation and adaptation activities and climate-change impacts (Continued)

Adaptation supports Mitigation	Efforts to increase rainwater infiltration on the land to improve water security and reduce the risk of sewer overflows and flooding during extreme rainfall events also reduces the need for energy-intensive sewage treatment and pumping.	Borel 2009; Waterfall 2006; PWA 2010; DeLaune and White 2011
	Coastal seagrass bed and wetland restoration increases carbon uptake and increases coastal protection against storms [1].	
Adaptation undermines Mitigation	Desalinization of seawater to increase local water security during drought years is a highly energy-intensive adaptation options, thus increasing CO_2 emissions (unless the desalination plant is solar-powered).	DOE 2006; Stokes and Horvath 2006; Lofman, Petersen, and Bower 2002
	Increased pumping for groundwater and increased recharge of depleted groundwater aquifers is energy-intensive and thus, typically, increases CO_2 emissions.	Biesbroek, Swart, and van der Knaap 2009
	Relocation of residents out of floodplains in ways that increase the overall need for driving increases one-time relocation-and rebuilding-related emissions and possibly increases transportation-related emissions.	Boden, Marland, and Andres 2011
	Extensive fortification of coastlines against sea-level rise and coastal flooding with seawalls also increases CO_2 emissions from cement.	

Note: [1] Additional benefits and cost savings may arise if sediment trapped in nearby bays or channels is used to help wetlands build up vertically; carbon storage benefit may be smaller if coastal storms cause severe damage to wetlands.

For example, in a survey of over 600 federal public land managers in Colorado, Wyoming, and Utah (Dilling 2012), lack of funding and lack of information (including both the uncertainty of information and its usefulness) were both ranked highly as barriers in moving forward to plan or implement adaptation strategies for climate change. Lack of specific agency direction was also mentioned as a key barrier. Public perception, including the perceived lack of importance and lack of demand from the public to take action on climate change may also act as hurdles in preparing for climate change. A perhaps unique challenge for public lands and other resources governed by federal law such as interstate water compacts (i.e., the Colorado River Compact) is that they have a decision process and legal framework that was developed under an assumption of climate stationarity—the concept that patterns of past climate provide a reasonable expectation of those of the future—an assumption that is no longer valid (Milly et al. 2008; Ruhl 2008). The legal framework defining decision making on public lands is likely to be another barrier to making adaptive decisions.

Box 18.6

Case Studies of Climate Choices for a Sustainable Southwest

Energy and Climate in the Southwest

With all states in the Southwest implementing Renewable Portfolio Standards (RPS), the development of renewable energy sources is thriving across the region. Taking advantage of its unique position at the intersection of three of the country's ten major electrical grids as well as its natural resources, New Mexico has the potential to become a major hub for renewable energy with a proposed Tres Amigas "superstation" linking to the Electric Reliability Council of Texas, the Southwest Power Pool, and the Western Electricity Coordinating Council. New Mexico's RPS requires 10% of its energy to be generated from renewable sources by 2011, with an increase to 20% by 2020. The state is capitalizing on its diverse renewable energy potentials, including wind, solar, geothermal, and biofuels. To encourage the increased production and demand for alternative and renewable energy, New Mexico is implementing a variety of tax credits, tax deductions, and innovation funds. In addition, the state is expanding green-job training as well as research and development of clean technology across the state (as through the new North American Wind Research and Training Center, which partners with Sandia National Laboratories and New Mexico State University) (Thorstensen and Nourick 2010).

Box 18.7

Case Studies of Climate Choices for a Sustainable Southwest

Salt Lake City's Emission Reduction Efforts

Salt Lake City is striving to reduce GHG emissions from municipal operations by 3% per year for the next ten years. By 2040, the city aims to reduce emissions by 70% (EPA n.d.). EPA and DOE have awarded an ENERGY STAR Award for Excellence to the Utah Building Energy Efficiency Strategies (UBEES), a coalition of government agencies, members of the building industry, and stakeholders, for their energy efficiency and renewable energy goals (Energy Star Program n.d.). Utah aims to source 20% of its energy from renewable energy sources by 2025. The state also aims to improve energy efficiency 20% by 2015 (Energy Star Program n.d.). Utah's first commercial wind power project generates nearly 19 MW of energy through an urban wind turbine installation. Located in Spanish Fork, a city of 32,000 located fifty miles south of Salt Lake City, the project is a remarkable example of small-scale renewable energy production that faced many political, market, and social barriers and overcame them successfully through a transparent and patient stakeholder engagement process (Hartman, Stafford and Reategui 2011).

A detailed study on barriers to adaptation focused on four local coastal communities (two cities and two counties) and a regional process in San Francisco Bay (Moser and Ekstrom 2012). Its findings were extended through a survey of coastal communities along the entire California coastline (Hart et al. 2012), thus allowing for verification and generalization. The case study found institutional- and governance-related barriers to be the leading impediments to greater adaptation planning and implementation, followed by attitudinal and motivational barriers among the individuals and groups involved. Economic barriers mattered also, even in some of the wealthiest communities in that region (and the nation). Multiple lines of evidence confirmed the importance of institutional, individual, and economic barriers, which is also echoed in the broader literature. At the same time, the study revealed that communities have significant leverage over the barriers they face in the "here and now," as well as many important advantages, and assets that either help avoid barriers in the first place, or help overcome them if they are encountered. To move beyond barriers created through decisions made in the past or at other levels of governance, as well as to manage obstacles resulting from entrenched local political dynamics and pressures, communities need assistance from higher levels of governance (see also Chapter 9, Section 9.5).

To help overcome the barriers that prevent communities, organizations, and businesses from planning for a climate-altered future or that pose time-consuming and costly obstacles to those ready to implement mitigation and adaptation actions, several critical steps can be taken. Much of the adaptation activity to date can be characterized as building capacity (including gathering relevant information, assessing risks, educating decision makers and affected stakeholders, and improving communication and cross-sectoral and cross-scale collaboration) (Moser and Ekstrom 2010, 2012). Several categories of supporting activities can be broadly categorized into cooperation and collaboration (across scales, agencies, public/private), market mechanisms (e.g., trading systems, pricing, valuing ecosystem services), legal reforms, mandates and standards, education, information and decision support, and—to move any and all of these forward—both technical and political leadership. Framing responses in terms of water conservation or energy efficiency, for example, may be more effective than making explicit links to climate change for some Southwest residents who are confused by the debate over climate science (Nisbet 2009; Resource Media 2009).

18.8 Coping with the Risks of Rapid Climate Changes

There is a risk that climate change might bring unacceptably large, sudden, or abrupt changes to the Southwest (see Chapter 7) and elsewhere, such as multi-decadal droughts, shifts to significantly higher temperatures (e.g., +3°F) in less than ten years, sea-level rise that is much faster than what has historically occurred, dramatic shifts in ecosystems (crossing of local- or larger-scale tipping points), or significant increases in the incidence of climatic extremes (Lindenmayer et al. 2010; Park et al. 2011; Smith, Horrocks et al. 2011). Even if such changes prompted steep emission reductions globally, the lags in the climate response would make it difficult to immediately stabilize the climate. Should such a scenario unfold, the Southwest may need to consider more dramatic and transformational adaptations to a changed climate (Smith, Horrocks et al. 2011; Kates, Travis and Wilbanks 2012; O'Brien 2012) or push for large-scale manipulations of the climate (also called geoengineering).

In conditions of water scarcity, for example, choices would need to be made about water-allocation priorities that would challenge traditional water rights in the West. Agriculture and ranching might need to shift into different places or species. Desalination and water reuse might become much more viable and socially acceptable options and urban areas might need to transform water use (Larson et al. 2005). Coastal settlements and infrastructure, as well as valued ecosystems, might need to be relocated on short timescales and thus possibly at considerable cost. Southwestern residents would need to consider their positions and choices on geoengineering options, which involve intentional interventions in the carbon cycle or in solar radiation to cool the planet (Victor et al. 2009; Caldeira and Keith 2010).

Box 18.8

Case Studies of Climate Choices for a Sustainable Southwest

Private Sector Responses in the Southwest: Freeport McMoRan mining

Multinational mining corporation Freeport McMoRan, based in Phoenix, operates eight copper mines in Arizona, Colorado, and New Mexico and has responded to environmental concerns, including climate change, by developing solar energy facilities in two Arizona mining communities, Bagdad and Ajo, and completing GHG inventories. Most of its emissions are from materials transport and the company states it is focusing on improved fuel consumption. As a global business, Freeport McMoRan report to the Global Reporting Initiative and Carbon Disclosure Project. In 2010 the company reported worldwide emissions of 10 MMT; it is working on overall emission reduction plans, energy efficiency, and carbon offsets (http://www.fcx.com/envir/wtsd/pdf-wtsd/2010/WTSD_Bk_2010.pdf).

18.9 Research Gaps

A significant amount of general knowledge about mitigation and adaptation options is available to Southwest stakeholders. Few of these options have specifically assessed the costs, legal feasibility, or possible trade-offs of climate solutions with other policy goals. Thus, the practical basis for informed decision making is still relatively weak, even if much is known in general about possible climate responses. Tracking and evaluation of mitigation and adaptation activities is missing. Research on private sector actions is especially difficult and therefore largely missing. In addition, little has been done to evaluate plans and responses already underway and to assess the effectiveness of secondary actions that indirectly contribute to climate responses. For example, claims of climate action undertaken for other reasons such as energy or food security need to be assessed for their impacts. Other key research gaps include the analysis of trade-offs and of the long-term implications of choices on environmental impacts, vulnerability, and economic well-being.

The least developed or understood solutions are generally those that require deeper intervention in the various systems, such as through legal changes (for example, to water rights) or large-scale market mechanisms (for example, a functional regional carbon-trading scheme). Similarly, understanding the potential impacts of geoengineering interventions on many systems—regional climate, crop production, water availability, and human well-being—is a considerable challenge.

With key agencies, collaborative projects, and universities actively engaged in use-inspired[4] climate research, the Southwest is uniquely endowed with research centers that have considerable expertise in developing effective relationships with stakeholders and decision makers and in developing decision-relevant information (Table 18.4). A fair amount is understood about how to do this well, and the Southwest may well lead the nation in this regard. The demand for use-inspired research and decision support is growing rapidly, and there is a growing need to expand the expertise and capacity to deliver on this need. Scaling up the capacity-building efforts among decision makers to understand and meet the challenges involved in risk management in the face of rapid changes must also be a priority.

Table 18.4 Climate science and assessment example activities in the Southwest

Type of Organization	Specific Programs in the Southwest	Geographic Scope of Program	Description and Mission
Regional Integrated Sciences and Assessments (RISAs; funded by NOAA)	Western Water Assessment *wwa.colorado.edu*	CO, UT	Identifying regional vulnerabilities to and impacts of climate variability and change, and developing information, products, and processes to assist decision makers throughout the Intermountain West.
	Climate Assessment for the Southwest *climas.arizona.edu*	AZ, NM	Improving the region's ability to respond sufficiently and appropriately to climatic events and climate changes.
	California-Nevada Applications Program *meteora.ucsd.edu/cap*	CA, NV	Developing and providing better climate information and forecasts for decision makers in California, Nevada, and the surrounding region.
Climate Science Center (CSC; funded by Department of the Interior)	*doi.gov/csc/southwest*	Entire Southwest	Providing scientific information, tools, and techniques that land, water, wildlife, and cultural-resource managers and other interested parties can apply to anticipate, monitor, and adapt to climate and ecologically driven responses at regional-to-local scales.

Table 18.4 Climate science and assessment example activities in the Southwest (Continued)

Type of Organization	Specific Programs in the Southwest	Geographic Scope of Program	Description and Mission
Landscape Conservation Coopera-tives (LCCs; funded by Dept. of the Interior)	California LCC *californialcc.org*	Portions of CA	LCCs are public-private partnerships that complement and build upon existing science and conservation efforts—such as fish habitat partnerships and migratory bird joint ventures—as well as water resources, land, and cultural partnerships as part of the Department of the Interior's collaborative, science-based response to climate change.
	Desert LCC *usbr.gov/WaterSMART/lcc/desert.html*	Portions of AZ, CA, NM, NV	
	Southern Rockies LCC *doi.gov/lcc/Southern-Rockies.cfm*	Portions of AZ, CO, NM, UT	
	Great Plains LCC *greatplainslcc.org*	Portions of CO and NM	
	Great Basin LCC *blm.gov/id/st/en/prog/Great_Basin_LCC.html*	Portions of CA, NV, and UT	
	North Pacific LCC *fws.gov/pacific/Climatechange/nplcc/*	Portions of CA	
	Great Northern LCC *nrmsc.usgs.gov/gnlcc*	Portions of CO and UT	
NOAA Regional Climate Services	NOAA Western Region RCSD *noaaideacenter.org/rcsd/west/*	Entire Southwest	Building and strengthening regional partnerships to better assess and deliver regionally focused climate science and information products and services to help people make informed decisions in their lives, businesses, and communities.
Bureau of Reclamation	Colorado River Basin Water Supply & Demand Study *www.usbr.gov/lc/region/programs/crbstudy.html*	Colorado River Basin	Defining current and future imbalances in water supply and demand, and developing and analyzing adaptation and mitigation strategies to resolve those imbalances.

Table 18.4 Climate science and assessment example activities in the Southwest (Continued)

Type of Organization	Specific Programs in the Southwest	Geographic Scope of Program	Description and Mission
The Nature Conservancy	Southwest Climate Change Initiative *conserveonline.org/workspaces/climateadaptation/documents/southwest-climate-change-initiative-0/view.html*	AZ, CO, NM, UT	Providing guidance to conservation practitioners and land managers in climate change adaptation planning and implementation on more local scales.
Northern Arizona University Institute for Tribal Environmental Professionals	Southwest Tribal Climate Change Network *www4.nau.edu/itep/climatechange/tcc_SWProj.asp*	AZ, NM	Identifying existing tribal climate change efforts being undertaken in Arizona and New Mexico; assessing tribal research and information needs regarding climate change issues; and developing strategies for meeting those needs.
University of Arizona Institute of the Environment	Southwest Climate Change Network *southwestclimatechange.org*	AZ, NM	Fostering a dialog and exchange of science and policy information among climate experts, other scientists, natural resource managers, utility providers, policy and decision makers, community groups, the public, and the media about climate-change issues in the Southwest.
Desert Research Institute	Western Regional Climate Center *wrcc.dri.edu*	Entire Southwest	Tracking and disseminating high quality climate data and information for the Western United States; fostering better use of climate data in decision making; conducting applied climate research; improving the coordination of climate-related activities.
Multi-university	Southwest Climate Alliance *southwestclimatealliance.org*	Entire Southwest	Working with the Southwest Climate Science Center to help regional stakeholders meet the needs of climate variability and change.
Multi-agency	Western Mountain Initiative *westernmountains.org*	Entire Southwest	Scientists from USGS and U.S. Forest Service working to understand responses of Western mountain ecosystems to climate variability and change.
Arizona State University	Decision Center for a Desert City *http://dcdc.asu.edu/*	AZ	Conducting climate, water, and decision research and developing innovative tools to bridge the boundary between scientists and decision makers and put this work into the hands of those whose concern is for the sustainable future of Greater Phoenix.

References

Amundsen, H., F. Berglund, and H. Westskog. 2010. Overcoming barriers to climate change adaptation - a question of multilevel governance? *Environment and Planning C: Government and Policy* 28:276–289.

Archie, K. M., L. Dilling, J. B. Milford, and F. C. Pampel. 2012. Climate Change and Western Public Lands: a Survey of U.S. Federal Land Managers on the Status of Adaptation Efforts. Ecology and Society 17 (4):C7-20.

Averyt, K., J. Fisher, A. Huber-Lee, A. Lewis, J. Macknick, N. Madden, J. Rogers, and S. Tellinghuisen. 2011. *Freshwater use by U.S. power plants: Electricity's thirst for a precious resource.* A report of the Energy and Water in a Warming World Initiative. Cambridge, MA: Union of Concerned Scientists.

Barsugli, J., C. Anderson, J. B. Smith, and J. M. Vogel. 2009. *Options for improving climate modeling to assist water utility planning for climate change.* N.p.: Water Utility Climate Alliance.

Berck, P. and L. Xie. 2011. A policy model for climate change in California. *Journal of Natural Resources Policy Research* 3:37–47.

Biesbroek, G. R., R. J. Swart, and W. G. M. van der Knaap. 2009. The mitigation-adaptation dichotomy and the role of spatial planning. *Habitat International* 33:230–237.

Blate, G. M., L. A. Joyce, J. S. Littell, S. G. McNulty, C. I. Millar, S. C. Moser, R. P. Neilson, et al. 2009. Adapting to climate change in United States national forests. *Unasylva* 60:57–62.

Boden, T., G. Marland, and B. Andres. 2011. Global CO_2 emissions from fossil-fuel burning, cement manufacture, and gas flaring: 1751-2008. Oak Ridge, TN: Oak Ridge National Laboratory, Carbon Dioxide Information Analysis Center. http://cdiac.ornl.gov/ftp/ndp030/global.1751_2008.ems.

Borel, V. 2009. *Rain gardens.* N.p.: California Sea Grant Extension / University of California Cooperative Extension. Green Sheet Series No. 3. http://www-csgc.ucsd.edu/BOOKSTORE/Resources/GS3%20Rain%20Gardens_8-10-09.pdf.

Burch, S. 2010. Transforming barriers into enablers of action on climate change: Insights from three municipal case studies in British Columbia, Canada. *Global Environmental Change* 20:287–297.

Caldeira, K., and D. W. Keith. 2010. The need for climate engineering research. *Issues in Science and Technology* 27:57–62.

Carpe Diem West. 2011. *Watershed investment programs in the American West. An updated look: Linking upstream watershed health and downstream security.* Sausalito, CA: Carpe Diem West.

Center for Climate and Energy Solutions. 2012. State adaptation plans. http://www.c2es.org/us-states-regions/policy-maps/adaptation.

Chen, F., S. Miao, M. Tewari, J-W. Bao, and H. Kusaka. 2011. A numerical study of interactions between surface forcing and sea-breeze circulations and their effects on stagnation in the greater Houston area. *Journal of Geophysical Research* 116: D12105, doi: 10.1029/2010JD015533.

Chou, B. 2012. *Ready or not: An evaluation of state climate and water preparedness planning.* NRDC Issue Brief IB:12-03-A. Washington, DC: Natural Resources Defense Council.

Clawson, M. 1983. *The federal lands revisited.* Baltimore: Johns Hopkins University Press.

DeLaune, R. D., and J. R. White. 2011. Will coastal wetlands continue to sequester carbon in response to an increase in global sea level? A case study of the rapidly subsiding Mississippi river deltaic plain. *Climatic Change* 110:297–314.

Dilling, L. 2012. Climate adaptation barriers and opportunities in the United States: A focus on policy and decision making at the sub-national scale. Poster presented at *Planet Under Pressure International Conference, London, United Kingdom, March 26-29, 2012.* http://sciencepolicy.colorado.edu/news/announcements/2011-2012/dilling_pup_poster.pdf.

Ekstrom, J. A., S. C. Moser, and M. Torn. 2011. *Barriers to climate change adaptation: A diagnostic framework.* Public Interest Energy Research (PIER) Program final project report CED-500-2011-004. Sacramento: California Energy Commission. http://www.energy.ca.gov/2011publications/CEC-500-2011-004/CEC-500-2011-004.pdf.

Energy Star Program. N.d. Utah building energy efficiency strategies. http://www.energystar.gov/index.cfm?fuseaction=pt_awards.showAwardDetails&esa_id=4104 (accessed June 14, 2012).

Engel, D., and D. M. Kammen. 2009. *Green jobs and the clean energy economy.* Copenhagen Climate Council, Thought Leadership Series No. 4. Copenhagen: Copenhagen Climate Council. Also available at: http://www.climatechange.ca.gov/eaac/documents/member_materials/Engel_and_Kammen_Green_Jobs_and_the_Clean_Energy_Economy.pdf.

Farr, D. 2007. *Sustainable urbanism: Urban design with nature.* New York: Wiley.

Fthenakis, V., J. E. Mason, and K. Zweibel. 2009. The technical, geographical, and economic feasibility for solar energy to supply the energy needs of the U.S. *Energy Policy* 37:387–399.

Franco, G., D. Cayan, A. Luers, M. Hanemann, and B. Croes. 2008. Linking climate change science with policy in California. *Climatic Change* 87 (Suppl. 1): S7–S20.

Friedlingstein, P., R. A. Houghton, G. Marland, J. Hackler, T. A. Boden, T. J. Conway, J. G. Canadell, M. R. Raupach, P. Ciais, and C. Le Quéré. 2010. Update on CO_2 emissions. *Nature Geoscience* 3:811–812.

Garde, A. M. 2004. New urbanism as sustainable growth? *Journal of Planning Education and Research* 24:154–170.

Garnaut, R. 2008. *The Garnaut climate change review: Final report.* Melbourne: Cambridge University Press. http://www.garnautreview.org.au/2008-review.html.

Georgetown Law Center. N.d. State and local adaptation plans. http://www.georgetownclimate.org/adaptation/state-and-local-plans (accessed 1 October 2012).

Gelman, R. S. 2010. *2010 Renewable energy data book.* Washington, DC: U.S. Department of Energy.

Glenn, E. P., C. Lee, R. Felger, and S. Zengel. 1996. Effects of water management on the wetlands of the Colorado River Delta, Mexico. *Conservation Biology* 10:1175–1186.

Gifford, R., C. Kormos, and A. McIntyre. 2011. Behavioral dimensions of climate change: Drivers, responses, barriers, and interventions. *WIREs Climate Change* 2:801–827.

Giridharan, R., S. S. Y. Lau, S. Ganesan, and B. Givoni. 2007. Urban design factors influencing heat island intensity in high-rise high-density environments of Hong Kong. *Building and Environment* 42:3669–3684.

Hardt, S. W. 1994. Federal land management in the twenty-first century: From wise use to wise stewardship. *Harvard Environmental Law Review* 18:345–403.

Hart, J. F., P. Grifman, S. C. Moser, A. Abeles, M. Meyers, S. Schlosser, and J. A. Ekstrom 2012. *Rising to the challenge: Results of the 2011 California coastal adaptation needs assessment.* USCSG-TR-01-2012. Los Angeles: University of Southern California, Sea Grant.

Hartman, C. L., E. R. Stafford, and S. Reategui. 2011. Harvesting Utah's urban winds. *Solutions* 2 (3). http://www.thesolutionsjournal.com/node/930.

Headwaters Economics. 2010. *Clean energy leadership in the Rockies: Competitive positioning in the emerging green economy.* Bozeman, MT: Headwaters Economics.

Hundley, N., Jr. 1991. The Great American Desert transformed: Aridity, exploitation, and imperialism in the making of the modern American West. In *Water and arid lands of the western United States,* ed. M. El-Ahsry and D. Gibbons, 21-84. Cambridge: University of Cambridge Press.

Jantarasami, L. C., J. J. Lawler, and C. W. Thomas. 2010. Institutional barriers to climate change adaptation in U.S. national parks and forests. *Ecology and Society* 15 (4): article 33. http://www.ecologyandsociety.org/vol15/iss4/art33.

Jimenez, A., R. Gough, L. Flowers, and R. Taylor. 2009. Wind power across Native America: Opportunities, challenges, and status. Poster presented at Windpower 2009 Conference, May 4-7, 2009, Chicago, IL. http://www.nrel.gov/docs/fy09osti/45411.pdf.

Jones, V. 2009. *The green collar economy: How one solution can fix our two biggest problems.* San Francisco: HarperOne.

Kammen, D. M., K. Kapadia, and M. Fripp. 2004. Putting renewables to work: How many jobs can the clean energy industry generate? Report of the Renewable and Appropriate Energy Laboratory (RAEL). Berkeley: University of California, RAEL. http://rael.berkeley.edu/old-site/renewables.jobs.2006.pdf.

Karlstrom, S. 2010. Fort Collins, Colorado: 2010 Smarter City – Energy. N.d.: National Resources Defense Council, Smarter Cities Project. http://smartercities.nrdc.org/topic/energy/fort-collins-co-2010-smarter-city-energy.

Kates, R. W., W. R. Travis, and T. J. Wilbanks. 2012. Transformational adaptation when incremental adaptations to climate change are insufficient. *Proceedings of the National Academy of Sciences*, 109(19): 7156-7161, doi: 10.1073/pnas.1115521109.

Larson, E. K., N. B. Grimm, P. Gober, and C. L. Redman. 2005. The paradoxical ecology and management of water in the Phoenix, USA metropolitan area. *International Journal of Ecohydrology & Hydrobiology* 5:287–296.

Lempert, R. J., and D. G. Groves. 2010. Identifying and evaluating robust adaptive policy responses to climate change for water management agencies in the American West. *Technology Forecasting and Social Change* 77:960–974.

Lindenmayer, D. B., W. Steffen, A. A. Burbidge, L. Hughes, R. L. Kitching, W. Musgrave, M. Stafford Smith, and P. A. Werner. 2010. Conservation strategies in response to rapid climate change: Australia as a case study. *Biological Conservation* 143:1587–1593.

Lofman, D., M. Petersen, and A. Bower. 2002. Water, energy and environment nexus: The California experience. *International Journal of Water Resources Development* 18:73–85.

Loomis, J. 2002. *Integrated public lands management.* New York: Columbia University Press.

MacDonald, G. M. 2010. Water, climate change, and sustainability in the Southwest. *Proceedings of the National Academy of Sciences* 107:21256–21262.

McKinsey & Company (McKinsey). 2007. *Reducing US greenhouse gas emissions: How much at what cost?* U.S. Greenhouse Gas Abatement Mapping Initiative, Executive Report. http://www.mckinsey.com/Client_Service/Sustainability/Latest_thinking/Reducing_US_greenhouse_gas_emissions.

McMullen, C. P., and J. Jabbour, eds. 2009. *Climate change science compendium 2009*. N.p.: United Nations Environment Programme.

McNeeley, S. 2012. Examining barriers and opportunities for sustainable adaptation to climate change in Interior Alaska. *Climatic Change* 111:835–857.

Means, E., M. Laugier, J. Daw, L. Kaatz, and M. Waage. 2010. *Decision support planning methods: Incorporating climate change uncertainties into water planning*. WUCA White Paper. San Francisco: Water Utility Climate Alliance.

Measham, T., B. Preston, T. Smith, C. Brooke, R. Gorddard, G. Withycombe, and C. Morrison. 2011. Adapting to climate change through local municipal planning: Barriers and challenges. *Mitigation and Adaptation Strategies for Global Change* 16:889–909.

Milly, P., J. Betancourt, M. Falkenmark, R. M. Hirsch, Z. W. Kundzewicz, D. P. Lettenmaier, and R. J. Stouffer. 2008. Stationarity is dead: Whither water management? *Science* 319:573–574.

Moser, S. C. 2012. Adaptation, mitigation, and their disharmonious discontents: An essay. *Climatic Change* 111:165–175.

Moser, S. C., and J. A. Ekstrom. 2010. A framework to diagnose barriers to climate change adaptation. *Proceedings of the National Academy of Sciences* 107:22026-22031.

—. 2012. *Identifying and overcoming barriers to climate change adaptation in San Francisco Bay: Results from case studies*. White Paper CEC-500-2012-034. Sacramento: California Climate Change Center.

Nakićenović, N., and R. Swart, eds. 2000. *IPCC Special report on emissions scenarios: A special report of Working Group III of the Intergovernmental Panel on Climate Change*. Cambridge, UK: Cambridge University Press.

National Research Council (NRC). 2010a. *Adapting to the impacts of climate change*. Washington, DC: National Academies Press.

—. 2010b. *Advancing the science of climate change*. Washington, DC: National Academies Press.

—. 2010c. *Informing an effective response to climate change*. Washington, DC: National Academies Press.

—. 2010d. *Limiting the magnitude of future climate change*. Washington, DC: National Academies Press.

Nisbet, M. C. 2009. Communicating climate change: Why frames matter for public engagement. *Environment: Science and Policy for Sustainable Development* 51:12–23.

Nobriga, M. L., F. Feyrer, R. D. Baxter, and M. Chotkowski. 2005. Fish community ecology in an altered river delta: Spatial patterns in species composition, life history strategies, and biomass. *Estuaries* 28:776–785.

Nowak, D. J., D. E. Crane, and J. C. Stevens 2006. Air pollution removal by urban trees and shrubs in the United States. *Urban Forestry and Urban Greening* 4:115–123.

O'Brien, K. 2012. Global environmental change II: From adaptation to deliberate transformation. *Progress in Human Geography* 36:667–676.

Park, S., N. Marshall, E. Jakku, A. Dowd, S. Howden, E. Mendham, and A. Fleming. 2011. Informing adaptation responses to climate change through theories of transformation. *Global Environmental Change* 22:115–126.

Pataki, D. E., R. J. Alig, A. S. Fung, N. E. Golubiewski, C. A. Kennedy, E. G. McPherson, D. J. Nowak, R. V. Pouyat, and P. Romero Lankao. 2006. Urban ecosystems and the North American carbon cycle. *Global Change Biology* 12:2092–2102.

Peters, G. P., G. Marland, C. Le Quéré, T. Boden, J. G. Canadell, and M. R. Raupach. 2011. Rapid growth in CO_2 emissions after the 2008-2009 global financial crisis. *Nature Climate Change* 2:2–4, doi:10.1038/nclimate1332.

Pew Charitable Trusts. 2009. *The clean energy economy: Repowering jobs, businesses and investments across America*. http://www.pewcenteronthestates.org/uploadedFiles/Clean_Economy_Report_Web.pdf.

Philip Williams and Associates (PWA). 2010. *Preliminary study of the effect of sea level rise on the resources of the Hayward shoreline*. Report for the Hayward Area Shoreline Planning Agency. PWA REF. 1955.00. San Francisco: PWA.

Post, J. E., and B. W. Altman. 1994. Managing the environmental change process: Barriers and opportunities. *Journal of Organizational Change Management* 7:64–81.

Quay, R. 2010. Anticipatory governance: A tool for climate change adaptation. *Journal of the American Planning Association* 76:496–511.

Resource Media 2009. Water and climate change in the West: Polling and media analysis; March 2009. Carpe Diem—Western Water and Climate Change Project. San Francisco: Resource Media. http://www.carpediemwest.org/sites/carpediemwest.org/files/MediaAnalysis_0.pdf.

Roland-Holst, D. 2008. *Energy efficiency, innovation, and job creation in California*. Research Papers on Energy, Resources, and Economic Sustainability. Berkeley: University of California, Center for Energy, Resources, and Economic Sustainability (CERES). http://www.next10.org/energy-efficiency-innovation-and-job-creation-california.

Rosenzweig, C., S. Solecki, S. Hammer, and S. Mehrotra. 2010. Cities lead the way in climate-change action. *Nature* 467:909–911.

Ruhl, J. 2008. Climate change and the Endangered Species Act: Building bridges to the no-analog future. *Boston University Law Review* 88 (1); Florida State Univ. College of Law, Public Law Research Paper No. 275; FSU College of Law, Law and Economics Paper No. 07-18. http://papers.ssrn.com/sol3/papers.cfm?abstract_id=1014184.

Schultz, T. C., R. L. Shelby, and A. M. Agogino. 2010. Co-design of energy efficient housing with the Pinoleville-Pomo Nation. In *Proceedings of the ASME 2010 4th International Conference on Energy Sustainability (ES2010), May 17-22, 2010, Phoenix, Arizona, USA*, vol. 2, 925-934. Paper ES2010-90190. http://best.berkeley.edu/~aagogino/papers/ES2010-90190.pdf

Scott, C. A., and M. J. Pasqualetti. 2010. Energy and water resources scarcity: Critical infrastructure for growth and economic development in Arizona and Sonora. *Natural Resources Journal* 50:645–682.

Smith, J., Vogel, J., Carney, K. and C. Donovan. 2011. *Adaptation case studies in the western United States: Intersection of federal and state authority for conserving the greater sage grouse and the Colorado River water supply*. Washington, DC: Georgetown Climate Center.

Smith, M. S., L. Horrocks, A. Harvey, and C. Hamilton. 2011. Rethinking adaptation for a 4°C world. *Philosophical Transactions of the Royal Society A* 369:196–216.

Stokes, D. E. 1997. *Pasteur's quadrant: Basic science and technological innovation.* Washington, DC: Brookings Institution Press.

Stokes, J., and A. Horvath. 2006. Life cycle energy assessment of alternative water supply systems. *The International Journal of Life Cycle Assessment* 11:335–343.

Storbjörk, S. 2010. "It takes more to get a ship to change course": Barriers for organizational learning and local climate adaptation in Sweden. *Journal of Environmental Policy & Planning* 12:235–254.

Thorstensen, L., and S. Nourick. 2010. *Getting prepared: Economic development in a transforming energy economy*. Washington, DC: International Economic Development Council.

U.S. Bureau of Reclamation (Reclamation). 2005. Water 2025: Preventing crises and conflict in the West. http://permanent.access.gpo.gov/lps77383/Water%202025-08-05.pdf.

U.S. Department of Energy (DOE). 2006. *Energy demands on water resources: Report to Congress on the interdependency of energy and water*. Albuquerque: Sandia National Laboratories. http://www.sandia.gov/energy-water/congress_report.htm.

U.S. Environmental Protection Agency (EPA). 2012a. 2010 greenhouse gas emissions from large facilities. http://ghgdata.epa.gov/ghgp/main.do (data reported to EPA as of August 15, 2012).

—. 2012b. Greenhouse Gas Reporting Program 2010: Reported data. http://www.epa.gov/ghgreporting/ghgdata/reported/index.html (last updated October 4, 2012).

—. N.d. Climate change action plans: Utah. http://www.epa.gov/statelocalclimate/local/local-examples/action-plans.html#ut (accessed June 14, 2012).

Verbruggen, A., M. Fischedick, W. Moomaw, T. Weir, A. Nadaï, L. J. Nilsson, J. Nyboer, and J. Sathaye. 2009. Renewable energy costs, potentials, barriers: Conceptual issues. *Energy Policy* 38:850–861.

Victor, D. G., M. G. Morgan, F. Apt, and J. Steinbruner, J. 2009. The geoengineering option: A last resort against global warming. *Foreign Affairs* 88:64–76.

Waterfall, P. H. 2006. *Harvesting rainwater for landscape use*. Tucson: University of Arizona, Cooperative Extension Service. http://ag.arizona.edu/pubs/water/az1052/harvest.html.

Wei, M., S. Patadia, and D. M. Kammen. 2010. Putting renewables and energy efficiency to work: How many jobs can the clean energy industry generate in the US? *Energy Policy* 38:919–931.

Westley, F., P. Olsson, C. Folke, T. Homer-Dixon, H. Vredenburg, D. Loorbach, J. Thompson, et al. 2011. Tipping toward sustainability: Emerging pathways of transformation. *AMBIO: A Journal of the Human Environment* 40:762–780.

Wheeler, S. M. 2000. Planning for metropolitan sustainability. *Journal of Planning Education and Research* 20:133–145.

Wiek, A., F. Farioli, K. Fukushi, and M. Yarime. 2012. Sustainability science: Bridging the gap between science and society. *Sustainability Science* 7 (Suppl. 1): 1–4.

Wilkinson, C. F. 1992. *Crossing the next meridian: Land, water, and the future of the West*. Washington, DC: Island Press.

Willson, R. W. and K. D. Brown. 2008. Carbon neutrality at the local level: Achievable goal or fantasy? *Journal of the American Planning Association* 74:497–504.

Worster, D. 1992. *Rivers of empire: Water, aridity, and the growth of the American West*. Oxford: Oxford University Press.

Zimmerman, R., and C. Faris. 2011. Climate change mitigation and adaptation in North American cities. *Current Opinion in Environmental Sustainability* 3:181–187.

Zweibel, K., J. Mason, and V. Fthenakis. 2008. A solar grand plan. *Scientific American* 298:64–73.

Endnotes

i Observed global emissions have accelerated from an increase of 1.1% per year in the 1990s to 3.5% per year from 2000–2007 (see McMullen and Jabbour 2009). The global recession produced only a slight drop in emissions in 2009 with the overall trend now upward again (Friedlingstein et al. 2010; Peters et al. 2011).

ii Definition adapted from NRC 2010a.

iii See http://www.icleiusa.org/about-iclei/members/member-list.

iv The concept of "use-inspired" basic research was originally introduced by Stokes (1997); it refers to research that seeks basic understanding while considering social needs and potential usefulness.

v The company website is at http://www.fcx.com.

Chapter 19

Moving Forward with Imperfect Information

COORDINATING LEAD AUTHOR

Kristen Averyt (University of Colorado Boulder)

LEAD AUTHORS

Levi D. Brekke (Bureau of Reclamation), David E. Busch (U.S. Geological Survey)

CONTRIBUTING AUTHORS

Laurna Kaatz (Denver Water), Leigh Welling (National Park Service), Eric H. Hartge (Stanford University)

REVIEW EDITOR

Tom Iseman (Western Governors' Association)

Executive Summary

This chapter summarizes the scope of what is known and not known about climate in the Southwestern United States. There is now more evidence and more agreement among climate scientists about the physical climate and related impacts in the Southwest compared with that represented in the 2009 National Climate Assessment (Karl, Melillo, and Peterson 2009). However, there remain uncertainties about the climate system, the complexities within climate models, the related impacts to the biophysical environment, and the use of climate information in decision making.

Chapter citation: Averyt, K., L. D. Brekke, D. E. Busch, L. Kaatz, L. Welling, and E. H. Hartge. 2013. "Moving Forward with Imperfect Information." In *Assessment of Climate Change in the Southwest United States: A Report Prepared for the National Climate Assessment*, edited by G. Garfin, A. Jardine, R. Merideth, M. Black, and S. LeRoy, 436–461. A report by the Southwest Climate Alliance. Washington, DC: Island Press.

Uncertainty is introduced in each step of the climate planning-and-response process—in the scenarios used to drive the climate models, the information used to construct the models, and the interpretation and use of the models' data for planning and decision making (Figure 19.1).

There are several key challenges, drawn from recommendations of the authors of this report, that contribute to these uncertainties in the Southwest:

- There is a dearth of climate observations at high elevations and on the lands of Native nations.
- There is limited understanding of the influence of climate change on natural variability (e.g., El Niño–Southern Oscillation, Pacific Decadal Oscillation), extreme events (droughts, floods), and the marine layer along coastal California.
- Climate models, downscaling, and resulting projections of the physical climate are imperfect. Representing the influence of the diverse topography of the Southwest on regional climate is a particular challenge.
- The impacts of climate change on key components of the natural ecosystems (including species and terrestrial ecosystems) are ill-defined.
- The adaptive capacity of decision-making entities and legal systems to handle climate impacts is unclear. This creates a challenge for identifying vulnerabilities to climate in the Southwest.
- Regulation, legislation, and political and social responses to climate all play important roles in our ability to adapt to climate impacts and mitigate greenhouse gas (GHG) emissions.
- Climate change is one of multiple stresses affecting the physical, biological, social, and economic systems of the Southwest, with population growth (and its related resource consumption, pollution, and land-use changes) being particularly important.

19.1 Introduction

Climate assessments illustrate how natural resources and managed systems might fare under a variety of climatic and socioeconomic scenarios. Assessments take advantage of the best data and modeling tools and follow scientifically approved methodologies to develop projections of climate impacts to physical, biological, social, and economic systems associated with possible climate futures. Such climate projections are important to the success of adaptive measures (Millner 2012). This assessment of the climate of the Southwest takes a risk-based approach. The intention is to provide the decision-making public with information about the costs and benefits to society associated with different emissions scenarios. Although uncertain, scenarios can help identify risks and appraise our ability as a society to adapt to climate change. Science will never eliminate uncertainty. Even concepts as seemingly simple as gravity are subject to uncertainties in a scientific context. Scientists cannot eliminate uncertainties about climate and related risks. Nonetheless, climate observations and projections can provide useful information. For this reason, characterizing what is known and what is not known about the past, current, and future climate and related impacts is necessary to help decision makers identify appropriate mitigation strategies and adaptive measures.

This chapter summarizes the scope of knowledge and uncertainty about climate in the Southwest. Throughout this assessment, each chapter has outlined key findings about our regional climate. Included with each key finding is a statement of "confidence," i.e., a statement intended to convey the degree of knowledge based on evaluation of available data and scientific interpretations in the literature (Box 19.1). This chapter outlines the uncertainties that collectively present challenges in using climate information to inform decisions. It also highlights cases in the Southwest where climate information—imperfect as it may be—is successfully being incorporated into planning and management. Drawing upon these examples and on the literature pertaining to decision making under uncertainty, this chapter offers steps for moving forward with imperfect information.

19.2 Uncertainty Typologies

The "uncertainty continuum" in Figure 19.1 outlines the process through which the impacts of climate change are projected and indicates numerous points at which uncertainties are introduced. These include everything from the scenarios used to drive models, the information used to construct climate models, and the interpretation and use of the models' data for planning and decision making. Discussed here are three types of uncertainty that can impact climate change: scenario uncertainties, model uncertainties, and communication uncertainties.

Scenario uncertainties

POPULATION, TECHNOLOGY, PRODUCTION, CONSUMPTION AND GREENHOUSE GAS EMISSIONS. Population growth and economic trends are the critical components driving greenhouse gas (GHG) emissions. The scenarios that feed into climate models represent different combinations of assumptions about population change and economic conditions, and show their related trends in greenhouse gas emissions. As described in Chapters 2 and 6, the high-emissions (A2) and low-emissions (B1) scenarios used in this assessment are from the IPCC Special Report on Emissions Scenarios (SRES; Nakićenović and Swart 2000). Emissions scenarios illustrate a suite of possibilities to aid in planning, but they are not perfect. For example, none of the SRES trajectories developed in 2001 presented a scenario that captured the global economic downturn in 2008. The SRES trajectories also did not include the entire suite of social, economic, policy, and regulatory responses that affect adaptive response and ability to mitigate emissions (Hawkins and Sutton 2009). As climate projections move further into the future, particularly beyond the fifty-year mark, accurately capturing population trends, economic trends, and technological advances becomes more difficult. There is no broadly accepted method for quantifying the uncertainties associated with future emissions.

Model uncertainties

ATMOSPHERIC CONCENTRATIONS, RADIATIVE FORCING, TEMPERATURE CHANGE. General circulation models (GCMs, often called global climate models) integrate the components of climate based on observations (Hawkins and Sutton 2009,

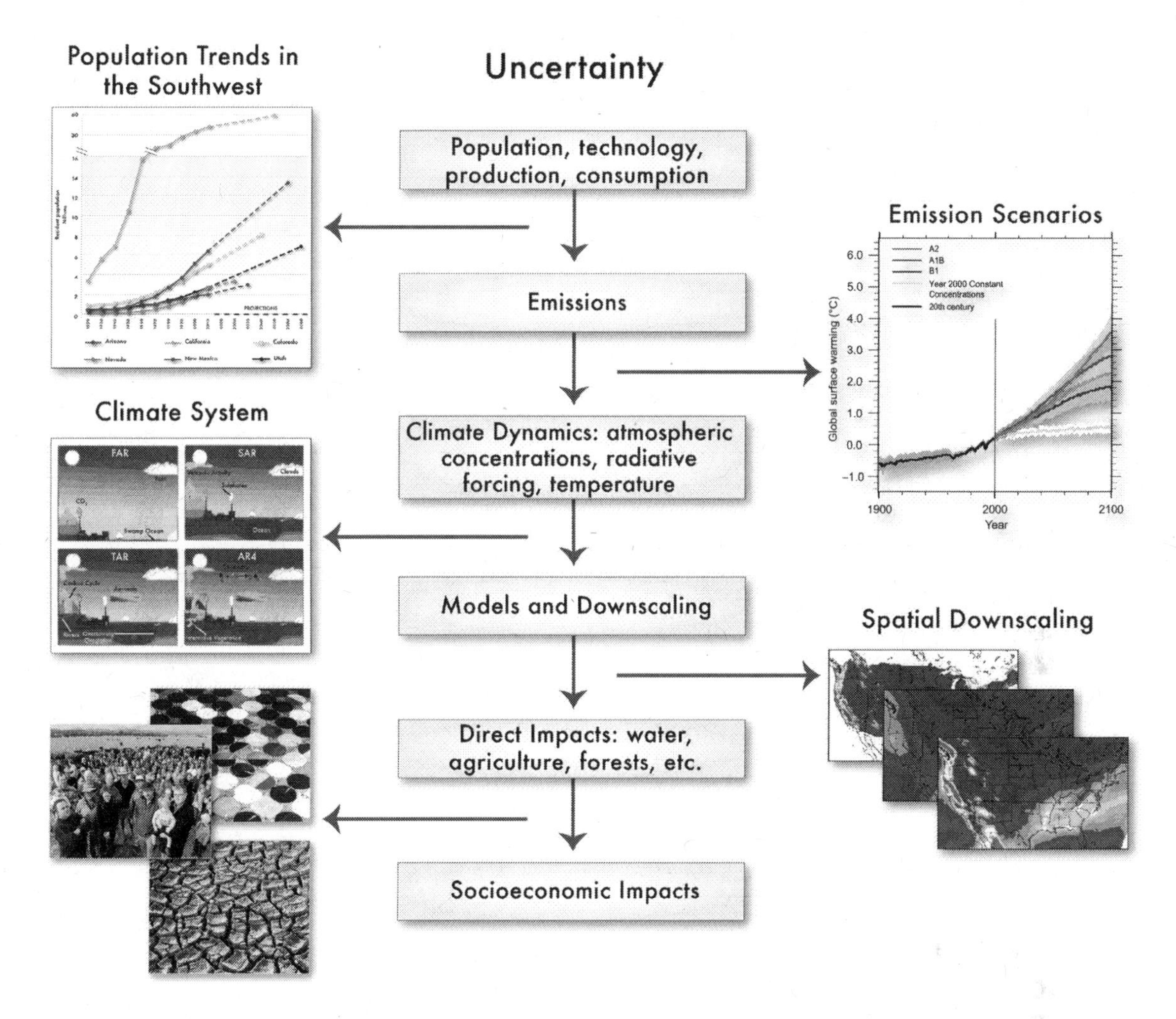

Figure 19.1 Working with uncertainty. Continuum of uncertainties, knowledge gaps and challenges related to projecting future climate changes and their impacts, and assessing vulnerabilities to future changes. See Tables 19.1 and 19.2 for syntheses of knowledge and uncertainties identified by authors of this assessment report. Adapted from Pidgeon and Fischhoff (2011).

2011). Although numerous emissions paths are represented in the GHG scenarios, they do not precisely translate into changes in radiative forcing (i.e., changes in the balance of radiated energy), which can warm or cool the climate system.

Observational data is a key research need that feeds into these uncertainties. Fewer observations make it difficult for scientists to tease out the information they need to accurately represent climate dynamics. In the Southwest, there are minimal climatic and meteorological observations for much of the region, especially at high elevations and on tribal lands—thus impeding our understanding of regional climate processes.

Model uncertainty can also be attributed to factors affecting climate that have yet to be identified (Risbey and O'Kane 2011). Consider the role of aerosols in moderating climate. Prior to 2003, the role of these particulates in the atmosphere and in regulating climate was unknown, and so they were not represented in GCMs. They were an "unknown uncertainty" discovered through scientific inquiry to be important components, even though considerable uncertainty remains about their precise influence on climate processes (IPCC 2007). This raises an important concept: discovering new parts of a climate system may add to the body of climate knowledge while introducing additional uncertainties (Trenberth 2010; Pidgeon and Fischhoff 2011).

GCMs have been shown to exhibit biases when trying to simulate historical climate. These biases vary locally, from wet to dry or warm to cool, and vary seasonally. Assessments adjust for these biases, but the approaches used to identify and correct them can vary. Bias correction can even affect projected climate trends and subsequently the impacts projected to occur to natural and managed systems (Pierce et al. 2012).

GCMs have a proven ability to simulate the influence of increased greenhouse gas emissions on global and continental temperature trends (IPCC 2007), demonstrating that climate models are doing pretty well at capturing the dynamics of the climate system despite the aforementioned uncertainties. However, climate models are less successful in simulating observations at smaller geographic scales.

DOWNSCALING. Because adaptation measures are often most successful at a regional level, global climate output from GCMs must be translated into regional terms to aid decision making. A key problem in applying global data to regional scales is that at smaller scales the internal (natural) variability in the climate system has a greater influence than climate change. As an example, in the mid-latitudes—which encompass the Southwest—this natural variability is especially pronounced and is greater than observed and projected precipitation signals (Hawkins and Sutton 2009).

Translating global climate data into regional information can be accomplished through the process of *downscaling*. Simply, downscaling merges large-scale climate information from GCMs with local physical controls (such as mountain ranges, deserts, water bodies, or large urban areas) on climate. The two methods of downscaling are statistical and dynamical, and both have different strengths and weaknesses (Fowler, Blenkinsop and Tebaldi 2007). Statistical downscaling relates the GCM temperature and precipitation output to the observed small-scale variability in a given grid cell. These techniques are computationally efficient and permit downscaling of many global climate projections at a given location, but assume that the relationship between large-scale circulation and local surface climate does not change through time, even as the large-scale climate changes. Dynamical downscaling uses regional climate models (RCMs) to simulate small-scale processes, and resolve data at a higher spatial resolution. The downside is that these techniques require significant computing power. Thus, the choice of which downscaling method to use in developing regional projections involves tradeoffs between model output that is meaningful for local impact assessment and yet can still be performed in a mathematically efficient manner, given computational limitations. (See further discussion of downscaling in Chapter 6, Section 6.1).

In the present assessment, different downscaling methods are referenced in different chapters. Thus, understanding the tradeoffs and inherent uncertainties associated with each technique, as they apply to the Southwest, is important. For example, while the Rocky Mountains reach elevations over 14,000 feet and play an important role in influencing regional and local climatology, in GCMs (such as the NCAR Community Climate System Model 3.0[i]), the elevation of the mountains is represented as about 8,000

Box 19.1

Treatment of Uncertainty in the Southwest Assessment Report

Critical questions or problems related to climate change are included in this report as "key findings." For each key finding, the scientific team evaluated the body of scientific information and described the type of information used, the standards of evidence applied (noting the amount, quality, and consistency of evidence), the uncertainty associated with any results, and the degree of confidence in the outcome. This process constitutes a "traceable account" of the authors' reasoning and evidence. The uncertainty and confidence associated with each finding is an important component in assessing risk.

For findings that identify outcomes with potential high consequences (see guidance on risk-based framing in Chapter 2), uncertainty is estimated probabilistically. Probabilities are expressed as the likelihood that a particular outcome could occur under a given condition or scenario. Likelihoods are based on quantitative methods—such as model results or statistical sampling—or on expert judgment. In some cases, authors used standardized ranges:

Qualitative Language	Quantitative Language
More than a 9 in 10 chance	Greater than 95% likely
More than a 6 in 10 chance	Greater than 66% likely
About a 5 in 10 chance	Between 33% and 66% likely
Less than a 4 in 10 chance	Less than 33% likely
Almost no chance	Less than 5% likely

Wherever possible, the authors used quantitative estimates and describe consequential outliers that may fall outside a statistical confidence interval of 90% (which increases the reliability of a dataset).

The authors also assessed the degree of confidence (high, medium-high, medium, medium-low, or low) by considering the quality of the evidence and the level of agreement among experts with relevant knowledge and experience (Mastrandrea et al. 2010; Mastrandrea et al. 2011). Confidence is a subjective judgment, but it is based on systematic, transparent evaluation of the type, amount, quality, and consistency of evidence, and the degree of agreement among experts.

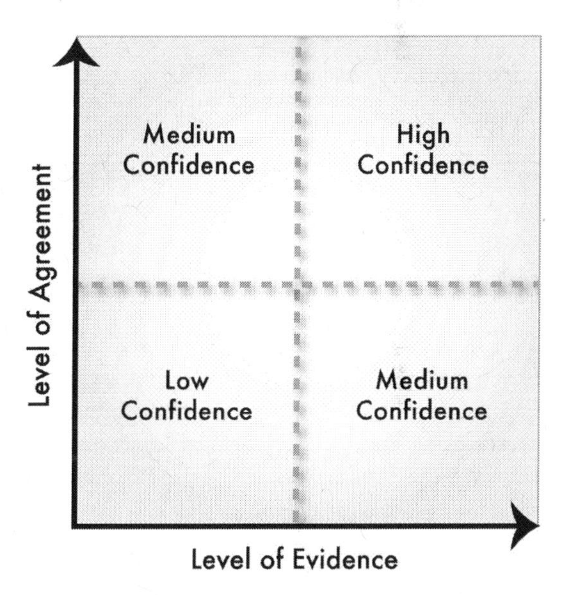

Figure 19.2 Summary evaluation of confidence, in terms of levels of evidence and agreement of the evidence. Adapted from Mastrandrea et al. (2011).

feet. In regional climate models (such as the Weather Research and Forecasting Model, or WRF[ii]) the mountains are represented as over 10,000 feet. The difference is because the topography must be simplified for global models and because of different model resolutions.[iii] Although the mountains are better represented in the RCMs, their higher resolution requires more intensive computational resources, which, in a practical sense, means that the RCMs are only able to utilize the inputs from a subset of the twenty-two available GCMs. Clearly, more data would be gained by using a larger suite (number) of GCMs, yet GCMs alone cannot account adequately for the important role of topography in the Intermountain West. The GCMs used in the IPCC's Fourth Assessment Report have a weak but systematic bias for overestimating the speed of upper-level westerly winds near 30°N and November-to-April precipitation in the Southwest. Of relevance is that the wettest models project the greatest drying in this region with climate change. As it turns out, all of these models have "subdued" topography that may contribute to the zonal wind bias and may also underestimate rain shadow effects, producing wet biases on the lee side of the mountains (McAfee, Russell, and Goodman 2011). Thus, in this case, the tradeoff between statistical and dynamical downscaling involves either a greater range of potential futures (which is valuable in planning and risk-based management) or potentially more accurate representation of climate.

DIRECT IMPACTS. Regional climate projections from downscaling are in turn used to drive other models of the physical environment. In the Southwest, water is a critical component of climate. Therefore, assessments typically must translate future climate projections into impacts on the region's hydrologic processes (such as precipitation, snowmelt runoff, streamflow, infiltration, groundwater recharge and discharge, evapotranspiration, and so on). Simulation models are often used for this task, with most of the effort spent characterizing future weather conditions that are consistent with climate projections. Those weather conditions are then used to simulate hydrologic processes. The hydrologic model itself is typically developed and verified under historical climate and watershed conditions. Uncertainty in projecting hydrologic processes arises from how the hydrologic model is structured, the way future weather over the watershed is characterized (which often requires some blending of historical weather observations and projected changes in climate), and assumptions about other features of a watershed that might change as climate changes and affects runoff. (See also the discussion presented in Chapter 10, especially in Section 10.3 and in "Planning Techniques and Stationarity" in Section 10.5.)

Despite limitations associated with such hydrologic models, outputs from these models are most influenced by the choice of GCM used to provide input, followed by the type of downscaling method used, then by the hydrologic model chosen (Wilby and Harris 2006; Crosbie, McCallum and Walker 2011). This suggests that GCMs and the level of understanding of large-scale processes are the largest source of uncertainties in the model uncertainty typology continuum discussed earlier. Given that outputs based on the averaging of results of numerous models are better than those based on the results of an individual model (Reichler and Kim 2008), impact studies that are informed by multiple global climate models will have a greater certainty than those based on a single global model.

Box 19.2

Case Study 1: Denver Water: Addressing Climate Change through Scenario Planning

Denver Water serves a growing population of customers and prepares long-range plans for meeting future water needs. Historical streamflow and weather records plus paleohydrologic data have been key information in projecting future water supply and demand conditions. Climate change fundamentally challenges the concept that the weather and hydrologic patterns of the past are the best representation of future conditions (Milly et al. 2008). But, there is a lot of uncertainty about how the climate will change. In addition to climate, other key uncertainties in long-range water planning include possible economic, regulatory, social, and demographic changes. Denver Water now uses scenario-planning techniques to try to prepare for these future uncertainties.

The "cone of uncertainty" (Figure 19.3) illustrates the growing uncertainty of future conditions over time. Scenarios are created to try to represent a plausible range of future conditions. Plans are created to meet each scenario, and common near-term strategies across plans are identified. "Decision points" note when strategy diverges from the common path. The goal is to take actions today that prepare for a range of future conditions. Maintaining flexibility and adaptability as well as identifying and preserving options are key elements in successfully preparing for future uncertainties such as climate change.

As a first step in climate change adaptation, Denver Water is testing the implications of a simple 5°F (3°C) temperature increase. Initial results show major supply losses and demand increases. Additional climate change conditions will be evaluated in an effort to develop a robust adaptation plan.

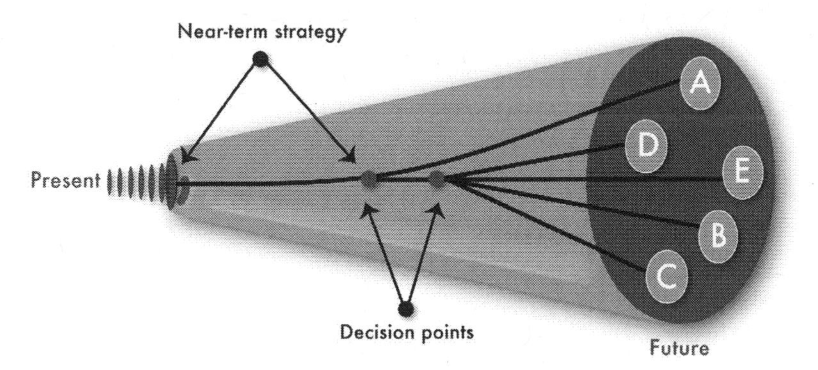

Figure 19.3 Cone of uncertainty used in Denver Water Scenario Planning Initiative. Uncertainties, due to knowledge or communication gaps or imperfect information increase as time progresses from present to future. The increase in uncertainties related to scientific understanding of the distant future (around 100 years hence), has prompted many resource managers and planners to consider multiple scenarios of the future, which can be evaluated at key decision points in the near or medium term (roughly 10-50 years into the future). Adapted from Waage and Kaatz (2011).

Box 19.3

Case Study 2: The National Park Service—Exploring Climate Futures and Decision Making in the Mojave Desert

Resource management decisions must be based on future expectations. However, in an era of rapid climate change, the future will be characterized by highly consequential and unprecedented changes that cannot be fully predicted. In February 2011, the National Park Service (NPS) convened a workshop to explore scenario planning as an approach for science-based decision making in the face of uncertainty for Southwestern parks and conservation areas.

Since 2007, the National Park Service has worked with other federal, state, and academic partners to develop a user-driven approach to build scenarios as a long-range planning tool for incorporating climate change into a range of NPS management processes and documents. The purpose is to better acquaint decision makers with climate complexity and uncertainty, evaluate management options, and ultimately implement effective, science-based decisions. The approach requires participation and transparency, and is structured in a way that encourages end-user input and ownership throughout the process. In addition to including climate-change information, the NPS scenario development process explores other external factors that define a park's operational environment, such as leadership and public values.

The February 2011 training workshop included scientists from the University of Arizona and other academic and governmental organizations, along with managers from the National Park Service, Bureau of Land Management, and Bureau of Reclamation. Participants explored how climate change could impact arid lands in the desert Southwest, using the Mojave Desert as a case study. Impacting factors that were considered to be uncertain but consequential included changes in precipitation, frequency of extreme storm events, extreme temperature events, duration and frequency of droughts, as well as societal concerns about these issues and leadership's capacity to implement adaptive measures. From these biophysical and sociopolitical drivers, participants created four plausible futures (scenarios) to test management and public response. Discussions centered on multiple pressures converging in the Southwest: public expectations for services such as water and renewable energy development, along with habitat connectivity (the interconnection of different habitats to allow species movement) and ecosystem resiliency as climate change forces species to move and adapt. Consensus emerged that future desert conservation efforts should be collaborations that are broad-based, landscape-scale, and multi-jurisdictional.

SOCIO-ECONOMIC IMPACTS. In a risk-based framework (planning based on the pros and cons of a given set of possibilities), decision makers are interested in the socio-economic impacts associated with different scenarios. However, socio-economic impacts encompass the entire sum of uncertainties in each step along the climate continuum (Figure 19.1). These impacts are also represented as being constant, whereas in reality, regulatory, institutional, and legislative policies change over time. In essence, decision making and the capacity to act are key elements of the uncertainty associated with socio-economic impact projections.

Communication uncertainties

COGNITIVE BARRIERS. The various uncertainties outlined above set up a number of analytic uncertainties and ultimately different interpretations about the results. Even if our understanding of climate science were 100% certain, science does not exist in a vacuum. Societal and individual perspectives are all molded by experiences and this affects the production of scientific information and its use to make decisions.

For example, climate scientists may choose from many different climate scenarios and models and tend to exhibit overconfidence in their results (see CCSP 2009). On the other hand, most people are psychologically distant from the concept of climate change. Not only must one sort through pervasive images of penguins and polar bears to rationally consider the problem, but the timeline for the onset of tangible impacts tends to be beyond most people's lifetimes. The decision-making public also often has many other interests—such as economic vitality, public health, and safety—that may have a higher value than concerns about climate change. Taken together, these factors can hinder the incorporation of climate information in planning and management.

The complexity of the connections and feedbacks in the climate system make bridging this gap difficult but not impossible. As examples, the nonlinear relationship between GHG emissions and atmospheric concentrations, or the reasons why a single winter storm does not invalidate the scientific perception that the global climate is warming, can be conveyed and understood through effective communication and mental models (Sterman 2008). Whether improved climate education will change perceptions about the utility of climate information is unclear (see, for example, Boykoff 2011; McCright 2011), but there are indications that improving understanding of the climate and the uncertainties inherent in climate projections may facilitate the inclusion of climate information in planning and management (Pidgeon and Fischhoff 2011).

19.3 Confidence and Uncertainty

Scientists use a variety of tactics to express scientific uncertainty. In general, people are familiar with probabilities and odds, which quantify the likelihood of an outcome. But uncertainty is more nuanced in an assessment where a large body of work is being represented. Unfortunately, the labels "likely" and "unlikely" to indicate the probability of occurrence of an event are interpreted very differently by different people and therefore do not always effectively communicate risk (see CCSP 2009). Recognizing this, in 2001 the IPCC implemented uncertainty guidelines for the use of such language into its assessment process. The intention of the guidelines is to convey the amount of evidence (uncertainty) and degree of consensus (confidence) about climate information (Moss and Schneider 2000). These uncertainty standards were modified slightly for the IPCC's Fourth Assessment Report (Manning et al. 2004; IPCC 2007). The 2000 U.S. National Climate Assessment adopted similar uncertainty standards and language to the IPCC (National Assessment Science Team 2001); the uncertainty language was altered again for the U.S. Global Change Research Program (USGCRP) synthesis and assessment products (CCSP 2009; Karl, Melillo and Peterson 2009). The IPCC has once again revamped its approach to uncertainty for its Fifth Assessment Report (Mastrandrea et al. 2010; Mastrandrea et al. 2011). The labeling conventions for uncertainty used in this report are modified from the current IPCC guidelines and outlined in Box 19.1.

Box 19.4

Case Study 3: Planning in the San Francisco Bay Using Sea-Level Rise Projections

The San Francisco Bay Conservation and Development Commission (SFBCDC), created in 1965 by the state of California, is "dedicated to the protection and enhancement of San Francisco Bay and to the encouragement of the Bay's responsible use." In an effort to update twenty-two-year-old sea-level data in the San Francisco Bay Plan, the SFBCDC commissioned a report to reevaluate sea-level-rise projections and its impact to the bay. The report concluded that sea level in the bay could rise 10 to 17 inches (26 to 43 cm) by 2050, 17 to 32 inches (43 to 81 cm) by 2070, and 31 to 69 inches (78 to 176 cm) by the end of the century (San Francisco Bay Conservation and Development Commission 2011). In October 2011, the SFBCDC approved these findings and incorporated the information into policies in the San Francisco Bay Plan, including future project designs, shoreline plans, and permit approvals. This new section details the impacts of climate change and, in particular, addresses issues regarding adaptation to sea-level rise. Policies in the plan specifically related to construction along vulnerable shorelines were changed to both promote habitat restoration and encourage building only in suitable regions of the bay.

19.4 What Is Known and Not Known About Climate in the Southwest

With few exceptions, there is now more evidence and more agreement among climate scientists about the physical climate and related impacts in the Southwest than there was in the 2009 National Climate Assessment (Karl, Melillo, and Peterson 2009) (Table 19.1). The body of research about processes affecting both global and regional climate is growing, as are some observational datasets, allowing for the detection of trends. Uncertainty and confidence about climate fluctuates with the ebb and flow of new data. Sometimes as scientists learn more, they become more confident in findings. This is particularly true of studies that rely on observational data. For example, the long and continuous time series of streamflow data has allowed scientists to document the early onset of the peak spring season pulse of streamflow in the region. On the other hand, additional data and new observations can sometimes muddy the works, drawing previously held conclusions into question. As scientists learn more about the climate system and the factors that naturally impact it, other parameters about which scientists know relatively little can factor more prominently in discussions of uncertainties in predicting future changes.

The synthesis of the evolution of knowledge regarding climate changes and their impacts in the Southwest (Table 19.1) is drawn from the judgment of the authors of this assessement report. Statements included in Table 19.1 were quoted from the Southwest section of the 2009 National Climate Assessment (Karl, Melillo, and Peterson 2009). The authors of this chapter made no attempt to correct or update the statements extracted from the 2009 National Climate Assessment. For each statement from the 2009 National Climate Assessment, the author team of this report identified the relative change in level of agreement among scientists about the statement, and changes in the level of evidence available to evaluate the statements. The table can be used as a coarse baseline for evaluating the evolution of knowledge since the 2009 National Climate Assessment.

Table 19.1 Evolution of knowledge about climate in the Southwest

2009 Southwest Assessment	This Assessment							
	← Agreement →					← Evidence →		
	Much Less	Less	Same	More	Much More	Same	More	Much More
Human-induced climate change appears to be well underway in the Southwest. Recent warming is among the most rapid in the nation, significantly more than the global average in some areas.				X			X	
Projected declines in spring snowpack and Colorado River flow					X			X
Projections suggest continued strong warming					X			X
Projected summertime temperature increases are greater than the annual average increases in some parts of the region, and are likely to be exacerbated locally by expanding urban heat island effects					X			X
Further water cycle changes are projected, which, combined with increasing temperatures, signal a serious water supply challenge in the decades and centuries ahead.					X			X
Water supplies are projected to become increasingly scarce, calling for trade-offs among competing uses, and potentially leading to conflict.			X				X	
Water supplies in some areas of the Southwest are already becoming limited, and this trend toward scarcity is likely to be a harbinger of future water shortages.			X			X		
Limitations imposed on water supply by projected temperature increases are likely to be made worse by substantial reductions in rain and snowfall in the spring months, when precipitation is most needed to fill reservoirs to meet summer demand.			X			X		
Increased likelihood of water-related conflicts between sectors, states, and even nations			X			X		
Increasing temperature, drought, wildfire, and invasive species will accelerate transformation of the landscape.				X			X	

Table 19.1 Evolution of knowledge about climate in the Southwest (Continued)

2009 Southwest Assessment	This Assessment							
	← Agreement →					← Evidence →		
	Much Less	Less	Same	More	Much More	Same	More	Much More
Competing demands from [Native] treaty rights, rapid development, and changes in agriculture in the region, exacerbated by years of drought and climate change, have the potential to spark significant conflict over an already over-allocated and dwindling [water] resource.			X			X		
Climate change already appears to be influencing both natural and managed ecosystems of the Southwest.				X			X	
Future landscape impacts are likely to be substantial, threatening biodiversity, protected areas, and ranching and agricultural lands.				X			X	
Record wildfires are also being driven by rising temperatures and related reductions in spring snowpack and soil moisture.			X			X		
How climate change will affect fire in the Southwest varies according to location. In general, total area burned is projected to increase.			X			X		
Fires in wetter, forested areas are expected to increase in frequency, while areas where fire is limited by the availability of fine fuels experience decreases			X			X		
Climate changes could also create subtle shifts in fire behavior, allowing more "runaway fires"—fires that are thought to have been brought under control, but then rekindle.			X			X		
The magnitude of fire damages, in terms of economic impacts as well as direct endangerment, also increases as urban development increasingly impinges on forested areas.					X			X
Increasing temperatures and shifting precipitation patterns will drive declines in high-elevation ecosystems such as alpine forests and tundra.			X			X		
As temperatures rise, some iconic landscapes of the Southwest will be greatly altered as species shift their ranges northward and upward to cooler climates, and fires attack unaccustomed ecosystems which lack natural defenses.		X				X		
Increased frequency and altered timing of flooding will increase risks to people, ecosystems, and infrastructure.			X			X		

Table 19.1 Evolution of knowledge about climate in the Southwest (Continued)

2009 Southwest Assessment	This Assessment: ← Agreement →					← Evidence →		
	Much Less	Less	Same	More	Much More	Same	More	Much More
Some species will move uphill, others northward, breaking up present-day ecosystems; those species moving southward to higher elevations might cut off future migration options as temperatures continue to increase.		X				X		
Potential for successful plant and animal adaptation to coming change is further hampered by existing regional threats such as human-caused fragmentation of the landscape, invasive species, river-flow reductions, and pollution.			X			X		
A warmer atmosphere and an intensified water cycle are likely to mean not only a greater likelihood of drought for the Southwest, but also an increased risk of flooding.			X			X		
More frequent dry winters suggest an increased risk of these [water] systems running short of water.			X			X		
A greater potential for flooding also means reservoirs cannot be filled to capacity as safely in years where that is possible. Flooding also causes reservoirs to fill with sediment at a faster rate, thus reducing their water-storage capacities.			X			X		
Rapid landscape transformation due to vegetation die-off and wildfire as well as loss of wetlands along rivers is also likely to reduce flood-buffering capacity.			X			X		
Increased flood risk in the Southwest is likely to result from a combination of decreased snow cover on the lower slopes of high mountains, and an increased fraction of winter precipitation falling as rain and therefore running off more rapidly.					X			X
Increase in rain on snow events will also result in rapid runoff and flooding.			X			X		
Impact of more frequent flooding is a greater risk to human beings and their infrastructure. This applies to locations along major rivers, but also to much broader and highly vulnerable areas such as the Sacramento–San Joaquin River Delta system.				X			X	
Projected changes in the timing and amount of river flow, particularly in winter and spring, is estimated to more than double the risk of Delta flooding events by mid-century, and result in an eight-fold increase before the end of the century.				X			X	

Table 19.1 Evolution of knowledge about climate in the Southwest (Continued)

2009 Southwest Assessment	This Assessment							
	← Agreement →					← Evidence →		
	Much Less	Less	Same	More	Much More	Same	More	Much More
Efforts are underway to identify and implement adaptation strategies aimed at reducing these risks [to the Delta and Suisun Marsh].			X			X		
Unique tourism and recreation opportunities are likely to suffer.			X			X		
Increasing temperatures will affect important winter activities such as downhill and cross-country skiing, snowshoeing, and snowmobiling, which require snow on the ground.				X			X	
Projections indicate later snow and less snow coverage in ski resort areas, particularly those at lower elevations and in the southern part of the region.				X			X	
Decreases from 40% to almost 90% are likely in end-of-season snowpack under a higher emissions scenario in counties with major ski resorts.			X			X		
Earlier wet snow avalanches—more than six weeks earlier by the end of this century under a higher emissions scenario—could force ski areas to shut down affected runs before the season would otherwise end.			X			X		
Ecosystem degradation will affect the quality of the experience for hikers, bikers, birders, and others.				X			X	
Water sports that depend on the flows of rivers and sufficient water in lakes and reservoirs are already being affected, and much larger changes are expected.			X			X		
Agriculture faces increasing risks from a changing climate.				X			X	
Urban areas are also sensitive to temperature-related impacts on air quality, electricity demand, and the health of their inhabitants.				X		X		
The magnitude of projected temperature increases for the Southwest, particularly when combined with urban heat island effects for major cities such as Phoenix, Albuquerque, Las Vegas, and many California cities, represent significant stresses to health, electricity, and water supply in a region that already experiences very high summer temperatures.				X		X		

Table 19.1 Evolution of knowledge about climate in the Southwest (Continued)

2009 Southwest Assessment	This Assessment							
	← Agreement →					← Evidence →		
	Much Less	Less	Same	More	Much More	Same	More	Much More
Rising temperatures also imply declining air quality in urban areas such as those in California which already experience some of the worst air quality in the nation.				X		X		
With more intense, longer-lasting heat wave events projected to occur over this century, demands for air conditioning are expected to deplete electricity supplies, increasing risks of brownouts and blackouts.				X			X	
Electricity supplies will also be affected by changes in the timing of river flows and where hydroelectric systems have limited storage capacity and reservoirs.				X			X	
Agriculture will experience detrimental impacts in a warmer future, particularly specialty crops in California such as apricots, almonds, artichokes, figs, kiwis, olives, and walnuts.				X			X	
Accumulated winter chilling hours have already decreased across central California and its coastal valleys. This trend is projected to continue to the point where chilling thresholds for many key crops would no longer be met.				X			X	
California's losses due to future climate change are estimated between 0% and 40% for wine and table grapes, almonds, oranges, walnuts, and avocadoes, varying significantly by location.			X			X		
Adaptation strategies for agriculture in California include more efficient irrigation, which has the potential to help compensate for climate-driven increases in water demand for agriculture due to rising temperatures.	X						X	
Adaptation strategies for agriculture in California include shifts in cropping patterns, which have the potential to help compensate for climate-driven increases in water demand for agriculture due to rising temperatures.				X			X	

Note: To construct this table, the authors of this chapter quoted statements from the Southwest section of 2009 National Climate Assessment (Karl, Melillo and Peterson 2009). For each statement, the authors of this report identified the relative change in level of agreement among scientists about the statement, and changes in the pertinent level of evidence, based on the current assessment of climate in the Southwest.

Table 19.2 presents an assessment of knowledge gaps and scientific challenges related to improving the understanding of physical and biological processes, impacts, vulnerabilities and societal responses to climate change. The authors of this report identified knowledge gaps and uncertainties, and the authors of this chapter evaluated and classified the information into key challenges. In each key challenge area, the knowledge gaps are divided into the three categories of uncertainty, as follows: model uncertainties (those related to understanding and modeling physical and biological processes and phenomena), scenario uncertainties (those related to identifying vulnerabilities, mitigation and adaptation choices), and communication uncertainties (those related to the effective exchange of knowledge between scientists and decision makers). Table 19.2 can be used as a coarse baseline for understanding sources of uncertainty related to climate and adaptation science challenges, and to inform future research priorities.

Table 19.2 Knowledge gaps and key challenges to improving understanding, reducing uncertainty, identifying and addressing vulnerabilities to climate changes in the Southwest

	Model Uncertainty				**Scenario Uncertainty**				
Knowledge Gaps Contributing to Key Challenges	Observational Data	Understanding Physical Climate Dynamics	Climate Models & Downscaling	Physical Climate Impacts to Biological & Human Systems	Social & Behavioral Factors	Economic Factors	Policy & Regulatory Factors	Communication & Education	**Chapter(s)**
KEY CHALLENGE: There is a dearth of climate observations at high elevations and on tribal lands in the Southwest.									
Changes in weather and climate observations, variability, and trends across mountain gradients and at variable elevations, including representation of topography in climate models	X	X	X						Present Weather and Climate: Average Conditions (4) Present Weather and Climate: Evolving Conditions (5) Water: Impacts, Risks, and Adaptation (10) Coastal Issues(9)
Weather and climate observations, variability, and trends on tribal lands	X								Present Weather and Climate: Average Conditions (4) Unique Challenges Facing Southwestern Tribes (17)

Table 19.2 Knowledge gaps and key challenges to improving understanding, reducing uncertainty, identifying and addressing vulnerabilities to climate changes in the Southwest (Continued)

	Model Uncertainty				Scenario Uncertainty				
Knowledge Gaps Contributing to Key Challenges	Observational Data	Understanding Physical Climate Dynamics	Climate Models & Downscaling	Physical Climate Impacts to Biological & Human Systems	Social & Behavioral Factors	Economic Factors	Policy & Regulatory Factors	Communication & Education	**Chapter(s)**
Measurements of precipitation amount and type	X								Present Weather and Climate: Average Conditions (4) Present Weather and Climate: Evolving Conditions (5) Future Climate: Projected Average (6)
KEY CHALLENGE: There is limited understanding of the influence of climate change on natural variability (e.g. ENSO, PDO), extreme events (droughts, floods), and the marine layer along coastal California.									
Ability to connect climate change and extreme events		X							Human Health (15)
Understanding of physical processes such as atmospheric convection, evapotranspiration, snow pack formation, and runoff production		X							Present Weather and Climate: Evolving Conditions (5)
Connections between modes of natural variability (ENSO and PDO) and climate change; including effect on SW Monsoon		X							Future Climate: Projected Extremes (7) Future Climate: Projected Average (6) Water: Impacts, Risks, and Adaptation (10)
Occurrence of compound high-impact extremes such as drought and heat waves		X							Future Climate: Projected Extremes (7)
Understanding of marine layer processes		X							Coastal Issues (9)

Table 19.2 Knowledge gaps and key challenges to improving understanding, reducing uncertainty, identifying and addressing vulnerabilities to climate changes in the Southwest (Continued)

	Model Uncertainty				Scenario Uncertainty				
Knowledge Gaps Contributing to Key Challenges	Observational Data	Understanding Physical Climate Dynamics	Climate Models & Downscaling	Physical Climate Impacts to Biological & Human Systems	Social & Behavioral Factors	Economic Factors	Policy & Regulatory Factors	Communication & Education	**Chapter(s)**
KEY CHALLENGE: Climate models, downscaling and resulting projections of the physical climate are imperfect. Representing the influence of the diverse topography of the Southwest on regional climate is a particular challenge.									
Downscaling methodologies and inconsistencies			X					X	Water: Impacts, Risks, and Adaptation (10)
Reproducibility of extreme high-frequency precipitation events by climate models			X						Future Climate: Projected Extremes (7)
KEY CHALLENGE: The impacts of climate change on key components of the natural ecosystem (including species and land regimes) are ill constrained.									
Links between impacts and climate change				X					Unique Challenges Facing Southwestern Tribes (17)
Impacts to tribal lands and societies				X					Unique Challenges Facing Southwestern Tribes (17)
Relationship between climate and distributions of species				X					Natural Ecosystems (8)
Connections between climate and disease systems				X					Human Health (15)
Response of individual species to changes in climate				X					Natural Ecosystems (8)
Extent to which individuals in different populations or species can observably change physical characteristics in response to climate				X					Natural Ecosystems (8)

Table 19.2 Knowledge gaps and key challenges to improving understanding, reducing uncertainty, identifying and addressing vulnerabilities to climate changes in the Southwest (Continued)

	Model Uncertainty				Scenario Uncertainty				
Knowledge Gaps Contributing to Key Challenges	Observational Data	Understanding Physical Climate Dynamics	Climate Models & Downscaling	Physical Climate Impacts to Biological & Human Systems	Social & Behavioral Factors	Economic Factors	Policy & Regulatory Factors	Communication & Education	**Chapter(s)**
Range of potential rates of evolution of individual populations or species				X					Natural Ecosystems (8)
Extent to which phenological events among species that interact will become asynchronous				X					Natural Ecosystems (8)
Effect of climate change on "dryland" production -- primarily dryland grain production in Colorado and Utah and forage production throughout the Southwest				X					Agriculture and Ranching (11)
Ecosystem responses (e.g., sensitivity, adaptive capacity) as water types (e.g. snow v. rain), water quantities, water quality, and water management practices change				X			X		Natural Ecosystems (8) Climate Change and U.S.-Mexico Border Communities (16)
KEY CHALLENGE: The adaptive capacity of decision-making entities and legal doctrines to handle climate impacts is unclear. This creates a challenge for identifying vulnerabilities to climate in the Southwest.									
Ability of the transportation system to manage large disruptions					X	X			Transportation (14)
Sensitivity and adaptive capacity of border communities to climate change impacts				X	X				Climate Change and U.S.-Mexico Border Communities (16)
Sensitivity and adaptive capacity of border agriculture and ranching sector to a range of stressors					X				Climate Change and U.S.-Mexico Border Communities (16)
Capacity of water infrastructure to address changes					X		X		Water: Impacts, Risks, and Adaptation (10)

Table 19.2 Knowledge gaps and key challenges to improving understanding, reducing uncertainty, identifying and addressing vulnerabilities to climate changes in the Southwest (Continued)

	Model Uncertainty				Scenario Uncertainty				
Knowledge Gaps Contributing to Key Challenges	Observational Data	Understanding Physical Climate Dynamics	Climate Models & Downscaling	Physical Climate Impacts to Biological & Human Systems	Social & Behavioral Factors	Economic Factors	Policy & Regulatory Factors	Communication & Education	**Chapter(s)**
Economic status of urban public works departments and ability to reduce flood risk						X			Urban Areas (13)
Fiscal capacity of cities to respond rapidly and effectively to climate change challenge						X			Urban Areas (13) Transportation (14)
Regulatory capacity to address climate adaptation and mitigation							X		Coastal Issues (9)
Capacity and flexibility of water and land regulations, agreements and legislation to accommodate climate adaptation and planning							X		Water: Impacts, Risks, and Adaptation (10) Agriculture and Ranching (11) Unique Challenges Facing Southwestern Tribes (17)
Financial risk to property						X			Coastal Issues (9)

KEY CHALLENGE: Regulation, legislation, political and social responses to climate all play an important role in our ability to adapt to climate impacts and mitigate greenhouse gas emissions.

Knowledge Gaps Contributing to Key Challenges	Observational Data	Understanding Physical Climate Dynamics	Climate Models & Downscaling	Physical Climate Impacts to Biological & Human Systems	Social & Behavioral Factors	Economic Factors	Policy & Regulatory Factors	Communication & Education	Chapter(s)
How the current and future fleet of power plants will evolve, particularly with respect to utilized fuel type and impacts on GHG emissions					X	X	X		Energy: Supply, Demand, and Impacts (12) Transportation (14)
The type and intensity of fuels used in the transportation sector and impacts on GHG emissions					X	X	X		Energy: Supply, Demand, and Impacts (12) Transportation (14)
Social and political responses to climate change; including market incentives					X	X	X	X	Coastal Issues (9)
Communication between planners and academics					X			X	Coastal Issues (9)

Table 19.2 Knowledge gaps and key challenges to improving understanding, reducing uncertainty, identifying and addressing vulnerabilities to climate changes in the Southwest (Continued)

	Model Uncertainty				Scenario Uncertainty				
Knowledge Gaps Contributing to Key Challenges	Observational Data	Understanding Physical Climate Dynamics	Climate Models & Downscaling	Physical Climate Impacts to Biological & Human Systems	Social & Behavioral Factors	Economic Factors	Policy & Regulatory Factors	Communication & Education	**Chapter(s)**
Extent of upper-level and/or grass roots leadership to effect change					X		X		Urban Areas (13)
Socio-economic and political conditions						X	X		Unique Challenges Facing Southwestern Tribes (17)
City-scale decisions about adaptation and regulatory frameworks					X		X		Urban Areas (13)
Environmental and economic impacts of extensive water transfers and effect on agriculture						X	X		Agriculture and Ranching (11)
Agricultural and environmental policies							X		Agriculture and Ranching (11)
Effect of water availability (physical and legal) on agriculture output				X			X		Agriculture and Ranching (11)
National policies related to air quality standards							X		Human Health (15)
Understanding of how adaptation to climate change develops and functions is limited, as is the role played by institutions in promoting effective adaptation							X	X	Climate Change and U.S.-Mexico Border Communities (16)
KEY CHALLENGE: Climate change is a multi-stressor problem, and many factors are at play. In the Southwest, population growth is particularly important.									
Future demand for energy; including temporal and spatial shifts					X	X	X		Energy: Supply, Demand, and Impacts (12)
Age distribution in the population					X				Transportation (14)

Table 19.2 Knowledge gaps and key challenges to improving understanding, reducing uncertainty, identifying and addressing vulnerabilities to climate changes in the Southwest (Continued)

	Model Uncertainty				Scenario Uncertainty				
Knowledge Gaps Contributing to Key Challenges	Observational Data	Understanding Physical Climate Dynamics	Climate Models & Downscaling	Physical Climate Impacts to Biological & Human Systems	Social & Behavioral Factors	Economic Factors	Policy & Regulatory Factors	Communication & Education	**Chapter(s)**
Global and U.S. economic outlook						X			Transportation (14)
Global and U.S. manufacturing and industrial patterns						X			Transportation (14)
The extent to which heat-related morbidity and mortality are a multi-stressor problem						X			Human Health (15)

Note: To construct this table, the authors of each chapter in this report identified key knowledge gaps and uncertainties. For Chapters 3–8, authors, outlined the major elements needed to improve confidence in observed and projected climate trends. For Chapter 9–18, author teams identified factors and knowledge gaps that need to be addressed in order to improve the ability of the respective sector to identify vulnerabilities and/or adaptive responses. The author team for this chapter identified Key Challenges based on common themes in the compilation of inputs from different chapters.

19.5 Moving Forward

Climate projections can provide information for understanding risks associated with physical, biological, and social impacts. Although model projections are imperfect given the uncertainties outlined above, entities in the Southwest are moving forward and using innovative strategies to incorporate climate information in their planning and management schemes.[v] Both public and private planners are employing strategies that run the gamut from iterative risk management frameworks (which adapt management strategies to new information and changing circumstances) to resilience strategies (which enhance the capacity to withstand and recover from emergencies and disasters) to approaches that optimize for a particular desired set of conditions (NRC 2011). Case studies from the Southwest are highlighted throughout this chapter.

Box 19.5

Case Study 4: Transmission Planning in the Western States

Resource management decisions must be based Western governors have long identified clean, diverse, and reliable energy as a regional and national priority. But access to transmission lines is a significant impediment to increasing renewable resources as a portion of the overall energy portfolio. There is also a broad recognition of the need to consider water, land use, and wildlife when planning and developing energy supplies in the West. To address these issues, the Western Governors' Association and the Western States Water Council are collaborating with the Department of Energy and the National Laboratories on the Regional Transmission Expansion Project (RTEP). A major focus of the project is to seek generation and transmission options that are compatible with reliable water supplies and healthy wildlife communities in the West (Iseman and Schroder 2012).

Electricity generation and reliability of the grid are dependent on availability of water resources. Most of the power generated in the West requires water, and in order to move electricity to population centers, transmission lines need to be sited near these power plants. Even low-carbon electricity portfolios require additional water supplies. Consequently, the reliability of the grid and electricity supplies depends on the availability of water.

To make better decisions on energy and water, risks associated with a variable water supply must be considered. Drought has always been a fact of life in the arid West. Thus, considering drought in this planning effort is prudent in order to minimize risks to both the grid and water supplies. However, projections of drought in the short term (less than fifty years) are uncertain. Long-term climate projections indicate there will potentially be more severe drought events; but in the short term, natural variability trumps climate change.

To address risks posed by drought, the RTEP team is using past droughts to test the vulnerability to dry conditions of proposed transmission systems. These droughts are not necessarily those recorded in the observation records, but rather paleodroughts that occurred up to 1,000 years ago, as evidenced by tree rings in the region.

References

Boykoff, M. 2011. *Who speaks for the climate? Making sense of media reporting on climate change.* Port Melbourne, Australia: Cambridge University Press.

Crosbie, R. S., J. L. McCallum, and G. R. Walker. 2011. *The impact of climate change on dryland diffuse groundwater recharge in the Murray-Darling Basin.* Waterlines Report No. 40. Canberra, Australia: National Water Commission.

Fowler, H. J., S. Blenkinsop, and C. Tebaldi. 2007. Linking climate change modelling to impacts studies: Recent advances in downscaling techniques for hydrological modelling. *International Journal of Climatology* 27:1547–1578.

Hawkins, E., and R. Sutton. 2009. The potential to narrow uncertainty in regional climate predictions. *Bulletin of the American Meteorological Society* 90:1095–1107, doi: 10.1175/2009BAMS2607.1.

—. 2011. The potential to narrow uncertainty in projections of regional precipitation change. *Climate Dynamics* 37:407–418.

Intergovernmental Panel on Climate Change (IPCC). 2007. *Climate change 2007: The physical science basis. Contribution of Working Group I to the Fourth Assessment Report of the Intergovernmental Panel on Climate Change*, ed. S. Solomon, D. Qin, M. Manning, Z. Chen, M. Marquis, K.B. Averyt, M. Tignor and H.L. Miller. Cambridge: Cambridge University Press.

Iseman, T., and A. Schroder. 2012. Integrated planning: Transmission, generation and water in the western states. In *The water-energy nexus in the American West*, ed. D. Kenney and R. Wilkinson, chapter 15. Williston, VT: Edward Elgar.

Karl, T. R., J. M. Melillo, and T. C. Peterson, eds. 2009. *Global climate change impacts in the United States*. Cambridge: Cambridge University Press. http://downloads.globalchange.gov/usimpacts/pdfs/climate-impacts-report.pdf.

Manning, M., M. Petit, D. Easterling, J. Murphy, A. Patwardhan, H-H. Rogner, R. Swart, and G. Yohe, eds. 2004. *IPCC workshop on describing scientific uncertainties in climate change to support analysis of risk and of options, National University of Ireland, Maynooth, Co. Kildare, Ireland, 11-13 May, 2004: Workshop report*. Geneva: Intergovernmental Panel on Climate Change (IPCC).

Mastrandrea, M. D., C. B. Field, T. F. Stocker, O. Edenhofer, K. L. Ebi, D. J. Frame, H. Held, et al. 2010. *Guidance note for lead authors of the IPCC Fifth Assessment Report on Consistent Treatment of Uncertainties*. Geneva: Intergovernmental Panel on Climate Change (IPCC). http://www.ipcc-wg2.gov/meetings/CGCs/Uncertainties-GN_IPCCbrochure_lo.pdf .

Mastrandrea, M. D., K. J. Mach, G-K. Plattner, O. Edenhofer, T. F. Stocker, C. B. Field, K. L. Ebi, and P. R. Matschloss. 2011. The IPCC AR5 guidance note on consistent treatment of uncertainties: A common approach across the working groups. *Climatic Change* 108:675–691.

McAfee, S. A., J. L. Russell, and P. J. Goodman. 2011. Evaluating IPCC AR4 cool-season precipitation simulations and projections for impacts assessment over North America. *Climate Dynamics* 37:2271–2287.

McCright, A. M. 2011. Political orientation moderates Americans' beliefs and concern about climate change. *Climatic Change* 104:243–253.

Millner, A. 2012. Climate prediction for adaptation: Who needs what? *Climatic Change* 110:143–167.

Milly, P. C. D., J. Betancourt, M. Falkenmark, R. M. Hirsch, Z. W. Kundzewicz, D. P. Lettenmaier, and R. J. Stouffer. 2008. Stationarity is dead: Whither water management? *Science* 319:573–574.

Moss, R. H., and S. H. Schneider. 2000. Uncertainties in the IPCC TAR: Recommendations to lead authors for more consistent assessment and reporting. In *Guidance papers on the cross cutting issues of the Third Assessment Report of the IPCC*, ed. R. Pachauri, T. Taniguchi, and K. Tanaka, 33–51. Geneva: World Meteorological Organization. http://www.ipcc.ch/pdf/supporting-material/guidance-papers-3rd-assessment.pdf.

Nakićenović, N., and R. Swart, eds. 2000. *Report on emissions scenarios: A special report of Working Group III of the Intergovernmental Panel on Climate Change*. Cambridge: Cambridge University Press. http://www.grida.no/climate/ipcc/emission/index.htm.

National Assessment Science Team. 2001. *Climate change impacts on the United States: The potential consequences of climate variability and change.* Report for the US Global Change Research Program. Cambridge: Cambridge University Press. http://downloads.globalchange.gov/nca/nca-2000-foundation-report.pdf.

National Research Council (NRC). 2011. *America's climate choices.* Washington, DC: National Academies Press.

Pidgeon, N., and B. Fischhoff. 2011. The role of social and decision sciences in communicating uncertain climate risks. *Nature Climate Change* 1:35–41.

Pierce, D. W., T. Das, D. R. Cayan, E. P. Maurer, N. L. Miller, Y. Bao, M. Kanamitsu, et al. 2012. Probabilistic estimates of future changes in California temperature and precipitation using statistical and dynamical downscaling. *Climate Dynamics*, published online, doi: 10.1007/s00382-012-1337-9.

Reichler, T., and J. Kim. 2008. Uncertainties in the climate mean state of global observations, reanalyses, and the GFDL climate model. *Journal of Geophysical Research* 113: D05106.

Risbey, J. S., and T. J. O'Kane. 2011. Sources of knowledge and ignorance in climate research. *Climatic Change* 108:755–773.

San Francisco Bay Conservation and Development Commission (SFBCDC). 2011. *Living with a rising bay: Vulnerability and adaptation in San Francisco Bay and on its shoreline.* San Francisco: SFBCDC. http://www.bcdc.ca.gov/BPA/LivingWithRisingBay.pdf.

Sterman, J. 2008. Risk communication on climate: Mental models and mass balance. *Science* 322:532–533.

Trenberth, K. 2010. More knowledge, less certainty. *Nature* 4:20–21.

U.S. Climate Change Science Program (CCSP). 2009. *Best practice approaches for characterizing, communicating, and incorporating scientific uncertainty in climate decisionmaking*, ed. M.G. Morgan, H. Dowlatabadi, M. Henrion, D. Keith, R. Lempert, S. McBride, M. Small, and T. Wilbanks. Synthesis and Assessment Product 5.2 Report by the U.S. Climate Change Science Program and the Subcommittee on Global Change Research. Washington, DC: Global Change Research Information Office. http://downloads.globalchange.gov/sap/sap5-2/sap5-2-final-report-all.pdf.

Waage, M. D., and L. Kaatz. 2011. Nonstationary water planning: An overview of several promising planning methods. *Journal of the American Water Resources Association* 47:535–540.

Wilby, R. L., and I. Harris. 2006. A framework for assessing uncertainties in climate change impacts: Low-flow scenarios for the River Thames, UK. *Water Resources Research* 42: W02419.

Endnotes

i See http://www.cesm.ucar.edu/models/ccsm3.0.

ii See http://wrf-model.org.

iii Grid boxes are 100 miles on each side in the GCM, compared with 30 miles square in the RCM (with more than a ten-fold increase in resolution).

iv See San Francisco Bay Plan, http://www.bcdc.ca.gov/laws_plans/plans/sfbay_plan. Since its original adoption in 1968, the plan has been amended as warranted by new data, including in October 2011, as explained in the text.

v Climate projections based on scenarios of future emissions are inherently uncertain. Climate models were initially built as experiments intended to facilitate understanding of the physical processes driving climate systems—not to predict specific, optimal outcomes. Rather, projections emerging from climate models can provide suites of potential futures. At this point, even significant investment in computational models may not significantly increase the certainty of climate projections. However, despite their uncertainties, climate model outputs are being incorporated into decision making processes in different sectors, at different geographic scales, across the Southwestern US. Simply, uncertainty related to future climate (whether physical, biological, or regulatory) is not impeding the use of climate information in decision making.

Chapter 20

Research Strategies for Addressing Uncertainties

COORDINATING LEAD AUTHOR

David E. Busch (U.S. Geological Survey)

LEAD AUTHORS

Levi D. Brekke (Bureau of Reclamation), Kristen Averyt (University of Colorado, Boulder), Angela Jardine (University of Arizona)

CONTRIBUTING AUTHOR

Leigh Welling (National Park Service)

REVIEW EDITORS

Karl Ford (The Bureau of Land Management, Retired), Gregg Garfin (University of Arizona)

Executive Summary

There is an immense volume of information pertaining to research needs for addressing climate change uncertainties and resolving key information gaps. Fortunately, multiple independent efforts to establish research *priorities* have yielded similar results. Input on research needs is being used to craft national scientific priorities and strategies that are being implemented regionally by agencies and organizations. A number of regionally based efforts are already underway to aggregate and synthesize climate-related management needs and research priorities. Landscape Conservation Cooperatives and Climate

Chapter citation: Busch, D. E., L. D. Brekke, K. Averyt, A. Jardine, and L. Welling. 2013. "Research Strategies for Addressing Uncertainties." In *Assessment of Climate Change in the Southwest United States: A Report Prepared for the National Climate Assessment*, edited by G. Garfin, A. Jardine, R. Merideth, M. Black, and S. LeRoy, 462–482. A report by the Southwest Climate Alliance. Washington, DC: Island Press.

Science Centers, funded by the Department of the Interior, are conducting strategic syntheses of common resource-management priorities and related science needs across the Southwest, and many of these priorities and needs are related to climate variability and change.

The present Assessment includes many examples of the types of research that are needed to address key climate science uncertainties. The Assessment also includes examples of information needs related to understanding climate effects on systems (human, biophysical, ecosystems, and others) in the Southwest. Implementation of research strategies will increase understanding and improve the ability of the scientific community to anticipate the direction or magnitude of future climate-related change in these systems. The assembly of experts for the other chapters of this document provided a unique opportunity to draw upon the authors' collective expertise to share knowledge about priority research strategies. The peer-reviewed information sources assessed in this chapter highlight research strategies and priorities established by the research community. Other sources cover priorities that are based primarily on management and policy needs. In the latter type, the sources represent the consensus of senior leaders of organizations, generally with substantial input and advice from the organizations' technical and scientific experts.

20.1 Introduction

This chapter examines research strategies that aim to reduce uncertainty associated with climate drivers and their effects on systems in the Southwest. It also identifies scientific approaches that are being considered for implementation in programs of adaptive responses to climate change. This chapter was written collaboratively with Chapter 19, which outlines some of the most important uncertainties related to climate variability and change in the Southwestern United States. In these chapters the uncertainty derives from both our presently imperfect capability to model climate and other earth systems and from our inability to adequately characterize social, economic, policy and regulatory responses in the form of adaptation and mitigation.

Sources of information on research strategies infrequently utilize a risk-based perspective and usually do not incorporate a formal statistical definition of risk (Raiffa and Schlaiffer 2000). Although in this Assessment we utilize some sources that have undergone scientific peer review, others are heavily influenced by policy. In such cases, the products (which are often unpublished papers) represent the consensus of senior leaders of agencies and organizations (including inter-organizational and inter-agency collaboratives) who generally have received substantial input and advice from the organizations' and agencies' technical and scientific experts. In policy-influenced products, formal evaluations of confidence employing levels of evidence and agreement (e.g., Mastrandrea et al. 2010) are typically not undertaken or reported. Instead, the level of confidence expressed in these collaborative technical-policy products represents the judgment of high-level decision makers regarding alternative management approaches to adapt effectively to climate change. Thus, this chapter offers an opportunity for the findings of the *Assessment of Climate Change in the Southwest United States* that are associated with lower confidence or higher uncertainty (e.g., Moss and Yohe 2011) to be prioritized in planning future climate-effects research.

20.2 Developing Research Strategies from Information Needs

Raising confidence in research findings by increasing evidence and consensus (Chapter 19; Mastrandrea et al. 2010) often involves iterative and circuitous pathways that can affect both the amount of evidence and level of agreement, but may only infrequently address both together. Because formal determinations about confidence (see Figure 20.1) are rarely undertaken, uncertainty is generally only implicitly considered in determining information needs for climate-effects research. Nonetheless there are many fine examples linking information needs to research strategies. At a national scale, the U.S. Forest Service has used research needs to craft a scientific strategy and implementation plan to organize its climate-effects research along three themes: ecosystem sustainability (climate-change adaptation), carbon sequestration (climate-change mitigation), and decision support (USFS 2010). In the Southwest, the Forest Service Research and Development program is using these broad themes to implement its climate-change research program through its Pacific Southwest and Rocky Mountain Forest and Range research stations.

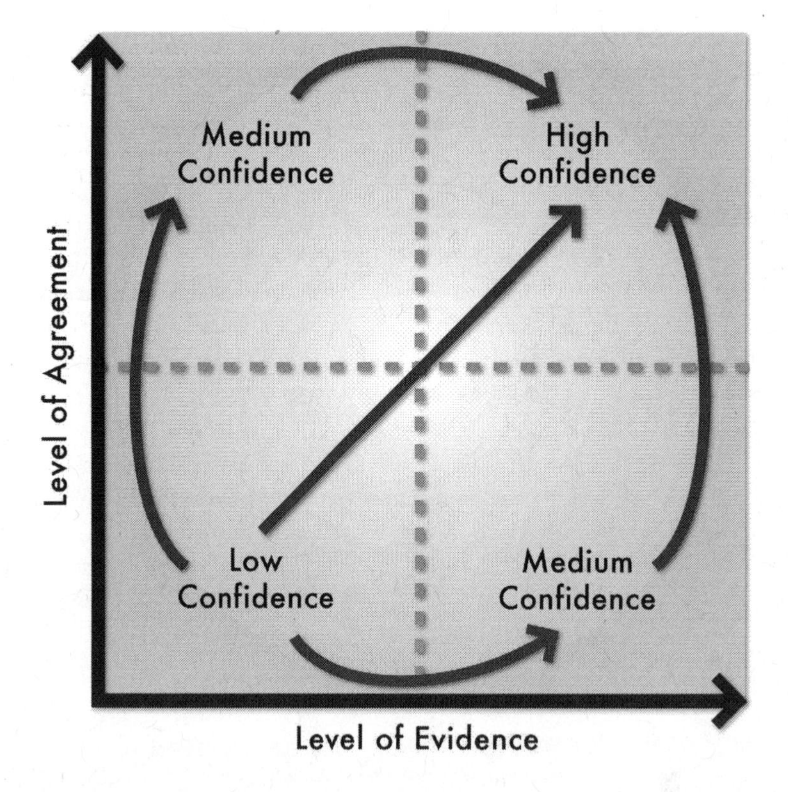

Figure 20.1 Research strategies for addressing uncertainties. Trajectories show how the level of evidence presented in climate effects research and agreement about the research, independently and collectively generate greater confidence in the research findings. Source: Mastrandrea et al. (2010), Mastrandrea et al. (2011).

Other national information needs and research priorities related to climate change have been developed (Lucier et al. 2006; DOE 2010; IWGCCH 2010; NRC 2010) along with syntheses of natural resource management-related gaps in scientific information pertaining to various regional resources and lands (California Coastal Commission 2008; BLM 2011; Brekke et al. 2011). However, climate-oriented scientific needs assessments are currently at an embryonic stage of development for the Southwest as a distinct region.

Within the Southwest, a number of efforts have been aggregating and synthesizing management needs and research priorities. For example, the Great Basin Research and Management Partnership[i] has been particularly active in developing momentum toward collaborative management and research (Chambers, Devoe and Evendon 2008). Climate variation is among the drivers of change considered in its work.

The Regional Integrated Sciences and Assessments (RISA) program, funded by the Climate Program Office at the National Oceanic and Atmospheric Administration (NOAA), was designed in part to enable NOAA to work with constituents to further its mission of climate science, monitoring, and data management. One of the Southwestern RISAs, the Western Water Assessment, has identified and characterized key people, projects, and documents related to climate in the state of Utah to provide guidance for research needs there.[ii]

The Department of the Interior's development of a nation-wide system of Regional Climate Science Centers (CSCs) represents a new approach to evaluate needs for scientific information about climate influences on natural and human resources at the regional scale, and to address such needs through collaborations of university and government research institutions (Salazar 2009). CSCs such as the Southwest Climate Science Center[iii] work with stakeholders throughout the Southwest to identify key scientific needs at the regional scale. Partner institutions such as NOAA's RISA program and the Forest Service Research and Development program will be important collaborators for the CSCs. The CSCs also are working with management-oriented inter-organizational groups, notably the Landscape Conservation Cooperatives (discussed below), which are being developed concurrently with the CSCs to help address the impacts of climate change on the nation's natural and cultural resources (Figure 20.2).

A number of factors can improve the dialogue about climate-effects research findings with those who are implementing climate mitigation and adaptation programs. Chief among these factors are:

- communication networks, science translation, and capacity for ongoing assessment;
- elimination of possible duplication and insufficient coordination of efforts among federal, state, and local agencies;
- improved access to climate change data and information; and
- improved understanding of the impact of laws and regulations on adaptation policy and implementation.

Research products will have more impact if such translational factors are considered as an essential part of strategy development rather than as an ancillary component. As a changing climate necessitates novel demands on decision processes (NRC 2009), the process of strategy development can be optimized if institutions charged with making decisions about climate adaptation and mitigation are involved early on. Pilot efforts by the National Park Service to test scenario planning related to climate futures (Box 19.3) are an example of one approach that is proactively integrating managers' perspectives.

20.3 Research Strategies Derived from the Southwest Climate Assessment

The assembly of experts for the *Assessment of Climate Change in the Southwest United States* presented a unique opportunity to draw upon the authors' collective expertise to outline uncertainty (see Chapter 19) and evaluate research-strategy priorities from a scientific perspective. A summary of strategies to address gaps in knowledge and data, monitoring needs, and modeling and other deficiencies that are outlined throughout this Assessment are presented in Table 20.1.

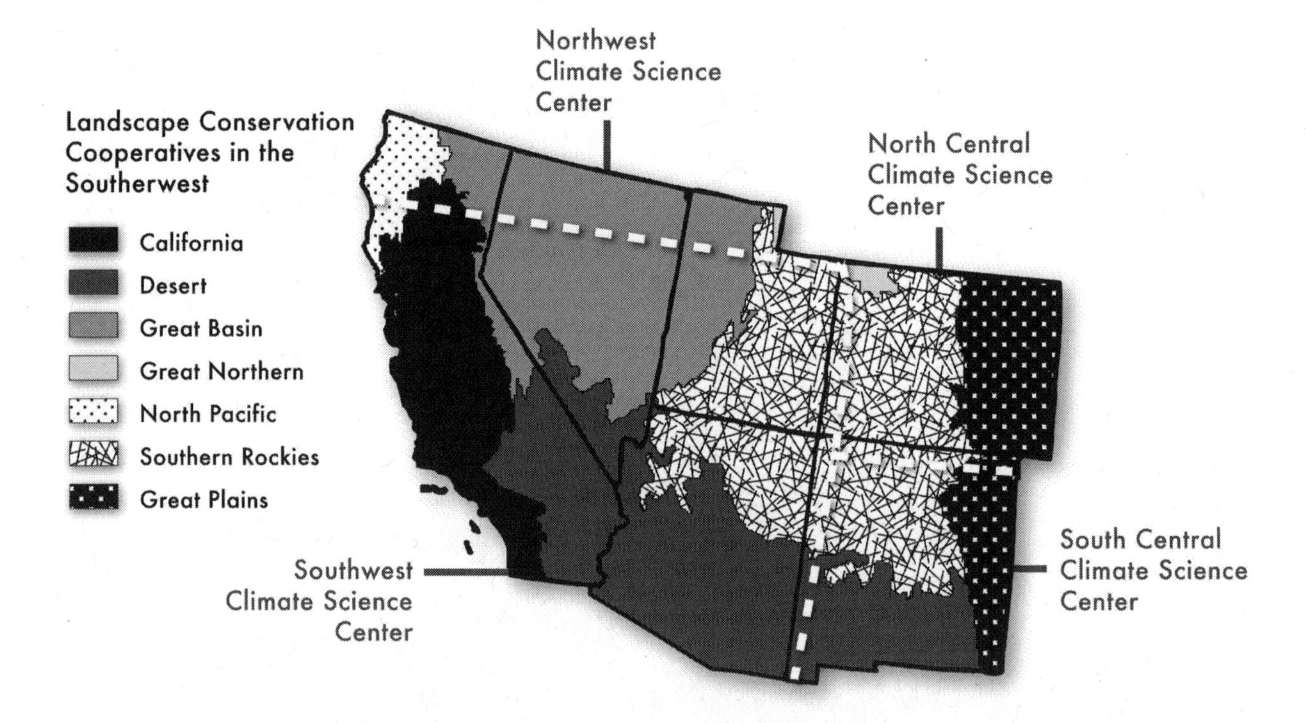

Figure 20.2 Federal climate-science and landscape-management initiatives. Shading indicates Landscape Conservation Cooperatives (http://www.doi.gov/lcc/index.cfm) in the Southwest, overlaid by the approximate research regions of the Department of the Interior's Climate Science Centers (http://www.doi.gov/csc/index.cfm), indicated by the dashed line boundaries.

Table 20.1 Research needs and strategies associated with themes in the *Assessment of Climate Change in the Southwest United States*

Uncertainty theme	Research need	Research strategy
	CLIMATE (CHAPTERS 4–7)	
Distinguishing long-term climate change from interdecadal and interannual climate variability at the regional scale.	Detection and attribution studies, supported by improved observations and data interpolation methods.	Increased emphasis on detection and attribution, supported by rigorous research protocols. Establish and maintain high quality weather and climate stations, prioritizing the largest data voids. Improved representation, in models, of physical processes such as atmospheric convection, evapotranspiration, snowpack formation, and runoff production. Improved models and modeling techniques for multi-year to decade prediction. Increased collaboration between modelers and scientists whose research focuses on observations.
Uncertainty in evaluating trends and variability in mountainous areas and montane environments.	Improved observations across mountain gradients and at a range of elevations.	Increased emphasis on mountain climate analyses, including studies that link climate, hydrology, soil science. Augmented capabilities should address the occurrence of heavy precipitation during winter storms and summer convection, rain versus snow and rain-on-snow events, snowpack formation and melt-off, and basin-scale runoff efficiency. Development of improved techniques for the automated measurement of precipitation in mountainous areas, especially remote locations.
Inadequate confidence in estimates of variations in current and future local climate conditions.	Enhanced meteorology and hydrology observations to better monitor at scales consistent with terrain. Improved modeling for studies of local variability.	Improved climate and hydrological modeling at scales consistent with Southwest terrain.
Assumption that study of past climate variations can provide a representation of future climate that is adequate to estimate future risks.	A suitable replacement for the stationarity principle.	Focused research on non-stationarity. Investigation into statistical approaches for dealing with time-varying climate and hydrological baselines.

Table 20.1 Research needs and strategies associated with themes in the *Assessment of Climate Change in the Southwest United States* (Continued)

Uncertainty theme	Research need	Research strategy
Few scientific studies have considered future projections of climate and hydrologic extremes. Even fewer have focused on regional extremes. Lack of studies increases uncertainty with regard to common claims that the magnitudes and frequencies of some extremes will increase.	Definition of the most impact-based indicators of environmental extremes that are relevant to society. The necessary cross-sector relationships are still in their infancy.	Define extremes by first understanding their impacts in key sectors. Spur and nurture close collaborations between science and the public-private policy sectors, in order to define policy and impact-relevant extremes.
	ECOSYSTEMS (CHAPTER 8)	
Despite the clear responses of the distributions of some species to climate, the relationship between changes in climate and recent changes in the geographic distribution of species is highly uncertain. Considerable uncertainty remains on how species and the communities and ecosystems they form will respond to projected changes in climate.	Projections of the effects of climate variability on geographic ranges, accounting for multiple factors affecting species persistence and distribution.	Projections that take into account species' environmental stress tolerances, and ability to adapt. Consideration of environmental change, land-use change, and management interventions. Elucidation of mechanisms of change in the interactions among factors such as climate variation and species fitness. Re-estimation of the probabilities of persistence of species. Knowledge gaps may be partly filled by identification of biotic and abiotic drivers of genetic change and selection, identifying which traits (or combinations of traits) will be targets of this selection, and determining how genetic change and phenotypic plasticity affect selection of potentially correlated traits.
Most projections of current or future distributions of species are based on their current climatic niches, which unrealistically assume that niches are static and uniform.	Species range projections that account for uncertainties due to climate projections that fall outside the ranges of data used to build the models.	Develop robust methods for accounting for changing niche delineations. Characterization of uncertainties associated with extrapolations beyond observations used in constructing niche definitions. Estimating likely temporal and spatial changes in these drivers, given multiple scenarios of climate change.
	COASTAL (CHAPTER 9)	
Local jurisdictions vary considerably in their technical expertise and capacity to conduct effective coastal land use management.	Improved understanding of persistent adaptation barriers that inhibit preparedness and active implementation of climate-change adaptation in coastal California.	Social science and communication research aimed at characterizing adaptation barriers. Definition and development of best practices for building capacity for implementation of alternative land-use management practices.

Table 20.1 Research needs and strategies associated with themes in the *Assessment of Climate Change in the Southwest United States* (Continued)

Uncertainty theme	Research need	Research strategy
Extent of ocean acidification and its regional impacts.	Improved understanding of the causes of ocean acidification and its effects on ecosystems and their constituent species.	Monitoring of ocean acidification and diagnostic studies of acidification effects. Analyses to identify sensitive ecosystems, locations, and species.
	WATER (CHAPTER 10)	
Twentieth-century water management was based on the principle that the future would look like the past. Statistical downscaling implicitly preserves stationarity in existing large-scale synoptic patterns.	A suitable replacement for the stationarity principle is needed to reduce inhibitions to the process of adaptation and the search for solutions.	Investigation into statistical approaches for dealing with time-varying baselines.
There is a mismatch between the temporal and spatial scales at which climate models produce useful outputs, and the scales of output needed by water decision makers.	Improvements in statistical downscaling methods that currently produce substantially different results. Improved depiction of factors related to fine-scale topography in climate models.	Reconciliation of downscaling methods. Guidance on best practices for interpreting the output of different downscaling methods. Improved model topography and resolution, and validation of output that uses the improved topography in order to address potential mismatches between observed and projected climate variability.
Differing responses across models, especially with respect to precipitation, lack of realistic topography, lack of realistic monsoon simulation, and lack of agreement about ENSO all provide uncertainty, which is difficult to reduce.	Improving models to better simulate modes of climate variability that have important effects on the region, such as ENSO and the North American monsoon.	Conduct intensive modeling studies using models with the best representations of ENSO and North American monsoon dynamics and regional effects. Develop focused initiatives on these key processes.
	AGRICULTURE AND RANCHING (CHAPTER 11)	
It is not known how much information private or public intermediaries use to transfer or interpret climate change science and projections and their implications for farmers and ranchers	Investigation of climate change knowledge transfer to farmers and ranchers.	Assess the sector-specific availability and use of information. Evaluate the effectiveness of sector-specific climate change communication and extension strategies, electronic media, and technologies, such as the Internet and phone applications.
	ENERGY (CHAPTER 12)	
Climate-change influence on projected peak loads. Estimates exist for California, but not the entire Southwest.	Rigorous projections of peak loads, peak demand, and associated impacts for the entire Southwest region.	Integrated assessment of climate change, demand, peak loads across the region, using common assumptions, model ensembles, socio-economic factors.

Table 20.1 Research needs and strategies associated with themes in the *Assessment of Climate Change in the Southwest United States* (Continued)

Uncertainty theme	Research need	Research strategy
Temporal aspects of future power production.	Simultaneous spatial and temporal assessment of evolution of power production, plant performance, fuel type and mix.	Estimate temperature impacts on different types of electricity-generating capacity and power-plant performance throughout the Southwest. Improve projections of spatial shifts in wind regimes; investigate probability and duration of no-wind conditions on hot days. Improve projections of extreme surface temperatures and their effects on photo-voltaic and concentrated solar power production. Estimate evolution of current and future power, with respect to fuel type.
Vulnerability of power plants to flooding.	Flood risk is site-specific, and relatively few studies directly quantify potential climate change impacts on hydropower production.	Conduct site-specific studies of power plant vulnerability to flooding. Synthesize and assess individual studies, in order to evaluate vulnerability and risk for the region.
Exposure and vulnerability of thermoelectric power production to drought and climate change.	Rigorous assessment of the exposure and vulnerability of thermoelectric power production to drought and climate change.	Assessment of hydrologic conditions, combined with operational characteristics and institutional factors. Development of a definitive measure of aggregate impact. Inventory and rigorously assess the robustness of existing contingency plans for individual plants.
A lack of accurate projections of future ground-level solar radiation adds to uncertainty in regional energy potential and production projections.	Improved simulation of regional cloud cover, directly from physical principles.	Increase spatial resolution in climate models, and improve model physics for estimation of ground-level solar radiation. In the meantime, ground-level radiation generated by global and regional climate models should be interpreted cautiously.
Uncertainty in future transportation sector fuel types and use.	Examinations of the type and intensity of future fuel use in the transportation sector.	Projections of the type and intensity of fuels used in the transportation sector, given projected future climate changes.
Implications of climate change across the entire Western interconnection.	Comprehensive assessment of impacts of climate change on multiple modes of power production, in conjunction with projections of shifts in power demand, and risks to the transmission grid.	Evaluate concurrent impacts of climate on West-wide hydropower production, coincident impacts on thermoelectric production and induced shifts in electric power demand. Include studies of fire risk to the transmission grid, given projected increases in regional fire frequency. Identify loss minimizing operation practices.

Table 20.1 Research needs and strategies associated with themes in the *Assessment of Climate Change in the Southwest United States* (Continued)

Uncertainty theme	Research need	Research strategy
	URBAN AREAS (CHAPTER 13)	
Determination of the adaptive capacity of urban communities.	Integrated evaluation of urban adaptive capacity and institutional complexity.	Evaluation of the connections between municipal agency funding and how municipal services capabilities might be used to implement mitigation and adaptation strategies. Studies of the fiscal capacity of localities in relation to their ability to monitor and act on climate change challenges.
Urban area observation collection, data amount, availability, and format are not sufficiently standardized to fully assess the ways in which cities may contribute to climate change through their urban metabolisms, such as flows of water, energy, materials, nutrients, and air.	Inventory and evaluation of data collection and analysis practices.	Urban metabolism studies can improve understanding of a host of climate-change vulnerabilities such as water use and waste generation. Improve collection of information on urban vegetation cover, to evaluate urban water consumption patterns in the Southwest. Improve quantification of the contribution of urban areas to climate change for sectors other than transportation. Couple household-level energy use with land-use data, to reveal important aspects of urban activities and mechanisms. The expected increase in climate-driven urban-fringe fires will call for improved post-fire monitoring, management, and treatment of stormwater runoff to reduce impacts to city water supplies and downstream ecosystems.
There is a paucity of observations on land and built-environment interactions with the atmosphere in the Southwest's urban centers.	Better information is needed on building technologies, particularly impermeable surfaces and surface albedo (the portion of solar energy reflected back from a surface to the atmosphere).	Improved use of information on urban vegetation cover to evaluate albedo patterns in the Southwest.
	TRANSPORTATION (CHAPTER 14)	
The magnitude of potential impacts to the transportation system for a particular system or location is too uncertain to be reliably estimated. Depending on the levels and types of investments now, the effects of climate change could be significantly increased or reduced.	Studies of the impact of climate change on passenger travel. Quantification of key variables for evaluating economic impacts of changing climate.	Focus studies on key economic variables, such as loss of time, loss of money, loss of productivity (and wages), relocation costs, and loss of life. Prioritize future research to quantify long-term effects of climate change on transportation systems. Refine studies to simultaneously examine multiple scenarios of climate changes of various magnitudes in conjunction with multiple scenarios of levels and types of investments and the effect of multiple timelines for implementation of investments.

Table 20.1 Research needs and strategies associated with themes in the *Assessment of Climate Change in the Southwest United States* (Continued)

Uncertainty theme	Research need	Research strategy
	PUBLIC HEALTH (CHAPTER 15)	
Limited understanding of the associations between meteorological factors and health impacts contributes to uncertainty, and limits the capacity of current statistical models to predict future health impacts.	High-quality, high-resolution long-term health and climate data are necessary to fully characterize their relationship and adequately estimate future impacts to health from climate change. Deficiencies in the quality of health data limit our ability to characterize linkages between climate change and health and develop predictive models for climate-related health impacts.	Incorporate physiological, societal, and behavioral effects to reduce uncertainty in predicting health outcomes. Improved data collection, combined with exploratory analyses that make use of sparse data, and hypothesis-driven diagnostic analyses can help build experimental predictive capacity.
Uncertainty in the future occurrence of allergies and asthma, in association with future pollen production, which may be influenced by increases in atmospheric CO_2 concentration.	Examination of linkages between pollen production, phenology, and public health.	Systematic focus on allergies and asthma, in conjunction with field observations, greenhouse experiments, and modeling. Develop interdisciplinary research initiatives and multidisciplinary research teams.
Uncertainty due to possible multiple causes of mortality in the case of heat-related deaths.	Criteria for heat-related mortality are not standardized. It is often difficult to identify where and when cases were exposed to infectious diseases.	Carefully explore questions of exposure source and location at the time of diagnosis, to identify factors other than temperature (e.g., socio-economic or other environmental issues) that influence the relationship of climate change and mortality.
Limitations to public health data.	Data on the spatial attributes of vector-borne diseases are required to estimate future infectious disease impacts. Criteria for identifying cases are not always consistent, suggesting a need to standardize diagnostic criteria. Multi-year data sets with high spatial and temporal resolution from multiple locations are needed to allow us to assess risk changes over time and estimate future impacts at a regional scale.	Develop standards for diagnostic criteria associated with vector-borne diseases. Invest in consistent, long-term monitoring networks, with sufficient resolution to answer research and public health professionals' needs.

Table 20.1 Research needs and strategies associated with themes in the *Assessment of Climate Change in the Southwest United States* (Continued)

Uncertainty theme	Research need	Research strategy
Insufficient understanding of the physiological, societal, and behavioral factors that affect human health, and the interactions between these factors.	Predictive modeling of climate impacts on health that accounts for the complexity introduced by non-climatic factors.	Assess the linkages between climate and human health at city, state, and regional levels, and develop action plans for each level that reflect differences in current and predicted climate conditions and vulnerabilities. Characterize vulnerability and future risk not only in terms of the impacts of climate change on health, but also with attention to demographics, local geographical differences, and socioeconomic factors.
TRIBAL COMMUNITIES (CHAPTER 17)		
Lack of scientific studies that have examined climate-change impacts on Native lands within the Southwest region. Lack of data and observations is a substantial source of uncertainty in documenting changes and attributing observed changes to anthropogenic climate change.	Indian reservations need improved monitoring of climatic conditions to provide tribal decision makers the necessary data to quantify and evaluate the changes taking place for planning and management of their resources. Reservations lack the data necessary for accurate downscaling of climate models, because meteorological monitoring is sparse over areas of significant size. Additional studies are needed, in order to make comprehensive assessments of observed changes.	In addition to commonly used land-based observations, a research program to evaluate climate change impacts on Native lands in the Southwest might include: increased use of remotely sensed observations to detect environmental changes, ethnographic studies of traditional environmental knowledge, and citizen-science observations. An inventory of climate-related observed changes could provide the first step in developing a comprehensive assessment.

20.4 Research Strategies from Southwestern Ecoregional Initiatives

Each of the Landscape Conservation Cooperatives[iv] (LCCs, Figure 20.2) is conducting strategic syntheses of common resource management priorities and related science needs across the Southwest. Although LCCs are not exclusively directed toward research nor toward climate adaptation and mitigation, climate change information needs are a major part of the LCCs' agenda to work at a landscape scale to protect natural and cultural resources. We queried four LCCs that cover the majority of the Southwest assessment area—California, the Desert region, the Southern Rockies, and the Great Basin[v]—regarding the status of their syntheses, as well as for information on the primary sources used to develop these syntheses. Responses indicated that the LCCs are in the earliest stages of conducting comprehensive science-needs assessments. They anticipate they will be able to make available in the near future a full compilation of these sources, the criteria for developing priorities among science needs, and approaches for applying the criteria. These assessments will be of immense value in the structuring of climate change-related research in the Southwest.

Additionally, a number of federal, state, and local agencies, universities, and inter-organizational cooperatives in the Southwest have started to consider research needs pertaining to climate adaptation to help them achieve their missions. Federal and state agencies, along with their university partners, have conducted several workshops over the past four years to acquaint land and water management agency leadership and staff with the state of climate-change research in the Southwest, and to provide opportunities for participants to articulate what they consider important uncertainties and gaps in our scientific understanding. Table 20.2 lists a sample of recent workshops held on climate-effects science in the Southwest. Incorporating information from such efforts in climate-change-related research needs assessments will be important in creating climate-effects-oriented research programs that are effectively integrated with land and water managers.

20.5 Strategies to Improve Characterization in Climate and Hydrology

Given the Southwest's aridity and climatological variability, water resources managers in the region face many challenges in predicting when they will have too much or too little water. Water resource managers have issued several "requirements" surveys of their water users' perceived needs, including data, methods, tools, and agency capacity. Many of these surveys were developed to address implications of a changing climate for water resources (Milly et al. 2008; Karl et al. 2009) and concerns about how water managers can adequately plan for and manage such changes (Brekke et al. 2009). Whether they are focused on preparing for longer-term climate change or shorter-term weather and climate variations, a common theme among these surveys is the promotion of research and capacity-building that leads to:

- better-quality predictions;
- better use of existing predictions while we wait for quality to increase; and
- better communication of risk and uncertainty during decision-support processes.

Requirements surveys tend to be better barometers of the relevance of research and development efforts than of research feasibility. The following are some requirements surveys that have been completed in recent years, all of which offer insight on user needs related to water-cycle science and prediction research:

- "Addressing Climate Change in Long-Term Water Resources Planning and Management" (Brekke et al. 2011) outlines the various types of analyses necessary to assess climate-change implications for water-resources management. It offers a technical discussion of desired capabilities, current capabilities, and gaps, and is aimed at science/management research collaborations.
- "Water Needs and Strategies for a Sustainable Future" (WGA 2006) offers states' perspectives regarding needs related to several water management aspects,

Table 20.2 Recent workshops that have included discussion about strategies for climate-effects science in the Southwest

Workshop Title	Sponsors	Publication of results
Natural Resource Mitigation, Adaptation and Research Needs Related to Climate Change in the Great Basin and Mojave Desert	Fish and Wildlife Service, US Geological Survey, National Park Service, Desert Research Institute, Bureau of Land Management, Environmental Protection Agency, University of Nevada Las Vegas	Hughson et al. 2011
Effects of Climate Change on Fish, Wildlife and Habitats in the Arid and Semiarid Southwestern United States: Putting Knowledge and Science into Action	Fish and Wildlife Service, US Geological Survey, Climate Assessment for the Southwest	Guido, Ferguson and Garfin 2009
Climate Change, Natural Resources, and Coastal Management: a Workshop on the Coastal Ecosystems of California, Oregon, and Washington	Fish and Wildlife Service, US Geological Survey	USFWS 2009
The Climate and Deserts Workshop: Adaptive Management of Desert Ecosystems in a Changing Climate	Desert Managers Group, George Wright Soc, University of California, University of Arizona, University of Nevada, Great Basin Cooperative Ecosystem Studies Unit, The Wildlife Society, Climate Assessment for the Southwest, The Nature Conservancy, Bureau of Land Management	Desert Managers Group 2008
Research and Development Workshop: Roadmap – Managing Western Water as Climate Changes	Bureau of Reclamation, National Oceanic and Atmospheric Administration, US Geological Survey	Brekke et al. 2009
Southwest Climate Summit	US Geological Survey, Climate Assessment for the Southwest (CLIMAS)	Southwest Climate Science Center 2012

inviting both research and capacity-building activities. The report emphasizes the need for both enhanced hydrologic data collection to track changing climatic conditions and improved capabilities in the areas of hydrologic prediction, modeling, and impact assessment. An associated "Next Steps" report was subsequently issued (WGA 2008), offering a more technical discussion of needs, including those related to managing during drought and other shorter-term weather variations and developing locally relevant (downscaled) long-range projections of climate and hydrology needed to support climate-change vulnerability and adaptation assessments in the Western United States.

- "Options for Improving Climate Modeling to Assist Water Utility Planning for Climate Change" (Barsugli et al. 2009) discusses water utilities' perspectives on global and regional climate projections in relation to their planning activities. It reviews the state of science on developing global to regional climate projections and prospects for improving this science.
- "Decision Support Planning Methods: Incorporating Climate Change Uncertainties into Water Planning" (Means et al. 2010) serves as a companion to Barsugli and others (2009), providing a review of methods for making decisions under climate-change uncertainty, discussing research needs in relation to probabilistic information on data and modeling uncertainties.
- "The Future of Research on Climate Change Impacts on Water—A Workshop Focusing on Adaptation Strategies and Information Needs—Subject Area: Water Resources and Environmental Sustainability" (Raucher 2011) focuses on needs and potential research directions in five areas, including flooding and wet weather, water supply and drought, and the water-energy nexus.

Next-generation climate models should be evaluated for their credibility in making climate projections useful in environmental risk assessments (e.g., Brekke et al. 2008), as must assessments of Southwestern ecosystems involving hydroclimatological drivers. As an initial step in this assessment, the Southwest Climate Science Center[vi] is evaluating the characteristics of California coastal zone influences, the sharp topographic gradients characteristic of the mountainous Southwest, and the North American monsoon.

Although their occurrence is highly variable, intense storms associated with narrow currents of concentrated water vapor ("atmospheric rivers" or ARs) make landfall on the California coast. ARs often make the difference between floods and plentiful water supply versus drought, and are therefore important to planning for integrated water resources and flood planning (see Chapter 4). ARs have been shown to penetrate into the interior Southwest, so the influence of climate change on their frequency and magnitude is in need of additional attention (Dettinger et al. 2011).

There is considerable uncertainty about how the Southwest's major river systems will evolve in the future. For example, in the Sacramento-San Joaquin drainages and San Francisco Bay, differences in projected futures arise from the different sensitivities of global climate models to the range of greenhouse gas emission trajectories. Such uncertainties propagate further into other drivers of transformative change such as landscape modification, water development, and pollutant loads (Cloern et al. 2011).

20.6 Strategies to Improve Characterization of Impacts and Vulnerabilities

Widespread drought has affected large areas of the Southwest for the last decade (see Chapter 5). Concern about persistence of this drought—and of longer, more severe droughts—due to projections for continued variability in precipitation amounts, decreased streamflows, and increased temperatures (see Chapters 6 and 7), is generating new research approaches that will improve policy prescriptions for wildland and urban/suburban systems (MacDonald 2010). However, the availability of scientific information regarding climate change and the capacity to adapt to or mitigate it are uneven across the Southwest. California is a leader among states in assessing climate change impacts associated with natural and managed systems, having completed its second integrated assessment, including thirty-nine individual studies (Franco et al. 2011). Other Southwestern states have not implemented programs to conduct assessments at this depth or breadth.

In addition, the importance of traditional knowledge of indigenous communities is beginning to be acknowledged in ecosystem management (see, for example, the Traditional Knowledge Bulletin[vii] of United Nations University). In other work, interviews with seventy Navajo elders were used to catalog changes in weather, vegetation, location of water sources, and the frequency of wind and dust storms, helping to corroborate research on sand dune movement and growth (Redsteer, Bogle and Vogel 2011). Inclusion of traditional knowledge from the roughly 180 Southwestern tribes could improve both climatic analyses and climate adaptation. Native Americans in the Southwest are thought to be particularly vulnerable to climate change (see Chapter 17). Resiliency can be affected by multiple climate-related threats, and because of tribal communities' close reliance on reservation resources for sustenance, economic development, and the maintenance of cultural traditions, they are particularly vulnerable (National Wildlife Federation 2011). Further assessment of such threats to Native American communities appears to be a pressing need across Southwestern landscapes.

Projections of the potential impacts of coastal flooding in California due to sea-level rise are presently imperfect due to information needs in a number of areas (see Chapter 9). Among these are (1) the capability to model factors such as flood duration and velocity; (2) economic analysis of transportation risk, health issues, and habitat loss; (3) integration of coastal development scenarios; and (4) better characterizations of coastal zone policies (Hebeger et al. 2011).

Climate-driven changes in stream temperature and hydrology are affecting aquatic ecosystems and fishes throughout the West (see Chapter 8), reinforcing the need to synthesize trends in monitoring data, form cross-disciplinary collaborations, and develop alternatives for climate adaptation across river basins (Rieman and Isaak 2010).

Few environmental studies covering the Southwest have explicitly considered dust flux and wind erosion, yet these factors are important drivers of ecosystem processes, can cause human health impacts, and act as a source of uncertainty in climate models (Field et al. 2010). Additional information could help determine ways to reduce the dust layer derived from human activities that is accumulating in the Colorado River Basin snowpack. Dust has increasingly reduced the capacity of the snow to reflect solar radiation, hastening and increasing snowmelt and causing early runoff in this important source of Southwestern water supply (Painter et al. 2010) (see Chapter 4, Box 4.1).

Climate significantly impacts ecosystem structure and plant-animal interactions, such as in plant and bird communities in montane Arizona, where Martin and Maron (2012) demonstrated that declining snowfall indirectly affects both plant and bird populations by allowing more extensive grazing by elk. It is becoming increasingly important for managers to consider such interactions as they struggle to achieve natural resource goals in a changing environment.

As in other regions, keystone species may be affected by climate change and this may have consequences for entire ecosystems. One such Southwestern icon, *Yucca brevifolia*, lends its name to Joshua Tree National Park. A study of past and future shifts in its distribution indicates that only a few of its populations appear to be sustainable, while barriers to dispersal may limit its potential to expand its range (Cole et al. 2011). This is but one example of how climate and land use change will affect species' capacity to migrate. Additionally, since the pace of shifting climate is itself variable, climate changes can affect biodiversity and species that are found only in the Southwest (endemic species); thus, species in many marine and terrestrial environments in the Southwest are likely to be affected (Burrows et al. 2011; Sandei et al. 2011). However, phenotypic plasticity and evolutionary potential could provide a degree of resilience and enhance probabilities of persistence for populations and species in the face of a changing climate (Reed et al. 2011). These findings point to the need for new approaches to integrate climate variability in population biology research in the Southwest.

As a consequence of such climate-species interactions, the nature of geographic boundaries for Southwestern deserts likely will change due to a number of factors affecting vegetation composition, diversity and productivity, water availability and evapotranspiration, and soil erosion (Archer and Predick 2008; Gonzales 2011; Munson et al. 2012). Southwestern forests and woodlands that are sensitive to fire and insect infestations appear to be increasingly vulnerable to rapid conversion to novel vegetation types (Williams et al. 2010). Climate is a principal driver for wildfire frequency, intensity, type, extent, and seasonality, and fire regimes in the Southwest are affected by the invasion of non-native species such as *Bromus tectorum* (cheatgrass) in the Mojave Desert (Brooks et al. 2004). Increases in wildfire related to climate change are projected for future decades due to changes in ignitions, fuel condition, and volume; a new generation of dynamic vegetation models appears necessary to help assess fire severity (Hessl 2011).

A growing awareness of the effects of ongoing and impending climate change on Southwestern ecosystems, urban areas, and socio-economic structures is creating a need to review management approaches to evaluate what lines of research are needed to fill information gaps. As new climate-driven natural and human community structures and relationships develop, it will be important for research strategies relating to climate effects to be tailored to address not only scientific uncertainty, but also to address our need to manage adaptively.

References

Archer, S. R., and K. I. Predick. 2008. Climate change and ecosystems of the southwestern United States. *Rangelands* 30:23–28.

Barsugli, J., C. Anderson, J. B. Smith, and J. M. Vogel. 2009. *Options for improving climate modeling to assist water utility planning for climate change.* N.p.: Water Utilities Climate Alliance. http://www.wucaonline.org/html/actions_publications.html.

Brekke, L., K. White, R. Olsen, E. Townsley, D. Williams, F. Hanbali, C. Hennig, C. Brown, D. Raff, and R. Wittler. 2011. *Addressing climate change in long-term water resources planning and management: User needs for improving tools and information.* U.S. Army Corps of Engineers Civil Works Technical Series CWTS-10-02. Washington, DC: U.S. Bureau of Reclamation / U.S. Army Corps of Engineers. http://www.usbr.gov/climate/userneeds/.

Brekke, L. D., M. D. Dettinger, E. P. Maurer, and M. Anderson. 2008. Significance of model credibility in estimating climate projection distributions for regional hydroclimatological risk assessments. *Climate Change* 89:371–394.

Brekke, L. D., J. E. Kiang, J. R. Olsen, R. S. Pulwarty, D. A. Raff, D. P. Turnipseed, R. S. Webb, and K. D. White. 2009. *Climate change and water resources management—A federal perspective*. U.S. Geological Survey Circular 1331. http://pubs.usgs.gov/circ/1331/.

Brooks, M. L., C. M. D'Antonio, D. M. Richardson, J. B. Grace, J. E. Keeley, J. M. DiTomaso, R. J. Hobbs, M. Pellant, and D. Pyke. 2004. Effects of invasive alien plants on fire regimes. *BioScience* 54:677–688.

Burrows, M. T., D. S. Schoeman, L. B. Buckley, P. Moore, E. S. Poloczanska, K. M. Brander, C. M. Brown, et al. 2011. The pace of shifting climate in marine and terrestrial ecosystems. *Science* 334:652–655.

California Coastal Commission. 2008. *California Coastal Commission climate change and research considerations*. N.p.: California Coastal Commission. http://www.coastal.ca.gov/climate/ccc_whitepaper.pdf.

Chambers, J. C., N. Devoe, and A. Evendon. 2008. *Collaborative management and research in the Great Basin - examining the issues and developing a framework for action*. General Technical Report RMRS-GTR-204. Fort Collins, CO: U.S. Forest Service, Rocky Mountain Research Station.

Cloern, J. E., N. Knowles, L. R. Brown, D. Cayan, M. D. Dettinger, T. L. Morgan, D. H. Schoellhamer, et al. 2011. Projected evolution of California's San Francisco Bay-Delta-River system in a century of climate change. *PLoS ONE* 6: e24465.

Cole, K. L., K. Ironside, J. Eischeid, G. Garfin, P. B. Duffy, and C. Toney. 2011. Past and ongoing shifts in Joshua tree distribution support future modeled range contraction. *Ecological Applications* 21:137–149.

Desert Managers Group (DMG). 2008. The Climate and Deserts Workshop: Adaptive Management of Desert Ecosystems in a Changing Climate, April 9–11, 2008, Laughlin, NV. http://www.dmg.gov/climate/index.html.

Dettinger, M. D., F. M. Ralph, T. Das, P.J. Neiman, and D. R. Cayan. 2011. Atmospheric rivers, floods and the water resources of California. *Water* 3:445–478.

Field, J. P., J. Belnap, D. D. Breshears, J. C. Neff, G. S. Okin, J. J. Whicker, T. H. Painter, S. H. Ravi, M. C. Reheis, and R. L. Reynolds. 2010. The ecology of dust. *Frontiers in Ecology and the Environment* 8:423–430.

Franco, G., D. R. Cayan, S. Moser, M. Hanemann, and M-A. Jones. 2011. Second California assessment: Integrated climate change impacts assessment of natural and managed systems. *Climatic Change* 109 (Suppl. 1): S1–S19.

Gonzales, P. 2011. Science for natural resource management under climate change. *Issues in Science and Technology* Summer 2011:65–74.

Guido, Z., D. Ferguson, and G. Garfin. 2009. *Putting knowledge into action: Tapping the institutional knowledge of U.S. Fish and Wildlife Service Regions 2 and 8 to address climate change; A synthesis of World Café discussion sessions during the FWS, USGS, and UA sponsored Climate Change Workshop, August 18–20, 2008, Tucson, Arizona*. http://www.fws.gov/southwest/Climatechange/docs/Knowledge_into_Action_FINAL[1].pdf.

Hebeger, M., M. Cooley, P. Herrera, P. H. Gleick, and E. Moore. 2011. Potential impacts of increased coastal flooding on California due to sea-level rise. *Climatic Change* 109 (Suppl. 1): S229–S244.

Hessl, A. E. 2011. Pathways for climate change effects on fire: Models, data, and uncertainties. *Progress in Physical Geography* 35:393–407.

Hughson, D. L., D. E. Busch, S. Davis, S. P. Finn, S. Caicco, and P. S. J. Verburg. 2011. *Natural resource mitigation, adaptation and research needs related to climate change in the Great Basin and Mojave Desert: Workshop summary.* U.S. Geological Survey Scientific Investigations Report 2011-5103. http://pubs.usgs.gov/sir/2011/5103.

Interagency Working Group on Climate Change and Health (IWGCCH). 2010. *A human health perspective on climate change: A report outlining the research needs on the human health effects of climate change.* Research Triangle Park, NC: Environmental Health Perspectives / National Institute of Environmental Health Sciences. doi:10.1289/ehp.1002272. http://www.niehs.nih.gov/health/assets/docs_a_e/climatereport2010.pdf.

Karl, T. R., J. M. Melillo, and T. C. Peterson, eds. 2009. *Global climate change impacts in the United States.* Cambridge: Cambridge University Press. http://downloads.globalchange.gov/usimpacts/pdfs/climate-impacts-report.pdf.

Lucier, A., M. Palmer, H. Mooney, K. Nadelhoffer, D. Ojima, and F. Chavez. 2006. *Ecosystems and climate change: Research priorities for the U.S. Climate Change Science Program; Recommendations from the scientific community.* Special Series No. SS-92-06. Solomons: University of Maryland Center for Environmental Science, Chesapeake Biological Laboratory. http://www.usgcrp.gov/usgcrp/Library/ecosystems/eco-workshop-report-jun06.pdf.

MacDonald, G. 2010. Water, climate change, and sustainability in the Southwest. *Proceedings of the National Academy of Sciences* 107:21256–21262.

Martin, T. E., and J. L. Maron. 2012. Climate impacts on bird and plant communities from altered animal plant interactions. *Nature Climate Change* 2:195–200, doi:10.1038/nclimate1348.

Mastrandrea, M. D., C. B. Field, T. F. Stocker, O. Edenhofer, K. L. Ebi, D. J. Frame, H. Held, et al. 2010. *Guidance note for lead authors of the IPCC Fifth Assessment Report on consistent treatment of uncertainties.* Geneva: Intergovernmental Panel on Climate Change (IPCC). http://www.ipcc.ch/pdf/supporting-material/uncertainty-guidance-note.pdf.

Mastrandrea, M. D., K. J. Mach, G-K. Plattner, O. Edenhofer, T. F. Stocker, C. B Field, K. L. Ebi, and P. R. Matschoss. 2011. The IPCC AR5 guidance note on consistent treatment of uncertainties: A common approach across the working groups. *Climatic Change* 108:675–691, doi: 10.1007/s10584-011-0178-6.

Means, E., M. Laugier, J. Daw, L. Kaatz, and M. Waage. 2010. *Decision support planning methods: Incorporating climate change uncertainties into water planning.* WUCA White Paper. San Francisco: Water Utilities Climate Alliance. http://www.wucaonline.org/html/actions_publications.html.

Milly, P. C. D, J. Betancourt, M. Falkenmark, R. M. Hirsch, Z. W. Kundzewicz, D. P. Lettenmaier, and R. J. Stouffer. 2008. Stationarity is dead: Whither water management? *Science* 319:573–574.

Moss, R. H., and G. Yohe. 2011. Assessing and communicating confidence levels and uncertainties in the main conclusions of the NCA 2013 report: Guidance for authors and contributors. N.p.: National Climate Assessment Development and Advisory Committee (NCADAC).

Munson, S. M., R. H. Webb, J. Belnap, J. A. Hubbard, D. E. Swann, and S. Rutman. 2012. Forecasting climate change impacts to plant community composition in the Sonoran Desert. *Global Change Biology* 18:1083–1095, doi: 10.1111/j.1365-2486.2011.02598.x.

National Research Council (NRC). 2009. *Informing decisions in a changing climate.* Washington, DC: National Academies Press.

—. 2010. *Advancing the science of climate change.* Washington, DC: National Academies Press.

National Wildlife Federation (NWF). 2011. *Facing the storm: Indian tribes, climate-induced weather extremes and the future for Indian Country.* Boulder, CO: NWF, Rocky Mountain Research Center.

Painter, T. H., J. S. Deems, J. Belnap, A. F. Hamlet, C. C. Landry, and B. Udall. 2010. Response of Colorado River runoff to dust radiative forcing in snow. *Proceedings of the National Academy of Sciences* 107:17125–17130.

Raiffa, H., and R. Schlaiffer. 2000. *Applied statistical decision theory*. New York: Wiley.

Redsteer, M. H., R. C. Bogle, and J. M. Vogel. 2011. Monitoring and analysis of sand dune movement and growth on the Navajo Nation, southwestern United States. U. S. Geological Survey Fact Sheet 2011-3085. Washington, DC: USGS.

Reed, T. E., D. E. Schindler, and R. S. Waples. 2011. Interacting effects of phenotypic plasticity and evolution on population persistence in a changing climate. *Conservation Biology* 25:56–63.

Rieman, B. E., and D. J. Isaak. 2010. *Climate change, aquatic ecosystems, and fishes in the Rocky Mountain West: Implications and alternatives for management*. General Technical Report RMRS-GTR-250. Fort Collins, CO: U.S. Forest Service, Rocky Mountain Research Station.

Salazar, K. 2009. Order No. 3289: Addressing the impacts of climate change on America's water, land, and other natural and cultural resources. Washington, DC: Secretary of the Interior. http://www.doi.gov/whatwedo/climate/cop15/upload/SecOrder3289.pdf.

Sandei, B., L. Arge, B. Dalsgaard, R. G. Davies, K. J. Gaston, W. J. Sutherland, and J.-C. Svenning. 2011. The influence of Late Quaternary climate-change velocity on species endemism. *Science* 344:660–664.

Southwest Climate Science Center. 2012. Southwest climate summit. http://swcsc.arizona.edu/content/southwest-climate-summit-0.

U.S. Bureau of Land Management (BLM). 2011. Rapid ecoregional assessments. http://www.blm.gov/wo/st/en/prog/more/Landscape_Approach/reas.html (last updated May 25, 2012).

U. S. Department of Energy (DOE). 2010. *Climate Research Roadmap Workshop: Summary report; May 13–14, 2010*. DOE SC-0133. Washington, DC: U.S. Department of Energy, Office of Science, Office of Biological and Environmental Research.

U.S. Fish and Wildlife Service (USFWS). 2009. *Climate Change, Natural Resources, and Coastal Management: A Workshop on the Coastal Ecosystems of California, Oregon and Washington, January 29–30 2009, San Francisco, California.* http://www.fws.gov/pacific/Climatechange/meetings/Coastal.cfm.

U.S. Forest Service (USFS). 2010. *Forest Service global change research strategy, 2009–2019: Implementation plan; May 2010*. FS-948. N.p.: USFS. http://www.fs.fed.us/research/publications/climate/GlobalChangeStrategy_7.7.pdf.

Raucher, R. S. 2011. *The Future of Research on Climate Change Impacts on Water: A workshop focusing on adaptation strategies and information needs; jointly sponsored by Water Research Foundation, National Oceanic and Atmospheric Administration, U.S. Environmental Protection Agency, Water Environment Research Foundation, and Universities Corporation for Atmospheric Research.* Denver: Water Research Foundation.

Western Governors' Association (WGA). 2006. *Water needs and strategies for a sustainable future*. Denver: WGA.

—. 2008. *Water needs and strategies for a sustainable future: Next steps*. Denver: WGA.

Williams, A. P., C. D. Allen, C. I. Millar, T. W. Swetnam, J. Michaelson, C. J. Still, and S. W. Levitt. 2010. Forest responses to increasing aridity and warmth in the southwestern United States. *Proceedings of the National Academy of Sciences* 107:21289–21294.

Endnotes

i See http://greatbasin.wr.usgs.gov/GBRMP/

ii Western Water Assessment, personal communications.

iii See http://www.doi.gov/csc/southwest/

iv See http://www.doi.gov/lcc/

v See http://californialcc.org/, http://www.usbr.gov/WaterSMART/lcc, and http://www.blm.gov/id/st/en/prog/Great_Basin_LCC.html

vi See http://www.doi.gov/csc/southwest/

vii See http://tkbulletin.wordpress.com/

Glossary

A2 – a greenhouse gas emissions scenario used as input into global climate models to project climate changes in the IPCC Fourth Assessment Report, as described in its *Special Report on Emissions Scenarios*. This scenario assumes a future with a high population growth rate, slow economic development rate, slow technological change and fossil fuels use at rates slightly lower than observed in historical records. This combination results in higher GHG emissions and substantially increased global temperatures.

accrete – build up through gradual accumulation, can refer to the increase in size of a tectonic plate by addition of material as well as process where coastal sediments return to the beach following storm erosion

acequia – shared system of irrigation ditches used by farmers in New Mexico

acidification, ocean acidification – increase in the pH of the oceans from the absorption of atmospheric CO_2. Some scientific evidence shows that ocean acidification affects shellfish, such as oysters and clams.

acre foot – the amount of water required to cover one acre of water one foot deep, equal to 43,560 cubic feet. This is approximately the amount of water used by two and a half households in one year.

adaptation – increasing the readiness and resilience of sectors to reduce the impacts of climate change; preparing and planning for climate change, minimizing those impacts that cannot be avoided and turning expected climate changes into opportunities wherever possible

adaptation barriers – barriers that prevent or limit response to change; these include institutional, economic, regulatory, or attitudinal barriers

adaptive capacity – the ability (of a household, community, or other unit of organization) to reduce its vulnerability to climate-related risks through coping strategies such as application of social, technical, or financial resources

adaptive pathway – intentionally adaptive operations or strategies responsive to climatic change

aerosol – "a collection of airborne solid or liquid particles, with a typical size between 0.01 and 10 micrometer (a millionth of a meter) that reside in the atmosphere for at least several hours. For example, dust, or small particles ejected from a volcano. Aerosols may be of either natural or anthropogenic origin. Aerosols may influence climate in several ways: directly through scattering and absorbing radiation, and indirectly through acting as cloud condensation nuclei or modifying the optical properties and lifetime of clouds." (From *IPCC Technical Paper—Climate Change and Water*)

agricultural drought – a period of low soil moisture sufficiently long or severe that it affects crops

air gap – the space between the bottom of a bridge and the top of a ship sailing under it

airshed – a geographical space wherein the air normally flows or is contained, so that air pollution conditions are relatively uniform within it

albedo – The percentage of light reflected by an object. Snow-covered areas have a high albedo (0.9 or 90%) due to their white color. Human activity has changed the albedo of various regions around the globe.

ambient temperature – temperature of the surroundings; background temperature

anadromous fish – fish species that spend most of their lives in the ocean but hatch and spawn in fresh water; for example, salmon are anadromous

annual plants – plants that live only for a single season or year; some ornamental plants, like zinnias, are annual, as are some crops, such as peas, corn and wheat

anomaly – a deviation from the norm; climate data are often expressed in terms of anomalies from a long-term average, such as "the 2002 annual temperature was a 3°F positive anomaly from 1971-2000" (in other words the 2002 annual temperature was 3°F higher than the 1971-2000 average)

anthropogenic – human-caused or human-induced

anthropogenic waste heat – heat produced by human activities for which there is no useful application, such as the heat generated to cool a structure

anticyclone – the circulation of winds around a high pressure center, traveling clockwise in the Northern Hemisphere and counterclockwise in the Southern Hemisphere; anticyclones are associated with high atmospheric pressure and dry conditions.

arbovirus – a virus spread by arthropod vectors such as mosquitoes, lice and ticks

armoring – the artificial reinforcement of natural areas, as by the erection of hard structures, such as a seawall, to protect a shoreline

assessment – critical evaluation of information for purposes of guiding decisions on a complex issue

atmospheric rivers – narrow corridors of atmospheric moisture typically found near or ahead of cold front storms, which deliver enormous amounts of water vapor in low level (<2 km [1.2 mi] above sea level), long (>2000 km [1243 mi]), and narrow (less than about 500 km [311 mi] wide) corridors from over the Pacific Ocean

B1 – a greenhouse gas emissions scenario used as an input into global climate models to project climate changes in the IPCC Fourth Assessment Report, as described in its *Special Report on Emissions Scenarios*. This scenario assumes a future in which global population peaks in the year 2050 and then declines, with economies that shift rapidly toward the introduction of clean and resource-efficient technologies and an emphasis on global solutions to economic, social, and environmental sustainability. This combination results in lower GHG emissions and smaller increases in global temperatures.

backshore – the coastal area above the high-water line

basin scale – denotes a larger-scale area with the extent of a watershed or series of watersheds

beach nourishment – replenishment of beach sand (that has been lost through erosion or drift) from an external source, usually to increase the area that can be used for recreation or to protect the shoreline from coastal storms

bias correction and spatial downscaling method (BCSD) – a method of creating climate projection results at spatial scales that are meaningful for analysis of regional impacts, by taking GCM results and statistically relating them to the historic climate of the region. Also known as bias-correction and spatial disaggregation.

bioallergens – pollens and molds that induce allergic reactions in some individuals

biomass – biological mass from living organisms or recently living organisms

biophysical process – combination of physical processes in biological organisms or systems, such as the uptake of water and nutrients in plants

C3 plants – plants that use the C_3 carbon fixation pathway, which grow and lose water during the day; these include most broadleaf and temperate zone plants

C4 plants – plants that use the C_4 carbon fixation pathway in their metabolism, which lose little water during the day and makes them well-adapted to hot and dry areas; examples include sugarcane and maize

CAM plants – plants with crassulacean acid metabolism, which store carbon dioxide at night and thus minimize water loss during the day; examples include orchids, cactuses and agaves

calcium carbonate – the component of marine shell material that dissolves in more acidic waters

carbon sequestration, carbon storage – the removal of carbon dioxide from the atmosphere, by natural or other means, and storage of carbon in, for example, plants, soil, oceans

carbon uptake – the absorption of carbon by soils and plants

chill time – the accumulation of hours during which temperatures are between 32°F–45°F during bud dormancy in plants; some plants require a certain amount of chill time before their buds open and growth can occur

climate change – ways in which systematic trends in some climate factors, such as increases in heat-trapping gases in the atmosphere (greenhouse gases) and associated increases in temperature, alter the climate system and its variations

climate regime – type of climate; climate classification, or an extended period of certain climatic characteristics, such as a "wet regime" or a "dry regime"

climate variability – the inherent variability of climate, for instance, from year to year or decade to decade

climatology – the study of climates and their phenomena

co-benefits – multiple benefits of a program, policy, or intervention; for example, a benefit of bicycling to work is reduced greenhouse gas emissions, and a co-benefit of bicycling is improved cardiovascular health

cold spell index – an index that reflects frequency, intensity, duration, and spatial extent of wintertime cold spells over a region. In this report the index was derived from observations and the CNRM-CM3 model.

cold wave – a prolonged period of unusual cold weather, variously defined as a four-day period colder than the threshold of a one-in-five year frequency or the coldest five percent of the wintertime daily temperature distribution, aggregating degree days below the local 5th percentile thresholds from November to March, averaged over the region

colonia – a U.S.-Mexico border-region residential community that is economically distressed and usually underserved by infrastructure

Colorado delta – the region where the Colorado River flows into the Gulf of California

concentrating solar – a type of solar energy production method that concentrates a large area of sunlight onto a small area, using lenses or mirrors. The concentrated light is converted to heat, which in turn drives a heat engine, often a turbine, connected to a power generator.

cone of uncertainty – a schematic device to illustrate the growing uncertainty of future conditions over time

confidence – a subjective judgment of the reliability of an assertion, based on systematic evaluation of the type, amount, quality, and consistency of evidence, and the degree of agreement among experts

conjunctive use – Conjunctive use of surface and groundwater consists of combining the use of both sources of water in order to minimize the undesirable physical, environmental, and economic effects of each solution and to optimize the balance between water demand and supply.

consumption-based emissions – emissions measurement that takes into account net imports and exports of goods and services rather than just emissions production

consumptive use – water that is not returned to a water system after use; for example, water lost through evapotranspiration of crops. Consumptive use makes the water unavailable for other uses, usually by permanently removing it from local surface or groundwater storage as the result of evaporation and/or transpiration.

convection – the movement of relatively warm air upward into the atmosphere, which cools, forms clouds, and often causes the downward movement of cooler air. In the Southwest, convection is associated with the development of summer thunderstorms.

cooling degree days – a measurement that reflects the amount of energy needed to cool a home or structure. This index is derived from daily temperature records. The "cooling year," during which cooling degree data are accumulated extends from January 1 to December 31. An average daily temperature of 65°F is the base for cooling degree day computations. Cooling degree days are summations of positive differences from 65°F. For example, cooling degree days for a station with daily mean temperatures during a seven-day period of 67, 65, 70, 74, 78, 65 and 68, are 2, 0, 5, 9, 13, 0, and 3, for a total for the week of 32 cooling degree days. (Adapted from NOAA Climate Prediction Center, www.cpc.ncep.noaa.gov/products/analysis_monitoring/cdus/degree_days/ddayexp.shtml)

coupled models – this phrase refers to the coupling of individual components in climate models, which allows for interaction between different parts of the climate system within the models. This phrase often refers to the coupling of individual atmosphere and ocean model components.

crassulacean acid metabolism – refers to plants that store carbon dioxide at night and thus minimize water loss during the day

cyclone – the circulation of winds around a low pressure center, traveling counterclockwise in the Northern Hemisphere and clockwise in the Southern Hemisphere. Cyclones are usually associated with precipitation-generating conditions.

decision analysis – a planning tool wherein uncertainties are throughly described and decision trees can be used to represent different decision pathways and find optimal solutions

decoupling – when previously linked phenomena or processes cease to be connected. For example when cold air drains down from mountain ranges and pools in mountain valleys, the cold air pool can become decoupled from the large-scale atmospheric flow. In such cases, local conditions can differ substantially from regional conditions, which is important in assessing the impacts of climate change in mountainous areas.

deficit irrigation – irrigation that aims to achieve greater crop output per unit of water applied rather than maximizing crop output per acre

detection and attribution study – a study to demonstrate the likelihood that an observed change (as in climate) is occurring (detection), that it is statistically significantly different from what could occur from natural variability, and the most likely cause for that change (attribution) in terms of a defined level of confidence

diapause – a period in which growth or development is suspended

dieback – tree mortality noticeably above usual mortality levels. In the Southwest, dieback often refers to the death of large numbers of conifer trees, as a result of drought, high temperatures, and insect pest outbreaks.

discharge – flow; the volume of water (and suspended sediment, if surface water) that passes a given location within a given period of time.

disturbance/ecological disturbance – a cause; a physical force, agent, or process, either abiotic or biotic, causing a perturbation (which includes stress) in an ecological component or system; examples are fire (abiotic) and outbreaks of disease (biotic)

downscaling – a method providing finer spatial detail of climate model (GCM) results. Often scientists refer to two methods: statistical downscaling, which uses mathematical relationships between the GCM data and historical data to adapt GCM projections to local conditions, and dynamical downscaling, in which GCM output is used as input to a regional-scale model which can then merge large-scale climate information from GCMs with local physical controls on climate (such as small mountain ranges).

driver – something that creates and fuels activity, or gives force or impetus; for example, energy from the Sun is one driver of the climate system

dry cooling – a cooling system in which air rather than water is used to cool a power plant's working fluid

dry warming scenario – a climate scenario in which the climate becomes drier (less precipitation) as well as warmer

dust flux – The flow of dust particles through a given area within a certain amount of time; for example the transport, by wind, of loose soil from a valley bottom to the top of a mountain range would be called a flux of dust to the mountain top

dynamical downscaling – a method of modeling climate by use of a limited-area, regional climate model (RCM) which uses the output from a global climate model as input to model climate at a finer spatial scale. The method uses a physically based process model at a grid spacing of tens (rather than hundreds) of miles (or kilometers) and so better represents complex topography. The method is very expensive to use, thus there are fewer dynamically downscaled data available.

ecoregion – large area of land and water characterized by distinctive plant and animal communities and other environmental factors

ecosystem-based adaptation – an adaptation approach that seeks to achieve the preservation and sustainability of biological resources and protect the ecosystem services that these resources provide humans

ecosystem services – the benefits provided by natural systems, such as flood protection, water treatment by filtering through soils, recreation, the storage (or sequestration) of carbon in plant matter, biodiversity, and wildlife habitat

ecotone – boundary between ecological systems

eddy covariance – a micrometeorological technique designed to measure turbulent exchanges of mass, momentum, and heat between an underlying surface and the atmosphere

El Niño – a disruption of the ocean-atmosphere system in the tropical Pacific that impacts weather around the globe. El Niño is characterized by a large scale weakening of the trade winds and warming of the surface layers in the eastern and central equatorial Pacific Ocean. In the southern parts of the Southwest, El Niño winters often deliver above–average precipitation, and there is more tropical storm activity in the eastern Pacific Ocean. In coastal areas of California and Baja California, El Niño often results in warmer ocean water, higher sea levels, more rainfall, and flooding.

El Niño Southern Oscillation (ENSO) – A term used to describe large climate disturbances that are rooted in the tropical Pacific Ocean and occur every 3 to 7 years. El Niño refers to warming of the tropical eastern Pacific Ocean and the Southern Oscillation refers to changes in atmospheric circulation across the Pacific Ocean basin. ENSO includes the full range of variability observed in these interactions between ocean and atmosphere, including both El Niño and La Niña episodes. (Compare **La Niña**, below).

elevational gradient – changes in elevation or slope; for example, changes in elevation along a mountain range. It is useful in many scientific studies to contrast changes in plant life or temperature at different elevations.

endemic species – species that are native to and occur only in a particular location

ensemble – multiple simulations used to construct a possible distribution of climate change; often, the average of multiple climate model runs will be referred to as the ensemble average

ephemeral streams or streamflow – surface water flow in streams and drainages that occurs only temporarily after precipitation or snowmelt

evapotranspiration – the combination of evaporation of water from the earth's water surfaces and soil and the transpiration of water by plants. Evapotranspiration is frequently measured in two ways: (1) the amount that occurred; and (2) the potential amount that would have occurred if enough water had been present to meet all evaporation and transpiration needs. In arid areas the actual amount is frequently less than the potential amount.

extreme events – (from IPCC AR3 WGI): "an event that is rare at a particular place and time of year. Definitions of rare vary, but an extreme weather event would normally be as rare as or rarer than the lowest or highest 10 percent of all weather events that have ever been observed in a location or region. By definition, the characteristics of what is called extreme weather may vary from place to place in an absolute sense. Single extreme events cannot be simply and directly attributed to anthropogenic climate change, as there is always a finite chance the event in question might have occurred naturally. When a pattern of extreme weather persists for some time, such as a season, it may be classed as an extreme climate event, especially if it yields an average or total that is itself extreme (e.g., drought or heavy rainfall over a season)."

exurban land – land on the outer ring of suburbs, here defined as having a housing density of one unit per 2.5 to 40 acres

feedback, natural system feedback – Feedbacks are interactions in which outputs from a process have an effect on the inputs to that same process. Sometimes feedbacks can offset or inhibit a change (negative feedback), and sometimes they can amplify a change (positive feedback). An example of a positive feedback is when the atmosphere heats up it melts ice. Ice reflects a lot of incoming energy from the sun, so when it melts and is replaced by heat-absorbing water, soil or vegetation, the land surface warms more quickly and warms the atmosphere, making it easier for more snow and ice to melt in a sustained manner that can eventually cause extensive snowmelt. An example of a negative feedback is when carbon dioxide increases, it causes plants to grow faster, which allows them to absorb more carbon dioxide, which eventually can lead to a large reduction of carbon dioxide.

forb – a herbaceous flowering plant that is not a grass

forcing – (n.) From IPCC AR4 WGI: "The climate system can be driven, or "forced" by factors within and external to the system. Processes within the system include those related to the atmosphere, the cryosphere (ice-covered parts of the Earth), the hydrosphere, the land surface, and the biosphere. Volcanic eruptions, solar variations and human-caused changes in the composition of the atmosphere and land use change are external forcings."

forward contract – a bilateral agreement to buy and sell an asset at a specified time and price in the future, with both terms agreed upon today

freeze-free season – the period between the last frost of spring and first autumn frost; the length of the season equals the number of consecutive days during which minimum daily temperatures are above freezing. An increase in the length of the season may be considered a proxy for potential increased evaporative or heat stress on plants in the arid Southwest.

futures contracts – a standardized contract, traded on a commodity exchange, to buy and sell an asset at a specified time and price in the future, with both terms agreed upon today traded on a commodity exchange

global climate model (GCM) – a computer-driven model of global climate, used to project climate change based on mathematical equations that represent key physical processes; also called general circulation model

greenspace – protected and reserved areas of undeveloped land

GWh – gigawatt hour; the unit of energy representing one billion watt hours or one million kilowatt hours; a gigawatt of power, depending on local factors, can provide enough energy to power several hundred thousand homes for one year

grassland – land dominated by grasses rather than large shrubs or trees

growing season – the period of each year in which native plants grow and crops can be grown; in the United States this is usually defined as the days between the last overnight frost or freeze and the first occurrence in the fall

habitat connectivity – the interconnection of different habitats to allow species movement. Habitat connectivity is important because some species can die out if they are impeded from migrating, but habitat connectivity can also facilitate the migration of invasive species.

haboob – severe and extensive dust storm

Hadley Cell – the tropical atmospheric circulation that moves warm moist rising air, from near the equator, poleward; the so-called "descending limb" of the Hadley Cell contains warm dry descending air and often defines the locations of arid and semi-arid regions; the Hadley Cell connects closely with the trade winds and jet stream in the tropics and subtropics

hard freeze – a freeze sufficiently long and severe to destroy seasonal vegetation and lead to ice formation in standing water and hard ground

heat flux – heat-energy transfer

heat island effect, urban heat island effect – The term "heat island" describes built-up urban areas that are hotter than nearby rural areas. Heat islands occur because the built environment, including impervious surfaces, such as roads, parking lots, and buildings, retain more heat than soils and vegetation. Consequently, minimum daily temperatures (which measure the degree to which the city cools off at night) increase substantially compared to surrounding naturally vegetated areas. Heat islands can affect communities by increasing summertime peak energy demand, air conditioning costs, air pollution and greenhouse gas emissions, heat-related illness, and mortality.

heat stress – the physiological response to excessive heat

heat wave – a period of abnormally elevated temperature, defined in relation to local normal conditions and thresholds deemed relevant by researchers and stakeholders

heat wave index – magnitude of a heat wave; the difference between the actual maximum daily temperature or minimum daily temperature and its corresponding 95th percentile threshold and summed over the consecutive days of the heat wave

heating degree days – a measurement that reflects the amount of energy needed to heat a home or structure. With a baseline of 65°F, heating degree days are the sum of the temperature differences of the daily mean temperature subtracted from 65°F, for all days when the mean temperature is less than 65°F. (See also **cooling degree days**)

host – an organism that harbors another organism in or on itself

Hueco Bolson – a groundwater basin in the El Paso/Ciudad Juarez region (shared among New Mexico and Texas in the United States, and Chihuahua in Mexico)

hydroclimatology – refers to key physical processes that connect climate and hydrology; a broad definition suggests the study of moisture in the atmosphere and water in and on the surface of the earth; thus, hydroclimatology includes processes such as transfers of moisture between the atmosphere and the surface, the connections between climate and soil moisture, and studies of the influence of climate upon water

hydrograph – a chart that shows the rate of change of a hydrologic variable through time, such as rate of flow in a stream past a particular location or a change in the temperature or pH of water through time

hydrologic drought – extended period of low water supply; below-normal streamflow, lake, and groundwater levels, due to a decline or deficit in precipitation

hydrologic parameters – examples are precipitation, snowpack, and temperature

hydrology – the study of the properties of water in all its forms—liquid, solid, and gas—especially its movement and distribution on and below the earth's surface and in the atmosphere

hypoxia, hypoxic events – the occurrence of dangerously low oxygen levels that can lead to widespread die-offs of fish or other organisms

IPCC – (from Bureau of Reclamation Climate Technical Work Group—Appendix U): "The Intergovernmental Panel on Climate Change (IPCC) established by World Meteorological Organization (WMO) and United Nations Environmental Programme (UNEP) provides an assessment of the state of knowledge on climate change based on peer-reviewed and published scientific/technical literature in regular time intervals."

impervious – resistant to penetration by water or plant roots

infiltration – the flow of water through the ground surface that percolates through soil and layers of geologic material in a generally downward direction

informal housing – unplanned residential areas where a group of housing units has been constructed on land to which the occupants have no legal claim, or where housing is not in compliance with current planning and building regulations. Many other terms and definitions have also been devised for informal human housing, for example: unplanned settlements, squatter settlements, marginal settlements, and non-permanent structures.

in-situ – in the natural environment; a term that refers to data and measurements taken in the field as opposed to in a laboratory

instrumental record – observed data using a variety of weather instruments, such as thermometers or rain gauges

interannual – year-to-year

interdecadal – decade-to-decade

intraseasonal – applies to time scales from a few days to less than a season

iterative management – management that adapts its strategies to new information and changing circumstances

just-in-time manufacturing – a business management strategy that strives to produce goods that meet exact customer demand with minimal waste in time and resources

keystone species – a species with a large effect on its environment by playing a critical role in maintaining the structure of an ecological community and whose removal will cause a dramatic ecosystem shift

La Niña – years in El Niño-Southern Oscillation (ENSO) that have below-normal temperatures in the central and eastern equatorial Pacific Ocean, and enhanced trade winds. In the southern part of the Southwest region, La Niña is associated with reduced winter and spring precipitation and drought.

latent heat – the quantity of heat absorbed or released by a substance undergoing a change of state, such as ice changing to water or water to steam, at constant temperature and pressure; for example, during the development of a thunderstorm, when water vapor rises in the atmosphere, and then cools and condenses to form ice pellets or rain droplets, latent heat is released and helps the thunderstorm grow.

lean supply chain – a supply chain management strategy that aims to streamline production processes by eliminating waste in time, supply, or inventory

Lower Rio Grande – the stretch of Rio Grande from Fort Quitman, Texas, to the Gulf of Mexico

macroscale – on a very large scale

maquiladora – in Mexico, a duty-free, foreign-owned assembly plant or factory

marine layer or **marine inversion** – persistent low-level clouds that hug the coast in the summer

Medieval Climate Anomaly – a period of warm climate in the Northern Hemisphere from ca. 900–1350 AD

megadrought – an extremely severe and sustained drought

mesic habitat – a habitat characterized by a moderate amount of moisture, such as temperate hardwood forest. This is in contrast to a xeric habitat (low moisture) or hydric habitat (saturated in water).

Mesilla Bolson – a groundwater basin in the El Paso/Ciudad Juarez region (shared among New Mexico and Texas in the United States, and Chihuhua in Mexico)

meteorological drought – extended period of low precipitation

microclimate – an atmospheric zone ranging from a few square feet to many square miles, where the climate differs from the surrounding area. Examples include areas near bodies of water and urban heat islands.

misery days – days where people feel strongly impacted by temperature, which occur when the temperature maximum is greater than or equal to 110°F or when the temperature minimum is less than 32°F

mitigation – reducing the causes of climate change; often this refers to reducing greenhouse gas emissions. Examples of mitigation measures include reduction of greenhouse gas emissions through improved home energy conservation, improved automobile gas mileage, or development of low-emission alternative energy power plants.

mixing height – the level of an inversion layer. It is like an atmospheric ceiling that limits the volume of air into which air pollution can mix. High air pollution is associated with a low mixing height/inversion layer, while a high mixing height is associated with better air quality.

model calibration – used in Box 10.3, this term refers to the adjustment of a model (for example a climate or hydrologic model), to insure that the model produces a realistic simulation of current conditions; the aforementioned "adjustment" usually requires improving the equation that represents a physical process

morbidity – illness

mortality – death

municipio – municipality (in Mexico)

National Climate Assessment – a report issued every four years to the President and Congress, as authorized by the Global Change Research Act of 1990. The report examines current trends in global change (both human-induced and natural) and projects major trends for the subsequent 25 to 100 years.

naturalized flow, naturalized streamflow – the "natural" amount of water in the stream in the absence of human activity: Raw streamflow data are often ill-suited for scientific studies, because they are affected by diversions of water for irrigation; naturalized flow data are developed to account for water diversions and give a more realistic picture of the flow of a river, if no water were diverted.

nonconsumptive use – water use that leaves the water available for other uses and is not lost to the atmosphere through transpiration or evaporation, such as water that is retained in water systems through drainage pipes or the portion of irrigation water left as return flow to a surface or groundwater basin

non-governmental organization – a legally constituted organization that is independent of a government and usually is not a for-profit business, and which usually pursues larger social aims that may or may not be partly political

nonlinear relationship – any relationship that is not linear: Most physical processes can be approximated by linear relationships; however, some key processes are non-linear and can increase rapidly. For example, as atmospheric temperature increases, the capacity of the atmosphere to take on water increases non-linearly.

non-market value – the value of goods not traded in markets, such as clean air and water

North American dipole – a situation in which relative conditions of precipitation occur in opposition simultaneously for the Pacific Northwest and for the southern Southwest;

a clear example of the dipole is during El Niño winters, when the Pacific Northwest tends to be dry and the Southwest tends to be wet.

North American monsoon – a shift in the large-scale atmospheric circulation that brings moisture originating from the Gulf of Mexico, Gulf of California, and Pacific Ocean into the Southwest from around July to September

one-hundred-year events – climatic events that have a 1% probability of occurrence in any given year and a 100% chance of occurring in 100 years

orographic precipitation – precipitation that occurs on the windward side of a mountain, caused by drafts of moist air forced upward along the ridge

ozone – a form of oxygen that forms naturally in the stratospheric portion of the atmosphere, where it absorbs most of the sun's UV radiation. In the lower atmosphere it is considered an air pollutant, where it is produced through chemical reactions between nitrogen oxides and hydrocarbons typically produced by the burning of fossil fuels. Ozone in the lower atmosphere is an important factor that worsens some respiratory ailments.

Pacific-Decadal Oscillation – a pattern of climate variability in the Pacific Ocean that shifts phases approximately every twenty or thirty years; the positive phase exhibits more El Niño-like conditions; the negative phase brings on more La Niña-like conditions

paleoclimate – (from IPCC AR4 WGI): "Climate during periods prior to the development of measuring instruments, including historic and geologic time, for which only proxy climate records are available. Proxy climate records include tree rings, pollen cores, and ice cores."

paleoclimate reconstruction – reconstruction of paleoclimate from indirect (proxy) evidence such as tree rings, pollen, and sediment layers

paleodrought – drought that occurred before humans began collecting instrumental measurements of weather, as determined through environmental proxy data such as tree rings and lake sediments

parameter – a constant in a mathematical equation (itself part of a model) that must be defined; a parameter is usually representative for some inherent property of the system described by the model. Sometimes the name of a type of data (e.g., precipitation, temperature, soil moisture) is referred to as a parameter of interest.

particulate – a small amount of solid matter suspended in a gas or liquid; air pollution studies often refer to dust in the atmosphere as particulate matter

Paso del Norte – the border region that includes the Ciudad Juárez municipality in Chihuahua, Mexico, El Paso County in Texas, and Doña Ana County in New Mexico

pathogen – a microorganism that causes disease in its host; can include viruses, bacteria and fungi

peak flow, peak pulse – the maximum instantaneous discharge of a stream or river at a given location. It usually occurs at or near the time of maximum amount of flow.

peaks over threshold approach – analysis of climatic conditions using location-specific definitions (or thresholds) of extreme temperature, precipitation, humidity, or wind

peer-review – the review process of an author's scholarly research by experts in the same field prior to its publication in order to critically evaluate the research, maintain academic standards and credibility

perennial flow or perennial streamflow – surface water flow in streams and drainages that occurs year-round

perennial plants – plants that have a life cycle of greater than two years; examples include woody plants, like trees and shrubs

peridomestic birds – species of birds that live around human habitation

peri-urban area – the area adjoining an urban area, a transition zone characterized by both urban and rural activities

persistence – the likelihood that a species will occupy and reproduce at a level that will not lead to local extinction in a certain geographic area for a certain number of years

phenology – the timing of seasonal events in the life cycle of plants and animals, such as the development of leaves, blooms of flowers, spawning of fish, and migrations of birds; phenology is important, because as climate changes, these life cycle events may change in different ways for different species, such that, if temperature increases significantly, a crop's flower may bloom and shrivel in advance of the arrival of an important pollinator (such as a bird, whose migration is timed with changes in the length of day), which would result in a lack of plant reproduction, and less food (possibly starvation) for the pollinator

phenotypic plasticity – the ability of an organism to change its phenotype in response to environmental changes; phenotype includes the physical and biochemical characteristics of an organism, such as its stature or blood type

photovoltaic solar (PV) – technology for conversion of sunlight into electricity through the photoelectric effect, a lower-cost solar technology than concentrated solar power (CSP)

photosynthetic pathway – metabolism type of a plant; examples are CAM, C3, and C4 (see definitions above)

PM2.5 – particulate matter (airborne dirt and dust) smaller than 2.5 microns in diameter; pm2.5 particles are also referred to as "fine particles" and are believed to pose very large health risks, because they can easily lodge deeply into the lungs

primary production – production of green plant tissue in a given time period, accomplished through conversion of the sun's energy to chemical energy (organic compounds) by photosynthetic plants; primary production is an important measure of ecosystem output and health

prior appropriation doctrine – a legal water right that assigns priority of use to the first person or entity to put the water to "beneficial" use," granting them right to the full amount from available supplies before a junior appropriator (one who came later) can use his: "first in time, first in right."

proxy – physical evidence from the past that provides an indication of prehistoric climatic conditions in place of direct measurements; examples are tree rings, ice cores, glacier size and movement, sand dunes, lake sediments, and cave speleothems

radiative forcing – changes in the balance of radiated energy between different layers of the atmosphere. Positive forcing warms the climate system while negative forcing cools the system.

resilience strategies – management strategies that enhance the capacity to withstand and recover from emergencies and disasters

return flow – water left over from irrigating a crop that does not evaporate but returns to a surface flow (such as an acequia or canal) or to a groundwater source

revetment – facings of masonry or other hardened surface built to protect an embankment

risk – likelihood of harm plus the consequence

risk-based framing – planning based on the pros and cons of a given set of possibilities. For the National Climate Assessment, risk-based framing includes assessment of a risk in terms of the likelihood of its occurrence and the magnitude of the impact associated with the risk.

robust – can refer to a scientific finding or a method that can stand up to a wide range of critique; in the context of climate adaptation planning, a robust adaptation strategy will be successful across a wide range of possible future conditions

runoff efficiency – the ratio of precipitation that infiltrates as groundwater as compared to runoff; a high runoff efficiency means that the precipitation will soak into the soil and percolate down into the groundwater, thus saving water within a region, as opposed to a low runoff efficiency, which would result in precipitation being quickly diverted out of a region, as occurs during a high intensity rainfall and flash flood

salmonid – species of fish that spawn in freshwater but may spend a portion of their life in the ocean

Santa Anas – strong, hot, gusty, and dry winds in Southern California that periodically blow from the inland deserts during the otherwise cool, moist fall and winter there, contributing to fire risk

scenario – A scenario is a coherent, internally consistent and plausible description of a possible future state of the world. It is not a forecast; rather, each scenario is one alternative image of how the future can unfold.

scenario planning – a process designed for managing into the future under conditions of high uncertainty and lack of control. The objective of scenario planning is to develop and test decisions under a variety of plausible futures.

seasonality – with reference to climate, the characteristic weather and climate attributes of a particular time of year. Scientists often refer to shifts in seasonality, such as a frequently recurring delay in the onset of the first winter snowfall

sea-surface temperature (SST) – the temperature of the water close to the ocean's surface, a measurement which can vary between 1 mm (.04 inches) and 20 m (3.3 feet) below the ocean surface

sediment – Transported and deposited particles derived from rocks, soil, or biological material, that forms in layers on the earth's surface. Also the layer of soil, sand, and minerals at the bottom of surface water, such as streams, lakes, and rivers.

sediment load – the amount of sediment a stream or river can carry; this is a function of the river's flow and speed. If a river is supplied with less sediment than it can carry, it will erode its bed to supply the missing sediment; if it gets more sediment from the landscape than it can carry, the sediment will be deposited on the riverbed, causing it to rise (aggradation).

sensitivity – the degree to which a vulnerable system responds to the climate phenomenon or stimulus

sheet sands – a thin accumulation of coarse sand or fine gravel having a flat surface

shrubland – land dominated by shrub vegetation

sink – something that acts as a reservoir to absorb a greenhouse gas on a short- or long-term basis, such as forests, which can absorb carbon dioxide, serve as a land-based sink for greenhouse gases; similarly, carbon dioxide is absorbed by the oceans, so the oceans serve as a "sink"

snowpack – an accumulation of snow on the ground, generally in high altitudes in the West; snowpack serves as an important water resource and the gradual melt of snowpack is important for providing surface water supplies that will last through the summer

snow water equivalent – the amount of water that would be obtained if the snowpack were melted, usually expressed in inches

soil moisture – water diffused in shallow soil and potentially available to plants.

soil-moisture balance – a method of accounting for the addition, removal, and change in storage of water within some volume, ranging from a soil sample to a watershed, over a specific period of time

soil water deficit – the amount of available water removed from the soil within the active root depth of plants

solar irradiance forcing – see **radiative forcing**

source – a process or activity through which a greenhouse gas is released into the atmosphere

spatial resolution – the level of geographic detail of data or model grid size

stakeholder – natural resource managers, decision makers, and other parties with an interest in a particular outcome, or in the way a climatic phenomenon might affect them, such as the way a water resource manager might be interested in the way that drought might affect their operations, thus they would be referred to as stakeholders, with regard to the topics of climate variation and change

stationarity, principal of stationarity – the idea that the future will look like the past; i.e., that statistical relationships developed in a historical period are applicable to a future period

statistical (or empirical) downscaling – a method providing finer spatial detail of climate model (GCM) results; statistical downscaling uses mathematical relationships between the GCM data and historical data to adapt GCM projections to local conditions

storm surge – "an abnormal rise of water generated by a storm's winds. Storm surge can reach heights well over 20 feet and can span hundreds of miles of coastline" (from NOAA National Hurricane Center)

stressor – interactions of economic and cultural globalization, demographic change, and climate change

subtropics – regions bordering the tropics; the southern part of the Southwest can be considered part of the Northern Hemisphere subtropical region. The climate of the subtropics is usually characterized by semi-arid conditions, and many of the world's deserts are located in subtropical climate zones.

sun kink – misalignment of a railroad from heat stress

supply chain – the movement and storage of raw materials, manufacturing, and finished goods from point of origin to point of consumption Supply chains are components of the urban metabolism energy flow.

sustainable – meeting the needs of the present without compromising the wellbeing of future generations

synoptic circulation – regional atmospheric pressure patterns and their associated surface winds

tailwater – excess surface water draining from an irrigated field

teleconnection – a linkage between weather or climate changes occurring in widely separated regions of the globe. The most well-known teleconnection is the El Niño Southern Oscillation, in which persistent climate changes in the eastern and central tropical Pacific Ocean can affect the weather and climate of places as far away as East Africa, Alaska, Florida, and so on.

temperature gradient – the rate at which temperature changes with depth or height

thermocline – an abrupt temperature gradient in a body of water extending from a depth of about 100m (328 ft) to 1000m (3281 ft); the temperature of the thermocline in the eastern Pacific Ocean is a determinant of future El Niño Southern Oscillation activity

tidal prism – the volume of water leaving an estuary at ebb tide

time series – data taken at fixed intervals at successive points in time; for example, scientists might refer to a 100-year record of precipitation in the Southwest as the region's precipitation time series

traceable account – in this document, this consists of (1) the reasoning behind the conclusion, (2) the sources of data and information contributing to the conclusion, (3) an assessment of the amount of evidence and degree of agreement among sources of evidence, (4) an assessment of confidence in the finding, and (5) an assessment of uncertainty associated with the finding.

transparency – openness and accountability

transpiration – the process by which plants take up and use water for cooling and for the production of biomass

uncertainty – estimating uncertainties is intrinsically about describing the limits to knowledge and for this reason involves expert judgment about the state of that knowledge. Two primary types of uncertainty are 'value uncertainties' and 'structural uncertainties'. Value uncertainties arise from the incomplete determination of particular values or results, for example, when data are inaccurate or not fully representative of the phenomenon of interest. Structural uncertainties arise from an incomplete understanding of the processes that control particular values or results, for example, when the conceptual framework or model used for analysis does not include all the relevant processes or relationships. Value uncertainties are generally estimated using statistical techniques and expressed probabilistically. Structural uncertainties are generally described by giving the authors' collective judgment of their confidence in the correctness of a result.

upwelling – the rising of a layer of water to the surface; upwelling is important in the El Niño Southern Oscillation, where upwelling of cool or warm water can change the state of surface water temperature and trigger an El Niño or La Niña episode; it is also important to fish species, who often migrate to nutrient-rich upwelled coastal waters

urban domes – pockets of increased ozone, carbon dioxide, and particulate matter present in air above urban spaces

urban land – developed land where the housing density is greater than one unit per 2.5 acres

urban metabolism – the total urban system flow of materials, resources, energy, and outputs in the form of waste

urban-wildland interface – transitional zone between developed land and unoccupied land; this interface is important in the start and spread or impact of wildfires near urban areas

valley fever – *coccidioidomycosis*, a disease that occurs almost exclusively in Arizona and California, caused by the inhalation of a soil-dwelling fungus

vector – an organism such as a mosquito that transmits disease from one host to another

VIC model – a macroscale, distributed, physically based hydrologic model that balances both surface energy and water over a grid mesh; the VIC model is used in many hydrologic studies of the potential impacts of climate change

vulnerability – "a function of character, magnitude and rate of climate change to which a system is exposed, as well as the system's sensitivity and adaptive capacity" (NRC 2010).

water cycle – The natural transport of water in all its states from the atmosphere to the earth and back to the atmosphere through various processes. These processes include: precipitation, infiltration, percolation, evaporation, transpiration, and condensation.

water transfer – the transfer of water to different uses, different sectors, or across jurisdictional lines

wave run-up – the maximum vertical extent of the rush of a wave onto a beach or a structure above the still water level; wave run-up is an important factor in determining the extent of beach erosion

wet warming scenario – a climate scenario in which the climate becomes wetter (more precipitation) as well as warmer

wildland-urban interface – transitional zone between developed land and unoccupied land; this interface is important in the start and spread or impact of wildfires near urban areas

wind stress – the dragging force of air moving over a surface

Winters doctrine – doctrine established by the U.S. Supreme Court in 1908 in Winters v. United States, which held that when the U.S. government establishes a reservation, it also implicitly reserves water rights sufficient to meet the current and future needs of the tribe and the purpose for which the reservation was set aside (including fisheries, where applicable). The priority date for tribal water rights under the Winters doctrine is thus the date the reservation was established.

Authors and Review Editors

CHAPTER 1: SUMMARY FOR DECISION MAKERS

Jonathan Overpeck (University of Arizona), Gregg Garfin (University of Arizona), Angela Jardine (University of Arizona), David E. Busch (U.S. Geological Survey), Dan Cayan (Scripps Institution of Oceanography), Michael Dettinger (U.S. Geological Survey), Erica Fleishman (University of California, Davis), Alexander Gershunov (Scripps Institution of Oceanography), Glen MacDonald (University of California, Los Angeles), Kelly T. Redmond (Western Regional Climate Center and Desert Research Institute), William R. Travis (University of Colorado), Bradley Udall (University of Colorado)

CHAPTER 2: OVERVIEW

Gregg Garfin (University of Arizona) and Angela Jardine (University of Arizona).

Review Editor: David L. Feldman (University of California, Irvine)

CHAPTER 3: THE CHANGING SOUTHWEST

David M. Theobald (National Park Service), William R. Travis (University of Colorado), Mark A. Drummond (U.S. Geological Survey), Eric S. Gordon (University of Colorado).

Review Editor: Michele Betsill (Colorado State University)

CHAPTER 4: PRESENT WEATHER AND CLIMATE: AVERAGE CONDITIONS

W. James Steenburgh (University of Utah), Kelly T. Redmond (Western Regional Climate Center and Desert Research Institute), Kenneth E. Kunkel (NOAA Cooperative Institute for Climate and Satellites, North Carolina State University and National Climate Data Center), Nolan Doesken (Colorado State University), Robert R. Gillies (Utah State University), John D. Horel (University of Utah), Martin P. Hoerling (NOAA, Earth System Research Laboratory), Thomas H. Painter (Jet Propulsion Laboratory).

Review Editor: Roy Rasmussen (National Center for Atmospheric Research)

CHAPTER 5: PRESENT WEATHER AND CLIMATE: EVOLVING CONDITIONS

Martin P. Hoerling (NOAA, Earth System Research Laboratory), Michael Dettinger (U.S. Geological Survey and Scripps Institution of Oceanography), Klaus Wolter (University of Colorado, CIRES), Jeff Lukas (University of Colorado, CIRES), Jon Eischeid (University of Colorado, CIRES), Rama Nemani (NASA, Ames), Brant Liebmann (University of Colorado, CIRES), Kenneth E. Kunkel (NOAA Cooperative Institute for Climate and Satellites, North Carolina State University, and National Climate Data Center).

Review Editor: Arun Kumar (NOAA)

CHAPTER 6: FUTURE CLIMATE: PROJECTED AVERAGE

Dan Cayan (Scripps Institution of Oceanography, University of California, San Diego and U.S. Geological Survey), Mary Tyree (Scripps Institution of Oceanography, University of California, San Diego), Kenneth E. Kunkel (NOAA Cooperative Institute for Climate and Satellites, North Carolina State University and National Climate Data Center), Chris Castro (University of Arizona), Alexander Gershunov (Scripps Institution of Oceanography, University of California, San Diego), Joseph Barsugli (University of Colorado, Boulder, CIRES), Andrea J. Ray (NOAA), Jonathan Overpeck (University of Arizona), Michael Anderson (California Department of Water Resources), Joellen Russell (University of Arizona), Balaji Rajagopalan (University of Colorado), Imtiaz Rangwala (University Corporation for Atmospheric Research), Phil Duffy (Lawrence Livermore National Laboratory).

Review Editor: Mathew Barlow (University of Massachusetts, Lowell)

CHAPTER 7: FUTURE CLIMATE: PROJECTED EXTREMES

Alexander Gershunov (Scripps Institution of Oceanography), Balaji Rajagopalan (University of Colorado), Jonathan Overpeck (University of Arizona), Kristen Guirguis (Scripps Institution of Oceanography), Dan Cayan (Scripps Institution of Oceanography), Mimi Hughes (National Oceanographic and Atmospheric Administration [NOAA]), Michael Dettinger (U.S. Geological Survey), Chris Castro (University of Arizona), Rachel E. Schwartz (Scripps Institution of Oceanography), Michael Anderson (California State Climate Office), Andrea J. Ray (NOAA), Joe Barsugli (University of Colorado/Cooperative Institute for Research in Environmental Sciences), Tereza Cavazos (Centro de Investigación Científica y de Educación Superior de Ensenada), and Michael Alexander (NOAA).

Review Editor: Francina Dominguez (University of Arizona)

CHAPTER 8: NATURAL ECOSYSTEMS

Erica Fleishman (University of California, Davis), Jayne Belnap (U.S. Geological Survey), Neil Cobb (Northern Arizona University), Carolyn A.F. Enquist (USA National Phenology Network/The Wildlife Society), Karl Ford (Bureau of Land Management), Glen MacDonald (University of California, Los Angeles), Mike Pellant (Bureau of Land Management), Tania Schoennagel (University of Colorado), Lara M. Schmit (Northern Arizona University), Mark Schwartz (University of California, Davis), Suzanne van Drunick (University of Colorado), Anthony LeRoy Westerling (University of California, Merced), Alisa Keyser (University of California, Merced), Ryan Lucas (University of California, Merced).

Review Editor: John Sabo (Arizona State University)

CHAPTER 9: COASTAL ISSUES

Margaret R. Caldwell (Stanford Woods Institute for the Environment, Stanford Law School), Eric H. Hartge (Stanford Woods Institute for the Environment), Lesley C. Ewing (California Coastal Commission), Gary Griggs (University of California, Santa

Cruz), Ryan P. Kelly (Stanford Woods Institute for the Environment), Susanne C. Moser (Susanne Moser Research and Consulting, Stanford University), Sarah G. Newkirk (The Nature Conservancy, California), Rebecca A. Smyth (NOAA, Coastal Services Center), C. Brock Woodson (Stanford Woods Institute for the Environment).

Review Editor: Rebecca Lunde (NOAA)

CHAPTER 10: WATER: IMPACTS, RISKS, AND ADAPTATION

Bradley Udall (University of Colorado).

Review Editor: Gregory J. McCabe (U.S. Geological Survey)

CHAPTER 11: AGRICULTURE AND RANCHING

George B. Frisvold (University of Arizona), Louise E. Jackson (University of California, Davis), James G. Pritchett (Colorado State University), John P. Ritten (University of Wyoming).

Review Editor: Mark Svoboda (University of Nebraska, Lincoln)

CHAPTER 12: ENERGY: SUPPLY, DEMAND, AND IMPACTS

Vincent C. Tidwell (Sandia National Laboratories), Larry Dale (Lawrence Berkeley National Laboratory), Guido Franco (California Energy Commission), Kristen Averyt (University of Colorado), Max Wei (Lawrence Berkeley National Laboratory), Daniel M. Kammen (University of California-Berkeley), James H. Nelson (University of California- Berkeley).

Review Editor: Ardeth Barnhart (University of Arizona)

CHAPTER 13: URBAN AREAS

Stephanie Pincetl (Institute of the Environment and Sustainability, University of California, Los Angeles), Guido Franco (California State Energy Commission), Nancy B. Grimm (Arizona State University), Terri S. Hogue (Civil and Environmental Engineering, UCLA), Sara Hughes (National Center for Atmospheric Research), Eric Pardyjak (University of Utah, Department of Mechanical Engineering, Environmental Fluid Dynamics Laboratory), Alicia M. Kinoshita (Civil and Environmental Engineering, UCLA), Patrick Jantz (Woods Hole Research Center).

Review Editor: Monica Gilchrist (Local Governments for Sustainability)

CHAPTER 14: TRANSPORTATION

Deb A. Niemeier (University of California, Davis), Anne V. Goodchild (University of Washington), Maura Rowell (University of Washington), Joan L. Walker (University of California, Berkeley), Jane Lin (University of Illinois, Chicago), Lisa Schweitzer (University of Southern California).

Review Editor: Joseph L. Schofer (Northwestern University)

ER 15: HUMAN HEALTH

eidi E. Brown (University of Arizona), Andrew C. Comrie (University of Arizona), Deborah M. Drechsler (California Air Resources Board), Christopher M. Barker (University of California, Davis), Rupa Basu (California Office of Environmental Health Hazard Assessment), Timothy Brown (Desert Research Institute), Alexander Sasha Gershunov (Scripps Institution of Oceanography), A. Marm Kilpatrick (University of California, Santa Cruz), William K. Reisen (University of California, Davis), Darren M. Ruddell (University of Southern California).

Review Editor: Paul B. English (California Department of Public Health)

CHAPTER 16: CLIMATE CHANGE AND U.S.-MEXICO BORDER COMMUNITIES

Margaret Wilder (University of Arizona), Gregg Garfin (University of Arizona), Paul Ganster (Institute for Regional Studies of the Californias, San Diego State University), Hallie Eakin (Arizona State University), Patricia Romero-Lankao (NCAR), Francisco Lara-Valencia (Arizona State University), Alfonso A. Cortez-Lara (Colegio de la Frontera Norte), Stephen Mumme (Colorado State University), Carolina Neri (National Autonomous University of Mexico), Francisco Muñoz-Arriola (Scripps Institution of Oceanography).

Review Editor: Robert G. Varady (University of Arizona)

CHAPTER 17: UNIQUE CHALLENGES FACING SOUTHWESTERN TRIBES

Margaret Hiza Redsteer (U.S. Geological Survey), Kirk Bemis (Zuni Tribe Water Resources Program), Karletta Chief (University of Arizona), Mahesh Gautam (Desert Research Institute), Beth Rose Middleton (University of California, Davis), Rebecca Tsosie (Arizona State University).

Review Editor: Daniel B. Ferguson (University of Arizona)

CHAPTER 18: CLIMATE CHOICES FOR A SUSTAINABLE SOUTHWEST

Diana Liverman (University of Arizona), Susanne C. Moser (Susanne Moser Research and Consulting, Stanford University), Paul S. Weiland (Nossaman Inc.), Lisa Dilling (University of Colorado), Maxwell T. Boykoff (University of Colorado), Heidi Brown (University of Arizona), Eric S. Gordon (University of Colorado), Christina Greene (University of Arizona), Eric Holthaus (University of Arizona), Deb A. Niemeier (University of California, Davis), Stephanie Pincetl (University of California, Los Angeles), W. James Steenburgh (University of Utah), Vincent C. Tidwell (Sandia National Laboratories).

Review Editor: Jennifer Hoffman (EcoAdapt)

CHAPTER 19: MOVING FORWARD WITH IMPERFECT INFORMATION

Kristen Averyt (University of Colorado Boulder), Levi D. Brekke (Bureau of Reclamation), David E. Busch (U.S. Geological Survey), Laurna Kaatz (Denver Water), Leigh Welling (National Park Service), Eric H. Hartge (Stanford University).

Review Editor: Tom Iseman (Western Governors' Association)

CHAPTER 20: RESEARCH STRATEGIES FOR ADDRESSING UNCERTAINTIES

David E. Busch (U.S. Geological Survey), Levi D. Brekke (Bureau of Reclamation), Kristen Averyt (University of Colorado, Boulder), Angela Jardine (University of Arizona), Leigh Welling (National Park Service).

Review Editor: Karl Ford (The Bureau of Land Management, Retired), Gregg Garfin (University of Arizona)

.ewers

Lee Alter*
Tim Barnett*
Julio Betancourt*
Evelyn Blumenberg*
Bethany Bradley*
Levi Brekke*
Maria Brown*
Tim Brown*
Jeanne Chambers*
Mike Crimmins*
Theresa Crimmins*
J.E. de Steiguer*
Henry Diaz*
Doug Eisinger*
Andrew Ellis*
Miriam Ellman-Rotkin*
Denise Fort*
Janet Franklin*
Randy Fuller
Chris Funk*
Patricia Gober*
David Gochis*
Dave Graber*
Tammy Greasby*
Jeff Greenblatt*
Pasha Groisman*
Zack Guido
W. Michael Hanemann*
Sharon Harlan*
Brian Holland*
Debra Hughson*
Joe Jojola*
Jeanine Jones
Sarah Krakoff*
Jeremy Lowe*
Bradfield Lyon*
Jordan Macknick*
Bob Maddox*
Neil Maizlish*
Paul Miller*
Dan Osgood*
David Pierce*
Ray Quay*
Margarito Quintero-Nuñez*
Dave Revell*
Roberto Sanchez*
Chris Scott*
Nancy Selover*
Jill Sherman-Warne*
Brian Smith*
Paul Starrs*
Court Strong*
Nick Sussillo
Will Travis*
Frank Ward*
Robert Stabler Webb
Arnim Wiek*
John Wiener
Bob Wilkinson*
Deborah Young*

* Invited expert reviewers